Y0-BRQ-118

Basic
Solid-State
Electronics
COMPLETE COURSE
VOLUMES 1 THROUGH 5

Basic
Solid-State
Electronics

The Configuration and Management of Information Systems

COMMON-CORE

VAN VALKENBURGH, NOOGER & NEVILLE, INC.

COMPLETE COURSE
VOLUMES
1 THROUGH 5

PROMPT.
PUBLICATIONS
An Imprint of
Howard W. Sams & Company
Indianapolis, Indiana

BRASWELL MEMORIAL LIBRARY
727 NORTH GRACE STREET
ROCKY MOUNT, NC 27804

© 1982, 1983, 1992 by Van Valkenburgh, Nooger & Neville, Inc.

REVISED EDITION, 1992

PROMPT® Publications is an imprint of Howard W. Sams & Company, 2647 Waterfront Parkway, East Drive, Indianapolis, IN 46214-2041.

For permission and other rights under this copyright, write to Van Valkenburgh, Nooger & Neville, Inc.

All rights reserved. No part of this book shall be reproduced, stored in a retrieval system, or transmitted by any means, electronic, mechanical, photocopying, recording, or otherwise, without written permission from the publisher. No patent liability is assumed with respect to the use of the information contained herein. While every precaution has been taken in the preparation of this book, the author, the publisher or seller assume no responsibility for errors or omissions. Neither is any liability assumed for damages resulting from the use of information contained herein.

The words "COMMON-CORE," with device and without device, are trademarks of Van Valkenburgh, Nooger & Neville, Inc.

International Standard Book Number: 0-7906-1042-6

Cover Design by: *Sara Wright*

Acknowledgements

All photographs not credited are either courtesy of Author, Van Valkenburgh, Nooger & Neville, Inc., or Howard W. Sams & Company.

All terms mentioned in this book that are known or suspected to be trademarks or service marks have been appropriately capitalized. Howard W. Sams & Company cannot attest to the accuracy of this information. Use of a term in this book should not be regarded as affecting the validity of any trademark or service mark.

Printed in the United States of America

9 8 7 6 5

ACKNOWLEDGMENTS

Van Valkenburgh, Nooger & Neville, Inc., acknowledges with gratitude the assistance of the following organizations that have supplied large amounts of materials and granted permission for their use in this series. All of these materials come under the heading of *proprietary information,* and were prepared by staff scientists and engineers working at the highest level of their particular specialties. Dedicated people, they contributed greatly to the sum total of data in this series to make it a reservoir of accurate, important, and up-to-date information on solid-state electronics.

Acopian Corp.; American Microsystems, Inc.; Ampex Corp.; Bailey Instrument Corp.; Bell Laboratories; Columbus Instruments; Crown International; Datametrics, Inc.; The Devon-Adair Corp.; Digital Equipment Corp.; Fairchild Semiconductor Division; General Electric Co. — Semiconductor Products Dept.; Globe Union, Inc. — Battery Division; Gulton Industries; Honeywell, Inc. — Solid State Electronics Center; IBM, Inc.; E.F. Johnson Co., LND, Inc.; P.R. Mallory & Co., Inc. — Battery Company; Monroe Calculator Co.; Motorola Semiconductor Products; National Semiconductor Corp.; North American Philips Controls Corp.; RCA Solid-State Division; Signetics Corp.; Union Carbide Corp. — Battery Products Division; and Zenith Radio Corp.

For their assistance, the authors also particularly thank Dr. Anthony P. Uzzo, Director, Radar Systems, Airborne Instruments Laboratory, Eaton Corp.; and Nathan Buitenkant, former Chief Engineer and Editor in charge of the original *Basic Electronics* development.

PREFACE

This series is essentially a redevelopment of our earlier work, *Basic Electronics,* originally developed as part of the COMMON CORE® Program —*Basic Electricity, Basic Electronics, Basic Synchros and Servomechanisms,* etc. —for the U.S. Navy during the years 1950-1954. At that time we were concerned with the prerequisite knowledge and skills for the vacuum tube technology of that day as applied primarily to radio communications equipment, radar, and sonar.

Technology wise, although electronics technology over the intervening years has changed *drastically* and *dramatically* via LSI, VLSI, etc., the concepts, purposes, and system building blocks functionally are *essentially the same* — that is, for *communications or information transfer* — with the notable exceptions being the advent of digital/logic switching circuits and their high-density packaging.

Systems wise, the really new aspects have been the internal incorporation of *management* and *control* functions — via digital formatting, processing, storage, retrieval, etc. — into *"intelligent" electronic systems* by means of computers and microprocessors.

Education wise, we faced ironically the same *fractionation* of subject matter as we did back in 1950-1954! Again, there appears to be no one place where a student can go to get a relatively *simple, clear overview* of what electronics is now all about. Again, we have tried to meet this need and challenge by presenting electronics in terms of an *Overall Information Management System.*

Format wise, we continue to use the original, innovative, basic text-format, system-design elements of the COMMON-CORE® Program — the Program by means of which over 100,000 U.S. Navy technicians have been trained along with hundreds of thousands more civilian students and technicians here and in South America, Europe, the Middle East, Asia, Australia, and Africa. This format of proved effectiveness — now incorporating individual learning/testing features and techniques within the texts themselves, and in the accompanying interactive student mastery tests — has withstood the test of time.

This, then, is our second effort in over 25 years to put the basics of electronics back together again by presenting electronics in terms of an *Overall Information Management System,* and in the form of the original innovative basic text-format, system-design elements so that an average person can learn what the basics of electronics — solid-state electronics — are all about.

VAN VALKENBURGH, NOOGER & NEVILLE, INC.

New York, N.Y.

CONTENTS

VOLUME 1 — INFORMATION SYSTEM BUILDING BLOCKS

Basic Concepts; Semiconductor Devices; Discrete Components and Integrated Circuits; Passive Components R/L/C; Active Diode and Transistor Components; Simple Power Supplies/Rectifiers/Filters; Regulated Power Supplies; Inverters and Converters; Battery-Type Power Supplies; Servicing/Learning System Power Supply

BASIC CONCEPTS

INTRODUCTION TO ELECTRONICS OVERVIEW

SEMICONDUCTOR DEVICES

DISCRETE SEMICONDUCTOR DEVICES

BIPOLAR DEVICES

MOS DEVICES

SEMICONDUCTOR DEVICES OVERVIEW

SIMPLE POWER SUPPLIES/RECTIFIERS/FILTERS

REGULATED POWER SUPPLIES

TYPICAL POWER SUPPLIES

INVERTERS AND CONVERTERS

INVERTERS AND CONVERTERS

REGULATED POWER SUPPLY/INVERTER AND CONVERTER OVERVIEW

POWER SUPPLIES — SERVICING AND APPLICATION

SERVICING AC POWER SUPPLIES

BATTERY—TYPE POWER SUPPLIES

BATTERY POWER SUPPLIES

SERVICING/LEARNING SYSTEM POWER SUPPLY

INTRODUCTION TO SERVICING SOLID-STATE CIRCUITS

RADIO TRANSMITTER — RECEIVER LEARNING SYSTEM POWER SUPPLY

SERVICING AND POWER SUPPLY OVERVIEW

VOLUME 2 — AUDIO INFORMATION SYSTEMS

Amplification; Audio Amplifiers; Audio Systems; Sound Transducers; Public Address Systems; Hi-Fi Audio Systems; Video Amplifiers; IF Amplifiers; RF Amplifiers; Oscillators; Troubleshooting Amplifiers and Oscillators

BASIC AMPLIFICATION

INTRODUCTION TO AMPLIFICATION

BASIC AMPLIFIER CHARACTERISTICS

AUDIO AMPLIFIERS

INTRODUCTION TO AUDIO AMPLIFIERS

AUDIO AMPLIFIERS OVERVIEW

BASIC AUDIO SYSTEMS

INTRODUCTION TO AUDIO SYSTEMS

SOUND TRANSDUCERS

PUBLIC ADDRESS SYSTEMS

BASIC AUDIO SYSTEMS OVERVIEW

HIGH-FIDELITY AUDIO SYSTEMS

HIGH-FIDELITY AUDIO SYSTEMS OVERVIEW

VIDEO/IF/RF AMPLIFIERS AND OSCILLATORS

VIDEO AMPLIFIERS

IF AMPLIFIERS

RF AMPLIFIERS

OSCILLATORS

VIDEO/IF/RF AMPLIFIERS AND OSCILLATORS OVERVIEW

TROUBLESHOOTING/APPENDIX

TROUBLESHOOTING AMPLIFIERS AND OSCILLATORS

APPENDIX

VOLUME 3 — INFORMATION TRANSMISSION

Radio Wave Propagation; Radio Transmitters; Transmission Lines; Antennas; Amplitude Modulation (AM); Single-Sideband AM Transmission; Frequency Modulation (FM); Pulse Modulation; TV Transmission

PROPAGATION AND TRANSMITTER RF SYSTEMS

INTRODUCTION TO INFORMATION TRANSFER

INTRODUCTION TO RADIO WAVE PROPAGATION

INTRODUCTION TO RADIO TRANSMITTERS

THE RF SECTION OF A TRANSMITTER

PROPAGATION AND TRANSMITTER RF OVERVIEW

ANTENNA SYSTEMS AND
AMPLITUDE MODULATION

TRANSMISSION LINES

ANTENNAS

ANTENNA SYSTEMS AND
AMPLITUDE MODULATION OVERVIEW

AMPLITUDE MODULATION SYSTEMS

AMPLITUDE MODULATION

SINGLE-SIDEBAND AM TRANSMITTER

AMPLITUDE MODULATION OVERVIEW

FREQUENCY MODULATION SYSTEMS

INTRODUCTION TO FREQUENCY MODULATION

FREQUENCY MODULATION OVERVIEW

INTRODUCTION TO TELEVISION SYSTEMS

INTRODUCTION TO PULSE MODULATION

INTRODUCTION TO TELEVISION TRANSMISSION

INTRODUCTION TO TELEVISION OVERVIEW

VOLUME 4 — INFORMATION RECEPTION

Audio/Video/Data/Sensory Reception; Receiving Antennas; AM Receivers; FM Receivers; Communication Receivers; TV Receiver Fundamentals; Black-and-White TV Receivers; Color TV Receivers; Video Recording; Video Display Terminals; Troubleshooting/Alignment

RADIO RECEPTION AND RECEIVER CHARACTERISTICS

INTRODUCTION TO INFORMATION RECEPTION

INTRODUCTION TO RADIO RECEPTION

FM RECEIVERS

CHARACTERISTICS OF FM RECEIVERS

FM RECEIVERS OVERVIEW

COMMUNICATION RECEIVERS

CHARACTERISTICS OF COMMUNICATION RECEIVERS

COMMUNICATION RECEIVERS OVERVIEW

TELEVISION RECEPTION

TELEVISION RECEPTION FUNDAMENTALS

TELEVISION RECEPTION OVERVIEW

BLACK-AND-WHITE TV RECEIVERS

INTRODUCTION TO BLACK-AND-WHITE TV

BLACK-AND-WHITE TV RECEIVERS OVERVIEW

COLOR TV RECEIVERS AND VIDEO RECORDERS

INTRODUCTION TO COLOR TV/VIDEO RECORDERS

COLOR TV/VIDEO RECORDERS OVERVIEW

EPILOGUE

VOLUME 5 — INFORMATION MANAGEMENT

Digital Systems; System Elements; Digital Arithmetic; Digital Timing and Counting; Alphanumeric Display; Computers and Microprocessors; Input/Output Devices; Computer/Microprocessor Applications; Digital Communication Systems; Troubleshooting Digital Circuits/Systems

DIGITAL SYSTEMS — BACKGROUND INFORMATION

INTRODUCTION TO DIGITAL SYSTEMS

DIGITAL SYSTEMS — BACKGROUND INFORMATION OVERVIEW

DIGITAL SYSTEM ELEMENTS

BASIC DIGITAL SYSTEM ELEMENTS

DIGITAL SYSTEM ELEMENTS OVERVIEW

DIGITAL ARITHMETIC

INTRODUCTION TO DIGITAL ARITHMETIC

DIGITAL ARITHMETIC OVERVIEW

BASIC DIGITAL SYSTEM FUNCTIONS AND APPLICATIONS

BASIC DIGITAL SYSTEM FUNCTIONS

DIGITAL TIMING AND COUNTING CIRCUITS

ALPHANUMERIC DISPLAY DEVICES

APPLICATION OF COUNTERS

BASIC DIGITAL SYSTEM FUNCTIONS
AND APPLICATIONS OVERVIEW

COMPUTERS AND MICROPROCESSORS —
BACKGROUND INFORMATION

COMPUTERS AND MICROPROCESSORS

COMPUTERS AND MICROPROCESSORS —
BACKGROUND INFORMATION OVERVIEW

COMPUTER INPUT/OUTPUT DEVICES

INTRODUCTION TO INPUT/OUTPUT DEVICES

COMPUTER INPUT/OUTPUT DEVICES OVERVIEW

COMPUTER/MICROPROCESSOR APPLICATIONS, DIGITAL COMMUNICATIONS, AND TROUBLESHOOTING

COMPUTER/MICROPROCESSOR APPLICATIONS

Basic
Solid-State
Electronics
VOL. 1

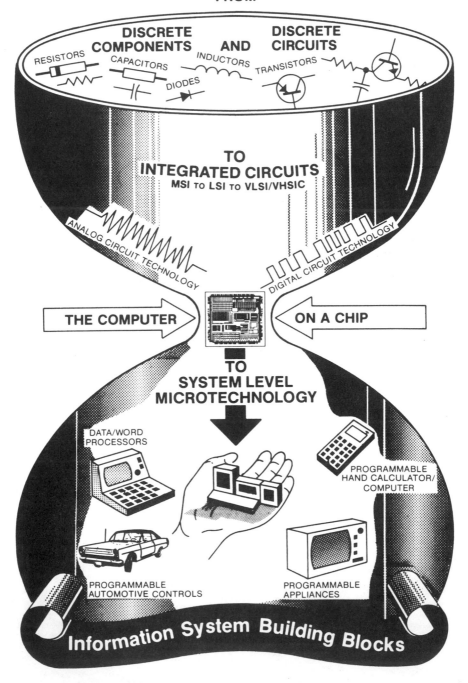

ELECTRONIC TECHNOLOGY
FROM

DISCRETE COMPONENTS **AND** **DISCRETE CIRCUITS**

RESISTORS   CAPACITORS   DIODES   INDUCTORS   TRANSISTORS

TO
INTEGRATED CIRCUITS
MSI TO LSI TO VLSI/VHSIC

ANALOG CIRCUIT TECHNOLOGY

DIGITAL CIRCUIT TECHNOLOGY

THE COMPUTER   **ON A CHIP**

TO
SYSTEM LEVEL
MICROTECHNOLOGY

DATA/WORD PROCESSORS

PROGRAMMABLE HAND CALCULATOR/ COMPUTER

PROGRAMMABLE AUTOMOTIVE CONTROLS

PROGRAMMABLE APPLIANCES

Information System Building Blocks

Basic
Solid-State
Electronics

COMMON CORE

The Configuration and Management of Information Systems

INFORMATION SYSTEM BUILDING BLOCKS
VOL. I

BASIC CONCEPTS/CIRCUITS
ANALOG/DIGITAL BUILDING BLOCKS
DISCRETE/IC SEMICONDUCTOR DEVICES
RADIO TRANSCEIVER LEARNING SYSTEM
SIMPLE/REGULATED/BATTERY POWER SUPPLIES
SERVICING/TROUBLESHOOTING

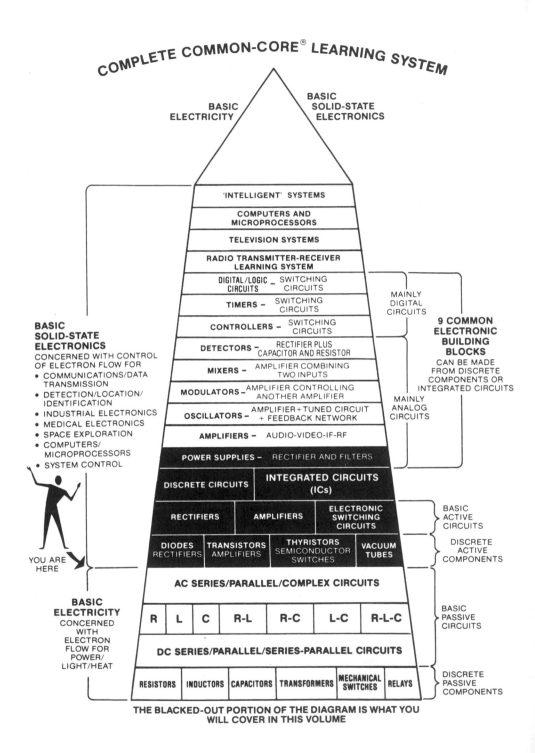

COMPLETE COMMON-CORE® LEARNING SYSTEM

BASIC ELECTRICITY

BASIC SOLID-STATE ELECTRONICS

'INTELLIGENT' SYSTEMS

COMPUTERS AND MICROPROCESSORS

TELEVISION SYSTEMS

RADIO TRANSMITTER-RECEIVER LEARNING SYSTEM

DIGITAL/LOGIC CIRCUITS – SWITCHING CIRCUITS

TIMERS – SWITCHING CIRCUITS

CONTROLLERS – SWITCHING CIRCUITS

DETECTORS – RECTIFIER PLUS CAPACITOR AND RESISTOR

MIXERS – AMPLIFIER COMBINING TWO INPUTS

MODULATORS – AMPLIFIER CONTROLLING ANOTHER AMPLIFIER

OSCILLATORS – AMPLIFIER+TUNED CIRCUIT + FEEDBACK NETWORK

AMPLIFIERS – AUDIO-VIDEO-IF-RF

POWER SUPPLIES – RECTIFIER AND FILTERS

DISCRETE CIRCUITS | **INTEGRATED CIRCUITS (ICs)**

RECTIFIERS | **AMPLIFIERS** | **ELECTRONIC SWITCHING CIRCUITS**

DIODES RECTIFIERS | **TRANSISTORS** AMPLIFIERS | **THYRISTORS** SEMICONDUCTOR SWITCHES | **VACUUM TUBES**

AC SERIES/PARALLEL/COMPLEX CIRCUITS

| R | L | C | R-L | R-C | L-C | R-L-C |

DC SERIES/PARALLEL/SERIES-PARALLEL CIRCUITS

| RESISTORS | INDUCTORS | CAPACITORS | TRANSFORMERS | MECHANICAL SWITCHES | RELAYS |

BASIC SOLID-STATE ELECTRONICS
CONCERNED WITH CONTROL OF ELECTRON FLOW FOR
- COMMUNICATIONS/DATA TRANSMISSION
- DETECTION/LOCATION/ IDENTIFICATION
- INDUSTRIAL ELECTRONICS
- MEDICAL ELECTRONICS
- SPACE EXPLORATION
- COMPUTERS/ MICROPROCESSORS
- SYSTEM CONTROL

YOU ARE HERE

BASIC ELECTRICITY
CONCERNED WITH ELECTRON FLOW FOR POWER/ LIGHT/HEAT

MAINLY DIGITAL CIRCUITS

MAINLY ANALOG CIRCUITS

9 COMMON ELECTRONIC BUILDING BLOCKS
CAN BE MADE FROM DISCRETE COMPONENTS OR INTEGRATED CIRCUITS

BASIC ACTIVE CIRCUITS

DISCRETE ACTIVE COMPONENTS

BASIC PASSIVE CIRCUITS

DISCRETE PASSIVE COMPONENTS

THE BLACKED-OUT PORTION OF THE DIAGRAM IS WHAT YOU WILL COVER IN THIS VOLUME

Overall Information Management System

We live in an *information society,* and the *flow* and *management* of that information is fundamental. But if this is a course on the fundamentals of solid-state electronics, *why* are we talking about *information?* Why? Because *this is what electronics is all about!*

Modern electronics technology/microtechnology is concerned with the management of information in all of its aspects—generation, transmission, reception, storage, retrieval, processing, display, and control. And the chart on the following two pages gives you an overall picture so you can see how it all fits together into one such overall system.

Both electrical and electronic equipment and systems are concerned with the effects of controlled electron flow, so that most of what you learned in *Basic Electricity* is directly applicable to your study of *Basic Solid-State Electronics* (and *Basic Electronics,* Vacuum-Tube Edition). We usually associate electricity and electrical equipment and systems with the control of electron flow for the functions of *generating heat* and *light,* the *generation* and *transmission* of electric *power,* and the *conversion* of *electrical energy* to *mechanical energy* (or vice versa). As you know, these electrical functions are usually accomplished through the use of discrete circuit elements. (See Complete COMMON-CORE® Learning System chart in Preface.)

In electronic equipment and systems, on the other hand, we are concerned primarily with the control of electron flow for the functions of *generating* or *sensing, transmitting, receiving, storing, retrieving, processing, displaying,* and *managing information.* These various electronic functions are accomplished through the use of discrete and integrated analog and digital circuits to form various electronic systems—audio, radio, TV, teletext, videotex, data and word processing, radar, sonar, scientific instrumentation, etc.

So that you can learn and understand this Information Management System one step at a time, this course is divided into the following areas:

Volume 1: INFORMATION SYSTEM BUILDING BLOCKS—
Semiconductor Fundamentals, Power Supplies

Volume 2: AUDIO INFORMATION SYSTEMS—Amplifiers, I/O Devices, Hi-Fi, Stereo Systems—Oscillators and IF, RF and Video Amplifiers

Volume 3: INFORMATION TRANSMISSION—Radio and TV Transmitters

Volume 4: INFORMATION RECEPTION—Radio and TV Receivers

Volume 5: INFORMATION MANAGEMENT—Digital Circuits, Digital Systems, Computers/Microprocessors, 'Intelligent' Systems

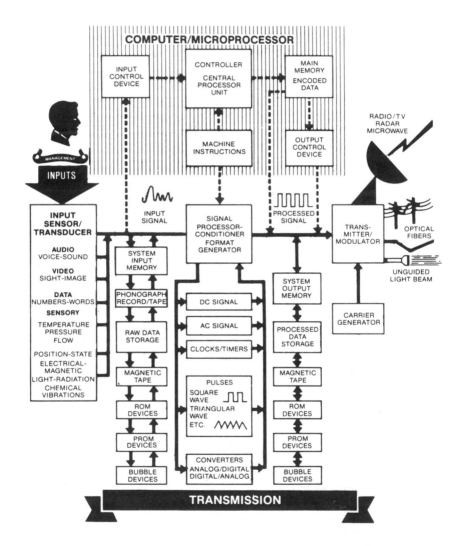

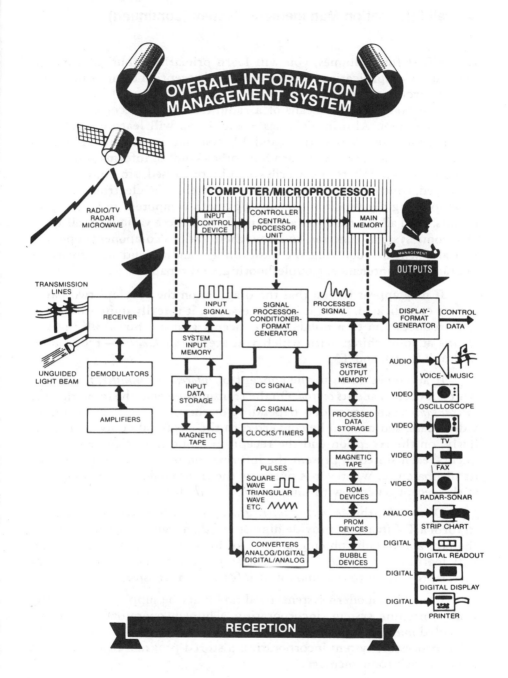

OVERALL INFORMATION MANAGEMENT SYSTEM

COMPUTER/MICROPROCESSOR

RADIO/TV RADAR MICROWAVE

INPUT CONTROL DEVICE

CONTROLLER CENTRAL PROCESSOR UNIT

MAIN MEMORY

MANAGEMENT

OUTPUTS

TRANSMISSION LINES

RECEIVER

INPUT SIGNAL

SIGNAL PROCESSOR-CONDITIONER-FORMAT GENERATOR

PROCESSED SIGNAL

DISPLAY-FORMAT GENERATOR

CONTROL

DATA

SYSTEM INPUT MEMORY

UNGUIDED LIGHT BEAM

DEMODULATORS

INPUT DATA STORAGE

DC SIGNAL

AC SIGNAL

CLOCKS/TIMERS

SYSTEM OUTPUT MEMORY

AUDIO

VOICE-MUSIC

AMPLIFIERS

MAGNETIC TAPE

PROCESSED DATA STORAGE

VIDEO

OSCILLOSCOPE

VIDEO

TV

PULSES
SQUARE WAVE
TRIANGULAR WAVE
ETC.

MAGNETIC TAPE

VIDEO

FAX

ROM DEVICES

VIDEO

RADAR-SONAR

PROM DEVICES

ANALOG

STRIP CHART

CONVERTERS
ANALOG/DIGITAL
DIGITAL/ANALOG

BUBBLE DEVICES

DIGITAL

DIGITAL READOUT

DIGITAL

DIGITAL DISPLAY

DIGITAL

PRINTER

RECEPTION

Overall Information Management System (continued)

In the first four volumes, you will learn primarily about information generation, amplification, transmission, and reception using principally analog circuit technology—in the form of audio (sound) information inputs and outputs—by means of a radio transmitter-receiver communication system. Also in Volumes 3 and 4 you will learn about video information inputs and outputs and TV system transmission and reception. In Volume 5, you will learn how other kinds of information inputs and outputs—numbers and words—can be processed, stored, retrieved, displayed, controlled, and used in various 'intelligent' electronic systems by employing digital circuit technology with computers and microprocessors. And all the time you are learning how these various 'intelligent' electronic systems function and how their various components operate, you also will be acquiring the basic understanding and skill for their operation, maintenance, troubleshooting, and repair.

It may appear to you that the diagram on the previous two pages is very complicated. Well, it isn't. In essence, it is really quite simple for it introduces you to a number of concepts that are but *another way of looking* at some things with which you are *already familiar*—radio, phonograph (hi-fi), and TV.

It all comes down to the concept of *information transfer*, the *rate* at which it occurs, and its *control*. On the transmission end, the input device, or *transducer*, can vary widely depending on the *type of information*—audio, video, data, sensory—to be converted to the electrical equivalent. Similarly, on the reception end the reception input goes to *different* transducers, each suited to present the information in a form most suited to its ultimate use. Whatever is put in at the transmission end will appear at the reception end, with some *acceptable level* of *distortion*.

Obviously, the *amount of information per unit time* (second) required to form a TV image or provide high-speed data transfer is much greater than for audio (radio/phonograph) or for low-speed data transfer. As you will learn, the *bandwidth* of the transmit/receive channel increases in direct proportion to the amount of information transfer.

Control in modern systems need no longer be supplied *externally* by a man-machine system (incorporating a human operator) but can be provided *internally* and *automatically* by an electronic machine (computer/microprocessor) system incorporating a stored program of instructions from an electronic memory.

By *stressing similarities rather than differences*, you will begin to see how all the apparent sophistication of the various modern electronic systems *can be related*, and related to electronic systems that you use every day—telephone, radio, stereo phonograph, and TV.

Electronic Systems

You will learn in these volumes that most electronic equipment or systems are made up of a relatively small number of basic circuits or building blocks. It is *how* these are *interconnected* that make for *different* functions. When you know how these electronic building blocks work, you will be prepared for work with any system. Because the field of electronics is so great, it is best subdivided—like the human body—into branches based on function. As with the definition of electricity and electronics, these branches are indistinctly divided and with much overlap. Remember, however, as in electricity, a relatively *few* basic principles and building blocks can give you the foundation to understand them *all*. Some of the major branches of electronics are shown in the illustration.

ELECTRONIC SYSTEMS

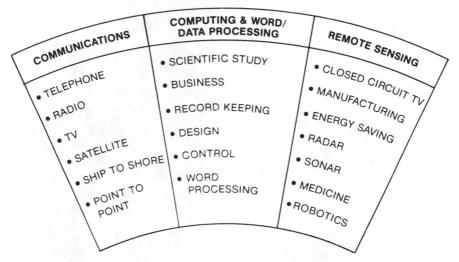

Communications is probably the best-known branch of electronics. (And this was the basis for the original edition of *Basic Electronics*, Vacuum-Tube Edition). In communication systems—like the human nervous system—we arrange our electronic building blocks to transmit and receive information from one place to another.

Computing and word/data processing, another branch, has become an indispensable part of our modern world. As you will learn, these systems—like the human mind—while very complicated in design, again are made up of relatively simple building blocks like all other electronic systems. More is being done in this area nowadays than in any other branch of electronics.

Another branch of electronics that is of great importance involves *remote sensing*—an extension of the human sensory system—where our electronic building blocks are arranged into systems used for *detecting*, *locating*, and *identifying distant* objects, or *gathering data* from a remote point.

A Brief History of Electronics

Electronics as we usually think of it began in 1896 when Guglielmo Marconi transmitted and received radio signals for a distance of 2 miles. In 1883, Thomas Edison discovered the flow of electrons between a hot filament and a nearby electrical conductor in a vacuum. John Fleming made use of this discovery in 1904 when he used it to convert an alternating current (ac) to a pulsating direct current (dc). By adding a third element to this elementary device in 1906, Lee De Forest produced the vacuum tube (thermionic valve) that could amplify a small voltage to a duplicate voltage many times larger in amplitude. This major improvement was the basis for vacuum-tube electronics, which was the foundation for electronic progress for many years.

The next major step came in 1948 when the transistor was invented by Shockley, Bardeen, and Brattain of Bell Laboratories. This device of solid materials that could perform electronic amplification was much smaller, consumed much less power, became much cheaper and more reliable than the vacuum tube. Since these solid-state devices were *much smaller* and used *much less power* but could do the job of vacuum tubes, it became possible to include *many more functions* in a small space. In addition, the improved reliability of transistors made it possible to start building really complex systems that could be kept in operation. In addition, solid-state devices operate very well as switches, making it easier to develop solid-state digital circuits. This, in turn, spurred the development of digital computers that has continued to the present time.

A Brief History of Electronics (continued)

The age of solid-state electronics really began in the late 1950s when solid-state devices started to become available commercially. New circuits were developed to be used in place of existing vacuum-tube circuits. The size and weight of electronic equipment decreased rapidly while reliability increased enormously. The low-cost transistor portable radio became an actuality in the late 1950s.

At this time electronic equipment consisted of resistors, capacitors, transistors, etc. used as *separate* (*discrete*) *components* connected together in circuits to perform various functions. In 1958 Texas Instruments and the Fairchild Corporation developed *single units* containing the equivalent of *several* transistors, resistors, and capacitors that worked together to do a complete circuit function. These *integrated circuits*, also known as *ICs*, were rapidly developed to the point where they soon saved electronic equipment manufacturers costs in assembling, while *again reducing* equipment size, weight, and power required, with improved reliability. Also, the development of low-power semiconductor technology allowed the packing of more and more functions into a single integrated circuit chip. This also led to the rapid development of *digital circuit technology*, which requires great numbers of simple circuits.

MICROCOMPUTERS

SPACE TRAVEL

CB RADIOS

POCKET CALCULATORS

SOLID-STATE ELECTRONICS NOW BRINGS YOU

PORTABLE TV

COMPUTERS FOR BUSINESS, INDUSTRY, AND SCIENCE

As more and more functions were packed into ICs, their variety increased enormously. These *medium-scale integration* (*MSI*) devices started the electronics revolution as we see it today. As new needs developed, it became apparent that *entire major functions* could be put on *one chip*. However, development of these *special* chips is quite expensive. Therefore, interest developed in the production of single devices that when properly programmed (instructed) could perform a wide range of functions using the same basic device.

A Brief History of Electronics (continued)

This led the Intel Corporation in 1971 to introduce the *microprocessor chip*, which was an integrated circuit that contained *all* that was required to make up (1) a *control system* and (2) the appropriate *external circuits* that could perform a great number of *control functions* merely by changing the instructions (programing).

The introduction of *large-scale integration* (LSI) made it possible for Hewlett-Packard in 1973 to produce the first hand calculator (computer-on-a-chip) that could perform almost any desired calculation with great speed. These first calculators contained three ICs, each equivalent to about 5,000 transistor circuits. Computers made up of such elements are thousands of times smaller and faster in operation than the most advanced equivalent vacuum-tube computer and hundreds of times smaller than *discrete* component systems.

EXAMPLE OF MICROMINIATURIZATION

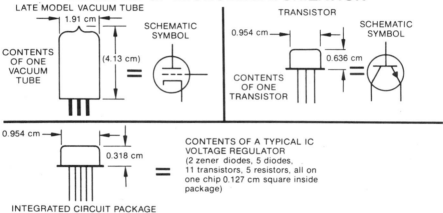

We now use sophisticated ICs containing the equivalent of thousands of transistors in microprocessors for control of things from complex industrial applications to the cooking time in home microwave ovens. These microprocessors are all *basically the same*, with their function determined by *how they are programmed* to do a particular function. In spite of their internal complexity, they are very inexpensive to manufacture, costing in some cases less than a U.S. dollar. You can buy a hand calculator that essentially consists only of a single IC chip, with display and keyboard, for approximately five U.S. dollars. However, that *single* IC chip is the equivalent of *thousands* of *discrete circuits* and actually is a *very sophisticated* electronic device.

We are now approaching the era of *very large-scale integration* (VLSI) and *very high-speed integrated circuits* (VHSIC) where *entire systems* are contained on a single chip that may hold the equivalent of a hundred thousand transistor circuits. Because of the greatly reduced cost of ICs, programmable personal computer systems for home and personal use are becoming common. These small systems have the advantage that they can be programmed (instructed) to do almost anything from personal record keeping and calculation to playing games.

Controlling Current (Electron) Flow

Controlling current (electron) flow is the basis of operation for all electronic equipment. From your knowledge of *Basic Electricity,* you already know some of the basic ways to control current flow:

1. With a battery, generator, or other source of power to start it and keep it flowing.
2. With a switch to interrupt current flow or direct it along a new path.
3. With an impedance (Z), resistor, capacitor, and/or inductor to control the magnitude of current flow.
4. With a capacitor or inductor to store it.
5. With a transformer to change its amplitude.

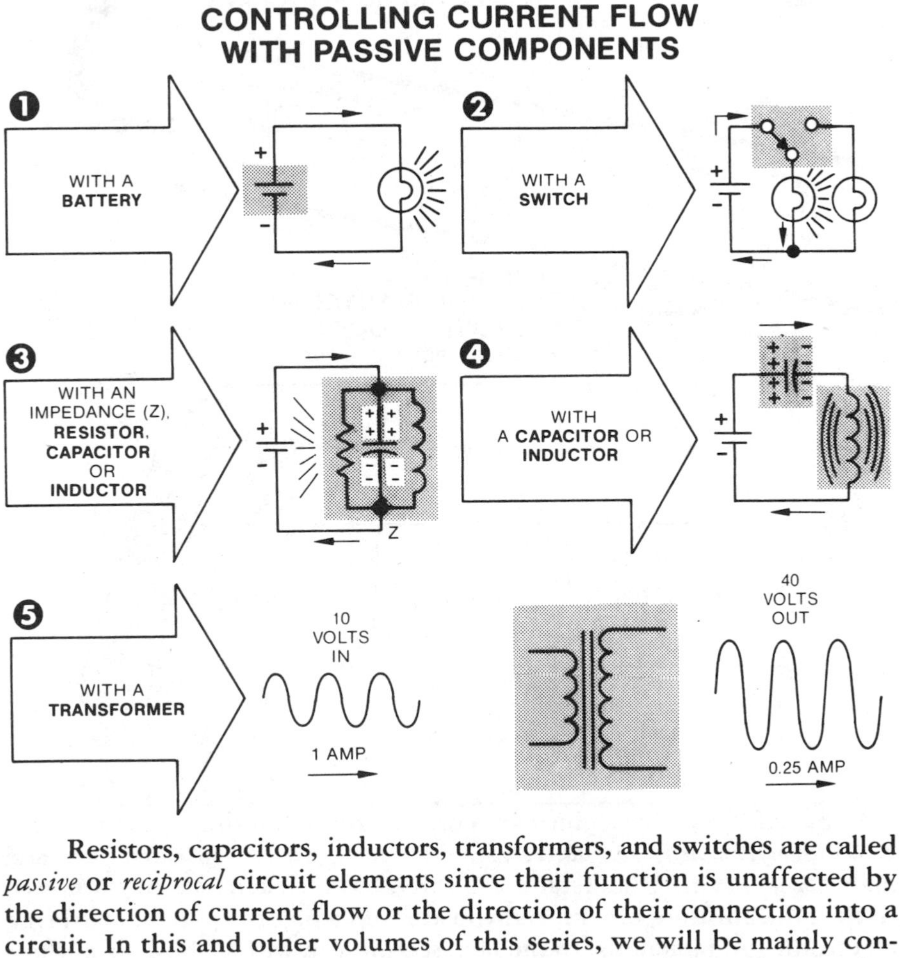

CONTROLLING CURRENT FLOW
WITH PASSIVE COMPONENTS

❶ WITH A **BATTERY**

❷ WITH A **SWITCH**

❸ WITH AN IMPEDANCE (Z), **RESISTOR**, **CAPACITOR** OR **INDUCTOR**

❹ WITH A **CAPACITOR** OR **INDUCTOR**

❺ WITH A **TRANSFORMER**

10 VOLTS IN

1 AMP

40 VOLTS OUT

0.25 AMP

Resistors, capacitors, inductors, transformers, and switches are called *passive* or *reciprocal* circuit elements since their function is unaffected by the direction of current flow or the direction of their connection into a circuit. In this and other volumes of this series, we will be mainly concerned with *active* or *nonreciprocal* circuit elements—that is, devices that have distinctly different properties as the *direction* or *magnitude* of current flow *changes*. These elements must always be connected in a specific way to properly function.

The Three Basic Electronic Circuits/Nine Common Building Blocks

As indicated earlier, although there is a great variety of electronic equipment available today, the vast majority make use of only *three basic circuits*. These are the *rectifier*, *amplifier*, and *electronic switching circuits*. Used in various circuit arrangements (*configurations*) with additional resistors, capacitors, inductors, and transformers, these three basic circuits make up the *nine common electronic building blocks* that are used in nearly all electronic equipment. As you will learn, in many cases, these functionally are done within an IC. However, you will study *discrete* solid-state electronic circuits initially so you can *understand better* these ICs as well as be able to understand systems using discrete electronic components.

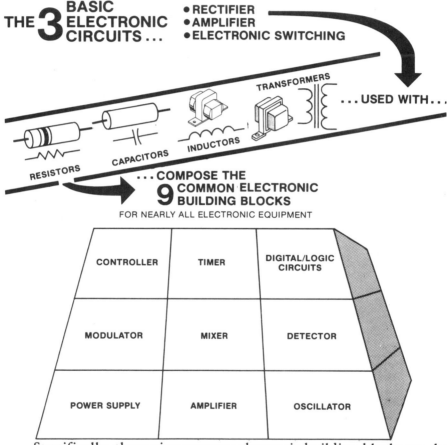

Specifically, these nine common electronic building blocks are the *power supply*, *amplifier*, *oscillator*, *modulator*, *mixer*, *detector*, *controller*, *timer*, and *digital/logic circuits*. Understand the three basic circuits as they are used in the nine building blocks, and you are well along the path to mastering electronic fundamentals. In the next section you will see how six of these common circuits are used as building blocks to form a complete radio transmitter-receiver system. This is part of the fundamental plan of this course to give you an *overview* of what is ahead and then to proceed from the *simple* to the *complex* in presenting it to you.

The Block Diagram—System Function

So that we can better understand an electronic system's functions, it is usual to use a highly simplified representation called a *block diagram*. The block diagram is a drawing of a set of boxes, each containing a *circuit function* described by the box label. These boxes are *interconnected* by lines showing important *information-signal flow* paths. By using the block diagram, it is possible to understand *overall system function* without all the complication that the circuit details introduce. When the block diagram is understood, we can identify the various circuit functions and then proceed to the individual circuit diagram that gives details on how the individual circuit elements are configured in detail.

While you will not understand all of the terms and symbols used below until later, you will see the utility of block diagrams for providing a simple way to discuss *how systems work*. The first block diagram shown is for a simple two-stage radio frequency (RF) amplifier system coupled to an antenna. As shown, the low-level signal input is amplified in the low-power amplifier that feeds the high-power amplifier that in turn produces the high-level signal output that drives the antenna. The power supply (to feed these amplifiers) is shown supplying the voltages and currents necessary to operate these amplifiers.

THE BLOCK DIAGRAM

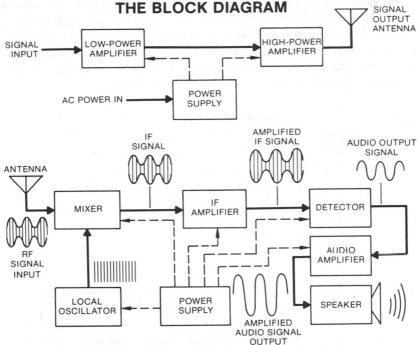

The second diagram shows a simple radio receiver. The antenna picks up a radio signal and supplies it (the RF signal input) to the mixer circuit. The RF signal and local oscillator signal are mixed, and the output signal from the mixer (the IF signal) is amplified and detected to produce an audio output signal. This audio output signal is amplified in the audio amplifier, which, in turn, drives the speaker. The power supply provides the necessary voltages and currents to operate the receiver.

Radio Transmitter-Receiver Learning System

Shown below is a block diagram of a complete radio transmitter-receiver system. The arrangement is typical of that used when a radio station transmits a program to a radio receiver in your home. It also applies to the arrangement used when one amateur radio station transmits a message to another, when a Citizens Band (CB) operator in one automobile talks to another operator several miles away or to any communication system. Although these various systems *differ* in size, cost, operating range, and information sent and received, they *all use* the common building blocks shown here.

Comparison of the diagram with the nine common building blocks listed earlier shows that the transmitter-receiver uses six of the nine common building block types. Electronic controllers, timers, and digital/logic circuits are not used here, although they may be incorporated for accessory and control purposes. As you will learn, the function of one or more building blocks can often be done within a *single* IC.

This transmitter-receiver system will be the *learning system* that will help take you from here to a mastery of electronic fundamentals. The next few pages will review the fundamental purpose of each of the six common building blocks illustrated. In the volumes that follow you will see how these circuits operate, what their common configurations are, and how these building blocks are connected together to accomplish vital functions in communications, industrial electronics, computer technology, and the other areas mentioned earlier. The remaining three building blocks are covered in detail in Volume 5.

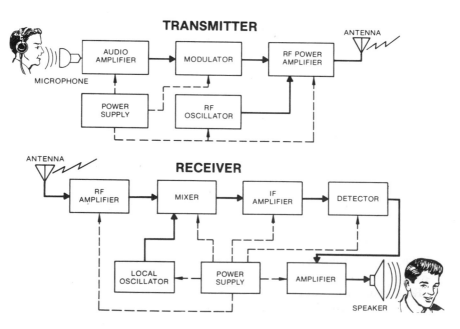

Radio Transmitter-Receiver Learning System—System-to-Parts Approach

By using the transmitter-receiver system as the fundamental learning vehicle that takes you through these volumes, you will have a good picture of how the details all fit together in a practical working relationship to accomplish a given system function. Electronics is often studied by the *parts-to-system* approach—by starting with one particular component and learning every detail about it before going on to the next step of connecting it into an elementary circuit. It is far better to learn systems and the function of the blocks within the system *before* you learn the details, so you can more easily and better understand the relationship between circuit design and function.

Therefore, these volumes use the *system-to-parts approach*, so you will always have a clear understanding of how the circuit you are learning about at the moment can be used to serve a *real, practical purpose* in a system. As each new area of subject matter is introduced, this approach will be employed and highlighted to show exactly how the new material *relates* to this transmitter-receiver learning system.

To get started, we will summarize the purposes of the various electronic building blocks in our transmitter-receiver system and show how they work together to transmit and receive information.

THE SYSTEM-TO-PARTS APPROACH

THE TRANSMITTER-RECEIVER LEARNING SYSTEM IS THE
"VEHICLE" THAT WILL TAKE YOU FROM HERE TO A
WORKING KNOWLEDGE OF:

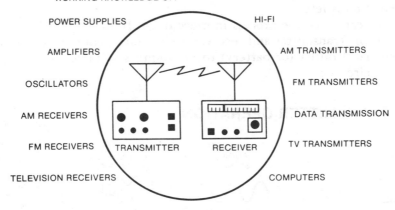

POWER SUPPLIES — HI-FI — AMPLIFIERS — AM TRANSMITTERS — OSCILLATORS — FM TRANSMITTERS — AM RECEIVERS — DATA TRANSMISSION — FM RECEIVERS — TRANSMITTER — RECEIVER — TV TRANSMITTERS — TELEVISION RECEIVERS — COMPUTERS

In accordance with the learning plan that has just been mentioned—getting the *big picture* before becoming concerned with the *details* (the system-to-parts approach)—there will be a brief review of the overall process of using radio signals for long-distance message communication. You will see that this explanation is highly simplified and that the contents of four of these volumes will be required to present the details.

Summary of Radio Transmitter-Receiver Operation

All the explanation on this page refers to the transmitter-receiver block diagrams on this and the next page. Note that we are describing an *amplitude modulation* (*AM*) system—that is, a system where information is conveyed by changing the *amplitude* of the signal. Later, we will describe the configurations that describe *frequency-modulated* (*FM*) systems where information is conveyed by changing the *frequency* of the signal.

As you will learn shortly, low-frequency signals such as audio signals travel very easily through wires. However, they can be radiated through space to a distant point only with very great difficulty. For electrical signals to radiate into space it is usual to use much *higher* frequencies called *radio frequencies*. For example, the frequencies used in U.S. commercial AM broadcasting are between 535,000 Hz (535 MHz) and 1,605,000 Hz (1.605 MHz). Frequencies of millions of hertz (megahertz) are usually used in many communication systems.

Because of this problem it is necessary to generate an electrical signal at a frequency high enough to travel easily through space to a distant point and to *superimpose* the *information* upon this signal. This high-frequency signal is known as the *carrier*, and the process of modifying the amplitude (or frequency) of the carrier to superimpose information is known as *modulation*. In an AM transmitter an RF carrier signal is generated and amplitude-modulated by voice signal; for example, this amplitude-modulated carrier is radiated into space via the antenna. The receiver picks up this signal and *demodulates* it by removing the carrier signal so that only the electrical equivalent of the original voice signal information is left.

We can trace the process in more detail from the block diagrams. In both the transmitter and receiver a *power supply* converts power from an electrical outlet (or battery) to dc voltages suitable for use by the various circuits.

AM TRANSMITTER OPERATION

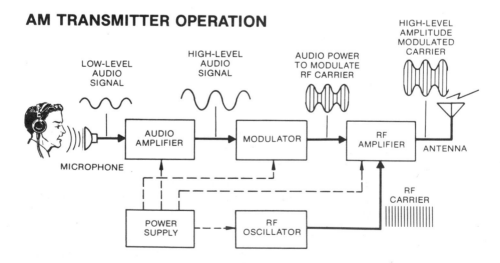

Summary of Radio Transmitter-Receiver Operation (continued)

In the transmitter a radio-frequency (RF) carrier wave is generated by the *RF oscillator*. A *microphone* is used to convert the pressure of the voice sound waves to a low-level audio electrical signal, and an *audio amplifier* steps up this signal to a high level. The audio voice signal is fed into the *modulator*, which produces a very high-level audio signal that is fed to the RF amplifier (modulates it) so that the resulting peaks and valleys of the carrier *duplicate* the variations in the voice signal. The RF oscillator produces a constant-level carrier that drives the *RF amplifier* where modulation takes place. This modulated RF carrier is then fed to the *antenna* where it is radiated over long distances.

At the receiver it is necessary to remove the carrier signal from the received modulated RF signal to recover the electrical equivalent of the voice signal. The problem is that the received signal is *very weak*. The weak signal picked up by the receiver *antenna* is usually stepped up by an *RF amplifier* (like the transmitter RF amplifier except the power levels are completely different) to raise it to a level suitable for further processing to begin. Because amplification at RF frequencies is relatively inefficient, the modulated carrier is converted to a lower *intermediate frequency* (IF) at which it is much easier to amplify. This is accomplished by the *mixer* that takes in both the modulated RF signal and a constant-level RF carrier at an offset frequency that is produced by the RF *local oscillator*. The amplitude variations in the mixer output signal duplicate those of the signal input carrier, but at a frequency—the *intermediate frequency*—equal to the difference between the RF signal and RF local oscillator frequencies. Further amplification is now done by the *IF amplifier*. The *detector* (or *demodulator*) removes the intermediate-frequency component from the output of the IF amplifier, and an audio electrical duplicate of the original voice signal remains. Now, an *audio amplifier* boosts this audio signal to a level sufficient for the *speaker* to *convert it back* to sound duplicating the original voice message.

AM RECEIVER OPERATION

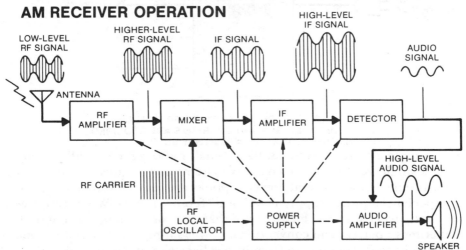

What Power Supplies Do

The power supply converts the available power source, usually 120 volts of ac at 60 Hz or a battery, to a form that is needed by the electronic circuits to perform the various tasks required for equipment operation. In general, all electronic circuits must be supplied with a proper dc voltage in order to perform their designated functions.

Electronic equipment that plugs into an electric outlet of 120 volts of ac requires a power supply to change available line voltage to the dc voltages required by the circuits inside. Although most power supplies convert ac to dc, there are power supplies that convert ac to ac, dc to dc, and dc to ac. You will study some of these later in this volume.

Why are there different types of power supplies? The answer is the power supply must meet the requirements of the functional circuits it is supplying. Some of the more obvious requirements may be the need for:

1. one or several different dc voltages;
2. dc voltages of both positive and negative polarity;
3. dc voltage with extremely low residual ac content;
4. very high voltages or currents.

Of equal importance to the above requirements is the current demand on the power supply. Every power supply is designed for some maximum current output. A portable transistor radio may require no more than a few milliamperes of current. Large transmitters may require tens of amperes. So power supplies must be designed to produce the current required by the functional circuits it supplies.

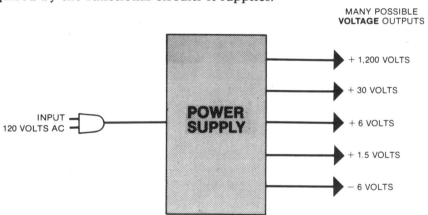

**POWER SUPPLIES ARE DESIGNED TO PRODUCE
WHATEVER VOLTAGES AND CURRENTS ARE REQUIRED
BY THE FUNCTIONAL CIRCUITS THEY SUPPLY**

Simple electronic systems, like a small radio receiver, will operate from a single voltage. More complex systems may require several voltages. In many cases, several power supplies are required to take care of the specialized power supply needs of different circuits. When batteries are used, every effort is made to see that only one voltage is required. Sometimes a battery and conventional power supply are used in the same equipment so it can operate from either batteries or the ac power line.

What Amplifiers Do

Of the three basic types of electronic circuits, amplifiers are by far the most widely used. In just about every type of electronic equipment, it is almost certain that several amplifiers will be included. What do amplifiers do? Amplifiers take a smaller signal and *increase* the signal amplitude to a higher level. In most cases, the signals of interest are not at the right level to be used, so they must be increased in level so that we can use them to drive other circuits or devices.

Amplifiers are used in a wide range of applications. The most common and important amplifiers are *audio*, *RF*, and *video* amplifiers. At this point, the discussion of amplifiers will be limited to those used for audio, RF, and video amplification. These amplifiers are characterized by the frequency range they are intended to amplify; however, you should understand that these classifications are very general.

AUDIO AMPLIFIERS 10 Hz to 20–30 kHz	RF AMPLIFIERS 100 kHz to 10,000 MHz	VIDEO AMPLIFIERS 30 Hz to 10 MHz
• RADIO • COMMUNICATIONS • HI-FI SYSTEMS • SONAR	• RADIO • TV • COMMUNICATIONS • RADAR	• TELEVISION • RADAR • DIGITAL SYSTEMS • COMMUNICATIONS

Audio Amplifiers: These are used to amplify a band of frequencies from about 10 Hz to about 20 to 30 kHz. This is the range of frequencies that the ear can hear in the form of sound waves—therefore the name *audio*.

Radio-Frequency Amplifiers: RF amplifiers usually amplify a *narrow* band of frequencies, but this narrow band may be anywhere within the wide range of frequencies from 100 kHz to several billion hertz. These frequencies are used in radio, communications and television receivers, and transmitters. RF amplifiers are used in receivers to amplify weak signals from the receiving antenna and also in transmitters to provide high power for driving the transmitting antenna.

Video Amplifiers: These are similar to audio amplifiers. The frequency band of operation, however, is very much expanded, typically covering frequencies from 30 Hz to 10 MHz or more. Video amplifiers are used primarily to amplify signals in television, radar, and digital systems.

What Oscillators Do

An oscillator is a variation of the amplifier circuit that can be powered by a dc voltage and produces an ac signal at a desired frequency. Oscillators are used to generate ac signals at almost any frequency. For example, they can generate audio-frequency and radio-frequency carrier signals as well as timing and television sweep signals. Basically, all oscillators consist of an amplifier and a special feedback circuit. Oscillators can be designed to produce sinusoidal, sawtooth, square waveshapes, or almost any waveshape desired.

The general requirements for an oscillator circuit are as follows:

1. It must have a frequency-determining element, such as a tuned circuit, which is resonant at the desired frequency of oscillation. The frequency-determining element can be an inductance-capacitance network (LC tank), resistance-capacitance (RC network), or a quartz crystal, depending on the frequency and waveshape desired.

2. It must contain an amplifier to compensate for power losses in the frequency-determining element.

3. It must have a positive feedback network to return the signal from the amplifier to the frequency-determining element in the phase required to sustain the oscillation.

OSCILLATOR OPERATING PRINCIPLES

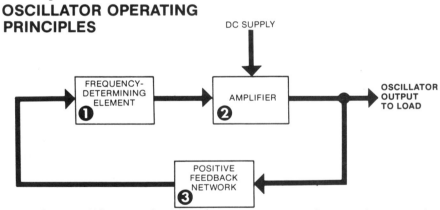

The amplifier can be just about any type as long as its operating frequency range covers the desired frequency of oscillation. The amplifier also must have enough *gain* (amplification) to compensate for losses in the feedback circuit. If the gain is too low, oscillation will not occur. On the other hand, if it is too high, spurious oscillations may occur that produce output signals at other frequencies, as well as the desired one.

The positive feedback network must be such that the phase of the oscillation is shifted so that it is returned to the frequency generator in phase to reinforce the oscillations. This means that a total of 360 degrees of shift is required. Since oscillators produce a steady train of signals, you can see that they can be used in timing circuits as well.

What Modulators, Mixers, and Detectors Do

The modulators and mixers to which you were introduced in the review of the transmitter-receiver learning system are only special variations of an amplifier circuit. The detector is a special variation of the rectifier circuit that you learned about in *Basic Electricity*.

A modulator is an amplifier circuit with two inputs. Its most common application is in a transmitter where one input is an audio signal that varies in amplitude and frequency in accordance with the information that is being transmitted. The other input is the constant-amplitude, radio-frequency carrier signal. The output is a signal at the frequency of the carrier, and the *amplitude* of this carrier rises and falls to *duplicate* the input information signal.

The mixer also is a special type of amplifier circuit with two inputs. It is commonly used in radio receivers to convert the received frequency to another more convenient frequency for amplification. Here the amplitude-modulated RF carrier is fed into one input, and a constant amplitude local oscillator L.O. carrier at an offset frequency is fed into the other input. The output from a mixer is the input signal shifted to a new frequency equal to the difference (or the sum of) between the two input frequencies (carrier + L.O. or carrier − L.O.).

An AM detector is a half-wave or full-wave rectifier with an RC (resistive-capacitive) filter connected across its output. Each peak of the carrier signal is rectified to a pulsating dc voltage that is applied to the capacitor and leaks away through the resistor. Leakage is slow enough to retain the charge between the peaks of the high-frequency RF signal but fast enough to follow the rises and falls in the low-frequency audio signal. Consequently the output is a *close duplicate* of the audio signal originally superimposed upon the carrier at the transmitter.

FUNCTION OF MODULATORS, MIXERS, AND DETECTORS

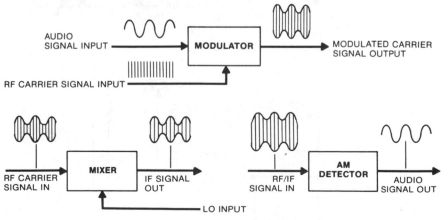

What Controllers, Timers, and Digital/Logic Circuits Do

Controllers, timers, and digital/logic circuits use a common element—the switching circuit. They are reviewed together here because of their close relationship. We will defer detailed discussion of these devices until later in these volumes.

As the basis for identification, timers are defined here as switching circuits whose primary function is to introduce a *time delay* or to *control* the *timing* of events. Controllers are similarly defined as switching circuits whose primary function is to control the *magnitude* or *speed* at which some function is being performed. Digital/logic circuits are essentially switching circuits that are confined to two states—on and off—which are used to *convey information* in computers, microprocessors, etc.

Timers are frequently used to apply or remove power from a load at a preselected time after a *start* signal is applied. Such an activating signal could be the throwing of a switch, the activation of a relay, or the application of a rapid rise and fall of voltage.

An example of a common type of controller is the speed control that you learned about in *Basic Electricity*. It operates by turning power on and off at a rate of 120 times per second, as shown in the diagram. By operating at this rate and controlling the duration of the *on* period that ac power is applied to the load, you have an efficient method of controlling the average power applied to such loads as motors, heaters, lamps, and power supplies.

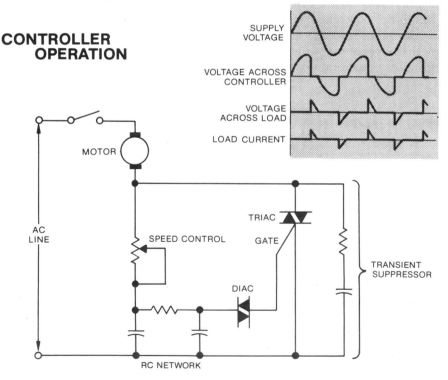

CONTROLLER OPERATION

SUPPLY VOLTAGE

VOLTAGE ACROSS CONTROLLER

VOLTAGE ACROSS LOAD

LOAD CURRENT

MOTOR

AC LINE

SPEED CONTROL

TRIAC

GATE

DIAC

TRANSIENT SUPPRESSOR

RC NETWORK

Discrete Electronic Components

As you know, all of the common electronic building blocks that have been reviewed can be made from *discrete components* or *integrated circuits*. Basic circuit components such as resistors, diodes, capacitors, and transistors are *individual* units that are interconnected to form complete electronic circuits and systems. These components are *separately* (*discretely*) packaged devices, *unlike integrated circuits* (*ICs*),in which *many* components are *manufactured together* in a *single compact package* that contains all or most of the vital components. It must be understood that ICs are *not* just discrete components packaged together but are designed quite differently even though their *external function* may be the *same* as for a circuit made up of discrete components. These differences will be discussed in detail as you proceed with your study of electronics. In comparison to ICs, discrete component circuit assemblies are much larger in size and are more costly and time consuming to manufacture. In addition, the characteristics of two identical discrete components are not *matched as well* as two identical components in a *single IC.*

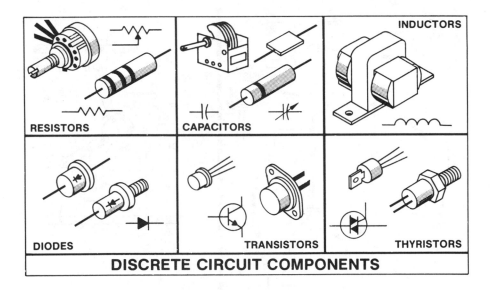

DISCRETE CIRCUIT COMPONENTS

Because of the *superior characteristics* and low cost of ICs, they have replaced, to a great extent, discrete component circuits. It is, however, necessary and important for you to know *how discrete component circuits work* because they are still widely used today, and you must become familiar with them in order to better understand how the electronic building blocks operate and to better understand how ICs are used. The diagram shows some common discrete components and their schematic symbols. The pages that follow review them in greater detail. Resistors, capacitors, transformers, and inductors are described in considerable detail in *Basic Electricity.* So only essential points concerning them will be reviewed.

Discrete Electronic Circuits

Discrete electronic circuits use the discrete, individual components described on the previous page. Usually, these circuits are not hand wired but are assembled on a printed circuit board. You should refer to *Basic Electricity* for the characteristics of resistors, capacitors, and inductors since these components are used and have the same properties as for electricity. The *major* difference in electronic circuits is that they are associated with *active discrete* semiconductor devices. Most solid-state electronic circuits use small components in an attempt to keep the equipment compact. Thus resistors are generally ⅛ and ¼ watt typically; capacitors are low-value ceramic types; and for larger capacitor values, electrolytic (aluminum or tantalum) are commonly used. The components are often specially designed to be mounted on a *printed-circuit (PC) board* and held in place by soldering to the conductors on the PC board.

DISCRETE COMPONENT CIRCUIT

PRINTED CIRCUIT (PC) EQUIVALENT

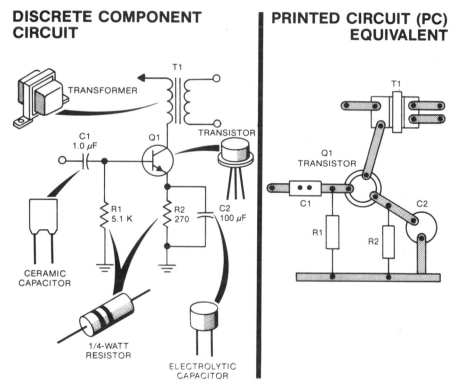

As shown in the figure, the individual components must be assembled and wired together—an often costly approach. The use of PC boards reduces this cost considerably because the conductors are part of the board; therefore, only soldering of components need be done. In the drawing, the conductors are shown as crosshatched areas. The wider areas where the component lead goes through the board via a drilled hole is called a *land*. Except for experimental equipment, essentially all solid-state circuits, discrete or IC, are mounted on PC boards.

Diodes/Transistors/Thyristors—In Discrete Circuits

Diodes are solid-state rectifier units that come in a wide variety of types and packages. As you know from *Basic Electricity*, diodes conduct well in one direction and poorly in the other. Shown on the next page are outline drawings of some of the available types. Details of their operation will be presented later in these volumes in connection with their circuit applications. Packaging can include ceramic jackets and plastic and metal cases. Primary diode characteristics include forward voltage drop, reverse breakdown voltage, and maximum current. These characteristics will be considered in the section on the principles and operation of semiconductor devices.

Transistors are solid-state components fundamentally capable of amplification. Transistors are interconnected with resistors, capacitors, and other discrete circuit components to form most of the fundamental electronic building blocks. Significant characteristics of transistors are gain, voltage rating, and power rating. Discrete transistors come in a variety of types and packages, some of which are shown on the next page. Note that there is no specific relationship between type and package appearance—that is indicated by markings. Details of transistor operation will be introduced in this and the next volume.

As you learned in *Basic Electricity*, thyristors are a family of semiconductor switches that can be made to direct an electric current to flow in either one of two directions. Basic thyristors include the silicon-controlled rectifier (SCR), triac, and diac, whose symbols and general appearance are shown in the diagram on the next page. Significant characteristics of these devices include maximum current capacity and turn-on voltage. The operation of these devices will be introduced later in this volume.

Most discrete transistors are *bipolar* as are IC analog devices like amplifiers. Simple digital ICs also are often bipolar in design, while many modern digital ICs use *MOS* (*metal-oxide semiconductor*) *technology*. Almost all MSI, LSI, and VLSI devices use MOS technology because it allows very high packaging density. You will learn about ICs later in this volume.

Standard nomenclature has been agreed to such that the two terminal semiconductor devices (diodes) are designated by an *1N* prefix followed by a two- to four-digit number; e.g., 1N914, 1N4568. Three terminal devices such as transistors have a *2N* prefix; e.g., 2N3055, 2N4560, 2N2222, etc. There are no standard ways to designate ICs except by type.

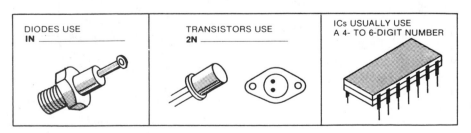

DIODES USE
IN _____

TRANSISTORS USE
2N _____

ICs USUALLY USE
A 4- TO 6-DIGIT NUMBER

TYPICAL DISCRETE DIODES

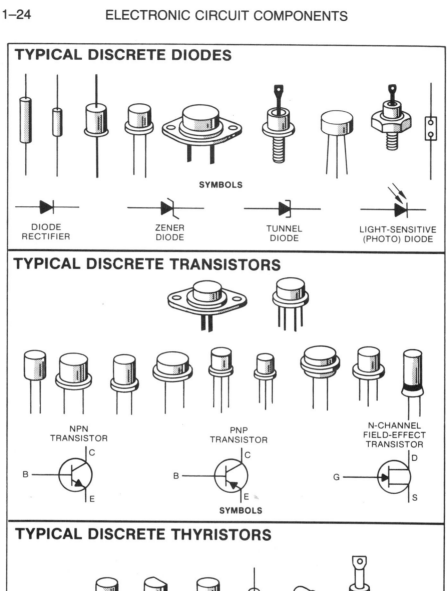

SYMBOLS

DIODE
RECTIFIER

ZENER
DIODE

TUNNEL
DIODE

LIGHT-SENSITIVE
(PHOTO) DIODE

TYPICAL DISCRETE TRANSISTORS

NPN
TRANSISTOR

PNP
TRANSISTOR

N-CHANNEL
FIELD-EFFECT
TRANSISTOR

C

B

E

C

B

E

D

G

S

SYMBOLS

TYPICAL DISCRETE THYRISTORS

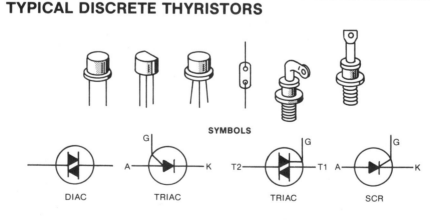

SYMBOLS

DIAC

G

A   K

TRIAC

G

T2   T1

TRIAC

G

A   K

SCR

Discrete and Integrated Circuit Components

Integrated circuit designers use all their design ingenuity to avoid using anything but semiconductor elements because these can easily be made simultaneously in one manufacturing step for a great number of units. These circuits are made by deposition of appropriate layers of material, in sequence, so that the circuit and the semiconductors are integral parts. Discrete capacitors and inductors require special additional processing or adding on as a separate, often expensive, operation. Resistors can be fabricated as a thin film of resistance material or as biased semiconductors, since semiconductor devices can function as resistors *when properly biased.* They also can be used as impedance transformers and as voltage step-down devices. Therefore, the circuit in discrete form shown earlier may look, in IC form, as shown in the figure below. While the two circuits may function the same, the IC circuit is all semiconductors and resistors and hence can be laid down in very compact form in a single operation.

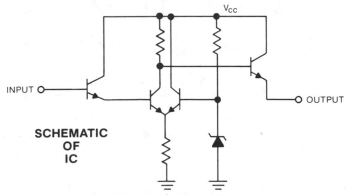

In some cases, an analog IC device will require one or two discrete external components such as capacitors or resistors, and in this case terminals are available on the IC device to which these components can be connected.

Digital ICs usually are mainly semiconductors with a few resistors for required biasing. A schematic for an IC amplifier is shown on the following page. All of the components are deposited on a single chip with lead wires brought out to the case to make connections to the IC. The amplifier shown is an example of MSI (medium-scale integration) technology since it is not very densely packed. LSI (large-scale integration) is quite different than MSI because the packaging densities are much greater. In addition, they use different semiconductor technology, which you will learn about a little later. Devices like the microprocessor shown are too complicated to show in schematic form, since there are several thousand transistors involved. Yet the entire circuit may be on a single chip about 1½ × 3 cm! The chip shown is a complete microprocessor having all of the elements necessary to form a microcomputer, except the memory and input/output circuits.

MSI IC AMPLIFIER

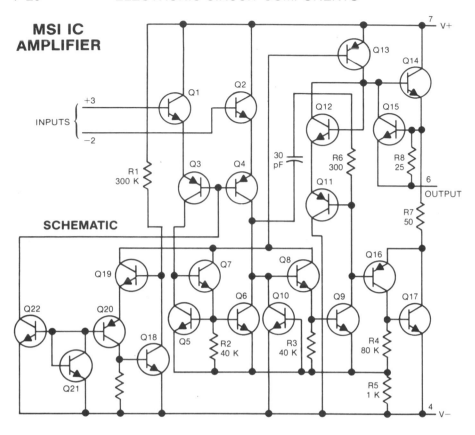

SCHEMATIC

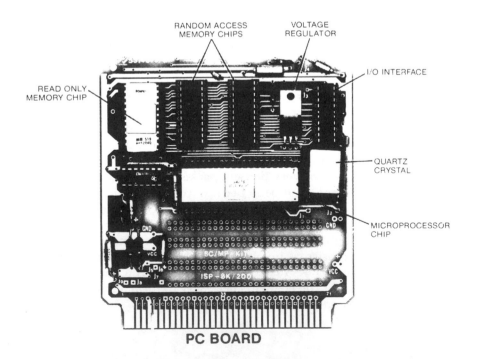

PC BOARD

Integrated Circuits—MSI, LSI, VLSI, VHSIC

As you know, an integrated circuit is a complete assembly of electronic components designed to perform the function of one or more electronic circuits. Integrated circuits are made by using the techniques of microphotography to form masks through which layers of semiconductor materials are deposited. By means of its vacuum deposition and selective chemical etching, transistor elements and metallic conductors are formed. The chips are designated as MSI (medium-scale integration), LSI (large-scale integration), or VLSI (very large-scale integration), depending on how much circuitry is on a single chip. More recently, VHSIC (very high-speed integrated circuits) are being developed to provide for wide-band and high-speed operation of ICs.

The completed IC is a package that can take many forms. Typical examples of packaging are shown in the diagram below. Thousands of different ICs are available for general and special analog and digital devices.

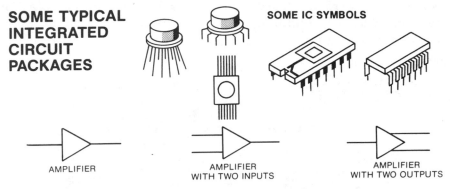

SOME TYPICAL INTEGRATED CIRCUIT PACKAGES

SOME IC SYMBOLS

AMPLIFIER

AMPLIFIER WITH TWO INPUTS

AMPLIFIER WITH TWO OUTPUTS

All of the nine common electronic circuit building blocks can be made from ICs. The manner in which this is accomplished is described in these volumes. *In general, each electronic circuit will be described (1) first in terms of its construction from discrete solid-state elements* and *(2) then in terms of its integrated circuit version, when a useful IC is available.* Integrated circuits using large-scale integration (LSI) will be considered in the volume on digital/logic circuits at the end of this series.

Most analog ICs contain a single function or group of closely related functions; for example, amplifier, FM stereo decoder, voltage-controlled oscillator, etc. Digital ICs can be similar in containing a single or small group of related functions. But now techniques such as MSI, LSI, and VLSI are used to have a single IC chip that performs an *entire system* function. For example, a calculator, microwave oven controller, small computer, or automobile engine controller. As stated earlier, because of high labor costs and the low price of ICs, discrete circuits are rapidly disappearing from most equipment except for a few special applications such as for RF amplification. Therefore, although you will use discrete circuits to learn how electronic systems work, you also will learn about ICs since you will probably have more contact with IC circuit devices in your work.

Vacuum Tubes

Vacuum tubes (or valves) comprise a large family of electronic circuit components that have been replaced by solid-state components in nearly all new electronic equipment. Vacuum tubes operate on the principle that electron flow can be initiated and controlled in a container usually made of glass or metal from which the air has been pumped. In this vacuum, free electrons are made available by using an electrically heated filament, like that in a small light bulb, to heat a surrounding cathode coated with materials that have loosely bonded electrons.

As this *cathode* is heated, electrons are freed and collect in a cloud around the cathode. If a metal plate is placed in the vacuum and charged positively with respect to the cathode, the electrons will flow to the *plate*. Now, a meshlike metal *grid* can be placed between the cathode and the plate, and it will be found that a very small change in grid-to-cathode voltage can cause a very large change in cathode-to-plate current. This is the basis of vacuum-tube amplification, and thousands of types of vacuum tubes were developed for general and special purposes. All of the basic electronic building blocks have been developed by using vacuum tubes.

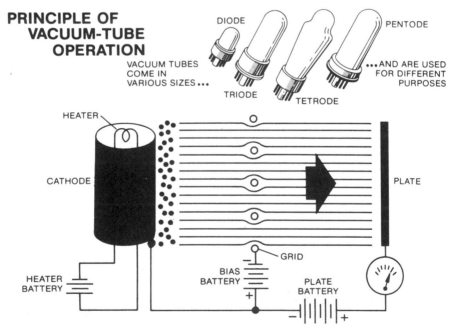

PRINCIPLE OF VACUUM-TUBE OPERATION

DIODE   PENTODE

VACUUM TUBES COME IN VARIOUS SIZES ...

...AND ARE USED FOR DIFFERENT PURPOSES

TRIODE   TETRODE

HEATER

CATHODE   PLATE

GRID

HEATER BATTERY   BIAS BATTERY   PLATE BATTERY

Although vacuum tubes are not usually used in new electronic equipment except for special purposes, they are still found in equipment manufactured before and during the perfection of solid-state devices. Electronic system operation based on vacuum tubes is described in the separate, self-contained series entitled *Basic Electronics*. The descriptions of circuits and equipment included here is restricted to those based on the use of solid-state devices. However, it should be noted that the basic operating principles of these circuits are *essentially unchanged*.

Review of Basic Concepts

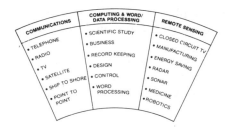

1. ELECTRONIC SYSTEMS are concerned primarily with the control of electron flow for generating or sensing, storing, processing, controlling, transmitting, receiving, and displaying *information*.

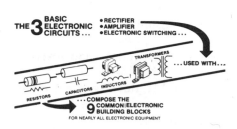

2. ELECTRONIC SYSTEMS are made up from three basic circuits— *rectifier*, *amplifier*, and *electronic switching*—which make up nine common building blocks. These are *power supply, amplifier, oscillator, modulator, mixer, detector, controller, timer*, and *digital/logic circuits*.

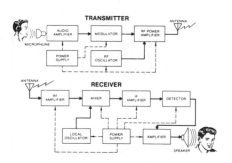

3. A RADIO TRANSMITTER/RECEIVER LEARNING SYSTEM will be used as the basic learning system to give you a view of how electronic systems are made up from the basic building blocks.

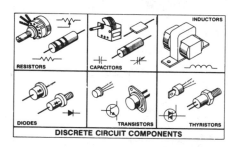

4. DISCRETE COMPONENT ELECTRONIC DEVICES are used in combinations to perform a given function. These consist of semiconductor devices, resistors, capacitors, and inductors usually assembled on a printed circuit board.

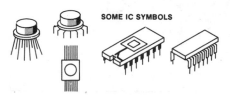

5. INTEGRATED CIRCUIT (IC) ELECTRONIC DEVICES are more common today and include entire functions or systems on a single chip. Clearly, space, power, and cost are greatly decreased.

Self-Test—Review Questions

1. What are the major changes that have taken place recently in electronics that have totally changed the use of electronic equipment in our daily lives?
2. List some of the current major uses of electronic systems.
3. Briefly list the major points in the development of modern electronics.
4. What are the fundamental differences between discrete and integrated electronic circuit technology?
5. List the three basic circuits of electronics.
6. List the nine building blocks that are derived from the three basic circuits.
7. Draw a simple block diagram of the radio transmitter-receiver system and identify the function of each block.
8. Why are integrated circuits more commonly used today than discrete circuits?
9. Draw outlines and symbols of some discrete and IC devices. What is the difference between IC, MSI, LSI, and VLSI?
10. How are diodes and transistors designated (usually)?

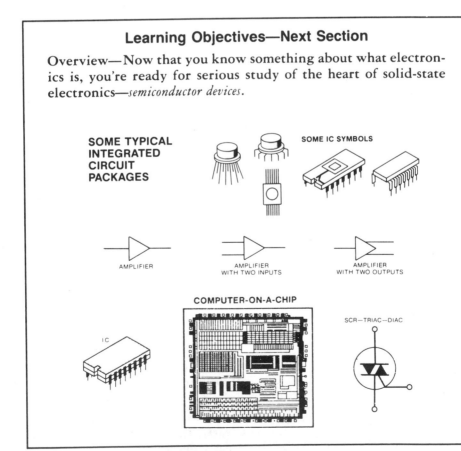

Learning Objectives—Next Section

Overview—Now that you know something about what electronics is, you're ready for serious study of the heart of solid-state electronics—*semiconductor devices*.

SOME TYPICAL
INTEGRATED
CIRCUIT
PACKAGES

SOME IC SYMBOLS

AMPLIFIER

AMPLIFIER
WITH TWO INPUTS

AMPLIFIER
WITH TWO OUTPUTS

COMPUTER-ON-A-CHIP

IC

SCR—TRIAC—DIAC

Semiconductor Materials

All semiconductor devices like diodes, transistors, and ICs are made of semiconductor materials. As you know from *Basic Electricity*, all materials may be roughly classified as either *conductors*, *semiconductors*, or *insulators*. This classification depends on the ability to conduct an electric current, which in turn depends on the number of free or loosely held electrons in the material. Good conductors such as silver, copper, and aluminum have large numbers of free electrons. Insulators such as mica, glass, paper, rubber, and plastics have few free electrons. Semiconductor materials have characteristics of *both* conductors and insulators and lie somewhere between the extremes of insulators and conductors—not conducting as well as conductors—but better than insulators.

The semiconductor materials most used in solid-state devices are silicon and to a much lesser extent, germanium. In their pure crystalline form, germanium and silicon are poor conductors because their structure has few free electrons. The outer electrons of individual atoms are shared by adjacent atoms to form a symmetrical arrangement, as shown in the diagram.

To provide free electrons, the pure crystal is modified by adding controlled amounts of impurities (for example, arsenic, antimony, aluminum) in a process called *doping*. These materials are added in extremely small but controlled amounts in the order of one part in ten million. (A larger amount would make conductivity too high.) These atoms enter into the structure, replacing a few of the silicon (or germanium) atoms, but not altering the basic crystal structure. Such a small change seems of no consequence, yet in terms of conductivity, the results are *dramatic*. For example, conductivity of silicon increases by a factor of *30,000* times when processed in this way!

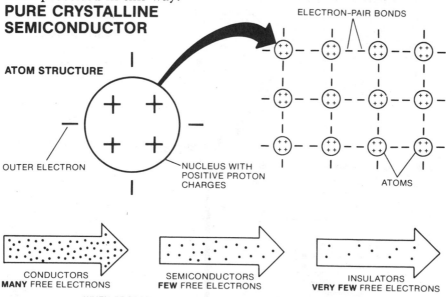

PURE CRYSTALLINE SEMICONDUCTOR

ELECTRON-PAIR BONDS

ATOM STRUCTURE

OUTER ELECTRON

NUCLEUS WITH POSITIVE PROTON CHARGES

ATOMS

CONDUCTORS
MANY FREE ELECTRONS

SEMICONDUCTORS
FEW FREE ELECTRONS

INSULATORS
VERY FEW FREE ELECTRONS

WHEN PROPERLY MADE AND USED, SEMICONDUCTORS ARE LIKE CONDUCTORS IN ONE DIRECTION

N-Type Semiconductor Material

If the impurity atoms added to the crystalline structure have *one more* outer electron than the pure crystal atom, this extra electron is held *very loosely* because it does not fit in the crystalline structure and becomes a loosely bound electron. The resulting crystal structure is shown in the illustration. It is the presence of these loosely bound electrons that results in the great increase in the conductivity of the semiconductor material when the doping material is added.

If a dc voltage is connected across the ends of a piece of such material, electrons flow through the crystal structure toward the *positive* terminal. The total number of free electrons in the crystal always remains the same; each electron that leaves the crystal at the positive terminal is replaced by one that enters at the negative terminal. As a result, there is a *continuous current* flow. Note that there is *no specific polarity* to the material. If the battery terminals are reversed, the electrons will flow in the opposite direction. All that has happened at this point is that the presence of the doping atoms has made the semiconductor more like a conductor by allowing electrons to move through the material.

Since the current in this material consists of excess *negative* charges (electrons), the material is known as *n-type* semiconductor material. Impurities added to silicon to form n-type semiconductor material include arsenic and antimony.

N-TYPE SEMICONDUCTOR MATERIAL

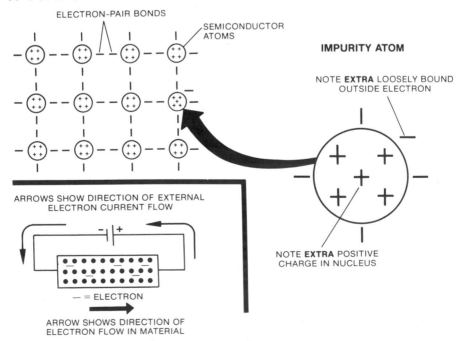

ELECTRON-PAIR BONDS

SEMICONDUCTOR ATOMS

IMPURITY ATOM

NOTE **EXTRA** LOOSELY BOUND OUTSIDE ELECTRON

ARROWS SHOW DIRECTION OF EXTERNAL ELECTRON CURRENT FLOW

NOTE **EXTRA** POSITIVE CHARGE IN NUCLEUS

− = ELECTRON

ARROW SHOWS DIRECTION OF ELECTRON FLOW IN MATERIAL

P-Type Semiconductor Material

Another way to modify the pure basic crystalline material is to add impurity atoms having *one less* outer electron than the basic material. These impurity atoms enter into the crystalline structure, as shown in the diagram. Thus, the modified structure has a *missing* electron for each impurity atom. The space in the structure caused by the missing electron is known as a *hole*.

Electrons from adjacent semiconductor atoms can move through the crystal to fill the hole. The presence of holes, therefore, encourages the flow of electrons in the semiconductor material.

If a dc voltage is connected across the ends of a piece of material doped in this way, the holes appear to flow toward the *negative* terminal of the voltage source. The total number of holes in the crystal does not change, since each time a hole reaches the end of the crystal connected to the negative terminal, an electron leaves the negative terminal and enters the crystal and fills the hole. An electron is then discharged to the positive voltage terminal to create another hole. The new hole moves toward the negative terminal, and the result is the continuous flow of holes through the crystal and a continuous flow of electrons through the external circuit.

Since the current flow is caused by the flow of holes in the crystal structure and these holes are *positive* charges, the material is known as *p-type* semiconductor material. Impurities that are added to form p-type semiconductor material include aluminum, gallium, and indium.

P-TYPE SEMICONDUCTOR MATERIAL

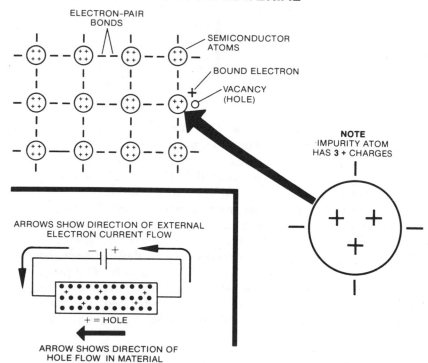

The PN Junction

As you learned earlier, current will flow through both p-type and n-type semiconductor materials, although n-type conducts via *electron* flow and p-type conducts via *hole* flow. Suppose you joined these materials together by alloying the junction (melting the materials at the junction together) or by some similar means so that the two materials formed a continuous circuit. When this is done, a junction forms between the p and n material called a *pn junction*, and an interaction takes place at the junction that is fundamental to transistor operation. When the junction is formed, some of the free electrons from the n-type material move across the junction and recombine with holes in the p-type material; similarly, some holes in the p-type material move across the junction and combine with free electrons in the n-type material. Thus, when the junction is

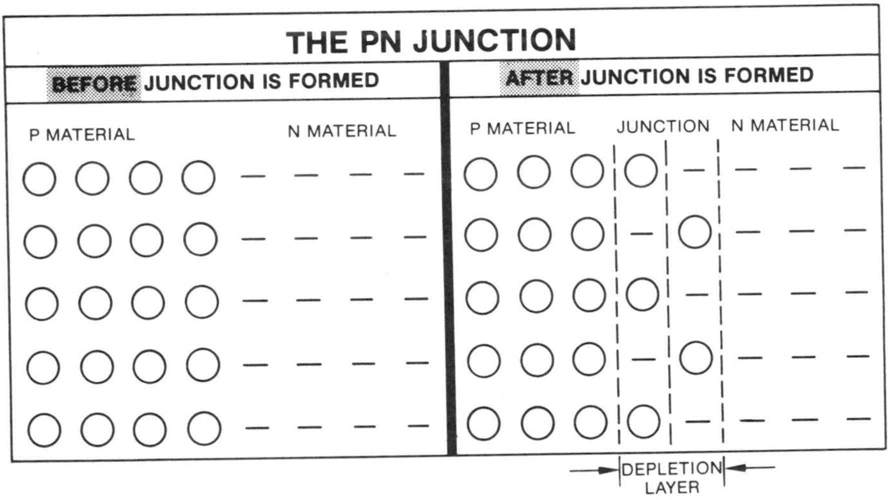

THE PN JUNCTION

BEFORE JUNCTION IS FORMED	AFTER JUNCTION IS FORMED

formed, an initial current flows. The initial flow of electrons from n material to the p material and holes from the p material to the n material at the junction places a slight negative charge on the p-type and a slight positive charge on the n-type material in the area of the junction. The initial current flow is called the *diffusion current* and it occurs during the time that the junction is being formed. After a very brief interval, the current flow ceases leaving the junction with a defined distribution of charges. Diffusion of free electrons and holes does not affect the entire semiconductor block but is confined to a well-defined area around the junction called the *space-charge region* or *depletion layer*. The processes are shown in the figure above. The left side of the figure shows the charge distribution immediately *before* the junction is formed. The right side of the figure shows the charge distribution immediately *after* the junction is formed.

Reverse-Biased PN Junction

When a battery is connected across a pn junction, the amount of current that flows depends on which way the battery is connected (polarity) and the amount of battery voltage.

When the battery *positive* terminal is connected to the *n-type* material end (cathode), and the negative terminal is connected to the p-type material end, as shown in the diagram, the pn junction is *reverse-biased*. Electrons in the n-type material are attracted toward the positive terminal of the battery away from the junction. Holes from the p-type material are attracted toward the *negative* terminal of the battery and away from the junction. Thus, the depletion region at the junction becomes wider. This widening of the depletion region at the junction creates a region with few electrons or holes. This widening holds the current flow to very low values because electrons and holes are pulled in *opposite* directions, as shown in the diagram. Thus, if the pn junction is reverse-biased, current flow is *extremely small* because the resistance across the junction is very high.

If the reverse voltage is made too high, a condition is created where the junction resistance drops very rapidly, and a *very large* current increase takes place with no further increase in voltage. This happens because the voltage is high enough to cause the junction to break down. The voltage at which this effect takes place is called the *reverse breakdown voltage* or *peak inverse voltage.*

A little later in this volume, you will learn about devices that operate in the reverse breakdown region as a proper operating condition. Normally, however, it is important to keep this from happening because the semiconductor junction can be destroyed by attempts to operate in this manner.

THE REVERSE-BIASED PN JUNCTION

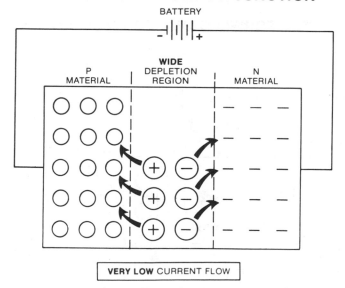

Forward-Biased PN Junction

If you take the pn junction and *reverse* the battery connections so that the positive terminal of the battery is connected to the p material and the negative terminal is connected to the n material, you now have a *forward-biased* junction. When the pn junction is forward-biased, outer electrons in the p-type material near the positive battery terminal enter the battery, creating new holes. Simultaneously, electrons from the negative battery terminal enter the n-type material and diffuse toward the junction. Thus, the depletion region becomes effectively much narrower. The excess electrons from the n-type material are no longer confined to the n-type material by the wide depletion layer barrier. Electrons can now readily flow across the junction and move toward the positive terminal of the battery by way of the holes in the p-type material. Thus, current flows easily across the junction.

As higher forward-bias voltages are applied, current flow *increases*, being limited essentially by the resistance of the material. Excessive forward-bias currents should be avoided without current-limiting resistance because they can cause permanent damage to the junction due to overheating of the junction. In a forward-biased pn junction, little current will flow until the depletion layer is very narrow. Since the depletion layer has a potential across it, as discussed earlier, it is necessary to overcome this bias potential (narrow the depletion layer) before significant amounts of current will flow. It should be noted that all semiconductor devices operate on the principles described here. Discrete devices like diodes, transistors, etc., as well as ICs use pn junctions in various configurations to do their job.

THE FORWARD-BIASED PN JUNCTION

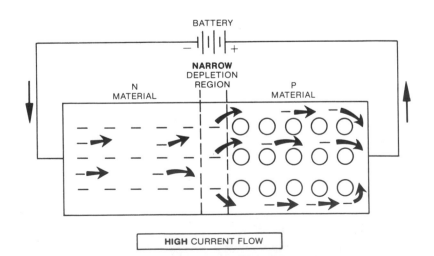

The Semiconductor Diode

One of the first uses of semiconductor materials was to join a p-type material with an n-type to form a simple but important semiconductor device called a *diode*. The *n-type* material of the diode is the negative electrode and is called the *cathode*. The *p-type* material is the positive electrode and is called the *anode*. The arrow symbol used for the anode (see next page) represents the *direction* of *conventional current flow*. The *electron current* flows in a direction *opposite* to the arrow.

> *Note:* In these volumes we will use *conventional current flow* because you will find that this is most generally used in electronics. Actually, it doesn't matter which direction you use, *provided you* are *consistent.*

As shown in the illustration for pn junction or diode, a high current can flow (low-voltage drop) when the diode junction is forward-biased and essentially no current flows when the diode junction is reverse-biased—until the reverse breakdown voltage is reached, at which point the current flow becomes very high and the diode will be destroyed.

DIODE CHARACTERISTICS

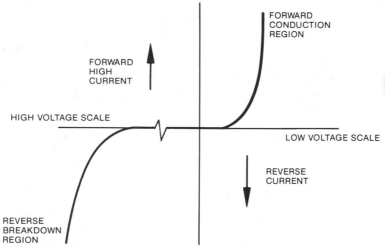

The diode junctions can be viewed as an automatic switch that is closed when the p-material is relatively positive and open when the p-material is relatively negative. Remember, it is the *voltage across the diode* that is important—*not* the voltage to another part of the circuit. Most diodes in use today use silicon semiconductor material; however, for some applications, germanium also is used. Because the depletion region never is completely absent, it takes a small voltage across a diode to make current flow, even when the polarity is correct for high current flow. This amounts to about 0.5 volt for silicon and 0.25 volt for germanium type diodes.

The Semiconductor Diode (continued)

The symbol for the semiconductor diode is shown below. The arrow points in the direction of *conventional current flow*. Diodes have *polarity* and therefore must be placed in circuits in the *correct direction*. Small diodes are marked with a band to indicate the cathode end. Larger diodes use the diode symbol to indicate their polarity. As shown earlier, diodes come in many package configurations, depending on their ratings. Small diodes that are used for low-current application in circuits up to 1 ampere of average current are generally packaged in a simple glass or plastic tube, as shown in the drawing. Diodes to handle larger currents are packaged in metal containers and are often mounted by studs on metal plates to conduct heat away.

DIODE MARKINGS

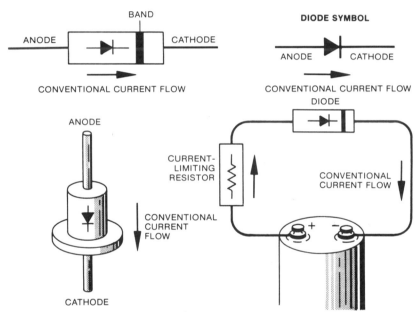

Diodes are rated in terms of their peak and average forward current-carrying capability and in terms of their reverse voltage or peak inverse voltage ratings. None of these values can be exceeded; otherwise the diode will be destroyed. For the pn diode junction, the reverse voltage rating applies to the instantaneous peak reverse voltage across the diode.

Some special diodes are of the point-contact type. These were the earliest diodes and involved the formation of a junction around the contact point by suitable processing after the diode is assembled. These diodes are usually used for special application in high-frequency circuits.

The Semiconductor Diode (continued)

Some diodes are designed to operate in the reverse breakdown region. These diodes are called *zener diodes*. When zener breakdown occurs, the strong potential across the depletion region causes direct rupture of electron bonds, thereby freeing large numbers of electrons so a current can flow. The potential difference across the depletion region is constant and depends on how the diode is fabricated. It can be between a few volts and greater than 10 volts. Zener diodes are connected into circuits with their *cathode* as the *positive* terminal because they operate in the reverse-bias mode. Zener diodes are packaged in the same way as conventional semiconductor diodes. Small zener diodes are marked the same as conventional diodes with a cathode band. The large diodes are marked with the zener diode symbol shown in the illustration below. In addition to having a voltage rating (reverse breakdown), zener diodes have a power rating that must not be exceeded.

ZENER DIODE

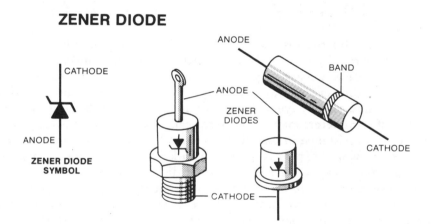

Zener diodes are widely used as voltage stabilizers in circuits. This is because the voltage drop across the zener diode is relatively constant. Zener diodes behave like regular diodes when connected in the normal way; however, they are not usually used this way. When connected in reverse, and sufficient voltage is applied (with a suitable current-limiting resistor), the diode breaks down in the zener mode and holds the voltage across it constant. You will learn more about the zener diode later when you learn how they are used in electronic circuits.

Experiment/Application—Observing and Testing Diode Operation

You can check the properties of a semiconductor diode very easily with only an ohmmeter and a diode. As you remember, the ohmmeter works by applying a potential across the resistance to be measured. This potential is sufficient to forward-bias or reverse-bias a semiconductor diode. Suppose you had a diode and an ohmmeter.

OBSERVING/TESTING DIODE OPERATION

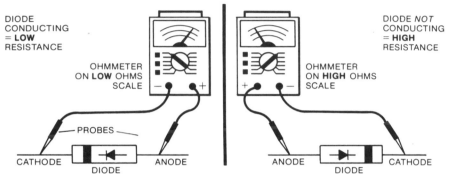

If you did the test shown above, you would find that the diode conducted readily (low resistance) with the leads connected one way; and the diode was essentially an open circuit (very high resistance) when connected the reverse way. You might have to experiment to find out which connections gave you high or low resistance because different makes of ohmmeters may not have the same polarity. However, you will get a high resistance in one direction and low resistance in the other—just as you would expect from what you know about diodes. Obviously, this is a good way to *test* semiconductor diodes, too. A low resistance or high resistance in *both* directions means that the diode is defective.

You can carry out other simple experiments to show diode operation. Suppose you hooked up the circuit shown below. With the circuit connected, the diode will conduct, as shown by the meter reading about 1.0 volt (about 0.5-volt drop in the diode). If the battery polarity is reversed (or the diode is reversed), the meter will read zero, showing that the diode is not conducting.

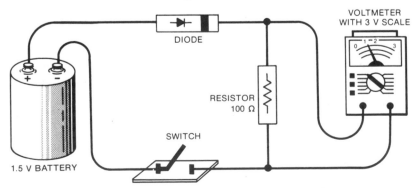

The Bipolar Transistor

A third semiconductor region (p-type or n-type) can be added to the diode semiconductor pn junction to form three regions (two pn junctions). The resulting device, called a *bipolar transistor*, provides *amplification*. The three regions of the device shown are called the *emitter*, the *base*,and the *collector*. In this two-junction construction, the arrangement can be either npn or pnp, as shown in the diagram. These devices are called *bipolar transistors* because, as you shall see, there are *two components* of current flowing in the emitter circuit that are in *opposite* directions.

For normal operation, one pn junction (emitter-base) is forward-biased and behaves very similarly to a forward-biased diode. The second pn junction (base-collector) is reverse-biased and behaves very similary to a reverse-biased diode. Internally, the current through one junction is almost equal to the current in the other junction. The reason for this effect in the second pn junction is that the transistor's pn junctions are constructed differently than for a diode.

In the diode, the p- and n-regions are similar in size and numbers of holes or electrons. In the reverse-bias condition, the depletion layer widens and the diode stops conducting. In the forward-bias condition, conduction continues because there is a constant flow of electrons from the n-type material. As each electron enters the p-type material, there is an equivalent number of holes to accept the electrons. These electrons then move through the p-type material by way of the holes.

In the transistor, on the other hand, the construction of the pnp or npn regions consists of two thick end regions (emitter and collector) with a large number of holes or electrons and a very thin center region (base) with few surplus charges. Thus, there is an imbalance of surplus charges between the center portion and the two outer portions. This creates the condition necessary for the transistor to function as a transistor and not like two diodes.

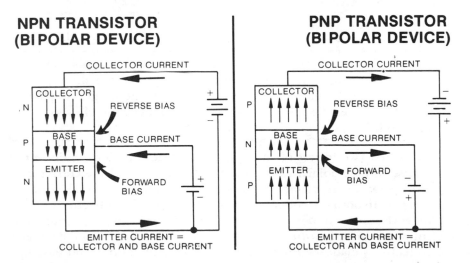

NPN TRANSISTOR (BIPOLAR DEVICE)

COLLECTOR CURRENT
COLLECTOR
N
REVERSE BIAS
+
BASE
P
BASE CURRENT
EMITTER
N
FORWARD BIAS
+
EMITTER CURRENT = COLLECTOR AND BASE CURRENT

PNP TRANSISTOR (BIPOLAR DEVICE)

COLLECTOR CURRENT
COLLECTOR
P
REVERSE BIAS
−
BASE
N
BASE CURRENT
EMITTER
P
FORWARD BIAS
−
+
EMITTER CURRENT = COLLECTOR AND BASE CURRENT

The Bipolar Transistor (continued)

In operation, the transistor's forward-biased pn junction (emitter-base) conducts constantly like a forward-biased diode. However, because the base region is so narrow and lacking in surplus charges, it can absorb only a few of the oppositely charged carriers that enter from the emitter. These charges therefore can flow across the base junction in large numbers to the collector junction. On the other hand, the current flow through the base-emitter junction is restricted because of the relatively few carriers available.

Because of all these excess charges, the transistor reverse-biased pn junction (base-collector) cannot act like a reverse-biased diode with a wide depletion region to inhibit current flow. Instead, these excess charges continue across to the collector.

TRANSISTOR SYMBOLS

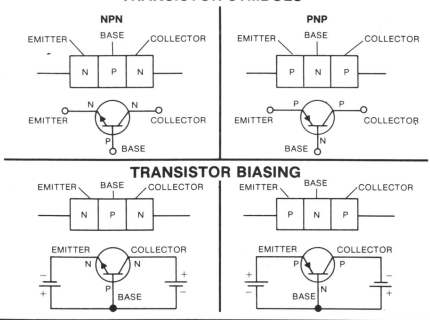

Note: In the transistor symbol it is important to remember that the direction of *electron flow* is always *opposite* to the direction indicated by the emitter arrow, as shown in the diagram. The arrow indicates the direction of conventional current flow as in the diode.

The first two letters of the npn and pnp designations indicate the respective voltage polarities applied to the emitter and the collector in normal operation. In an npn transistor, the emitter is biased *negatively* with respect to both the collector and the base, and the collector is biased *positive* with respect to both the emitter and the base. In a pnp transistor, the emitter is biased *positive* with respect to both collector and base, and the collector is made *negative* with respect to both emitter and base.

Transistor Current Flow

Remember that the *forward-biased* junction (p-type material positive) provides high current flow and is therefore a *low-resistance* circuit. The *reverse-biased* junction (n-type material positive) provides low current and is therefore a *high-resistance* circuit.

The npn and pnp transistor circuits illustrated here show current values that illustrate the relative magnitudes of the currents involved. For the npn transistor, current flows easily from collector to emitter even though this junction is reverse-biased. Most of this current moves directly through the thin base region toward the negative battery terminal. A small amount of current does not make it across the junction to the collector but instead recombines with a hole in the base region, and an equivalent current flows into the transistor base. These few electrons are the base current of the transistor. For the pnp transistor, the same thing happens except that the current flows are reversed.

In most transistors, 95 to 99.9% of the current flow caused by the forward-biased, emitter-base junction reaches the collector output. This high percentage of current is defined in terms of the *current gain* designated by the symbol α (called *alpha*). Although the base is a forward-biased junction, the current flow is small because of the way transistors are made. Note that α is always *less than one* because the collector current can never equal the emitter current since some emitter current must flow in the base circuit. The transfer of current from low input resistance to high output resistance with almost no loss is the reason the device is known as a *transfer resistor* or *transistor*.

CURRENT FLOW IN COMPLETE TRANSISTOR CIRCUITS

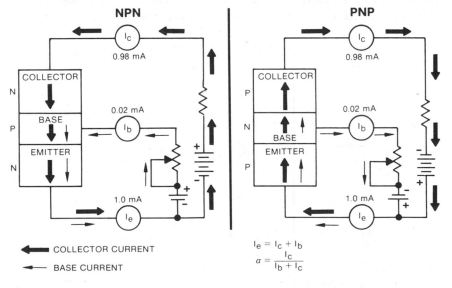

$$I_e = I_c + I_b$$
$$\alpha = \frac{I_c}{I_b + I_c}$$

Transistor Circuit Configurations

Before you learn how transistors amplify—that is, provide greater signal output voltage, current, or power, by controlling current flow from a power source—it is necessary to know about the ways transistors can be connected. There are three basic configurations (arrangements) known as the *common emitter*, *common base*, and *common collector*. Each configuration is used to suit particular circuit needs, as you will learn later in these volumes. The most commonly used configurations are the common-emitter and common-collector circuits. The circuits below show the three npn transistor circuit configurations. With proper biasing, pnp transistors can be used in the same way, as shown in the lower figure.

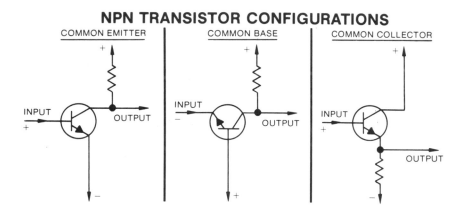

As you can see, the circuits are named for the terminal of the transistor that is not part of the input or output circuit—i.e., the common terminal. Remember that npn transistors require that the collector and base must be relatively positive with respect to the emitter for proper operation. Pnp transistors require that the collector and base must be relatively negative with respect to the emitter for proper operation.

Transistor Current Gain

In the review of transistor operation, it was seen that the base current generally amounts to a fraction of a percent of the total current through the emitter and collector circuits. By increasing or decreasing the current (ΔI_b) through the base circuit alone (as shown in the npn transistor diagram), it becomes possible to obtain much larger current changes (ΔI_c) in the collector circuit. The reason for this is that the flow of current through the base-emitter junction depends upon the forward bias in that junction, and the amount of current flow is proportional to the base resistance R_b. The base resistance R_b is the sum of the internal base-emitter resistance plus any external series resistance in the base circuit. The internal portion of the base resistance is called the *intrinsic base resistance*. If there is no current flow (I_b) in the base circuit, there can be no forward bias at the base-emitter junction, and there will be no current flow in the collector-emitter circuit. When the current in the base circuit increases, this leads to an increasing forward bias at the base-emitter junction; and the emitter-collector current will increase greatly. Thus, *small* changes in base current lead to *large* changes in collector current. Except for the change in the polarity of the collector and base bias, the same conditions apply exactly to the pnp transistor.

TRANSISTOR CURRENT AMPLIFICATION

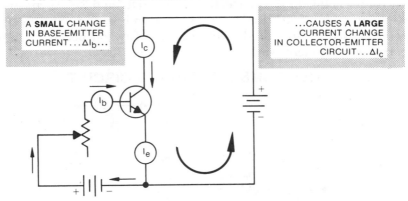

A SMALL CHANGE IN BASE-EMITTER CURRENT...ΔI_b...

...CAUSES A LARGE CURRENT CHANGE IN COLLECTOR-EMITTER CIRCUIT...ΔI_c

$$\text{COMMON EMITTER CURRENT GAIN} = \beta = \frac{\text{CHANGE IN COLLECTOR CURRENT}}{\text{CHANGE IN BASE CURRENT}} = \frac{\Delta I_c}{\Delta I_b}$$

$$I_e = I_b + I_c ; \quad \Delta I_c = \Delta I_b + \Delta I_c$$

$$\alpha = \frac{\beta}{1+\beta} ; \quad \beta = \frac{\alpha}{1-\alpha}$$

The current gain obtained in this way is called the *beta* (β) of the transistor. β is equal to the change in collector current (ΔI_c) divided by the change in base current (ΔI_b). Typical values for beta fall in the region from 25 to over 300. High beta values are usually associated with high alpha (α) values. The relationship between β and α is shown in the illustration above.

Transistor Voltage Gain

The common-emitter circuit can provide voltage gain by taking advantage of the current amplification:

$$\text{Voltage gain} = \frac{\text{voltage}_{out}}{\text{voltage}_{in}}$$

From Ohm's Law,

$$E = IR$$

Therefore,

$$E_{out} = I_c R_c \quad \text{and} \quad E_{in} = I_b R_b$$

Or, in differential form,

$$E_{out} = \Delta I_c R_c \quad \text{and} \quad E_{in} = \Delta I_b R_b$$

Therefore,

$$\text{Voltage gain} = \frac{E_{out}}{E_{in}} = \frac{\Delta I_c R_c}{\Delta I_b R_b} = \beta \frac{R_c}{R_b}$$

For the simple common-emitter circuit illustrated, the voltage gain is easily calculated if R_b and R_c and β are known. Note that the R_b and R_c are the total circuit impedances including both the internal and external components. Thus, R_b is equal to R_b' and R_{bi} where R_b' is the external source resistance and R_{bi} is the intrinsic base resistance. Similarly for the collector resistance, R_c is equal to R_c' and R_{ci}. As you proceed with your study of electronics, you will learn how the common-emitter circuit is used to provide the voltage amplification necessary to make electronic equipment work.

THE COMMON-EMITTER CIRCUIT

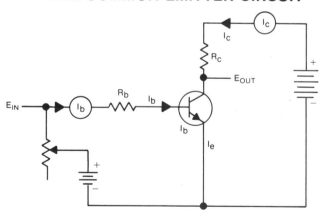

$$\text{VOLTAGE GAIN} = \frac{\text{OUTPUT VOLTAGE}}{\text{INPUT VOLTAGE}} = \frac{I_{OUT} \times R_{OUT}}{I_{IN} \times R_{IN}}$$

$$\text{SINCE } I_{OUT}/I_{IN} \text{ (CURRENT GAIN)} = \beta \qquad \text{VOLTAGE GAIN} = \beta \times \frac{R_{OUT}}{R_{IN}}$$

$$\text{FOR A TYPICAL TRANSISTOR, VOLTAGE GAIN} = 150 \times \frac{5,000}{250}$$
$$= 150 \times 2,000$$
$$= 3,000 \text{ TIMES}$$

Transistor Power Gain

Upon preliminary examination there may seem to be no advantage to transistor operation in the common-base configuration since 1.0 mA of current change in the input circuit (emitter to base) is required to produce 0.98 mA of current change in the output circuit (collector to base). In fact, the common base current gain (α) is less than unity. For example, a transistor with a collector current of 0.98 mA (milliampere) and emitter current of 1 mA has a current gain of 0.98 ($\alpha = 0.98/1 = 0.98$).

Although no useful current amplification is produced by the common-base connection for a transistor, significant *voltage* and *power gain* are available. To see why, it is only necessary to compare the current and resistance conditions in the input and output circuits. As you know, for our example transistor, the currents in the input (emitter) and output (collector) circuits are almost identical. However, the resistance in the input and output circuits can be very different. The forward bias across the emitter and base junction makes this junction have a low resistance typically ranging from 40 to 800 ohms. On the other hand, the bias across the base and collector is in the reverse direction, giving the junction between them a high resistance. Such resistances generally range from 10,000 to 100,000 ohms.

In *Basic Electricity* you learned from Ohm's Law that $E = IR$ and that the power developed is equal to I^2R. Since almost identical currents flow in the input and output circuits and since the output circuit resistance is much higher than the input circuit resistance, it can be seen that voltage and power gains are available, as shown in the illustration.

THE COMMON-BASE CIRCUIT

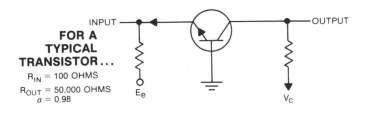

FOR A TYPICAL TRANSISTOR...
INPUT — OUTPUT
$R_{IN} = 100$ OHMS
$R_{OUT} = 50,000$ OHMS
$\alpha = 0.98$
E_e V_c

$$\text{POWER GAIN} = \frac{\text{OUTPUT POWER}}{\text{INPUT POWER}} = \frac{I^2_{OUT} \times R_{OUT}}{I^2_{IN} \times R_{IN}} = \alpha^2 \times \frac{R_{OUT}}{R_{IN}}$$

$$\text{POWER GAIN} = (0.98)^2 \times \frac{50,000}{100} = 480$$

$$\text{VOLTAGE GAIN} = \frac{E_{OUT}}{E_{IN}} = \alpha \frac{R_{OUT}}{R_{IN}}$$

$$\text{VOLTAGE GAIN} = 0.98 \times \frac{50,000}{100} = 490$$

Translstor Power Gain (continued)

Thus, for a typical transistor connected in the manner shown in the drawing, a voltage gain also can be achieved. This gain is not due to current amplification, as it was previously. Instead, it is entirely due to the high resistance in the output circuit as compared with the low resistance in the input circuit. Amplification is achieved because the semiconductor arrangement has transferred a current, with almost no loss, from a low-resistance circuit to a high-resistance circuit. Similarly, power gain is achieved by the transistor because of the fact that in the relationship $P = I^2R$, although the current has not changed, the resistance has, thus providing power gain.

The common-collector configuration in some ways resembles the common-base circuit, using the current gain of the transistor from base to emitter. A small base-current change results in a large current change in the collector-emitter circuit as before. However, since the base-emitter voltage is almost constant (because of forward bias), the voltage difference between them is small. Thus, the output from the emitter is almost the same as applied to the base. Because the currents are widely different, the apparent resistances are widely different as well. For example, if 1 mA is flowing in the collector-emitter circuit, then only $1/\beta$ mA flows in the base circuit. Since $R = E/I$, then the resistance seen at the base is β times as great as seen at the emitter. Thus, the common-collector configuration is often used to connect a high-resistance (impedance) circuit to a lower-resistance (impedance) circuit without loading of the input circuit.

IMPEDANCE CHARACTERISTICS OF
COMMON-COLLECTOR CIRCUIT

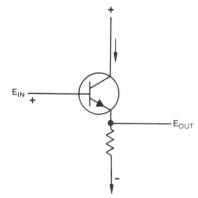

BASE IMPEDANCE $= \beta /$ EMITTER IMPEDANCE

EMITTER IMPEDANCE $\frac{1}{\beta}$ BASE IMPEDANCE

VOLTAGE GAIN $= \alpha$

Transistor Operating Characteristics

The common-emitter configuration can be used to draw up a series of *characteristic curves* that show the relationships among the various currents in basic transistor circuits. These curves plus other data are made available from the transistor manufacturer and describe each device.

To supply a constant input current, an approximation can be obtained by using a higher voltage dc supply with the output connected in series with a large value of fixed resistance—at least 100 times greater than the input resistance of the transistor base. Thus the current in the circuit is determined by the series resistor rather than by the base resistance. As shown, a six-volt battery with a 100,000-ohm series resistor is satisfactory for most applications.

DETERMINING TRANSISTOR CHARACTERISTICS

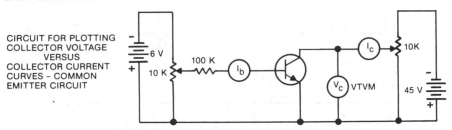

CIRCUIT FOR PLOTTING
COLLECTOR VOLTAGE
VERSUS
COLLECTOR CURRENT
CURVES - COMMON
EMITTER CIRCUIT

To plot the curves shown, a fixed value of base current is selected and the collector voltage is varied over a predetermined range. For each value of collector voltage, the collector current is recorded. When the collector current begins to show signs of saturation, or leveling off to a fixed value, another base current is selected and the process is repeated.

Examination of the family of curves shows that small changes in base current produce large changes in collector current. It is not unusual for a 10-μA (microampere) change in base current to produce a 1-mA change in collector current, a beta (β) gain of 100. Since $\alpha = \beta/(1 + \beta)$, $\alpha = 0.99$.

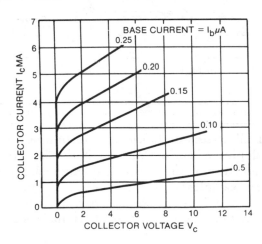

COLLECTOR VOLTAGE
VERSUS
COLLECTOR CURRENT
CURVES FOR TYPICAL
NPN JUNCTION
TRANSISTOR
COMMON - EMITTER
CIRCUIT

BRASWELL MEMORIAL LIBRARY
727 NORTH GRACE STREET
ROCKY MOUNT, NC 27804

Design and Fabrication

Transistors are designed and fabricated to meet widely different operating characteristics such as current capacity, gain, stability, and operating frequency. Different fabricating techniques are used to optimize those particular characteristics for which there is a defined circuit requirement. Today, there are literally thousands of transistor types to choose from. The design and fabrication of a transistor always involves trade-off decisions. Usually, not all of the desirable characteristics can be achieved in one device. Design and fabrication techniques are continually being improved or modified.

DETAILS OF TRANSISTOR CONSTRUCTION

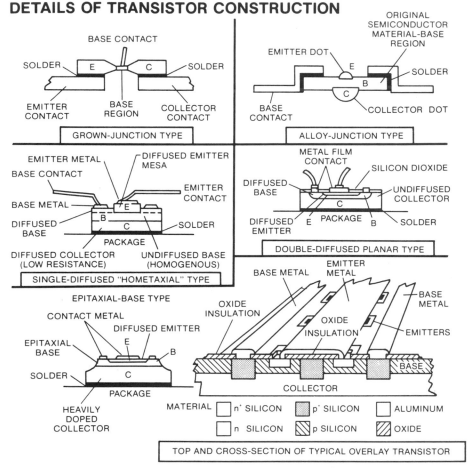

In general, the aim of all transistor fabrication techniques is to control the size, shape, and doping of various pn junctions. These techniques include various alloy, diffusion, and crystal-growing techniques. The sketches shown above are examples of transistor construction. Since you will use the completed device, it is the *external* characteristics that interest you. The structural details shown here are only included for your general reference.

Experiment/Application—Transistor Characteristics—Common Collector

Suppose you connected a transistor in the common-collector configuration as shown.

CHECKING THE COMMON-COLLECTOR CIRCUIT

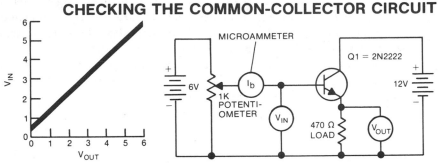

As you vary V_{in} with the potentiometer, you can see that the output voltage at V_{out} the emitter follows the input voltage at the base. In fact, this configuration is usually called an *emitter follower.* The difference in dc voltage between the base and the emitter is the emitter-base junction voltage drop and is constant at about 0.6 volt. As you know, this configuration of a transistor circuit does not provide any voltage gain, but it does provide *power gain.* If we set the dc operating point (usually called the *quiescent operating point*) to 3 volts (base), the output voltage V_{out} will be about 2.4 volts. The current through the load resistor can easily be calculated as

$$I = \frac{E}{R} = \frac{2.4}{470} = 0.005 \text{ ampere} = 5 \text{ mA}$$

Thus, the power in the load is

$$P = VI = 2.4 \times 0.005 = 0.012 \text{ watt} = 12 \text{ mW}$$

As you learned, the base current is about $1/\beta$ times the emitter current since we can usually ignore the base current contribution to the total emitter current. Thus, for the 2N2222 transistor with a $\beta = 150$ (nominal value), the base current is $Ie/\beta = 0.005/\beta = 0.33\ \mu A$. A sensitive micrometer in the base leg, as shown, will show this current flow when the emitter current is 5 mA. (Actually, when you do this experiment, the base current may be considerably different since the manufacturer's data on the 2N2222 shows that the β can be between 75 and 375, but for a given transistor it will be constant.) Assuming $\beta = 150$, the power in the base circuit is

$$P = VI = 3 \times 33 \times 10^6 = 0.0001 \text{ watt} = 0.1 \text{ mW}$$

As you see from this experiment, there is a power gain of 12 mW/0.1 mW = 120. That is, you can get 120 times the power from the emitter circuit than you put into the base. Circuits like the common-emitter configuration are very useful when it is necessary to convert from a circuit that will deliver little current to one that will deliver considerable current, without appreciably changing the signal level.

Experiment/Application—Transistor Characteristics—Common Collector (continued)

In the earlier part of this experiment, you saw how the common-collector circuit has power gain but no voltage gain. You also saw how the dc operating point for the circuit (the quiescent operating point) is established by proper biasing. In electronics, we are usually concerned with ac signals. This part of the experiment will show how ac signals have effects similar to dc on the emitter-follower circuit.

Suppose you had the same basic circuit as before, but modified as shown below to allow the introduction of audio signals. The 4.7 volts on the input simulates a high-resistance signal source.

VOLTAGE GAIN OF COMMON-COLLECTOR CIRCUIT

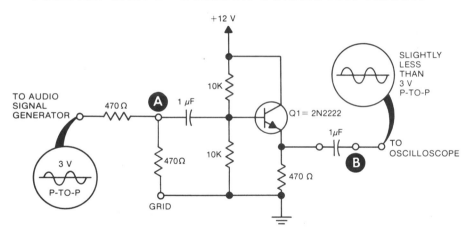

As you can see, we have fixed-biased the base to about +6 volts. The emitter will be at about 5.4 volts (base-emitter junction drop = 0.6 volt). An ac audio signal of 3 volts peak-to-peak on the input as observed on an oscilloscope at point A also will be observed to be slightly less than 3 volts on the emitter at point B. The capacitors are included so that the ac can be added without disturbing the dc operating point. You can see the effects of current gain by placing a 470-ohm resistor across the input at point A. As you can see, the signal level drops enormously. However, when the 470-ohms resistor is placed across point B (to ground), the output drops comparatively little. Thus, it is clear that the emitter can deliver much more power into the load than was present at the input.

Experiment/Application—Transistor Characteristics—Common Emitter

In this experiment you will see how the common-emitter configuration acts as a voltage amplifier. This is the most commonly used configuration for transistor amplifiers, as you will see as your study of *Basic Solid-State Electronics* proceeds.

Suppose you connected the circuit shown below:

THE COMMON-EMITTER CIRCUIT

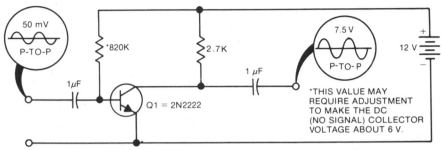

As you remember about transistor characteristics, the base current is about 1/β of the collector-emitter current. For a collector current of about 2.2 mA, we need a base current of about 2.2/β = 2.2/150 = 15 μA. This is supplied by the 820-K resistor to the + 12-volt supply. As noted, you may have to adjust this resistor to get the proper transistor quiescent collector current because the β of the transistor varies from about 75 to 375 for the 2N2222, according to the manufacturer's data.

Unlike the emitter follower, this circuit has voltage gain, as you learned in your study of how transistors work. Thus, you will need a very small input to obtain a larger output. For example, a 50-mV peak-to-peak input signal produces about a 7.5-volt peak-to-peak output (for β = 150) or a voltage gain of

$$\text{Voltage gain} = \frac{E_{out}}{E_{in}} = \frac{7.5}{0.05} = 150$$

As you will see later in your study of *Basic Solid-State Electronics*, there are many variations of this basic common-emitter configuration, but the operating principle is unchanged.

Note that with this basic amplifier circuit, results may vary widely from transistor to transistor, so the data shown is only nominal. Most of the variations of this circuit are used to stabilize the gain so that different transistors do not greatly affect the amplification obtained. One of the most common variations is to put a resistor in the emitter line. This greatly reduces the gain but stabilizes it so that variations in β do not affect the overall gain significantly. You will learn more of this when you study feedback amplifiers in Volume 2 of this series. To check this, put a 100 ohms resistor in the emitter line. You will see that the gain is now about 27.

Experiment/Application—Transistor Characteristics—The Transistor as a Switch

In this experiment you will see how a transistor operates as a switching device. Suppose you wired up the circuits shown below.

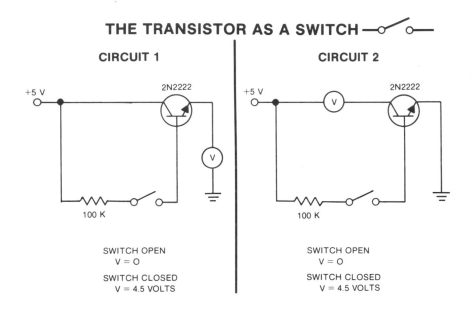

THE TRANSISTOR AS A SWITCH

CIRCUIT 1

CIRCUIT 2

Circuit 1: When the switch is open, the transistor is cut off (does not conduct) because there is no forward bias on the base-emitter junction. Therefore the output voltage V is 0 because the transistor is an open circuit. When the switch is closed, the base becomes strongly biased on, and the voltage drop across the transistor drops to a very low value (essentially saturated) so that almost the entire +5 volts appears at the emitter (about 4.5 volts). Thus, as you can see, the transistor is effectively switching the 5-volt supply.

Circuit 2: No voltage appears across the voltmeter when the switch is open since the transistor is an open circuit and no current can flow. When the switch is closed, sufficient current will flow in the base-emitter junction to saturate the transistor, so it looks like a short circuit except for the voltage drop across the junctions and, therefore, the voltmeter will read about 4.5 volts. Again, as you can see from the experiment, the transistor operates as a switch.

As you can see, the current switched in the base circuit is small compared to the current switched in the output circuit. Therefore we can control larger amounts of power with transistor switches.

Experiment/Application—Checking Transistors with an Ohmmeter

As you learned earlier, an ohmmeter can be used to test a diode because the potential across the ohmmeter leads biases the diode so that a check for forward-bias operation (low resistance) and reverse-bias operation (high resistance) can be made. We can use the *same idea* to test transistors. While the analogy below is not very useful for describing transistor operation, it is very useful in understanding *how to test them*.

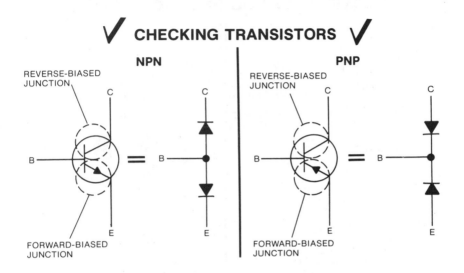

We can consider a transistor as being made from two diodes connected in series as shown in the figure. From what you know about checking diodes with an ohmmeter, you can see what will happen when you put an ohmmeter terminal across a transistor terminal. For example, the resistance from collector to emitter should always be high, regardless of the ohmmeter connections for any transistor since one of the diodes is always nonconducting. For npn transistors, when the positive probe of the ohmmeter is on the base, the resistance to both the collector and the emitter will be *low*. When the *negative* probe of the ohmmeter is on the *base*, the resistance to both the collector and emitter will read *high*. For pnp transistors, the reverse will be true. Thus, when the *positive* ohmmeter probe is on the *base*, the resistance to the collector and the emitter will be *high*. When the negative ohmmeter probe is on the base, the resistance to the collector and the emitter will be *low*.

The test of a transistor by the method mentioned above only checks for proper functioning of the *junctions* and is by no means a *complete test* of the device. However, the most common failures of transistors involve failure of the junctions, so you will find this a useful procedure.

Thyristors—The SCR

Thyristor is the name for a family of bistable semiconductor switching devices made up from multiple pn junctions. Thyristors include the *bidirectional trigger diode (diac)* and two-gate controlled thyristors: *the reverse-blocking silicon-controlled rectifier (SCR)* and the *bidirectional triode (triac)*.

The silicon-controlled rectifier (SCR) is the most widely known thyristor. This gate-controlled unidirectional switch has three terminals (anode, cathode, gate) and is made up of a pnpn semiconductor structure.

A SCR CONSISTS OF TWO PN JUNCTIONS IN SERIES

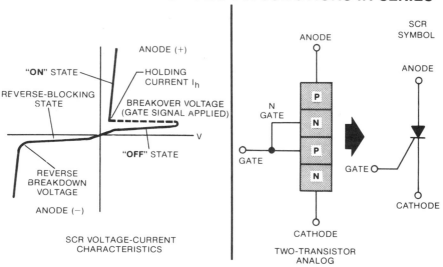

SCR VOLTAGE-CURRENT CHARACTERISTICS

TWO-TRANSISTOR ANALOG

For reverse-biased conditions, the SCR reverse-blocks current flow like a diode. For forward-biased conditions, the SCR has two states. In the first state, the SCR exhibits high impedance with no gate signal and except for a small forward current, the device is *off*. If a gate signal is applied, the current increases rapidly and the voltage across the SCR drops to a very low value. The SCR is now in its second forward-biased state, the *on* or conducting state.

The breakover voltage point is controlled by application of a current pulse to the gate, as shown by the voltage-current characteristic curve.

The SCR is usually operated with current signals of sufficient magnitude and duration to assure good switching action and gate control of switching. After switching to the *on* state, the SCR remains on independently (self-sustaining) of the gate input until the anode/cathode current falls to a low value. The current necessary to keep the SCR conducting is called the *holding current*. When the holding current drops below the threshold value, the SCR turns off and must be triggered again for turn-on. SCRs are used to control current flow by acting as electrically controlled switches.

Thyristors—The Triac and Diac

The triac has three electrodes designated as terminal 1 (T1), terminal 2 (T2), and gate. The triac exhibits the same forward-blocking, forward-conducting characteristics as the SCR, but for *both* directions. Thus, it is comparable to two SCRs connected in parallel with one SCR inverted with respect to the other and with the gates connected together. Therefore, there is *no* reverse-bias condition. For gate biasing above the threshold in either direction, the triac is in a forward-conducting condition. As for the SCR, the breakover voltage point is controlled by application of a gate current pulse. However, a gate pulse of either polarity will trigger the triac. It is usually used as a triggering device for SCRs.

The diac is a two-electrode, bidirectional diode that can be switched from *off* to *on* for either polarity of applied voltage, whenever the voltage level exceeds the breakover voltage point.

The diac is similar in construction to the pnp bipolar transistor junction. It differs from the bipolar transistor in that the base has no electrode and the impurity element concentrations at the two pn junctions are about the same in order to obtain symmetrical bidirectional switching at the breakover point. The diac is used as a triggering device in triac-SCR phase-control circuits used for motor speed control, and home heating, ventilation, cooling, and lighting control.

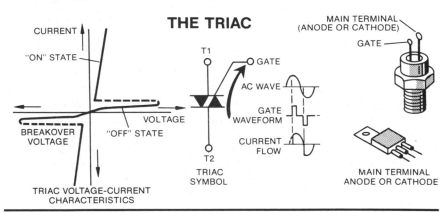

THE TRIAC

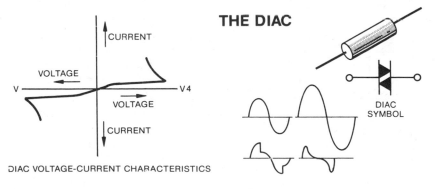

THE DIAC

Field-Effect Transistors—FET/MOSFET/MOS

Like the bipolar transistor, the *field-effect transistor (FET)* is basically a three-terminal device that makes use of pn junctions and can be used as an amplifier or switch. However, unlike the bipolar transistor, whose performance depends on the interaction of two types of charge carriers (holes and electrons), the FET is unipolar and relies on only one type of charge carrier—either holes in p-channel FETs or electrons in n-channel FETs. The FET functions more like a vacuum tube than a bipolar transistor in that it is a voltage-controlled device that has a very high input resistance when properly biased, so it can be used in circuits where it is important not to load down output from the device that drives the FET.

The three terminals of the FET are called the *gate, source,* and *drain.* These terminals are somewhat analogous to the bipolar transistor terminals—base, emitter, and collector, respectively. Like the bipolar transistor base, the FET gate is the controlling terminal, except that the gate controls the FET output by its voltage rather than current.

The first FETs were the *junction FET(JFET)* that were packaged as a single FET device. For this type of FET, the gate is insulated from the channel by a pn junction as shown.

JUNCTION-GATE FIELD-EFFECT TRANSISTOR (JFET)

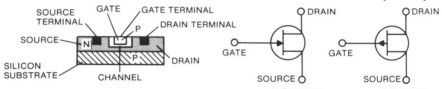

A newer type of FET is of the metal-oxide-semiconductor type called *MOSFET*, or just *MOS*. This type of FET is packaged as a discrete device or in IC circuits. MOS devices have a metal control *gate* separated from the semiconductor *channel* by an insulating oxide layer. Since the gate is not a junction that must be biased, the very high input resistance of MOS transistors is not affected by the polarity of the bias on the gate.

MOSFET TRANSISTORS - CONSTRUCTION DETAILS

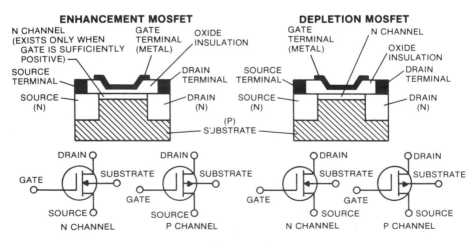

Field-Effect Transistors—FET/MOSFET/MOS (continued)

All FETs operate under the charge control concept. That means that the voltage placed on the gate induces a field effect on the semiconductor channel located beneath the gate. The field effect in turn alters the conductivity of the channel. Since the channel is the path taken by output current between the source and drain, the gate voltage thereby controls the FET output.

In metallic-oxide-semiconductor FETs (MOSFETs), the p-channel (holes) and n-channel (electrons) can be either of the *depletion* type or *enhancement* type as shown. The direction of the arrowhead in the symbol identifies the device as n-channel or p-channel. The channel line is solid to identify the normally *on* depletion type or is a dashed line to identify the normally *off* enhancement type.

In a depletion-type MOSFET, the channel between the source and drain is inherently as conductive as both the source and drain. Therefore, this channel provides substantial drain current even when no gate bias voltage is applied. A reverse gate voltage depletes the charge in the channel and thereby reduces the channel conductivity. A forward gate voltage draws more charge carriers into the channel and thereby increases the channel conductivity.

In a enhancement-type MOSFET, the gate must be forward-biased to produce carriers necessary for conduction through the channel. No useful channel conductivity exists at zero or reverse gate bias. This configuration is used in switch applications since its channel conductivity can be readily changed from off to on by a simple gate voltage change.

MOSFET OPERATING CHARACTERISTICS
ENHANCEMENT TYPE　　DEPLETION TYPE

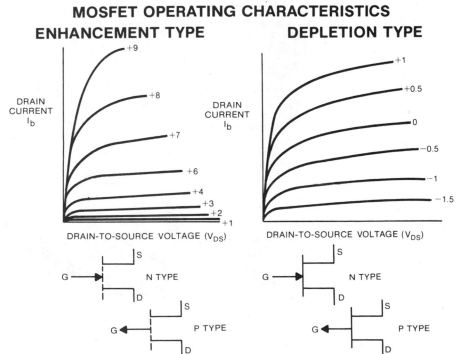

IC MOS Devices—MSI, LSI, VLSI

The early development of IC technology was restricted because bipolar technology was used. This technology was used in MSI devices that are extensively used in solid-state electronics. However, the need to provide isolation between devices in an IC restricts the number of devices that can be included in one IC. In addition, the power consumption of bipolar devices caused heat dissipation problems when the packing density became too great.

The development of MOS technology has been the key factor in development of LSI and VLSI devices. The important point is that an MOS device is *self-isolating*; that is, MOS devices can be placed very close together without requiring any special isolation between them. In addition, *very-low-power* MOS devices are quite *practical*. Therefore, they can be packed together without too much concern about heat dissipation becoming a problem as it would if the devices consumed appreciable power. Furthermore, MOS devices are *symmetrical*; that is, the source and drain can be *interchanged* so that not only can they be operated in *series* but also can be operated *bilaterally*. Thus, *complex circuits* and *modes* of *operation are practical* with *MOS*, while *not* practical in bipolar form. All these factors allow very high *packing density* as well as very high *functional density*, which in turn allows for complex functions on a single small chip. LSI technology allows for the construction of a calculator on a few chips. The VLSI technology provides a computer on a chip. Some typical Dual-In-Line (DIP) IC packages are shown below.

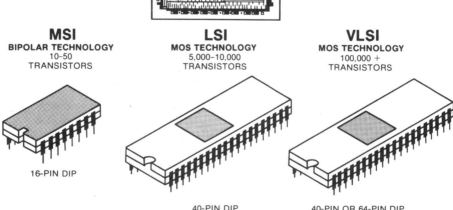

MAC-4
COMPUTER-ON-A-CHIP

MSI	**LSI**	**VLSI**
BIPOLAR TECHNOLOGY	MOS TECHNOLOGY	MOS TECHNOLOGY
10–50	5,000–10,000	100,000 +
TRANSISTORS	TRANSISTORS	TRANSISTORS

16-PIN DIP

40-PIN DIP

40-PIN OR 64-PIN DIP

Review of Semiconductor Devices

N-TYPE SEMICONDUCTOR MATERIAL

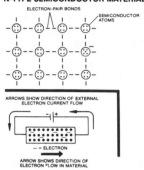

1. SEMICONDUCTOR MATE-RIALS are available in two types: n type (excess electrons) and p type (excess holes).

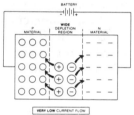

2. PN JUNCTIONS are formed when n- and p-type material are joined. Interaction takes place here to create space charge or depletion layer regions. This produces special properties of the pn junction: its resistance to current flow depends on the polarity of applied voltage.

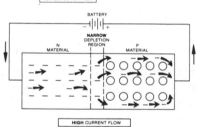

3. The SEMICONDUCTOR DIODE was the first use of semiconductor materials and consists of a pn junction in one of several configurations.

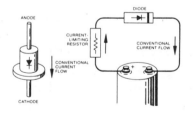

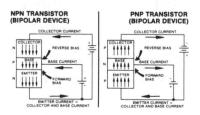

4. The BIPOLAR TRANSISTOR has two junctions of pn material—either pnp (pn-np) or npn (np-pn). These are called bipolar transistors because *two* components of current flow in them.

Review of Semiconductor Devices (continued)

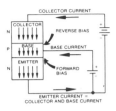

5. Bipolar transistor current flow and biasing junction (called base emitter) is forward-biased, while the second junction (called base collector) is reversed-biased.

CURRENT FLOW IN COMPLETE TRANSISTOR CIRCUITS

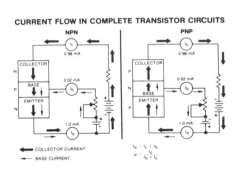

6. Bipolar transistor current flow is in the base-emitter circuit and in the base-collector circuit. Because of construction, most electrons (or holes) reach collector output and only few reach base itself. Thus, major current flow is in emitter-collector circuit. The ratio of collector current to emitter current is called α or current gain.

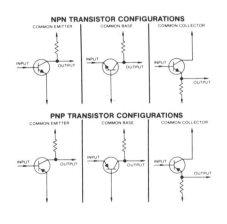

7. Bipolar transistor circuit configurations are common emitter, common base, and common collector. All are useful.

$$\beta = \frac{\Delta \text{ collector current}}{\Delta \text{ base current}}$$

8. TRANSISTOR CURRENT GAIN (β) is defined as the change in collector current for a given charge in base current.

$$V_g = \text{voltage gain} = \beta \, \frac{R_c}{R_b}$$

9. VOLTAGE GAIN depends on β and is typically $\beta(R_c/R_b)$.

$$\text{Power gain} = \alpha^2 \, \frac{R_c}{R_b}$$

10. TRANSISTOR POWER GAIN depends on α and the ratio of R_{in} to R_{out}.

Review of Semiconductor Devices (continued)

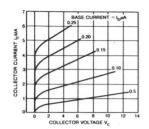

COLLECTOR VOLTAGE
VERSUS
COLLECTOR CURRENT
CURVES FOR TYPICAL
NPN JUNCTION
TRANSISTOR
COMMON - EMITTER
CIRCUIT

11. TRANSISTOR CHARACTERISTICS describe the transistor operating condition change in response to outside stimulus.

SCR—TRIAC—DIAC

12. SCRs, TRIACs, and DIACs are special multifunction semiconductor devices that have special properties for power control.

DRAIN

GATE

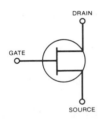

SOURCE

13. FET-MOSFET unipolar devices have only one type of carrier and are hence unipolar. FETs are voltage controlled rather than current controlled.

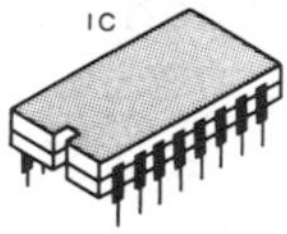

IC

14. MOS TECHNOLOGY represents a major step forward in semiconductor technology. Confined current flow allows for dense packing, leading to development of integrated circuits.

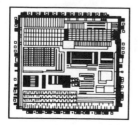

15. MSI, LSI, and VLSI are extremely important technologies that led to whole functions on a chip (microtechnology).

Self-Test—Review Questions

1. Describe p- and n-type materials. How are they made? Describe their properties as compared to the pure materials?
2. Describe what occurs at a pn junction when p and n materials are joined. What occurs when a bias is applied to a pn junction (both polarities)?
3. Discuss the semiconductor diode: (a) What are its characteristics? (b) Draw a typical voltage-current characteristic and describe each region. (c) Draw the diode symbol and explain the meaning of the marking. (d) Sketch some typical diode packages. (e) How are diodes designated?
4. Describe the junction structure in bipolar transistors. How is it biased? Show the current flow for npn and pnp bipolar transistors. Why are they called bipolar?
5. Define α and β in terms of transistor operation. From a practical standpoint, what do they mean? How are these data obtained from transistor operating curves?
6. Draw the three bipolar transistor configurations for both npn and pnp transistors. Show the polarity of the biasing voltage on each terminal.
7. For each circuit configuration of question 6, describe the available gain (voltage, current, and power) and how it is obtained. How does a transistor function as a switch? What are the essential differences between SCRs, TRIACs, and DIACs and transistors?
8. Describe and sketch some transistor cases. How are transistors usually designated?
9. Describe the difference between MOSFET and bipolar semiconductors. Why are MOSFETs called unipolar?
10. Define MSI, LSI, and VLSI. Why are these important today? What particular properties of MOS makes LSI and VLSI possible?

Learning Objectives—Next Section

Overview—Now that you know something about semiconductor devices, you are ready to see how these can be used to make the first of the nine common electronic building blocks—*power supplies*—that provide the voltages and currents required by electronic systems.

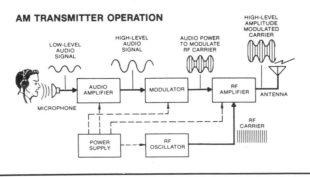

Purpose of the Power Supply

Earlier in this volume, you learned that *power supplies* are one of the nine common building blocks used in all electronic equipment. To become familiar with these nine building blocks, you were introduced to a radio transmitter-receiver system. This system is to be the *aid* for learning how the various common building blocks work together to form a complete electronic system. The power supply converts power from an available source of electric power to a voltage and current suitable for use by the various circuits in electronic equipment.

Most electronic equipment requires specific dc voltages and currents to operate efficiently. The power supply provides these specific voltages and currents to the circuits in the equipment. While power supplies may differ greatly in terms of their voltage and current outputs, they all operate on the *same basic principles*. So when you learn about one type of power supply, you will know what is necessary to understand all power supplies.

Initially, we will start our study of power supplies with simple units. Later we will learn about power supplies that produce multiple voltage outputs, have voltage regulators, protective circuits, etc.

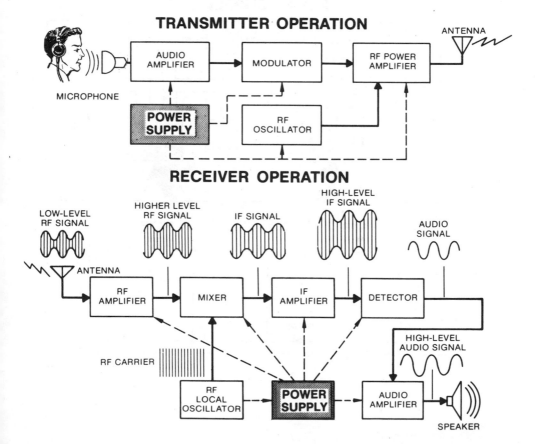

Basic Components of a Simple Power Supply

The basic components of a simple power supply are the *transformer*, *rectifier*, and *filter*. These are used together in the circuit diagram shown on this page. The power supply illustrated takes electric power from the ac line at 120 volts of ac and supplies a dc output voltage.

As shown, 120 volts of ac is supplied to the *transformer* that raises or lowers the voltage as required in accordance with the ratio of turns in the primary and secondary windings, which you learned in *Basic Electricity*. Note that there may be two or more secondary windings to supply a number of secondary voltages, if this is required. The *rectifier* accepts the ac voltage from the secondary winding and changes it to a pulsating dc voltage. As you learned, diodes only pass current well in *one* direction, and as you can see, diodes are used as rectifiers. Finally, the *filter* smooths out the pulsations in the rectifier output and provides a dc voltage and current with a residual ac component (ripple) suitable for the electronic circuits being used. Again note that there may be several rectifiers and filters connected to different transformer secondaries if the electronic circuits in the equipment require several different outputs, each with its own voltage, current, and ripple characteristics.

BASIC COMPONENTS OF AN AC-TO-DC POWER SUPPLY

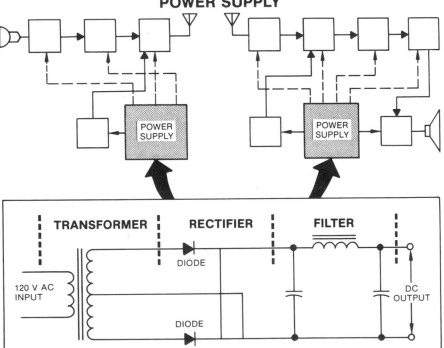

Voltage Regulators

As you have seen, the basic dc power supply is made up of a transformer, rectifier, and filter circuit. This is all that is basically required to provide the high- or low-voltage dc outputs required to operate electronic circuits. However, when current is drawn from the power supply, the voltage drops. Also, as the load current changes with different load conditions, the voltage again changes. These effects are due to the internal resistance of the power supply. In addition, the voltage output will change as the ac supply voltage changes.

This voltage change is not serious for many electronic circuits, and they will go right on working. However, there are many electronic circuits that cannot operate properly if the voltage varies *significantly*. These circuits require that the power supply have a *voltage regulator* circuit added to it to stabilize the output voltage by holding the voltage change to an acceptable amount. In complex solid-state systems, several regulated voltages may be required for different circuits. In this case, several regulators of different types may be used. Each is suited to the voltage regulation requirements for the particular circuit. When a power supply has a voltage regulator, only those circuits that require a constant voltage are generally connected to the voltage regulator. Other circuit elements are connected directly to the unregulated dc output terminal.

There are many different types of regulators. Which regulator is used depends, as you might suspect, on the voltage and current requirements of the circuits operated by the power supply. You will learn about these after you learn about the basic power supply.

ADDING A VOLTAGE REGULATOR

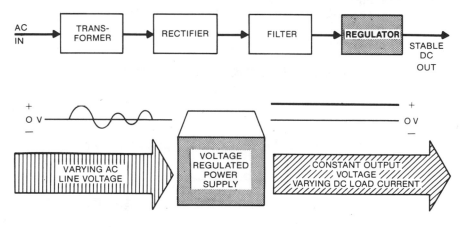

Purpose of Rectifiers

The most commonly used source of power for electronic equipment is the ac power line, usually 120 volts of ac at 60 Hz. As you know from *Basic Electricity*, transformers are used to convert the ac line voltage to a suitable level for the particular power supply operation. This transformed ac must be changed to dc to be used in electronic equipment. Devices that convert ac to dc are called *rectifiers*. Circuits that use rectifier devices are called *rectifier circuits*.

The diode described earlier is almost universally used as a rectifying device since diodes allow current to flow through them in only one direction. The rectifier diode accomplishes this by acting as a conductor for current flow in one direction and as an insulator for current in the reverse direction. For example, when placed in an ac circuit, the rectifier will pass current only on all positive half cycles of the ac voltage or only on all negative half cycles. The choice of either positive cycle rectification or negative cycle rectification depends only on how the rectifier diode is connected into the circuit. Thus, the polarity of the pulsating dc output from a rectifier circuit depends on how the rectifier diodes are connected into the circuit.

Regardless of the direction chosen to pass current, the current flow in a simple rectifier circuit is pulsating dc corresponding to the chosen positive (or negative) half cycles of ac. The point to remember is that the current output of the rectifier circuit is not a steady dc. It is a pulsating flow in one direction. As you will learn soon, the pulsations occur at the line frequency or twice the line frequency, depending on whether the rectifier circuit is *half wave* or *full wave*. Thus, for a 60-Hz ac supply, the half-wave rectifier circuit produces a half sine-wave pulsation of $1/120$-second duration every $1/60$ second, while the full-wave rectifier circuit produces a half sine-wave pulsation of $1/120$-second duration every $1/120$ second.

RECTIFIER CIRCUITS CHANGE
AC TO DC

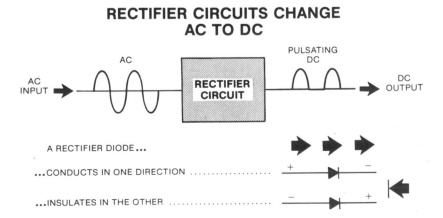

The Rectifier

The rectifiers used in power supplies are usually high-current-capacity diodes made from silicon, germanium, or selenium. In certain special applications, SCRs (silicon-controlled rectifiers) also are used. Each rectifier device has a *peak inverse voltage rating*, which limits the maximum reverse voltage that can be applied across the diode in the nonconducting state. However, by stacking several diodes in series, the inverse voltage rating can be increased. Each diode can pass a limited amount of current in the conducting direction. Rectifiers are not usually connected in parallel for higher currents because current generally divides unequally, causing one of the devices to be overloaded and damaged.

Silicon rectifiers are generally used today because they are very rugged and can operate at higher temperatures than other types. On the other hand, the conducting voltage drop across most silicon rectifiers is between 0.5 and 1 volt. This leads to a considerable power loss in high-current applications. Thus, in some cases for very high-current use, germanium is used because of the lower conducting voltage drop of about 0.25 volt, even though the germanium diodes cannot operate at as high temperatures.

The type of rectifier circuit to be used in electronic equipment depends on the requirements of the equipment. Characteristics to be considered include dc voltage and current requirements and amount of ripple that can be tolerated. There are three basic types of rectifier circuits in general use—the *half-wave*, *full-wave*, and *full-wave bridge types*. Operation of these types will be described on the following pages.

HALF-WAVE RECTIFICATION

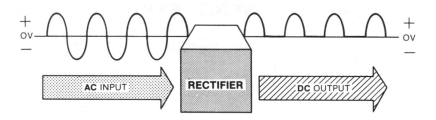

FULL-WAVE RECTIFICATION

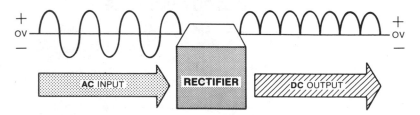

The Half-Wave Rectifier Circuit

The half-wave rectifier circuit consists of a single rectifier diode connected in series with the ac voltage source and the circuit load resistance. The rectifier permits current only during the half cycle of the applied ac voltage that makes the voltage across it right for conduction. The circuit current then is pulsating dc. In the half-wave rectifier circuits shown, the frequency of the ac input is 60 Hz, and current flows only for one-half of each cycle. Thus, the current flow through the load is in pulses at the rate of 60 pulses per second.

These simple circuits illustrate the basic method used to change ac to dc. When connected as shown in the first circuit, the dc voltage across the load resistor is positive at the end that connects the rectifier and negative at the grounded end.

To reverse the polarity of the dc voltage obtained, the rectifier is reversed as in the second circuit. This allows current to flow only on the negative half cycles as compared to the previous circuit. This circuit is used to obtain a negative dc voltage with respect to ground.

The advantage of the half-wave rectifier is its extreme simplicity. A major disadvantage of the half-wave rectifier is that no current is produced for one-half of the ac cycle. This results in a very large amount of ac ripple (residual ac component) that requires a very effective filter unless the current drain is minimal. There are, however, a number of applications where limited current capability and large ripple are of little importance, and the economy offered by the half-wave rectifier makes it useful.

HALF-WAVE RECTIFIER CIRCUIT

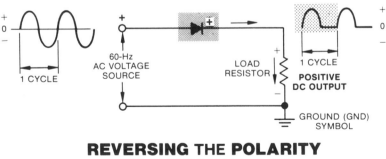

REVERSING THE POLARITY
OF OUTPUT VOLTAGE

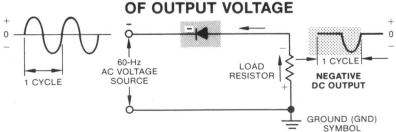

The Full-Wave Rectifier Circuit

The full-wave rectifier requires a center-tapped transformer so that current can be made to flow in the load resistor on both halves of the ac wave. In a full-wave rectifier circuit, a diode rectifier is placed in series with each half of the transformer secondary and load. Effectively, you have two half-wave rectifiers working into the same load.

On the first half cycle the transformer's ac voltage wave makes the upper diode rectifier anode *positive* so that it conducts. As a result, current flows through the load, causing a pulse of voltage across the load. Notice that while the upper diode is conducting, the lower diode anode is negative with respect to its cathode so that it does *not* conduct.

On the second half cycle the anode of the upper diode is *negative* so that it cannot conduct, whereas the anode of the lower diode is positive so that current flows through it and through the load. Since both pulses of current through the load are in the *same* direction, a pulsating dc voltage now appears across the load. The full-wave rectifier has changed *both* halves of the ac input into a pulsating dc output. Therefore, *two* pulses of current are provided for *each* cycle of the ac wave.

The advantage of a full-wave rectifier is that it produces output current for both half cycles of the line voltage. This results in an ac ripple characteristic much easier to filter. The disadvantage is that a center-tapped transformer is required, adding an appreciable cost. This is partially offset by the fact that the current is divided between two halves of the transformer secondary, and hence smaller wire can be used. The cost of the extra diode is usually of no great consequence. As before, reversing *both* diodes will reverse the polarity of the full-wave rectifier. Moving ground from the centertap to the other side of the load also reverses the polarity. This is convenient when large stud-mounted devices are used since cathodes do not have to be insulated from the metal frame or chassis.

FULL-WAVE RECTIFIER CIRCUIT

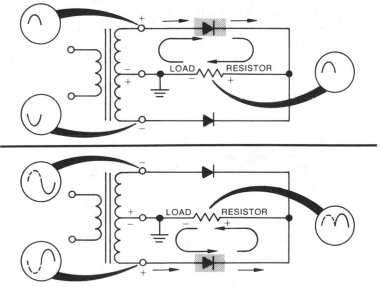

Experiment/Application—The Diode Rectifier

You can demonstrate the rectifying action of the semiconductor diode by a simple experiment. Suppose you connected up the circuit shown below using a transformer and rectifier diode.

DIODE RECTIFYING ACTION

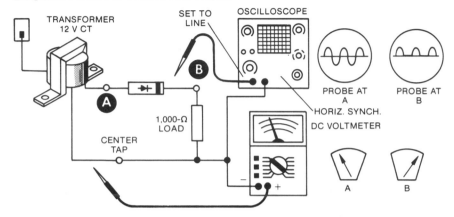

If you put the oscilloscope probe at point A, you will see the ac waveform at the power line frequency as you might expect. If the oscilloscope probe is moved to B, you will see the pulsating dc that results from the diode only conducting when the ac waveform is positive. Similarly, if a dc voltmeter is put across the ac source (use 30-volt scale) at point A, there will be no significant meter deflection, but if the voltmeter probe is moved to B, there will be a voltage of about 10–12 volts measured. Obviously, if the diode is reversed, everything reserves.

You can convert this circuit easily to a full-wave rectifier by using the other half of the transformer secondary and a second diode as shown below. As you can see from the oscilloscope trace, the pulsating dc is delivered on both halves of the ac wave, showing that both diodes are conducting alternately. By placing the scope probe at point B, you can see that when A is positive and C negative, diode CR1 conducts. During the other half cycle when A is negative and C positive, CR2 conducts. Thus, one diode conducts for each half cycle in a full-wave circuit.

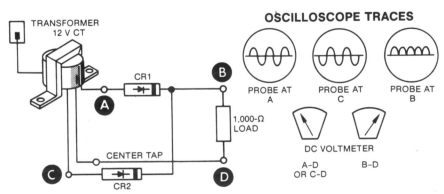

The Full-Wave Bridge Rectifier

To avoid the necessity and expense for a center-tapped transformer, the bridge rectifier circuit can be used to provide full-wave rectification. The bridge rectifier, just like the other rectifiers you have studied, changes ac voltage to dc voltage. Four diode rectifiers are hooked together with the ac input and the load as shown. As the upper terminal of the transformer secondary swings positive with respect to the bottom terminal, current flows from the upper secondary terminal through rectifier CR3, through the load, and then through rectifier CR2 back to the other side of the input.

Then, when the lower terminal of the transformer input swings positive, current flows through CR4 to the load and to the transformer via CR1. Notice that the current flow through the load is in the same direction during both half cycles of the input wave and that two current pulses are produced for each cycle. Thus, this is a full-wave rectifier. Therefore, the voltage developed across the load can be filtered just as the pulsating dc output from the full-wave rectifier circuit described on the previous page.

Because rectifier diodes are inexpensive, the full-wave bridge rectifier is very commonly used with modern solid-state electronic equipment. In many cases, a special package of four diodes can be obtained to avoid the extra wiring required in this circuit. Note that either polarity of voltage can be obtainable from the same diode configuration, depending on which of the load terminals are grounded. This is often convenient when large stud-mounted diodes are used, since the cathodes do not have to be insulated from the metal frame or chassis. Many stud-mount diodes are available with the anode and cathode leads reversed to allow operation with the stud at ground potential.

THE FULL-WAVE BRIDGE RECTIFIER

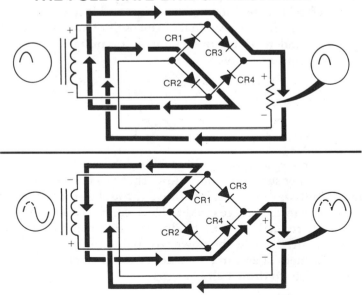

Experiment/Application—The Full-Wave Bridge Rectifier

You noticed that with the full-wave circuit demonstrated previously, only half of the secondary is active at a given time. The full-wave bridge, as you know, does not require a center-tapped transformer and still uses both halves of the ac wave. Thus, using the same transformer, we can get about twice the voltage that we got from the full-wave circuit by using the bridge circuit. Suppose we hooked up the bridge circuit as shown below using four diodes and the same transformer.

THE FULL-WAVE BRIDGE RECTIFIER

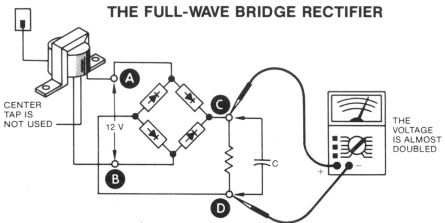

As you can see, ac voltage input (A–B) is doubled, and dc voltage (C–D) also is almost doubled as would be shown by a voltmeter test. Since we go through two diodes per half-cycle (instead of one as in the full-wave bridge) the diode drop is twice as great (or about 2 × 0.5 = 1 volt). Therefore dc output (filtered) is about (12 − 2 × 0.5) = 11 volts × 1.4 = 15.4 volts dc, or almost twice the previously measured voltage.

By using the oscilloscope we can examine waveforms as before and also see the effects of the filtering capacitor and load changes. Note that we must move the ground terminal of the oscilloscope to B for observing ac waveforms and to D for observing dc or pulsating dc waveforms. These would appear as shown below.

| SCOPE ACROSS TERMINALS A–B | SCOPE ACROSS TERMINALS C–D WITHOUT FILTER CAPACITOR | SCOPE ACROSS TERMINALS C–D WITH FILTER CAPACITOR |

You can't get something for nothing, however, since the current in the transformer secondary only flows through half the winding at one time in the full-wave circuit, but flows through the entire winding for both half-cycles in the bridge circuit. Thus, current available from a given transformer is reduced for the bridge circuit to about half that of the full-wave circuit. That is, the power available is about the same from a given transformer. Therefore, if we get twice as much voltage, we would expect to be able to draw half the current.

The Purpose of Filters

For most applications, a relatively pure dc voltage is required. However, the output from any rectifier circuit is pulsating dc. This pulsating dc has two components—the dc value and superimposed ac component called the *ripple*. If you put a dc meter across the load resistor of the output from a rectifier, you will get a voltage reading corresponding to the dc component. If you now use an ac meter through a blocking capacitor to keep the dc out of the meter, you will get a reading corresponding to the ac component or *ripple*. Ripple is usually undesirable in power supplies for electronic circuits and is reduced by filters and in many cases further reduced by voltage regulators. As you know, voltage regulators suppress output variations that result from input changes and hence will suppress ripple, too.

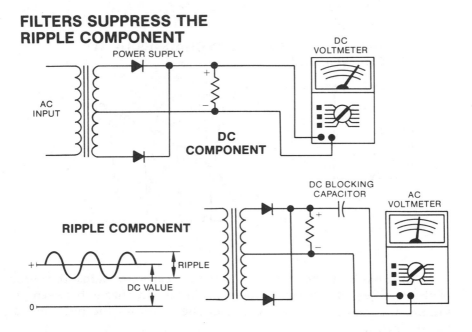

FILTERS SUPPRESS THE RIPPLE COMPONENT

The purpose of a filter then is to attenuate (reduce) the ac component while leaving the dc component unaffected. Filters have many configurations, and it is not unusual to have several filters on the same power supply feeding different circuits that can tolerate differing amounts of ripple. All filters use energy storage devices (capacitors and sometimes inductors) to provide filtering action. Because most solid-state systems use low voltages at high current, most filters use only large capacitors rather than LC filters, which you may have seen in vacuum-tube equipment that draw low currents at high voltages. Large, low-voltage capacitors are relatively inexpensive for this application.

The Capacitor in the Filter Circuit

If you remove the load resistor from the output of the rectifier and replace the resistor with a large capacitor, essentially pure dc will appear across the capacitor.

You know from *Basic Electricity* that, when a capacitor is placed across a battery, it eventually charges up to the battery voltage. The same is true when a capacitor is placed across the output of a rectifier. The rectifier starts charging up the capacitor every time it conducts. If the capacitor does not have time to charge up to the peak of the pulsating dc wave on the first half cycle, it will do so during the next few half cycles. After a few cycles have passed, there will be pure dc across the capacitor. Because current can flow in only one direction through the rectifier, the capacitor will not discharge between the peaks of the pulsating dc voltage (the rectifier diodes cannot conduct in reverse).

A CAPACITOR WITH NO LOAD CHARGES UP TO PEAK DC VOLTAGE

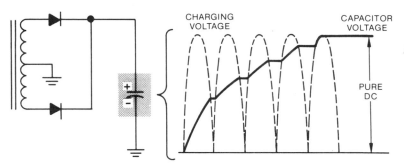

What has been the effect of placing the capacitor across the output of the rectifier? By charging up and holding this charge between pulsations, the capacitor has filtered out the ripple in the pulsating dc, leaving *pure* dc.

If a power supply did not have to supply current to circuits, pure dc voltage could ·be obtained simply by connecting a capacitor across the rectifier output. However, the operating circuits connected to the power supply output draw *load current*. The effect of this load current can be simulated by connecting a load resistor across the capacitor.

You know from your study of RC circuits in *Basic Electricity* that when a resistor load is placed across a charged capacitor, the capacitor will discharge through the load. The rate of the discharge will depend on the relative sizes of the capacitor and the load. The lower the resistance of the load, the more current will be drawn from the capacitor, and the faster will be the discharge. Conversely, for a given load, the larger the capacitor, the slower will be the discharge.

The Capacitor in the Filter Circuit (continued)

As soon as a load is connected across the capacitor of the filter circuit, the capacitor will begin to discharge and the voltage will start to drop. The voltage, however, will not drop to zero because a new voltage peak is provided by the rectifier output (60 per second for the half-wave rectifier and 120 per second for a full-wave rectifier).This voltage peak will recharge the capacitor that will then proceed to discharge through the resistor until the next voltage peak comes along. The result will be increased ripple. However, the ripple can be very much less than with no capacitor, depending on the size of the capacitor.

As you will learn, most solid-state electronic circuits use *low* voltages (5 to 50 volts), and very large capacitors in this voltage range are readily available (as high as 10,000 µF). Thus, the single capacitor is the most commonly used filter in solid-state equipment. For low-power circuits where very low ripple is required, additional filtering components are added to further remove the ripple.

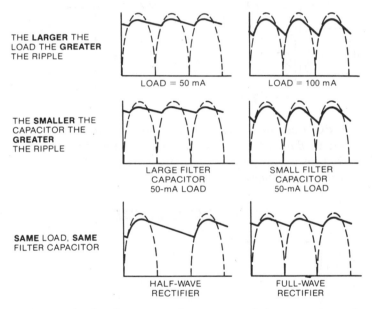

THE **LARGER** THE LOAD THE **GREATER** THE RIPPLE

LOAD = 50 mA   LOAD = 100 mA

THE **SMALLER** THE CAPACITOR THE **GREATER** THE RIPPLE

LARGE FILTER CAPACITOR 50-mA LOAD   SMALL FILTER CAPACITOR 50-mA LOAD

SAME LOAD, **SAME** FILTER CAPACITOR

HALF-WAVE RECTIFIER   FULL-WAVE RECTIFIER

The larger the load current drawn out of the capacitor, the larger will be the voltage drop between pulses and the larger will be the ripple. For a 60-Hz ac supply, half-wave rectifiers charge the capacitor 60 times per second, and there will be more time for the capacitor to discharge through the load than with a full-wave rectifier that charges the capacitor 120 times per second. Thus, the ripple will be greater for a half-wave rectifier than for a full-wave rectifier because the voltage will drop more between charging pulses. Thus, half-wave rectifiers are not usually used.

For special applications like aircraft and space ac systems, a 400-Hz supply frequency (or even higher) is used. This of course reduces the ripple proportionally for a given size capacitor and load.

The RC Filter

As you have learned, the larger you make the filter capacitor the lower will be the ripple. The usual filter in power supplies for solid-state electronic equipment is one or more large filter capacitors connected to the rectifier output. This is true for high-power circuits where large currents (amperes) at low voltages (5 to 50 volts) are delivered to the load, and the power-supply circuit resistance must be held to a minimum to avoid excessive power dissipation (I^2R loss) and voltage drop (IR loss). For the low-power portion of the electronic circuit (which also are usually more sensitive to ripple), additional filtering often is required. A filter commonly used is the *RC filter*. It consists of an additional resistor and capacitor arranged as shown in the figure below.

THE RC FILTER

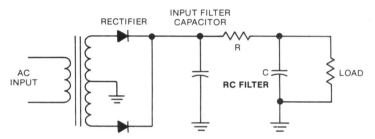

To understand how the RC filter works, we have to review what you learned in *Basic Electricity* about the voltage divider principle. Remember that in a voltage divider, the voltage divides in inverse proportion to the resistance values.

THE VOLTAGE DIVIDER

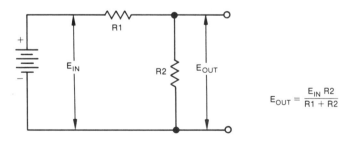

$$E_{OUT} = \frac{E_{IN}\,R2}{R1 + R2}$$

In the circuit shown, the output voltage is given by

$$E_{out} = \frac{E_{in} \times R2}{R1 + R2}$$

If you don't remember this, review voltage dividers in Volume 2 of *Basic Electricity*, which you must understand before you can understand the RC filter. The voltage divider principle can be applied to the RC filter to show how ripple is reduced by this circuit, as described on the next page.

The RC Filter (continued)

Consider the circuit shown below.

THE RC POWER SUPPLY FILTER

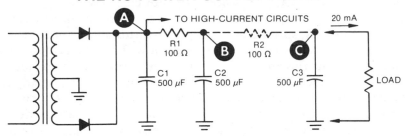

Assume for the moment that the high-current load at point A is such that 1.5 volts of ac ripple (120 Hz) is present at this point. Applying the voltage divider principle (for ac circuits), the ripple at point B is reduced by the R_1C_2 combination as follows:

$$E_{out} = \frac{E_{in}X_C}{(R + X_C)} \qquad X_C = \frac{1}{2\pi + C}$$

$$= \frac{1}{2 \times 3.14 \times 120 \times 0.0005} = 2.65 \text{ ohms}$$

Remember that to find the total impedance (Z) of an R and C in series we must use the relationship

$$Z = \sqrt{R^2 + X_C^2} = \sqrt{100^2 + 2.65^2} = \text{approx. 100 ohms}$$

For our circuit

$$E_{out} = \frac{E_{in}X_C}{Z} = \frac{1.5 \times 2.65}{100} = 0.04 \text{ volt}$$

Thus, the ripple at point B is 0.04 volt, an improvement of almost 38 to 1. If we were to add another section of RC filter, we would get a 38:1 further improvement for a ripple output of about 0.001 volt at point C. The major disadvantage of RC filters is that the series resistor has a voltage drop across it for the dc load as well. Thus, if the load current to point C is too high, the dc output voltage at C will be appreciably less than at point A. For example, if the dc load current is 20 mA, the drop across each 100-ohm resistor will be 2 volts (E = IR = 0.02 × 100 = 2 volts) or 4 volts total, which is not too great. However, if the load current were 200 mA, the voltage drop across each 100 ohms would be 20 volts (E = IR = 0.2 × 100 = 20 volts) for a total of 40 volts. This could be a serious problem for a single-section filter and disastrous for a two-section filter. Thus, the RC filter is pretty much confined to circuits that draw *little* current. Sometimes, the voltage drop in RC filters is used to advantage to lower the supply voltage to circuits that operate at less than the supply voltage. A major disadvantage of the RC filter is the fact that the load voltage is highly dependent on the load current. In some circuit applications, this poor regulation of voltage can cause difficulty.

The Inductor in the Filter Circuit

To reduce the large output voltage variations due to changing load demands, a filter requires a component that has a high ac impedance but a low dc resistance. As you know, an inductor has these properties. Inductors used in filters are called *filter chokes*. Filter chokes are usually iron core inductors, specifically designed for this application.

From your study of ac circuits, you know that a choke opposes any change of current flowing through it. Thus, a filter choke or inductor has an ac impedance $X_L = 2\pi f l$. Since a choke is made up of many turns of copper wire wound around an iron core, it can be made to have a low resistance to dc. Thus, a choke has the qualities that are needed in a filter circuit series element (replacing the resistor) to reduce ripple and yet have minimal effect on the dc output voltage with varying load demands because of the choke's low dc resistance.

ADVANTAGES OF THE FILTER CHOKE

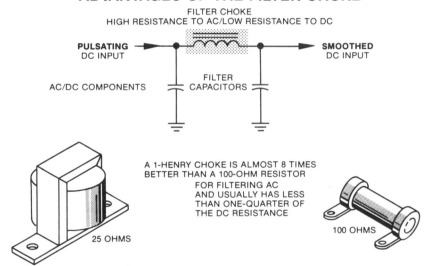

FILTER CHOKE
HIGH RESISTANCE TO AC/LOW RESISTANCE TO DC

PULSATING DC INPUT

SMOOTHED DC INPUT

AC/DC COMPONENTS

FILTER CAPACITORS

A 1-HENRY CHOKE IS ALMOST 8 TIMES BETTER THAN A 100-OHM RESISTOR FOR FILTERING AC AND USUALLY HAS LESS THAN ONE-QUARTER OF THE DC RESISTANCE

25 OHMS

100 OHMS

A filter choke can be employed in place of the series resistor used in the RC filter circuit to provide even better filtering. For example, a 1-henry choke will present a reactance of about 754 ohms to 120-cycle ripple from a full-wave rectifier and will typically have a dc resistance of only about 25 ohms. Thus, the variation in voltage output is less than with an RC filter.

Because of these excellent qualities, you will find that chokes are sometimes used in filter circuits of some electronic power supplies that are unregulated, but must supply high currents. Since filter inductors are relatively expensive, you will find they are rarely used in power supplies for solid-state electronic equipment. Instead, voltage regulators are used as a better method of reducing ripple and stabilizing the output voltage in solid-state equipment.

Filter Configurations—Capacitor Input Filters

The filter circuits used in solid-state electronic power supplies are configured in several different ways. The simplest type of power supply filter consists of a single capacitor across the load. This filter is used in most applications because high-capacitance low-voltage capacitors are readily available. For improved filtering in low-current application, the RC filter often is used to feed portions of the electronic circuits that require lower ripple. Filters with a capacitor on the input side are called *capacitor input filters* and provide dc output equal to the peak ac voltage. Thus, a transformer with a 10-volt rms (root-mean-square) secondary (total voltage for half wave, or bridge voltage from each secondary half for full wave), will produce a dc voltage of 1.4 times the rms value or 14 volts (less the rectifier drop that is typically about 0.5 volt). When equipment with capacitor input filters is used for high-current applications, the initial surge of current to charge the capacitor when the equipment is first turned on may be excessive. Sometimes, you will see a special resistor called a *thyristor* connected between the rectifier and the filter capacitor or in series with the transformer primary. The thyristor is a voltage-sensitive resistor that has a relatively high resistance when cold and a very low resistance when hot. Thus, it initially reduces the surge current when the equipment is initially turned on but does not interfere with circuit operation once the circuit is on.

CAPACITOR INPUT FILTERS
FILTER CONFIGURATIONS

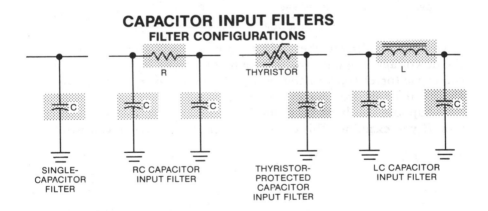

SINGLE-CAPACITOR FILTER

RC CAPACITOR INPUT FILTER

THYRISTOR-PROTECTED CAPACITOR INPUT FILTER

LC CAPACITOR INPUT FILTER

As described earlier, additional RC or LC filter sections can be cascaded to improve filtering. RC and LC capacitor input filters are sometimes called *pi filters* because their circuit diagram is shaped like the symbol for pi (π). Occasionally for special applications you will find filters with a choke or resistive input (no input capacitor). These are, however, almost never used in solid-state equipment.

Experiment/Application—The Filter Circuit

The effects of filtering on pulsating dc can be demonstrated simply by using the circuit shown earlier for the diode rectifier. Suppose you had the circuit shown below hooked up as before.

POWER SUPPLY FILTERS

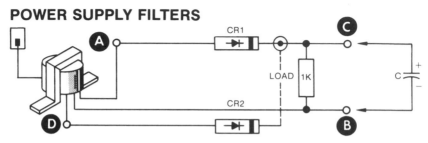

If CR2 is not connected, we have a half-wave rectifier circuit. The pulsating dc can be verified as before. If we put a 10-μF, 35-volt capacitor across terminals C to B (observe polarity!), we would observe that there is some filling in of the area between conducting half cycles. If we increased the capacitor to 100 μF, this filling in would be much more complete, showing that there is less ripple present as the capacitance increases, providing better filtering action. The scope trace under these conditions is shown below.

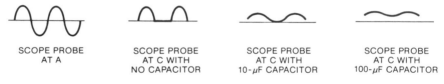

If you put a voltmeter between terminals C and B, you will find that the dc voltage is about 8-volts for a 6-volt rms ac input. This is because the capacitor charges to the peak voltage of $1.4 \times 6 = 8.4$ volts less about a 0.5-volt drop across the diode.

Suppose you hook up diode CR2 to make a full-wave rectifier system. If you examined this circuit with the oscilloscope, you would see:

As you can see, the ripple is greatly reduced with the full-wave circuit. Indeed, it is essentially gone with the 100-μF capacitor in the circuit. As you know, the greater the load, the greater the ripple for a given filter capacitor. You can see this by increasing the load. If you add another 1-K resistor in parallel with the 1-K resistor already there, to make the total load resistance 500 ohms, and then repeat these experiments, you will see that the ripple has increased. You also can do the reverse experiment by decreasing the load to 5 K and noting that the ripple is greatly reduced.

The Voltage Multiplier

Unlike vacuum-tube systems, solid-state electronic systems essentially never use voltage multipliers or transformerless ac power supplies because solid-state devices are almost invariably *low-voltage devices*. Therefore, a step-down transformer is required. An exception is the high-voltage section of a TV receiver or oscilloscope where there is a need for high voltage at very low current. These will be considered in Volume 4 of this series in some detail. In this unit, we will describe how voltage multipliers work (just for completeness), and we will restrict our discussion to the voltage doubler. However, voltage triplers and quadruplers, etc., can be configured using the *same* principles.

All voltage multipliers operate on the same principle; that is, capacitors are charged in parallel from the source and discharged in series across the load. With this basic idea in mind, you can figure out how any voltage multiplier works, regardless of configuration. Basically, one diode and one capacitor are required for each additional voltage step up. Thus, *two* diodes and *two* capacitors are required for a *doubler*, *three* diodes and *three* capacitors for a *tripler*, etc.

THE VOLTAGE DOUBLER

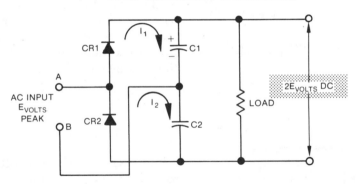

As shown in the circuit diagram above, for a simple half-wave voltage doubler during one half cycle, when terminal A of the ac input is positive, current I_1 flows through diode CR1 and changes capacitor C1 to the peak input voltage. During the other half cycle, when terminal B of the input is positive, current I_1 flows through diode CR2 and charges capacitor C2. If you trace the charging current through C1 and C2, you will find that they are charged so that their outputs add across the load. As you can see from the diagram, the capacitors cannot discharge through the source because of the diodes; however, they can supply current to the load resistance. Since each is charged to the peak input voltage, the output voltage is twice the peak input voltage. Actually, the voltage is not quite twice the input voltage because the charging is not 100% efficient, but it is close if the load currents are small.

Review of Simple Power Supplies/Rectifiers/Filters

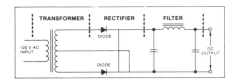

1. A SIMPLE AC POWER SUPPLY consists of a transformer, rectifier diode, and filter circuit.

2. The HALF-WAVE RECTIFIER uses a single-diode rectifier and produces a pulsating dc at the line frequency.

3. The FULL-WAVE RECTIFIER uses two diodes and a center-tapped transformer. Full-wave rectifiers produce a pulsating dc at twice the line frequency.

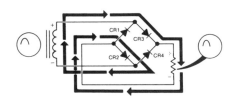

4. The BRIDGE RECTIFIER is used to avoid the complication of a center-tapped transformer. It uses four diodes.

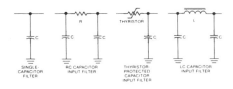

5. FILTER CIRCUITS are used to reduce ripple. For solid-state equipment, the filter capacitor is usually used alone and is used with RC filter circuits to reduce ripple further in low-power circuits.

Self-Test—Review Questions

1. List the essential elements of a simple power supply.
2. For the elements listed in question 1, describe the function of each.
3. Draw a schematic diagram of a half-wave rectifier power supply and describe how it operates.
4. Draw a schematic diagram of a full-wave rectifier power supply and describe how it operates.
5. Draw a schematic diagram of a bridge rectifier power supply and describe how it operates.
6. Compare half-wave and full-wave power supplies in terms of cost, complexity, and performance.
7. Most solid-state power supplies use a single large capacitor as the filter element. Why?
8. Draw an RC filter. Where are they used? Why?
9. If an RC filter has a resistance of 200 ohms and a capacitor of 200 μF and is connected to a power supply rectifier that has a 5-volt, 60-Hz ac ripple, how much will the ripple be reduced?
10. Draw the schematic diagram for a complete full-wave power supply with a capacitor output filter. If the rms ac voltage on each half of the transformer is 15 volts, what is the output voltage?

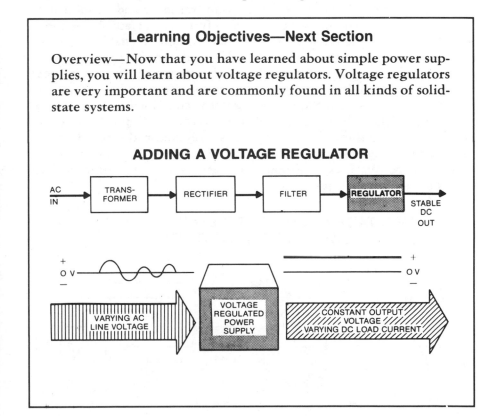

Learning Objectives—Next Section

Overview—Now that you have learned about simple power supplies, you will learn about voltage regulators. Voltage regulators are very important and are commonly found in all kinds of solid-state systems.

ADDING A VOLTAGE REGULATOR

Purpose of Voltage Regulators

You now are going to learn about voltage-regulated power supply circuits, which are used to provide better regulation and lower ripple than can be obtained from an unregulated power supply. Like other circuits you will learn about, voltage-regulator circuits range from very simple circuits using only one or two parts to very complex circuits requiring many components. However, all of these circuits operate in the same fundamental way as the basic regulator circuits.

You already know the two most important factors that affect the voltage output in a conventional power supply. The first factor is variation in the line voltage. When the ac line voltage goes up, the power supply output voltage goes up; similarly, when the line voltage goes down, the output voltage goes down.

The second factor that affects the voltage output is variation in output loading placed on the power supply by the operating circuit. When the operating circuit draws a small amount of current, the supply voltage is higher than when there is a large current demand.

Ripple also can be considered as an input voltage variation, and voltage regulators can provide for very high ripple reduction as well as line and load regulation. Because some of the examples of power supply regulators presented in this section contain transistor and IC solid-state amplifier circuits, you may not be able to fully appreciate all the interactions involved. Such examples are presented here to enable you to see real power supplies in operation, and you will appreciate the sophisticated concepts more as you learn about solid-state amplifiers in the next volume. As you will learn, many regulators now are available as complete integrated circuits with the entire regulator contained within a single case. Many of these devices have only three terminals (input, output, and ground) and contain, in addition to a basic regulator, the necessary circuits to protect against overcurrent, overvoltage, and short circuits. These regulators are very inexpensive and are very commonly used in modern solid-state electronic equipment.

VOLTAGE REGULATORS STABILIZE THE POWER SUPPLY DC OUTPUT

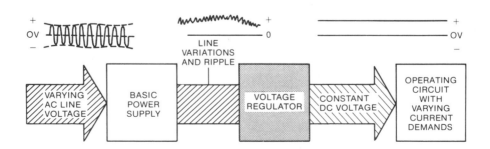

Simple Hand-Operated Voltage Regulator

A simple, hand-operated voltage regulator can be made by connecting a potentiometer across the power supply output terminals.

Assume that you have a 100-ohm potentiometer and a power supply with an output of 10 volts. Also assume that you want a constant output voltage of 5 volts. You first adjust your potentiometer so that the arm is at the point that provides 5 volts of output. If the output voltage rises due to an increase in ac line voltage or a decrease in current demand, all you do is move the center tap closer to ground (*decrease the resistance between tap and ground*) until you get 5 volts again. If the output voltage *falls* due to a decrease in the ac line voltage or an increase in current demand, all you do is move the center tap away from ground (*increase the resistance between tap and ground*) until you get 5 volts again.

You can see that the hand-operated voltage regulator works as a voltage divider, as you learned in *Basic Electricity*. You increase or decrease the resistance between ground and the output voltage tap to increase or decrease the output voltage back to the desired value whenever the supply voltage falls or rises for any reason.

The main fault with this method is that it requires a human operator, and he or she will always be *too slow*. First, the output voltage must change. Then you must notice that it has changed, and then you must increase or decrease the resistance between the voltage tap and ground to get back the desired voltage output. So what you really need is an *automatic* voltage regulator, one that automatically and quickly increases or decreases its internal resistance as the supply voltage falls and rises so as to maintain a constant voltage to the operating circuit.

HAND-OPERATED VOLTAGE REGULATOR

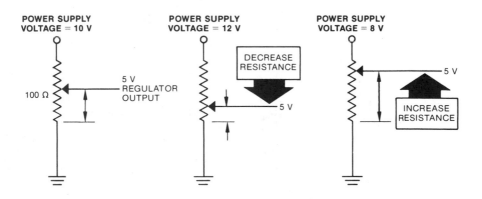

Diode Regulators

The constant formed (conducting) voltage drop of diodes can be used to regulate a dc voltage. Very inexpensive and simple regulator circuits can be obtained with simple diodes where the voltage required is very low, since the forward drop across a diode is only about 0.5 to 0.75 volt but is relatively constant irrespective of the current through it. In low-voltage applications, several diodes can be placed in series across the unregulated power supply output as shown. The diode characteristics and a number of diodes in series are selected so that the total voltage drop across all diodes is equal to the desired regulated output voltage. The series resistor R_s limits the total current (diode and load).

In this circuit configuration, the regulated dc output voltage will remain relatively constant for small variations in both unregulated voltage and operating circuit current demand.

Since the voltage drop across the diode string is constant, as more current is diverted to the load, the diode current decreases and vice versa. Thus, the apparent resistance (R) of the diode string (R = E/I) varies inversely with the current (I) through it since the voltage (E) is constant. It thus acts as an automatically varying resistance to keep the load voltage constant.

Some of the disadvantages of this circuit are as follows:

1. Diode characteristics are not perfect, and there will be some variations of forward voltage drop across diodes, therefore producing some variation in the regulated output voltage.
2. The regulated output voltage must be considerably lower than the unregulated supply.
3. The regulating circuit can consume a large portion of available power. Powerwise, the circuit is very inefficient.
4. For voltages higher than 3 or 4 volts, an impractical number of diodes is required since each diode has only about a 0.6-volt drop (about a 0.3-volt drop for germanium) across it in the forward-biased direction.

TYPICAL DIODE REGULATOR
(FORWARD BIASED)

Zener Diode Regulator

The zener diode regulator is very commonly used in solid-state electronic power supplies to eliminate some of the disadvantages of the forward-biased diode regulator. In this type of regulator, the diode is reverse-biased and operates in the reverse-breakdown *zener region* with a relatively fixed voltage, even with relatively large variations of reverse current. The figure below shows the voltage-current characteristics of a typical 15-volt zener diode. Single zener diodes can be obtained for operation from a few volts to over 100 volts. They also can be operated in series to provide for almost any voltage desired.

The simple diode regulator that has been discussed regulates the dc voltage output by controlling the amount of current passed through it to ground. This diode regulator is placed in parallel with the operating circuit load. Therefore, we may say that this regulator maintains a constant voltage output by shunting appropriate amounts of current away from the operating circuit. They are therefore known as *shunt-type voltage regulators*.

BASIC ZENER DIODE SHUNT REGULATOR

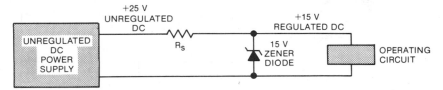

CHARACTERISTIC OF 15-VOLT ZENER DIODE

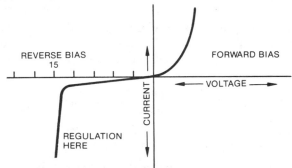

The zener diode regulator also is a shunt-type regulator. For the moment focus your attention on the operation of this simple zener diode shunt regulator. You will see from the circuit analysis on the next page just how a zener diode shunt regulator maintains a constant voltage to the operating circuit load by shunting appropriate amounts of current to ground.

The shunt-type voltage regulator is connected between the unregulated dc power supply via a series resistance R_s and the operating circuit, as shown in the diagram.

Zener Diode Regulator (continued)

The size of the resistor R_s is chosen to limit current through the zener regulator in the maximum current condition (no load). This must be done so that the current through the regulator will not exceed the zener diode's maximum rated current, should the operating circuit load become disconnected and all the current from the unregulated power supply flow through the regulator. We will assume that the zener diode is rated at 50-mA maximum current at 15 volts. Zener diodes come in various power ratings from a few hundred milliwatts to hundreds of watts. A 50-mA, 15-volt zener diode is rated as a 750-mW unit. Obviously, you must choose a zener diode with an appropriate power rating.

BASIC ZENER DIODE WITH NO LOAD

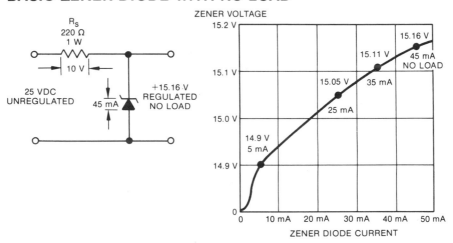

As shown, the output of the power supply is 25 volts, so the voltage across the resistor will have to be 25 volts − 15 volts = 10 volts. For these current and voltage conditions, the size of the resistor is calculated from Ohm's Law:

$$R = \frac{E}{I} = \frac{10 \text{ volts}}{0.05 \text{ ampere}} = 200 \text{ ohms}$$

The power dissipated in the resistor is found from the power formula

$$P = EI = 10 \text{ volts} \times 0.05 \text{ ampere} = 0.5 \text{ watt}$$

The resistance you want according to the above results is a resistor of 200 ohms rated at 0.5-watt minimum. The nearest 10% value of resistance available is 220 ohms so this has been chosen for use. This resistor would allow 45 mA to flow through the regulator with no load. A 0.5-watt resistor could be used, but a 1-watt resistor would probably be best since the size and cost are not much more and the danger of overheating will be reduced.

When no load is connected to the 15-volt output, 45 mA will flow through the zener diode. We can see from the voltage-current (VI) characteristics of the zener diode that the voltage output is 15.16 volts.

Zener Diode Regulator (continued)

Suppose you attached a 10-mA load (1.5 K) across the zener diode. Of the 45 mA flowing through the series resistor R_s, 10 mA flows through the load and 35 mA flows through the regulator. Referring to the VI characteristic of the zener diode, we see that the output voltage is now 15.11 volts.

If you now increase the load on the output to 20 mA (750 ohms), 25 mA will flow through the regulator, and the output voltage will be 15.05 volts. As long as the current flowing through the regulator is within its regulating range of about 5 mA to 50 mA, the regulator is able to keep the output voltage very close to 15 volts.

You can increase the load on the output terminals until the current through the load reaches about 40 mA. At this load, only 5 mA will flow through the regulator, which is about the minimum current that may flow through the zener diode and still keep the output voltage near 15 volts. Any further increase in load current will cause less than 5 mA to flow through the regulator, and it will cease to regulate. From this point on, the regulator will have little effect on the output voltage, and the output voltage will be determined only by the resistance of the dropping resistor R_s and the resistance of the operating circuit. For example, if the load current were to increase to 60 mA, the zener diode current would be zero, and the zener diode would no longer regulate the output because the voltage from the dividers R_s and R_{load} would be below the zener breakdown voltage. In many cases, a filter capacitor is put across the zener diode. This provides additional filtering in conjunction with the series resistor R_s.

ZENER DIODE REGULATOR WITH VARYING LOADS

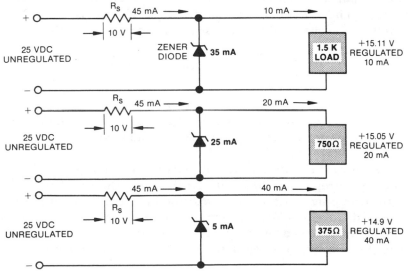

Zener Regulator with Supply Voltage Decrease

Zener diode regulators also regulate the output when the supply voltage changes. The input voltage rises as the line voltage rises, and falls as the line voltage falls. In addition, there often are other circuits connected to the supply before the regulator. When these circuits draw more current, the unregulated voltage drops; and when these other circuits draw less current, the voltage rises. The voltage-regulator circuit must provide a regulated voltage output in spite of these changes in unregulated voltage input.

ZENER REGULATOR WITH SUPPLY VOLTAGE DECREASE

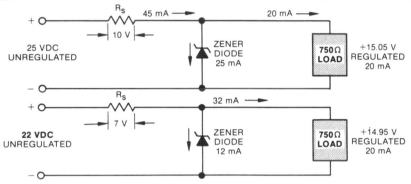

Under the operating conditions shown in the figure, 45 mA flows through the dropping resistor R_s, 25 mA flows through the regulator, and 20 mA flows through the load.

If the unregulated input voltage were to fall to 22 volts, the regulator would have to adjust its internal resistance so that the output voltage remains near 15 volts. Let's see if the regulator can make this adjustment. The voltage across the dropping resistor is now 22 volts − 15 volts = 7 volts. The current through the dropping resistor is

$$I = \frac{E}{R} = \frac{7}{220} = 0.032 \text{ ampere} = 32 \text{ mA}$$

The load draws 20 mA, and the remainder of the current (32 mA − 20 mA = 12 mA) flows through the regulator. The regulator will do its job as long as the current flow through it remains within its regulating range of from 5 to 50 mA.

In order for the regulator to fail in its job, the unregulated voltage would have to drop below 20.5 volts. At this point there will be 5.5 volts across the dropping resistor and 25 mA total current through this resistor. The load current would be 20 mA, and the regulator current would be 5 mA. Any further drop in unregulated voltage will cause less than 5 mA to flow through the regulator, and it will stop regulating.

Zener Regulator with Supply Voltage Increase

Now that you have examined what happens when there is a *fall* in the unregulated voltage supplied to the voltage-regulator circuit, suppose you find out what happens when this voltage *rises*.

ZENER REGULATOR WITH SUPPLY VOLTAGE INCREASE

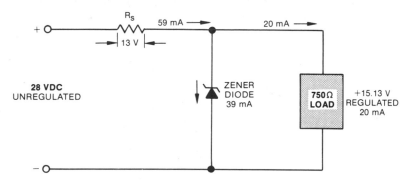

Suppose the input voltage increased to 28 volts. Under these conditions, there will be 13 volts across the resistor and the current flow through that resistor will be determined by Ohm's Law as follows:

$$I = \frac{E}{R} = \frac{13}{220} = 0.059 \text{ ampere} = 59 \text{ mA}$$

Since the load draws 20 mA at 15 volts, the remainder of the current (59 mA − 20 mA = 39 mA) must flow through the regulator. The regulator is designed to do its job of regulating if the current flow through it remains between 5 and 50 mA. The regulator can adjust for this unregulated voltage change and still maintain the regulated voltage output near 15 volts.

For the regulator to fail in its job, the unregulated voltage would have to go up to over 30.4 volts. At this point there would be 15.4 volts across the dropping resistor and 70 mA total current through this resistor. The load current would be 20 mA and the regulator current would be 50 mA. Any further increase in unregulated voltage would cause over 50 mA to flow through the regulator, and it could be damaged by heating from the excessive current.

As you might suspect, the zener diode regulator also will act as a filter. The ripple voltage can be considered variations in the input voltage and can be treated like those described above. The filtering of the supply via the zener diode regulator can be further improved by use of a capacitor across the zener diode as described earlier. This is a common practice because zener diodes produce considerable noise. The capacitor reduces noise and prevents it from getting into sensitive circuits.

Summary of Zener Regulator Operation

You have examined the principles behind the operation of a zener diode shunt-type voltage-regulator circuit. You have seen that the regulated voltage output will remain essentially constant as long as the current operating range of this regulator is not exceeded. By using a voltage-regulator circuit of this type, you can get a constant voltage output in spite of fairly large changes in unregulated power supply voltage and in spite of sizable changes in current drain by the operating circuit. As shown below, the zener diode can be used to regulate either a positive or negative supply.

THE ZENER REGULATOR
IS SIMPLE AND EFFECTIVE

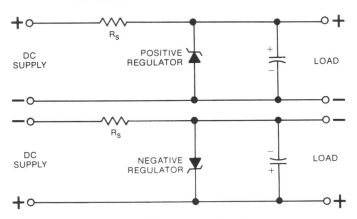

In many cases, you will find zener diode regulators are used with a higher voltage supply (even regulated) to provide special voltages for operation of specific circuits in electronic equipment. For example, if the basic regulated supply voltage for most of the circuits were 15 volts and you needed 5 volts for a specific circuit, a zener regulator would be very appropriate to use.

Zener diodes are sometimes used for overvoltage protection. A zener diode directly across the supply will act as an open circuit until the zener voltage is reached, and then it will conduct heavily. Therefore, if you want to protect circuits that operate at 5 volts from excess voltage, you could place a 7.5-volt zener diode directly across the supply. If the supply voltage should get to 7.5 volts by accident, the zener diode current can blow a fuse or open a circuit breaker, before the equipment is damaged.

Some disadvantages of zener diode regulators are:

1. The regulated output is limited to a fixed voltage depending on the particular zener diode chosen.
2. The regulated output voltage must be lower than the supply voltage.
3. The regulating zener diode can consume a large portion of the available power and can handle only limited amounts of power.

Experiment/Application—The Simple Zener Diode Regulator

You can see the effect of adding a regulator to a power supply by a simple experiment. Initially, suppose you checked the regulation of the full-wave bridge that you experimented with earlier. It might be advisable to mount the components on an appropriate board, called a *breadboard* because wooden boards, like breadboards, were used to hold components in the early days of electronics.

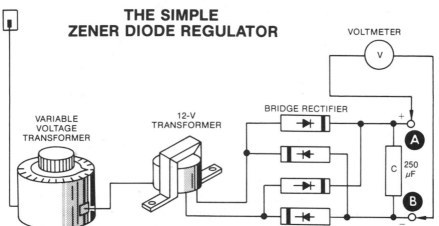

THE SIMPLE ZENER DIODE REGULATOR

When you turn on this simple supply, the output voltage will be about 16–18 volts dc when the line voltage is 120 volts of ac. As you change the line voltage, you will see dc output voltage change proportionately. Thus, if the line voltage is reduced 10% (to 108 volts), the dc output will be affected similarly and will drop 10% (14.4–16.2 volts). The reverse happens as the line voltage is increased above normal.

You can determine the dc regulation by setting the input at 120 volts of ac and measuring the dc terminal voltage. This is the no-load output voltage. Let's assume that the rated output current from the supply is 250 mA. This current will flow when a load of about 75 ohms (use a 10-watt resistor) is placed across terminals AB.

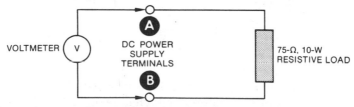

The voltage indicated on the voltmeter will be less because the supply regulation is not perfect. Suppose the voltage dropped from 18 volts to 15 volts, the load regulation of the supply would then be

$$\text{Percent regulation} = \frac{18-15}{18} \times 100 = 16.66\%$$

As you can see, the regulation is only fair, and while this might be alright for some users, it could be poor for feeding circuits that need stable voltage.

Experiment/Application—The Simple Zener Diode Regulator (continued)

You can see the improved regulation obtained by the addition of a simple zener diode regulator. Suppose you had a 10-watt, 12-volt zener diode such as the type 1N2976. You could use it to construct a simple zener diode regulator for the bridge rectifier power supply described in the earlier experiments. The 1N2976 is a stud-mounted zener diode that is available in several tolerances. 1N2976 has a 20% tolerance, the 1N2976A has a 10% tolerance, and the 1N2976B has a 5% tolerance. The 1N2976 also is available with the case as the anode (designated as 1N2976R).

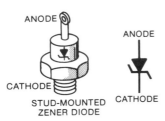

ANODE
ANODE
CATHODE
STUD-MOUNTED   CATHODE
ZENER DIODE

ZENER DIODE SPECIFICATIONS

1N2976 CHARACTERISTICS

NOMINAL ZENER VOLTAGE	= 12 VOLTS
TYPICAL CURRENT	= 210 mA
MAXIMUM DC CURRENT	= 720 mA
MAXIMUM REVERSE CURRENT	= 5 mA

For a full-load current of 250 mA at 12 volts, you can choose a total supply current of 300 mA (250 mA load + 50 mA minimum zener diode current). Assuming that supply voltage is 15 volts under load, you can calculate the series resistor value and wattage from Ohm's Law as follows:

$$R = \frac{E}{I} = \frac{(15 - 12)}{3} = \frac{3}{3} = 10 \text{ ohms}$$

$$P = I^2R = (0.3)^2(10) = 0.09 \times 10 = 0.9 \text{ watts}$$

Therefore, you would use a 2-watt resistor. The circuit can now be set up as shown below.

IMPROVED REGULATION FROM A ZENER DIODE

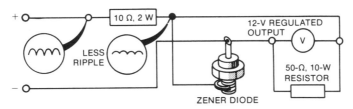

The no-load voltage will be about 12 volts. If you increase the load to about 250 mA (by placing a 50-ohm, 10-watt resistor across the output) you will find that the output voltage will change much less than before— typically about 0.25 volt (to 11.25 volts). The new regulation is now

$$\text{Percent regulation} = \frac{12 - 11.25}{12} \times 100 = \frac{0.025}{12} \times 100 = 2\%$$

If you changed the input voltage via the variable voltage transformer at the supply ac input, you would see little change at this regulated output. Also, as shown in the oscilloscope traces above, the ac ripple improves proportionately to the regulation improvement.

Zener Diode—Transistor Regulator

For many applications, it may not be desirable to use a zener diode-resistor combination because the load will vary widely, a high-current output is desired, or somewhat better stability is sought. In this case, a transistor can be added to the zener circuit to provide this capability. The circuit is configured as shown.

ZENER DIODE—TRANSISTOR REGULATOR

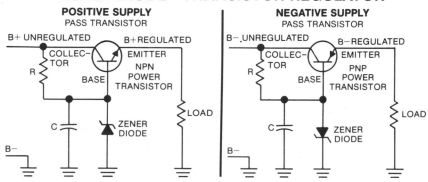

As you can see, the circuit closely resembles the zener diode regulator you studied earlier, except that the zener diode regulator is connected to the base of a transistor in the common-collector configuration. Since the base-emitter junction is forward-biased, it has a relatively constant voltage drop (e.g., 0.6 volt for a silicon transistor). As you know, for the emitter follower, the emitter voltage follows the base voltage very closely. If the base voltage is held constant, then the emitter voltage also will be constant.

From the schematic diagram you can see that the base voltage is held constant by the zener diode regulator circuit. Thus, the voltage at the emitter is regulated by following the constant base voltage.

In the common-collector circuit, the current required by the base is only $1/\beta$ times the current passing through the transistor from collector to emitter. Thus, a low-power zener diode can regulate the base voltage of a transistor that can pass many times this current. For example, if the transistor has a β of 100 and is to supply 1 ampere, the base current is approximately $\frac{1}{100}$ of an ampere or 10 mA. As you can see, a low-power zener diode working in a circuit with 10 mA flowing can regulate 1 ampere of current in the pass transistor. The transistor is called a *pass transistor* because it passes the total circuit current. Note that there is no improvement in power supply efficiency in this configuration since the power dissipated in the pass transistor is the same as it would be in the resistor of the simple zener diode regulator. Thus, it is usually necessary to heat-sink the pass transistor to eliminate the heat generated by the drop across it. The capacitor C can be added to reduce the noise from the zener diode as well as improve the ripple since it operates with resistor R to form an RC filter.

Zener Diode—Transistor Regulator (continued)

In some cases, the zener diode circuit cannot supply enough base current. In this case, a second transistor can be added to the circuit. This transistor acts as a current amplifier and also is in the common-collector configuration.

ZENER DIODE—TRANSISTOR REGULATOR

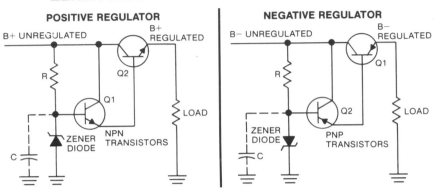

As you can see from the circuit diagrams, the pass transistor Q1 is as before but is driven at the base with a second transistor Q2 in the common-collector configuration. Transistor Q2 has its base connected to the zener regulator. Suppose transistor Q2 has a β of 10 and Q1 has a β of 100. If the load current is 10 amperes, then the base of Q2 needs $20/\beta = 0.5$ ampere into its base. The transistor Q1 can supply this current from its emitter with a base current of $0.5/\beta = 0.5/100 = 5$ mA. Thus, the zener regulator need only provide 5 mA for a load current of 10 amperes. The capacitor functions with R, as described on the previous page. It is apparent that the transistor is in series between the source and load. Thus, this type of regulator is called a *series* regulator, in contrast to the simple zener regulator, in which the regulator *shunts* the load and hence is called a *shunt regulator*.

The circuits shown here are quite practical and you will see them often in your work. While they are not the best regulators from a voltage-regulator standpoint, they can provide good regulation for many applications. Notice that for all these circuits (including those on the previous page) the unregulated voltage has to be higher by at least 1 or 2 volts for regulation to take place. This condition must be maintained at *all* times so that the lowest voltage (at the lowest part of the ripple cycle) must be above the minimum value for regulation to take place. You should note that the pass transistor dissipation is proportional to the voltage drop across it and the load current, rather than being a constant value as in the shunt regulator.

Note that the transistors are designated with the symbol *Q*. You will find this to be a consistent practice in your study of transistor circuits. In the next pages, you will learn about more elaborate *feedback* regulators. These are almost invariably of the series (pass transistor) type.

Experiment/Application—Voltage Regulator with Pass Transistor

As you learned, you can greatly improve the simple zener diode regulator by adding a pass transistor. When you do this, it is possible to use a much lower power zener diode because the zener diode does not have to handle all of the current. In addition, as you shall see, the voltage regulation is better and remains better at lower input supply voltages.

CHARACTERISTICS OF THE 2N3055 TRANSISTOR
POWER DISSIPATION 117 WATTS MAX. AT 25°C CASE TEMPERATURE

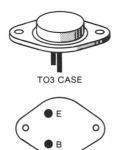

NPN SILICON HIGH-POWER AUDIO TRANSISTOR

COLLECTOR—BASE VOLTAGE (V_{cb})	100 VOLTS DC MAX.
COLLECTOR—EMITTER VOLTAGE (V_{ce})	60 VOLTS DC MAX.
EMITTER—BASE VOLTAGE (V_{eb})	7.0 VOLTS DC
MAX. COLLECTOR CURRENT (I_c)	15 AMPERES DC
MAX. BASE CURRENT (I_b)	7.0 AMPERES DC

TO3 CASE

COLLECTOR = CASE

The pass transistor we will use in this experiment is the 2N3055. As you can see from the above list of characteristics of the 2N3055, this transistor is capable of handling large voltages and currents, it is inexpensive and readily available and quite rugged for this application. Suppose you hooked up such a regulator as shown below.

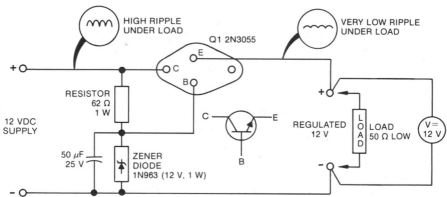

As shown in the diagram, the voltage will be about 12 volts and will change very little under load. Usually, the regulation will be better than 1%. If you connect an oscilloscope probe to the regulator input, you will find that the ripple is quite high under load, but the regulated output is almost pure dc. This is due not only to the regulating action but also to the additional filtering on the base of Q1, due to the 50-μF capacitor. You can run regulation tests on this supply as in earlier experiments/applications, and you will see that line and load variations have little effect on this regulator.

Feedback Regulators

In the previous pages you learned about the operation of simple shunt and series regulators using zener diodes as the regulating or reference elements. It should be pointed out that these simple regulators can regulate only at the operating voltage of the particular diode chosen; however, zener diodes of many various voltage ratings are available. These simple regulators do not provide very good voltage regulation. Because of their excellent characteristics, the feedback regulator is most commonly used today. While these have more complicated circuits, the availability of ICs that include most of the regulator circuitry has made these feedback-regulated power supplies very commonly used. In feedback regulators, excellent power supply regulation is obtained by the use of a *feedback circuit* that senses the regulated output and develops a control signal to oppose changes, usually via a pass transistor in series with the regulated output. Most feedback regulators can be classified as either *linear regulators* or as *switching regulators*. The type depends on the operating mode of the regulating element (usually a transistor). If the regulating element operates strictly in an on or off mode (transistor operating as a switch), the regulator is a *switching* regulator. If the regulating elements operate in an *active mode* (transistor operating linearly, as we saw earlier for the pass transistor), the regulator is a *linear* regulator.

As you will learn, switching-type regulators can be made very efficient and are therefore used mainly where large currents must be regulated or where high efficiency is most important. However, linear regulators are most commonly used when the load requirements are not too great.

BASIC SERIES FEEDBACK REGULATOR

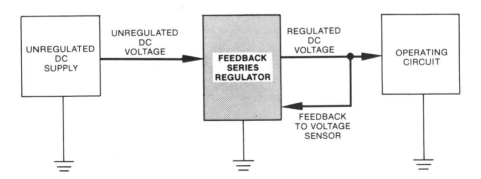

In addition, as you know, all linear regulators are of two basic operating types: *shunt* and *series*. Shunt regulators are in parallel with the load and regulate by shunting appropriate amounts of supply current to ground. On the other hand, series regulators are in series with the load and regulate by controlling the voltage across the series-regulating element so that the voltage to the load is maintained at a constant.

Linear Series Regulator

The linear series voltage regulator consists of a series (pass) transistor, a control element (dc amplifier), a dc reference voltage, and an output-sensing network.

From the diagram of the basic series regulator, you can see how regulation is achieved. The output-sensing network, made up of a resistance divider across the output load, constantly senses the voltage across the operating circuit and feeds this information to the control amplifier.

The control amplifier, usually a discrete transistor or an IC dc amplifier, compares the feedback voltage to a fixed voltage reference. The difference between these two voltages is amplified and applied to the base of the pass transistor. The polarity of the feedback difference voltage is such that if the voltage from the network starts to become greater than the reference (indicating that the load voltage is high), the current into the base of the pass transistor decreases, increasing the voltage drop across the pass transistor. Thus, the output voltage is reduced. If the load voltage starts to decrease, the opposite happens, thus stabilizing the voltage. This feedback control operates as soon as a small change begins to occur in either direction. Thus, the regulator is continuously and automatically adjusting to keep the output voltage *constant*. The basic improvement in this regulator over the simple regulators described earlier is that the control signal is obtained directly from the output of the regulator and the error signal (the difference voltage) is amplified before it is applied to the base of the pass transistor, thus making the feedback regulator very sensitive to small voltage changes.

LINEAR SERIES FEEDBACK VOLTAGE REGULATOR

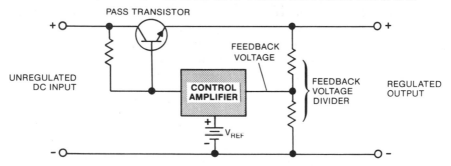

As you learned in *Basic Electricity*, load regulation is

$$\text{Percent load regulation} = \frac{\text{no-load voltage} - \text{full-load voltage}}{\text{no-load voltage}} \times 100$$

$$\text{Percent line regulation} = \frac{\text{input voltage change} - \text{load voltage change}}{\text{input voltage change}} \times 100$$

Linear series regulators commonly have voltage regulation better than 0.1%. Regulators of this type are very good at ripple reduction because of the fast response of the feedback loop.

A Simple Feedback Series Regulator

A simple practical feedback series regulator is shown below. In this circuit, the series-pass element is transistor Q2. The control amplifier is a single discrete transistor (Q1) that is operating in the common-emitter mode and is therefore a current amplifier. The emitter of Q1 is held at a constant voltage by the zener diode CR1. The base of Q1 is fed from the voltage divider R2–R3. As you can see, the base-emitter current of Q1 (dependent on the base-emitter voltage) increases if the load voltage wants to increase and decreases as the load voltage wants to decrease. The current in the collector circuit of Q1 increases as the base-emitter current increases and decreases as the base-emitter current decreases. The current for the base of Q2 and the collector of Q1 is obtained through resistor R1.

Suppose the voltage at the output of the regulator wants to increase. When this happens, the base current of Q1 increases. The current through R1 supplies the base of Q2 and the collector of Q1. Thus, when the base current of Q1 increases, more current is drawn in the collector circuit of Q1 and less is available for the base of Q2. This decreased current at the base of Q2 raises the voltage drop across Q2 and thus reduces the output voltage proportionately. The reverse happens when the load voltage drops.

Changes in line voltage are similarly sensed as changes in output voltage and hence are similarly controlled.

The reference voltage in this circuit is derived from a zener diode. The reference voltage circuit can be more complicated. The complexity of the reference voltage circuit depends on the required accuracy of the output voltage to the load, which is highly dependent on the accuracy of the reference supply.

As you also can see, the regulation depends on how sensitive the control amplifier is to voltage changes; therefore, the greater the amplification in Q1, the better the regulation. As you will see a little later in this volume, the control amplifier may be a much more complicated amplifier to give very good regulation.

SIMPLE SERIES REGULATOR

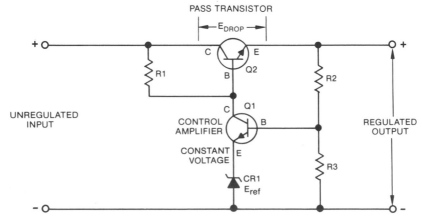

Switching Regulators

In the previous pages you learned about linear regulators. Now you will learn about another kind of regulator—the *switching regulator*. The switching regulator is much more efficient than the series linear regulator.

The efficiency of a regulator is its ability to use as little power as possible inside the regulator in performing its function. In linear regulators, the regulating element (usually a series-pass transistor) operates in the region between fully on (saturation) and fully off (cutoff). The pass transistor can be looked upon as a variable resistor, and the power dissipated by it is equal to the voltage across it and the current through it ($P = EI$). For even modest operating conditions, the power dissipation can be considerable. For example, a 5-volt drop in a 10-ampere supply will require the pass transistor to dissipate 50 watts.

TRANSISTOR SWITCHING REGULATOR

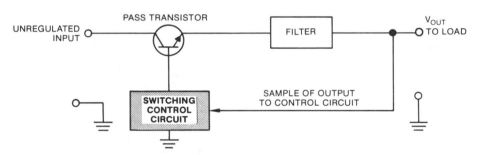

In switching regulators, the regulating element (usually a transistor or SCR) operates in the *on* or *off* condition. In these two operating modes, the control element is very efficient since it is either off (current = 0) or fully on, where the voltage drop is a few tenths of a volt and therefore the power dissipated in the transistor is minimized. The switching regulator operates on the same basic idea as the vibrating regulator you learned about in *Basic Electricity*. You will remember that the average voltage from a rapidly switched dc source is proportional to the ratio of on to off time. If this switch is closed (on) all of the time, you get 100% of the dc voltage. If it is open (off) all of the time, you get 630% of the dc voltage. If it is on half the time and off half the time, you get an *average* voltage of 50%, etc. As you can see, the voltage can be varied over its complete range by adjusting the ratio of on to off time. Thus, a switching regulator varies the ratio of on to off time, in accordance with the line or load voltage changes to keep the average voltage at the output constant.

In the switching regulator the control circuit generates a signal that switches the pass transistor on and off at a very high rate with the ratio of on to off being controlled so that the output is held constant. Since the output consists of a series of dc voltage pulses, these must be smoothed to get an average dc voltage. This is accomplished in the filter circuit as shown above.

Switching Regulators (continued)

A basic switching transistor regulator is shown below. In this circuit, you recognize the series switching transistor and the feedback network resistor divider because they are exactly the same as described previously in linear regulators. What you may not recognize are the elements comprising the filter circuit and the control circuit.

The filter shown consists of an inductor and capacitor. When the pass transistor switch is on, current flows through the inductor to the load. When the switch turns off, the energy stored in the filter circuits L and C keeps the current flowing in the same direction, through the *catch diode* that completes the circuit. The inductor-capacitor-diode filter provides a steady dc output, and removes the switching frequency component of ripple just as any filter will.

THE BASIC SWITCHING REGULATOR

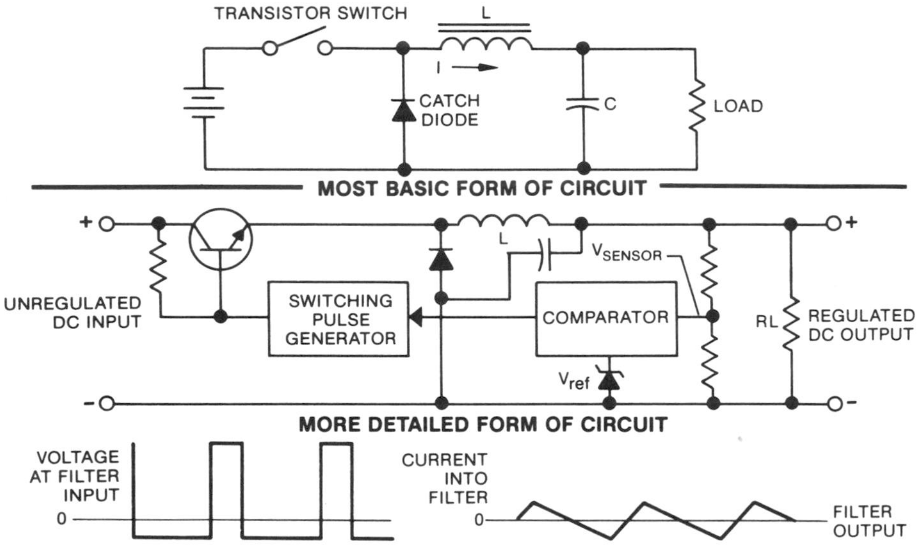

TRANSISTOR SWITCH   L

I →

CATCH
DIODE

C   LOAD

— MOST BASIC FORM OF CIRCUIT —

+   +

L

V$_{SENSOR}$

UNREGULATED
DC INPUT

SWITCHING
PULSE
GENERATOR

COMPARATOR   RL   REGULATED
DC OUTPUT

V$_{ref}$

–   –

MORE DETAILED FORM OF CIRCUIT

VOLTAGE
AT FILTER
INPUT
0

CURRENT
INTO
FILTER
0

FILTER
OUTPUT

The control circuit is made up of a voltage-reference circuit, a comparator circuit, and a switching pulse generator that drives the base of the pass transistor. The comparator circuit and voltage-reference circuit are similar to that used in the linear regulator. Their purpose is to generate a voltage proportional to the error in regulated output voltage.

The pulse generator accepts the comparator voltage and produces *turn-on* pulses of variable width to control the on-off ratio of the series transistor. It is important to note that on-off pulse width is variable because this is the key to the switching regulator's control, as described previously. The switching frequency is usually between 10 kHz and 20 kHz to minimize the filter requirements. As mentioned earlier, the switching regulator is used in cases where high efficiency of the regulator is required to conserve power when the total current drawn is so high that linear series regulation losses would be excessive.

This section reviews typical power supplies of increasing complexity that also make use of the rectification, filtering, and voltage regulation techniques you learned about in previous sections.

Simple Half-Wave Rectified Zener-Regulated Battery Charger

This simple zener-diode-regulated charger is very popular as a low-power, low-cost battery charger. It is practical for up to about 120 mA output. Although battery-charging time is relatively slow, it is preferred over constant-current trickle charging. The 1,000-ohm rheostat adjusts charging current and the 250-ohm rheostat adjusts voltage output.

BATTERY CHARGER POWER SUPPLY

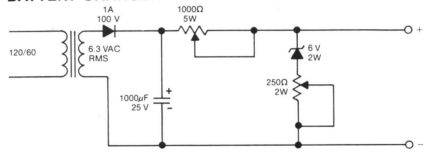

Adjustable Voltage-Regulated Power Supply

This power supply produces regulated dc voltage output of up to 30 volts. With components shown, up to 1 ampere can be supplied to operate electronic circuits. The circuit uses a full-wave bridge rectifier, a capacitor filter, and a series-pass regulating transistor. The 10-K potentiometer can be adjusted to alter the controlling voltage to the series-pass transistor; this in turn will adjust the regulated dc output voltage.

ADJUSTABLE VOLTAGE-REGULATED POWER SUPPLY

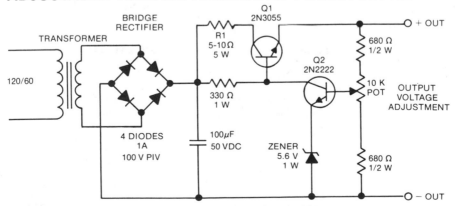

Power Supply Using Three-Terminal IC Regulator

Three-terminal integrated circuit regulators were originally introduced as voltage regulators for use on computer logic cards. Now they also are used in many regulated power supplies. These devices are available in a wide range of current and voltage ratings, including both positive and negative regulators. They are very simple to use and are the *preferred regulator* for most low- and medium-power requirements. Most of these devices provide internal current-limiting, short circuit protection, over-voltage protection, and automatic thermal shutdown build into the IC. In addition, they have very good voltage regulation.

POWER SUPPLY USING THREE-TERMINAL IC REGULATOR

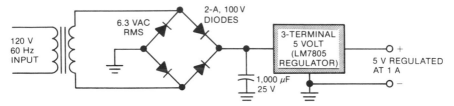

The circuit shown is suitable for producing 5-volt regulated dc using a National Semiconductor LM7805 regulator IC and supplying a continuous load of up to 1 ampere. The simplicity of these devices is apparent from the schematic diagram. Regulators of this type are available typically to provide $+5$, -5 $+12$, -12, $+15$, -15, $+22.5$, and -22.5 volts. Because of this built-in protective circuitry, they are almost indestructible. For example, a direct short across the output will not destroy them.

Low-Power IC Supplies

The schematic diagram below shows a simple, regulated power supply using an integrated circuit regulator for adjustable voltage output. Unregulated dc is supplied to the IC regulator (an RCA CA3085). The IC provides the necessary reference, amplifiers, and protective circuits and has built-in current limiting. The IC alone has output-current capability of up to 100 mA. If more current is needed, the IC regulator can be used to drive a pass transistor.

BASIC INTEGRATED CIRCUIT POWER SUPPLY

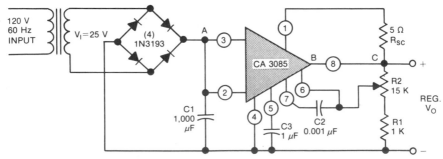

Experiment/Application—An IC Regulator

As you know, there are simple IC regulators that can provide excellent regulation, and short circuit and other protection within an integrated circuit package. One such IC device is the XX7812 IC regulator. (The two-letter prefix before the IC number is determined by the manufacturer of the device; for example, MC7812, LM7812, UX7812, etc.) These are all essentially identical units with the *12* in the 7812 giving the output voltage. Thus, a 7805 is a 5-volt unit, a 7815 is a 15-volt unit, etc. You will experiment with a positive regulator, but negative regulators also are available. Some characteristics of the 7812 are shown below.

CHARACTERISTICS OF THE 7812

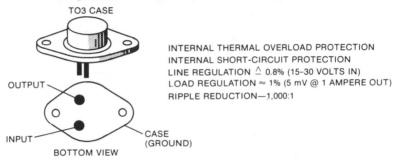

INTERNAL THERMAL OVERLOAD PROTECTION
INTERNAL SHORT-CIRCUIT PROTECTION
LINE REGULATION △ 0.8% (15-30 VOLTS IN)
LOAD REGULATION ≈ 1% (5 mV @ 1 AMPERE OUT)
RIPPLE REDUCTION—1,000:1

You can connect the 7812 to the basic power supply as shown below.

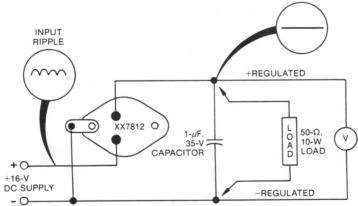

As you can see, the circuit connections are very simple since most of this has been done inside the IC package. This points up one of the major advantages of IC devices as well as their extreme compactness, since the components necessary to do the same job with *discrete* circuits would require an *entire printed circuit board*.

Suppose you test the regulation of the IC regulator. You would find that for a ±10% change in line voltage, there would be less than 0.1 volt change in output; for a load change of about 250 mA, the output voltage would change less than 0.25 volt. When you check for ripple reduction, you will find that there is more than *1,000:1 reduction* in ripple because of the complex feedback circuit in the IC. You can put a direct short on the output of this regulator, leave it there for a time, remove it, and the voltage will return immediately to 12 volts without harm.

The Need for Inverters and Converters

Many times, an ac line source is not available; other times, the desired value of dc voltage is not available from the primary dc power source. Thus, there are many instances in which dc-to-dc *converters* or dc-to-ac *inverters* may be used to provide the desired voltage. An *inverter transforms dc to ac*. If the ac output is rectified and filtered (and regulated sometimes) to provide dc again, the overall circuit is called a *converter*. The converter therefore is a combination of an inverter and an ac power supply.

Converters are needed to change *voltage level* from a dc source. Because these two circuits use an inverter as the common element, they are covered together in this section.

COMPARING INVERTERS AND CONVERTERS

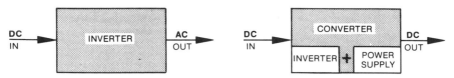

Both inverters and converters use a *chopper* to convert dc into pulsating dc or ac. In the simple chopper circuit shown, a switch (the chopper) is connected between the load and a dc voltage source. If the switch is alternately closed and opened for equal intervals, an ac-like waveform is produced and the average voltage across the load is equal to half the supply voltage. Waves of this type are called *square waves*.

BASIC CHOPPER CIRCUIT

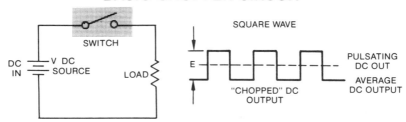

The pulsating square-wave dc can be applied to a transformer so that its voltage can be stepped up or down. When an appropriate rectifier, filter, (and possibly regulator) are added, then a dc output can be obtained. Several windings can be used for multiple voltage outputs.

Inverters are used to drive ac equipment such as motors, electric shavers, fluorescent lighting, or electronic equipment such as radio and television receivers originally designed for ac operation, where only a dc source is available. Converters are used to provide the operating voltages for electronic equipment such as radio and television equipment, communication sets, etc., where the dc voltages required are different than those available from the dc source.

The Driven Inverter

The simplest inverter circuit to understand is the *driven inverter*. The driven inverter is usually used only when it is necessary to precisely control the frequency of the ac wave. In these inverters, the square wave is generated in an external circuit. We will learn how this is done in later volumes of this series. For our present purposes, let's assume that such a source is available.

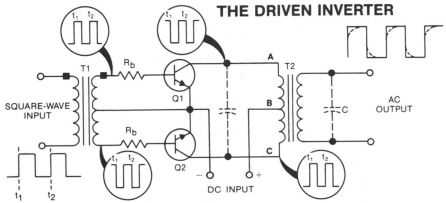

THE DRIVEN INVERTER

As shown, the input square wave is supplied to the transformer that is center-tapped to provide an ac wave of *opposite* polarities to the bases of Q1 and Q2. For the npn transistors, the transistors Q1 and Q2 conduct during alternate halves of the square wave when the wave is positive. Thus, for the condition shown, Q1 conducts when its base is driven positive. Q2 is off since its base is negative. During this interval, current flows from the positive (+) terminal of the supply, through half of T2 (B to A), through saturated transistor Q1 to the negative (−) terminal of the supply. Obviously, on the alternate half cycle, the situation reverses and Q2 conducts. Since the total supply voltage is applied to each half of the transformer primary, on alternate half cycles, the total ac voltage (peak-to-peak) across the transformer is equal to twice the dc supply voltage. This alternating waveform is supplied to the output at the proper level via the secondary of transformer T2.

Resistors in series with the base of transistors Q1 and Q2 limit the base current to a safe value.

A square wave has in it many harmonics of the fundamental ac frequency. Thus, a 60-Hz square wave will have components at 120 Hz, 180 Hz, 240 Hz, etc. These can be reduced somewhat in magnitude by putting a capacitor (C) across the output (or from collector to collector) at the transformer primary. The capacitor is chosen so that it partially shorts out these harmonic frequency components without affecting the fundamental frequency component. In addition, this capacitor suppresses high-voltage transient pulses that can occur when current is switched abruptly in an inductor. These large voltage transients can destroy semiconductor devices by exceeding their peak-inverse rating.

The Push-Pull Self-Excited Inverter

The *push-pull self-excited inverter* is probably the most widely used inverter. As an inverter the circuit provides an ac square-wave output. As before, rectifiers, filters, and regulators also can be used with the inverter to make a dc-to-dc converter.

A schematic diagram of a typical self-excited push-pull inverter is shown below. As shown, the inverter is similar to the driven inverter, except that the drive signal for the bases of the inverter transistors is taken from the output transformer itself. These inverters are called *push-pull* because the transistors work in sequence.

THE SELF-EXCITED INVERTER

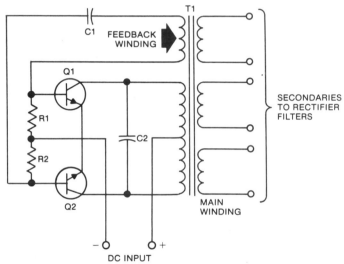

In this inverter, the operation starts when power is applied. One transistor starts to conduct (because of noise or circuit imbalance) and current is drawn through the transformer primary on that side. For example, suppose transistor Q1 starts to conduct. The current will increase through the inductance, and the changing field induces a voltage into the feedback winding that is connected to the bases of the transistors Q1 and Q2. The polarity of this winding with respect to the main winding is such that as transistor Q1 conducts and current flows from the transformer center tap (+ input) to the collector of Q1, the base of Q1 is driven more positive by the voltage induced in the *feedback winding*. Thus, the more current change occurring in the collector circuit of Q1, the harder its base is driven. This *feedback*, as it is called, is positive or *regenerative* because it chases itself to build up higher and higher unless stopped by some limitation in the feedback path. During this interval, the base of Q2 is negative, and hence Q2 is nonconducting.

The current drawn by Q1 through the main winding of the transformer magnetizes the core. This increases as the current increases through Q1, but finally the core reaches saturation. As you know, when this happens no more flux lines are created and the voltages induced in the other windings of the transformer will drop to zero.

The Push-Pull Self-Excited Inverter (continued)

When these voltages drop to zero, Q1 also loses the positive drive on its base and stops conducting. As the magnetic field in the transformer collapses, it induces a reverse-polarity voltage in the windings that turns on Q2, and the current through the collector circuit of Q2 induces a voltage in the feedback winding of the transformer that keeps Q2 on, and the cycle repeats as described previously.

As you can see from the previous explanation, the *transformer* plays a critical role in the operation of the self-excited inverter. Indeed, its characteristics determine the operating frequency of the inverter to a major extent. Thus, self-excited inverters are designed around a particular transformer and will have different operating characteristics if the transformer type is changed. The resistors R1 and R2 are bias returns for the bases of Q1 and Q2. The capacitor C2 is a *despiking* capacitor, and C1 provides dc isolation of the bases of Q1 and Q2. You might wonder why the feedback winding isn't center-tapped. Although sometimes it is, it has been found that Q1 and Q2 drives are better balanced (as transistors are changed) if this configuration is used.

There are some disadvantages to using the two-transistor, one-transformer self-excited inverter at high-power levels (over 50 watts). The most important of these is that peak collector current is independent of the load and depends on the transformer and transistor characteristics. This disadvantage can be partially overcome by adding a second transformer to the circuit as shown below.

THE PUSH-PULL SELF-EXCITED INVERTER

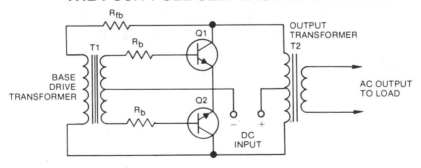

With two transformers, the base-drive transformer T1 controls the inverter switching at base-circuit power levels rather than collector-circuit power levels. A separate conventional output transformer T2 transfers power to the load. The peak collector current in this configuration is now determined by the load impedance. Thus, this circuit provides higher circuit efficiency than single-transformer configuration.

DC-to-DC Converters

As you learned earlier, the dc-to-dc converter is nothing more than an inverter and a power supply hooked together. The main function of these devices is to convert dc at one voltage to dc at another voltage. These converters do this with high efficiency. It doesn't matter what the inverter configuration is so long as it produces an ac waveform that can be transformed to the proper dc voltage for the power supply to operate. One typical converter is shown below to convert a 12-volt (nominal) input from an automobile battery to +5 volts at 1 ampere, and ±12 volts at 100 mA each.

The main inverter is self-excited using inverter transistors Q1 and Q2 in a push-pull configuration. As you studied earlier, capacitors C3 and C4 with inductor L1 form an input filter to keep the switching transients generated by the inverter from getting back into the dc input line and possibly causing interference with other equipment using the same dc input source.

One secondary winding (lower) is used to provide ac to the bridge rectifier that has a capacitor filter (C7) at its output. The rectified-filtered output is about 8 volts. The 8 volts of dc is applied to a three-terminal IC regulator (an LM7805) that provides 5 volts at 1 ampere.

The other secondary winding (upper) is center-tapped for full-wave rectifier operation. As shown, there are two sets of full-wave rectifiers connected to this winding (for economy) so that a positive and negative voltage are obtained from a single winding. Obviously, two separate windings also could have been used. The positive and negative rectifier outputs are filtered by capacitors C5 and C6 respectively. These provide 15 volts of dc to the zener diode shunt regulators made up of R5 and CR9 for positive supply and R6 and CR10 for negative supply.

As you can see, the converter is made up of components that you have already studied. You will find that while there are many variations in inverters and converters, all operate with the principles outlined here.

THE DC-DC CONVERTER (AUTOMOBILE)

Review of Regulated Power Supplies/Inverters and Converters

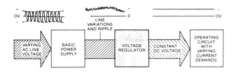

VOLTAGE REGULATORS STABILIZE THE POWER SUPPLY DC OUTPUT

1. **VOLTAGE REGULATORS** are devices associated with power supplies to keep the load voltage constant with changes in the line voltage or load current.

PERCENT LOAD REGULATION =

$$\frac{\text{NO-LOAD VOLTAGE} - \text{FULL-LOAD VOLTAGE}}{\text{NO-LOAD VOLTAGE}} \times 100$$

PERCENT LINE REGULATION =

$$\frac{\text{INPUT VOLTAGE CHANGE} - \text{LOAD VOLTAGE CHANGE}}{\text{INPUT VOLTAGE CHANGE}} \times 100$$

2. **PERCENT REGULATION** is a measure of how well the output voltage remains constant as the input voltage or load current changes.

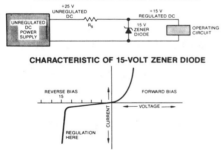

BASIC ZENER DIODE SHUNT REGULATOR

CHARACTERISTIC OF 15-VOLT ZENER DIODE

3. **SIMPLE ZENER DIODE SHUNT REGULATORS** use the constant voltage characteristics of zener diodes to regulate against load and line changes.

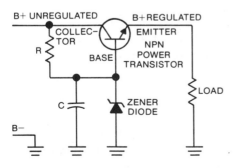

4. **ZENER DIODE PASS-TRANSISTOR REGULATORS** use a zener diode for the base reference of a pass transistor, thus providing improved regulators.

Review of Regulated Power Supplies/Inverters and Converters (continued)

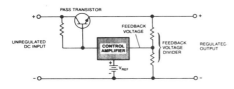

5. FEEDBACK SERIES REGULA-TORS use feedback of the output voltage to adjust the circuit for constant output. They are the best type of regulator and provide high stability against line and load variation.

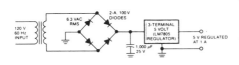

6. IC REGULATORS contain all of the required circuits for a feedback regulator in one package along with protective circuitry. They are very commonly used today for regulator circuits either with or without pass transistors.

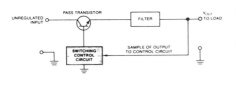

7. SWITCHING REGULATORS use transistors as choppers to provide a variable-duty-cycle pulsating dc output that is averaged by filtering the dc. The regulation is obtained by adjusting the duty cycle for constant output.

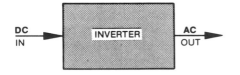

8. INVERTERS are devices that convert dc to ac by using the switching action of transistors.

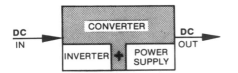

9. CONVERTERS are devices that convert dc to dc by using an inverter plus a power supply.

Self-Test—Review Questions

1. Draw a simple zener diode regulator circuit. Describe how it works as the line and load change.
2. Calculate the series resistor required for a zener diode regulator that will handle load currents between 50 and 200 mA at 15 volts. The 15-volt zener diode has a maximum current rating of 400 mA. The power supply has a nominal output voltage of 19 volts. What is the voltage rating of the resistor?
3. Draw a zener regulator with a pass transistor. How would you change the circuit above if the transistor had a β of 20?
4. Draw a feedback regulator circuit block diagram. Explain the function of each element.
5. Why would you prefer to use a feedback regulator rather than a zener regulator?
6. What role have ICs played in making feedback regulators more popular? If an IC regulator has a maximum current capability of 250 mA, show how it can be used to regulate a load current of 2 amperes.
7. Describe how a switching regulator differs from a conventional feedback regulator. Where would these regulators be used?
8. Draw a diagram of a self-excited inverter. Briefly describe how it works and where it is used.
9. Add the necessary elements to the diagram you drew for question 8 to make a converter. Briefly describe how it works and where it is used.
10. If you have to specify a regulated power supply to deliver 12 volts at 1 ampere, list some of the features you would want to see included.

Learning Objectives—Next Three Sections

Overview—Now that you know how power supplies work, you will be introduced to their servicing procedures. You also will be introduced to battery-type power supplies and their servicing procedures.

SERVICING AC POWER SUPPLIES

VOLTMETER

ELECTRONIC
UNIT

EXTERNAL LOAD

Types of Failures

If a power supply fails, it may be due to a problem *internal* to the supply or a problem *induced* by the operating circuit *external* to the supply. Therefore, as a first troubleshooting step in diagnosing the problem, the power supply should be electrically disconnected from the operating circuit if possible. If the problem originally was not the power supply and there has been no damage to supply components, the power supply output voltage should now return to normal. An external load should then be placed on the output to be sure the supply functions properly under load. If the supply functions normally under these test conditions, then the original problem lies in the operating circuit, which you will learn to service in later volumes.

If the power supply is defective, you can proceed to check components, as you already have learned in *Basic Electricity* for transformers, capacitors, resistors, and inductors. You can check diodes and transistors with an ohmmeter, as you learned earlier in this volume. ICs are very difficult to check, even for an expert, and the usual procedure is to *change* these if it becomes likely that these are the source of difficulty. This should be done only as a *last resort* since this replacement is difficult to do mechanically.

If the power supply contains a regulator with protective elements, the chances are that there probably will be no component damage due to current overload. Most three-terminal IC regulators are so well protected internally that they are almost indestructible except when too much input voltage occurs, which is unlikely in most cases. If there is no supply voltage due to overload, the supply output should immediately return to normal when the external failure condition is removed. If the regulator incorporates a thermal overload control that is tripped because of excessive power dissipation in the supply, the supply will require a cooling-down period before returning to normal output after the failure condition is removed.

SERVICING AC POWER SUPPLIES

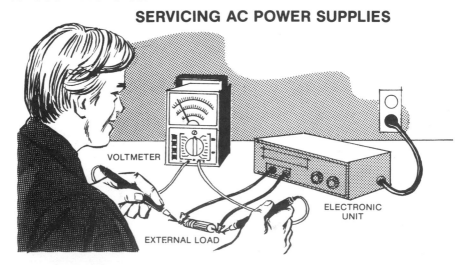

VOLTMETER

ELECTRONIC UNIT

EXTERNAL LOAD

Types of Failures (continued)

Other possible power supply failures include (1) loss of filtering due to capacitors that have lost their capacity and (2) loss of output voltage because of a shorted output capacitor or open-circuited resistor or choke. A more costly failure could occur in the power transformer. Such failures, such as a shorted winding, would produce reduced output dc voltage or no voltage at all; an open winding also would produce no output voltage.

The main reason for IC regulator failures is overdissipation in the internal series-pass transistors. If this happens, there will be no output voltage but the input voltage will be normal. Be cautious here because a shorted output will cause the same conditions to occur—that is, no voltage output because the regulator may have been shut down by the short-circuit protection provided internally.

Another cause of IC failures is filter capacitors. If the input filter capacitor is defective and operating with excessive ripple across it, near its maximum dc voltage rating, it will appear to have a low impedance momentarily. When this happens, the output capacitor of the regulator is discharged back through the reverse-biased pass transistors or the control circuitry, frequently causing destruction. Obviously, capacitor replacement will solve the problem.

ZENER CIRCUIT PROTECTION

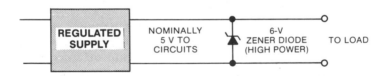

Another cause of problems with IC regulators can be due to severe transient voltage increases on the unregulated input. If severe enough, they can exceed the maximum voltage difference allowable between the regulator input and output, and cause failure of the regulator. In some cases, regulator failure puts the full unregulated voltage on the load circuit. ICs are very susceptible to failure due to excessive voltage and can fail in large numbers in complex circuits as a result. In some cases, you will find a zener diode across the regulator output to cause a short if the supply voltage becomes excessive, thus shutting the supply down without damaging the system.

You should know your equipment so that you can look first at the areas that will fail first. Also, look at the most obvious causes of problems *first*—ac or dc line input, power line cord, fuse or circuit breaker, series resistive fuse, and thermal overload circuit. After you have exhausted these simple checks, then you can look more deeply into the circuitry.

Step-by-Step Troubleshooting Procedure

For reference, the troubleshooting information that was presented on the previous pages is summarized here in the form of a step-by-step procedure:

STEP 1. If the equipment is completely inoperative, first check to see that all power switches are on and that the source of electrical power is functioning. When an ac electric outlet is used, check to see that the power cord is not damaged and that the plug prongs are clean and fit well into the outlet. For a battery power supply, make sure that the battery is in good condition (check with a voltmeter) and that its terminals are making good electrical connections with the power supply input leads.

STEP 2. Check to see if the power supply fuse or circuit breaker is open and replace, or reset, if necessary. If the fuse or circuit breaker blows again when power is restored, continue on to step 3.

STEP 3. If normal operation has not been restored, it will be necessary to perform other tests on the electronic circuits inside. As a safety precaution while gaining access to these circuits, you should *disconnect* the equipment from its power source. Just turning off the switch is *not* a safe alternative.

In the case of a battery power source, disconnect or remove the batteries. Although a low-voltage, low-current battery is not dangerous to you, inadvertent short circuits made while gaining access might cause damage. Shorting of a storage battery with high-current capacity can be *dangerous*. The high current can cause severe arcing and could overheat the battery, causing the acid electrolyte to be sprayed over you and the equipment. *So be careful!* (Also see page 1–132 on servicing rechargeable battery supplies, including the *Caution*.)

STEP 4. If the means for gaining access to the interior circuits is not obvious, consult the manufacturer's instruction manual and follow the directions.

Step-by-Step Troubleshooting Procedure (continued)

STEP 5. Make a thorough visual inspection. Look for disconnections or breakage that could have been caused by dropping solder and discolorations that may be a sign of overheating. Make any necessary repairs or replacements before going to the next step.

STEP 6. Restore the electric power disconnected in step 3, and turn on the power switch if you think you have found and repaired the difficulty. As long as the electric power is on, it is a vital safety precaution to touch the circuit only with *insulated* instrument test probes. Disconnect the power if it is necessary to rearrange the equipment or to use hand tools in making connections or disconnections.

STEP 7. With a voltmeter trace the path of electric power from the power source input to the transformer (if any) input terminals, transformer output terminals, rectifier input, rectifier output, filter input, and filter output. If there are voltage-regulator circuits with which you are not familiar, consult the instruction manual to learn the electrical path from input to output. In this voltage-tracing procedure, if the voltage fails to appear at any test point or is significantly different in amplitude than expected, the trouble is probably between this point and the last point where a correct voltage appeared. Look for open or short circuits and for signs of physical damage or overheating. Make any repairs or replacements and recheck for normal operation.

STEP 8. If this does not reveal the source of the problem, or if the trouble is not a component burnout, disconnect the power supply output from the circuits that follow. From the manufacturer's manual determine the normal load current and use Ohm's Law to determine the size of the load resistor that will draw that current. Install the resistor, apply power, and repeat step 7. If normal operation can now be restored, the trouble is not in the power supply but in the circuits that follow. You will learn how to find and make necessary repairs in these circuits later in these volumes.

STEP 9. Once normal operation is restored, disconnect the power source, and reassemble the equipment. Check for normal operation.

Features and Applications

The battery offers two very distinct advantages over the standard ac-dc power supply. The major advantage is *portability*, and the other advantage is *independence* from the possible interruptions of service.

From your study of *Basic Electricity* you remember that an electric *cell* consists of a container, two *electrodes* of different materials, and a liquid, paste, or gel *electrolyte* between the electrodes. The chemical action that takes place causes one of the electrodes (called the *anode*) to become positively charged (deficient in electrons) and the other (called the *cathode*) to become negatively charged (excess electrons). Since a single cell has a limited output voltage, a number of cells are often connected together to produce a unit called a *battery* that has the desired voltage and current characteristics.

It was difficult to achieve portability in the days of vacuum-tube electronics because these circuits required *sufficiently high* voltages and currents to require the use of several bulky batteries. The *low*-voltage and current requirements of most solid-state electronic circuits makes it quite practical to obtain many hours of operation with only small low-cost batteries.

As you know, there are two types of batteries used today. The first type is *primary*; these batteries are discarded after they are worn out. They wear out because the chemical process that produces electricity is essentially *irreversible*, and hence they *cannot be recharged. Secondary* batteries, on the other hand, *are rechargeable* since the chemical reaction that produces electricity in these batteries *can be reversed* by applying current in the opposite direction to recharge the battery. Thus, these secondary cells can be used repeatedly.

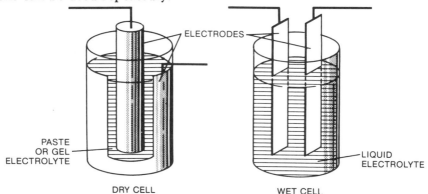

DRY CELL   WET CELL

Dry cells can be either rechargeable or nonrechargeable, but wet cells are almost always rechargeable. Dry cells are not really dry but have the electrolyte in the form of a paste or gel. They also are sealed (or semisealed) so that they can be operated so that no gas is produced during their use. Thus, no special operating instructions are necessary.

In general, dry batteries are used for small highly portable equipment. Some of these have rechargeable batteries built in. Often the charger is also built in.

Nonrechargeable Batteries

For small electronic systems such as portable radios and TVs, tape players, and some hand calculators, the nonrechargeable dry battery is used because it provides good life without the bother and relative expense of a rechargeable battery system. Some of these also have an ac power supply for power line operation, but these supplies are independent and used separately. Although the dry cell is not new, many new dry cell designs have been developed recently with improvements in the areas of peak power output, compactness, power longevity, freedom from maintenance, and in some cases, very limited rechargeability.

Recently different types of dry cells and the variety of sizes and shapes have grown enormously to suit a wide range of applications.

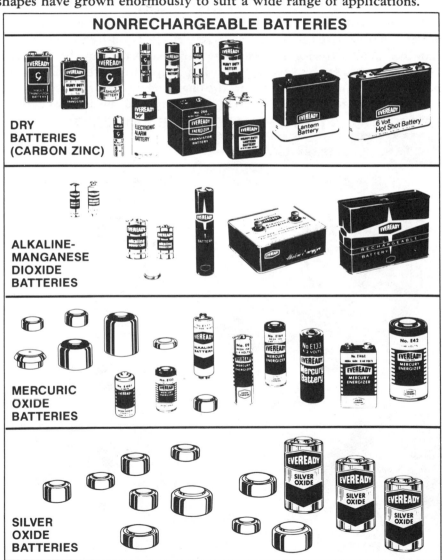

NONRECHARGEABLE BATTERIES

DRY BATTERIES (CARBON ZINC)

ALKALINE-MANGANESE DIOXIDE BATTERIES

MERCURIC OXIDE BATTERIES

SILVER OXIDE BATTERIES

Rechargeable Batteries

The lead-acid battery that is part of automotive electrical systems is probably the most commonly used rechargeable battery power source for electronic equipment since almost every vehicle now has at least a radio in it. The nature of the electrical system used to keep this battery charged has been covered in *Basic Electricity*. Care and maintenance of these batteries consists of periodic checking and replacement of water lost during charging by electrolytic action and being certain that the battery is kept charged. An uncharged battery deteriorates very rapidly.

Most modern portable or semiportable electronic equipment that use self-contained rechargeable batteries use *dry* nickel-cadmium cells to make up the battery. These are available as sealed (or semisealed) units, which use a paste electrolyte so that operation in any position is possible, and are characterized by being relatively rugged and lighter in weight than lead-acid batteries. In some cases, the batteries must be removed from the equipment and recharged using a separate battery charger. In most modern equipment, however, the charging circuits are built in and recharging can take place merely by plugging into the ac power line.

APPLICATIONS FOR RECHARGEABLE BATTERIES

HAND COMPUTER

AUTOMOBILE WITH CB EQUIPMENT
AND FM-AM RADIO

WALKIE-TALKIE

RADIO-CONTROLLED
MODEL AIRPLANE

Types of Dry Batteries

The rest of this section will describe *six types* of dry cells and give their general construction and performance characteristics. As an introduction to these particular dry cells, the graph shows a comparative summary of their discharge characteristics. From the graph you will notice that of the four nonrechargeable types, the lithium and mercury/Nicad® have a much higher energy output than the more popular but less expensive carbon-zinc and alkaline/manganese dioxide cells.

Of the rechargeable types, nickel-cadmium (Nicad®) is the most widely used. However, the new lead dioxide gel-electrolyte and lead dioxide sealed cells are being used more frequently because of their superior recharge characteristics. These cells are designed for up to five years of continuous trickle charging in standby applications and up to 500 normal discharge-charge cycles in use. Some alkaline batteries are designed to be recharged on a limited basis.

COMPARISON OF DISCHARGE CHARACTERISTICS OF 5 TYPES OF DRY BATTERIES

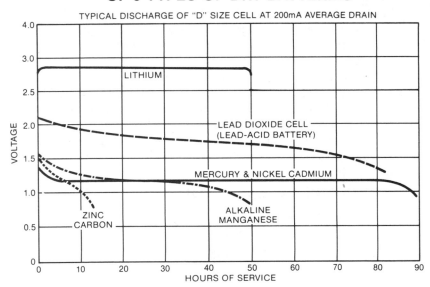

TYPICAL DISCHARGE OF "D" SIZE CELL AT 200mA AVERAGE DRAIN

As you can see, the output voltage from batteries is far from constant in most cases. As a result, most equipment designed for operation from batteries must be capable of meeting desired performance under varying supply voltage conditions. Voltage regulators are generally not used with battery supply equipment because of the inefficiency of voltage regulators that would dissipate battery power. In some cases, a small part of the circuit will be regulated with a simple pass transistor regulator. When stabilized voltage is required from a regulator, a switching regulator provides the highest efficiency.

Carbon-Zinc Cell

The voltage of a standard single carbon-zinc cell is 1.5 volts. Present carbon-zinc batteries made of many cells are available in voltages ranging from 1.5 volts to several hundred volts. Cells and batteries may be connected in series to obtain higher voltages or in parallel to achieve greater current capacity.

The carbon-zinc battery uses a zinc cathode, a manganese dioxide anode, a carbon-rod current collector, and an electrolyte of ammonium chloride and zinc chloride dissolved in water. The standard carbon-zinc battery is not designed for recharging.

The working voltage of the cell falls gradually as the cell is discharged. It is important that the proper size be used in replacement to get maximum service life from the battery. It is equally important that the electronic circuit must be able to operate not only at the full-rated battery voltage but also down to a voltage considerably less than the rated peak voltage of the battery; the so-called *cut-off* voltage of the operating circuit should be as *low* as possible (relative to the battery) to make maximum use of the battery as it discharges and its terminal voltage goes down.

Carbon-zinc cells are designed to operate at room temperature. Service life at *low* temperatures is reduced because of *decreased* chemical activity in the cell. Low temperatures are not harmful to these cells and full life will be restored on return to room temperature. Low-temperature storage (typically 10° to 0°C) extends cell shelf life. Since the shelf life of carbon-zinc batteries is limited, they should be used reasonably soon after purchase. The shelf life is decreased because the electrolyte and the zinc terminal interact, even when no current is being drawn, thus eventually exhausting the battery. This process is accelerated at high temperatures. These batteries also are subject to corrosion and should be removed from equipment that is to be stored for any period of time.

CARBON-ZINC CELL CONSTRUCTION (SIZE D)

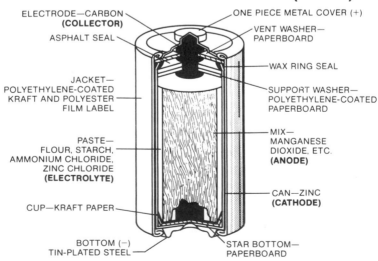

Alkaline Cell

Both nonrechargeable and rechargeable alkaline cells are made of a zinc anode, a manganese dioxide cathode, and an alkaline potassium hydroxide electrolyte. The alkaline manganese dioxide-zinc battery is capable of heavier and more continuous current drain than the carbon-zinc battery at both normal and low temperatures.

The alkaline cell is used in applications requiring more power or longer life than can be provided by carbon-zinc cells. Alkaline cells contain 50 to 100% more total energy than a carbon-zinc cell of the same size. The alkaline cell also can deliver high currents for short intervals. Because of their efficient construction, these cells operate well over a wide temperature range ($-20°C$ to $+70°C$). They also can withstand severe conditions such as mechanical shock and high pressures.

Alkaline cells have excellent shelf life and can retain more than 90% of their capacity after one year when stored at room temperature. Although these cells are superior to zinc-carbon cells in all applications, they are more costly. It is always possible to substitute alkaline batteries for carbon-zinc batteries. In some cases, carbon-zinc batteries cannot be substituted for alkaline batteries without reduction in equipment performance. This is true because the voltage of carbon-zinc batteries drops rapidly with use, and they are not able to supply high currents for short intervals without serious loss of voltage output. Alkaline batteries are less prone to corrosion than carbon-zinc ones but also should be removed when the equipment is to be stored for any extended period (more than a month).

Alkaline batteries are recommended for applications where there is a relatively high peak-current requirement for a short period of time, which is the case in portable transmitters for short communication and flash equipment used in photography.

CUTAWAY VIEW OF ALKALINE CELL

ONE-PIECE COVER—(+) PLATED STEEL
CAN—STEEL
ELECTROLYTE—POTASSIUM HYDROXIDE
CURRENT COLLECTOR—BRASS
CATHODE—MANGANESE DIOXIDE, ETC.
ANODE—POWDERED ZINC
SEPARATORS—NONWOVEN FABRIC
JACKET—TIN-PLATED LITHOGRAPHED STEEL
INSULATING TUBE—POLYETHYLENE-COATED KRAFT
SEAL—NYLON
INNER CELL BOTTOM—STEEL
METAL WASHER
INSULATOR—PAPERBOARD
PRESSURE SPRING—PLATED SPRING STEEL
METAL SPUR
RIVET—BRASS
OUTER BOTTOM (−) PLATED STEEL

Mercury Cell

The mercury cell consists of a mercuric oxide cathode, an anode of pure zinc, and an electrolyte of potassium hydroxide or sodium hydroxide. These cells are not rechargeable and are generally produced in the flat or cylindrical containers shown in the diagram.

The flat pellet cells are often used in hearing aids and watches, while the gelled anode cells are designed for improved low-temperature performance in cameras. The cylindrical cells are found in many general uses where appreciable power is required periodically. Mercury cells are usually used for very low-current, extended-life operation.

The components are housed in a sealed double-can structure. This structure allows gas to escape while trapping electrolyte in the paper sleeve between the cans, preventing the leakage of liquid that might damage components. This gas *venting* mechanism also prevents the buildup of internal pressure that can occur under abusive conditions such as reverse currents or short circuits.

CROSS-SECTIONAL VIEW OF FLAT- AND CYLINDRICAL-TYPE MERCURY CELLS

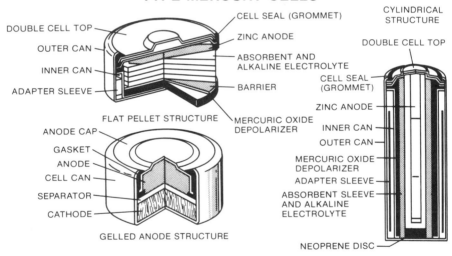

During discharge, voltage declines slightly at first and then remains constant until the end of useful life, where a sharp drop-off of voltage occurs. Their almost constant discharge voltage permits reliable long-term operation of watches, camera exposure meters, heart pacemakers, and other similar devices. The output voltage is stable over wide temperature ranges, and special versions of mercury cells are available for very low-temperature and high-temperature operation.

Mercury cells can retain 90% of their rated capacity after two years storage at room temperature. They sometimes are used as reference sources in regulated power supplies because of their stable voltage output.

Lithium Cell

The lithium cell is nonrechargeable and is made of a lithium foil anode, a separator, and a carbon-compound cathode, as shown. These elements are spirally wound together and the entire assembly is contained in a steel case. Lithium cells are comparatively new and expensive but are being used more and more because of their superior characteristics. Because the electrolyte contains no water there is no gas evolved during discharge. The top assembly contains a vent to prevent the buildup of high internal pressure due to excessive heat that can result with improper use.

Lithium cells offer reduced size and weight for a given output capability. Lithium batteries can provide energy densities of up to 330 watt-hours per kilogram (150 watt-hours per pound), nearly *three* times that of *mercury* cells, and *four* times that of *alkaline* manganese cells.

Nominal terminal voltage is 2.95 volts per cell, almost twice that of most other cells. During discharge, the voltage is extremely stable. This cell operates efficiently over a range of temperatures from −40°C to +70°C. In addition, higher discharge rates can be achieved at lower temperatures than are possible with other dry cells. Moreover, the cells are very rugged, and lithium cells are an excellent source for high currents over short intervals.

The shelf life extends for several years with storage over a wide range of temperatures. One advantage of lithium cells is high voltage output per cell that permits use of fewer cells for a given voltage output.

CONSTRUCTION OF LITHIUM CELL

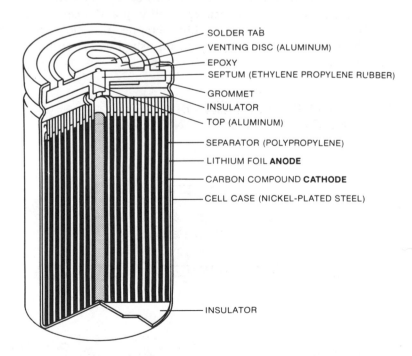

SOLDER TAB
VENTING DISC (ALUMINUM)
EPOXY
SEPTUM (ETHYLENE PROPYLENE RUBBER)
GROMMET
INSULATOR
TOP (ALUMINUM)
SEPARATOR (POLYPROPYLENE)
LITHIUM FOIL **ANODE**
CARBON COMPOUND **CATHODE**
CELL CASE (NICKEL-PLATED STEEL)
INSULATOR

Nickel-Cadmium (Nicad®) Cells

Any rechargeable cell is a combination of chemically active materials in which the negative electrode is oxidized while the positive electrode is reduced as electric current is taken from the cell. In a *rechargeable* cell these changes can be *reversed* by passing an electric current through the cell in the *opposite* direction.

In the uncharged condition the positive electrode of a nickel-cadmium cell is nickel hydroxide; the negative electrode is cadmium hydroxide. In the charged condition the positive electrode is nickel hydroxide; the negative electrode is metallic cadmium. The electrolyte is potassium hydroxide. The average cell voltage under normal conditions is about 1.2 volts and is usually quite constant during discharge, even under varying load and temperature conditions.

During the latter part of a recommended charge cycle and during overcharge, nickel-cadmium cells generate some gas. Oxygen is generated at the positive (nickel) electrode and hydrogen gas is formed at the negative (cadmium) electrode when the cell reaches full charge.

These gases must be vented from the conventional nickel-cadmium system. To have a sealed cell that can withstand overcharging, the generation of gases is prevented by having provisions to chemically recombine the oxygen and hydrogen within the cell container to form water. Nickel-cadmium cells also are used in the conventional wet-cell configuration for high-power applications. These cells are very rugged, can stand great abuse, and will not freeze when left in a discharged state. Some nickel-cadmium cells have life expectancies of 25 years or more. "Dry" nickel-cadmium cells are made in two types: the *button* cell and *cylindrical* cell. Button-type cells utilize a cell cup and a cell cover. The electrodes consist of pressed powder tablets wrapped in nickel wire gauze, kept apart by a separator. Sealing is accomplished by crimping the rim of the cell cup over the rim of the cell cover.

CONSTRUCTION OF
NICKEL-CADMIUM BUTTON CELL

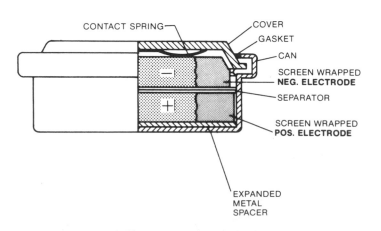

Nickel-Cadmium (Nicad®) Cells (continued)

The cylindrical cell type incorporates a different electrode arrangement than the button cell. The positive electrode consists of thin, highly porous nickel plates impregnated with active material. The negative electrode is a powder pressed into an expanded metal carrier. It is this electrode configuration with its large surface area that gives cylindrical nickel-cadmium cells outstanding cycle life and ruggedness.

Like all nickel-cadmium cells, sealed rechargeable nickel-cadmium cells have relatively constant discharge voltages. They can be recharged many times. The cells have very low internal resistance and are rugged and highly resistant to shock and vibration.

Nickel-cadmium cells can operate over a wide temperature range. However, they can be harmed by use at high temperatures, by charging at higher than recommended rates, or by repeated complete discharge. Leaving nickel-cadmium cells in a discharged condition for a long period of time also will ruin them.

In summary, nickel-cadmium batteries can be recharged many times and have a relatively constant voltage output during discharge. They will stand much abuse and have good low-temperature performance characteristics. They are widely used in portable electronic equipment where the load is relatively high and the batteries require frequent recharge, so nonrechargeable batteries would be too great an expense.

CONSTRUCTION OF NICKEL-CADMIUM CELL— CYLINDRICAL TYPE

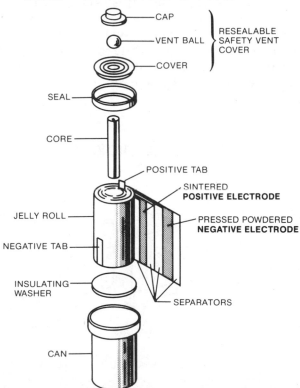

Lead Acid–Lead Dioxide Cell

The lead dioxide cell is the *dry battery* version of the lead-acid storage battery. This cell has lead dioxide positive plates, a gelled sulfuric acid electrolyte, and lead negative plates. Like the conventional lead-acid storage battery, it is rechargeable. When the cell delivers power to a load, the positive and negative plates become converted to lead sulfate, and the electrolyte is converted to water. The open circuit voltage of each cell is approximately 2.12 volts. As the battery is discharged, the terminal voltage slowly decreases.

When a charger is connected to the discharged cell, the water in the gel and the lead sulfate covering the plates combine to reform the gelled sulfuric acid electrolyte. In the process the lead oxide and lead of the plates is reformed and is again available for repeated power production. When too low a charging voltage is used, the current flow essentially stops before the battery is fully charged. This allows some of the lead sulfate to remain on the battery plates, which can eventually reduce the available capacity of the battery.

This cell can be used in a wide variety of typical portable applications as the primary power source. However, one excellent use of the lead-acid rechargeable cell is in continuous-charging standby applications such as intrusion (burglar) alarms, smoke detectors, uninterruptible power systems, computer memory standby power, and emergency lighting. In many of these applications, conventional lead-acid storage batteries are used if the power required is substantial.

For applications when the cell is operated with continuous charging, the charging voltage must be *slightly higher* than the cell working voltage (2.12 volts per cell). For a 6-volt battery, 6.75 volts to 6.90 volts should be the charging voltage; while for a 24-volt system, the charging voltage should be 27.0 or 27.6 volts. This type of voltage ratio relationship between cell working voltage and charging voltage will result in maximum service life of the cell.

LEAD ACID–LEAD DIOXIDE CELLS

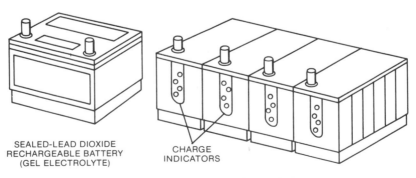

SEALED-LEAD DIOXIDE
RECHARGEABLE BATTERY
(GEL ELECTROLYTE)

CHARGE
INDICATORS

LEAD-ACID BATTERY
(HEAVY DUTY—WET ELECTROLYTE)

Nonrechargeable Battery Supplies

One of the major problems with nonrechargeable battery-powered equipment is *corrosion*. This is usually caused by leaving dead batteries in equipment for long periods of time. Another problem involves either dead or weak batteries or batteries that are improperly installed.

Corrosion can be taken care of by removing the defective batteries and cleaning the battery compartment carefully with a noncorrosive cleaner like carbon tetrachloride (*caution—use in a well-ventilated location*). The electrodes should be checked and, if necessary, should be cleaned with very fine sandpaper to remove corrosion without removing the corrosion-resistant plating. Interconnections are often only riveted in, and corrosion may cause an intermittent connection. If this is suspected, a solder joint should be made to ensure a tight connection.

Dry cells, even when essentially dead for equipment operation, will give a normal voltage indication (about 1.5 volts) when checked with a voltmeter alone. Therefore, they must be checked *under load*. A flashlight can be used to check conventional cells for normal light intensity. Alternately, a 10-ohm (1-watt) resistor can be placed across the battery when the voltage is measured. It should be greater than about 1.3 volts under load to be considered good. The commonly used 9-volt radio transistor battery can be checked under load by placing a 100-ohm resistor (2 watts) across its terminals when the voltage is measured. Of course, the simplest thing to do is to replace the battery. In any case, it is important to make sure that the cells are *properly installed*. If one cell is put in backward, the voltage across the set of cells will be reduced by twice the cell voltage. Thus, a set of four cells (6 volts nominal) will only produce 3 volts output if *one* cell is reversed. Check connectors and terminal voltage with the equipment on. If the battery is good but the voltage drops excessively under load and/or the batteries get warm, it is due to an equipment malfunction that must be remedied.

CHECKING BATTERY UNDER LOAD

CLEANING CONTACTS

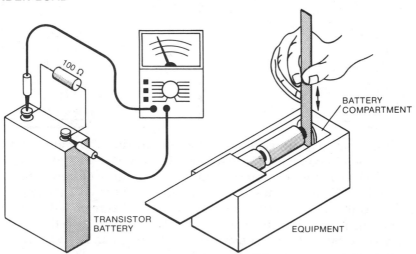

Rechargeable Battery Supplies

Rechargeable battery supplies can be divided into those that use an external charger and those that have the charger built in. Equipment that has nonrechargeable batteries and also an ac mode of operation often has a switch that enables one supply or the other to be used. Equipment with a built-in charger usually operates from the battery at all times; the charger is used to charge the battery simultaneously when the ac line is connected to the charger.

Common problems with rechargeable batteries are *corrosion* and *poor connections*. These can be remedied as described on the previous page. To test a cell under load, choose a resistor for the load so that about 200 mA will be drawn for low-power equipment and about 1 ampere for heavier duty equipment. Make sure you choose a proper wattage rating. *Caution: Some equipment should not be operated from the ac line with the battery removed; make sure of this before you operate equipment without the battery.*

One way to check the operation of the charger is to put the batteries on charge and note the terminal voltage under charge. It should be higher than the noncharging battery voltage by about 10%. The same scheme can be used to check batteries using a separate charger.

CHECKING BATTERY TERMINAL VOLTAGE WITH INTERNAL CHARGER

CHECKING A BATTERY WITH SEPARATE BATTERY CHARGER

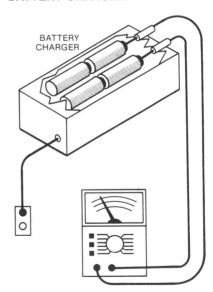

RECHARGEABLE BATTERIES

ELECTRONIC EQUIPMENT

BATTERY CHARGER IS INTERNAL

BATTERY CHARGER

If the battery voltage does not increase by the proper amount under charge, the charger may be defective. Battery chargers are simple dc supplies with a series resistance to limit the charging current, so you can troubleshoot them like any ac power supply. If the charger appears to be operating and the batteries either will not charge or will not hold a charge, then there may be a weak or defective cell, or cells, and these will need to be replaced. Sometimes the battery can be revived by dis-charging it completely and then recharging it.

Circuit Boards

When you remove a unit of modern electronic equipment from its cabinet, you will nearly always see that the electronic components are mounted on one or more boards called *printed-circuit boards*. Sheets of thin copper are firmly bonded to sheets of semirigid plastic to form a *copper clad* board of insulating material. The copper is coated selectively with a thin layer of light-sensitive material and exposed to a source of intense light through a photographic negative on which there is an image of the desired interconnections. The portions of the coating that have been exposed to the light are made resistant to acid, while the unexposed portions (covered by the black areas of the negative) are not made resistant to acid. When the board is acid-dipped, the unprotected copper is eaten away, leaving the desired interconnecting strips. After the acid-resisting material is removed, holes are drilled through the ends of the conductive strips to permit the component leads to be fed through for electrical and mechanical attachment.

The copper is sometimes coated with solder or flashed with a metal-like rhodium to resist corrosion. Sometimes gold plating is used. Now the components are mounted with their leads inserted into the holes, and the leads are soldered to the metal surrounding the holes. In some cases, the holes are plated through to provide contact to printed wiring on the other side of the circuit boards. These are called *double-sided* circuit boards. For very complex circuits, where extreme compactness is necessary, there may be many printed layers. These boards are called *multilayer boards*. When a piece of equipment uses a number of circuit boards, they are often mounted in standard racks and have special connectors to permit rapid and positive insertion and removal for servicing. For some applications, the ICs and other components may be mounted in sockets.

CIRCUIT BOARD CONSTRUCTION

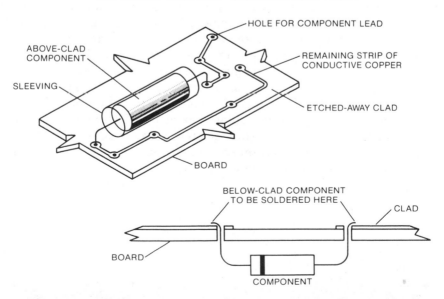

Block and Schematic Diagrams/Special Markings

Solid-state electronic equipment is very rugged and reliable. However, eventual malfunction or breakdown may occur. To locate the specific cause of the malfunction you must be able to *identify* the various components, identify their interconnections, and then perform a logical troubleshooting and adjustment procedure. There are a number of servicing aids that can assist you in doing this, and some or many of these are usually supplied by the equipment manufacturer via instruction and service manuals.

The simplest aid is the *system block diagram* (that you learned about earlier), which was used to describe the radio transmitter-receiver system and power supplies. This diagram is a shorthand description of the major steps in the electronic processing that takes place and is the fastest aid to understanding the overall operation of the system.

The most important aid to servicing is the *schematic diagram*. This shows the symbols for the various circuit components and specifically shows how they are interconnected, their value or type designation, and their part number. As you know from your earlier studies, schematic diagrams are arranged to show most clearly the sequence of electronic processing from the input to the output. They are most often arranged to be *read* from *left to right* and from *top to bottom*. The schematic diagram shown on the next page is for a discrete component solid state phonograph preamplifier, which is designed to accept the output from a magnetic phonograph pickup at terminals one and two, and deliver a high level output at terminal nine. You will learn more about these amplifiers in Volume 2.

Schematic diagrams are usually marked to show the value of the various components and also the identification numbers of the various transistors, diodes, and ICs.

When there is a distinction between leads or terminals coming out of a component, the part and schematic are marked appropriately. Where significant, electrolytic or tantalic capacitors and diodes are marked with a plus sign to show how the component is connected in the circuit. In many cases, these markings are also included on the printed circuit board.

Parts such as transistors and integrated circuits that have many leads or terminals are identified by basing diagrams. As seen from the bottom, the leads or terminals are numbered in sequence in a clockwise direction starting from a tab, notch, dot of colored paint, or other obvious marking. The schematic diagram shows these leads or terminals identified by the same numbers, but they are arranged in a manner to show the connections most clearly and with minimum crossovers.

Block and Schematic Diagrams/Special Markings (continued)

Often included are other markings to indicate circuit operation voltages and waveforms that appear at important points in the circuit. When more than one circuit board is required to mount all the components shown in the schematic, the diagram is often marked (with dotted lines for example) to indicate which components are mounted on each board.

Without doubt, the schematic diagram is the single most important aid to servicing. It contains all the most vital information and, in general, is sufficient to permit logical troubleshooting and repair. All the other aids that are described here are supplements that make the servicing process easier.

CIRCUIT BOARD LAYOUT (SCHEMATIC DIAGRAM)

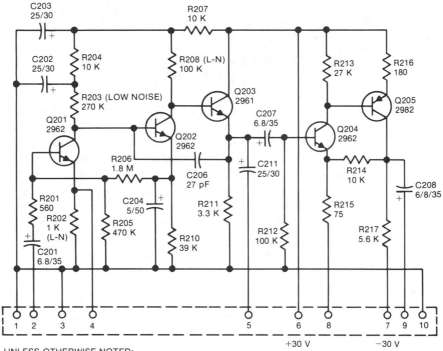

UNLESS OTHERWISE NOTED:
ALL RESISTANCES IN OHMS
ALL CAPACITORS IN MICROFARADS

Layouts, Photos, Wiring Diagrams, and Other Aids

Although schematic diagrams show the circuit components arranged in a sequence that best illustrates *how* the circuits are interconnected from a functional standpoint, they do not show *where* each component is physically located in the equipment. While you could compare the equipment to the schematic and eventually locate all components, this is a very time-consuming procedure. To simplify this process, other aids to servicing also are often supplied.

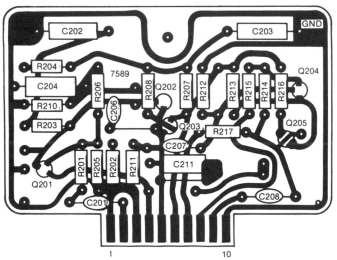

Component layout diagrams show a drawing of the circuit board with the outlines of the various components shown on it in a close representation of their relative sizes, shapes, and locations as shown above for the preamp shown earlier. Also included is the component designation so that parts on the schematic can be identified. Sometimes these diagrams are made with the parts shaded to make them appear three-dimensional, and they are known as *pictorial layout diagrams*. Some manufacturers strive for even greater realism and include a photograph of the circuit board with superimposed names, code numbers, and arrows to identify the various parts—this is a *labeled photograph*. Also frequently supplied is the *wiring diagram*, which is a component layout diagram to which lines have been added to show the interconnections.

Some larger equipment contains a number of circuit boards that are stacked one above the other or side-by-side in a frame or rack. This equipment also may contain rows of front panel controls and meters and a variety of electromechanical components. In addition, adjustments, alignment points, test points and other pertinent information is shown pictorially. *Equipment layout diagrams* are often supplied to show the outlines of the various miscellaneous components and circuit boards as well.

Instruction manuals supplied by the manufacturer describe special test and trouble localization procedures. The facing illustration shows a typical layout diagram for the CB radio that you studied earlier.

EQUIPMENT LAYOUT DIAGRAM OF CB SET SHOWING ALIGNMENT POINTS

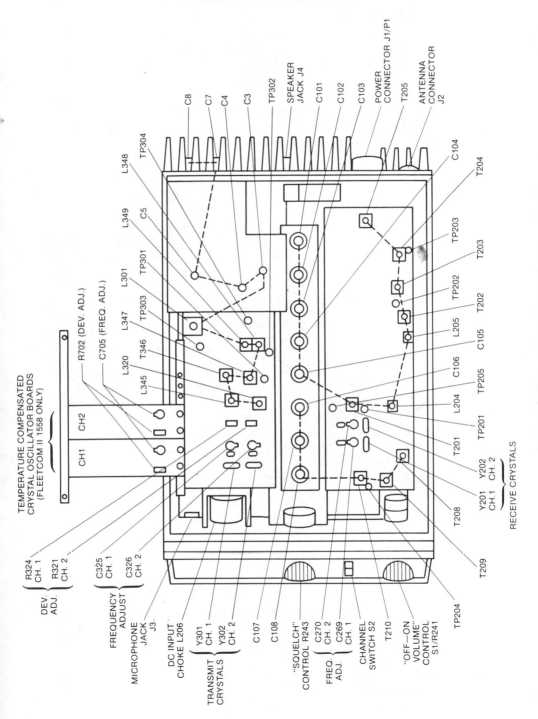

Component Removal and Replacement

As you gain experience and do actual troubleshooting and repair, you will find it necessary to remove components from circuit boards. This *isn't very easy* usually, and you have to be *very careful* to minimize damage to the circuit board. The simplest thing to do is to cut the component from the board and then remove the leads that are still soldered in by careful application of heat and gently tugging with a pair of long-nose pliers. Another approach is to gently apply heat to each terminal and use a *solder puller* (a rubber suction bulb that pulls the molten solder away from the connection so it is free). In this way each lead can be freed and then the component removed.

COMPONENT REMOVAL AND REPLACEMENT

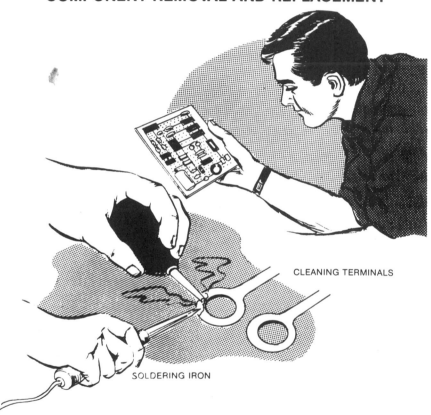

CLEANING TERMINALS

SOLDERING IRON

In either event, once the component has been removed, it is necessary to clean the board area of excess solder (use the solder puller), and if there is excess flux, this can be removed with alcohol. The new component can now be inserted and carefully, *using as little heat as possible*, soldered into place. It is wise to carefully examine the connections with a magnifying glass since those that appear to be good connections often are not.

Other Aids to Servicing

To further help the service technician, who may be confronted with all kinds of equipment at different times, some manufacturers give in-service courses on the specifics of servicing their hardware. These are extremely useful because you get the benefit of all of the manufacturer's experience with a particular line of equipment. If you have reason to believe that you will be dealing with much equipment from a particular manufacturer, it is a good idea to look into these short courses.

The manufacturer of the Citizens Band learning system for your study of solid-state electronics has provided a new approach to servicing. Information about particular equipment is provided in a JETPAC® (Johnson Educational Training Package) that includes not only schematic diagrams but a set of colored slides (or microfiche) that are keyed to a set of audio cassettes. You can learn about the equipment firsthand by following the schematics, diagrams, and photographs as you are instructed by the audio cassette. This new approach to learning about specific systems is very useful for providing training and servicing information particularly for those who cannot attend formal training courses and also can serve as a supplement to these. As video cassette machines have become more available hardware training programs of many types have been worked out using this medium. These video cassette learning systems now are becoming common as training aids in many areas and allow for group or individual instruction.

Learning System Power Supply

Earlier in this volume, you were introduced to the function of a power supply in a radio transmitter and a radio receiver. You also will remember that these would be the basis for learning about electronic systems using a candidate radio transmitter-receiver system. The equipment used for this candidate system is a Citizens Band (CB) radio transmitting-receiving set that embodies most of the nine common electronic building blocks in electronic systems.

The CB radio set is normally designed for mobile operation directly from a 12-volt (normally 13.5–14 volts) automobile battery. For base (fixed) station operation (at a fixed location where ac power is available), an ac supply, exactly like the ones you've been studying, is available as a separate accessory. This supply plugs into the 120-volt ac outlet and provides about 13.5–14 volts dc regulated for CB radio operation.

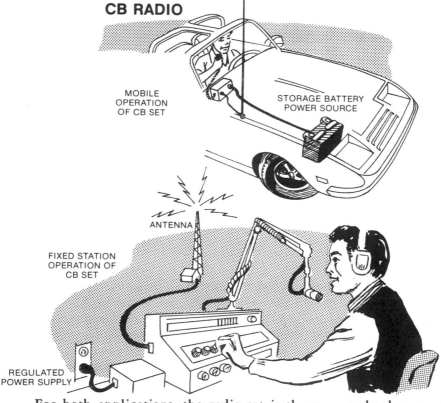

CB RADIO

MOBILE OPERATION OF CB SET

STORAGE BATTERY POWER SOURCE

ANTENNA

FIXED STATION OPERATION OF CB SET

REGULATED POWER SUPPLY

For both applications, the radio set is the *same*—only the power supply is changed. As you will see, the supply is no different than those you have already studied. Two versions of the power supply are available. One uses a zener-regulated pass transistor arrangement (discrete circuit) regulator, while the other uses an IC regulator. In both cases, a transformer-bridge rectifier-capacitor filter is used to provide basic unregulated dc. Both supplies provide about 13.8 volts dc at 1.2 amperes for operation of the CB radio. Good regulation is required because current drain on the power supply is three times greater in *transmit* than *receive*.

Learning System Power Supply—Discrete Component Version

Shown below is the schematic diagram of the power supply using discrete components. Note that component values and operating voltages are specified as an aid for servicing. The similarity to the power supplies you have already studied is immediately apparent.

CB RADIO AC POWER SUPPLY

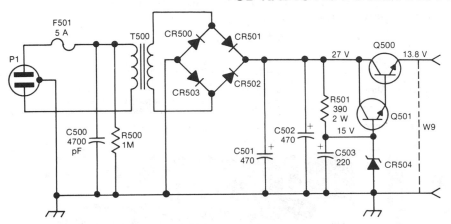

NOTES:
1. ALL RESISTOR VALUES ARE OHMS AND ALL CAPACITOR VALUES ARE MICROFARADS UNLESS OTHERWISE SPECIFIED

2. VOLTAGE READINGS ARE WITH RESPECT TO CHASSIS WITH A 500-mA LOAD.

As you can see, the transformer, bridge rectifier, and capacitor filter are exactly as you have studied. The operation of the regulator is exactly the same as the one you have studied. The output from the bridge rectifier is applied to the series-pass transistor Q500. The base of Q500 is driven from the emitter-follower (common-collector circuit) Q501 that acts as a current amplifier for the current applied to its base from the zener regulator circuit. As you remember, the base current of a transistor is about $1/\beta$ times the collector current. Therefore, to have a current of 1.2 amperes flow in the collector-emitter circuit of Q500, the base current of Q500 must be $1/\beta \times 1.2$ amperes. If the β of Q500 is about 20, then its base current must be $1.2/20 = 0.06$ ampere or 60 mA. Thus, Q501 must supply at least 60 mA from its emitter. If Q501 has a β of 50, then its base must be supplied with a current of $0.06/50 = 0.0012$ or 1.2 mA to allow about 60 mA to be supplied from its collector-emitter circuit. Thus, the current required into the base of Q501 is about $1/\beta$ (Q500) $\times$ $1/\beta$ (Q501) times the output current. The zener reference diode CR504 is in the base leg of Q501 and holds it at constant potential.

Learning System Power Supply—Discrete Component Version (continued)

Shown below are the parts list data and the printed-circuit layout so that you can see the arrangement of individual circuit components. These are very useful servicing aids and in conjunction with the schematic diagram and voltage data provide essentially all of the information necessary to service this power supply. Other useful specification data provided state that the ripple should be less than 0.1 volt and the regulation better than 5% so that you know what to expect from a normally operating power supply.

PARTS LIST

C500	4,700-pF, ±20% 1.4 KV Z5U	510-3001-472
C501	470-µF, 40-V ALUMINUM	510-4009-001
C502	SAME AS C501	
C503	220-µF ALUMINUM	510-4125-221
CR500	IN4818 200-V, 1.5-A RECTIFIER	523-0013-201
CR501	SAME AS CR500	
CR502	SAME AS CR500	

CR503	SAME AS CR500	
CR504	15-V, ±5% 10W ZENER	523-2003-150
F501	FUSE 0.5 A, 200 V FB AGC	534-0003-017
Q500	SILICON NPN 40 V, 4 A, 60 W	576-0002-026
Q501	SILICON NPN 40 V, 4 A, 40 W	576-0002-029
R500	1 M, ±10% 1/2 W	569-1004-105

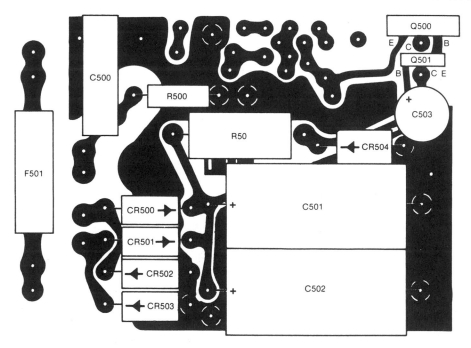

POWER SUPPLY BOARD (SOLDER SIDE VIEW)

You might wonder how you can have a 13.8-volts supply with a 15-volts zener diode. The answer is that there is a potential difference (as you learned earlier) of about 0.5–0.6 volt across the base-emitter junction. Thus, to get 13.8-volts output (at the emitter), you need 14.4 volts on the base of Q500; and to get 14.4 volts on the emitter of Q501 you must have 15 volts on its base. As you can see from the schematic, 15 volts on the base of Q501 provides 13.8 volts out at the emitter of Q500.

Learning System Power Supply—IC Version

The newer version of the power supply for the CB radio set we have been discussing uses an IC regulator. As you might suspect, this is much simpler in *outward appearance*, although the internal construction of the IC is quite complicated. A schematic diagram of the IC regulated power supply is shown below. Note that component values and pertinent voltages again are shown for reference.

As you can see, the circuit diagram is essentially identical to the power supply unit that you studied and experimented with earlier. The regulator (IC, U101), shows the symbol *commonly used* to designate *amplifier ICs*, and you will see more of this symbology in later volumes. The diode CR105 protects IC U101 from reverse voltage, and the series resistor R102 (1.5 ohms) limits the transient current when switching from *transmit* to *receive* and vice versa.

CB RADIO AC POWER SUPPLY WITH IC REGULATOR

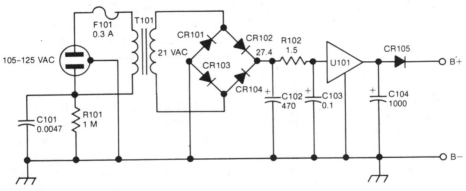

NOTES:
1. CAPACITANCE VALUES IN MICROFARADS, RESISTANCE VALUES IN OHMS UNLESS OTHERWISE SPECIFIED
2. ALL VOLTAGE MEASUREMENTS TAKEN WITH RESPECT TO CHASSIS COMMON, USING A VACUUM-TUBE VOLTMETER

It is apparent that the IC power supply is much simpler in construction than the discrete circuit version. Furthermore, the cost for assembly of the IC supply is less because of the reduced number of components. Obviously, most of the circuitry is included in the IC. In addition, the IC regulator version of the CB radio ac power supply provides for much better performance because it provides better regulation and power supply stability.

Review of Servicing and Power Supplies

1. SERVICING POWER SUPPLIES is a step-by-step logical procedure. Make sure you know what you're doing. Don't take it apart unless you have good reason to believe it's the problem.

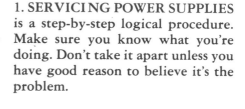

DRY BATTERIES (CARBON ZINC)

2. NONRECHARGEABLE CELLS are used once and then discarded. The most typical of these is the common carbon-zinc dry cell.

SEALED-LEAD DIOXIDE RECHARGEABLE BATTERY (GEL ELECTROLYTE)

CHARGE INDICATORS

LEAD-ACID BATTERY (HEAVY DUTY—WET ELECTROLYTE)

3. RECHARGEABLE BATTERIES are often used with portable equipment because they can be recharged and used again. The most common example is the lead-acid storage battery in an automobile. Nickel-cadmium (ni-cad) is very popular for portable equipment.

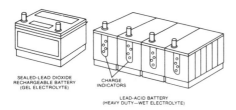

CHECKING BATTERY UNDER LOAD

CLEANING CONTACTS

100 Ω

BATTERY COMPARTMENT

TRANSISTOR BATTERY

EQUIPMENT

4. MAINTENANCE OF BATTERY POWER SUPPLIES involves keeping the contacts clean (free of corrosion); removing dead batteries before they corrode; removing all batteries when equipment is to be stored; and keeping rechargeable batteries charged.

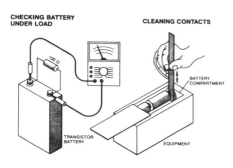

5. BLOCK AND SCHEMATIC DIAGRAMS—Most equipment is supplied with servicing information that gives block and schematic diagrams, wiring information, and component location and voltage levels and component values. These are essential to do rapid servicing work.

Self Test—Review Questions

1. List the step-by-step procedure you would use to service a power supply that you believe is defective.
2. How would you determine that the power supply was at fault?
3. List three types of nonrechargeable batteries and briefly describe their characteristics and uses.
4. List two types of rechargeable batteries and briefly describe their characteristics and uses.
5. Outline preventive maintenance procedures for nonrechargeable batteries.
6. Outline preventive maintenance procedures for rechargeable batteries.
7. List the major aids to servicing that you would expect a manufacturer to supply.
8. Outline the procedure for removing and replacing defective components from PC boards.
9. Briefly describe the differences between the discrete component and IC power supply for the CB learning system.
10. What is immediately apparent from an examination of both of these power supply schematics?

Learning Objectives—Next Volume

Overview—Now that you have learned some fundamentals of solid-state electronics and learned about power supplies, you're ready for the next step in your study of solid-state electronics— the study of amplifiers and audio information systems in Vol. 2.

**TYPICAL PRESENT-DAY
HIGH-FIDELITY AUDIO SYSTEM**

- MAGNETIC TAPE RECORDER
- RECORD PLAYER
- STEREO TUNER
- STEREO AMPLIFIER
- TWO LOUDSPEAKERS

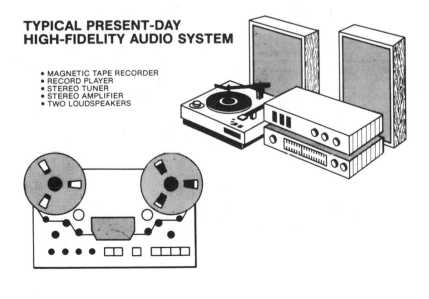

Basic
Solid-State
Electronics
VOL. 2

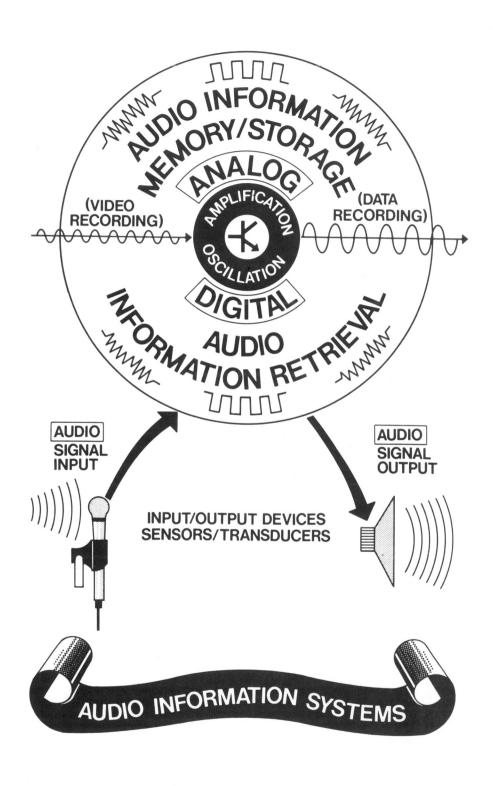

Basic Solid-State Electronics

COMMON·CORE

The Configuration and
Management of
Information Systems

AUDIO INFORMATION SYSTEMS
VOL. 2

BASIC AMPLIFICATION/AUDIO SYSTEMS
AUDIO/VIDEO/IF/RF AMPLIFIERS
HI-FI/PUBLIC ADDRESS AUDIO SYSTEMS
SOUND TRANSDUCERS/OSCILLATORS
TROUBLESHOOTING AMPLIFIERS/OSCILLATORS

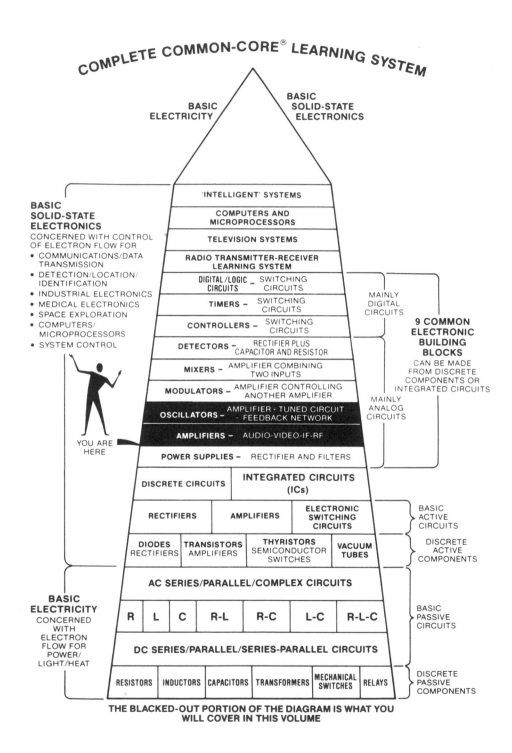

COMPLETE COMMON-CORE® LEARNING SYSTEM

BASIC ELECTRICITY

BASIC SOLID-STATE ELECTRONICS

BASIC SOLID-STATE ELECTRONICS
CONCERNED WITH CONTROL OF ELECTRON FLOW FOR
• COMMUNICATIONS/DATA TRANSMISSION
• DETECTION/LOCATION/ IDENTIFICATION
• INDUSTRIAL ELECTRONICS
• MEDICAL ELECTRONICS
• SPACE EXPLORATION
• COMPUTERS/ MICROPROCESSORS
• SYSTEM CONTROL

YOU ARE HERE

BASIC ELECTRICITY
CONCERNED WITH ELECTRON FLOW FOR POWER/ LIGHT/HEAT

'INTELLIGENT' SYSTEMS

COMPUTERS AND MICROPROCESSORS

TELEVISION SYSTEMS

RADIO TRANSMITTER-RECEIVER LEARNING SYSTEM

DIGITAL/LOGIC CIRCUITS – SWITCHING CIRCUITS

TIMERS – SWITCHING CIRCUITS

CONTROLLERS – SWITCHING CIRCUITS

DETECTORS – RECTIFIER PLUS CAPACITOR AND RESISTOR

MIXERS – AMPLIFIER COMBINING TWO INPUTS

MODULATORS – AMPLIFIER CONTROLLING ANOTHER AMPLIFIER

OSCILLATORS – AMPLIFIER + TUNED CIRCUIT · FEEDBACK NETWORK

AMPLIFIERS – AUDIO-VIDEO-IF-RF

POWER SUPPLIES – RECTIFIER AND FILTERS

MAINLY DIGITAL CIRCUITS

MAINLY ANALOG CIRCUITS

9 COMMON ELECTRONIC BUILDING BLOCKS
CAN BE MADE FROM DISCRETE COMPONENTS OR INTEGRATED CIRCUITS

DISCRETE CIRCUITS	INTEGRATED CIRCUITS (ICs)	
RECTIFIERS	AMPLIFIERS	ELECTRONIC SWITCHING CIRCUITS
DIODES RECTIFIERS	TRANSISTORS AMPLIFIERS	THYRISTORS SEMICONDUCTOR SWITCHES	VACUUM TUBES

BASIC ACTIVE CIRCUITS

DISCRETE ACTIVE COMPONENTS

AC SERIES/PARALLEL/COMPLEX CIRCUITS

R	L	C	R-L	R-C	L-C	R-L-C

BASIC PASSIVE CIRCUITS

DC SERIES/PARALLEL/SERIES-PARALLEL CIRCUITS

RESISTORS	INDUCTORS	CAPACITORS	TRANSFORMERS	MECHANICAL SWITCHES	RELAYS

DISCRETE PASSIVE COMPONENTS

THE BLACKED-OUT PORTION OF THE DIAGRAM IS WHAT YOU WILL COVER IN THIS VOLUME

Audio Information Systems

In this volume you are going to take your first step in information management and control and see how certain basic electronic functions are incorporated into an electronic system—an *audio information system.* (See the Information Management System chart in Volume 1 on pages 1-2 and 1-3.) In Volume 1 you learned about semiconductor fundamentals and how simple, discrete components and integrated circuits (ICs) can be used to build power supplies that provide the necessary operating potentials for solid-state electronic equipment. Now you are going to learn how the concepts of *amplification* and *oscillation* are utilized to build amplifiers—audio, video, IF, and RF—and oscillators. You also will learn about your first information system *input/output (I/O) devices* —microphones and loudspeakers. You will discover how these power supplies, amplifiers, and I/O devices are used in the construction of hi-fi and stereo sound systems.

You also will learn how *information* can be *recorded*, *stored*, and *retrieved*, such as audio information on records and tapes. This will be your prelude to the *electronic storage* and *retrieval* of other types of information, such as video and data. In addition, you will be introduced to the first use of *digital techniques* for the reduction of noise in sound recording.

Amplifiers are used to increase the amplitude of a signal (voltage or power). Thus, they boost the signal level in an electronic system to a suitable level for use or further processing. Amplifiers are classified on the basis of their intended use and the band of frequencies over which they amplify.

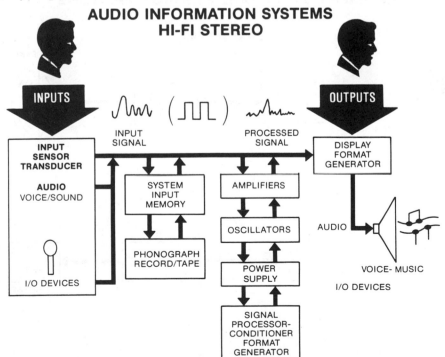

Amplification

In addition to learning about amplifiers, you also will learn about *oscillators*. Oscillators are a special form of amplifier; they represent the next step in the electronic learning ladder, and are the subject of the last part of this volume.

Why are amplifiers so important and so widely used? An amplifier of any type (audio, video, RF, or IF) takes a small *voltage* or *current change*—an input from the sensor/transducer of the Overall Information Management System—and *increases* the size or amplitude and/or power level of that change. The output, in turn, after transmission/reception, can be used to make sound (audio) come out of a loudspeaker; form a picture (video) on the face of a TV tube or radar screen; control the speed of an electric motor; transfer data to control the temperature of a mixing vat in a plastics factory; or guide a space vehicle in its flight—and so on.

In all of these uses, the *original electrical change*—whether it is the result of sound waves striking a microphone, light falling on the sensitive face of a TV camera tube, or a variation in the temperature of the sensitive element in a mixing valve—is *too small* to do the job required. What the amplifier does is to *increase* the amplitude and/or power level of the original change until it is large enough to drive the loudspeaker, TV picture tube, electric motor, heating element, or rocket motor control.

AMPLIFIERS INCREASE SIGNAL AMPLITUDE

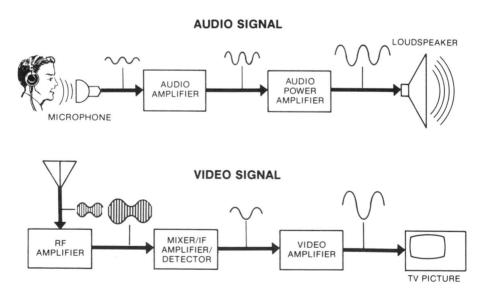

Radio Transmitter-Receiver Learning System

In Volume 1 of this series, you also were introduced to the block diagram of a complete radio transmitter-receiver system. You learned that this would be the *learning system* that would take you from basic concepts and circuits to a working knowledge of complete radio transmitter-receiver systems.

The arrangement of this learning system is typical of that used when a radio station transmits a program to a radio receiver in your home. It also applies to the arrangement used when one amateur radio station transmits a message to another or when a Citizens Band (CB) radio (or walky-talky) operator in one automobile talks to another operator several miles away. Although these various systems differ in size, cost, operating range, and sound quality, they all employ the basic arrangement shown here.

In Volume 1 you learned about the basics of semiconductors and how they are used in power supplies. Now that you are about to learn about amplifiers, it is worthwhile to review the operation of the radio transmitter-receiver learning system. This will enable you to see how widely different types of amplifiers are used, and your understanding of them will establish a reference upon which to base details of this volume.

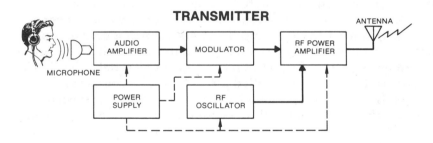

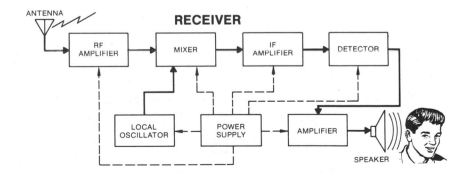

Summary of Radio Transmitter-Receiver Operation

Although the electrical signals that represent sound waves travel very easily through a conducting wire, they are incapable of radiating through air and space to a distant point. For electrical signals to radiate easily into space and thus travel without the use of a wire, their *frequency* usually must be *considerably higher*. The lowest frequencies used in commercial broadcasting in the United States are about 535 kHz (kilohertz), and frequencies of thousands of megahertz (MHz) are commonly used in communications. Because of this, it is necessary to generate an electrical signal at a frequency high enough to travel readily without wires to a distant point and to superimpose the message or information on this signal. This high-frequency signal is known as the *carrier*, and the process of modifying the amplitude of the carrier to superimpose the information is known as *amplitude modulation* (or *AM*). Later we will consider another technique for superimposing information on a carrier called *frequency modulation* (or *FM*).

In the transmitter, a constant-level carrier signal is generated and modulated by the voice signal. The receiver picks up this signal and *demodulates* it by removing the carrier signal so that only the electrical equivalent of the original sound (voice) signal is left.

You can trace the process in the block diagrams. In both the transmitter and receiver, a *power supply* converts power from the electric outlet to a voltage and current suitable for use by the various circuits.

In the transmitter, a radio-frequency (RF) carrier wave is generated by the *oscillator* (a special type of amplifier). A *microphone* is used to convert the pressure of the voice (sound) information to an equivalent electrical signal, and an *audio amplifier* steps up this signal to a suitable level. The high-power audio amplifier is called a *modulator*. The constant-level RF carrier signals are fed into the modulated amplifier (another special type of amplifier) that also is fed with the audio signal from the modulator.

AM TRANSMITTER OPERATION

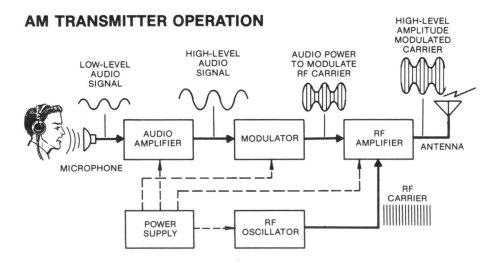

Summary of Radio Transmitter-Receiver Operation (continued)

The modulated amplifier raises and lowers the amplitude of the carrier so that the resulting peaks and valleys duplicate the variations in the voice signal. The modulated RF carrier from the modulated amplifier is fed into the *antenna* to radiate it for long distances.

At the receiver, it is necessary to remove the carrier signal from the received modulated RF signal and leave only the electrical equivalent of the voice signal. The problem is that the received signal is *very weak*, often only a few millionths of a volt. The weak signal picked up by the receiver *antenna* is first stepped up by an *RF amplifier* to raise it to a level suitable for processing to begin. Because amplification at this frequency is inefficient, the modulated carrier is reduced to an *intermediate frequency (IF)* at which it is much easier to amplify. This is accomplished by the *mixer* (another special form of amplifier), which takes in both the modulated RF carrier and a constant-level carrier at a fixed frequency difference from the signal input produced by the receiver (local) *oscillator* (a special form of amplifier). The variations in the mixer output signal duplicate those of the modulated RF carrier, but the new carrier is at an intermediate frequency (IF) equal to the sum or difference between the incoming RF signal and local oscillator frequencies. Considerable amplification can now be accomplished by the *IF amplifier*. The *detector* (or *demodulator*)—a special form of rectifier—removes the intermediate frequency component from the output of the IF amplifier, and only an electrical duplicate (audio) of the original voice signal remains. Now, an *audio amplifier* boosts this audio signal to a level sufficient for the *speaker* to convert it back to sound duplicating the original voice message.

As you can see, amplifiers are used in a wide range of applications to increase the amplitude of various signals. The pages that follow will review the outstanding features of amplifiers in much greater detail. The final part of this volume will describe oscillators.

AM RECEIVER OPERATION

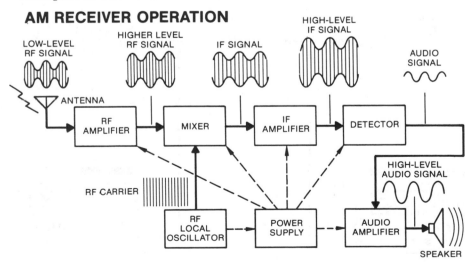

The Electromagnetic Frequency Spectrum

On the previous pages that reviewed the radio transmitter-receiver learning system, you were introduced to the fact that *different frequencies are used to carry the information* contained in the voice of a CB operator in one automobile to the ear of another CB operator in another automobile several miles away. You learned that in the transmitter the sender's audio-frequency voice signals were superimposed, or modulated, on a radio-frequency carrier signal, to enable the antenna to radiate a wave that could travel readily through space. Then, in the receiver, the modulated radio-frequency carrier was reduced to a lower intermediate frequency to simplify the process of amplification. Next, in the receiver's detector, the audio signal is extracted from the intermediate-frequency signal and amplified to drive a loudspeaker that reproduces the sound waves of the original voice signal.

As you learned in *Basic Electricity, the number of complete cycles of change per second* in an electrical signal is known as its *frequency*. Frequency is measured in *hertz (Hz)*, and the number of hertz equals the number of cycles per second. This periodic voltage or current change can actually be seen by connecting an *oscilloscope* to the circuit in which the change is taking place. The oscilloscope provides a *visual image* of the electrically varying signal, so the actual waveform can be seen at selected points in the system. You will learn later how the oscilloscope works. For now it is sufficient to understand that by the use of the oscilloscope it is possible to see a *time-varying waveform*.

THE OSCILLOSCOPE SHOWS THE FREQUENCY, WAVESHAPES, AND AMPLITUDE OF ELECTRICAL SIGNALS AT ANY SELECTED POINT IN AN ELECTRONIC SYSTEM

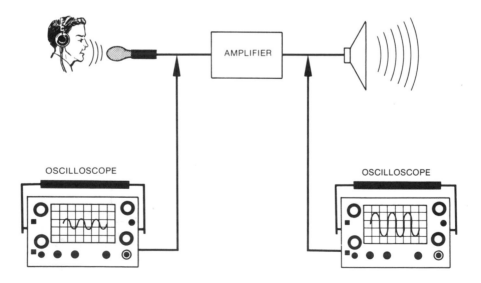

The Electromagnetic Frequency Spectrum (continued)

Sound is the motion of pressure waves in air. Any device that changes the pressure of the surrounding air produces sound. That is what your vocal cords do when you talk. When you increase the tension of these cords, you increase the frequency at which they vibrate. This produces a tone that is higher in frequency. Many musical instruments make use of this principle by making some part, such as a taut string, a reed, or a stretched membrane vibrate and produce varying pressure waves in air.

In the piano, when a key is struck a taut string is set into vibration. The string vibrates on both sides of its resting point and compresses and expands the surrounding air. As shown in the illustration, when the string moves from its resting point to the right, the air to the right of the string is compressed, or increased, in pressure. When the string moves to the left, the air at the right is reduced in pressure. If a sound-detecting (or sensing) device such as the human ear is located in the vicinity of the vibrating string, the varying-pressure waves will strike the eardrum and produce the sensation of sound. The number of complete vibrations of the string taking place per second defines the frequency of the resulting sound wave. The *loudness* or *amplitude* of the resulting sound wave is determined by the amount of *displacement* of the string from its resting place. The harder the string is struck to displace it, the louder the sound will be. The sound of the human voice generally includes the frequency range from about 60 Hz to over 5,000 Hz and has intensity variations of greater than 10,000 to 1. In music the frequency range is from about 40 Hz to above 20,000 Hz, and the intensity variations may be greater than 100,000 to 1. Frequencies in the range from about 16 Hz to 20 kHz are called *audio frequencies*.

SOUND WAVES PRODUCED BY VIBRATING STRING

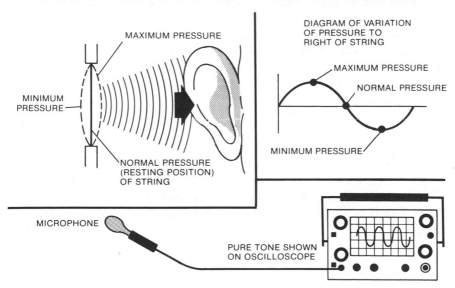

DIAGRAM OF VARIATION OF PRESSURE TO RIGHT OF STRING

MAXIMUM PRESSURE

MAXIMUM PRESSURE

NORMAL PRESSURE

MINIMUM PRESSURE

MINIMUM PRESSURE

NORMAL PRESSURE (RESTING POSITION) OF STRING

MICROPHONE

PURE TONE SHOWN ON OSCILLOSCOPE

The Electromagnetic Frequency Spectrum (continued)

As indicated earlier in the review of the transmitter-receiver learning system, audio frequencies are not the only ones in use. After the microphone converts the sound waves of the voice into audio-frequency electrical signals, the transmitter uses this audio signal to modulate an RF (radio-frequency) carrier. The antenna receives the carrier signal and propagates it in the form of *electromagnetic waves*—waves that have the characteristics of both electrical and magnetic fields—which travel through space to reach the receiving antenna.

ELECTROMAGNETIC WAVE

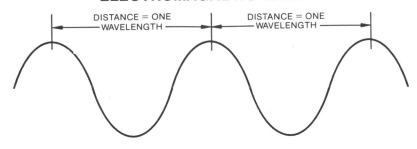

The presently used range of electrical and electromagnetic frequencies extends from the direct current (dc) range on through the ultraviolet and even beyond. This entire band of known frequencies is called the *electromagnetic spectrum*. The details of the most widely used frequencies and their applications are shown on the next page.

The best way to become familiar with these applications is to examine these diagrams in detail. As you progress through this course, you will learn more details about various portions of the spectrum and the associated equipment shown here.

The diagram on the next page indicates the various portions of the spectrum in terms of both frequency and wavelength. The wavelength (in meters) for any frequency can be found by dividing the speed of light (300,000,000 meters per second) by the frequency (in hertz):

$$\lambda = \frac{3 \times 10^8}{f}$$

where λ (the Greek symbol lambda) is the wavelength in meters and f is the frequency in hertz. Thus, the wavelength of a 300,000,000-Hz (300-MHz) signal is 1 meter, and the wavelength of a 30-MHz signal is 10 meters.

ELECTROMAGNETIC FREQUENCY SPECTRUM WITH ASSOCIATED EQUIPMENT SYSTEMS

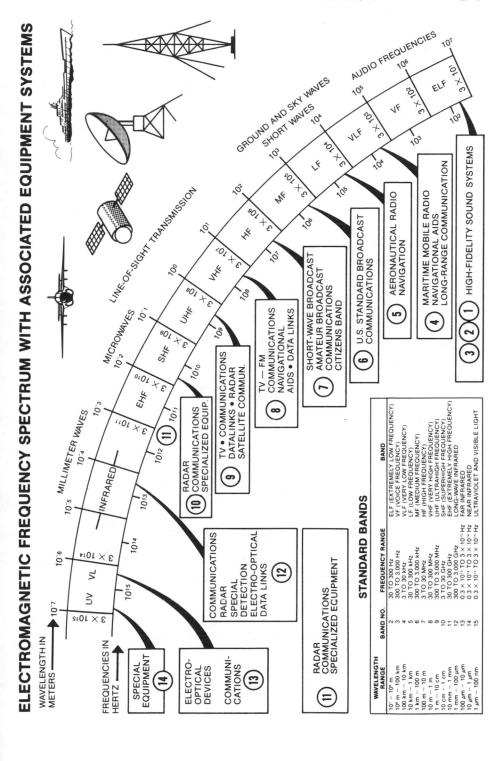

STANDARD BANDS

WAVELENGTH RANGE	BAND NO.	FREQUENCY RANGE	BAND
$10^7 - 10^6$ m	2	30 TO 300 Hz	ELF (EXTREMELY LOW FREQUENCY)
10^6 m – 100 km	3	300 TO 3,000 Hz	VF (VOICE FREQUENCY)
100 km – 10 km	4	3 TO 30 kHz	VLF (VERY LOW FREQUENCY)
10 km – 1 km	5	30 TO 300 kHz	LF (LOW FREQUENCY)
1 km – 100 m	6	300 TO 3,000 kHz	MF (MEDIUM FREQUENCY)
100 m – 10 m	7	3 TO 30 MHz	HF (HIGH FREQUENCY)
10 m – 1 m	8	30 TO 300 MHz	VHF (VERY HIGH FREQUENCY)
1 m – 10 cm	9	300 TO 3,000 MHz	UHF (ULTRAHIGH FREQUENCY)
10 cm – 1 cm	10	3 TO 30 GHz	SHF (SUPERHIGH FREQUENCY)
10 mm – 1 mm	11	30 TO 300 GHz	EHF (EXTREMELY HIGH FREQUENCY)
10 mm – 100 μm	12	300 TO 3,000 GHz	LONG-WAVE INFRARED
100 μm – 10 μm	13	0.3×10^{13} TO 3×10^{13} Hz	FAR INFRARED
1 μm – 100 nm	14	0.3×10^{14} TO 3×10^{14} Hz	NEAR INFRARED
	15	0.3×10^{15} TO 3×10^{15} Hz	ULTRAVIOLET AND VISIBLE LIGHT

Types of Amplifiers—Audio, Video, RF, IF

A number of different methods have been used for classifying and describing amplifiers. Amplifiers can be described in terms of the relationship between the *signal level* and the *bias voltage* placed on the active elements—that is, their *bias conditions* identified in terms of their types of operation. These types are called *Class A*, *Class B*, and *Class C*. Amplifiers also can be described in terms of the *power level* of the input and output signals. Other methods of description include the range of signals they amplify or by the particular way they are interconnected with resistors and capacitors.

A common method of describing amplifiers is by their *function*; this often also happens to define the *frequency range* of the signals that they amplify. You will initially learn about amplifiers in terms of the four categories listed in the illustration: audio, video, RF, or IF. The other methods of classification also will be considered later.

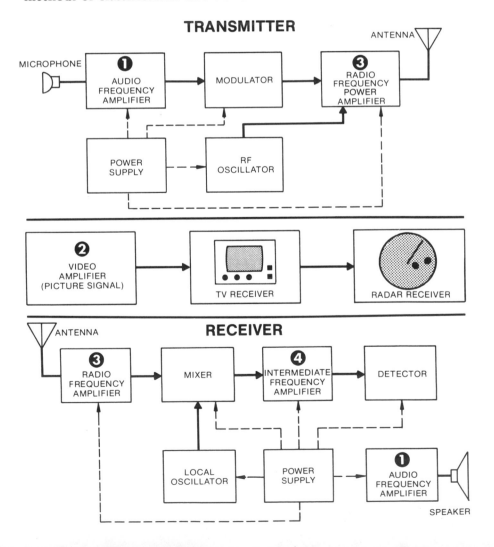

Types of Amplifiers—Audio, Video, RF, IF (continued)

Summarized below are the major features of the four types of amplifiers you will learn about.

TYPES OF AMPLIFIERS

TYPE	FREQUENCY RANGE	APPLICATIONS
AUDIO AMPLIFIERS	16 Hz-20,000 Hz	HI-FI, INTER-COMMUNICATION, SOUND ON FILM, SOUND ON TAPE, SONAR, TELEPHONE
VIDEO AMPLIFIERS	DC-10MHz	PICTURE DISPLAYS ON TV, RADAR, SONAR, FACSIMILE, OSCILLOSCOPES
RADIO-FREQUENCY AMPLIFIERS	30,000 Hz-300 GHz	TRANSMITTERS AND RECEIVERS FOR RADIO, TV, RADAR
INTERMEDIATE-FREQUENCY AMPLIFIERS	FIXED BAND BETWEEN 400 kHz AND 50 MHz	RECEIVERS FOR RADIO, TV, RADAR

1. *Audio Amplifiers*: These amplify a band of frequencies from about 16 Hz to 20,000 Hz. This is the range of frequencies that the human ear usually can hear—hence, the name *audio*. These amplifiers are used in the audio sections of radio transmitter-receiver systems, in hi-fi systems, in intercommunication and telephone systems, in sound-on-film and tape equipment, in sonar, and so on.

2. *Video Amplifiers*: These are similar in design and general arrangement to audio amplifiers, but they amplify a much broader band of frequencies, from the dc range to a few hundred kilohertz or up to 10 MHz and higher. Video amplifiers are used to amplify the signals that create the image on oscilloscopes, TV, facsimile, radar, sonar, and other visible displays formed on TV-type picture tubes. They also are used in pulsed circuits where broadband operation is necessary.

3. *Radio-Frequency Amplifiers*: Unlike the two previous types, RF amplifiers usually amplify only a *specific band* of frequencies. This specific band, however, may be anywhere within the wide range of frequencies from 30,000 Hz up to microwave frequencies. Radio frequencies are used as carriers in radio, TV, and communication systems. When you tune this equipment, you are changing the specified band of radio frequencies that the RF amplifier will amplify.

4. *Intermediate-Frequency Amplifiers*: These are a special type of RF amplifier used in nearly all types of receivers. They are designed to amplify *one particular band* of RF frequencies, and this band always *remains the same* even though the RF amplifier associated with it may be tuned through a wide range. In most AM broadcast band receivers, for example, the IF amplifier amplifies a band of frequencies centered around 455 kHz.

The dB Measurement Scale

You probably have heard the term *decibel* used and may have seen it abbreviated as *dB* on equipment specification sheets and other data. The decibel is ¹⁄₁₀ of a bel, and is a unit for measuring gain or loss in voltage, current, or power. It also is the usual unit for measuring the relative loudness of sounds. It is approximately the smallest degree of difference of loudness detectable by the human ear. The decibel is based on the logarithmic scale and allows us to express very large numbers with ease. It also allows us to calculate system power gains and losses by simple addition and subtraction, which you will find particularly useful with amplifiers.

The power formula for calculating the decibel is

$$dB = 10 \log_{10}\left(\frac{P1}{P2}\right)$$

where P1 is the signal power at some terminal (usually the output) and P2 is the signal power at another terminal (usually the input) or the signal power at the reference level.

Generally, you do not have to calculate. Instead, you can use the chart below to give you the dB reading directly or the voltage, current, or power ratio if you know the dB value. To use the chart, simply take the gain (or loss) and look across to the power- or voltage-current curves, then down to the dB value. By reversing the procedure, you can determine gain when you know the dB value. If the dB value is a loss, then the ratio will be less than 1; invert the ratio (to make it greater than 1), proceed as before, and put a minus sign in front of the dB value. For calculating dB values, see the Appendix. A dB-value drill with answers also is included in the Appendix.

VOLTAGE, CURRENT, OR POWER VS. DECIBEL GAIN

THIS CHART IS USED FOR ATTENUATION BY INVERTING THE RATIO AND MAKING THE DECIBELS NEGATIVE

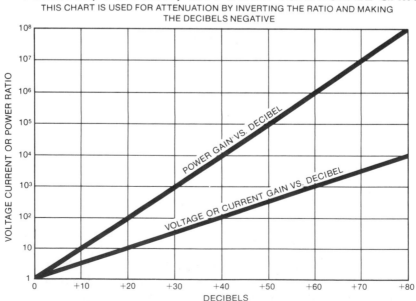

Review of How Transistors Amplify

Solid-state amplifiers have a number of features that are common to all types—whether they are audio, video, IF, or RF amplifiers. The fundamental features include the manner in which a transistor provides amplification, basic circuit configurations, biasing methods, and classes of operation. In this section, basic amplifier circuits will be described. This information can then be used as a basis for describing the special types of amplifiers in later sections, since all amplifiers share these common properties. Some of the material presented here is a review and should reinforce what you learned in Volume 1.

Remember that a transistor is an assembly of semiconductor materials, usually silicon, arranged in the manner shown in the illustration. Semiconductor material with impurities added to make it have an *excess of electrons* is known as *n-type* material. Semiconductor material with impurities added to make it have a *shortage of electrons*—an *excess of positive charges*, or *holes*—is known as *p-type* material.

Transistors are made of three sections of semiconductor material in an npn or pnp configuration, as shown in the diagram. Also shown are the schematic symbols for each type. The three parts of the transistor are the *emitter*, *base*, and *collector*. The significant electronic effects that take place in a transistor occur primarily at the junctions between the p and n materials.

TRANSISTOR CROSS-SECTIONAL DIAGRAMS AND SCHEMATIC SYMBOLS

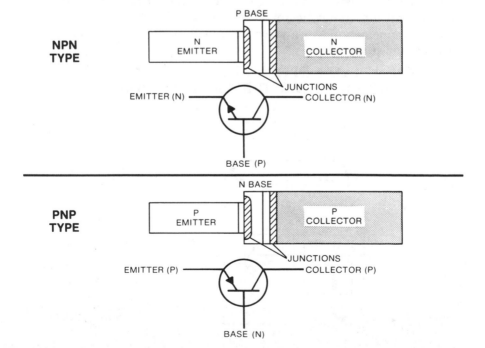

Review of How Transistors Amplify (continued)

You learned in Volume 1 that in a pn or diode junction, when a negative voltage is applied to n material and a positive voltage is applied to p material, current flows readily from the p material through the junction into the n material fusing conventional current flow. This particular polarity of the voltages applied across the pn diode junction is known as *forward bias*. If the polarity of the applied voltage is reversed, there is only an *insignificant* current flow. This polarity is known as *reverse bias*. As you know, if the reverse-bias voltage is made large enough, reverse current flow will take place, as in zener diodes.

If bias is applied across the junctions of an npn transistor, as shown in the diagram, current will flow in the paths indicated. Note that there may seem to be a contradiction to the statement in the first paragraph when you examine the condition at the base-collector junction. At this junction, you see current flowing from n material to p material under what actually is a reverse-bias condition. The reason for this is that the base is only several thousandths of an inch (or millimeter) thick. The electrons leaving the emitter penetrate right through the base, where they are attracted by the strong positive voltage of the collector material. Consequently, only a small percentage of the current flowing out of the emitter actually flows into the emitter-base junction. The relationship between the currents is that the collector current (I_c) and the base current (I_b) add up to equal the emitter current (I_e). The ratio I_c/I_e is called α *(alpha)*. Also important is parameter β *(beta)*, which is the ratio of I_c/I_b.

As you can see, for the pnp transistor (again using conventional current flow) the applied voltages are reversed. The current flow inside the transistor consists of a flow of positive holes from the emitter through the base and into the collector. The diagram for this condition shows the equivalent current flow (instead of hole flow) to maintain consistency with the npn diagram.

CURRENT FLOW IN NPN AND PNP TRANSISTORS UNDER FORWARD-BIAS CONDITIONS

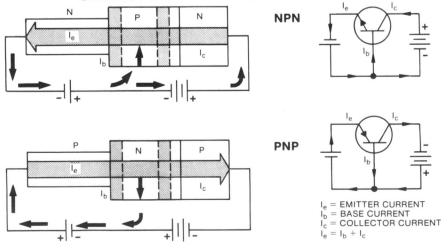

I_e = EMITTER CURRENT
I_b = BASE CURRENT
I_c = COLLECTOR CURRENT
$I_e = I_b + I_c$

Review of How Transistors Amplify (continued)

There might seem to be no useful result from the current flows that have been described. Not all the current that left the emitter actually reached the collector because a small percentage of it flows in the base circuit. Where is the amplification?

Transistor amplification results from the fact that a *small change* in emitter current can produce a *large change* in collector current. Amplification produces a large change from a small change.

You saw on the previous page that except for the very small base current (I_b), the collector current (I_c) is equal to the emitter current (I_e). That is, the input and output currents are almost equal. There is, however, a great difference in the *resistance* in the input and output circuits. Because of the forward bias, the emitter-base junction resistance has a low resistance. Because the base-collector junction is reverse-biased, its junction resistance is high. Since $E = IR$, the voltage developed across the base-collector junction is much larger than the voltage developed across the emitter-base junction. This represents voltage and power gain. It is the *nature* of the *input/output circuits* that determines whether an amplifier is an audio, video, RF, or IF amplifier. For audio and video amplifiers, these are generally resistance-capacitance (RC) coupling networks, or audio transformers are sometimes used in audio power amplifiers. RF/IF amplifiers use tuned circuits in their input/output circuits to select the desired band of frequencies that are to be amplified.

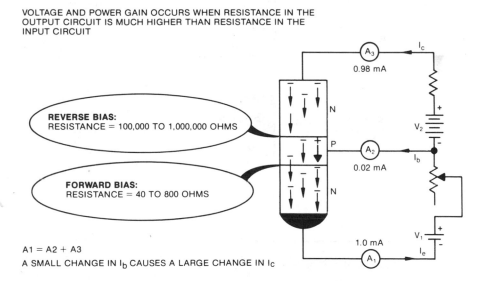

VOLTAGE AND POWER GAIN OCCURS WHEN RESISTANCE IN THE OUTPUT CIRCUIT IS MUCH HIGHER THAN RESISTANCE IN THE INPUT CIRCUIT

REVERSE BIAS:
RESISTANCE = 100,000 TO 1,000,000 OHMS

FORWARD BIAS:
RESISTANCE = 40 TO 800 OHMS

A_3
I_c
0.98 mA

V_2

A_2
I_b
0.02 mA

V_1

1.0 mA
I_e
A_1

A1 = A2 + A3
A SMALL CHANGE IN I_b CAUSES A LARGE CHANGE IN I_c

The diagrams on the next page show the relationships that enable you to calculate voltage and power gain. The next few pages will show you how the small change in *input* signal can produce large changes in the *output* signal for various transistor circuit configurations.

TRANSISTOR VOLTAGE AND POWER GAIN

THE VOLTAGE GAIN IN ANY AMPLIFYING DEVICE IS

$$\text{VOLTAGE GAIN} = \frac{\text{OUTPUT VOLTAGE}}{\text{INPUT VOLTAGE}}$$

$$= \frac{I_{OUT} \times R_{OUT}}{I_{IN} \times R_{IN}}$$

SINCE I_{OUT}/I_{IN} (CURRENT GAIN) HAS ALREADY BEEN DEFINED AS ALPHA (α),

$$\text{VOLTAGE GAIN} = \alpha \times \frac{R_{OUT}}{R_{IN}}$$

FOR A TYPICAL TRANSISTOR,

$$\text{VOLTAGE GAIN} = 0.98 \times \frac{500,000}{250}$$

$$= 0.98 \times 2,000$$

$$= 1,960 \text{ TIMES}$$

FOR A TYPICAL TRANSISTOR,

$$\text{POWER GAIN} = \frac{\text{OUTPUT POWER}}{\text{INPUT POWER}}$$

$$= \frac{I^2_{OUT} \times R_{OUT}}{I^2_{IN} \times R_{IN}}$$

$$= \alpha^2 \times \frac{R_{OUT}}{R_{IN}}$$

$$\text{POWER GAIN} = (0.98)^2 \times \frac{500,000}{250}$$

$$= 0.9604 \times 2,000$$

$$= 1,920.8 \text{ TIMES}$$

Common-Emitter Configuration

You were introduced to the subject of transistor circuit configuration in Volume 1. The circuit shown here was identified as the *common-emitter circuit* because the emitter is included in (is common to) both the input and output circuits.

This common-emitter arrangement is the most frequently used transistor amplifier configuration. The reason is that this configuration usually produces higher voltage gain than the common-base or common-collector configurations to be described later.

Consider what happens in the common-emitter configuration with an npn transistor that has a low-level ac signal applied to its base. When the ac input voltage is zero (time T_0), there is a steady flow of current in both the base-emitter and the collector-emitter circuits. Collector voltage is at its average voltage (time T_0).

Now, when the input voltage starts to become positive, the base becomes more positive. This increases the forward bias across the base-emitter junction. In response to this, more current flows in the base-emitter junction, and, as described in Volume 1, there is an increase in collector current. As more collector current flows through the resistor in the collector circuit, the voltage drop across that resistor increases in accordance with Ohm's Law. As a result of this voltage drop, the voltage at the collector decreases, or becomes less positive.

When the input voltage reaches its maximum positive level (time T_1), the forward current in the base-emitter junction reaches its maximum. In response, the collector current reaches a maximum, and the voltage drop across the collector circuit resistor reaches a maximum and the collector voltage reaches a minimum (time T_1).

DC BIAS CURRENT FLOW IN NPN COMMON-EMITTER CONFIGURATION

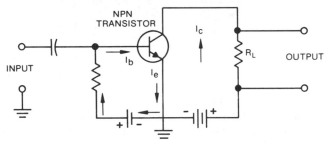

FIRST ¼ CYCLE OF SIGNAL AT BASE AND COLLECTOR OF TRANSISTOR VOLTAGE OUTPUT

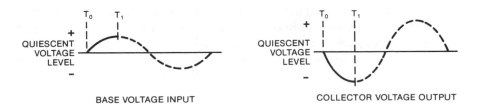

BASE VOLTAGE INPUT COLLECTOR VOLTAGE OUTPUT

Common-Emitter Configuration (continued)

Now the input voltage begins to decrease. As it does, the current in the base-emitter junction decreases, collector current decreases, voltage drop across the collector resistor decreases, and the collector voltage gradually rises. When the input voltage reaches zero (time T_2), the collector voltage is at the same level as it was before the input voltage began to rise (time T_0).

When the input signal starts to become negative, the base-emitter current decreases from the quiescent value. This causes a decrease in the forward bias across the base-emitter junction. As a result, fewer electrons flow from the emitter into the base. The result is a decrease in collector current through the resistor. The voltage drop across the collector resistor decreases even more, and the collector voltage rises.

When the input voltage reaches its most negative level (time T_3), the current in the base-emitter junction reaches its minimum. Now the minimum number of electrons flow in the base-emitter junction. Collector current reaches a minimum, the voltage drop across the resistor reaches its minimum, and collector voltage is at its peak (time T_3).

Now the input voltage becomes less negative and begins its return to zero. As it does, the bias across the base-emitter junction begins to increase. Collector current begins to increase again, causing the voltage drop across the resistor to increase and the collector voltage to begin to decrease. When the input voltage reaches zero (time T_4), the collector voltage is at the same average level as at the beginning (time T_0).

COMPLETE AC CYCLE SIGNAL AT BASE AND COLLECTOR OF TRANSISTOR VOLTAGE OUTPUT

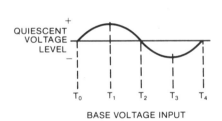

BASE VOLTAGE INPUT

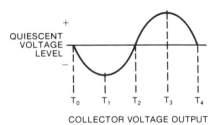

COLLECTOR VOLTAGE OUTPUT

Examine what has been described on the previous pages, and you will see that a *small* change in input signal voltage has produced a *large* change in collector voltage. This is voltage amplification, and the common-emitter circuit configuration has operated as a voltage amplifier.

Because the signal of interest is ac, a capacitor can be used to couple the signal into the base. This capacitor separates any dc bias in the input signal source from the dc bias deliberately set on the base. Similarly, a capacitor at the collector permits the ac part of the collector voltage to be removed as an output signal without disturbing the dc bias on the collector.

Common-Emitter Configuration (continued)

An important fact to notice is that the change in collector voltage is always in a direction *opposite* to the change in the input voltage. This means that there is a 180-degree phase reversal between the input and output signal. As you will see, the presence of this phase reversal is important in a number of applications.

COMPLETE SIGNAL CYCLE OF COMMON-EMITTER CONFIGURATION

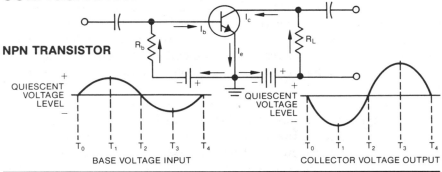

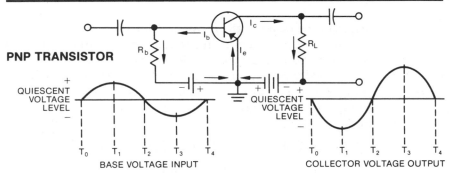

If a pnp transistor is used in the same circuit configuration, the dc bias voltages must be reversed. Except for this difference, the same results are produced. When a positive-going signal makes the base less negative, fewer holes flow from the emitter into the base. This makes the collector current decrease, and the collector voltage increases (becomes more negative). A negative-going signal makes the base more negative. Now more holes flow from the emitter into the base, the collector current increases, and the collector voltage decreases (becomes less negative). Again notice that there is a 180-degree phase shift between the input and output signals.

Common-Base Configuration

In the common-base circuit configuration shown here, the transistor base is common to both the input and output circuits. Both the pnp and npn transistor forms of this circuit are shown. They are identical except for the reversal of the polarity of the applied bias. Because you now know that the npn and pnp transistor variations of the same circuit operate in an *identical* manner, only the npn version will be described here.

When the input signal is zero, the bias in the base and collector circuits produces a steady current flow in these circuits. When an ac input signal on the emitter becomes positive, part of the forward bias across the emitter-base junction is canceled. (The net positive voltage between base and emitter is reduced.) This causes a decrease in emitter current and, consequently, a decrease in collector current. Less current flowing through the load resistor in the collector circuit produces less voltage drop across that resistor, and the collector voltage becomes more positive.

When the input signal on the emitter becomes negative, there is an increase in the forward bias across the emitter-base junction. As a result, there is an increase in the emitter current and, consequently, in the collector current. More current flows through the load resistor, there is an increase in the voltage drop across it, and the collector becomes less positive.

A comparison of the peaks of the input and output waveforms shows that they are in phase. Amplifiers using the common-emitter configuration that produce a 180-degree phase shift between input and output are called *inverting amplifiers*. Amplifiers using the common-base configuration are called *noninverting amplifiers*. Compared to the common-emitter configuration, this circuit has *lower* voltage and power gains, and a current gain of less than 1.

THE COMMON-BASE CONFIGURATION

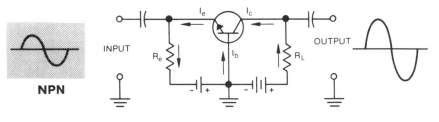

NPN

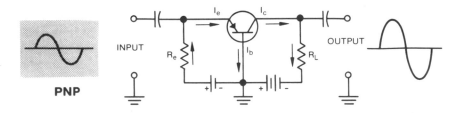

PNP

Common-Collector Configuration

The common-collector circuit configuration shown here in both npn and pnp transistor versions can have high current gain but always has a voltage gain of less than 1. Because of your growing familiarity with these circuits, only the npn version will be described.

When the input signal becomes positive, the forward bias across the emitter-base junction is increased. The result is an increase in emitter current and an increase in collector current. The voltage drop across the load resistor (the emitter resistor) increases, and the emitter voltage becomes more positive.

When the input signal becomes negative, the forward bias across the emitter-base junction is reduced. The result is a decrease in emitter current and a decrease in collector current. The voltage drop across the load resistor decreases, and the emitter voltage becomes less positive (more negative).

Because the emitter voltage rises and falls in time with the input signal, there is no phase reversal between the input and output signals. The outstanding characteristic of this circuit configuration is that the voltage gain is nearly 1, although considerable power gain can be achieved. Another notable feature is that the circuit can have very high input impedance and low output impedance. The input impedance is approximately $\beta \times R_L$, while the output impedance is R_c, which is typically less than 50 ohms. This makes the circuit configuration useful for matching high-impedance sources to low-impedance outputs with little loss. Because the emitter voltage follows the base voltage so closely, this configuration is very often called an *emitter-follower*.

THE COMMON-COLLECTOR CONFIGURATION

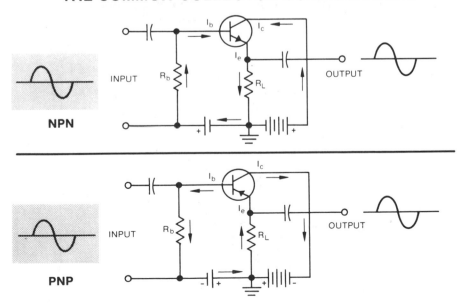

Transistor Operating Point

The family of characteristic curves shown here should be familiar to you. These curves can be used to determine the operating characteristics of a transistor amplifier. For any value of base current on any of the curves, you can find the associated collector current by looking at the I_c scale directly to the left, and you can find collector voltage by looking at the V_{ce} scale directly below.

When the transistor is fully turned on (maximum collector current for a given value of R_L flows), it is operating in its region of *saturation*. When the transistor is fully turned off (no collector current flows) it is operating in its region of *cutoff*. The region between cutoff and saturation is known as the *active* or *transition region*, which is the region in which amplification is accomplished.

For typical amplifier operation, the base current is set so that when no signal is applied, the operating point is approximately at point B. Under these conditions, the base voltage is generally about 0.5 volt greater than emitter voltage for silicon transistors. Now, if the base voltage is gradually increased, the base current also will increase, causing collector current to increase, which increases the voltage drop across R_L (the load resistor) and produces a drop in collector voltage. Consequently, an increase in collector current causes the collector voltage to move along the straight line shown through point A. When it reaches the saturation region, its value will be about 4.0 mA. Similarly, if the base voltage is decreased, the collector current will decrease along the straight line through point C and into the cutoff region.

TYPICAL COMMON-EMITTER COLLECTOR CHARACTERISTICS

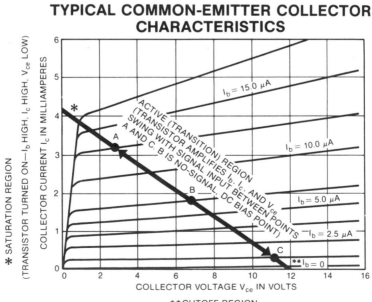

Transistor Operating Point (continued)

The straight line shown in the graph on the previous page is known as the *load line*. This name is appropriate because the line shows the collector voltage and current relationships for any fixed value of load resistor and voltage source.

For example, from the load line you can see that the voltage source is 12 volts. You can determine that it is 12 volts from the fact that this is the voltage on the collector when no collector current is flowing. This is the maximum voltage available in this circuit and therefore must be the source voltage.

You can then determine the value of the load resistor by dividing the source voltage (12) by the current at the $V_{ce} = 0$ point (about 4.25 mA). Under these conditions there is a voltage drop of 12 volts across the resistor and about 4 mA flowing through it. Thus

$$R_L = \frac{E}{I} = \frac{12}{0.004} = 3,000 \text{ ohms (approx.)}$$

The amplifier designer selects the load line according to the type of amplification desired and according to the voltage source available. The desired load line can then be drawn on the characteristic curve to show the operating region for the transistor. In this particular load line, amplification with minimum distortion is obtained by setting the dc base bias at point B where the base current is about 6.5 µA. This is approximately midway between the cutoff and saturation regions.

With this bias, the input signal can drive the collector current through a swing from point A (I_c about 3.3 mA and V_{ce} about 2.25 volts) to point C (I_c about 0.5 mA and V_{ce} about 10.75 volts). With a collector current swing from A to C, the signal output waveshape will be a reasonable reproduction of the input signal waveshape.

COLLECTOR CIRCUIT BIAS ADJUSTMENT

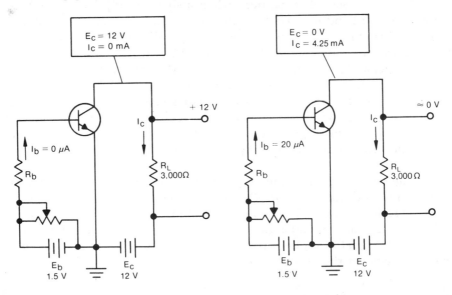

Transistor Operating Point (continued)

Up to this point you have mainly considered the voltage and current in the *collector* circuit. To visualize the conditions in the *base* circuit more easily, suppose you simplify the arrangement by giving the base its own 1.5-volt source and a single resistor to regulate current flow in the base circuit.

Now, for a dc (no signal) operating point at B and a bias current of 6.5 µA, the value of the base resistor is determined from Ohm's Law as follows:

$$R_b = \frac{E_b}{I_b} = \frac{1.5}{0.0000065} = 230 \text{ K (approx.)}$$

If another voltage source were selected for the base, you would simply divide that voltage by 6.5 µA to find the base bias resistor. This alternate resistor would not change conditions in the collector circuit or the load line.

BASE BIAS ADJUSTMENT

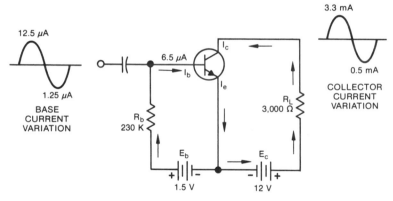

Now, let's see what happens when a signal is applied to the base. Suppose that the positive peak of the input signal increases the collector current so that it reaches point A, where the collector current I_c is about 3.3 mA and the base current I_b is about 12.5 µA. Also suppose that the negative peak of the input signal decreases the collector current so that it reaches point C, where I_c is about 0.5 mA and I_b is about 1.25 µA.

For these conditions a 1.25- to 12.5-µA change (11.25 µA) in base signal current has produced a 0.5- to 3.3-mA change (2.8 mA) in collector signal current. This represents a current gain ($\beta = I_c/I_b$) = 240. Since this is the gain produced by a varying or *dynamic* input, this is the *dynamic current gain*.

The voltage gain of the amplifier is equal to beta times the ratio between the output resistance (R_L = 3,000 ohms) and the internal base-emitter resistance, which is about 4,000 ohms for this type of transistor. Thus, the dynamic voltage gain is 24 × 2,800/4,000 = 168. Therefore, if a 10-mV signal is applied to the input, a 1,680-mV signal (1.68 volts) would appear at the output.

Transistor Operating Point (continued)

Because the β of a transistor is very variable, the operating point (and gain) are very variable as well. The operating point can be stabilized (at some sacrifice in gain) by adding a resistance in the emitter circuit. As you know, the emitter voltage is almost the same as the base voltage since the base-emitter junction is forward-biased. Suppose you had the circuit shown in Fig. (A) below.

The base is fixed-biased at a voltage determined by the voltage divider R_{e1} and R_{e2}. Suppose this were set at 2 volts ($R_{e1} = 2R_{e2}$, neglecting base current). You know that the emitter must be about 0.5 volt below the base, or $2.0 - 0.5 = +1.5$ volts. Thus, the total transistor current can be set by setting the value of R_e to maintain the operating point at B, as shown earlier. The value of R_e (for Z mA) is $R_e = \dfrac{E}{I} =$

$\dfrac{1.5}{0.002} = 0.75$ K (or 750 ohms).

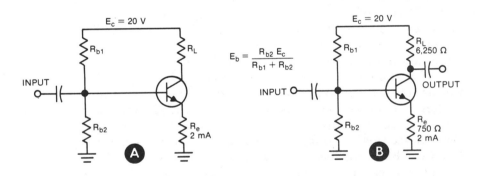

$$R_e = \frac{E}{I} = \frac{1.5}{0.002} \, 0.75 \text{ K (or 750 ohms)}$$

Let's assume a supply voltage of 20 volts. To have +6 volts on the collector (with respect to the base) or a total 8.5 volts, the load resistor is:

$$R2 = \frac{E}{I} = \frac{12.5}{0.002} = 6,250 \text{ ohms}$$

The complete schematic is shown above in Fig. (B). The gain and operating point are now essentially free of β variations. The gain is simply calculated approximately as

$$\text{Gain} = \frac{R_L}{R_e}$$

Experiment/Application—Measuring Transistor Amplifier Parameters

You can see the way transistor amplifiers operate by setting up simple transistor amplifiers, as shown on the following page, using the npn transistor 2N2222.

In circuit A, the gain will be about $R_L/R_e = 2,000/390 = 5x$, and the output will be inverted from the input. You can check this by using the oscilloscope to measure the input and output voltages. For this measurement, if you set the input audio signal to 1 volt peak-to-peak at about 1 kHz, the output voltage from the amplifier will be about 5 volts peak-to-peak. If you bypass the emitter resistor, as shown, you will realize the maximum gain of this configuration, since the signal current feedback will be eliminated by doing this. The gain will go up to a few hundred (the increase depending on the internal impedance of the signal generator and the transistor). If you put the emitter-bypass capacitor across the emitter resistor, you can see how the low frequency gain will decrease at frequencies below the point where the bypassing is adequate (about where X_c is equal to 390 ohms). As you lower the frequency, the gain will remain constant until X_c is nearly equal to 390 ohms, and then the gain will drop off gradually because more and more signal is appearing in the emitter circuit corresponding to more feedback. Eventually, a frequency will be reached where the bypassing reading will be negligible, and the gain will be the same as it was for the unbypassed case.

If you connect up the common-collector circuit (circuit C), you will see that the voltage gain is almost unity, as you learned in your earlier studies. But you know that the current gain is high so that a load resistor placed across the output will not materially affect the output voltage. You can see this by noting that the output does not change appreciably when you halve the load resistor by adding the 3.3-K resistor across the output. The capacitor coupling is used as in the other circuits to allow for coupling the ac signal into and out of the circuit without disturbing the dc operating point. As you can see by increasing the input voltage, the common collector circuit will deliver a peak-to-peak output almost as great as the supply voltage.

The common-base circuit (circuit B) is shown for completeness, but it is rarely used today. If you connected it into your test circuit, you would find it had a gain of about 10, depending on the generator impedance. You also would note that it is a noninverting amplifier; that is, the phase of the input and output are the *same*. This also is true of the common-collector circuit, but you will notice that the common-emitter circuit has a phase inversion (180-degree phase change) between the input and the output.

Experiment/Application—Measuring Transistor Amplifier Parameters (continued)

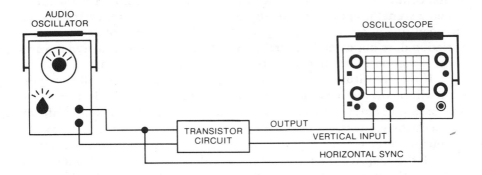

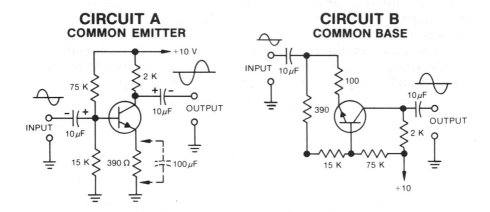

Signal Clipping

In the linear type of operation that has been considered, the no-signal condition produces a collector current at point B on the load line. When a signal is applied to the input, the collector current varies along the load line above and below point B. As long as the collector current variation does not extend beyond points A and C, the output-signal waveform is a good reproduction of the input-signal waveform.

If the amplitude of the input signal is increased, the collector current will be driven into the regions *above* A and *below* C. In these regions, the base current versus collector current variations are not uniform. Consequently, the peaks of the output signal will be limited, causing the output waveform to be a greatly distorted reproduction of the input.

If the amplitude of the input signal is increased even more, the peaks of the input signal will overdrive the collector current into the region of saturation or into the region of cutoff. In either of these extremes, the collector current cannot change further, and the output signal will have a waveform with *clipped positive* and/or *negative peaks*.

For linear amplifiers, such clipping and distortion is highly undesirable. However, as you will see, there are some circuits (FM receivers, for example) where such limiting or clipping is desirable.

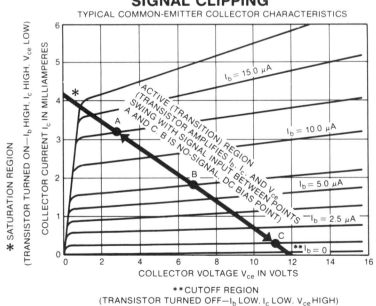

SIGNAL CLIPPING

TYPICAL COMMON-EMITTER COLLECTOR CHARACTERISTICS

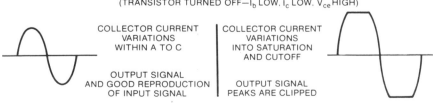

COLLECTOR CURRENT VARIATIONS WITHIN A TO C	COLLECTOR CURRENT VARIATIONS INTO SATURATION AND CUTOFF
OUTPUT SIGNAL AND GOOD REPRODUCTION OF INPUT SIGNAL	OUTPUT SIGNAL PEAKS ARE CLIPPED

Classes of Operation

There are some applications in which nonlinear biasing is desirable. For the present, you should know about the biasing conditions that are most generally useful in amplifier applications. These are defined as *classes of amplifier operation*, usually called A, AB, B, and C.

In Class A operation the output signal is a linear representation of the input signal without clipping or limiting; that is, the peaks and valleys of the collector current are never driven into either saturation or cutoff, and collector current flows for the entire 360 degrees of the input signal cycle.

In Class AB operation the zero-signal bias is set closer to the region of current cutoff. The bias is set so that the negative peak of the collector current is partially clipped. Collector current flows for less than 360 degrees of the signal cycle but for more than 180 degrees of the input signal.

In Class B operation the zero-signal bias is set just at cutoff. Collector current flows for 180 degrees of the input signal. The collector current is clipped for the entire 180 degrees of the negative peak of the input signal.

In Class C operation (not usually used in solid-state devices), the zero-signal bias is set well below the point of cutoff. Now the input signal must rise somewhat above zero before collector current will flow. Consequently, collector current flows for less than 180 degrees of the input signal, and more than half the output signal waveform is clipped.

Class A amplifiers are least efficient but provide greatest *linearity* (*fidelity*) of amplification. Class C amplifiers are the most efficient, but their use is confined to RF power amplifiers because of the extreme distortion. This distortion is reduced because of the presence of the tuned circuits. Class AB and Class B amplifiers are intermediate in both efficiency and distortion and are widely used for audio power amplifier work.

CLASSES OF AMPLIFIER OPERATION

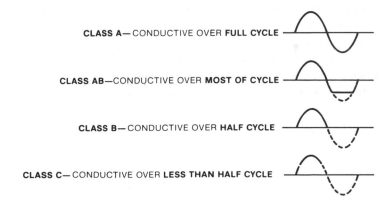

CLASS A— CONDUCTIVE OVER **FULL CYCLE**

CLASS AB—CONDUCTIVE OVER **MOST OF CYCLE**

CLASS B— CONDUCTIVE OVER **HALF CYCLE**

CLASS C— CONDUCTIVE OVER **LESS THAN HALF CYCLE**

Integrated Circuit Elements

As you learned in Volume 1, many circuit elements such as amplifiers occur as complete ICs (integrated circuits) that, with a few external connections, can perform entire circuit functions. In this and later volumes, we will discuss these as their use arises in each circuit or circuit configuration.

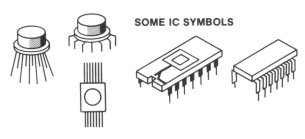

SOME IC SYMBOLS

MAC-4
COMPUTER-ON-A-CHIP

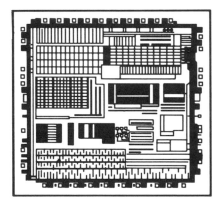

More and more analog functions are being done with ICs, and, of course, essentially all digital circuits are ICs. As you will see, the simplicity and stability of ICs make them most attractive since all of the difficult design work and testing have been done by the manufacturer. In addition, it's so easy for the manufacturer to include everything necessary for good operation that almost invariably ICs perform better than discrete circuits. The major exception is in power-handling circuits where the dissipation necessary prohibits making the circuits too compact.

Review of Basic Amplification

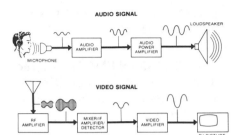

1. AMPLIFIERS are used to increase the signal voltage or power level.

TYPE	FREQUENCY RANGE
AUDIO AMPLIFIERS	16 Hz - 20,000 Hz
VIDEO AMPLIFIERS	DC - 10 MHz
RADIO-FREQUENCY AMPLIFIERS	30,000 Hz - 300 GHz
INTERMEDIATE-FREQUENCY AMPLIFIERS	FIXED BAND BETWEEN 400 kHz AND 50 MHz

2. AMPLIFIER CLASSIFICATION
   a. AUDIO AMPLIFIERS typically amplify signals at frequencies between 16 Hz and 20,000 Hz.
   b. VIDEO AMPLIFIERS typically amplify signals at frequencies between dc and 10 MHz or more.
   c. RF/IF AMPLIFIERS amplify signals in a narrow band from a few hundred kilohertz to above 20 GHz.

$$dB = 10 \log \frac{P1}{P2}$$

$$dB = 20 \log \frac{V1}{V2}$$

3. LOGARITHMIC SCALE (the dB)—A decibel (dB) is $\frac{1}{10}$ of a bel and is used as a convenient measure of gain or loss.

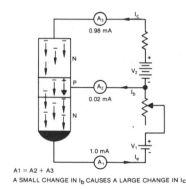

A1 = A2 + A3

A SMALL CHANGE IN I_b CAUSES A LARGE CHANGE IN I_c

4. TRANSISTOR AMPLIFIERS— The transistor amplifies because a small change in base current causes a large change in collector current.

Review of Basic Amplification (continued)

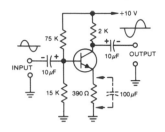

5. TRANSISTOR AMPLIFIER CONFIGURATION

a. The COMMON EMITTER is the most common voltage amplifier. It has the emitter common to input and output signal.

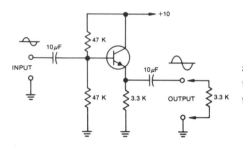

b. The COMMON COLLECTOR, also called the emitter follower, is used to provide current gain—high-to-low impedance transformation.

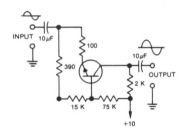

c. The COMMON BASE is used to provide power gain—low-to-high impedance transformation.

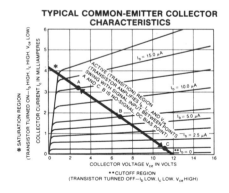

TYPICAL COMMON-EMITTER COLLECTOR CHARACTERISTICS

6. TRANSISTOR OPERATING POINT—Transistor operation can be described in terms of the characteristic transistor-operating curves.

Self Test—Review Questions

1. Briefly draw the block diagram for and describe the elements of a transmitting system.
2. Briefly draw the block diagram for and describe the elements of a receiving system.
3. Briefly define the characteristics of the four major groups of amplifiers.
4. Convert the following, using the dB chart on page 2–189:
   a. Power gain of 2, 3, 10, 50, 33, 100, 10,000, 45,000, 10^6, 3×10^6, to decibels
   b. Voltage gain of 2, 4, 10, 30, 100, 350, 500, 1,000, 5,000, 10,000, to decibels
   c. Decibel (voltage) gain of 3, 6, 10, 35, 46, 52, 60, 75, to voltage ratio
   d. Decibel (voltage) loss of 3, 6, 10, 35, 46, 52, 60, 75, to voltage loss ratio
5. Define p and n semiconductor materials. How are these configured to make npn and pnp transistors?
6. Describe briefly how transistors amplify.
7. Draw the three basic configurations of transistor amplifier circuits.
8. Describe the major characteristics of each transistor circuit configuration.
9. Define α and β and show how they are used to calculate circuit gain.
10. Define the Class A, Class AB, Class B, and Class C types of operation.

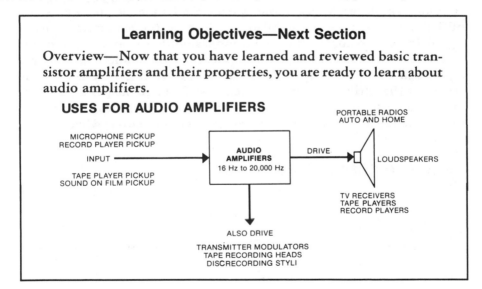

Learning Objectives—Next Section

Overview—Now that you have learned and reviewed basic transistor amplifiers and their properties, you are ready to learn about audio amplifiers.

USES FOR AUDIO AMPLIFIERS

PORTABLE RADIOS
AUTO AND HOME

MICROPHONE PICKUP
RECORD PLAYER PICKUP

INPUT ⟶ **AUDIO AMPLIFIERS** 16 Hz to 20,000 Hz ⟶ DRIVE ⟶ LOUDSPEAKERS

TAPE PLAYER PICKUP
SOUND ON FILM PICKUP

TV RECEIVERS
TAPE PLAYERS
RECORD PLAYERS

ALSO DRIVE

TRANSMITTER MODULATORS
TAPE RECORDING HEADS
DISCRECORDING STYLI

Applications of Audio Amplifiers

In your review of the electromagnetic frequency spectrum earlier in this volume, you learned that the audio-frequency range typically extends from about 16 to 20,000 Hz (20 kHz). Sometimes the audio band is extended to below 16 Hz to dc and may be considerably beyond 20 kHz. In many cases, audio amplifiers have a deliberately restricted range to improve voice communication, etc. In any case, amplifiers that operate in this general frequency range are called *audio amplifiers*.

You have had contact with audio amplifiers because it is their output that reaches you from automobile, portable, and home radios, from the loudspeakers in TV receivers, from tape recorders and record players, and from public address systems. You are familiar with some of the faults of inadequate audio amplifiers—those that cut off the low and high frequencies in voice and music and those that hum or are noisy because they pick up interfering electrical signals. You may be familiar with how well a good audio amplifier can reproduce sound. You learned this when you heard high-fidelity sound systems in homes, theaters, and auditoriums. In general, you will find audio amplifiers are coupled by transformers, resistance/capacitance (RC) or may be directly coupled, without intervening dc-blocking elements.

There are audio amplifiers whose output you do not hear directly. You know that in the radio transmitter-receiver learning system there are audio amplifiers whose input is from a microphone and whose amplified output is used to modulate the level of the transmitter carrier signal. In a tape recorder, there is an audio amplifier whose output drives a coil that impresses a magnetic duplicate of the audio signal on the tape's iron oxide coating. In a disk recording system, there is an audio amplifier that drives the cutting needle (stylus) that shapes a duplicate of the audio signal in the form of a groove in the record's surface. In fact, there are audio or audio-type amplifiers in use everywhere.

USES FOR AUDIO AMPLIFIERS

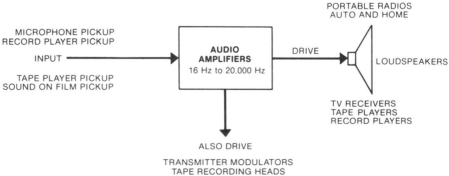

Microphones and Loudspeakers

Although microphones and loudspeakers are described in detail later in this volume, a brief review will be presented here to enable you to understand something about the input and output devices most commonly used with audio amplifiers. The principles of sound production were discussed earlier in this volume. A microphone is a device that converts the varying pressures from sound waves in air into an electrical signal whose *electrical variations duplicate* those of the sound. All microphones operate on the principle that the varying pressure from sound waves causes an element to vibrate mechanically; the element, in turn, is connected to or is part of an electromechanical system that generates electrical signals. Microphones, like other devices that convert non-electrical inputs to electrical signals, are called *transducers*. All devices that perform this function are called *transducers*. One of the most popular microphones is the dynamic microphone.

In the dynamic microphone a coil is attached to a diaphragm that is suspended in an air gap between the poles of a strong permanent magnet. When sound waves strike the diaphragm, the energy in the sound waves moves the diaphragm back and forth, thus moving the coil in the magnetic field. As you know, moving a coil in a magnetic field creates an electric current. Since the coil motion is produced by the sound waves, the electrical signal represents or is the *electrical analog* of the sound wave. The frequency of this current produced is the *same* as the frequency of the sound waves, and the amplitude of the current is *proportional* to the sound waves' pressure changes in the air.

The dynamic microphone output is usually fed to an IC or transistor amplifier by means of a transformer that matches the very low impedance of the microphone's coil to the impedance of the amplifier input.

THE DYNAMIC MICROPHONE

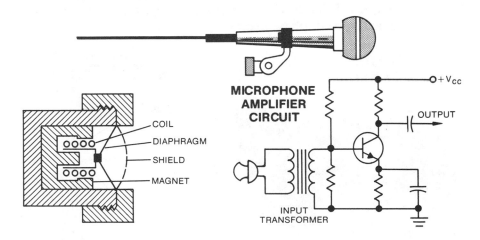

Microphones and Loudspeakers (continued)

Shown here is the construction of the dynamic loudspeaker universally used today. In many ways, it is very similar to the dynamic microphone. In fact, this type of speaker is sometimes used in intercommunications systems as *both* speaker and microphone.

In the dynamic loudspeaker a coil (called the *voice coil*) with relatively few turns is suspended around the pole piece of a very strong permanent magnet. The voice coil is fastened directly to a cone made of a special type of paper or plastic. The voice coil is held in position by a *spider* that is usually a flexible corrugated disk cemented to the frame, which allows the cone to move back and forth, but *not* sideways. The outer edge of the cone is attached to the metal speaker frame via a flexible corrugated rim. The corrugations allow the cone to move back and forth freely within the frame. Thus, the cone is held at the edges and at the center where the voice coil is attached by means of the corrugated rim and spider. In this way, the entire cone can move in and out but is restrained so that it cannot move from side to side.

The voice coil is connected to the output of the audio amplifier either directly from a low-output impedance amplifier or through a transformer that matches the low impedance of the voice coil to a higher impedance amplifier. Amplified audio signal current flows through the voice coil and generates a magnetic field whose polarity and amplitude varies with the changes in the audio signal. When the voice coil field has a polarity that aids the field in the gap, the voice coil and cone are pulled inward, reducing the air pressure in front of the cone. When the voice coil field is in opposition to the field in the air gap, the voice coil and cone are pushed out, compressing the air in front of the cone. In this manner, the voice coil and cone convert *electrical* waves into *sound* waves. It is a curious fact of our hearing that we do not sense the *phase* of an audio signal. Thus, reversing the speaker or microphone connections, etc., which would reverse the phase of the signal, causes *no audible difference*. As you will learn later, however, if more than one speaker or microphone is involved, phase must be preserved.

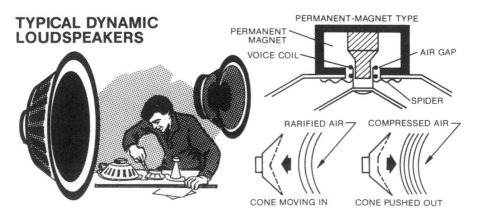

TYPICAL DYNAMIC LOUDSPEAKERS

PERMANENT-MAGNET TYPE

PERMANENT MAGNET

VOICE COIL

AIR GAP

SPIDER

RARIFIED AIR

COMPRESSED AIR

CONE MOVING IN

CONE PUSHED OUT

Multistage Amplifiers—Discrete and IC

A single transistor in a common-emitter configuration can produce an amplification or gain of several times to several hundred times. This may be insufficient for some applications. In addition, for stability reasons, it may not be wise to try for maximum stage gain, as you will see later. To achieve the voltage and power amplification required, most audio amplifiers contain a number of individual amplifiers coupled together in series. The second amplifier receives the output of the first, the third amplifies the output of the second, and so on. Each such amplifier, discrete transistor or IC, is known as a *stage*, and an amplifier consisting of many stages is known as a *multistage amplifier*.

The gain of an amplifier connected in such a series or *cascade* arrangement is equal to the numerical gain of the first stage multiplied by the gain of the second stage, multiplied by the gain of the third, etc. If dB values are used, the dB stage gains are added. Thus, the gain of three amplifier stages with individual voltage gains of 20 is $20 \times 20 \times 20$ or 8,000. A voltage gain of 20 is about 26 dB ($20 \log_{10} 20$), and the total voltage gain is $26 + 26 + 26$ or 78 dB.

MULTISTAGE AMPLIFIER

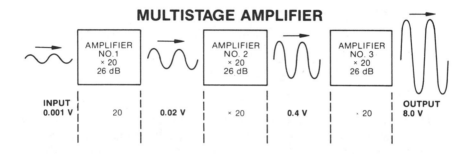

The selection of circuit configuration and methods of coupling are determined by the available signal input level, the required output signal level, the nature of the load to be driven, the range of frequencies to be amplified, the acceptable distortion level, the available voltage source, size, weight, cost, and operating environmental factors such as temperature, shock, vibration, etc. All of these problems must be solved by the electronic design engineer who makes the best possible compromise between performance, cost, size, weight, etc. that is possible at that time.

The most widely used discrete audio amplifier circuit arrangements are described on the pages that follow. Although there are many variations for each basic circuit, it should be obvious that the basic circuit configurations are very simple and conform to the basic principles that you know. Later, when you study real circuit configuration, you will see that discrete circuit audio amplifiers have been almost entirely replaced by IC amplifiers, at least for all but power amplifier stages.

Coupling Methods

There are three basic methods in widespread use for coupling the various stages in an IC or transistor amplifier. These methods are *transformer, resistance–capacitance,* and *direct coupling.*

Transformer coupling is simple and permits matching the output impedance of one stage to the input impedance of the next stage, thus providing most efficient power transfer. However, it is the most expensive method and can limit frequency response. It is usually used only in amplifiers of limited performance where transformer deficiencies can be tolerated. The widely used common-emitter amplifier configuration has a low input impedance and a high output impedance. A transformer at the input and output of each such stage can be used to match these widely different impedances efficiently.

TRANSFORMER-COUPLED COMMON-EMITTER STAGE

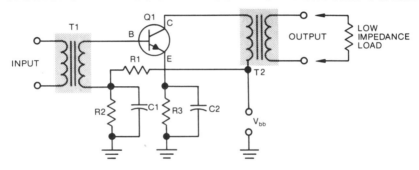

In the transformer-coupled circuit shown here, operating bias for the transistor base is supplied by R1 and R2, which are connected as a voltage divider across the dc supply. Capacitor C1 bypasses the T1 secondary without changing the dc operating point. This is necessary to prevent part of the input signal from appearing across R2. Resistor R3 is the bias-stabilizing resistor in the emitter circuit and is bypassed for audio signals by C2 to avoid loss in amplifier gain. The transistor output is coupled to the next stage by transformer T2 as the collector lead.

One significant disadvantage of transformer coupling is that the inductive reactance of the transformer increases with frequency. This can result in nonuniform amplification of different frequencies. Perhaps the most important disadvantage is that the cost of a transformer is often many, many times more than a transistor, and it is much more economical to use transistors as impedance transformers (e.g., common collector) to couple the stages. Other disadvantages are that the size and weight of transformers are many times more than transistors. Therefore, transformers are used in those special applications where their advantages overcome these shortcomings. Because IC amplifiers almost invariably have a high input impedance and a low output impedance, transformers are not used as coupling devices.

Coupling Methods (continued)

One of the most common methods of coupling in discrete circuit audio amplifiers is *resistance-capacitance (RC) coupling*. In this method, which is shown in the diagram, R4, C3 and R5, R6 replace the transformer between the stages. The method of biasing is similar to that used in the transformer-coupled amplifier shown earlier. The value of C2 is usually high, on the order of 2 to 10 μF, because of the low input impedance of Q2. Miniature electrolytic capacitors are usually used for this purpose. Although the RC coupling method shown does not efficiently match the output and input impedances, this method is widely used for reasons of economy, size, and weight. Such amplifier configurations are most useful for low-level audio amplifiers.

The *direct-coupled* amplifier offers maximum economy because it uses the fewest components. In this circuit, R3 serves as both the collector load resistor for the first stage and the bias resistor for the second stage. Resistors R1 and R2 set the operating bias of Q1. Nowadays, most IC audio amplifiers are direct-coupled and offer the significant advantage of being self-contained and having impedance matching integrally with the amplifier—that is, high input impedance and low output impedance. As shown below, the 741 IC amplifier provides amplification from dc through the audio range when connected as shown with the gain set by the ratio of R_f to R_{in}. This is a feedback amplifier, which you will learn more about later.

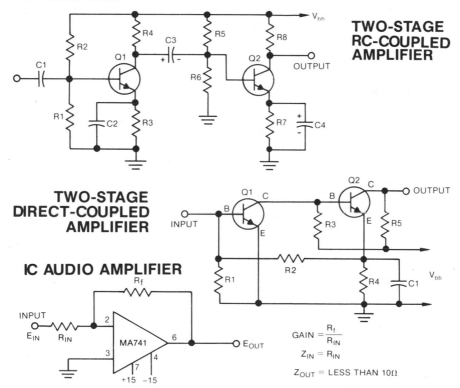

TWO-STAGE RC-COUPLED AMPLIFIER

TWO-STAGE DIRECT-COUPLED AMPLIFIER

IC AUDIO AMPLIFIER

$$GAIN = \frac{R_f}{R_{in}}$$

$$Z_{IN} = R_{IN}$$

$$Z_{OUT} = LESS\ THAN\ 10\Omega$$

Impedance Matching

Most low-level audio amplifiers use the popular common-emitter configuration. Examples typically show common-emitter stages connected in series, ignoring the poor impedance match that results on the basis that gain is easy and economical to obtain. There are, however, many cases where it is very useful to use the common-collector configuration in the coupling circuit in order to obtain a better impedance match. (The common-base configuration is rarely used today but is illustrated here for completeness.) The diagram below shows the input and output impedances of the three basic circuit configurations and thus suggests how they can be used together.

The common-collector circuit is often used as a coupling device between common-emitter stages since it has a high input impedance and lightly loads the prior stage, and a low output impedance that will drive the next stage adequately. As you can see from the diagram, Q1 is in a common-collector configuration to provide high input impedance and low output impedance to drive Q2. The input impedance of Q1 is approximately (R1 in parallel with R2) in parallel with the β of Q1 × R3. The output impedance is a few ohms. The input impedance of Q2 is approximately (R4 in parallel with R5) in parallel with β of Q2 × R6. If R6 is bypassed at audio frequencies by a suitable capacitor, then the input impedance of Q2 is $\beta \times R_e$, where R_e is the intrinsic resistance in the emitter circuit. Q3 and Q5 are common-collector impedance transformers, and Q4 is a second audio amplifier stage similar to Q2. The utility and simplicity of common-collector impedance-matching stages is evident from the diagram.

IMPEDANCE MATCHING IN TRANSISTOR AMPLIFIERS

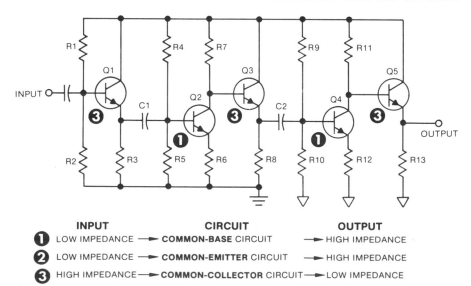

	INPUT	CIRCUIT	OUTPUT
❶	LOW IMPEDANCE ⟶	**COMMON-BASE** CIRCUIT	⟶ HIGH IMPEDANCE
❷	LOW IMPEDANCE ⟶	**COMMON-EMITTER** CIRCUIT	⟶ HIGH IMPEDANCE
❸	HIGH IMPEDANCE ⟶	**COMMON-COLLECTOR** CIRCUIT ⟶	LOW IMPEDANCE

Power Amplifiers

The usual purpose for coupling amplifier stages together is to raise the input signal to a level at which it can perform the desired function. In the case of an audio amplifier, the input is usually from a microphone, from the detector stage of a radio receiver, or from the pickup of a phonograph record or magnetic tape player. The output of an audio amplifier is most generally a loudspeaker, although audio amplifiers also are used to drive a cutting stylus for engraving the sound groove on a recording disk or to drive the magnetic coil for recording sound on a magnetic tape.

If the final stage of an audio amplifier is to be used for driving a load that consumes appreciable power, that final stage is known as the *power amplifier*. It is designed not to provide maximum voltage or current gain but to deliver *maximum power output* to the *load*. The diagram shows a common-emitter power amplifier that might be used in a small record player, tabletop radio, or portable TV. (Actually, such a simple design is rarely used.) Because it uses only a single transistor that feeds a transformer with a single primary winding, it is known as a *single-ended power amplifier*. The amplifier shown is a Class A-type and therefore is not very efficient. As a result, this simple, single-ended configuration is useful only for low-power applications. In modern low-power amplifiers, a single IC is used, and the impedance matching to the output is done without transformers, as will be discussed later.

The input RC coupling network is formed by C1 and R1 in parallel with R2. Bias is obtained from the voltage divider consisting of R1 and R2. An emitter–resister (R3) provides dc stability, and C2 bypasses R3 at audio frequencies. As you learned earlier, the gain of an amplifier with an emitter–resistor (R3) is equal to R_L/R_E, which means the dc gain will be low if R3 is large. Since it is bypassed at audio frequencies by the low reactance of C1, the full gain is obtained at audio frequencies. Thus, good stabilization of the dc operating point is obtained without sacrificing audio gain.

Transformer T1 couples the transistor output to the loudspeaker load for maximum power output with maximum efficiency. The impedance of most small loudspeakers ranges from 2 to 8 ohms, and a transformer provides a good impedance match to such a low output impedance.

SINGLE-ENDED POWER AMPLIFIER

Frequency Response

Now that you know some of the basics of audio amplifiers, you are ready to learn some of the details. As you know, audio amplifiers are designed to amplify audio signals and transfer impedance. It is necessary, however, that audio amplifiers provide this capability in a way suited to their intended use. Thus, practical audio amplifiers must process a *range* of frequencies *appropriate* to the application and must have controls for adjusting the *volume* (amplification) and *tone* (frequency response) of the output. For example, the audio amplifier in a CB radio has no need for a frequency range of more than about 100 Hz to 3,000 Hz, since it is intended to provide only *intelligible* voice reception. In fact, a very wide frequency response is undesirable because it may allow interference to be heard without increasing the intelligibility of the desired signal. Thus, special response-shaping networks are used to hold the frequency response to this range. On the other hand, the audio amplifier in a hi-fi music system will be expected to have a frequency range from at least 20 to 20,000 Hz or more and have elaborate controls for adjusting the response according to the highly refined tastes of the individual listener.

The usual method of describing the *frequency response* (frequency range) of an amplifier is to draw a response curve such as the one shown in the illustration. A plot is made of the output at each frequency, with the highest output shown as 100% (0 dB), and outputs below that are shown as decibel losses relative to the maximum. There is generally an approximately flat region in the center of the curve (midband response), and the response drops off at the high- and low-frequency ends. Usually, response curves are plotted with a decibel scale on the vertical axis and frequency on the horizontal axis. A 3-dB difference (2:1 in power) is about the least discernible audio-output change that can be noticed, and these 3-dB points are taken as the end (limit) of response of an audio amplifier (or any other type of amplifier). If amplifiers are cascaded, the frequency response of each must be wider to achieve the desired overall response.

FREQUENCY-RESPONSE CURVES

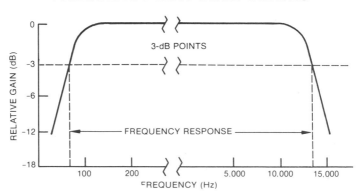

Frequency Response (continued)

The methods used for coupling audio amplifier stages do not produce flat-frequency response over a wide range. For RC coupling, the coupling capacitor causes a loss in gain at low frequencies. The reason is that the coupling capacitor and the input impedance of the next circuit (transistor input impedance) form a voltage divider across the signal voltage input to the stage. As a result, only part of the signal reaches the base of the transistor. Just how much of the signal goes to the base depends on the reactance of the coupling capacitor compared to the resistance of the input circuit.

In *Basic Electricity* you learned that capacitive reactance [$X_C = 1/(2\pi fC)$] increases as frequency decreases. The larger the reactance of the coupling capacitor, the larger the input signal drop across it, and the lower the signal voltage reaching the base. Thus, the amplifier output decreases as the frequency of the input signal decreases. Obviously, as the signal frequency increases, X_C becomes smaller, and this cause for poor response no longer exists.

For a single stage, the low-frequency, 3-dB point is reached when the coupling capacitor impedance equals the base circuit impedance. Thus, the reactance of the coupling capacitor should be made equal to the circuit impedance at the lowest frequency to be amplified, or $X_C = Z_{in}$. If there is more than one RC network in cascade, then the response of each must be adjusted so that the total response is down 3 dB. For example, if there were three networks involved, the low-frequency cutoff of each would be set so that the 1-dB loss point for each was at the frequency cutoff desired (1 dB + 1 dB + 1 dB = 3 dB). Conversely, one RC could be set at the desired cutoff and two set very much lower so that they would not interfere with the desired frequency response.

CAUSES OF FREQUENCY-RESPONSE LOSS
AT LOW FREQUENCIES

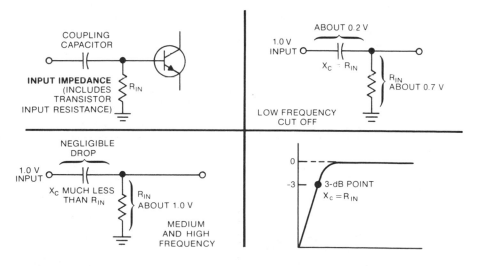

Frequency Response (continued)

At high frequencies (above midband), there is an effect that causes gain loss as frequency increases. This loss is due to capacitance that exists between the amplifier output and ground. This capacitance (called *circuit capacitance*) is made up of a number of components that together make up the total stray capacitance. One effect is due to the collector-to-emitter capacitance of the transistor in the previous amplifier stage. Another stray capacitance is due to the base-to-emitter capacitance in the stage that follows. Also, there is the capacitance between the ground and the signal-carrying wires and components called the stray capacitance. These capacitances are added together in the circuit and are in parallel with the load resistance. At low and midband frequencies, this small capacitance has high reactance and is too small to be of consequence. At higher frequencies, however, this reactance decreases and has a lower impedance to ground. Thus, the amplifier gain decreases as the frequency increases, because the amplifier gain is directly proportional to the load impedance.

The high-frequency cutoff point (− 3 dB) is the point where the total load resistance (output impedance of the previous stage in parallel with the input resistance of the succeeding stage) is equal to the reactance of the total circuit capacitance. Consequently, the amplifier gain is lower at high frequencies than at medium frequencies because of the circuit capacitance. To minimize the loss of gain at high frequencies, low values of load resistance can be used with wiring methods that reduce the capacitance between ground and the signal-carrying wires and components. In some cases, circuits include special provisions for extending the high-frequency response of amplifiers. You will learn about thes later.

CAUSES OF FREQUENCY-RESPONSE LOSS AT HIGH FREQUENCIES

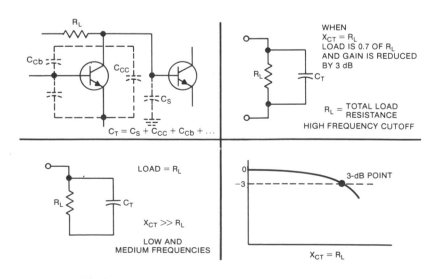

Describing Amplifier Characteristics

Describing audio amplifier characteristics in terms of human hearing is not easy. The reason for this is that the sound produced by an audio system is usually intended for the human ear, and human hearing does *not* correspond to easily described electronic measurements.

A simple method to describe amplifier gain is to state voltage input at a given impedance and output power at a given impedance for a specified range of frequencies. For example, an amplifier can be described as requiring a minimum input of 1 mV into a 50,000-ohm input impedance for a power output of 10 watts into an 8-ohm loudspeaker load. For some applications, the noise level is an important consideration. A more complete specification would state the frequency response (20 Hz to 25 kHz) and state that at 1-mV input, the noise level will be 65 dB below the signal level. While this is an adequate description in terms of electronic measurement, it will not define how this *will sound to you*.

The human ear does *not* hear different frequencies *equally well*. A young person can usually detect sound in the frequency range between 20 Hz to about 15 kHz. As a person ages, the high-frequency response decreases so that beyond 40 years of age this response is restricted to 5 kHz or less. The human ear is most sensitive in the region between 2 and 5 kHz. Also, the frequency response of the ear depends on the *loudness* of the sound. Electronic engineering loudness measurements are generally made in terms of the *decibel* defined earlier. If the threshold of hearing is defined as 0 dB, a quiet whisper at a distance of 5 feet is equal to 10 dB, while an average whisper at the same distance is equal to 20 dB, and average conversation at that range is about 65 dB. The level of a hi-fi system reproducing classical music is about 70 dB, while the same system used to play pop music is about 105 dB. A noise level of 130 dB can cause ear pain. Sustained exposure to levels above 70–75 dB can permanently impair hearing. (See the table of noise levels in decibels below.)

SOURCE OR DESCRIPTION OF NOISE	DISTANCE FROM NOISE	NOISE LEVEL IN DECIBELS
THRESHOLD OF PAIN		130
ROCK CONCERT OR DISCO	CLOSE UP	115-120
HAMMER BLOWS ON STEEL PLATE	2 ft	114
LISTENING LEVEL FOR HI-FI POP MUSIC		105
CAR		87
FACTORY		78
LISTENING LEVEL FOR CLASSICAL MUSIC		70
BUSY STREET TRAFFIC		68
LARGE OFFICE		65
ORDINARY CONVERSATION	3 ft	65
RESTAURANT		60
RESIDENTIAL STREET		58
MEDIUM OFFICE		58
SMALL STORE		52
THEATER (WITH AUDIENCE)		42
AVERAGE WHISPER	4 ft	20
QUIET WHISPER	5 ft	10
RUSTLE OF LEAVES IN GENTLE BREEZE		10
THRESHOLD OF HEARING		0

Additional Audio Amplifier Terminology

The development of *high fidelity (hi-fi) systems* has brought a new maze of terminology into popular use. In addition to frequency response, gain, etc., we are concerned in audio work with *signal-to-noise ratios*. As you will learn later, all electronic circuits produce *noise* (called *thermal noise*). Hum and other noise also may be present. The ratio of the signal to this noise is called the *signal-to-noise (S/N ratio)*, usually expressed in decibels. Typically, we expect an S/N ratio of better than 60 dB for adequate hi-fi performance. Much lower S/N ratios (20 dB) are suitable for communications systems.

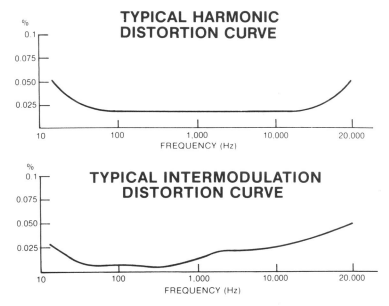

Another quality factor for audio systems is *distortion*, which is deviation of the output from the input (exclusive of amplification). Two types of distortion predominate—*harmonic distortion (HD)* and *intermodulation distortion (IMD)*—and both result from *nonlinearity* of amplification. Nonlinearity means that the output/input characteristic is not a constant.

Harmonic distortion occurs when signals appear at the output in multiples of the input signal frequency that were not present in the input signal (e.g., a 100-Hz tone input produces a harmonic distorted output of 200 Hz, 300 Hz, 400 Hz, etc.). Intermodulation distortion describes the amplifier's performance in terms of the production of sum and difference frequencies from two tone inputs (e.g., 1,000-Hz and 5,000-Hz simultaneous input signals produce tones at the output of 4,000 and 6,000 Hz, as well as at 1,000 and 5,000 Hz). From a listener's standpoint, harmonic distortion is much more tolerable than intermodulation distortion. In most modern solid-state amplifiers, both HD and IMD are negligible.

Push-Pull Amplifier

The power amplifiers that you have seen up to now have used only a *single* transistor. If you needed to increase the output from such a power amplifier, you could use a larger, more expensive transistor with greater power rating. You also could connect two of the lower-power transistors in parallel—base to base, emitter to emitter, and collector to collector— and carry twice the current as before. A better method is to use the *push-pull* arrangement of two transistors shown here. This also is known as a *double-ended* configuration, as compared to the *single-ended* configurations you have seen up to now.

In a push-pull amplifier, the input-signal waveforms to the bases of the two transistors are 180 degrees out of phase. One simple method of achieving this is by using an input transformer with a center-tapped secondary winding, as shown. Now, when one base becomes more positive, the other base becomes more negative (less positive) by the same amount. In this circuit, one base input signal *pushes* the collector current *upward*. At the same time, the 180-degree out-of-phase counterpart on the other base is *pulling* the collector current *downward* by exactly the *same* amount. This unique interaction between the two sections of the power amplifier explains why it is called a *push-pull* amplifier.

If an amplifier is biased for Class A operation, the quiescent current is the same in both transistors, and current flows through both transistors for the full input cycle. Thus, the sum of the two collector currents drawn from the supply remains a steady dc. The reason is that when one collector current increases, the other decreases by exactly the same amount. Across the primary winding of the output transformer, it is the *difference* between these two collector currents that creates the output signal. Thus, the total output at the secondary of the output transformer is the sum of the outputs from both transistors. Class A push-pull amplifiers are rarely used, except for low-level amplification, because of their low efficiency.

THE PUSH-PULL AMPLIFIER

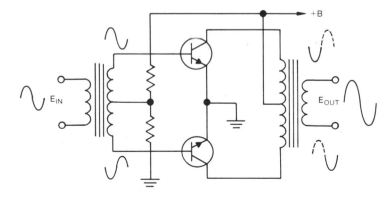

Push-Pull Amplifier (continued)

Although a push-pull amplifier can be biased for Class A operation, one real advantage to push-pull operation is that it can be operated as Class AB or Class B with high efficiency without significant distortion. Suppose both transistors are biased to cutoff. When an input signal is present, the input signal drives each of the transistors into conduction when its base is relatively positive (for npn transistors). Thus, one transistor is driven further into collector current cutoff during one half cycle of the input signal, while the other transistor collector current flows normally. On the other half cycle, the situation is reversed. Although

CLASS AB AUDIO AMPLIFIER

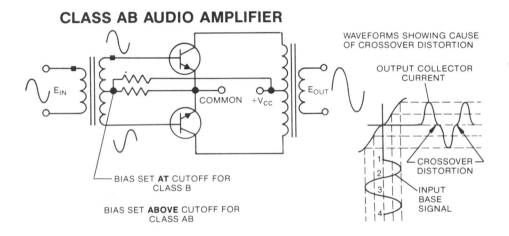

the individual collector current waveforms are slightly distorted, this distortion is greatly minimized when the two currents are combined in the transformer primary.

Class A amplifiers are inefficient because current flows continuously whether signal is present or not. However, because of the distortion reduction of the push-pull arrangement, two transistors can be biased for an approximation of Class B operation. Now each transistor produces collector current for about 180 degrees of the cycle. Pure Class B operation would be ideal because no collector current flows under no-signal conditions. However, at low-input signal levels the increase in collector current is not uniform with respect to the increase in base current. That is, the amplification is nonlinear near cutoff bias and low-level signal inputs. In such a case, the output signal will have *crossover distortion*, as shown in the above diagram. For this reason Class AB operation is often used for audio amplifier high-power applications, since in Class AB operation both transistors are conducting slightly. As you will see later, negative feedback is used to minimize this kind of distortion.

Phase Inverters

Push-pull amplifiers require that the signals to the two bases be 180 degrees out of phase and of equal amplitude. In the amplifiers described on earlier pages, this out-of-phase input was obtained from a center-tapped transformer. Because of the size, weight, cost, and relatively poor frequency response of transformers, it is desirable to use transistor circuits to supply the out-of-phase signals.

There are many types of phase inverters; however, the split-load phase inverter shown here accomplishes the job with a minimum of parts. In this arrangement, the transistor output current flows through both the collector load resistor R3 and the emitter load resistor R4. If R3 and R4 have equal resistance, the signals in the collector and emitter circuits are approximately equal. As you know, the emitter current is the sum of the base and collector current. Thus, the signal at the emitter is slightly greater if R3 and R4 are equal. Therefore, R3 and R4 are often chosen to be slightly different to make the signal level at the emitter and collector the same. The gain of this circuit obviously is a little less than 1. The collector signal is reversed with respect to the input, while the signal at the emitter is in phase with the input signal. Operating bias for the base is established by the voltage divider formed by R1 and R2. This circuit uses an npn transistor with a positive supply voltage, although the identical circuit could be assembled with a pnp transistor and a negative supply voltage.

Although R3 and R4 are made approximately equal to produce equal output voltages, the output impedances are unequal. They are unequal because the collector output impedance is higher than the emitter output impedance. This could produce distortion when large signal currents flow. The impedance can be equalized by placing a resistor between C2 and the emitter, as shown in the diagram. Normally, however, this circuit is designed to operate into a high impedance so that this source-impedance unbalance is not important. A disadvantage of the transistor split-load phase inverter is that it can supply little power to the driver stage. This problem can be overcome by using emitter followers after the phase inverter as impedance converters. In this way, the power delivered to the load can be greatly increased.

SPLIT-LOAD PHASE INVERTER (SPLITTER)

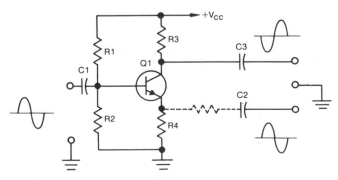

Emitter-Coupled Push-Pull Amplifier

The emitter-coupled push-pull amplifier shown can be used to provide phase inversion in addition to operating as a push-pull amplifier. As you will see, this configuration is often called a *differential amplifier*.

When the amplifier is operating as a push-pull amplifier, signals of equal amplitude and of opposite phase are applied to the bases of Q1 and Q2. If the transistors are identical and the input signals also are identical but 180 degrees out of phase, the currents through Q1 and Q2 are equal and opposite. Therefore, no signal appears at the junction of Q1–Q2 emitters and R3. Signals of equal and opposite phase appear at the collectors, as shown in the diagram, and provide a push-pull output. If the emitter resistor R3 is made large enough to approximate a constant current ($-V$ is much greater than the input signal), the current through the transistors must be matched so that as the current through Q1 increases, the current through Q2 must decrease in precisely the same way, and vice versa. In this way imbalance in the input amplitude can be corrected to balance the output.

EMITTER-COUPLED PUSH-PULL AMPLIFIER

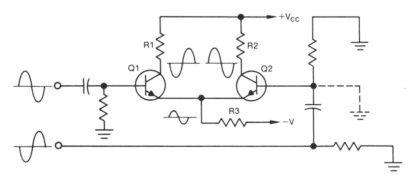

This circuit can be used as a phase inverter by simply omitting one input. Suppose the base of Q2 were grounded and an input supplied to the base of Q1. In this case, Q1 can be considered to be a common-emitter amplifier and Q2 a common-base amplifier. In this arrangement the signal voltage at the emitters is equal to half the input voltage. This is because the input impedance at the emitter of Q2 is exactly equal to the output impedance at the emitter of Q1. The gain of Q1 is half what it would be if its emitter were grounded because the base-emitter voltage is half of the input voltage. The gain of Q2 also is exactly half because its base-emitter voltage (base grounded) also is half the input voltage. Thus, two signals of equal amplitude, but opposite in phase, appear at the collectors of Q1 and Q2. An important advantage of this type of phase inverter is that the output signals are of approximately equal impedance and gain is obtained in the stage. Thus, in general, this circuit is to be preferred to the single-transistor phase inverter.

IC Differential Amplifier Circuits

The IC differential amplifier circuit shown here has many important applications in audio amplifiers and instrumentation, measurement, and control systems. It generally forms the input amplifier to most IC amplifiers because of its versatility. When the two transistor circuits are almost perfectly matched, as in integrated circuits, and the emitters are fed from a constant-current supply (a third transistor rather than a resistor), the circuit performs several useful functions: If the two input voltages V_{b1} and V_{b2} are either zero or equal in amplitude and polarity, the collector currents I_{c1} and I_{c2} are equal, and there is a zero voltage difference between the collectors of Q1 and Q2.

BALANCED DIFFERENTIAL AMPLIFIER

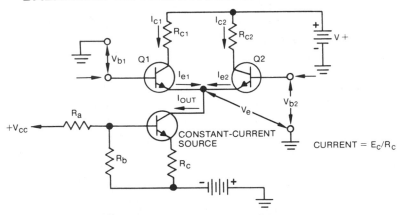

Note that the sum of I_{c1} and I_{c2} must always be equal to the constant-current source output I_{out}. Therefore, if either emitter current increases, the other emitter current must decrease by exactly the amount required to keep I_{out} constant.

Now, if the base of Q1 is made more positive than the base of Q2, the emitter current of Q1 must increase, and the emitter current of Q2 must decrease in order to keep I_{out} constant. Thus, I_{c1} must increase and I_{c2} must decrease in the same manner. Because of the voltage drop across the collector resistors R_{c1} and R_{c2}, the collector voltage of Q2 must be larger than the collector voltage of Q1, in proportion to the current change.

Thus, a difference in the voltages applied to the bases produces an equivalent difference in the voltages at either collector. *In other words, a voltage differential at the input produces an equivalent voltage differential at the output.* This is the reason that the circuit is called a *differential amplifier*.

The same effect takes place regardless of the polarity and amplitude of the inputs. As long as the circuit is not driven into cutoff or saturation, the collector voltage is an accurate means of measuring the difference between the base voltages.

IC Differential Amplifier Circuits (continued)

The constant-current source functions very simply. The base is held at a constant voltage by R_a and R_b. Therefore, the emitter is at a voltage about 0.5 volt less (constantly). The current flowing through the transistor is determined by R_c. Since the emitter current is almost entirely the collector current, as you can see, the collector circuit is a constant-current source.

If an alternating voltage signal is applied to the base of Q1, the output at the collector of Q1 is amplified and phase-inverted, as it would be with any single-ended, phase-inverting transistor amplifier. However, because of the fact that there is a constant-current source, the sum of the emitter currents must remain constant. Thus, if an alternating signal voltage is applied to the base of Q1, the output at the collector of Q2 is an amplified version of the input signal and is at the *same* phase.

The differential amplifier also produces a differential signal at either collector. Thus, if there is a difference between the voltages applied to the two bases, the collector voltage of either Q1 or Q2 will change in response to the difference between the two inputs. Thus, the circuit produces a single-ended output that is a measure of a differential input.

If the base of Q2 is connected to ground, the circuit can be operated as the emitter-coupled phase splitter described earlier. This perhaps seems to be a complex way of obtaining an emitter-coupled phase-splitter circuit. The differential amplifier is, however, very easily obtainable in the form of an IC. The great versatility of the differential amplifier has resulted in its extensive use and, hence, lower cost in IC form. Because it is widely used as a building block in a variety of ICs, the differential amplifier also is available in a variety of packages suitable for audio amplifier applications.

USING THE DIFFERENTIAL AMPLIFIER AS A PHASE INVERTER (SPLITTER)

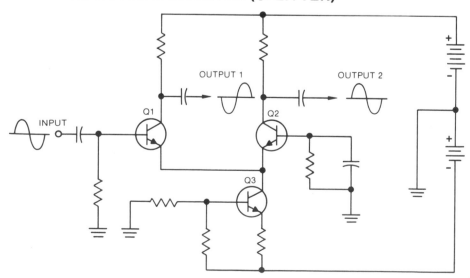

Negative Feedback

Negative feedback is a technique widely used to reduce distortion in amplifiers. It also can be used to make an amplifier's frequency response more uniform. Negative feedback involves feeding a portion of the output signal back to the input in such a manner that the signal fed back is 180 degrees out of phase with the signal present at the point of feedback. Although this reduces the gain of the stage, it significantly reduces the amount of distortion present in the output. In solid-state circuits where gain is easy and economical to obtain, even if it means adding extra stages, it is generally worth the small extra cost of using negative feedback to reduce distortion.

You are already familiar with one method of obtaining negative feedback. In your study of transistor bias methods, you learned that the use of an emitter-to-ground resistor creates an emitter bias that reduces the effects of drift in the average dc bias circuit. A capacitor is often placed across the emitter bias resistor to bypass the signal voltage and thereby prevent this effect from taking place with the ac signal voltage. Thus, the circuit has high ac gain and lower dc gain.

CURRENT FEEDBACK

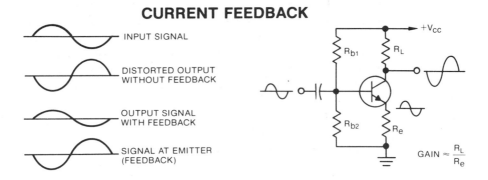

INPUT SIGNAL

DISTORTED OUTPUT WITHOUT FEEDBACK

OUTPUT SIGNAL WITH FEEDBACK

SIGNAL AT EMITTER (FEEDBACK)

$$\text{GAIN} \approx \frac{R_L}{R_e}$$

If the bypass capacitor is removed from across the emitter resistor, an ac signal voltage will develop across it. This voltage will be in phase with the input signal. Since the collector current flow is proportional to the base-emitter junction voltage, the higher the emitter resistor, the less signal current will flow in the collector circuit. As an extreme, if the emitter voltage was exactly equal to the base voltage, no current change would take place. As you know, the emitter resistor reduces or degenerates the amplifier gain. The emitter resistor also reduces distortions because the emitter signal also includes the collector current. If the collector current is not a faithful reproduction of the input signal, the base-emitter signal that results tends to correct the collector current so that it is a more faithful reproduction of the input signal.

When this type of degeneration is deliberately introduced, the distortion in the output signal is reduced. The feedback is obtained from emitter current flow and hence is known as *current feedback*.

Negative Feedback (continued)

Negative feedback is usually used around a multistage amplifier because, as you will see, the utility of feedback in distortion reduction increases as the amplifier gain increases. As you will learn, many high-gain IC amplifiers are designed to use negative feedback to set the amplifier gain with low distortion. The amplifier gain with the feedback loop open is called *the open-loop gain*. The amplifier gain with the feedback loop closed is called *the closed-loop gain*. The open-loop gain typically is from 10 to 60 dB greater than the closed-loop gain to obtain good distortion reduction.

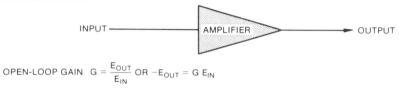

INPUT ————————————— AMPLIFIER ————————→ OUTPUT

OPEN-LOOP GAIN $G = \dfrac{E_{OUT}}{E_{IN}}$ OR $-E_{OUT} = G\,E_{IN}$

The minus sign indicates that E_{out} is 180 degrees out of phase with E_{in}.

Negative feedback (sometimes called *inverse feedback*) is added as shown below. The amplifier gain is now controlled by R1 and R2. If the open-loop gain A is large, then E_c must be very small. For example, if A is 1,000 and E_{out} is 10 volts, the input signal at E_c can only be 10 mV. This is a signal that is essentially zero. Therefore, if an input signal of 1 volt is applied at E_{in}, the current flow through R1 is $E_{in}/R1$. By the same token, the current flow must be essentially matched by current in R2 (opposite polarity) if E_c is to remain near zero. The only way this can happen is if the currents through R1 and R2 are equal and opposite.

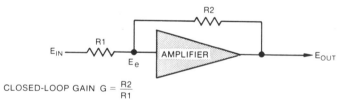

CLOSED-LOOP GAIN $G = \dfrac{R2}{R1}$

If the current through R2 is not the same as the current through R1, the difference (error) will appear as part of E_E. This difference is opposite in sense to the output error and tends therefore to correct the error so as to make the currents equal and opposite at the junction of R1 and R2. If R2 is 10 times as large as R1, the output voltage has to be 10 times the input voltage to provide equal currents. Therefore, the gain is almost exactly equal to R2/R1 if the open-loop gain is sufficiently higher than the closed-loop gain. If the open-loop gain was 60 dB and the closed-loop gain was 30 dB, the distortions, etc. are reduced by 30 dB.

Just as negative feedback will reduce distortion and noise, it also will reduce distortion in the frequency response; that is, negative feedback will even out the frequency response curve of an amplifier.

Complementary Amplifier Stages

There are several kinds of npn and pnp transistors with essentially the same operating characteristics. The only significant difference between such transistors is that they have internal pn junctions of opposite polarity. This requires that their applied bias voltages have opposite polarity. By using such transistors together in an amplifier, the circuit can be simplified.

COMPLEMENTARY AMPLIFIER STAGES

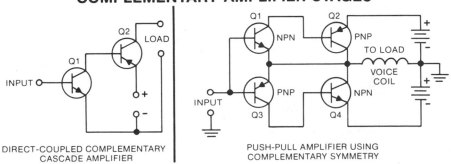

DIRECT-COUPLED COMPLEMENTARY
CASCADE AMPLIFIER

PUSH-PULL AMPLIFIER USING
COMPLEMENTARY SYMMETRY

As one example, a simple two-stage direct-coupled amplifier can be made without any resistors or capacitors. Only low gain is achieved with this arrangement because of the impedance mismatch, but the simplicity of the circuit compensates for this loss. In this circuit the base-to-ground circuit of Q2 is completed by the collector-to-emitter junctions of Q1. Consequently, the flow of Q2 base-emitter current supplies the bias current of just the right polarity required for the operation of Q1. In addition, the base-emitter junction of Q2 serves as the load across which the output signal of Q1 is developed. The bias and load conditions required for the correct operation of Q1 and Q2 have been supplied, and the output signal of Q1 is applied to the input of Q2. Thus, two-stage amplification has been obtained with extreme circuit simplicity.

The same techniques can be used to produce a simply constructed push-pull amplifier. In addition, the method provides the means for driving a low-impedance speaker voice coil from a much higher impedance source. In this circuit, Q1 and Q2 operate in the manner described above, and so do Q3 and Q4. In addition, Q1 and Q3 are interconnected so that their collectors produce 180-degree out-of-phase inputs to the bases of Q2 and Q4. As shown, the circuit is very simple even though two power supplies are required (positive and negative). For many applications, however, this complication results in an overall more simple system than one in which circuits require only one power supply.

This push-pull circuit usually requires a large amount of negative feedback to minimize distortion. It also is often operated as a Class B amplifier, producing maximum useful efficiency in audio amplification.

Experiment/Application—An IC Feedback Audio Amplifier

You can use the 741 IC operational amplifier (dc-coupled) shown on the following page as a low-level audio amplifier. As an audio amplifier, feedback is used to adjust the amplifier gain. This IC amplifier comes in several package configurations. The easiest to use is the 14-pin DIP (dual in-line package).

If you connect the 741 operational amplifier, you can see how varying the feedback resistor network will vary the gain of the amplifier. One advantage of this amplifier is that no external frequency-compensating components are necessary.

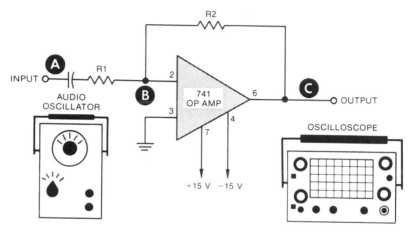

If you set up the circuit as shown, you can choose the gain you wish as the ratio of R1 to R2. Suppose you set R1 equal to 10 K and R2 equal to 100 K. If you check the gain at 1 kHz, you will find the signal at point C is 10 times that at point A. As you know, there is little signal at B because the amplifier has high gain, and the tendency is for the currents through both resistors to be equal and opposite.

If you set R2 equal to 10 K, the gain from A to C will be unity. When you set R2 equal to 220 K, the gain will be 22. Lastly, if you set R2 equal to 3.3 K, you will find the overall gain will be one-third.

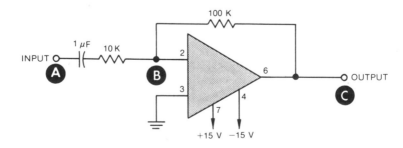

OPERATIONAL AMPLIFIERS 741/741C

GENERAL DESCRIPTION

The 741 and 741C are general purpose operational amplifiers that feature improved performance over industry standards like the 709. They are direct, plug-in replacements for the 709C, 201, 1439, and 748 in most applications.

The offset voltage and offset current are guaranteed over the entire common mode range. The amplifiers also offer many features which make their application nearly foolproof: overload protection on the input and output, no latch-up when the common mode range is exceeded, as well as freedom from oscillations.

The 741C is identical to the 741 except that the 741C has its performance guaranteed over a 0°C to 70°C temperature range, instead of -55°C to 125°C.

SCHEMATIC AND CONNECTION DIAGRAMS

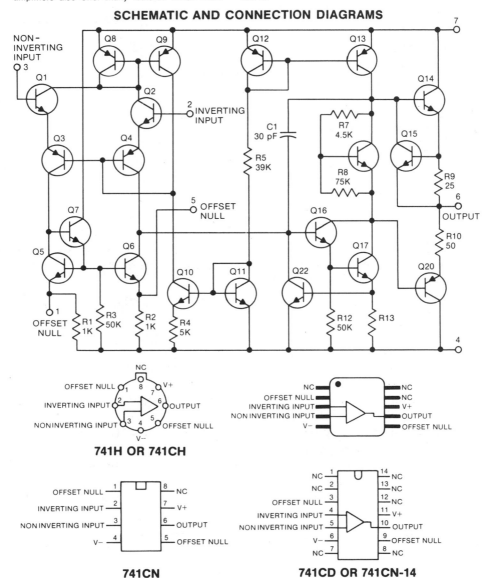

741H OR 741CH

741CN

741CD OR 741CN-14

Experiment/Application—Measuring Frequency Response of an Amplifier

A more practical use of the 741 operational amplifier as an audio amplifier employs RC coupling at the input and output, which limits the low-frequency response. The internal characteristics of the 741 determine the high-frequency response. (This response will depend on the gain.) The higher the gain, the poorer is the high-frequency response. Suppose you set up the 741 IC as shown below:

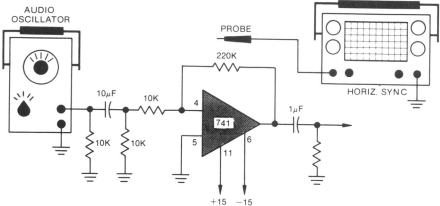

Now, if you measure the gain at midband, say at 2 kHz, you will find it equal to 10. Suppose you made a series of measurements at different frequencies. You could construct a table from which you could plot a graph of frequency response.

Frequency (Hz)	Gain	Frequency (Hz)	Gain
15	8	10,000	21
30	16	15,000	19
60	19	20,000	16
500	22	25,000	13
1,000	22	30,000	10
2,000	22	40,000	8
5,000	22		

The overall frequency response is from 30 Hz to 20 kHz. If you want to rapidly check the response, you can find the midband gain (typically at 1 kHz), and then you know the band ends are at 70% of the midband gain. For example, the band ends of an amplifier with a gain of 22 would be 22 × 0.7 or 15.5. The 741 makes a good general-purpose audio amplifier.

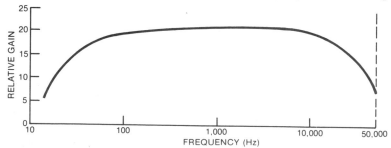

An IC Low-Level Preamplifier

In general, ICs are designed to provide either great flexibility of operation or are designed for a specific function. Since IC development is expensive, ICs are usually produced only for applications that require them in *large* quantities.

A useful general-purpose amplifier is the National Semiconductor LM381. This amplifier is very inexpensive and provides high gain and low-noise operation. The schematic and package configuration are shown on the next page.

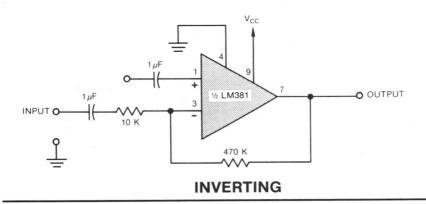

INVERTING

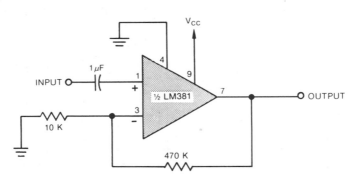

NONINVERTING

As you can see, two complete amplifiers are included in a single package. A typical amplifier with a closed-loop gain of 50 (34 dB) is shown schematically above. This amplifier will provide low-noise amplification with a minimum of components. While the IC was designed for a stereo system (two amplifiers in one chip), the two amplifiers can be cascaded if the gain of each amplifier is kept below about 40 dB. The amplifier has a differential input so that one input can be used as a noninverting input (+), while the other, inverting input (−) can be used for feedback control. This useful property of differential input is very convenient and is typical of this sort of amplifier design.

LM381/LM381A LOW-NOISE DUAL PREAMPLIFIER

GENERAL DESCRIPTION

The LM381/LM381A is a dual preamplifier for the amplification of low-level signals in applications requiring optimum noise performance. Each of the two amplifiers is completely independent, with individual internal power supply decoupler-regulator, providing 120-dB supply rejection and 60-dB channel separation. Other outstanding features include high gain (112 dB), large output voltage swing (V_{CC} −2 V) p-p, and wide power bandwidth (75 kHz, 20 V p-p). The LM381/LM381A operates from a single supply across the wide range of 9 to 40 V.

Either differential input or single-ended input configurations may be selected. The amplifier is internally compensated with the provision for additional external compensation for narrow band applications.

FEATURES

- Low noise—0.5 μV total input noise
- High gain—112-dB open loop
- Single supply operation
- Wide supply range 9-40 V
- Power supply rejection 120 dB
- Large output voltage swing (V_{CC} −2 V) p-p
- Wide bandwidth 15 MHz unity gain
- Power bandwidth 75 kHz, 20 V_{p-p}
- Internally compensated
- Short circuit protected

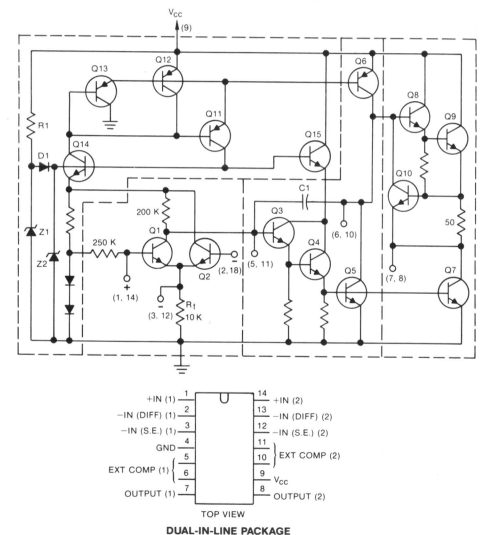

TOP VIEW

DUAL-IN-LINE PACKAGE

An IC Power Amplifier

A typical IC power amplifier that delivers sufficient output power to operate speakers on small phonographs, radios, TVs, and small public address systems is described below. Units of this type are available from many manufacturers. We will consider the LM380 made by National Semiconductor. The pertinent data is shown below. As shown, it is available as a dual amplifier or as a single-amplifier unit.

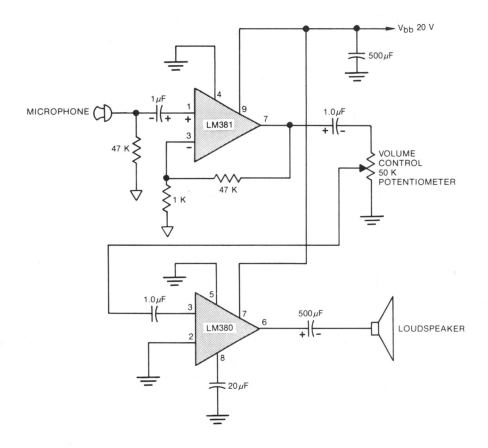

As described in the manufacturer's literature shown on the next page the amplifier gain is fixed at 34 dB (50). An amplifier like this, coupled with the simple amplifier shown previously, can easily make up a simple public address or paging system as shown below.

As you can see, the circuit configuration is simple and requires very few external components. The total gain is 68 dB (2,500), which can drive the amplifier to full output (2.5 watts) with a high-output microphone. A heat sink is necessary on the LM380 to keep its temperature within design limits.

LM380 AUDIO-POWER AMPLIFIER

GENERAL DESCRIPTION

The LM380 is a power audio amplifier for consumer application. In order to hold system cost to a minimum, gain is internally fixed at 34 dB. A unique input stage allows inputs to be ground referenced. The output is automatically self entering to one-half the supply voltage.

The output is short-circuit proof with internal thermal limiting. The package outline is standard dual-in-line. A copper lead frame is used with the center three pins on either side comprising a heat sink. This makes the device easy to use in standard p-c layout. A mini dual-in-line package version with reduced power capability also is available.

Uses include simple phonograph amplifiers, inter-

coms, line drivers, teaching machine outputs, alarms, ultrasonic drivers, TV sound systems, AM-FM radio, small verso drivers, power converters, etc.

FEATURES

- Wide supply voltage range
- Low quiescent power drain
- Voltage gain fixed at 50
- High peak current capability
- Input-referenced to GND
- High input impedance
- Low distortion
- Quiescent output voltage is at one-half of the supply voltage
- Standard dual-in-line package

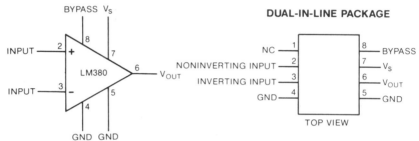

DUAL-IN-LINE PACKAGE

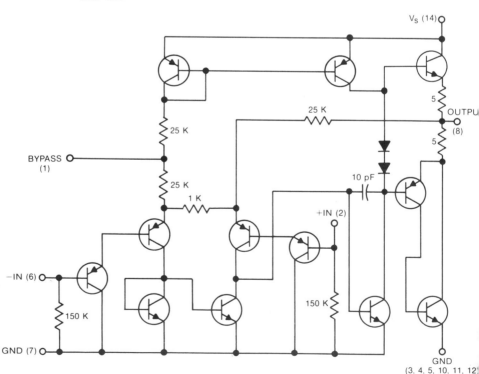

CB Transceiver Audio System

As you learned earlier, the CB transmitter-receiver learning system has an audio amplifier that takes the microphone input and amplifies it in an audio amplifier to a level sufficient to modulate the RF carrier generated in the transmitter. Another audio amplifier was shown taking the output from the detector and amplifying it sufficiently to drive the loudspeaker. An amplifier having high sensitivity and several watts of power output is required for both applications. In a transceiver the same audio amplifier is used for *both* functions. When transmitting, the microphone is connected to the amplifier input, and the output is used to modulate an RF amplifier stage in the transmitter RF section. When receiving, the amplifier input is switched to the output of the detector, and the output is connected to the speaker or headphones.

COMMON AUDIO AMPLIFIER FOR CB TRANSCEIVER

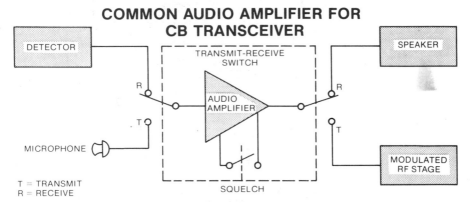

To improve the intelligibility and to assure a consistent high-modulation percentage, filters and compressors are included in the amplifier on both transmit and receive. In addition, a squelch circuit that shuts down the audio amplifier when no signal is present is used on receive. Thus, it isn't necessary to listen to receiver noise when there is no transmission. The squelch circuit senses the presence of a detected signal and enables the audio amplifier when this is so. Usually, a control is included to set the threshold value of signal that will trip the audio output-enabling circuit.

As you will learn, speech has many high peaks in output that are not necessary for intelligibility. If these are clipped, the transceiver has a greatly improved transmitting range without increasing the RF power. Clipping is used on transmit for this reason. On receive it is useful to minimize large noise pulses, which can be very annoying. As mentioned earlier, a restricted audio band is useful to minimize interference and to maximize intelligibility. Both these features are almost invariably included in communication equipment such as CB radios to provide improved performance. Obviously, these features would not be found in a broadcast receiver where high fidelity is desired.

CB Transceiver Audio System (continued)

The basic audio system for a 40-channel CB transceiver is shown below. The switch S5-B grounds the low microphone or speaker terminals on transmit and receive, respectively. In the transmit position, the audio input from the receiver detector is switched off so that there is no input from this source. The microphone is connected to audio preamplifier Q15, where the signal is amplified and coupled into the audio amplifier IC U3. Capacitor C78 (0.022 μF) in the collector of Q15 provides for restriction of the high-frequency response.

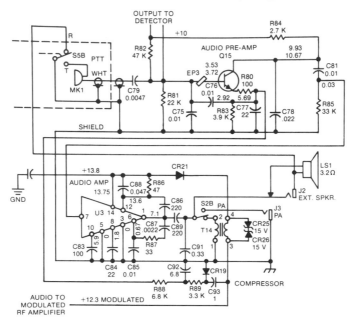

Audio amplifier U3 is a combined IC linear operational amplifier and power amplifier that can provide up to 4 watts of audio output at pin 1. This output is coupled to the speaker via capacitor C89 (220 μF) that also connects to the primary of modulation transformer T14. Frequency compensation of U3 is accomplished by the resistor-capacitor assemblies associated with U3. Compression to hold the modulating level constant for different levels of speech input, is provided by the feedback loop from of T14 via the network to the emitter circuit of Q15. When the signal level is excessive, part of it is fed back to reduce the audio gain or *compress* the audio modulating signal. The compressor does not function on receive because the secondary of T14 is an open circuit. When RF excitation is supplied to the RF power amplifier stages, the supply voltage to these stages is supplied via the T14 secondary. This voltage is modulated by the audio signal and thus varies the RF output (as you will learn later) to impress the audio signal on the RF carrier. Provisions are included to use an external speaker via switch S2 so that the amplifier can be used as a public address system.

Review of Audio Amplifiers

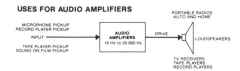

USES FOR AUDIO AMPLIFIERS

1. AUDIO AMPLIFIERS usually have frequency coverage in the range from 20 Hz to 20 kHz. They are used for amplification of electrical signals that represent audio or sound.

2. MICROPHONES AND LOUD-SPEAKERS—Microphones are transducers that convert sound pressure to electrical signals corresponding to sound pressure variations. Loudspeakers are transducers that convert these electrical signals into corresponding sound pressure variations.

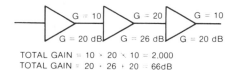

TOTAL GAIN = 10 × 20 × 10 = 2.000
TOTAL GAIN = 20 + 26 + 20 = 66dB

3. AMPLIFIER GAIN in a multistage amplifier is the producer of the individual stage gains or the sum of their gain in decibels.

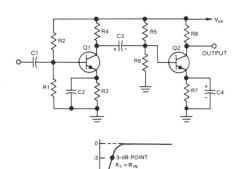

4. AUDIO AMPLIFIERS are usually either transformer-coupled or RC-coupled. Transformer coupling is simple but expensive and can limit frequency response. RC coupling is more popular because it is inexpensive and can have any desired frequency response.

5. FREQUENCY RESPONSE of an audio amplifier is defined as the band of frequencies between the lower and upper 3-dB points.

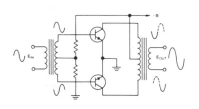

6. PUSH-PULL AMPLIFIERS use two transistors operating 180 degrees out of phase to provide output capability of two transistors.

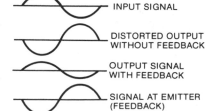

INPUT SIGNAL

DISTORTED OUTPUT WITHOUT FEEDBACK

OUTPUT SIGNAL WITH FEEDBACK

SIGNAL AT EMITTER (FEEDBACK)

7. NEGATIVE FEEDBACK reduces distortion in amplifiers by feeding back part of the output signal and comparing it to the input. The error represents defects in the output signal that is fed to the amplifier to reduce errors due to defects in the amplifier.

Self-Test—Review Questions

1. An audio amplifier consists of three stages, with voltage gain of 10 dB, 18 dB, and 27 dB, respectively. What is the numerical voltage gain?
2. List the coupling methods for amplifiers. Draw sketches showing each method.
3. Why are the 3-dB points taken as the limit of the frequency response of an audio amplifier?
4. What limits low- and high-frequency response in RC-coupled amplifiers?
5. Draw transformer-coupled, single-ended, and push-pull amplifiers. Describe their differences.
6. Draw a phase inverter coupled to a push-pull amplifier with an output transformer. Describe the operation of this configuration.
7. Draw the circuit of a differential amplifier and explain how it operates.
8. Discuss how negative feedback can reduce distortion.
9. Draw the circuit diagram of an IC feedback amplifier, and show the external resistor necessary to give a gain of 10.
10. Sketch the essential features of a complementary amplifier configuration. What are some of its advantages?

Learning Objectives—Next Section

Overview—You now know about audio amplifiers and some of their design characteristics. In the next section you will learn about the basic sound transducers, microphones, and speakers.

HI-FI LOUDSPEAKERS

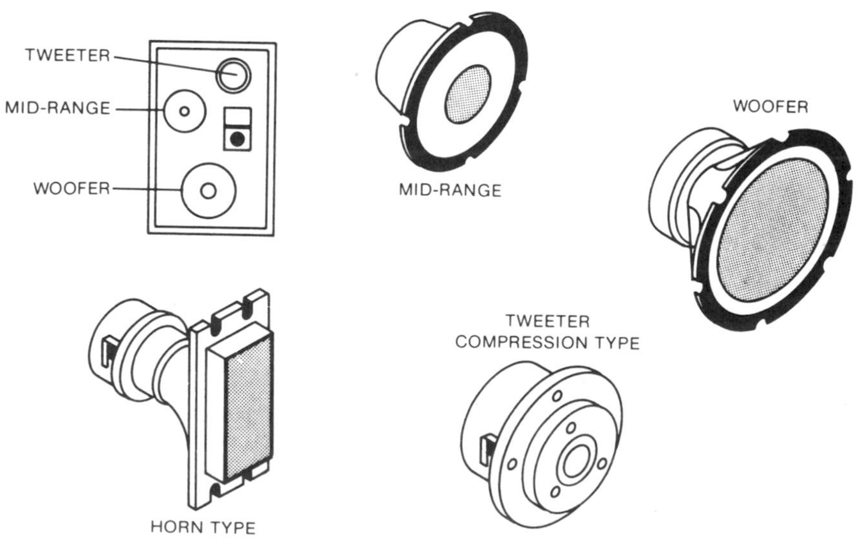

TWEETER

MID-RANGE

WOOFER

MID-RANGE

WOOFER

TWEETER
COMPRESSION TYPE

HORN TYPE

Modern Audio Systems

Sound plays a very important role in our everyday lives. Most of the communication that takes place between individuals is accomplished by voice communication. As you know, the CB transmitter-receiver learning system has audio amplifiers for both transmission and reception. Much of the general information that is distributed to large groups in schools, factories, railroad stations, airline terminals, sport events, etc. is delivered by voice announcements. Much entertainment and mood establishment is presented in the form of vocal and musical sound. All of the information and program material delivered by radio is in the form of sound. Without the sound track, motion pictures and TV would be much less flexible and attractive to audiences. Modern audio systems encompass the equipment that delivers sound from its source to the listener.

This section is devoted to descriptions of the basic components and arrangements of modern audio systems. Initially, we will be concerned with the most basic public address systems. The following sections will review the concepts of high-fidelity sound systems, including stereophonic and quadraphonic sound. All of this material is based on your previous knowledge of the basic electronic building blocks. The emphasis will be on how these building blocks work together in modern audio systems. As part of our study you will learn about microphones and loudspeakers—the first of many *input/output transducers* that you will encounter in your study of *Basic Solid-State Electronics*.

In this volume we will discuss *analog audio* recording and reproducing systems. Currently, *digital audio* recording is starting to be used for studio and other very exacting work. You will learn more about digital technology and its applications in Volume 5.

COMPONENTS OF A BASIC PUBLIC ADDRESS SYSTEM

Basic Principles of Sound

When sound is sent directly from one point to another, air molecules carry the sound. Sound is actually the motion of pressure waves of air. Therefore, any device that produces sound, such as human vocal cords or a loudspeaker, is a device for varying the *pressure* of the surrounding air. All musical instruments make use of this principle by having some part, such as a taut string, a reed, or stretched membrane, set into vibration to produce varying pressure waves of air. In the piano, when a key is struck, a taut string is set into vibration. The string vibrates on both sides of its resting point and compresses and expands the surrounding air. When the string shown in the figure moves from its resting point to the right, the air to the right of the string is compressed (increased in pressure). When the string moves to the left of its resting point, the air to the right of the string will be expanded (reduced in pressure). If a sound-detecting device, such as the human ear or a microphone, is located in the vicinity of the vibrating string, the varying pressure waves will strike the eardrum and produce the sensation of sound.

SOUND WAVES PRODUCED BY VIBRATING STRING

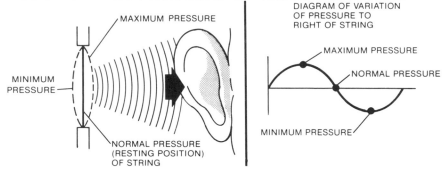

FREQUENCY = NUMBER OF VIBRATIONS PER SECOND
INTENSITY = AMOUNT OF DISPLACEMENT

The *number* of complete *vibrations* occurring per second determines the *frequency* or *pitch* of the resulting sound wave. The *intensity* or *amplitude* of the sound wave is determined by the *amount* of *displacement* of the string from its resting point.

The sound produced by the human voice may vary in intensity by more than 40 dB and cover a range of about 60 to 10,000 Hz (cycles per second). In music the intensity differences may be greater than 60 dB, and the frequency range typically is from 40 to 15,000 Hz.

Of course, sound will travel through many other media, such as water and solids. *Sonar* is an example of the use of sound to locate objects in water. Sound in the audio range and above is used for sonar, because radio waves used in *radar* do not travel through water.

Introduction to Microphones

The *sound waves* produced when you talk or sing can be converted into corresponding *electrical signals* by a microphone. An element inside the microphone body—the *transducer*—is actuated by the air pressure changes from the sound waves and produces an ac voltage that *corresponds* to the pressure changes from the original sound. The amplitude of this ac voltage is proportional to the intensity of the sound.

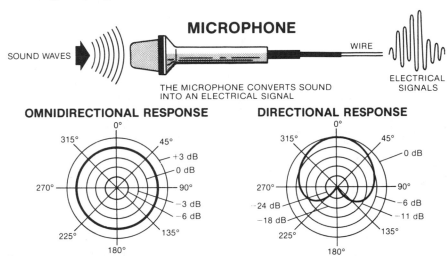

THE MICROPHONE CONVERTS SOUND INTO AN ELECTRICAL SIGNAL

The ratio of electrical output (voltage) to the intensity of sound input is the *sensitivity* of a microphone. This is often expressed in decibels with respect to a reference sound pressure of 1 dyne/cm². (Normal conversational speech at 2–3 ft corresponds to about 1 dyne/cm².) Thus, a microphone with an output rating of −70 dB will produce about 0.3-mV output with a pressure change of 1 dyne/cm². Sensitivity varies widely among different types of microphones.

The *frequency response* of a microphone is a measure of its ability to faithfully convert different acoustical frequencies into ac. For clear understanding of speech, only a limited frequency range is necessary, from about 100 to 3,000 Hz. For hi-fi applications the widest frequency response (from 20 to 20,000 Hz) is desirable.

Microphones also have different *directivity* characteristics. *Omnidirectional* microphones respond equally well in all directions; the *directional* type responds well in only a *limited* direction. Clearly, the choice of a microphone depends on its application.

If the output of a microphone shows only small changes in amplitude between its upper and lower frequency limits, it is said to have a *flat-frequency response*.

The Carbon Microphone

One of the earliest used microphones was the *carbon microphone*. Its use is restricted now almost entirely to telephone systems for the transmission of speech. This microphone is very rugged and supplies a much greater output for a given sound input than any other. This feature was important in earlier times when amplification was difficult and expensive to achieve.

THE CARBON MICROPHONE

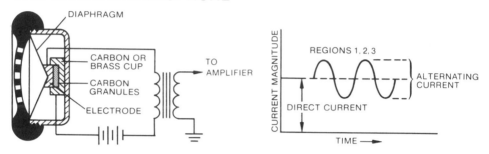

The carbon microphone operates by using the varying sound pressure waves to vary the resistance between loosely packed carbon granules. These carbon granules are enclosed in a brass or carbon cup with an electrode that is mechanically connected to a thin diaphragm. The electrode acts as a plunger by compressing the carbon granules in the cup, which is often called a *carbon button*. The carbon button is connected in series with a source of dc voltage. If a high-impedance input amplifier is used, a carbon button can be used with a step-up microphone transformer since it is a low-impedance device.

When no sound waves strike the diaphragm, the carbon granules are at rest; the resistance of the carbon button, between the cup and the electrode, is constant. The circuit current (region 1) also is constant. When the pressure waves of sound strike the diaphragm, the diaphragm and the attached electrode move in and out, varying the pressure on the carbon granules. An increase in air pressure moves the diaphragm in, compressing the carbon granules and lowering their resistance. This causes the current to increase (region 2). A decrease in air pressure causes the diaphragm to move out, which reduces the pressure on the granules, raising their resistance and decreasing the circuit current (region 3).

In this manner sound waves vary the circuit current in accordance with the sound pressure variations. Carbon microphones are not commonly used for radio and electronic systems because of their limited frequency response, the inconvenience of their external batteries, and their noisiness. The noise occurs because the loosely packed carbon granules create microscopic current changes that are heard as noise, even when no input sound signal is present.

Crystal and Ceramic Microphones

The carbon microphone has disadvantages: it requires an external source of dc voltage for operation, it is noisy, is relatively insensitive, and it has limited frequency response.

The *crystal microphone* eliminates many of the above difficulties because it operates on a somewhat different principle and requires no external source of voltage.

CRYSTAL/CERAMIC MICROPHONES

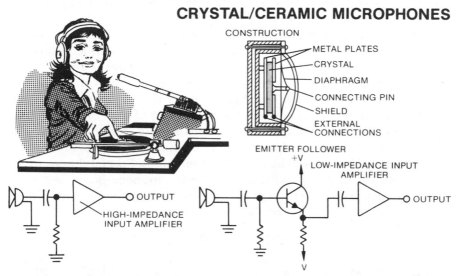

CONSTRUCTION

METAL PLATES
CRYSTAL
DIAPHRAGM
CONNECTING PIN
SHIELD
EXTERNAL CONNECTIONS

EMITTER FOLLOWER
+V
LOW-IMPEDANCE INPUT AMPLIFIER

OUTPUT
HIGH-IMPEDANCE INPUT AMPLIFIER

OUTPUT

Certain crystalline substances such as quartz, Rochelle salts, and barium titanate generate a voltage when pressure is applied. Known as the *piezoelectric* effect, this principle is used in the crystal microphone.

The construction of a crystal microphone is shown in the diagram. The flat crystal of Rochelle salts (used instead of quartz because it is more sensitive) is mounted between two metal plates that have external connections. A thin diaphragm is mechanically connected to the crystal through a hole in the front plate. When sound waves strike the diaphragm, varying pressure is applied to the crystal through the connecting pin, and a varying voltage is produced between the plates. Since the sound waves apply the pressure to the crystal, the output-voltage waveform will be a duplicate of the original sound. The output level is in the range of −50 to −65 dB, depending on the frequency response characteristic of the microphone.

In a more recent development of the same principles, ceramic materials such as barium titanate have been produced that incorporate the same piezoelectric properties as the crystal material just described. The result is *ceramic microphones*. They are considerably more rugged than earlier crystal types and are now more common, particularly for public address, communications, and amateur radio equipment. The output impedance of these microphones is very high and, therefore, they should be coupled to a high-impedance amplifier input, or coupled by an emitter follower to a low-impedance input.

Dynamic Microphones

The *dynamic microphone* is the most commonly used microphone at the present time. It is rugged, can be made with excellent frequency response, and is relatively inexpensive. It is widely used for communication, public address, hi-fi, recording, and broadcasting. Dynamic microphones also are available with a wide range of directivity characteristics.

In the dynamic microphone a coil of wire is suspended in a strong magnetic field produced by a permanent magnet. The coil is rigidly attached to a thin flexible diaphragm. When sound waves strike the diaphragm, the coil will vibrate in accordance with the impingent sound. Since the coil is in the magnetic field, an ac voltage will be induced in it. This ac voltage has the same form as the impingent sound wave; however, its amplitude will be proportional not only to the sound intensity but also to its frequency. This is so because the magnitude of the induced ac voltage is proportional to the *velocity* of the coil in the field. This means that the voltage output will increase as the frequency increases for a constant sound level, since the diaphragm must move faster to complete each cycle as the frequency increases. Special design of the diaphragm, tuned ports and ducts in the microphone body, and other design factors have solved this problem, leading to dynamic microphones that have very good, flat-frequency response.

TYPICAL DYNAMIC MICROPHONES

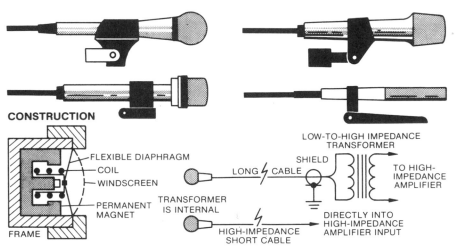

Dynamic microphones have a low impedance output. This is very convenient where a long connecting cable is necessary. Conversion to high impedance can be done at the amplifier end of the cable by using a transformer. Some dynamic microphones are designed with the transformer in the microphone body for high-impedance output only or have a switch so either high- or low-impedance outputs are available. Typically, the output is from −60 dB for voice frequency range models to −70 dB for high-fidelity units. The dynamic microphone generates its own signal and does not require an external voltage source.

Ribbon or Velocity Microphones

The *ribbon microphone* consists of a corrugated ribbon of aluminum alloy, which is suspended in a strong magnetic field so that the ribbon can be moved by sound waves. As sound waves move the ribbon back and forth, the ribbon cuts the lines of force between the poles of the magnets, and a voltage is induced in the ribbon. This voltage is very small but is stepped up by a transformer, which is usually enclosed in the microphone casing. In addition to stepping up the voltage, the transformer raises the very low output impedance of the microphone to a more reasonable impedance.

The microphones described previously were pressure-operated and had diaphragms that moved because the sound waves increased the air pressure on the front side of the diaphragm above the air pressure on the enclosed back of the diaphragm. The ribbon microphone has no diaphragm. Both front and back sides of the ribbon are exposed to the sound pressure, and the ribbon responds to the difference in pressure on each side. Thus, the voltage induced in the ribbon is determined not by the pressure of the air but by the velocity of the air particles striking the ribbon and, therefore, the output voltage is proportional to the ambient sound pressure. This type of microphone is a linear device without the compensation necessary for dynamic microphones. Velocity microphones have very good frequency response, particularly at lower frequencies. The output (after the first transformer) of velocity microphones is comparable to that of good dynamic microphones, about −70 dB.

A feature of the velocity microphone is its directivity characteristic. As you can imagine, it responds to sound from the front or back but has little response from the sides.

THE RIBBON MICROPHONE

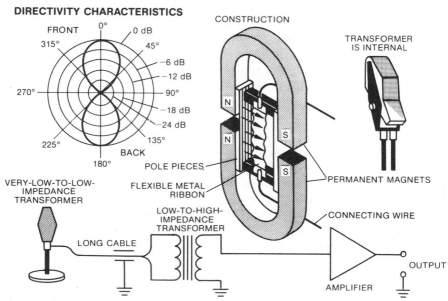

Directional Microphones

The special bidirectional characteristic of a velocity microphone makes it possible to construct a unidirectional microphone by combining elements from dynamic and velocity microphones into a single unit. The dynamic element and the velocity element are series-connected so that their outputs add or subtract, depending on the direction of the sound wave.

THE DIRECTIONAL MICROPHONE

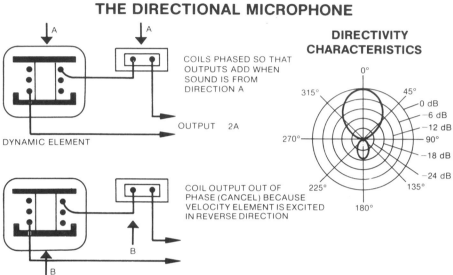

DIRECTIVITY CHARACTERISTICS

COILS PHASED SO THAT OUTPUTS ADD WHEN SOUND IS FROM DIRECTION A

OUTPUT   2A

DYNAMIC ELEMENT

COIL OUTPUT OUT OF PHASE (CANCEL) BECAUSE VELOCITY ELEMENT IS EXCITED IN REVERSE DIRECTION

If both elements have the same sensitivity, then sound coming from direction A generates the same voltage in each element. These voltages provide double the output from a single element. On the other hand, if sound comes from direction B, it still excites the dynamic element like sound from direction A, but the velocity element is excited in reverse phase, and the two outputs cancel. This produces the directivity pattern shown, which is commonly called a *cardioid* because it resembles the geometric figure called a cardioid (heart).

Directional microphones are very useful when you wish to exclude noise from undesired directions. As you will learn, they also reduce feedback from the loudspeakers in public address systems. In addition, they can reduce the effects of echo in large halls. On the other hand, they have the disadvantage that the speaker or performer must remain in a relatively fixed position.

Practical, modern directional microphones use common magnets, special diaphragm configurations, and other special arrangements to get the effect of a dynamic element and a velocity element in a single unit. The effect and the results, however, are basically the same. As you can see, if one element is switched off, the directional pattern will change. Some microphones are arranged so that they can be switched from unidirectional to omnidirectional.

The Condenser or Capacitor Microphone

The *condenser microphone* consists of a rigid metal plate and a very thin metal flexible or metallized plastic plate mounted parallel and very close together. The output signal is generated by the change in capacitance across the plates caused by spacing variations from impingent sound waves that make the flexible plate vibrate.

A dc voltage, called the *polarizing voltage*, is applied across the plates through a high resistance, charging the plates. Since the charge on the capacitor cannot change instantaneously, the change in capacitance produces a change in voltage according to $\Delta V = Q/\Delta C$, where ΔV is the voltage change, Q is the change and ΔC is the capacitance charge. Since capacitance changes directly with spacing, the voltage output is linearly proportional to the sound pressure. A very-low-noise amplifier with a very high input impedance is required to operate this type of microphone and to provide a low output impedance.

TYPICAL ELECTRET CONDENSER MICROPHONES

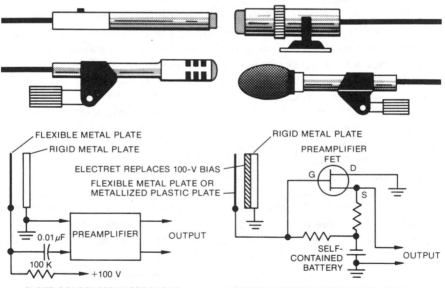

OLDER CONDENSER MICROPHONE   MODERN ELECTRET CONDENSER MICROPHONE

Condenser microphones can be awkward and expensive, usually requiring an amplifier built into the microphone and a source of high voltage. The development of plastics that can retain an electric charge indefinitely (the *electret*) and the development of the field-effect transistor (FET) to provide a compact, low-noise, very-high-impedance amplifier have made this type of microphone inexpensive and practical. The output from the condenser microphone preamplifier is about the same as that for other microphones (about −70 dB). Condenser microphones are very good in applications when very good high-frequency response is desired because the diaphragm can be made very light and responsive.

Conventional Headphones

The purpose of a microphone is to convert sound pressure waves into corresponding ac voltages. The purpose of any *sound-reproducing* device, such as a headphone or loudspeaker, is to change these ac voltages *back* into sound waves. To do this, the sound reproducer must be designed to vary the surrounding air pressure in accordance with the applied ac signal.

CONVENTIONAL HEADPHONES

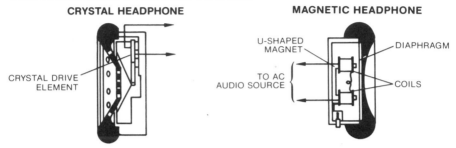

Until the advent of hi-fi systems, headphones were used most often in communication systems or in monitoring public address systems where information was to be received but high-quality sound was not necessary. Headphones of this type are still used extensively today. In some of these, crystal elements are used. These headphones make use of the piezoelectric effect and act like a microphone *in reverse*. The amplified ac signal voltages are applied to the metal plates of a crystal element, and these voltage variations cause the crystal to change its shape and produce pressure variations in the surrounding air, resulting in sound reproduction. Crystal headphones are light in weight and have excellent frequency response, but their sound output level is limited.

Most common headphones operate on magnetic principles, and a typical magnetic headphone is shown in the diagram. A coil of wire is wound on each pole of a U-shaped permanent magnet. These coils are connected in series and have external connecting leads. A soft iron diaphragm is held in place close to the pole ends.

With no signal applied to the coils, the permanent magnet exerts a constant pull on the diaphragm. When the audio-frequency currents (ac) flow through the coils, they become electromagnets that either aid or oppose the fixed permanent magnetic field. Thus they cause the diaphragm to move forward or back in accordance with the applied ac signal. The diaphragm, in turn, moves the air adjacent to it, producing the sound waves that will be heard.

In this way, the audio signal is converted into air pressure variations, or sound. These headphones are inexpensive and quite serviceable, but have limited frequency response—adequate for communications, but poorly suited to music. This has led to the development of high-fidelity (hi-fi) headphones, as described on the next page.

Hi-Fi Headphones

The advent of high-fidelity (hi-fi) audio systems has created a demand for high-quality headphones with high sound output levels that cannot be met with the conventional headphones described previously. To overcome this difficulty, the hi-fi headphone was developed. The basic construction of this headphone is like that of a dynamic microphone (and, as you will see, a loudspeaker).

HI-FI HEADPHONES

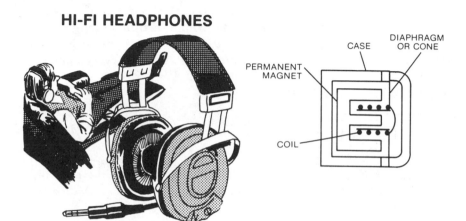

The internal construction consists basically of a permanent magnet-coil structure like the dynamic microphone described previously, except that the coil is driven by an electrical signal. The field of the coil interacts with the permanent magnetic field to generate a force that drives the diaphragm back and forth against the restraining force of the diaphragm at rest. Since this restraining force is linear (within limits), the diaphragm is capable of fairly wide excursions without distortion. Thus, high acoustical output is obtainable. In addition, the configuration is such that good to excellent fidelity is obtainable with care in design.

Hi-fi headphones are usually designed to have an input impedance of about 8 ohms, although high-impedance (600 ohms) units are also available. Because only limited power can be handled for headphone operation, they must not be connected directly to power amplifiers but are fed through resistive attenuators.

The development of hi-fi stereo, which you will learn about later in this volume, has been a major factor in the development of hi-fi headphones. Some units are actually loudspeakers in specially designed enclosures that are packaged to look like headphones. Hi-fi headphones with frequency response better or at least as good as the best loudspeakers are now widely used. Special attention to light-weight construction is necessary to provide for wearer comfort and to provide a good seal of the headphone to the ear.

Introduction to Dynamic Loudspeakers

The dynamic loudspeaker is almost universally used today. It is very similar in construction to the dynamic microphone. In fact, small dynamic speakers are used in intercommunication systems as *both* speaker and microphone.

As shown in the diagram, a strong permanent magnet made of Alnico® alloy produces a strong magnetic field in the air gap. A coil of wire with relatively few turns, called the *voice coil*, is suspended in the air gap and attached to a paper or fiber cone. The outer edge of the cone and the voice coil suspension are made from a flexible corrugated fiber and attached to the speaker frame. This allows the cone to be held in place laterally but allows it to be moved in and out freely against the restoring force of the corrugated supports.

THE DYNAMIC LOUDSPEAKER

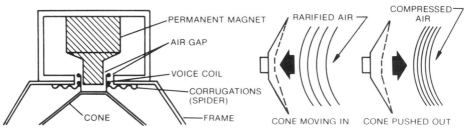

An audio amplifier is connected to the voice coil through an impedance-matching device. This is necessary because loudspeakers usually have impedances between 3 and 16 ohms. The ac audio signal current flowing through the voice coil causes it to generate a magnetic field whose polarity and amplitude are continuously varying. When the voice-coil field has a polarity that aids the field in the air gap, the voice coil and cone move in, reducing the air pressure in front of the cone. When the voice-coil field is in opposition to the field in the air gap, the voice coil and cone are pushed out, compressing the air in front of the cone. In this way, audio-signal currents are changed into sound pressure waves. While almost all speakers are based on this principle, they vary widely in details, particularly in the way the cone is restrained to move in the front-to-back direction. The power-handling capability of loudspeakers is limited by the heat-dissipation capability of the voice coil and the cone excursion possible without reaching the limits of the corrugated supports.

One of the major factors in speaker performance is called *baffling*. Baffles (or speaker cabinets) are designed specifically to improve the frequency response of speakers, particularly the low-frequency response. As you will learn, many hi-fi speaker systems design the speaker and baffle as integral assemblies. For hi-fi use, many speakers covering different frequency ranges are combined into a speaker system, as you will learn shortly.

Electrostatic Speakers

Electrostatic speakers enjoyed a popularity a few years ago which has now subsided. Their major problems have been poor low-frequency response and low efficiency. They are now used primarily in speaker systems for providing extended high-frequency response.

The electrostatic speaker is based on a principle that you learned about very early in your study of *Basic Electricity*—like electric charges attract each other and unlike charges repel. This type of speaker consists of a rigid plate and flexible plate mounted close together. A high dc voltage is maintained across the plates, keeping them oppositely charged. The plates are now attracted together, but their edges are held to keep them apart. The center of the flexible plate is pulled into the rigid plate to the point where its springlike resistance exactly counterbalances the electrostatic attraction.

ELECTROSTATIC LOUDSPEAKER | **ELECTRET HEADPHONE OR LOUDSPEAKER**

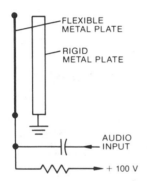

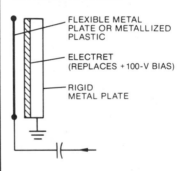

Now an audio amplifier output signal also is applied across the plates. This signal adds to and subtracts from the high dc voltage. When the charge on the flexible plate becomes more dissimilar to that on the rigid plate, the flexible plate center is pulled even more closely to the rigid plate. When the charge on the flexible plate becomes more similar to that on the rigid plate, the flexible plate is pulled less strongly, and its center moves slightly further away from the rigid plate. This motion of the center of the plate produces a sound wave.

The flexible plate of the electrostatic speaker can be connected to a cone in the same manner as the voice coil of a dynamic speaker. However, the flexible plate can itself be made very large and perform the same function as the cone. As in the case of the condenser microphone, the bias voltage can be replaced with an electret as shown in the diagram, thereby eliminating the need for the high bias voltage. The result is an electret headphone or loudspeaker.

Horn-Type Speakers

The moving cone of a loudspeaker is not the most efficient method of converting the movement of the voice coil into sound. The cone is used because it is compact, economical, and provides a very good frequency response.

The horn shown in the illustration is a considerably more efficient sound producer. The motion of the small diaphragm pushes and pulls small amounts of air in and out of the chamber at the throat of the horn, producing small increases and decreases in pressure. As these small pressure changes travel toward the mouth, they move increasingly larger amounts of air in front of them. At the mouth of the horn, considerably larger amounts of air are being moved than at the beginning of the throat. The result is that higher sound levels are produced for a given amount of energy fed into the voice coil. The effect of the horn is to match the impedance of the transducer to the physical impedance of the air in the room, thus producing the best energy transfer. Horn speakers are also directive, thus increasing the sound level in one direction at the expense of other directions.

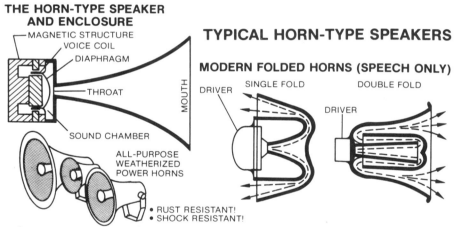

THE HORN-TYPE SPEAKER AND ENCLOSURE

MAGNETIC STRUCTURE
VOICE COIL
DIAPHRAGM
THROAT
MOUTH
SOUND CHAMBER
ALL-PURPOSE WEATHERIZED POWER HORNS
• RUST RESISTANT!
• SHOCK RESISTANT!

TYPICAL HORN-TYPE SPEAKERS

MODERN FOLDED HORNS (SPEECH ONLY)

DRIVER   SINGLE FOLD   DOUBLE FOLD
DRIVER

The disadvantage of horns is that they do not have wide frequency response, and they have to be large to be efficient, particularly for low frequencies, since the lowest frequency efficiently coupled is proportional to the size of the horn mouth. For example, the wavelength in air of a 100-Hz sound is 3.3 meters. To radiate effectively, therefore, the horn must have a mouth width of one-half wavelength, or 1.67 meters. Thus, you can see the disadvantages of simple horns for hi-fi reproduction. Since low frequencies are not usually important in many public address applications, horns are often used for public address work because they can easily be made compact and weatherproof for use in outdoor conditions by folding, as shown.

For home use, horns are sometimes used to get efficient output over the low-frequency audio range in spite of their size. For theater use, where space is often available, horn speakers are used for all sound reproduction.

Introduction to Hi-Fi Loudspeakers

Some large speaker systems are very efficient and require very little input power, while others are very inefficient and require much more power input than the actual sound power delivered. In many cases, the efficiency is deliberately reduced to improve the frequency response.

Ideally, a hi-fi speaker system should accurately reproduce the acoustical equivalent of the ac signal driving it. As a practical matter, this ideal is not achieved in even the most expensive speaker systems. Amplifier and recording technology have greatly outstripped speaker technology, with the result that speaker systems are usually the weakest link in a hi-fi system, regardless of expense.

You will note that speaker systems varying widely in price seem to have the same or very similar technical characteristics such as frequency response. The speaker characteristics that are often more important than frequency response are distortion, transient response, smoothness of response, dispersion of high frequencies, and listening angle over which high performance is possible. Today, hi-fi speakers generally are obtained assembled in an enclosure specifically matched to the speakers. This allows for optimized speaker performance.

HI-FI LOUDSPEAKERS

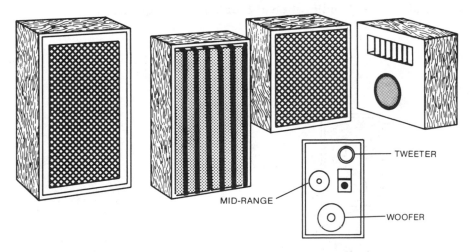

Since it is most difficult to design and construct a single speaker to perform well at all audio frequencies, the practice has been to use several different speakers for different parts of the audio range in all but the least expensive hi-fi systems. This allows optimization of design of speakers for different parts of the audio range, with the result that the overall performance is greatly improved when these speakers are combined into a speaker system. One of the most difficult areas has been in obtaining good, distortion-free low-frequency response. It is in this area that inexpensive speaker systems generally are deficient.

Introduction to Hi-Fi Loudspeakers (continued)

The need for separate speakers arises because the audio-frequency spectrum is so broad. For good low-frequency reproduction, a large speaker is needed. This is so because speakers have a *natural resonant frequency* (electromechanical) that depends on their size, cone shape, and material, etc. As in any resonant circuit, the electrical impedance becomes very high—little current can be driven through it—but the cone amplitude may be very large and uncontrolled. Essentially, no power can be radiated without distortion at the resonant frequency and below. Thus, to get the resonant frequency as low as possible, we use large speakers, special construction, etc. These features make these low-frequency speakers poor at higher frequencies, where the cone mass and shape are not suitable. This same reasoning applies for mid and high frequencies. Hence, mid- and high-frequency speakers are smaller and better suited to operation in their specific ranges.

Multiple-speaker systems, typically are made up of two, three, or four speakers. The *woofer* takes care of the low-frequency reproduction and may have a cone from 6 to 18 inches (15.25 to 45.75 cm) in diameter. The *mid-range speaker*, which handles the middle audio frequencies, is usually either a small (4-inch or 10.15-cm) dynamic speaker or a horn and driver. The *tweeter*, which handles the high audio frequencies, is either a direct radiating dome speaker or a horn and driver. Sometimes a second tweeter handles the very high audio frequencies.

HI-FI LOUDSPEAKERS

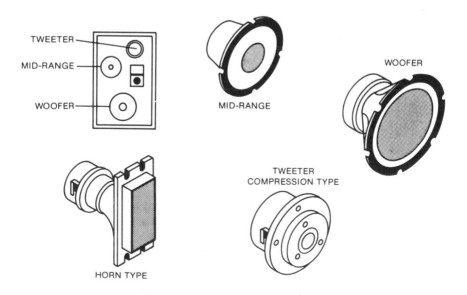

Coaxial and Triaxial Hi-Fi Speakers

To provide a compact assembly and to improve the fidelity of speaker systems, manufacturers frequently install a large *bass* speaker for low frequencies and a small *tweeter* for high frequencies on the same frame. This arrangement is called a *coaxial* speaker. The same principle is used to mount two successively smaller speakers on the frame of the low-frequency speaker, on the same axis; this arrangement is called a *triaxial* speaker.

As shown, in a coaxial arrangement two speakers are mounted concentrically on the same frame, with the small high-frequency speaker mounted within the cone of the larger bass speaker. These units typically are available in sizes from approximately 8-inch (20.30-cm) to 15-inch (38.10-cm) outside diameter. They are also available in 6- by 9-inch (15.25- by 22.85-cm) elliptical or oval shapes. The fidelity of these units is reasonable for the price, but is strongly influenced by the speaker cabinet design.

Construction of triaxial speakers is similiar to coaxial units. They consist of a large bass speaker, a centered mid-range speaker, and a small high-frequency range speaker in the center of the mid-range speaker. These speakers have better overall frequency response and higher power-handling capability. These speakers split the audio range into three regions for the three speakers. Again, these speakers require an appropriate cabinet, or baffle, for proper operation. In some triaxial and coaxial systems, the mid-range speaker is not a separate unit but is an auxiliary light-weight cone fastened to the voice coil assembly of the woofer.

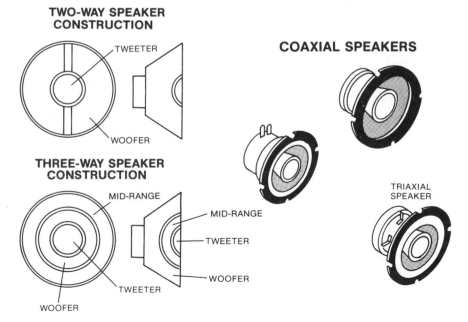

TWO-WAY SPEAKER CONSTRUCTION

TWEETER

WOOFER

THREE-WAY SPEAKER CONSTRUCTION

MID-RANGE

MID-RANGE

TWEETER

WOOFER

TWEETER

WOOFER

COAXIAL SPEAKERS

TRIAXIAL SPEAKER

Crossover Networks

In multiple-speaker systems, *crossover networks* are used to divide the power amplifier output properly so that each speaker receives electrical signals in the frequency range it can best reproduce.

In multiple-speaker systems where there are separate woofer, midrange, and/or treble speakers, the optimum crossover points are determined by the response of the individual speakers. The first criterion is that the frequency response range must properly overlap so that there will not be a *hole* in the reproduced range of frequencies.

The frequency range for crossover typically is between 1,000 and 2,000 Hz for a two-speaker system and 500 and 3,000 Hz for a three-speaker system.

The crossover networks usually consist of passive LC or capacitive filters. They are based on the fact that capacitive reactance decreases with increasing frequency and inductive reactance increases with increasing frequency.

Suppose you had a three-way speaker system in which each speaker had a nominal impedance of 8 ohms. For crossover frequencies of 500 and 3,000 Hz, the design is straightforward. You know that in a LR or CR divider circuit, the -3-dB point (half power) is when X_C or X_L equals R.

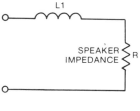

AT A FREQUENCY WHERE X_{L1} = R, RESPONSE IS DOWN 3 dB; BELOW THAT, RESPONSE IS FLAT.

AT LOW FREQUENCIES, X_{C1} IS LARGE COMPARED TO R AND THERE IS LITTLE RESPONSE; AT FREQUENCIES WHERE X_{C1} = R (ASSUMING X_L IS SMALL), RESPONSE IS DOWN 3dB; BUT ABOVE THAT FREQUENCY X_{C1} DECREASES AND RESPONSE IS **FLAT**. AT SOME HIGHER FREQUENCY, X_{L2} BEGINS TO INCREASE, SO RESPONSE IS DOWN 3dB AGAIN AND DECREASES AT HIGHER FREQUENCIES.

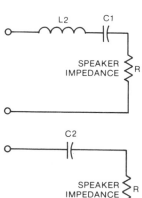

RESPONSE IS LOW AT LOW FREQUENCIES WHERE X_{C2} IS MUCH GREATER THAN R; RESPONSE IS -3 dB WHEN X_{C2} = R; ABOVE THAT FREQUENCY, RESPONSE IS FLAT.

Crossover Networks (continued)

RESPONSE OF TOTAL SPEAKER SYSTEM
(SUM OF EACH RESPONSE)

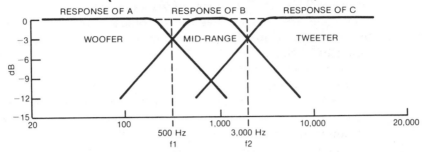

The system's electrical properties can be calculated easily.

For the woofer, X_{L1} = 8 ohms at 500 Hz.

Therefore, $L1 = \dfrac{8}{2\pi f} = \dfrac{8}{2 \times 3.14 \times 500} = 0.0025$ henrys = 2.5 mH.

For the mid-range, X_{C1} = 8 ohms at 500 Hz, X_{L2} = 8 ohms at 3 kHz.

Therefore, $C1 = \dfrac{1}{2\pi f \times 8} = \dfrac{1}{2 \times 3.14 \times 500 \times 8} = 0.000398$ farad or

40 μF, and $L2 = \dfrac{8}{2\pi f2} = \dfrac{8}{2 \times 3.14 \times 3000} = 0.42$ mH.

For the tweeter, X_{C2} = 8 ohms at 3 kHz.

Therefore, $C2 = \dfrac{1}{2\pi f2 \times 8} = \dfrac{1}{2 \times 3.14 \times 3000 \times 8} = 6.6$ μF.

The complete crossover network therefore is

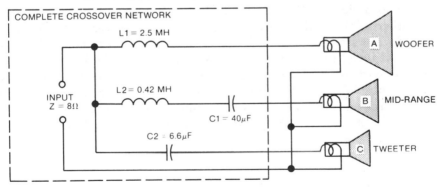

If you work out the input impedance to the speaker system, you will find that it is always equal to 8 ohms, which is what you want. As you can see, calculation of these networks is very easy. Often, an adjustable volume control is included in the mid-range and/or tweeter circuit to give some adjustment of relative level and to suit individual preference.

Loudspeaker Enclosures

A loudspeaker enclosure is not simply an attractive-looking box. There are two loudspeaker characteristics that the cabinet or *enclosure* can help to correct. The first is that sound waves generated by the back of the speaker cone are 180 degrees out of phase with sound waves generated by the front. At low frequencies, these two waves can cancel each other, and this is why an unbaffled or unenclosed speaker lacks bass response. This can be corrected in three ways: (1) by making the path from back to front very long, so that the difference in path length tends to bring the sound waves into phase; (2) by building a phase-reversing enclosure for the energy from the back of the speaker cone; or (3) by confining the sound from the rear of the cone so that it is suppressed.

The second major purpose of the enclosure is to attempt to correct acoustical-electrical resonance characteristics in speaker systems. This resonance is usually at a low frequency and can cause severe distortion if not properly handled. At the resonant frequency, the cone vibrates strongly with small inputs, so that severe distortion can occur.

THE INFINITE BAFFLE

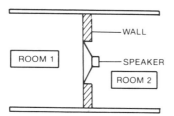

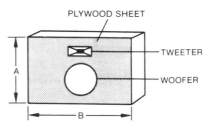

From one viewpoint, a good way to mount a speaker would be to place it in a rigid wall with the back and front regions isolated from each other. This would overcome the first problem very well. This so-called *infinite baffle* is not always easy to implement, but an approximation can be achieved by mounting the bass speaker on a large flat piece of plywood with or without shallow sides. With an infinite baffle, good performance is obtainable down to the bass resonant frequency. The basic limitation is that the distance around (from back to front) must be about equal to a half-wavelength at the lowest useful frequency (the resonant frequency). For good results, the dimension A or B (whichever is smallest) should be about

$$\text{a or b} = \frac{4,400}{f_R} \text{ ft}$$

where f_R is the bass-resonant frequency of the woofer. Mid-range and tweeter can be separately mounted on the baffle without significantly affecting performance. As a practical matter, neither an infinite baffle nor any simple surface provides good performance to bring out the best characteristics of a speaker. To do this, an enclosure specifically matched to the speaker characteristics must be used.

Bass-Reflex Loudspeaker Enclosures

The bass-reflex cabinet operates by damping the resonance and making use of both the front and rearward cone radiation. The bass-reflex cabinet consists of a solidly built closed box of the proper dimensions with a hole for the speaker and a second hole, or port, from which the energy from the back of the speaker cone is radiated. The box is designed so that the phase of the back-cone sound is reversed and thus comes through the port *in phase* with the radiation directly from the cone. Thus, the two sound waves *reinforce* each other, providing additional bass response. In addition, the box and vent act as a damper on the cone resonance, holding the impedance down to acceptable levels at resonance and significantly extending the bass response of the woofer.

SOUND TRANSDUCERS–BASS-REFLEX LOUDSPEAKER ENCLOSURES

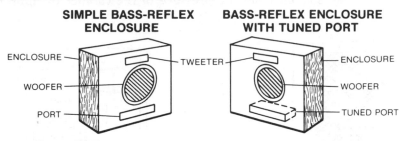

SIMPLE BASS-REFLEX ENCLOSURE

BASS-REFLEX ENCLOSURE WITH TUNED PORT

The rules for designing bass-reflex cabinets are somewhat empirical; typically, for a 12-inch (30.50-cm) speaker, a box volume of about 4 to 5 cubic ft with a port of about half the area of the speaker will give good results. It is usual to line the box with insulation to damp higher frequencies, since the port and enclosure are primarily to handle low frequencies.

A variation of the bass-reflex enclosure is the acoustical labyrinth. In this design, the rear of the speaker is loaded by a folded horn, as shown in the illustration.

ACOUSTICAL LABYRINTH

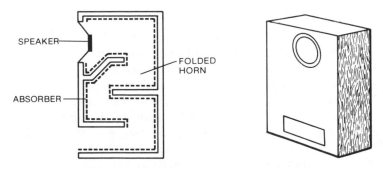

Acoustical-Suspension Speaker-Enclosure Systems

The enclosure of a woofer in a totally enclosed box of reasonable size raises the bass resonant frequency, which is undesirable; however, it does prevent front-rear speaker coupling. One way to permit the use of a totally enclosed, small enclosure is to make the resonant frequency of the speaker very low, so that the increase in bass resonant frequency in a total enclosure is not objectionable. To accomplish this, the speaker cone of the speaker is designed to be extremely free, with very flexible corrugated supports for the speaker cone at its rim and at the center. As a result, the cone is loosely suspended and does not reside at a specific point because of suspension stiffness as in a conventional speaker. A restoring force is necessary, but it is not supplied by the suspension as in a conventional speaker, but by the pressure of the air in the enclosed rear of the speaker. Mid-range speakers and tweeters in the same enclosure must be isolated from the pressure in the cabinet by appropriate rear covers.

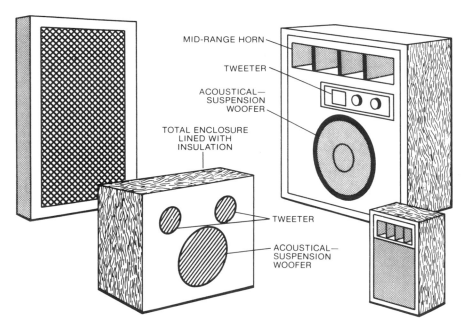

MID-RANGE HORN

TWEETER

ACOUSTICAL—
SUSPENSION
WOOFER

TOTAL ENCLOSURE
LINED WITH
INSULATION

TWEETER

ACOUSTICAL—
SUSPENSION
WOOFER

Acoustical-suspension speakers are capable of suprisingly good performance in very little space. They have made it possible to obtain real high fidelity in small bookshelf speaker systems and at modest cost.

One major disadvantage of acoustical-suspension systems is their low efficiency. With a high-efficiency bass-reflex system, an efficiency of 10–20% or greater is possible. With acoustical-suspension speakers, the efficiency is only a few percent, requiring a relatively high-power amplifier for even a modest volume level. For these speaker systems, a recommended minimum amplifier usually is specified.

Phasing of Components

To provide for proper operation when more than one microphone or more than one speaker are used, it is absolutely necessary to pay attention to the *phasing* of these devices. It should be obvious that two woofers operating out of phase, one with the cone pushing forward on a positive signal and one with the cone pushing backward on a positive signal, will result in cancellation of much of the resultant sound, particularly at low frequencies. In addition, the proper operation of multiple-speaker systems with crossover networks, etc., assumes that the woofer, mid-range, and tweeter speakers are properly phased.

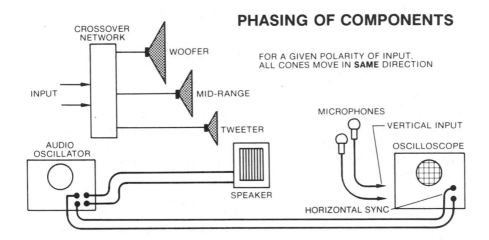

PHASING OF COMPONENTS

FOR A GIVEN POLARITY OF INPUT, ALL CONES MOVE IN **SAME** DIRECTION

Most of the time, the polarity of the terminals is marked with either a + sign or a dot of red paint, and if these terminals are parallel (through the crossover, etc.) the phasing will be correct. If the phasing is unknown, you can check it (if you can see the speaker cones) by using a flashlight battery and a small series resistor (4.7 ohms) across the speaker terminals momentarily and noting the direction in which the cone moves. The connections that give the same direction of deflection tell how the speakers should be phased when connected.

Phasing of microphones is more difficult if they are not marked. The simplest way is to use an audio oscillator (that generates sine waves) set at about 500 Hz connected to a speaker. Then an oscilloscope is *synched* (synchronized) to the audio signal. The vertical input to the oscilloscope is connected to the microphones under test. The microphone diaphragm should be at the same place for each microphone. The connections for the same polarity output, as seen on the oscilloscope, indicate that the microphones are phased.

Proper phasing of microphones and speakers becomes *critically important* in stereo and quadraphonic systems that you will study later in this volume. It also can affect the performance of public address systems.

Experiment/Application—Measuring Speaker Impedance and Phasing

Suppose you had a speaker but did not know its impedance or phasing. You could find these characteristics by means of a simple set of experiments.

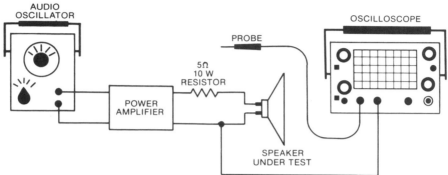

If you set the audio oscillator at 400 Hz (the frequency where most larger speaker impedances are defined), you can measure the voltage drop across the 5-ohm resistor. Using the voltage divider formula, you can calculate the speaker impedance. For example, suppose the total output voltage is 6.5 volts and 2.5 volts appears across the 5-ohm resistor. The speaker impedance is then given by the formula

$$Z = \frac{E_t - E_R}{E_R} R = \frac{6.5 - 2.5}{2.5} \times 5 = 8 \text{ ohms}$$

You can test the speaker impedance at any frequency; tweeters and small high-frequency speakers are usually checked above their resonant frequency (typically 1 kHz to 5 kHz). If you vary the oscillator frequency, you can find the speaker's resonant frequency simply by noting the frequency when the voltage across the resistor is maximum. This defines the resonant frequency, since this is also the point where the speaker impedance is maximum. This point will vary, depending on the enclosure (if any) that is used. To phase the speaker, use a small dry battery (1.5 volts) instead of an amplifier.

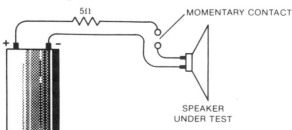

Momentarily touch the resistor lead to the speaker terminal and note the direction of cone movement. If it moves *out*, the speaker lead is *positive* and should be so marked. If it moves *inward*, the *other* speaker terminal is positive. If you are phasing a pair of speakers, it doesn't matter which way you phase them, as long as you are consistent.

Amplifier Requirements

You already know that a public address system is made up of a microphone (or several microphones) coupled into a preamplifier—a power amplifier combination—which, in turn, feeds loudspeakers. A logical question is the amount of power needed to satisfactorily fill a room, auditorium, or outdoor area. Questions like this are difficult to answer easily. However, curves have been generated to guide in selection of the necessary power level.

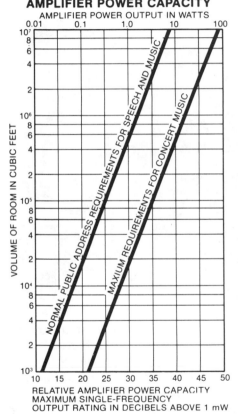

ROOM VOLUME AND RELATIVE AMPLIFIER POWER CAPACITY

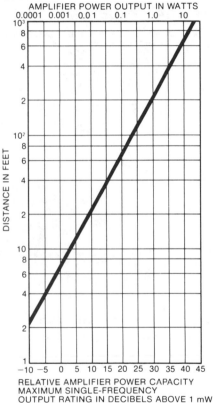

DISTANCE FROM LOUDSPEAKER AND RELATIVE AMPLIFIER POWER CAPACITY REQUIRED FOR SPEECH

As shown, the power requirements for indoor and outdoor use are quite different. Also, the requirements for speech and music are different. The actual amplifier capacity is dependent on speaker efficiency. The curves shown assume 100% efficiency speakers. For practical speakers, the power level should be increased by 6 dB (four times) for efficient horn speakers and by 10–20 dB for low-efficiency cone-type speakers. In any event, amplifier power is inexpensive, and the system should have a large reserve, particularly for music applications.

Special Problems

Almost everyone has heard loud squeals due to oscillation coming from a public address system indicating that a significant amount of sound power from the loudspeakers is reaching the microphone—*sound feedback*. When this happens, and the feedback signals are in phase (as you will learn later) oscillation will occur. To minimize acoustical (sound) feedback, the microphones should be directional and placed in back of the loudspeakers so that advantage is taken of loudspeaker and microphone directivity. Reducing the volume also is helpful. Reduced frequency range (by proper selection of amplifier settings) will be helpful in those situations where voice only is required.

SPECIAL PROBLEMS

SOUND FEEDBACK (LOUDSPEAKER TO MICROPHONE)
KEEP MICROPHONE BACK OF SPEAKERS
FOR MINIMUM FEEDBACK

ELECTRICAL NOISE PICKUP AND FEEDBACK

HORN SPEAKER

MICROPHONE

MICROPHONE

FAULTY SHIELD ON MICROPHONE CORD

HORN SPEAKER

Another problem with public address systems is that unwanted signal pickup can take place due to improperly grounded shielded cable or chassis. Because of the low-signal level in the microphone cable small stray electrical signals can cause noise or oscillation in the output. The outer shield of the microphone cable is normally to be connected to ground so that it can shield the internal wire or wires from stray electrical signals. Because this shield gets continuous rough handling, it is subject to fraying and breaking. This ungrounded shield and the unprotected wires inside then act like an antenna and can feed unwanted electrical signals to the high-gain stages of the system. Some public address systems overcome these problems by using a preamplifier at the master station and separate power amplifiers at each individual speaker station. This distributes the signal at higher levels. In all cases, low-impedance connection between the microphone and the input amplifier are most desirable since low-impedance circuits are much less prone to pickup and hum problems.

Review of Basic Audio Systems

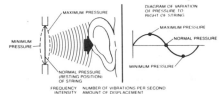

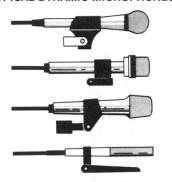

TYPICAL DYNAMIC MICROPHONES

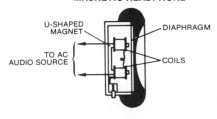

MAGNETIC HEADPHONE

HI-FI LOUDSPEAKERS

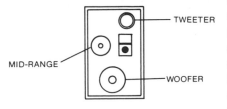

1. SOUND is the motion of air molecules that represent pressure waves which carry the sound from source to destination.

2. MICROPHONES are transducers that convert acoustical waves (sound energy) into electrical signals which have an amplitude/time relationship related to the acoustical waves. They operate by causing a diaphragm to vibrate in accordance with the acoustic wave and using this motion to generate the electrical signal. Commonly used microphones are the crystal microphone, dynamic microphone, ribbon microphone, and condenser microphone.

3. HEADPHONES are transducers that convert electrical signals into acoustical waves. In headphones, a diaphragm is forced to vibrate in accordance with the variation of electrical signals. The diaphragm makes the air move generating sound.

4. LOUDSPEAKERS do the same job as headphones except on a larger scale, so that much higher sound levels are obtainable. For proper operation, a loudspeaker must be properly baffled or enclosed. Enclosures can be bass reflex, horns, etc.

5. HI-FI SPEAKER SYSTEMS usually contain two or more speakers to cover different parts of the audio range. They are usually contained in a common enclosure. A typical hi-fi speaker system has a woofer for good bass response, a mid-range speaker to cover the middle audio range, and a tweeter to cover the very high audio frequencies.

Self-Test—Review Questions

1. List the basic elements of audio systems.
2. Define sensitivity and frequency response of microphones.
3. List, sketch, and describe briefly the operation of crystal, dynamic crystal, and ribbon microphones.
4. Describe, using a sketch, the operation of headphones. How do they differ from loudspeakers?
5. Compare dynamic and horn-type speakers. Where would each be used?
6. Why are multiple speakers used in high-fidelity systems?
7. Design a crossover network for a three-way speaker system having all 4-ohm units, with crossovers at 1,000 Hz and 4,000 Hz.
8. What are the essential differences between bass-reflex and acoustical-suspension speaker enclosures and speaker systems? Give some advantages and disadvantages of each.
9. Describe phasing of microphones and speakers. Why is it important?
10. Discuss some of the special problems associated with public address systems and describe how they are resolved.

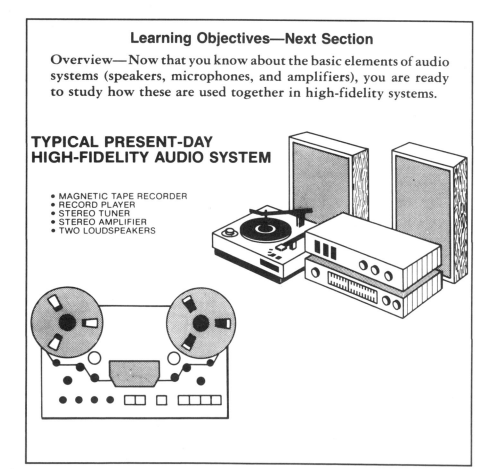

Learning Objectives—Next Section

Overview—Now that you know about the basic elements of audio systems (speakers, microphones, and amplifiers), you are ready to study how these are used together in high-fidelity systems.

**TYPICAL PRESENT-DAY
HIGH-FIDELITY AUDIO SYSTEM**

- MAGNETIC TAPE RECORDER
- RECORD PLAYER
- STEREO TUNER
- STEREO AMPLIFIER
- TWO LOUDSPEAKERS

Introduction to High-Fidelity Audio Systems

A *high-fidelity* audio system is one that *very closely duplicates* the original sound. Just exactly what *very closely* means is a subject of extensive discussion and only partial agreement. Although some insist that a sound system have a frequency response with very specific levels throughout the range from 20 to 20,000 Hz, most are satisfied with much less.

In the past, one or more microphones were used to pick up sound for processing through a single electrical channel. Even for multiple microphones all these sounds were combined or *mixed* in suitable proportions and then processed through a single electrical channel. Such single sound channeling is known as *monophonic* or *monaural*, because the sound is the same in both ears.

MONOPHONIC SOUND

ONE MICROPHONE-**ONE** LOUDSPEAKER

STEREOPHONIC SOUND

TWO MICROPHONES-**TWO** LOUDSPEAKERS

It has long been appreciated that people get a different impression of sound with each ear. This sensation enables a person to estimate the direction and distance of various different performers and instruments on a stage. When sound is picked up by two physically separated microphones, spaced like our ears with an absorber between, and then is processed through two separate electrical channels and reproduced through headphones into the listener's ears at the same volume as the original, the listener gets a sound sensation that is very close to the one he gets when actually sitting in an auditorium. This form of sound processing is called *binaural*, because it simulates sound heard with two ears. Obviously, this is a very inconvenient arrangement. Therefore, the microphones are widely separated, and the two channels are reproduced through speakers that are also widely separated. While this is not the same as binaural sound, it is quite satisfactory and convenient. This technique is used universally today and is known as *stereophonic*, because it gives the individual the effect of three dimensions, or depth, in sound. This idea has been extended to *quadraphonic* sound, which simulates the sound coming to the back of the listener's ears as well.

Introduction to High-Fidelity Audio Systems (continued)

Sound recordings have been available for over a century and were one of the first practical *memory* or *storage* devices. A major change occurred about 30 years ago with the introduction of *long-playing records* (vinyl pressings) that permitted true high-fidelity recording. The introduction of *stereo* about 20 years ago was made possible because of component improvements. Modern high-fidelity phonograph systems are capable of remarkably faithful reproduction. Such sound often does not resemble the original because of electronic processing of the sound during recording. However, the fidelity of playback of what is intended to be heard usually is very high.

High-fidelity magnetic tape recording is about as old as the long-playing record. Modern tape recorders are capable of truly remarkable fidelity; not only do they provide extended playing time and a durable recording medium, they also provide the individual with a most convenient and economical means for home recording either from live or recorded material.

Modern high-fidelity systems include either an AM/FM tuner or an AM/FM receiver. The tuner is simply a radio receiver without the audio amplifier portion and provides an economical method of adding high-fidelity radio reception to an existing audio system. Essentially, all present day tuners and receivers have stereo capability; some also have provision for quadraphonic sound. For many years the speaker has been the limiting factor in fidelity of sound reproduction. This continues to be a problem, although modern speaker systems are rather good in this regard. The interaction between the speaker system and the room in which it is located is a source of *coloration* of the sound. Coloration means that the sound has been given a quality that is different from the original, although not necessarily displeasing or distorted.

TYPICAL PRESENT-DAY HIGH-FIDELITY AUDIO SYSTEM

- MAGNETIC TAPE RECORDER
- RECORD PLAYER
- STEREO TUNER
- STEREO AMPLIFIER
- TWO LOUDSPEAKERS

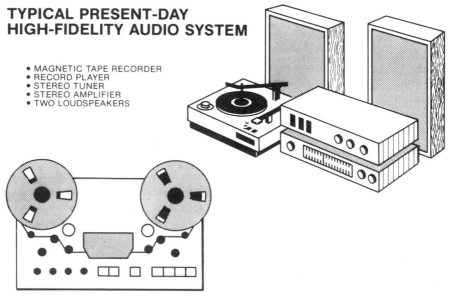

How Phonograph Records Are Made

In a typical case, the process of making a phonograph record starts with a suitable magnetic tape master that has been carefully made, edited, and modified electronically, as desired. In many cases, particularly with popular music, the recording is done on many tracks (8–24 or more) with separate instruments, vocals, etc. on different tracks. These then are mixed down usually onto stereo (two-track) tape for recording onto disk.

The tape is used to produce a master recording that is made by feeding the tape into a precision lathe with a flat master disk (it looks like an uncut phonograph record); the disk is rotated at the appropriate speed while a heated stylus driven by a cutting head (a magnetic motor that moves the stylus from side to side in accordance with the information from the tape master) cuts the grooves into the master disk. The finished master is then inspected for defects; if none are found, the master is ready for the next step. Because the peak-to-peak variations of the grooves depend on loudness, care is necessary to ensure that there is no over-cutting (one groove cutting into an adjacent groove). The recording process takes place at constant velocity of the cutter stylus, which means that the lower the tone, the larger the groove excursion. To avoid excessive groove excursion, the frequency response during recording is modified. Additional response modifications are also made to reduce noise. You will learn about this a little later.

The completed master is now copied in reverse by electroplating, and backed up to form a metal *stamper*. This stamper is reversed from the original recording. The stamper is used, with a ball of vinyl record material, in a press to stamp out a replica of what was originally impressed on the master. Two stampers (top and bottom) are used to produce the usual two-sided record. For a popular record, many replicas of the master are necessary, because the stampers become worn and must be replaced. The finished records are trimmed, labeled, jacketed, and boxed for shipment.

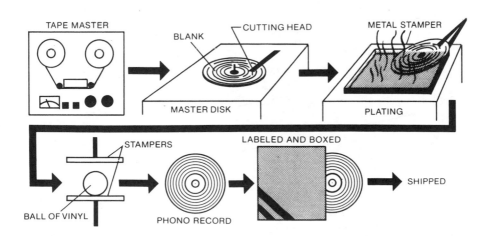

Record Players

A modern record player contains a turntable to rotate the record at constant speed plus a stylus and phono-cartridge to convert the groove variations into an electrical signal duplicating the recorded information. A *tone arm* holds the cartridge and enables the stylus to follow the groove from the outer edge of the recording to the inner limit of its travel. The resulting electrical signal from the cartridge is processed by preamplifiers, power amplifiers, and loudspeakers.

Early recordings used a rotational speed of 78 rpm. These were popular until modern manufacturing methods made it possible to increase the density of the information cut into the record groove, and long-playing 33⅓-rpm records were introduced mainly for classical music. These records were much quieter than the old 78-rpm records and had a greatly increased playing time. The 78-rpm recordings often required four to six or more records for a complete performance. Long-playing records have approximately 20 minutes (more in some cases) of playing time on each side of a 12-inch record and have totally replaced 78-rpm records for the past 25 years. To separate short recordings on such records, narrow unrecorded bands are spaced between individual performances. These bands can be seen by viewing the record at an angle. As a competitor to the 33⅓-rpm record, 45-rpm records were introduced. These are used currently mainly for single-play short selections.

With the variety of records available, automatic record players were improved with capabilities for playing any stack of records in sequence. They are also capable of handling records of various sizes at any rotation speed. However, 78-rpm capability is not available today on many modern machines, primarily to avoid the compromises in sound quality necessary to provide this capability and also because 78-rpm records have not been made for 20 years or so.

MODERN AUTOMATIC RECORD PLAYER

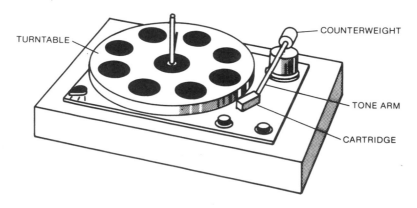

TURNTABLE

COUNTERWEIGHT

TONE ARM

CARTRIDGE

Record Player Turntables

The turntable rotates the record at the same speed that was used during recording, so that the playback stylus in the groove moves at the same speed as the recording stylus, faithfully reproducing the same frequencies. Repeated variation in turntable speed, due to imperfections in the drive system, cause repeated variations in the frequency of the reproduced sound. Speed variations that repeat slowly produce a repeating *wow* sound from the speaker; more rapid variations produce *flutter*. Flutter and wow are expressed in terms of percentage and usually must be less than 0.05% in a hi-fi system.

TURNTABLE DRIVE ARRANGEMENTS

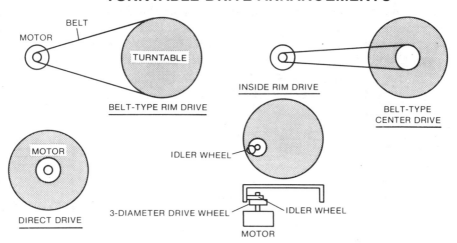

A turntable consists of a flat disk, typically 10–12 inches (25.40–30.50 cm) in diameter which rotates about a central shaft supported by a relatively heavy-duty bearing system. The heavier the turntable, the greater its mechanical inertia (due to its rotating mass) and the greater its resistance to changes in rotational speed (due to imperfections in the drive system). As a result, high-quality turntables use heavy castings and large bearings and drive components.

The most common turntable drive systems are shown in the illustration. The inside-rim drive is simple and widely used. Speed changes are accomplished by using a manual shift lever. To avoid turntable *rumble* (very-low-frequency sounds picked up from the mechanical rotation of the drive system), the belt drive has been developed. This decouples the motor and associated components from the turntable itself and reduces rumble proportionately. New, high-quality turntables use direct drive with a *servo* (feedback) control of motor speed. These are capable of providing the quietest turntables of all.

Automatic Record Changers

Early automatic record changers were developed to take care of the problem of the short playing time inherent in 78-rpm recordings. Record changers are still popular for use with modern recording, although the true hi-fi fan uses a single-play mechanism, eschewing the automatic changer's convenience for possibly better fidelity.

For most modern record changers, a stack of records is placed on an upper extension of a nonrotating spindle coming up through the center of the turntable. When the tone arm reaches the end of the record being played, the stylus encounters an offset inner circular groove present in nearly all records. The resulting large back-and-forth motion of the tone arm activates the changer mechanism. A simplified diagram of the changer cycle of operation is shown in the figure. Most modern records have the label area slightly raised so that the recorded portions of the records do not touch when stacked. In some designs, the turntable is stopped while records are being changed in order to minimize damage.

SIMPLIFIED OPERATING CYCLE
OF AUTOMATIC RECORD CHANGER
(DIMENSIONS DISTORTED FOR EASE OF ILLUSTRATION)

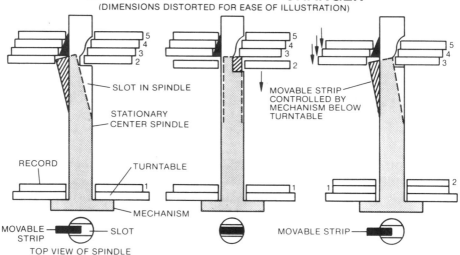

Although the weight of the tone arm is counterbalanced so that the force on the record is reduced to a few grams or less, the pickup cartridge can be damaged by frequent accidental dropping onto the record surface. For this reason, most hi-fi automatic changer systems use a *damper*. Fluid-type dampers are most common and act like a shock absorber. A piston with a small hole is pressed down through a fluid-filled cylinder by the weight of the tone arm. The slow flow of fluid through the hole permits only a slow fall of the tone arm if it's accidentally dropped. In order to simplify the process of selecting one specific band within a record, most changers and single-play turntables have a *cueing lever*. This lever allows the arm to be raised and lowered without touching the tone arm so that contact of the stylus with the record is controlled in the changer damping mechanism. This minimizes problems with accidental dropping of the tone arm or skidding it across the record.

Record Player Tone Arms

The tone arm and cartridge or phono pickup are not usually considered as one element, but in most systems each is selected independently. Tone arms have been made of various materials and in many shapes in order to prevent mechanical resonances in the audio range. The most popular materials are aluminum tubing and wood. The several common shapes include the S, L, straight, and combination shapes that tend to minimize tracking distortion.

As the stylus moves from the outer to the inner grooves, it is important that the tone arm hold the cartridge and stylus in a constant position with respect to the grooves to minimize distortion. The cartridge should be tangent to the record groove circle. This is most easily approximated by using a long tone arm. When such an arm is mounted so that it is tangential at mid-record, it provides a close approximation of the required tracking between the groove, stylus, and cartridge. The S, L, and similar tone arm shapes are used to provide a semblance of good tracking with reduced tone arm length.

Modern tone arms are very light in weight and use a complex gimbal-bearing pivot and counterbalance on the tone arm to require only the lightest pressure on the record to maintain tracking. As you can easily understand, if considerable force is necessary to get the tone arm to move, this force comes from the stylus in the groove and can result in excessive record wear.

TONE ARM SHAPES

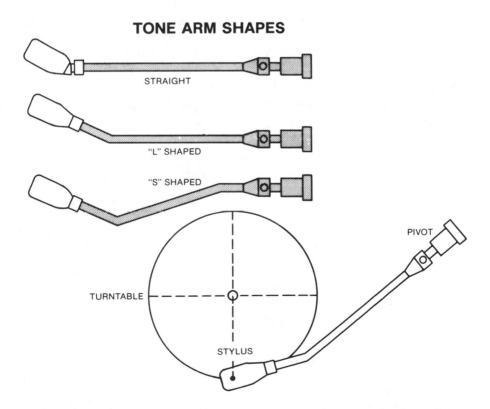

STRAIGHT

"L" SHAPED

"S" SHAPED

PIVOT

TURNTABLE

STYLUS

Record Player Cartridge and Stylus

The pickup cartridge and stylus are the elements that convert the variations in the record grooves to a corresponding electrical signal. All of the principles and devices that have been used in microphones described previously have also been used in record player pickups. The advantages and disadvantages of these various microphones also generally apply to the corresponding pickups.

Magnetic pickups are a frequent choice in the best hi-fi systems. There are two main types. In the most common, the stylus is attached to a small permanent magnet. As the stylus moves, the magnet moves in the vicinity of a coil. The motion of the magnet causes the generation of an electric current in the coil, and the amplitude and frequency of this current correspond to the variations in the record groove.

In the other type of magnetic pickup, the stylus drives a movable coil between the poles of a stationary U-shaped permanent magnet. A current is generated in the coil, and the amplitude and frequency variations of this current match the variations in the record groove. It doesn't really matter which scheme is used; both cartridge types are capable of excellent performance and are invariably used in high-quality systems. Their output is low (1–5 mV), so adequate preamplification is required. For inexpensive systems, a crystal cartridge is used. This consists of a crystal element that is mechanically stressed by the stylus/groove and produces an electrical output. These crystal pickups have high output (up to 1 volt), but have relatively poor fidelity.

The stylus (replaceable in some units) is usually made of diamond for long life, although sapphire is used in some inexpensive units. Typically, the stylus has a tip diameter of 1 mil (0.001 inch) for mono and 0.5–0.7 mil for stereo recordings. Some styli use an elliptical configuration (long axis parallel to the groove) for improved stereo performance.

SIMPLIFIED DIAGRAMS OF TWO TYPES OF MAGNETIC CARTRIDGES

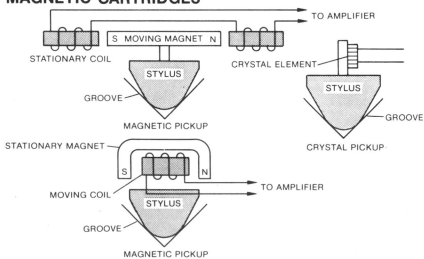

Magnetic Tape Recorders—Basic Components

Magnetic tape is a sound-recording medium that is subject to much less wear and accidental damage than disk recordings. While not as convenient as disk, the tape is a flexible medium that has no grooves to be damaged by a faulty stylus or rough handling. Moreover, magnetic tape offers a method whereby literally almost anyone can make good voice or music recordings from live or other sources. In addition, magnetic tape can be erased and reused. Magnetic tape is a strip of thin plastic material, one side of which is coated with a thin layer of very finely ground iron and other oxides mixed with an adhesive binder. Each of the microscopic bits of oxide can be considered to be made up of a large number of extremely small elements that are permanent magnets.

A mechanical tape-transport system moves the tape in close contact with an *erase*, *recording*, and *playback* head. These heads consist of a core of powdered iron in a binder. Such materials, called *ferrites*, are molded into a core, and the core is wound with many turns of wire. A single gap at one end of the core is precisely cut to provide a fine gap across which a magnetic field is produced when the coil is excited. It is this magnetic field that magnetizes the tape oxide material as it passes the heads. The narrower the gap, the better the quality of the sound produced. Because of possible wear from the tape, special very hard materials are used to make the ferrites (or coat them) to minimize wear.

BASIC COMPONENTS OF MAGNETIC TAPE RECORDER

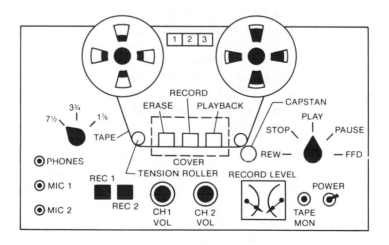

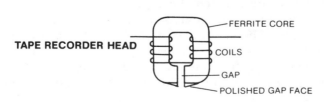

TAPE RECORDER HEAD

Magnetic Tape Recorders—Erasing and Recording

The erase and recording heads are used in recording. The erase head is fed with the output of an oscillator operating well above the audible range (50–100 kHz). (You'll find out about oscillators at the end of this volume.) This creates an alternating magnetic field across the erase-head gap. As the tape moves through this magnetic field, the minute permanent magnets in the tape are subjected to the ac field. As a result, the magnets are magnetized by the ac field and are randomly arranged when they leave the vicinity of the gap. Any fixed magnetic alignment on the tape existing from any previously recorded information is now erased. Thus, any earlier recorded material is deleted.

MAKING A MAGNETIC TAPE RECORDING

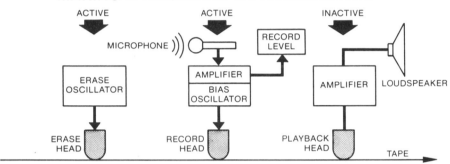

Now the tape moves past the gap in the recording head, which is fed with the amplified electrical signal of the sound to be recorded. This electrical signal produces a magnetic field across the gap of the record head, with the magnitude and direction of the field corresponding to the variations of the electrical signal. Now the minute magnets in the tape *orient* themselves in accordance with the magnetic field that exists across the gap at the moment those particular magnets happened to be passing the gap. Stronger magnetic fields succeed in orienting more of the magnets than weaker fields. Consequently, the organization of the minute magnets in the tape corresponds to the *frequency* and *amplitude variations* in the electrical signal.

A low-level signal from an oscillator operating at a frequency between 50 and 100 kHz is mixed with the signal to be recorded. This signal has the effect of overcoming the interactions between the magnetic domains on the tape and aids in making their orientation on the tape more precisely proportional to the amplitude of the signal being recorded, thereby reducing distortion of the recorded signal. This oscillator signal is known as *bias*, and many tape recorders have controls for adjusting bias to best match the type of oxide being used in the selected tape. Incorrect bias for a given type of tape results in increased noise and distortion. There is a maximum level that can be recorded, depending on how the magnetic materials saturate. To keep within this level, most tape recorders have meters (recording level) to allow adjustment of the recording level so that this maximum is not exceeded.

Magnetic Tape Recorders—Playback

To play back a tape recording that has been made it is necessary to move the tape across the gap in the playback head at the *same speed* and in the *same direction* as when it was recorded. What now travels across the gap are magnetic regions of various intensities, directions, and rates of change.

In your study of *Basic Electricity* you learned that a moving magnetic field causes an electric current to flow in a nearby conductor. Moreover, the changes in this magnetic field cause corresponding changes in the electric current. The playback head, like the record and erase heads, consists of a ferrite core with a very fine gap. The coil is wound with many turns of fine wire, so that the magnetic variations picked up by the core as a result of the tape moving across the gap are now induced in the coil.

PLAYING BACK A MAGNETIC RECORDING

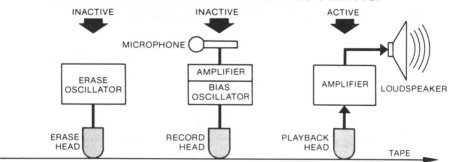

Consequently, the variation in the output current of the playback-head very closely corresponds to the electrical audio signal originally fed to the record head. Thus, when the signal in the playback head is fed to the amplifier and loudspeaker, the resulting sound is a close duplication of the sound that was recorded originally.

Inexpensive tape recorders use the same head for *both* recording and playback. They also use the *same* amplifier, with a few extra parts switched in and out as required, for both recording and playback.

In the more expensive recorders, a separate playback head is provided. This is done because the winding and gap configurations are *different* for record and playback if optimum performance is desired. Also, when the heads are separate, the playback head can be used while a recording is being made. Now the user can listen to or *monitor* the tape as the recording proceeds, and thus is able to judge the actual results being obtained during recording. To monitor using the playback head while recording requires not only a separate playback head but also another channel of amplification to drive the monitor.

Tape Transports

There are three methods in widespread use today for tape transport past the heads: (1) open reel-to-reel, (2) cassette, (3) cartridge. The oldest and most fundamental is the open reel-to-reel system. In this mode, the tape simply unwinds from an open supply reel, moves past the heads, and is rewound on an open takeup reel.

Tape is *not* pulled through the systems by the takeup reel. If that reel turned at constant speed, the tape would be pulled through with increasing speed as it collected on the takeup reel. Instead, the tape passes between a metal shaft (or *capstan*) and a rubber pressure wheel (or *idler*). The capstan is driven at constant speed by an electric motor, and this assures that the tape will always move at a constant speed past the heads regardless of the amount of tape on the reels. The takeup reel is turned by a slipping drive, and merely acts to take up slack in the tape as it comes out from between the capstan and the idler. Most tape transports include a tape counter so that positions along the tape can be determined.

BASIC COMPONENTS OF TAPE RECORDER TRANSPORT

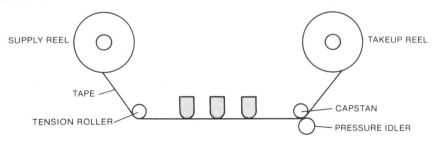

Most tape transports also have a selector (a knob or row of buttons) for control purposes. Typically, this selector has the following functions: *rewind, stop, play, pause,* and *fast forward*. The *record* function is accomplished by using the *play* position in conjunction with separate switches that are interlocked so that it cannot be operated unless the *play* control function is activated at the same time. This prevents accidental operation of the *record* function and thus avoids accidental erasure or rerecording. In most recorders, the operation of the selector cancels the record switch setting to prevent accidental recording over existing material.

The most popular tape width is ¼ inch (0.64 cm) for both mono and stereo systems. For very-high-quality multichannel recording work, tapes between ½ and 2 inches (1.27–5.08 cm) are used. Tape speeds between 1⅞ and 30 inches per second (4.76–76.20 cm per second) are used; the higher speeds of 15 and 30 inches per second (38.10 and 76.20 cm per second) are used on commercial equipment for professional recording and broadcasting. Home hi-fi tape recorders usually are operated at 7½ inches per second (18.05 cm per second) for best results.

Tape Transports (continued)

The *cassette* transport is based on exactly the same concepts and mechanisms as the open reel-to-reel system, except that both reels are packaged together and miniaturized. Thus, it is unnecessary to handle and thread the tape as with a conventional reel-to-reel transport. The cassette contains the supply reel and takeup reel and has an opening in the face into which the capstan and/or the heads fit whenever in an operating mode. Cassette tape is ⅛ inch (0.32 cm) wide and operates at 1⅞ inches (4.76 cm) per second (nonselectable). Typically, a cassette tape has four tracks (two stereo pairs). To load a cassette into the recorder you simply press the unit down into a depression in the top of the recorder or slide it into a slot in the front edge of the recorder. In either case, nothing else needs to be done except to select the desired function and proceed to play or record. As in the case of the reel-to-reel recorder, selecting the proper function also makes the required connections necessary to perform the selected function. By simplifying the problem of threading the tape through the transport, cassettes have succeeded in enormously popularizing tape recording.

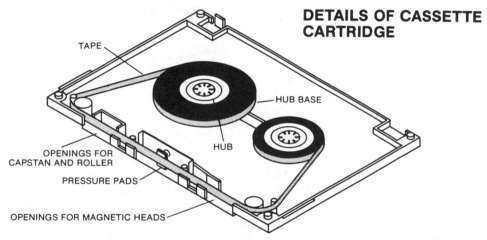

DETAILS OF CASSETTE CARTRIDGE

TAPE

HUB BASE

HUB

OPENINGS FOR CAPSTAN AND ROLLER

PRESSURE PADS

OPENINGS FOR MAGNETIC HEADS

Because of the narrow strip that is recorded and the low speed, early cassette tape systems were noisy and had poor high-frequency response. The development of new head materials and tape oxides greatly improved the performance of cassette tape systems; however, the development of noise-reduction systems (for example, Dolby®, to be described later) has made it possible to produce cassette systems having performance comparable to all but the best reel-to-reel systems.

In addition to their use for voice and music recording, tape cassettes have become very popular for recording digital data and programs. While special cassette tapes are used for these applications, techniques have been developed to use an ordinary inexpensive cassette system for this application. You will learn more about this in Volume 5 of this series.

Tape Transports (continued)

The eight-track tape *cartridge* uses a somewhat different method for simplifying threading the tape through the transport. The cartridge slips into a slot in the transport. Pressing the function selector moves the heads and capstan into position and makes the required electronic connections, just as for the cassette transport. The main difference is that the cartridge has only one reel. The end and beginning on the tape are joined into a continuous loop. During operation, the reel rotates continuously in one direction. The capstan pulls tape out of the center of the reel and the tape is rewound onto the outside of the reel. Although this may sound difficult to accomplish, it takes place very smoothly, provides for continuous play, and never needs rewinding the way open-reel or cassette tapes do. Typically, four pairs of stereo tracks are recorded on each cassette tape.

DETAILS OF TAPE CARTRIDGE

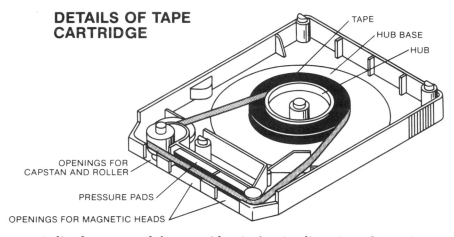

TAPE
HUB BASE
HUB
OPENINGS FOR CAPSTAN AND ROLLER
PRESSURE PADS
OPENINGS FOR MAGNETIC HEADS

A disadvantage of the cartridge is that its direction of travel *cannot* be reversed. When listening or when making a recording, it is very often desirable to re-record or listen to the same place on the tape. In the reel-to-reel or cassette systems, it is a simple matter to rewind and return to the beginning or to any point on the tape. To return to that spot on a cartridge it is necessary to go through the *entire* length of the tape. Thus, the cartridge has become primarily a convenient system for playing back commercially recorded tapes. The usual format for cartridge tapes is eight tracks (four stereo pairs) on ¼-inch (0.64-cm) tape operating at 1⅞ inches (4.76 cm) per second. The tape heads are arranged on tracks with a ratchet and solenoid so that any track pair can be selected. Like cassettes, cartridge tapes are almost invariably stereo. Because of the narrow width of the recorded area and the low speed, eight-track tape systems are somewhat noisier and have poorer fidelity than conventional open-reel tape systems. In addition, the pulling of the tape from the center of the reel results in a high rate of tape wear. In spite of these disadvantages, the eight-track system is convenient and easy to use and is commonly employed in automobile hi-fi systems.

Noise Reduction—Analog (Dolby®)

One of the problems with tape recording (and to a lesser extent with disk recording) is *noise*. This occurs as a result of the residual randomly oriented magnetic domains in the tape oxide coating, and hence is a fundamental problem. In conventional reel-to-reel tape machines, the problem is reduced by recording in a wide strip (two to four tracks on a ¼-inch (0.64-cm) or wider tape). The noise is reduced because the recording area on the tape (magnetic field) can be increased with wide tape. Cassette and eight-track systems are therefore inherently more noisy than reel-to-reel type systems. Because of this, early hi-fi systems did not include cassette or eight-track recorder/players. Another problem involves poor high-frequency response for tape systems operating at low speeds. Several new advances, however, have made the cassette system a competitor to the reel-to-reel system in hi-fi systems. These are chromium-oxide (and other exotic oxide) tapes and the patented Dolby® noise-reduction technique. The advantages of these tape materials are significantly increased high-frequency response with lower noise.

DOLBY® NOISE - REDUCTION SYSTEM

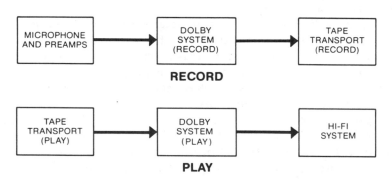

The Dolby® noise-reduction technique is based on the fact that tape noise and hiss are greatest at high frequencies and are only discernible during soft passages. While recording in the Dolby® system, the high-frequency as well as all portions of the music signal are boosted in level in a controlled, known way during the soft passages when noise would be the most audible; that is, the dynamic range is compressed. During playback, the process is reversed. This reduces the high frequencies and softer passages to normal levels but also reduces the tape noise to a much lower level than before. Unless Dolby®-recorded tapes are played back on Dolby® machines, the high frequencies will sound unnaturally loud, and the sound will have little variation in volume. Obviously, the Dolby® technique can be applied to any recording as well as to other tape systems. In addition, there are other noise-reduction systems that are normally used commercially. It is claimed that the Dolby® system can provide up to 20-dB noise reduction.

Noise Reduction— Digital

Digital tape recording techniques avoid the noise problems by converting the audio signal to digital form before recording, recording the digital information, and converting it back to analog form on playback. Because the recorded information is all digital (1,0), tape noise is not a factor in the output. Although you will not study digital systems until you reach Volume 5, you should have an understanding of how digital recordings work. While many forms of digital recording and processing exist, we will concentrate on the more conventional approaches.

In digital recording, the incoming analog signal can be treated in several ways. For example, it can be sampled at intervals and the result used to produce a *digital word* that describes its amplitude at a given instant. Alternatively, the width of a uniform amplitude series of pulses can be modulated in width, with the width proportional to the signal amplitude at the moment of sampling. For example, a weak signal produces a narrow pulse, and a loud signal produces a wider pulse. If these signals are recorded on tape, it is easy to see that amplitude noise becomes unimportant because we are only interested in the width or the number of pulses in a digital word. The recording amplitude is of no importance. Thus, amplitude noise from tape becomes relatively insignificant in digital tape systems.

On playback, the pulse-width-modulated signal (or the digital word, etc.) can be recovered with little loss of fidelity and converted to an amplitude or analog system, since this is what we hear. The important point is that we can suppress noise that originates in the recording medium by proper choice of system elements and are no longer dependent on having to put up with tape noise.

Digital tape recording is just now becoming popular for commercial use, so most noise reduction is of the analog type described earlier. This situation will change shortly to permit the hi-fi enthusiast to have the excellent acoustic properties of digital tape recording.

DIGITAL NOISE-REDUCTION SYSTEM

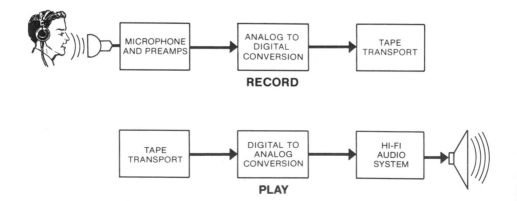

Preamplifiers and Mixers

Most high-fidelity systems use a preamplifier and/or mixer to join or select audio from various sources and provide for amplification of low-level signals. Mixer–preamplifiers are most commonly used in recording and public address work, while hi-fi systems use switched preamplifiers. Tape recorders usually have special preamplifiers in both the record and playback circuits. These circuits attenuate bass and boost high frequencies during record and reverse the procedure on playback, so the overall recording characteristics external to the tape recorder appears to be a flat frequency response, as shown in the diagrams. Thus, the system feeding or taking audio from the tape recorder should have flat frequency response. Similar circuits are used in disk recording systems.

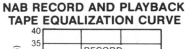

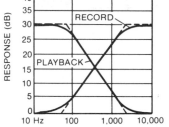

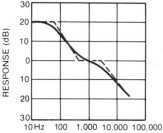

In audio systems, preamplification is used to raise the signal level, adjust frequency response, and provide impedance matching between the various audio signal sources and the power amplifier. Most modern hi-fi audio systems have the capacity to accept several inputs. Typically, these are for tape, phono-pickup, auxiliary, and possibly other functions. Some systems have additional input terminals for signal sources of different impedance and amplitude levels, such as from crystal and ceramic pickups.

The circuit of a typical preamplifier for high-fidelity equipment usually contains special circuits to compensate for the frequency response of the system elements. For example, disk recording circuits are made with the low frequencies attenuated and the high frequencies accentuated. The low frequencies are attenuated to limit the excursion of the stylus during recording, while the high frequencies are pre-emphasized to reduce high-frequency noise that occurs in the recording process. This occurs because the de-emphasis of high frequencies during playback also reduces noise introduced by the playback mechanisms. This must be compensated for, along with the characteristics of the phono-pickup. The diagram on the left shows the NAB (National Association of Broadcasters) standard equalization curves for magnetic tape recording. These equalization curves vary, depending on recording speed. The diagram on the right shows the RIAA (Recording Institute of America) curves for the equalization of records played with a magnetic pickup. This is the standard for all commercial phono records made today.

Preamplifiers and Mixers—Tone Control Circuits

Most hi-fi audio systems have a number of RC networks (filters) whose function is to alter the frequency response of the amplifier system. Moreover, when the amplifier has different inputs for tape, phono, and tuner, different types of frequency modification networks may be found at these inputs, as mentioned earlier, to provide an overall flat response.

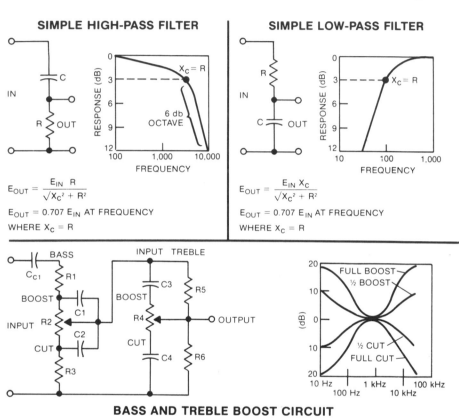

SIMPLE HIGH-PASS FILTER

$$E_{OUT} = \frac{E_{IN}\ R}{\sqrt{X_C{}^2 + R^2}}$$

$E_{OUT} = 0.707\ E_{IN}$ AT FREQUENCY
WHERE $X_C = R$

SIMPLE LOW-PASS FILTER

$$E_{OUT} = \frac{E_{IN}\ X_C}{\sqrt{X_C{}^2 + R^2}}$$

$E_{OUT} = 0.707\ E_{IN}$ AT FREQUENCY
WHERE $X_C = R$

BASS AND TREBLE BOOST CIRCUIT

In addition to these, there are adjustable frequency compensations to satisfy the personal tastes of listeners. The most common of these are the tone controls that vary the gain (either boost or cut) of the high and low frequencies. Some modern tone controls have a set of adjustments (up to 10 or more frequency bands) so that different parts of the audio spectrum can be boosted or cut separately across the entire band. In addition, there are scratch filters that attenuate high frequencies and rumble filters that attenuate very low frequencies associated with turntable rumble.

Preamplifiers and Mixers—Tone Control Circuits (continued)

All of these compensating circuits are based on the fact that capacitive reactance decreases as the frequency increases (as shown in the illustration on previous page). Variations of these basic arrangements are used to produce boost and attenuation of the bass and treble frequencies, as shown in the diagrams.

The signal from a record player with a magnetic pickup will rarely exceed a few millivolts. Signals from the tuner and tape recorders are usually about 0.25 volt. The diagram on the following page shows a typical discrete circuit preamplifier.

The magnetic phono signal is amplified by a two-stage, direct-coupled amplifier Q1 and Q2. A feedback network (from Q2 to Q1) contains a frequency-compensating network for the magnetic pickup. The output level of Q2, adjustable by R10, is coupled to the source selector switch S1 and volume control R12 with a compensation network switched by S2. When the loudness compensation is switched in, the lower frequencies are amplified more at low volume to compensate for the reduced low-frequency responses of the ear at low volume levels. Switch S1 also allows selection of other sources at higher level.

The selected level-controlled signal is applied to amplifier Q3, which has the bass and treble tone controls in the coupling circuit to Q4. A scratch filter, which can be switched in or out by S3, helps remove transient signals that are produced by scratches in the records. Treble and bass controls provide a boost of 10 dB and a reduction of 15 dB. Each control is independent, and the response is essentially flat with both controls set to their midpoints. The output from the tone control circuit feeds amplifier Q4 via C16. As shown, the tone controls are in a feedback loop around Q4. This configuration provides for bass and treble boost without midband attenuation, like the circuit shown previously.

The output of the magnetic phono preamplifier section or the selected tuner or tape input can be coupled to the tape output terminal. This signal line would be connected to a tape recorder input to allow taping a program coming from a record, tuner, or other tape recorder.

The output from the preamplifier is obtained from the Q4 collector. Typical inputs are 0.5-3mV for the phone input and between 0.1 and 0.5 volt for the other inputs. Inputs of at least the minimum levels specified will produce an output signal of at least 1 volt. This signal is adequate to drive almost any power amplifier to its maximum power output. Recently designed preamplifiers use ICs rather than discrete circuits because of the circuit simplification and ease of manufacture. An IC preamplifier is described later.

DISCRETE CIRCUIT HIGH-FIDELITY PREAMPLIFIER

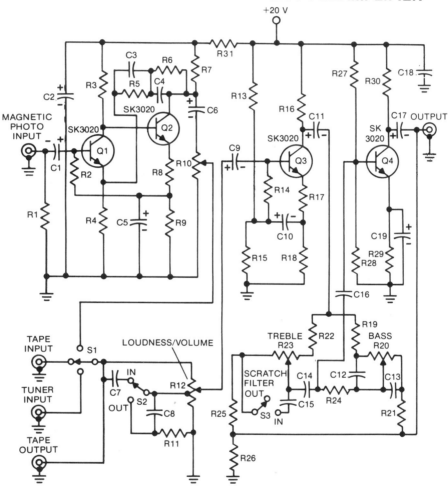

Experiment/Application—Checking the Tone Control Circuit

Most tone control circuits are designed to boost or cut high and low frequency response to suit a listener's taste. The circuit shown below is one of the most popular in use today. You can check its characteristics easily. Suppose you hooked up the circuit shown below.

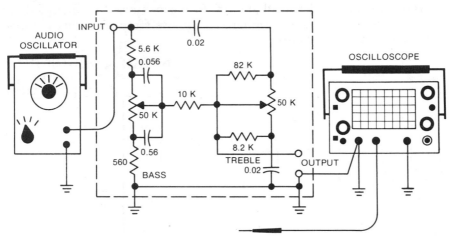

If you check the frequency response of this circuit at the full boost, ½ boost, flat, ½ cut, and full cut, using the point-by-point method used to check frequency response earlier, you will get a set of response curves as shown below.

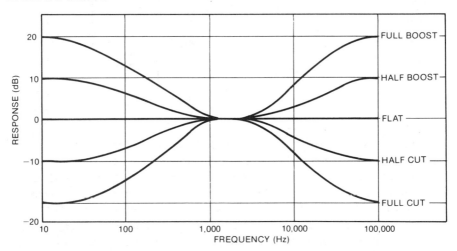

You can vary these response curves for a particular application by changing the resistors and capacitors. For example, you could change the bass boost by changing the capacitors, or the resistors. To keep the 20-dB (10:1) ratio between flat and boost/cut, you should keep the R/C ratios at 10:1 also. As you can see from this experiment, simple RC circuits are useful in adapting or correcting frequency response.

Preamplifiers and Mixers—IC Preamplifier

A single IC containing two low-noise preamplifiers (LM381, described previously) can be used as a very simple preamplifier-mixer, as shown below. Provisions are included for phono pickup (magnetic), tuner, tape, and auxiliary inputs. Volume and tone controls of the boost-cut type are included for all inputs. Provisions also are included for matching the phono input frequency response from a magnetic phono cartridge to the RIAA curve.

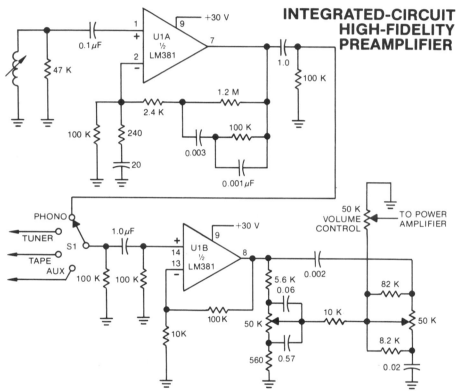

INTEGRATED-CIRCUIT HIGH-FIDELITY PREAMPLIFIER

As shown, amplifier U1A (one-half of the LM381 low-noise dual preamplifier) is used as a phono preamp to raise the level from the pickup from about 5 mV to about 1 volt at its output in the mid-band range. The feedback network sets the mid-band gain at about 400, with appropriate bass boost and treble cut as required to match the RIAA frequency-response requirements with a magnetic cartridge. Thus, the phono-preamp output is corrected for the pre-emphasized highs and the de-emphasized lows on recording. The output from this phono preamp is fed to the switch S1 for selection of the various inputs. The selected input is then amplified further (10 times) by the other half of amplifier U1 (U1B) in a noninverting feedback amplifier configuration. The output from this amplifier is applied to the tone control circuit and volume control to be fed to the power amplifier.

Preamplifiers and Mixers—IC Mixer

One typical IC mixer application allows any one or many microphones to supply the signal to a recording device or power amplifier. For example, when multiple sound pickups are used on several instruments and there is a soloist and background group, many simultaneous inputs must be open at all times. Any or all of these signals must be amplified without switching or selecting circuits. The mixer provides multiple inputs to a single amplifier. By using feedback, inputs can be added or removed without affecting other inputs. The input signals are usually set to suitable levels by variable resistors that can be used as gain controls for each individual input.

All unused input terminals must be grounded (or the volume control turned to zero), because each open lead can act as an antenna for low-frequency signals (such as power sources) and for wide-band noise generators (such as fluorescent lights and switching circuits).

The mixer shown can also be used to combine signals from different sources such as a microphone and tuner or tape deck. In fact, this configuration of the mixer is usually used to combine a number of signals that must be handled simultaneously and not switched independently. For example, one application might require a signal with a prerecorded tape or voice signal at the same time as background music or public address applications where several speakers each have individual microphones. Mixers of this type are also used in broadcast and recording studios.

IC MULTIPLE-INPUT AUDIO MIXER

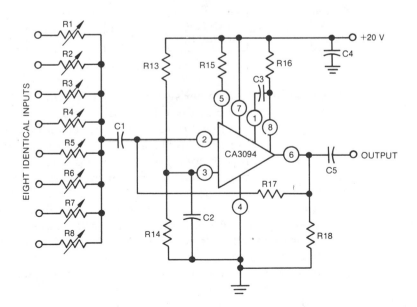

Power Amplifiers

The output of the preamplifier/mixer is usually sufficient to drive a power amplifier, which in turn drives the speaker. Most modern solid-state amplifiers do not use output transformers but couple directly (via a dc blocking capacitor) to the loudspeaker load. The actual power requirement depends on speaker type, number of speakers, and room characteristics. Obviously, you need more power for a large room than for a small one. In addition, acoustical-suspension speakers may require very high power for a modest acoustical output because of their low efficiency. Loudspeaker data will give you information on the appropriate power level for the power amplifier.

Only a few watts are required to drive an efficient speaker to produce enough sound to fill an average room. So why are high-powered systems commonly used? The basic reason is that the fidelity is better if the amplifier and speaker are operated *below* rated power rather than at full power to allow for high-energy transient peaks. A good phonograph recording may have a dynamic range of greater than 50 dB (300:1). Furthermore, as mentioned above, acoustical-suspension speakers are inefficient and may need 10–20 watts for comfortable listening even in a small room.

COMPLETE AUDIO AMPLIFIER
POWER AMPLIFIER SYSTEM

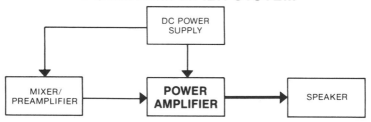

Except for inexpensive, low-power systems shown previously using ICs, power amplifiers are made up of discrete circuit components. Usually, the power amplifier uses feedback to reduce distortion and improve the bandwidth. Most power amplifiers use large power transistors in the output circuit with either a complementary or quasi-complementary configuration at their output so that no output transformer is required.

Important characteristics of power amplifiers are frequency response, output impedance, harmonic distortion (intermodulation and harmonic), and power output. An important part of a good power amplifier system is the power supply. It must be capable of delivering the peak currents without excessive voltage drop. In addition, it should be well filtered. A good power supply also has a large output filter capacitor so that it has a low output impedance at low frequency, since the power supply output impedance can affect the frequency response.

Power Amplifiers—Complementary-Symmetry Amplifier

The complementary-symmetry power amplifier is a very popular and convenient power amplifier configuration because it contains few parts but gives excellent performance. In addition, it can be built (using a dual power supply) so that the output is at dc ground, allowing the elimination of the speaker-coupling capacitor. This is desirable, because this capacitor is large and expensive.

40-WATT COMPLEMENTARY-SYMMETRY AMPLIFIER

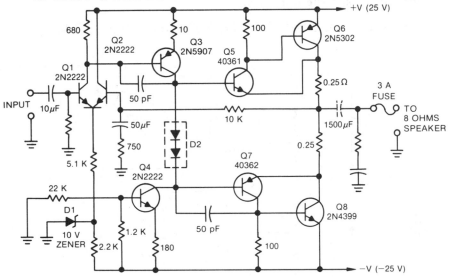

Transistors Q1 and Q2 function as a differential amplifier to conveniently sum the input (Q1 base) and feedback (Q2 base). Transistor Q3 acts as a common-emitter driver amplifier that drives the power amplifier pair Q5 and Q6. Transistors Q5 and Q6 are configured to provide high current gain with only unity voltage gain for the positive portion of the signal; Q7 and Q8 function similarly for the negative portion of the signal. The base of Q7 is driven from the Q3 collector through diode D2. Current through D2 is kept constant by constant-current source Q4, so that current flows through D2 under all signal conditions. The input to Q7 and Q8 is obtained from the collector of Q3 via D2 as a level shifter for the signal on the Q3 collector. As shown, each pair of output transistors (Q5–Q6 or Q7–Q8) is a complementary pair to permit direct coupling, as described previously.

The gain is such that a 1-volt rms input signal drives the output to 40 watts rms with a power supply of ± 25 volts. The amplifier can be modified for single power supply operation by grounding the negative terminal and placing all ground connectors at half the supply voltage. Some modification of the emitter-bias supply for Q1 and Q2 would also be necessary. As you can see, the power amplifier is very simple because direct coupling is used in the amplifier, thereby minimizing the need for capacitors.

Power Amplifiers—Quasi-Complementary-Symmetry Amplifier

Another very popular power amplifier configuration is the quasi-complementary-symmetry power amplifier. This resembles the complementary-symmetry power amplifier described previously except that it allows the use of the same transistor type in the power-output stages. In addition, this quasi-complementary-symmetry amplifier is well suited to operation from a single power supply. Because of this, the output is not at dc ground, and so a coupling capacitor is required between the amplifier output and the loudspeaker.

THE QUASI-COMPLEMENTARY AUDIO POWER AMPLIFIER

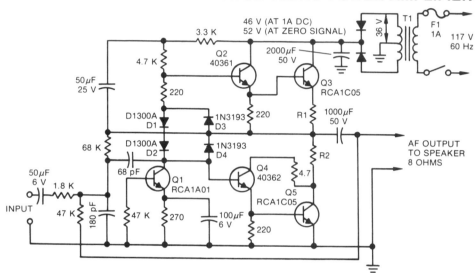

Transistor Q1 is a high-gain common-emitter amplifier. The input signal and the feedback signal are resistively mixed at the base of Q1 (ac feedback only). The gain is adjusted via feedback so that a 1-volt rms signal will provide full output. Transistors Q2 and Q3 operate as a dual emitter follower that functions as a high-gain, noninverting current amplifier with about unity voltage gain. (You may remember this configuration from your study of power supplies in this volume.) Transistor Q3 is an npn power transistor that drives the output on positive signal inputs. The base of Q2 is driven by a constant-current source from the collector of Q1. Constant current is maintained by the *bootstrap* capacitor (50 μF, 25 volts), which maintains both ends of the voltage divider feeding Q2 at the same ac potential. The collector of Q1 drives the unity-voltage-gain-inverting amplifier formed by Q4 and Q5, which provides high current gain on negative input signals. It is the combination of noninverting and inverting current amplifier configurations Q2–Q3 and Q4–Q5 that is the essential difference between complementary and quasi-complementary configurations. Like the complementary-symmetry amplifier, the quasi-complementary amplifier is very popular and capable of excellent performance.

Basics of Stereophonic and Quadraphonic Systems

When you sit in an auditorium and a live performance is taking place on the stage, you enjoy a *unique* sound experience. Instruments and performers on the left side of the stage sound as if they are on the left side of the stage. Similarly, instruments and performers at the right sound as if they are at the right. Moreover, reflected sounds coming off the left back and side walls and right back and side walls sound as if they were coming from those directions. These unique sound sensations result from the fact that your two ears have *directivity* and are more sensitive in a particular direction, the left being more sensitive to the left and the right to the right.

When the record player and the radio were developed, people immediately sensed that what they were hearing was *not* the same as in the auditorium. High-fidelity sound reproduction helped a lot, but the difference between left and right sound sensations still was not there. Your two eyes work together to give you the ability to estimate the direction and distance of objects. This is known as *stereoscopic vision*. Similarly, your two ears work together and enable you to estimate the direction and distance of a sound source. This hearing ability is known as *binaural hearing*.

STEREOPHONIC SOUND
LEFT-FRONT AND **RIGHT-FRONT** RECORDINGS PLAYED BACK
THROUGH **LEFT-FRONT** AND **RIGHT-FRONT** SPEAKERS

QUADRAPHONIC SOUND
ADD **LEFT-REAR** AND **RIGHT-REAR** RECORDINGS PLAYED BACK
THROUGH **LEFT-REAR** AND **RIGHT-REAR** SPEAKERS

The effects of binaural hearing are closely duplicated by using two microphones in order to duplicate the position and sensitivity pattern of the two ears. The two signals resulting from these microphones can be separately recorded and processed and then played back into left and right headphones, as mentioned previously. Because this is a very inconvenient arrangement, we use *stereophonic systems* that place the microphones considerably further apart than the ears so that more people can experience an enhanced sound sensation when two speaker systems are separated by some distance. This stereo arrangement does not, however, duplicate the reflected sounds coming from the *rear* of the studio or auditorium. To do that, two additional microphones and processing channels must be added. This four-source *quadraphonic* sound adds considerably to the cost of the system and has not been widely accepted, but it does add an additional element of realism, beyond stereo systems.

Basics of Stereophonic and Quadraphonic Systems (continued)

The simplest method of understanding how stereo and quad are recorded and played back is to study these processes on a tape recorder. The simplest reel-to-reel recording method is the *half-track monaural* technique. This also is used in monaural cassette recorders.

In this type of recording, the erase, record, and reproduce (playback) heads come in contact with only a narrow portion of the standard ¼-inch (0.64-cm) tape. The actual dimensions used in a professional-quality system are shown in the diagram. Thus, during recording a sound track is impressed only on a 0.082-inch (0.21-cm) strip on the upper part of the tape. The entire length of the tape is recorded in this manner. When the end of the tape is reached, the reels are reversed (turned over), and the recording continues on the previously unused portion of the tape. Recording time is thus doubled by using a magnetic track that is about half the width of the tape in both directions.

When the second side of the tape is completely recorded, the beginning end of the tape is now on the outside of the reel. Now the tape again can be placed on the left side of the recorder and replayed through its full length in the same manner as it was recorded. This is all accomplished without rewinding.

MAKING TWO HALF-TRACK MONAURAL RECORDINGS

DIMENSIONS OF HALF-TRACK RECORDING
HALF-TRACK MONAURAL

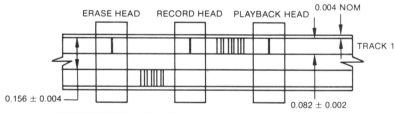

FIRST MOVEMENT OF TAPE PAST RECORD HEAD

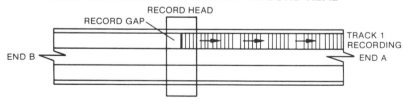

TAPE IS TURNED OVER AND MOVED PAST RECORD HEAD AGAIN

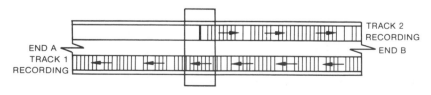

Basics of Stereophonic and Quadraphonic Systems (continued)

Using magnetic tape makes it an easy matter to add the second sound channel to produce stereophonic sound. The most basic method is shown here, and it is known as the *two-track stereo method*.

In this arrangement, the erase, record, and reproduce (playback) heads each contain two identical but independent magnetic units stacked one above the other. The upper and lower heads in each unit are separated by a 0.156-inch (0.40-cm) space. To make a recording, two independent electronic systems are required to feed the recording heads.

During recording, the output of the left microphone is fed through the *left* electronic channel and recorded on the *upper* track of the tape. In a similar manner, the output of the right microphone is processed by the *right* electronic channel and recorded on the *lower* tape track. When the end of the tape is reached, the tape is rewound onto the supply reel.

After rewinding, the beginning of the tape is on the outside of the supply reel. Now the tape can be replayed by threading it past the heads and switching the electronics and transport to the *play* mode. As the tape moves, the upper (left) magnetic track passes the upper (left) playback head. The resulting signal is amplified by the *left* audio amplifier and reproduces sound from the speaker on the left side of the room.

MAKING A TWO-TRACK STEREO RECORDING
DIMENSIONS OF TWO-TRACK STEREO RECORDING

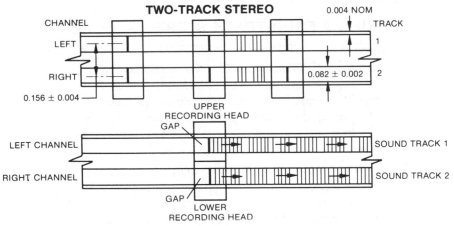

Simultaneously, the lower (right) magnetic track passes the lower (right) playback head. The resulting signal is amplified by the *right* audio amplifier and reproduces sound from the speaker on the right side of the room. Thus the requirements for stereoscopic sound reproduction have been met directly with the addition of another complete recording and reproduction channel. In the above discussion, it is assumed that the phasing of all components has been carefully done so that the proper phasing is maintained throughout the system, both during the recording and during the playback process. The maintenance of phase is essential if good performance is to be obtained.

Basics of Stereophonic and Quadraphonic Systems (continued)

For home tape use, a four-track stereo system has been devised which allows the tape to be used in both directions. This yields twice the amount of recording time for a given tape. The four-track (two stereo pairs) approach is made possible by improved manufacturing techniques that enable the width of the heads to be reduced by half without significant loss of quality and signal-to-noise ratio. The left and right heads are only 0.043-inch (0.109 cm) wide and are spaced apart as shown in the diagram. While making a recording, the left channel magnetically imprints track 1 while the right channel imprints track 3. When the end of the tape is reached, the reels are reversed (turned over), and recording continues with the left channel imprinting track 4 and the right channel working on track 2. Now when the end of the reel is reached, no rewinding is necessary, and playback can begin.

During playback, track 1 is in the correct position for moving past the upper (left) playback head, and track 3 is in the correct position for moving past the lower (right) playback head. When the end of the tape is reached, the reels are again reversed (turned over). Now the playback continues from tracks 2 and 4 in the same manner in which they were recorded. While this approach allows twice the recording time of the two-track system described on the previous page, it becomes impossible to *edit* the tape because an edit of one track pair affects the reverse track pair as well.

MAKING A FOUR-TRACK STEREO RECORDING

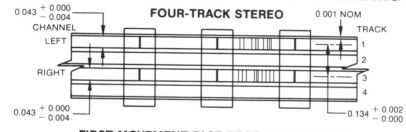

FIRST MOVEMENT PAST RECORDING HEADS

SECOND MOVEMENT PAST RECORDING HEADS

Basics of Stereophonic and Quadraphonic Systems (continued)

In cassette stereo recorders in contrast to reel-to-reel recorders, the left and right channels are recorded on adjacent tracks. Thus, any monophonic cassette player can play stereo cassettes, although the left and right channels will come out of a single loudspeaker. Similarly, any stereo cassette player can reproduce monophonic cassette recordings, although the same sound will come out of both loudspeakers.

For quadraphonic sound, four tracks are necessary. The four-track capability developed for stereo recordings also paved the way for making true quadraphonic recordings. To do this, four independent heads are required in the erase, record, and playback positions. Each of these heads has the same width as a four-track stereo head. Each head is fed by an independent electronic recording and reproduction channel; thus, four channels are required.

HEAD ARRANGEMENT ON STEREOPHONIC CASSETTES

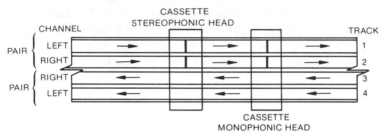

HEAD ARRANGEMENT ON QUADRAPHONIC CASSETTES

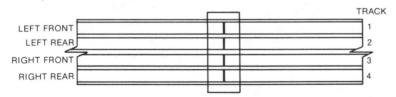

To accomplish this, four microphones are placed so as to receive sound from the left front, right front, left rear, and right rear. These signals are separately imprinted on each of the four tracks. When the end of the tape is reached, it is necessary to rewind the tape. Then the system is switched over to playback, and each magnetic track travels past the corresponding playback head, feeding signals to four independent audio amplifiers and speakers. Consequently, the sound coming from each of the four speakers will correspond in direction to the sound picked up by each of the four microphones if the speakers are placed in positions corresponding to the original positions of the microphones during recording. Thus, an additional element of realism has presumably been added to the system.

Basics of Stereophonic and Quadraphonic Systems (continued)

On a stereo phonograph record, the left-channel sound is engraved into the left wall of the groove. Right-channel sound is engraved into the right sidewall. The illustration shows how the sound of each channel can be recovered by a single stylus attached to a pickup with two magnetic coils. The construction shown is an extremely simplified view of a pickup with two moving magnets, but the concept applies to any of the pickup types.

Because the tip of the stylus is rounded, it is constantly in contact with both sidewalls, regardless of the up or down variations in either sidewall. For example, a rise in the left sidewall moves the stylus at right angles up from that sidewall, and the magnet marked L moves into its coil. Similarly, a fall in the left sidewall makes the same magnet move out of the coil. These in and out motions cause a duplicate of the left-channel signal to be generated at the output of coil L.

In a similar manner, variations in the depth of cut in the right sidewall cause a duplicate of the right-channel signal to be generated at the output of coil R. In this way, the two coils of the stereophonic phono-cartridge are separately excited by the appropriate recorded information on the disk. As in a tape system, the two cartridge outputs are applied to separate, identical audio systems for each channel.

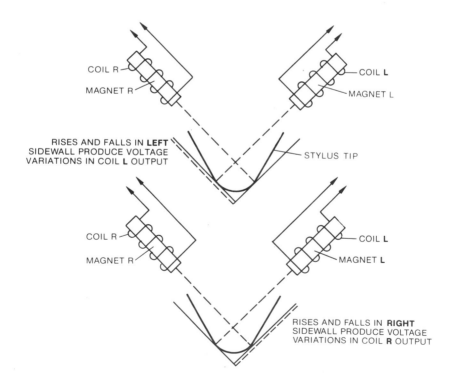

Basics of Stereophonic and Quadraphonic Systems (continued)

At least two major methods and a number of variations are in use for inscribing and reproducing quadraphonic sound on record player disks. The combination or mixing method is the simplest. As shown in the illustration, it employs a solid-state matrix to encode (by the phase-shift method to be described on the next page) the LF, RF, LR, and RR channels into the form of two sound channels—called the left combination or *left total* and the right combination or *right total*. These two channels are then engraved separately into the left and right sidewalls of the recording groove in the same manner as the left and right stereo signals considered previously.

The same stylus and pickup arrangement used for stereo can reproduce the variations in the walls into the form of separate left-total and right-total signals. These two signals are then fed into a solid-state decoding matrix, and four-channel sound is produced at the output.

ENCODING AND DECODING QUAD SOUND ON RECORD PLAYER

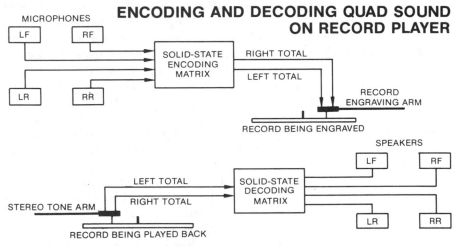

There are three variations of this system in general use today. These are identified as the *SQ®*, *QS*, and *RM* systems, and each is promoted by a different group of recording companies. All methods employ phase shifting and partial channel addition, and the decoder designed for any one system will operate for either of the others. A slightly different placement of the loudspeakers will be required for the different encoding methods.

What has been described here for record players is also true for *FM radio* and *stereophonic TV* broadcast reception. The left-total and right-total signals can be transmitted by FM radio or stereophonic TV in the same way as the left- and right-channel stereo signals (described in Volumes 3 and 4). Then, after stereo detection, the left- and right-total signals are fed into a decoder, and four-channel sound results. If the receiver does not have the decoder, the left-total signal, and the right-total signal are treated as two-channel stereo signals.

Basics of Stereophonic and Quadraphonic Systems (continued)

The *SQ*® decoding method used by CBS is shown here in block diagram form. Because the left-rear and right-rear signals are combined with the left-front and right-front signals, they must be distinguished in some manner that will permit later separation. The method used in the encoder is to shift the left-rear and right-rear signals by 90 degrees in phase and add these to the left-front and right-front signals to form left-total and right-total signals.

In the decoder, there are four preamplifier channels with equal gain and with either 0 or 90 degrees of phase shift, plus the two mixers indicated by circles. The separation achieved is only a compromise, because there is no stable reference present (such as the 38-kHz subcarrier in the FM stereo decoder that you will learn about in Volume 4) upon which to base the decoding.

In the left channel, the left-total signal passes through the 0-degree phase shift preamp and appears with no phase shift at the left-front output. The left-total signal also passes through the 90-degree phase shift preamp and is phase shifted by 90 degrees. Part of the right-total signal is mixed with this shifted signal and subtracts from the shifted left-total signal in such a manner as to produce a good approximation of the left-rear signal at the left-rear output of the mixer. What is really fed to the left-front speaker is the left-total signal, and what is fed to the left-rear speaker is a good approximation of the left-rear signal. Thus, all but the most exacting listeners are satisfied with the results. Similar processing takes place in the right channel, and the total result is one that pleases nearly all listeners.

In most systems, the circuits are all part of an integrated-circuit (IC) chip except for the RC portions of the phase shifters and the provisions for mixing. These portions are made external so that they can be adjusted for different encoding systems.

The mixing type of quad system that has been described is one designed for low cost and easy adaptability to existing stereo systems. Only an approximation of four-channel separation is achieved. In the CD-4 system to be described next, there is a carrier frequency to use as a reference for decoding, so accurate separation can be accomplished.

OPERATION OF TYPICAL FOUR-CHANNEL SQ® DECODER

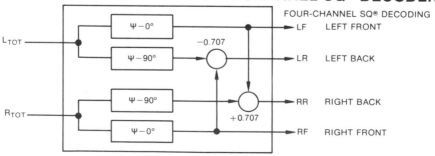

Basics of Stereophonic and Quadraphonic Systems (continued)

One successful method for recreating four discrete sound channels from a single groove in a record disk is the *CD-4 method* used by RCA and several other recording companies. The basics of the CD-4 method are shown in the diagram. Reference should also be made to the figure illustrating the broadcasting of stereo signals by FM radio in Volumes 3 and 4 of this series. The concepts of the two systems are the same.

In the left sidewall of the record groove there is engraved a complex of left-channel information. First, there is the left-front plus left-rear signal (LF + LR). In addition, there is a 30-kHz carrier signal modulated by the difference between the left-front and left-rear signals (LF − LR). This produces sidebands above and below the 30-kHz carrier. Sideband frequencies extend upward through 45 kHz, as shown. A similar complex of right-channel information is engraved in the right sidewall. Here again, sideband frequencies extending upward through 45 kHz are produced.

To reproduce these signals, the pickup must be capable of responding to frequencies up to 45 kHz. In addition, the pickup must be of the stereo type with two sensitive elements. Each of the two outputs of this pickup is fed to a separate decoder matrix that processes the signal in a manner similar to that described for stereo decoding. The difference is that each decoder separates front- and back-channel information from the incoming signal. There is little residual signal in any of the four reproduced channels.

BASICS OF CD-4 QUADRAPHONIC SOUND

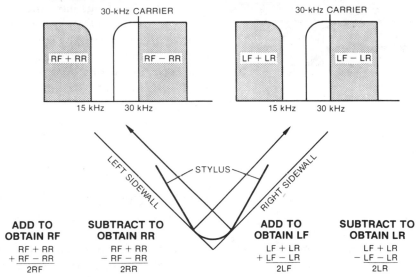

FOUR DISTINCT SIGNALS RECOVERED

Although various methods have been proposed for broadcasting CD-4 signals by FM radio, none has yet been approved officially. For the present, this system is thus only for home reproduction of CD-4 records.

Review of High-Fidelity Audio Systems

1. MONOPHONIC SYSTEMS use a single recorded or transmitted channel. This system requires *one* microphone and *one* loudspeaker.

2. STEREOPHONIC SYSTEMS use two channels to provide a spatial quality to the sound. This system requires *two* microphones and *two* loudspeakers.

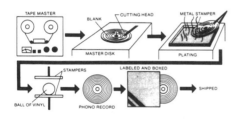

3. RECORDING METHODS—The most important recording methods involve either disk or tape recording. Disk recordings have the information stored as an undulating groove on a vinyl disk. Tape recordings have the information stored as magnetic variations on a tape covered with magnetic oxides.

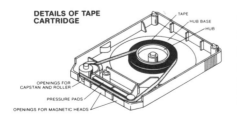

4. MAGNETIC TAPES are usually used in an open reel-to-reel configuration, a cassette configuration, or an eight-track cartridge configuration. Only the open-reel system can be edited conveniently.

Review of High-Fidelity Audio Systems (continued)

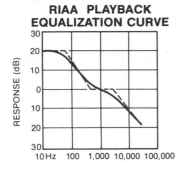

RIAA PLAYBACK
EQUALIZATION CURVE

5. PHONOGRAPH PREAMPLI-FIERS are used to amplify the weak signals from the phonograph pickup and correct the equalization (pre-emphasis and de-emphasis) deliberately introduced during the recording process. The most common type of equalization is the RIAA standard.

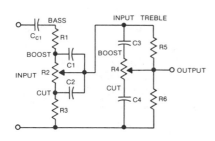

6. TONE CONTROL CIRCUITS are used to provide the listener with a means for adjusting the system frequency response to provide the most pleasing result.

STEREOPHONIC SOUND

LEFT-FRONT AND RIGHT-FRONT RECORDINGS PLAYED BACK THROUGH LEFT-FRONT AND RIGHT-FRONT SPEAKERS

7. STEREOPHONIC REPRODUC-TION requires two separate channels to reproduce the outputs from two separate input transducers.

QUADRAPHONIC SOUND

ADD LEFT-REAR AND RIGHT-REAR RECORDINGS PLAYED BACK THROUGH LEFT-REAR AND RIGHT-REAR SPEAKERS

8. QUADRAPHONIC REPRO-DUCTION uses two additional channels to reproduce the effects of sound heard from indirect sources (reflections from walls, etc.). It requires four channels that are encoded and decoded with the stereo signals to produce left- and right-front and left- and right-rear signals.

Self-Test—Review Questions

1. Define the terms high-fidelity, monophonic, stereophonic, quadraphonic, and coloration.
2. Describe briefly how modern phonograph records are made. How is sound reproduced by the phono-cartridge?
3. Sketch the basic components of a modern tape recorder/player. Describe the function of each component.
4. Sketch and describe tape schemes for utilization of cassettes and eight-track cartridges. How do they differ from the open-reel tape equipment? What are some of their advantages and disadvantages?
5. Describe briefly the idea behind noise-reduction schemes for tape recorders. Why are these noise-reduction schemes necessary?
6. How are preamplifiers and mixers used in audio systems?
7. How do preamplifiers for tape recorder/players and phonographs differ from a microphone preamp?
8. Sketch a typical bass and treble tone control circuit and describe how it works.
9. Describe (and/or sketch) the complementary and quasi-complementary power amplifier circuit. Describe the essential operation of each circuit.
10. Compare mono, stereo, and quad audio systems in terms of the amount of hardware (equipment) necessary, the principles of operation, and the performance obtained. Briefly describe how stereo and quad outputs are obtained on phonograph records.

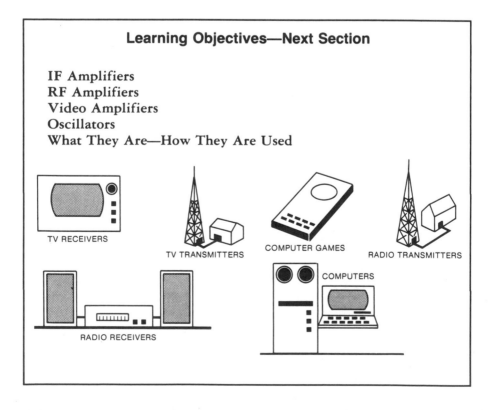

Learning Objectives—Next Section

IF Amplifiers
RF Amplifiers
Video Amplifiers
Oscillators
What They Are—How They Are Used

TV RECEIVERS

TV TRANSMITTERS

COMPUTER GAMES

RADIO TRANSMITTERS

RADIO RECEIVERS

COMPUTERS

Introducing the Video Amplifier

At the beginning of this volume you were introduced to the subject of video amplifiers. You learned that these are similar in design and general arrangement to audio amplifiers. The difference is in the frequency response, particularly at high frequencies. Video amplifiers work with signals covering the wide frequency range from the dc range up to many megahertz and higher, although a range of 30 Hz to 5 MHz is adequate for many applications. We will briefly discuss video amplifiers here, and will consider them in more detail when we study TV.

The term *video* is from latin and means *to see*. As the name suggests, video amplifiers are used in applications where the signal contributes to image formation. However, modern usage includes any wide-band amplifier of this general type. Video amplifiers are used in oscilloscopes, TV, radar, sonar, work data processing, and in any application where signals are shown on displays like a TV picture tube and on oscilloscope screens. They are also widely used to amplify pulsed waveforms.

VIDEO AMPLIFIERS PROCESS THESE KINDS OF SIGNALS

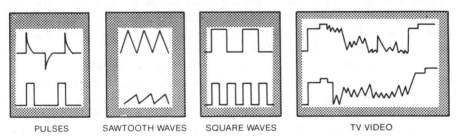

PULSES   SAWTOOTH WAVES   SQUARE WAVES   TV VIDEO

All that you have learned about audio amplifiers—circuit configurations, methods of biasing, classes of operation, methods of coupling, etc.—also applies to video amplifiers. Therefore, it is only the *broadbanding* of these amplifiers that make them *video* amplifiers rather than *audio* amplifiers. Video amplifiers must have a broad frequency response because they are usually required to process the pulse, sawtooth (triangular), and square waveforms shown here. It can be shown by mathematical analysis that such waves are actually the sum of many harmonically related sine waves. Some of these waves have frequencies that are 10 or more times as high as the fundamental frequency of the square, triangular, or pulse waveform itself. For narrow pulses, a good rule of thumb is that the high-frequency extent of the bandwidth of the video amplifier should be about 1/T, where T is the pulse length. Thus, for a 1-μ second pulse, the bandwidth should be about $1/10^{-6} = 10^6$ Hz = 1 MHz. At the low end of the response range, a good rule of thumb is that the response should be 10 times lower than the repetition rate of the waveform or the longest pulse involved. If the amplifier does not have the required broad-frequency response, the output waveform will be distorted.

Distortion Caused by Poor Frequency Response

Basically, most discrete circuit video amplifiers consist of RC-coupled, Class A, common-emitter stages. Suppose we fed an ideal pulsed wave into such an amplifier. If the amplifier had perfect response, the amplified square wave would also be perfect, with vertical sides, sharp corners, and a flat top and bottom. As you know, RC coupling restricts low-frequency response. If a dc voltage is applied to a series RC network, the voltage across the resistor equals the applied voltage and then decreases exponentially in accordance with the time constant RC. The voltage decreases to about 33% in one time constant. The same thing happens with the pulse, the top of which can be considered as the application of the dc voltage. To obtain a good reproduction of the square wave, then, we have to make the time constant sufficiently long so that no appreciable fall off (or droop) occurs across the square wave flat portions. Typically, we choose RC to be greater than 10 times lower than the frequency of the square wave. The droop cascades: thus for many stages, each RC may need to be much larger so that the overall effect will be 10 times the lowest frequency. Emitter-bypass capacitors can be a source of poor low-frequency response if the capacitor is insufficient to keep the emitter bypassed. Low-frequency alternating current (ac) signals in the emitter circuit will cause degenerative feedback and reduced gain at the frequency of these signals.

Poor high-frequency response is caused by a reduction in total load impedance as a result of the circuit capacitance. These capacitances, as you know, arise from the components and the circuit wiring. The effect of poor high-frequency response is to round the corners of the waveform and slow its rise time. These act in parallel with the load resistor and reduce gain as frequency increases, since capacitive reactance decreases with increasing frequency. Thus, it is important to keep circuit capacitance to a minimum when circuits are designed and to use lower values of load resistances so that the shunting capacitance will have less effect.

POOR FREQUENCY RESPONSE CAUSES DISTORTION

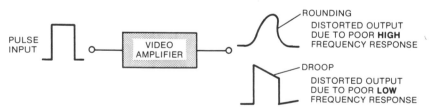

PULSE INPUT — VIDEO AMPLIFIER

ROUNDING
DISTORTED OUTPUT DUE TO POOR **HIGH** FREQUENCY RESPONSE

DROOP
DISTORTED OUTPUT DUE TO POOR **LOW** FREQUENCY RESPONSE

One very important cause of poor high-frequency response in solid-state devices has to do with properties of the semiconductor itself. Thus, it is important to know that amplification by a particular semiconductor at the desired high frequency is possible in the first place. Methods for compensating low- and high-frequency response are available. However, for most modern applications, amplifiers are dc coupled and use low enough load resistor values so that no compensation is necessary.

IC Video Amplifiers

Integrated circuits can effectively replace the discrete-circuit video amplifier. The extensive use of direct coupling between their internal stages and high input impedance greatly improves low-frequency response without the need for large coupling capacitors as in the discrete circuit amplifier. High-frequency response is aided by the fact that the very small size of the circuit greatly reduces the capacitances between the elements. The very small size of the interconnections greatly reduces their capacitance to ground, which also aids high-frequency response. Further, the semiconductor materials and fabrication methods are suited to providing high-frequency response.

The RCA CA3001 IC shown below is designed for video and other amplifier applications for frequencies up to 20 MHz. Examination of the circuit reveals that internal transistors Q3 and Q4 are connected in a differential amplifier configuration. The differential amplifier is fed by a built-in constant-current source formed by regulator diodes D1 and D2 and transistor Q7. Input transistors Q2 and Q5 provide for high input impedance, and output transistors Q1 and Q6 provide for low output impedance. As shown, Q1, Q2, Q5, and Q6 are in a common-collector configuration to accomplish this. As you can see, each stage (Q3 and Q4) has a gain of about 10 if driven with a single-ended input and a gain of about 20 if driven push-pull. The emitter followers Q1, Q2, Q5, and Q6 have slightly less than unity gain.

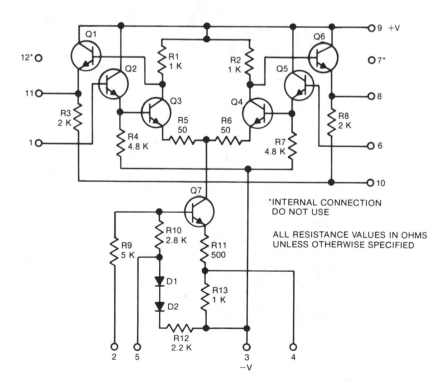

IC Video Amplifiers (continued)

The diagram on this page illustrates how three CA3001 ICs can be connected in cascade to form a stable gain of about 65 dB. Note that the circuit has a single-ended input (one input is grounded) and a double-ended (push-pull) output. The symmetrical arrangement of this cascade circuit makes it unnecessary to use elaborate decoupling to the power supply, because there are equal and out-of-phase currents in the supply leads.

THREE-STAGE CA3001 CASCADE AMPLIFIER AND FREQUENCY-RESPONSE CHARACTERISTICS

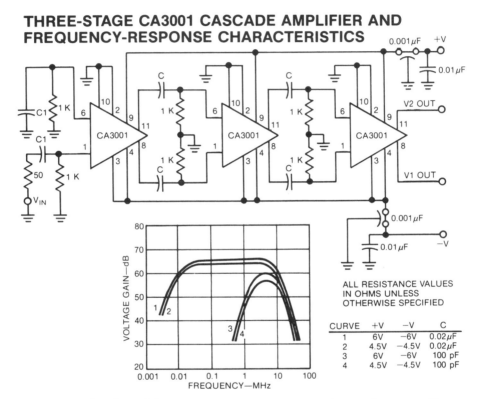

ALL RESISTANCE VALUES
IN OHMS UNLESS
OTHERWISE SPECIFIED

CURVE	+V	−V	C
1	6V	−6V	0.02 μF
2	4.5V	−4.5V	0.02 μF
3	6V	−6V	100 pF
4	4.5V	−4.5V	100 pF

The family of frequency-response curves shows the overall performance of the three stages with different supply voltages and different coupling capacitors (values in the column under C). Curves 1 and 2 are for wide-band video applications, and curves 3 and 4 are for intermediate-frequency (IF) amplifier applications, described in the next section.

The table illustrates other performance characteristics for two sizes of coupling capacitor and two levels of supply voltage. Note that the lower supply voltage provides the same bandwidth and almost the same gain with a considerable saving in power dissipation but with reduced voltage output capability. The values of C can be made considerably larger (up to 10 μF or more) and the 1-K resistors can be increased in value to provide better low-frequency response.

What IF Amplifiers Do

Intermediate-frequency (*IF*) amplifiers are a *simplified* form of radio-frequency (*RF*) amplifiers. Learning about the IF amplifier first makes it easier to understand about RF amplifiers in the next section.

IF amplifiers are used in superheterodyne radio receivers that you were introduced to in the transmitter-receiver learning system. In a superheterodyne receiver, the signal picked up by the antenna is first amplified by an RF amplifier. Because amplification at this frequency is inefficient, the amplified RF carrier is heterodyned via a mixer and local oscillator to convert it to a suitable fixed intermediate level at which frequency amplification can be accomplished more readily. The modulation of the mixer output signal duplicates that of the modulated RF carrier, but the frequency of the intermediate signal is constant and equal to the sum or difference between the RF and local oscillator frequencies, as shown. Considerable amplification is usually accomplished by the IF amplifier, whose output is fed to the detector, audio amplifier, and speaker to produce sound waves duplicating the original message. IF amplification is efficient because it is done at a fixed frequency.

In the superheterodyne receiver, the *main difference* between the RF and IF amplifiers is that the RF amplifier is tuned to the frequencies that the receiver is designed to pick up, whereas the IF amplifier is always tuned to the *same, fixed frequency.*

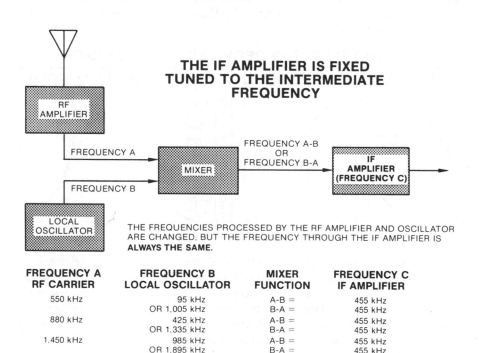

THE IF AMPLIFIER IS FIXED TUNED TO THE INTERMEDIATE FREQUENCY

RF AMPLIFIER

FREQUENCY A

FREQUENCY B

MIXER

FREQUENCY A-B OR FREQUENCY B-A

IF AMPLIFIER (FREQUENCY C)

LOCAL OSCILLATOR

THE FREQUENCIES PROCESSED BY THE RF AMPLIFIER AND OSCILLATOR ARE CHANGED, BUT THE FREQUENCY THROUGH THE IF AMPLIFIER IS **ALWAYS THE SAME.**

FREQUENCY A RF CARRIER	FREQUENCY B LOCAL OSCILLATOR	MIXER FUNCTION	FREQUENCY C IF AMPLIFIER
550 kHz	95 kHz	A-B =	455 kHz
	OR 1,005 kHz	B-A =	455 kHz
880 kHz	425 kHz	A-B =	455 kHz
	OR 1,335 kHz	B-A =	455 kHz
1,450 kHz	985 kHz	A-B =	455 kHz
	OR 1,895 kHz	B-A =	455 kHz

What IF Amplifiers Do (continued)

An IF amplifier is an amplifier that contains one or more frequency-selective devices that enable it to amplify a selected band of frequencies. All that you have learned about solid-state amplifiers—circuit configurations, methods of biasing, classes of operation, methods of coupling, etc.—*also applies* to IF amplifiers.

The frequency to which an IF amplifier is tuned and the band of frequencies above and below this *center* frequency are determined by the nature of the signal to be amplified. The band of frequencies (or bandwidth) of the IF amplifier is determined by the fact that when the information (message) is superimposed upon the carrier in the transmitter, several different frequencies are produced in a band adjacent to the carrier; these are called *sidebands*. In an amplitude-modulated (AM) system these frequencies are directly related to the modulating frequency. Thus, for a pure tone of 1 kHz, there will be two sidebands 1 kHz away from the carrier; if a 3-kHz tone is added, there will be 1-kHz and 3-kHz sidebands. Frequency-modulation (FM) sidebands are a little more complicated. You will learn more about AM and FM sidebands when you study these later. For now it is sufficient to understand that a *band* of frequencies centered around the carrier frequency must be amplified rather than a single carrier frequency. The band of frequencies the IF amplifier must process should be wide enough to enable the sidebands to get through to produce a final sound output of acceptable quality. This section is concerned with how IF amplifiers amplify the band of frequencies required.

AM AND FM RADIO RECEIVER IF BANDWIDTHS

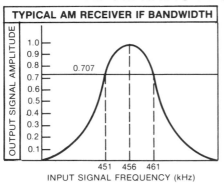

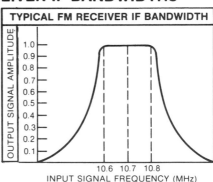

A receiver designed to receive AM stations in the broadcast band usually uses an IF amplifier tuned to a center frequency of 455 kHz with a bandwidth of about 10 kHz. A receiver designed to receive FM stations usually has an IF amplifier tuned to a center frequency of 10.7 MHz with a bandwidth of 200 kHz. The CB transceiver learning system has an IF amplifier center frequency of 4.3 kHz with a bandwidth of 10 kHz.

For TV receivers in the United States the sound carrier uses an IF center frequency of 41.25 MHz and the video carrier uses an IF center frequency of 45.75 MHz.

IF Amplifier Configuration

The schematic diagram shows a typical discrete-component transistor IF amplifier using transformer coupling. The main difference between this amplifier circuit and those you have learned about previously is the fact that it has a *tuned* transformer at its input and another at its output. The purpose of these tuned transformer circuits is to match impedance and produce selective amplification for the desired band of frequencies and to reject all other frequencies. As you can see, the circuit is parallel resonant. You should restudy the properties of parallel resonant circuits in *Basic Electricity*.

There are other frequencies in the output of the mixer stage, as you will learn about in Volume 4. Some of these are produced by stations in the channels adjacent to the one to which you are tuned, and which have frequencies very close to that being processed by the IF amplifier. If these signals were amplified, you would hear an interfering signal. The ability of the IF amplifier to process *only* the desired band of frequencies, is what gives the receiver *good selectivity*. It is the use of *tuned LC circuits* that enables the IF amplifier to amplify the desired band and sharply cut off signals with frequencies outside this band.

TYPICAL IF AMPLIFIER STAGE

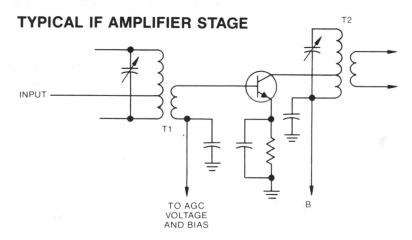

To achieve desired selectivity, a high impedance at parallel resonance is desired. In solid-state amplifiers, the input and output impedances are lower than are desirable across the tuned circuit. Therefore, the collector is tapped down on the primary so that the collector impedance is stepped up by autotrampower action by the square of the ratio of the total turns to the turns to the tap. The secondary winding has a few turns because it is desirable to match the low input impedance of the transistor base. In many cases, an automatic gain control (AGC) voltage is applied to the base to control the amplifier gain. In some modern IF amplifiers, crystal and/or ceramic filters are used to provide very high selectivity. These filter elements are especially useful in communication receivers where high selectivity is essential to minimize adjacent channel interference.

Review of Tuned Circuits

The tuned circuit in an IF amplifier consists of an inductor and capacitor connected in parallel, as shown in the illustration. Thus, the same voltage is present across both of them. Now suppose you adjust the frequency of the applied signal to one at which the inductor and capacitor have the same reactance. In theory, at this frequency the current in the inductor will be equal to and 180 degrees out of phase with the current in the capacitor. Under such conditions, the two currents will cancel and there will be no external current flowing through this combination. As you know, the frequency at which this occurs is known as the *resonant* frequency, which is equal to $\frac{1}{2\pi} \sqrt{LC}$.

Actually, there is a small current flowing through the external circuit. The reason is that the inductor and capacitor are not perfect—they have resistance and losses. As a result, at a band of frequencies above and below resonance, the impedance remains high and a low current flows through the external circuit.

In a radio receiver, signals at many frequencies are applied to the tuned circuit at the IF amplifier's input. For signals in the desired frequency band, the applied voltages "see" a high impedance and therefore are applied at full amplitude to the base of the transistor. These signals are amplified. Similarly, in the transistor's collector circuit, these signals are applied across another similar tuned circuit; again, they "see" maximum impedance, and a maximum collector output signal voltage is produced. The frequency difference between all points 30% below the peak (0.7 amplitude) and the peak itself is called the *bandwidth* of the circuit.

Signals at frequencies away from resonance "see" a low impedance at the input and a low impedance in the collector circuit, and thus the gain of the amplifier is significantly reduced at these frequencies.

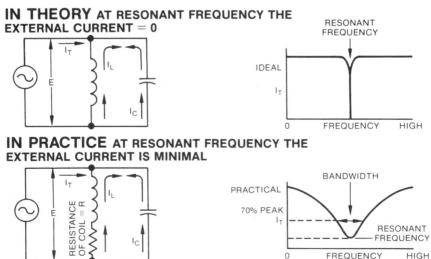

IN THEORY AT RESONANT FREQUENCY THE EXTERNAL CURRENT = 0

IN PRACTICE AT RESONANT FREQUENCY THE EXTERNAL CURRENT IS MINIMAL

Typical IF Amplifier Stages

The diagram on this page shows the circuit of a typical two-stage IF amplifier. This circuit is actually the second and third IF stages of the CB transceiver learning system, but it is typical of most solid-state discrete-circuit IF amplifiers. Note that there is a common-emitter circuit combination, and transformer-coupling to the previous and following stages. This form of coupling permits the use of tuned circuits at the inputs and outputs of both stages, thereby improving the selectivity (rejection of undesired signals).

In transformers of the type shown, tuning is not accomplished by using a variable capacitor across each of the coils as shown previously; instead, a fixed capacitor is used and a core made of compressed powdered iron (ferrite) is moved in and out of the coil to adjust the inductance. This provides a single convenient tuner arrangement at low cost. Such tuning is called *permeability tuning*.

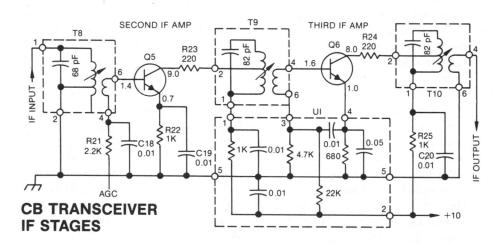

**CB TRANSCEIVER
IF STAGES**

As shown, the signal input from inductively tuned IF transformer T8 is applied to the base of Q5 via an impedance step-down in T8 to match the base impedance of Q5. Transistor Q5 operates a Class A amplifier biased via R22 and the AGC bias. Capacitor C19 provides an IF ground (low impedance) to the signal, but has no effect on the dc bias. The emitter circuit of Q5 provides for dc stabilization, but since C19 effectively grounds the emitter at the IF, full gain is obtained at the IF. The collector is connected to a tap on transformer T9 primary for impedance matching. The lower end of T9 primary is grounded at IF via C29 (0.01 μF) but allows for dc bias of the collector via R20 (1 K). The following IF amplifier stage Q6 is essentially the same, except that the biasing is somewhat different, so that no AGC is applied to the base of Q6. While this IF amplifier operates at 4.3 MHz, it is typical of IF amplifiers at other frequencies.

Automatic Gain Control

Automatic gain control (AGC) of IF amplifiers for AM reception leads to uniform volume between stations. In addition, portable units like auto radios, CB, and communication equipment show great changes in signal strength at different locations. For the listener to have to adjust the volume control constantly is inconvenient, so the answer is to *control* the *gain* of the receiver *automatically*, and this is usually done in the IF amplifier.

In an AGC system, the gain of one or more IF stages is varied inversely in accordance with the strength of the signal. Gain is reduced when the signal level is high and increased when signal level is low. The result is that receiver output volume remains fairly constant.

AGC is accomplished by means of a dc voltage derived from the receiver's detector stage. A diode rectifies the output from the IF amplifier output. The result is the audio signal as you know it plus a dc voltage proportional to the carrier level or signal strength. A low-pass filter bypasses the audio signal, and the remaining dc component is fed back to the IF as shown. The application of this bias to the base changes the gain of that stage. Assume that the detector produces a negative carrier voltage and the IF stage is an npn transistor. As the signal level increases, the negative output from the detector is added to the positive base bias, decreasing the net bias on the transistor base. This reduces the operating current, which in turn will reduce the stage gain. When the signal level decreases, the opposite happens. Thus, the system tends to maintain a constant level at the detector output. As you can see, we are dealing with a feedback system of a somewhat different sort than you have learned about before. In many cases, to improve the AGC action, an amplifier and sometimes a separate detector are used. In addition, a fixed bias must usually be added to the AGC voltage so that Class A operation of the transistor IF stages is maintained. You will learn more about AGC when you study receivers.

HOW AUTOMATIC GAIN CONTROL VOLTAGE IS DEVELOPED

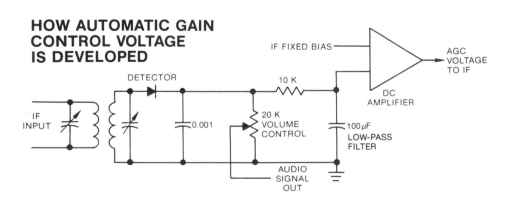

Selectivity

The previous pages should have shown you that there is nothing complex about IF amplifier circuitry. The major differences between RF and IF amplifiers are those related to the frequency to which they are tuned and the bandwidth. The IF is independent of the frequency to which the receiver is tuned. The bandwidth is related to the type of transmission being received. AM radio requires a bandwidth of only a few kilohertz, while FM and TV reception require a much wider bandwidth.

The selectivity of an IF amplifier is determined by its tuned circuits and is taken to be the 3-dB bandwidth point (70%). The measure of the selectivity is determined by the Q of the circuit. The Q of a coil is equal to its inductive reactance X_L divided by its resistance R_L. Since the leakage resistance of a well-constructed capacitor is extremely high, nearly all of the effects of lowered Q in the tuned circuit are due to the resistance of the coil and external circuit loading. The higher the Q in a tuned circuit, the sharper will be the peak in the impedance-versus-frequency curve, and the narrower will be the bandwidth. To make a coil with high Q, it is necessary to use wire of as large a diameter as is consistent with the application and to use high-quality (low-leakage) insulating material on that wire.

To broaden the bandwidth of a tuned circuit, a resistor or other circuit resistance such as transistor input, can be connected across the coil, to lower its Q. The lower the resistance, the lower the Q and hence the greater the bandwidth. For a single tuned circuit, the bandwidth is equal to the center frequency divided by the Q. When many stages are cascaded, the overall selectivity is the product of the individual selectivities. In some special cases, double-tuned overcoupled transformers are used to provide a wider response band with good selectivity, or other special tuning methods will be used to adjust the IF passband to a specific shape. These will be discussed in Volume 4 on receivers.

CASCADED TUNED CIRCUITS

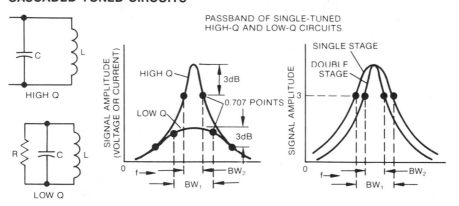

Bandwidth Control Using Crystal Filters

Quartz-crystal filters have been used in the IF amplifiers of communication receivers for many years. For reception of single-tone (code) signals, an IF bandwidth of less than 100 Hz is adequate, and crystal filters greatly enhance the selectivity of the receiver. A single-crystal filter connected as shown provides the desired narrow bandwidth. From your study of *Basic Electricity* you will recall that a *piezoelectric crystal* is one that develops a voltage across its large flat surfaces when compressed. Such a crystal also has the characteristics of an extremely high Q series-resonant tuned circuit. The resonant frequency of such a crystal is determined by its dimensions (primarily its thickness) and the type of material. In communication receivers, bandwidths of less than 100 Hz can be achieved with such an arrangement. Such bandwidths are not adequate for voice reception, where a broader band of frequencies ($\pm$ 3 kHz) is typically required.

In the illustrated circuit, the crystal filter is fed from a tuned IF transformer. The crystal filter acts as a very-high-Q series-tuned resonant circuit that presents minimum impedance to the desired very narrow frequency band, and permits only that narrow band to get through to the next stage. Capacitor C2 is a phasing capacitor used to cancel the capacitance across the crystal holder that would allow undesired signals to couple across the crystal filter.

A *crystal-lattice filter* such as that illustrated can be used to produce a broad frequency response with a sharp dropoff at the sides. In this arrangement, Y1 and Y2 are a matched pair and serve as series crystals. Y3 and Y4 are a matched pair having a higher resonant frequency than the first two and serve as the shunt or *lattice* crystals.

IF CRYSTAL FILTER   CRYSTAL LATTICE FILTER

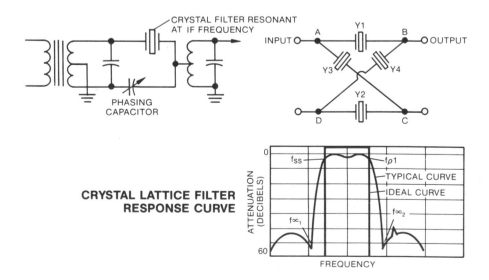

CRYSTAL FILTER RESONANT AT IF FREQUENCY

PHASING CAPACITOR

INPUT — Y1 — OUTPUT
A   B
Y3   Y4
Y2
D   C

CRYSTAL LATTICE FILTER RESPONSE CURVE

ATTENUATION (DECIBELS)

f_{ss} $f_{p}1$
TYPICAL CURVE
IDEAL CURVE
f_{∞_2}
f_{∞_1}

FREQUENCY

Bandwidth Control Using Crystal Filters (continued)

The bandpass of the arrangement shown on the previous page is determined by the difference in frequency between the two sets of crystals. In the diagram, the low-frequency cutoff f_{ss} is determined by the series crystals and the high frequency cutoff f_{p1} is determined by the parallel crystals. Frequencies $f_{\infty1}$ and $f_{\infty2}$ are those at which output is a minimum. A variety of different lattice arrangements can be used to produce almost any desired bandpass, and the low cost of crystal units makes it economical to do.

Modern high-performance receivers and communications equipment require very high selectivity for minimum interference; they therefore commonly use crystal filters in their IF stages. Typically, these are special assemblies that are packaged together in either a lattice or a series-resonant configuration to give the best passband control. The first IF amplifier of the CB transceiver learning system uses crystal filters. The circuit diagram is shown below.

CB TRANSCEIVER CRYSTAL FILTERS

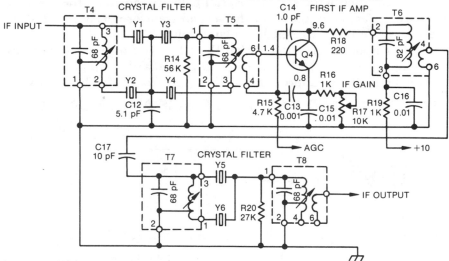

Transformer T4 and crystals Y1, Y2, Y3, and Y4 form a lattice filter coupled to T5 which in turn drives the first IF amplifier Q4. The output of Q4 is, in turn, coupled to the second IF amplifier Q5 via a second crystal filter Y5–Y6. These filters in combination with the tuned circuits give this receiver high selectivity. The overall selectivity is −6 dB at 6 kHz (± 3 kHz), and the response is down 60 dB at 30 kHz off resonance. This provides for very sharp rejection of signals in adjacent channels and insures minimum interference.

As you might suspect, such filtering is possible because the IF frequency is fixed. If these filters were required to be tunable over a band, their design would be impossible. It is the *fixed* frequency operation of IF amplifiers that makes high-selectivity, controlled-passband operation a *practical* thing to do.

Integrated Circuit IF Amplifiers

The RCA CA3011 and CA3012 ICs are designed especially for use in the IF amplifier sections of FM and communications receivers. Both circuits are supplied in the 10-terminal package shown here and are identical except that the CA3012 can be operated at higher supply voltages.

These ICs contains three differential amplifiers connected in cascade, plus a regulated dc power supply. The internal organization of these ICs is shown in block diagram form on this page, and the internal circuit connections are shown on the next page. Also shown on this page is the voltage gain versus frequency response characteristic, which shows that more than a 60-dB gain is available at 10 MHz. Amplitude limiting, of importance in FM receivers, is also provided.

The circuit diagram on the following page shows the details of an FM IF amplifier using the CA3012 IC. A double-tuned filter is used between the two ICs, and the frequency response curves for two levels of input signal are shown. The center frequency of the IF is 10.7 MHz. The gain for each stage and the overall gain are shown on the drawing. The circuits at the output and the second transformer are for FM detection, and will be described in a later volume.

The response curve shows that the selectivity of the double-tuned interstage filter is 200 kHz at the 3-dB points with an input of 10 μV and does not change appreciably with inputs from 500 μV to 0.5 volt.

RCA CA3011 AND CA3012 ICs FOR IF AMPLIFICATION

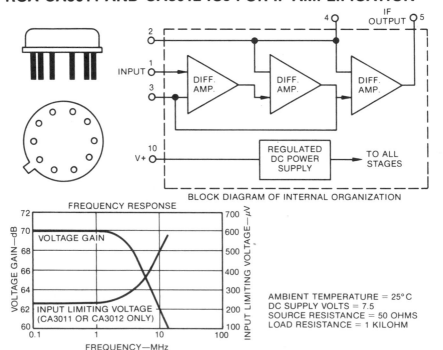

BLOCK DIAGRAM OF INTERNAL ORGANIZATION

AMBIENT TEMPERATURE = 25° C
DC SUPPLY VOLTS = 7.5
SOURCE RESISTANCE = 50 OHMS
LOAD RESISTANCE = 1 KILOHM

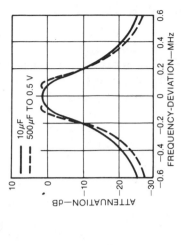

SELECTIVITY CURVE FOR DOUBLE-TUNED INTERSTAGE FILTER USED IN THE IF AMPLIFIER STRIP SHOWN ABOVE

10.7-MHz FM IF AMPLIFIER STRIP USING CA3012 IC

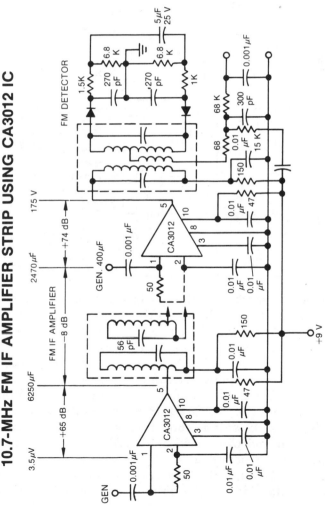

ALL RESISTANCE VALUES IN OHMS UNLESS OTHERWISE SPECIFIED.

Other Examples of IF Amplifiers

Shown here are examples of the AM and FM IF amplifier sections in the Zenith H-Line stereo systems.

Two AM IF amplifier stages are used, both with identical transistors Q202 and Q203. The stages use common-emitter configurations. A voltage-divider circuit consisting of R219, R220, and R214 establishes bias for Q202. A voltage divider consisting of R221 and R222 provides bias for Q203. AGC voltage from the detector is supplied to the base of Q202 and Q203 through R220 and R214.

AM IF AMPLIFIERS IN ZENITH H-LINE STEREO RECEIVER

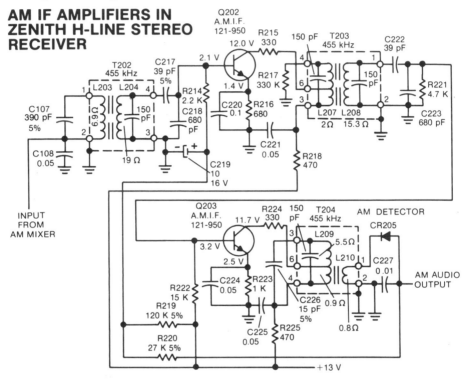

The schematic diagram for the FM IF amplifier section of the Zenith H-Line stereo systems is shown on the next page. IC 201 functions as an IF amplifier. IC 202 functions as an IF amplifier and FM detector (discriminator) so that these functions of IF amplification and detection are accomplished in two ICs. Also shown are parts of the muting circuit that quiets the receiver output when tuning from one station to another.

Transformer T201, at the input of the IF amplifier, is double-tuned with a center frequency of 10.7 MHz. In the second IC amplifier, there is a tuned inductor L205 with a center frequency of 10.7 MHz which is part of the detector and which will be described when we discuss FM receivers. Selectivity is obtained almost entirely by means of the two ceramic filters Y201 and Y202. The ceramic filters provide excellent frequency selectivity, and yet the basic circuits are quite simple and stable.

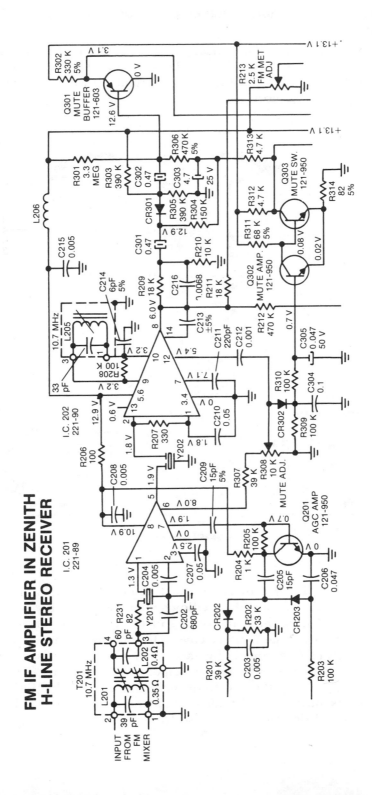

FM IF AMPLIFIER IN ZENITH H-LINE STEREO RECEIVER

Experiment/Application—Checking Characteristics of an IF Amplifier

IF amplifiers are a good example of a tuned RF amplifier. If the frequency is kept low (around 465 kHz), one can easily be set up in the laboratory. Suppose you hooked up the circuit shown below, keeping the leads reasonably short.

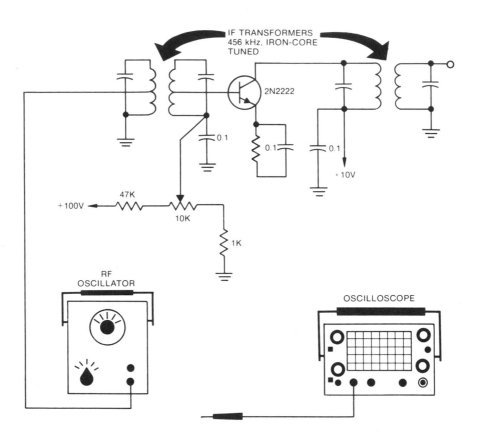

The RF oscillator provides a signal at 456 kHz, which drives the amplifier. You can connect the oscilloscope probe to the output to monitor it. You can now check the tuning of the input and output transformers by adjusting the tuning elements for maximum output. You can see the effects of gain control by varying the dc on the base of the IF amplifier transistor. The frequency response (or selectivity) can easily be checked by noting the two frequencies (on either side of 456 kHz) where the response is down to 70% of the maximum value. When you do this, you will find that the bandwidth is 10–20 kHz.

What RF Amplifiers Do

RF amplifiers have the same basic characteristics as IF amplifiers, however they are used in both transmitters and receivers. As you learned in the previous section, IF amplifiers are used in superheterodyne receivers, where they amplify the incoming radio signal at a frequency that is fixed to make the amplification more efficient and more selective.

In contrast, RF amplifiers are usually tunable and are tuned to the carrier frequency of the signal. By this time you are familiar with the fact that RF amplifiers amplify the carrier signal in the transmitter learning system and also amplify the incoming carrier signal in the receiver learning system. Since RF amplifiers are tunable, there must be a means for properly tuning them. While they are the same in operating principle, RF amplifiers for transmitters are usually *quite different* than for receivers. In transmitters, the object is to amplify to provide power to drive the antenna for transmission. In receivers, the object is to handle low-level receiver input signals and to amplify the signals with minimum additional noise. In either case, the RF amplifiers must be tunable or be sufficiently broad-banded to operate over the desired band of frequencies.

In this section you will learn about the differences between RF amplifiers that are used to produce low-level gain in receivers and those used to produce power gain in transmitters. You will also learn how such amplifiers are tuned and how the tuning is kept aligned. Remember that what you learned about IF amplifiers is also applicable to RF amplifiers. As with IF amplifiers, you will learn more about these in Volumes 3 and 4 when you study transmitters and receivers.

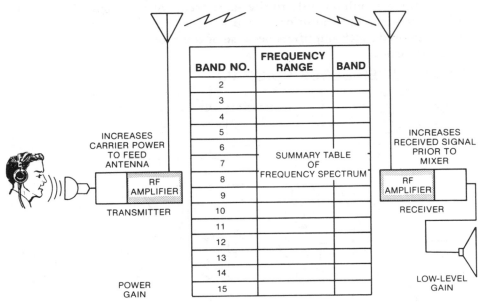

Typical Receiver RF Amplifier Stage

The RF amplifier circuit diagram shown here is typical of those used at the input of a radio receiver. A typical circuit configuration is the common-emitter arrangement that supplies maximum gain with simplicity of biasing arrangements. As you can see, the circuit closely resembles that for an IF amplifier.

Signals at a wide range of frequencies reach the antenna, but only those in the passband of the input tuned circuit reach the base. As for IF amplifiers, taps on the tuned circuit match the high impedance of the tuned circuit to the low impedance of the transistor base.

Base bias is set by the voltage divider formed by R1 and R2. Bias is selected for Class A operation because this is a small signal amplifier. Capacitor C4 bypasses the signal voltage around the emitter resistor R3, and thereby prevents degeneration that would lower gain. Capacitor C2 provides an ac (RF) ground for the base circuit of the RF amplifier. Capacitor C5 serves a similar function for the collector circuit.

The tuned circuit at the input consists of C1 and the secondary winding of T1. This parallel-resonant circuit presents maximum impedance to signals at the frequency to which it is tuned, and thereby transfers those signals at maximum amplitude to the base of Q1. Signals at other frequencies are met with a lower impedance and are reduced in amplitude. Similarly, C6 and the secondary winding of transformer T2 are tuned to that same frequency and present maximum impedance to it, so the amplifier has maximum gain at the signal or carrier frequency. The dashed line shown between C1 and C6 means that they are connected (ganged) together so that a single knob permits the operator to tune (track) them simultaneously to the same frequency. In some cases, two RF stages are used. In addition, as you will see, *field-effect transistors* (FETs) are often used as RF amplifiers because of their low-noise properties and their linearity.

RF AMPLIFIER BLOCK DIAGRAM

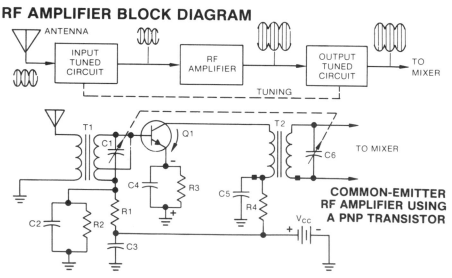

COMMON-EMITTER RF AMPLIFIER USING A PNP TRANSISTOR

Bandwidth and Selectivity

Although most of the selectivity in a modern radio receiver is determined by the IF amplifier, this factor is also important in the RF amplifier. In the standard U.S. commercial AM broadcast band, the signal bandwidth is about 10 kHz; in the standard U.S. commercial FM broadcast band, the signal bandwidth is about 200 kHz. It is very difficult to build an RF amplifier with this degree of selectivity that will also track in tuning over the entire band desired. Therefore, RF amplifiers act as *preselectors*, amplifying the RF signals in the vicinity of the carrier. Another important function of the RF amplifier is to provide discrimination against the *image frequency* in superheterodyne receivers. The image is generated during the mixing process because the mixer is equally sensitive to an input signal equal to the sum or difference between the local oscillator and the incoming carrier. The image is rejected by RF preselection, usually via an RF amplifier.

The bandwidth of an amplifier (or *bandpass* or *passband*, as it is also called) is the band of frequencies between the two 3-dB points on the response curve, as shown. Examination of the frequency response curve shows that the − 3-dB points are at 471 kHz and 529 kHz. The difference between these two frequencies is 58 kHz, and that is the bandwidth of this RF amplifier. As mentioned in the section on IF amplifiers, the bandwidth of a single-tuned circuit is equal to the resonant frequency divided by the Q of the tuned circuit. As you will remember, $Q = X_L/R$ or X_C/R. As the frequency increases, for a given Q it is apparent that the bandwidth will increase. For example, if a circuit has a Q of 10, then the bandwidth at 455 kHz is 45.5 kHz, while at 10.7 MHz, the bandwidth is 1.7 MHz.

Good selectivity and *poor selectivity* are terms that are meaningless unless the bandwidth required by the type of reception intended is also stated. What is good selectivity for a commercial broadcast FM receiver is poor (too wide) for a standard-broadcast band AM receiver. In many modern receivers where the overall received band is narrow, as, for example, in CB (26.965 to 27.405 MHz), the RF amplifier is fixed-tuned since an RF amplifier in this frequency range would have a bandwidth of at least this much in any event.

ILLUSTRATION OF THE DEFINITION OF BANDPASS

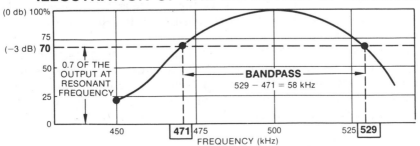

Ganged Tuning and Tracking

Ganged tuning becomes necessary when a number of tuned circuits are connected in cascade. For example, a receiver with one RF stage will have a tuned circuit at the RF input, the mixer input, and the local oscillator. The RF input and mixer interstage tuned circuits are tuned together in frequency, while the local oscillator is tuned so that there is a constant difference between its frequency (tuned circuit) and the RF and mixer frequency (tuned circuit). When this is done with a single knob, all the tuned circuits must change together so that all circuits remain tuned or the tuned circuits track each other to maintain tuning. Two methods are used commonly to accomplish the tuning.

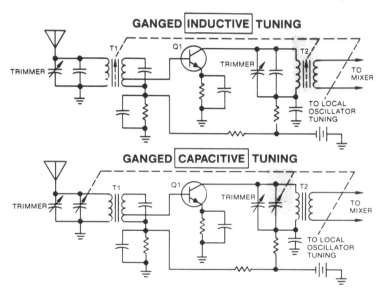

In the first method, each parallel-resonant circuit has a fixed tuning capacitor and is tuned by means of a core or *slug* of compressed powdered iron inside the inductor. The inductance of the coil (and therefore the tuning of the circuit) is changed as the core is inserted more or less deeply into the coil. Instead of having the depth of core insertion adjusted by means of a screw, as in IF transformers, all the cores are attached to a common crosspiece. When this crosspiece is raised and lowered, the cores are simultaneously moved in and out of their coils to simultaneously retune all the resonant circuits. Most tuners have an adjustment on the core of each tuned circuit to allow some small individual adjustment to equalize the tracking.

The other common arrangement for ganged tuning is to tune each resonant circuit by means of a variable capacitor. All of the variable capacitors are mounted on the same shaft and are tuned simultaneously by one knob. A small adjustable trimmer capacitor is connected across each tuning capacitor to equalize the tracking.

Sensitivity Improvement—Receiver Noise

The flow of electrons generates noise because the flow has a certain degree of nonuniformity or randomness. All circuits produce noise as a result. A mixer is much noisier than an RF amplifier; therefore an RF amplifier will amplify the signal with less noise than when a mixer is used alone, since the signal at the mixer is greater than it would have been without the RF amplification. Since the noise is inherent in the semiconductor device and its associated circuitry, it is very important to *match* the *input circuit* properly to the *amplifier* to get maximum signal in. The proper choice of the semiconductor device is very important since some are noisier than others and many are especially designed for low-noise applications. As stated above, an RF amplifier will produce less noise than a mixer. As you can see, if an RF amplifier has 0.1 μV of internal noise and a mixer has 1 μV of internal noise, and then if you have a signal of 5 μV, with an RF amplifier gain of 20, the signal-to-noise (S/N) ratio at the input to the RF amplifier is 50:1 (5/0.1) or 34 dB. Since the signal is amplified (also noise) by a factor of 20, the signal level is 100 μV or 40 dB greater than the mixer noise. The RF amplifier input noise is also amplified by 20 times from 0.1 μV to 2 μV, which masks the mixer noise. Therefore, from the above calculations, if the minimum useful signal level is at a signal-to-noise ratio of 10:1, it is necessary to have a 1 μV signal at the input to the RF amplifier. On the other hand, if the mixer only were used, the necessary input signal would be 10 μV for the same signal-to-noise ratio. Note that this is not due directly to gain but to the fact that the internal noise of an RF amplifier can be made lower than for a mixer. It should be noted that if a mixer is used directly, the noise introduced by the first IF stage can be significant because of the low signal output from the mixer.

RF AMPLIFIER NOISE REDUCTION

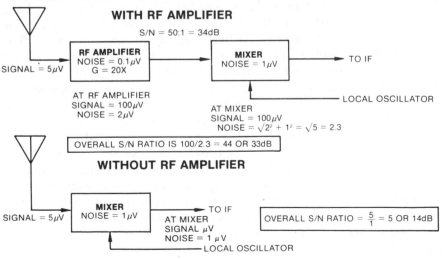

WITH RF AMPLIFIER

S/N = 50:1 = 34dB

SIGNAL = 5μV

RF AMPLIFIER
NOISE = 0.1μV
G = 20X

MIXER
NOISE = 1μV

TO IF

AT RF AMPLIFIER
SIGNAL = 100μV
NOISE = 2μV

LOCAL OSCILLATOR

AT MIXER
SIGNAL = 100μV
NOISE = $\sqrt{2^2 + 1^2} = \sqrt{5} = 2.3$

OVERALL S/N RATIO IS 100/2.3 = 44 OR 33dB

WITHOUT RF AMPLIFIER

SIGNAL = 5μV

MIXER
NOISE = 1μV

TO IF

AT MIXER
SIGNAL μV
NOISE = 1 μV

OVERALL S/N RATIO = $\frac{5}{1}$ = 5 OR 14dB

LOCAL OSCILLATOR

Typical RF Amplifier

The CB transceiver learning system described earlier has an RF amplifier to improve sensitivity. Since the band covered is very limited (26.965 to 27.405 MHz), the RF amplifier is fixed tuned. Specifications state that a signal-to-noise ratio of about 10 dB is available with a 0.5-μV signal; therefore, the noise generated in the RF amplifier is about 0.16 μV. Only a single RF stage is used. Image rejection is better than 12 dB. Another function of RF amplifiers is the reduction of radiation, via the antenna, of the local oscillator signal. This would cause considerable interference if not isolated. RF amplifiers provide this isolation.

CB TRANSCEIVER RF AMPLIFIER

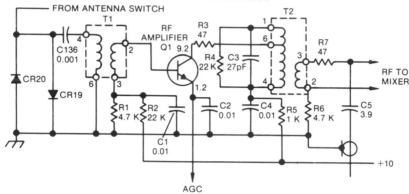

As shown, the signal input from the antenna is applied to transformer T1. Diodes CR19 and CR20, at the input, protect the input against strong signals. This protective circuit has a selectable bias level for normal (NOR) or local (LOC) receiving conditions. The RF amplifier stage (Q1) is conventional in design except for the fact that AGC bias is applied to the emitter rather than to the base of Q1. The base is fixed dc biased via resistance network R1–R2, and the transformer T1 secondary is ac grounded via C1. The emitter is bypassed (ac grounded) via C2. The output of Q1 is coupled to the mixer stage via transformer T2, which is also fixed-tuned to the RF frequency. Capacitor C4 provides a ground for transformer primary T2, while the collector voltage is applied via resistor R5. The secondary winding of T2 drives the mixer base circuit. The 47-ohm resistor in the collector circuit of Q1 is used to help stabilize the RF stage against unwanted spurious oscillation called *parasitic oscillations*. Resistor R4 (22 K) across the tuned circuit of T2 is used to lower the Q of T2 primary. This broadens its bandwidth (at some loss in gain) but allows the stage to be fixed-tuned—an important simplification.

Low-noise RF amplifiers can be made effectively using field-effect transistors (FETs). The metal-oxide semiconductor field-effect transistor (MOSFET), covered on the next page, provides stable gain and low noise, and operates efficiently at frequencies where standard transistors decline in performance.

MOSFET RF Amplifiers

The dual-gate MOSFET is well suited to RF amplifiers using automatic gain control, because the gain-control voltage can be connected to the second gate where it has minimum effect on amplifier operation except for gain control.

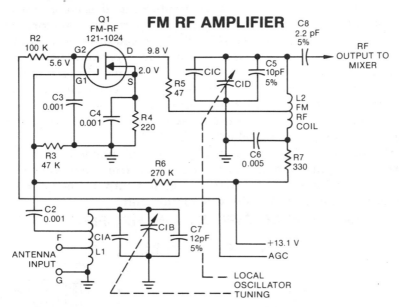

The RF amplifier shown here is from the FM receiver in the Zenith H-Line stereo system and employs a dual-insulated gate MOSFET (Q1). As shown in the detail diagram, back-to-back zener diodes are diffused into the MOSFET between each gate and the source. These protect the gate structure against high-voltage transients by bypassing any such voltage peaks around the gate, but are effectively open circuits under normal signal conditions.

Associated with the RF stage are an oscillator and mixer tuning capacitors that are ganged together with the tuning capacitor (C1B) in the RF stage. The tuning capacitor C1D is used to tune the interstage between the RF, amplifier, and the mixer. The voltage divider consisting of R3 and R6 supplies fixed bias to gate 1 (G1). Bias for gate 2 (G2) is supplied via an AGC amplifier not shown. Drain operating voltage is supplied from a + 13.1-volt source. The output of the RF stage is applied to the mixer (not shown), where it is mixed with the output of the local oscillator tuned by another section of C1 to produce a 10.7-MHz IF signal.

The tuned antenna circuit (consisting of coil L1, tuning capacitor C1B, trimmer C1A, and fixed capacitor C7) is coupled to G1 via C2 from a tap on L1. This circuit couples the FM RF signal into the gate of Q1. A similar tuned-circuit arrangement is used at the output where the drain is connected to a tap on L2 through suppressor R5. This in turn is capacitor-coupled to the mixer.

Integrated Circuit RF Amplifiers

Shown here is a IC amplifier designed specifically for use in RF amplifier applications. The internal circuit arrangement is shown together with the external outlines and connections as an RF amplifier circuit.

This IC is a general-purpose RF amplifier used as either a wide- or narrow-band amplifier. It also can be used as a mixer, detector, limiter, and modulator. The circuits consist of a balanced differential amplifier (Q1 and Q2) that is driven from a controlled constant-current source (Q3). These ICs are given special versatility by connecting useful internal points to outside terminals where they can be connected to external components to alter the basic configuration. By means of such external modifications, it is possible to use these ICs as push-pull amplifiers, or as single amplifiers in parallel channels.

Shown on this page is the arrangement for connecting this IC in a differential amplifier configuration operating from sources of positive and negative voltage. As shown, the input and output tuned circuits are essentially the same as for a discrete-circuit unit.

INTEGRATED-CIRCUIT RF AMPLIFIER

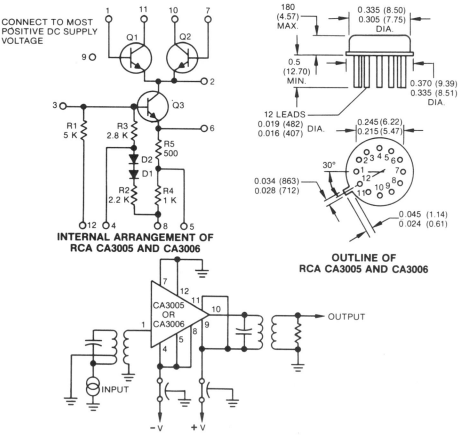

INTERNAL ARRANGEMENT OF RCA CA3005 AND CA3006

OUTLINE OF RCA CA3005 AND CA3006

RCA CA3005 AND CA3006 CONNECTED AS DIFFERENTIAL AMPLIFIER OPERATED FROM DUAL + AND − SUPPLIES

Introduction to RF Power Amplifiers

In a radio transmitter, an oscillator generates an RF carrier signal that is amplified and fed into an RF power amplifier, which steps it up to a power level sufficient to enable the antenna to radiate it. RF power amplifiers are used to do this. Usually, for maximum frequency stability it is necessary to operate the RF oscillator at low power levels. The power amplifier must usually put out a signal at considerably higher power level. In some cases, the oscillator operates at a lower frequency (a harmonic sub-multiple) so that not only amplifiers but frequency multipliers are necessary. The only difference between these is their biasing arrangement and the fact that a frequency multiplier has its output circuit tuned to a harmonic of the input signal. Thus, a doubler has its output tuned to twice the input frequency, a tripler is tuned to three times the input frequency, etc. Usually, these frequency multipliers operate at low power because they are relatively inefficient. Typically in familiar applications, the final power amplifier must deliver something between 0.25 watt and 100 kW. The amount of power radiated is usually set by the FCC. For a hand-held radio transmitter that will transmit a message about one-quarter of a kilometer, the power amplifier will put out about 0.25–1 watt of RF power. In CB radios where the signal is transmitted for perhaps up to 5 miles with some degree of reliability, the power amplifier will deliver up to 4 watts. In commercial, civil, and military communications intended for a small community or town and the immediate surroundings, the RF power amplifier output will range from 25 watts to several kilowatts. For commercial broadcasting over great distances, the RF power amplifier output may be up to 100 kW or more. High-power RF power amplifier stages use vacuum tubes because semiconductors cannot handle the necessary power.

To accomplish this, the RF amplifier must contain one or more *immediate* intermediate power amplifiers (IPA) to isolate or *buffer* the oscillator from the power amplifier and then multiply and/or amplify the oscillator output to that required for the application. Here we only touch on the subject of transmitters; we will consider it in great detail later in Volume 3 of this series.

BASIC THREE-STAGE TRANSMITTER

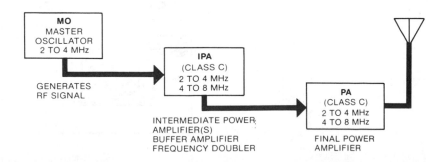

Introducing the Oscillator

Oscillators are circuits that generate an ac signal output when supplied with appropriate operating voltages. An oscillator generates an ac signal at a desired frequency. The important characteristics of these ac signals are the stability of the output frequency and the stability of the output signal level. The output waveform usually is in the form of a sine wave. Some oscillators produce a pulse, square, triangular, or other types of waveshapes.

Oscillators are amplifiers that have positive (in-phase) feedback from output to input at the frequency of oscillation. Oscillators are usually described by the frequency-control element. In many cases, the frequency of oscillation is determined by an LC resonant circuit. This is particularly convenient for tunable oscillators in receivers when a variable capacitor ganged to the other tuning capacitors can be used. These oscillators are called *LC oscillators*. In some oscillators (particularly audio types), the frequency of oscillation is determined by a filter composed of resistors and capacitors; such a circuit is known as an *RC oscillator*. For precise frequency control, quartz crystals are used as the frequency-control elements; such circuits are called *crystal oscillators*.

An important type of oscillator is the *frequency synthesizer*. These are based on analog or digital synthesis of the desired frequency either directly or via reference oscillators. Such systems are used in communications systems, TV receivers, etc., and will be discussed in Volumes 3 and 4. Here we will discuss basic oscillator configurations. As you know, the oscillator is the source of the RF carrier signal that is amplified, modulated, and coupled to the antenna for radiation into space. From your knowledge of receivers, you know that an oscillator is the source of the signal that is mixed with the received RF carrier signal so that an IF signal can be formed for more efficient amplification.

In this section, you will learn primarily about oscillators that generate sinusoidal waveforms. Oscillators generating other waveforms will be discussed in a later volume when data transmission is discussed.

OSCILLATORS ARE USED IN

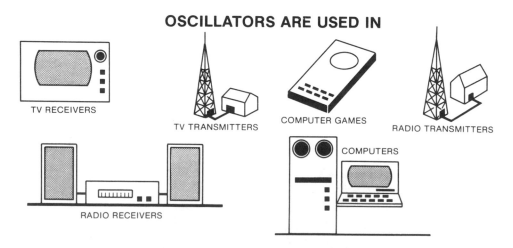

TV RECEIVERS

TV TRANSMITTERS

COMPUTER GAMES

RADIO TRANSMITTERS

RADIO RECEIVERS

COMPUTERS

What Oscillation Is

When something moves back and forth in a uniform manner, it is said to be *oscillating*. A violin string oscillates when a bow is drawn over it. A pendulum swinging in a grandfather clock oscillates. Perhaps the most familiar type of oscillator is a child on a swing or a swinging weight on a string. When the weight reaches the extreme left-hand side of its swing, it momentarily comes to rest, and all of its energy has been converted to *potential energy*—stored energy that is capable of doing work. Now the weight starts to swing back to the right. Its speed of motion is slow at first but gradually increases as the potential energy is converted to kinetic energy (moving). The speed of motion to the right is a maximum at the bottom of the arc. At this time all of the potential energy has been converted to *kinetic energy*—the energy of a mass in motion which is capable of doing work. This kinetic energy pushes the weight to the right and lifts it higher and higher against the force of gravity. As the weight swings upward, its speed of motion becomes slower and slower, because the kinetic energy is being expended in lifting it upward and stored as potential energy. Finally, all the kinetic energy has been converted back to the form of potential energy at the right-hand top of the arc.

If we plot the speed of motion of the weight against time, considering motion to the right as positive, we obtain the familiar positive half of a sine wave. If we now continue the plot, we complete the sine wave. As the weight moves back and forth, it traces out a series of sine waves or oscillations of constant frequency, as shown.

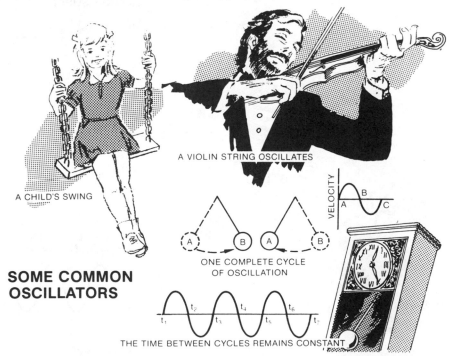

A VIOLIN STRING OSCILLATES

A CHILD'S SWING

ONE COMPLETE CYCLE
OF OSCILLATION

VELOCITY

**SOME COMMON
OSCILLATORS**

THE TIME BETWEEN CYCLES REMAINS CONSTANT

What Oscillation Is (continued)

You know from personal observation that the motion of the weight eventually runs down. That is, the weight swings through smaller and smaller arcs, each requiring the same time for a complete cycle, until all motion ceases. The reason for this is that not all the energy of one form is converted to the other form; a small bit of it is wasted in each cycle by the friction in the string, air resistance, etc. If you plot this motion, you see that the amplitude of successive sine waves gradually becomes smaller, although the time interval remains the same. This is called a *damped wave*, or damped oscillation.

You also know from observation that all you have to do is to give the weight a very slight push at the moment of direction reversal at the extreme right or left of its arc of motion. This push is *in phase* with the motion that is taking place and provides just enough extra energy to overcome the frictional losses.

Now you know that to *maintain* oscillation you must supply enough *additional* energy to overcome losses, and it must be supplied *in phase* with the motion. This is the basis on which all oscillators operate. The system must have the capability to overcome any losses so that the oscillation can be sustained. Continuous oscillations are therefore called *sustained oscillation*. In electronic systems, it is the amplifier that provides the gain necessary to overcome losses and sustain oscillation.

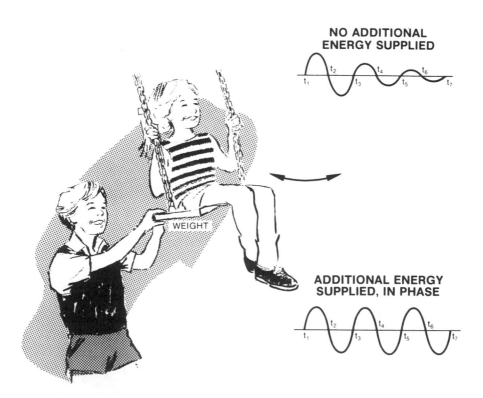

NO ADDITIONAL
ENERGY SUPPLIED

WEIGHT

ADDITIONAL ENERGY
SUPPLIED, IN PHASE

Electronic Circuit Oscillation

The action that takes place in an LC electronic oscillator is very similar to that which takes place with a swing. In a resonant circuit consisting of a capacitor and inductor connected in parallel, energy is alternately stored in the electrostatic field of the capacitor and in the magnetic field of the inductor. When a charge is stored in a capacitor, an electrostatic field is generated. Energy has to be expended to build up this charge, but this charge and field are capable of doing work in the form of forcing a current to flow into an external circuit. This energy is similar to the potential energy that is accumulated when a weight is lifted up against the force of gravity.

Similarly, when current flows through an inductance, it builds up a magnetic field, and energy is stored in this magnetic field as its magnitude is increased. The magnetic field resists change in current flow. It retards the buildup of current as it is being formed, but it tends to maintain the flow of a current that is decreasing. Energy stored in the magnetic field is similar to the kinetic energy referred to previously.

Suppose we had a circuit such as that shown in the illustration. Initially, the capacitor is charged from the battery by placing the switch arm to the left. Then the switch arm is thrown to the right and the charged capacitor is connected across the inductor. As shown at point 1 in the diagram, the current cannot rush into the coil, because as soon as a current starts to flow, a magnetic field begins to form around the coil. The magnetic field induces a voltage in the coil that opposes the current flow. From your study of *Basic Electricity*, you recognize this as a counter-emf. As the capacitor continues to discharge, the magnetic field becomes stronger. When the capacitor is discharged, its negative voltage has declined as shown at point 2 on the diagram. By the time the capacitor has completely discharged, all of its energy has been converted into magnetic-field energy.

STARTING THE CYCLE OF OSCILLATION

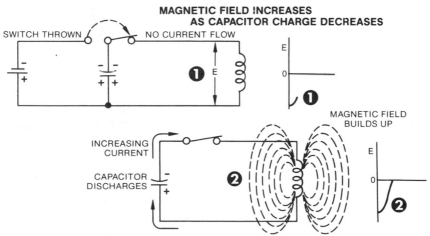

MAGNETIC FIELD INCREASES
AS CAPACITOR CHARGE DECREASES

SWITCH THROWN   NO CURRENT FLOW

MAGNETIC FIELD
BUILDS UP

INCREASING
CURRENT

CAPACITOR
DISCHARGES

Electronic Circuit Oscillation (continued)

Now the current through the coil begins to decrease, and the magnetic field begins to collapse around the coil. The collapsing magnetic lines cut across the turns of the coil and induce a voltage across the inductor. This counter-emf is opposite in polarity to the original voltage across the capacitor, and the voltage on the upper plate starts to become positive, as shown at point 3 on the diagram.

The energy of the collapsing magnetic field is used to produce a negative charge on the lower capacitor plate. By the time the field has completely collapsed, all the magnetic-field energy has been returned to the capacitor as an electrostatic field, and voltage across the capacitor is now the reverse of the original, as shown at point 4 on the diagram.

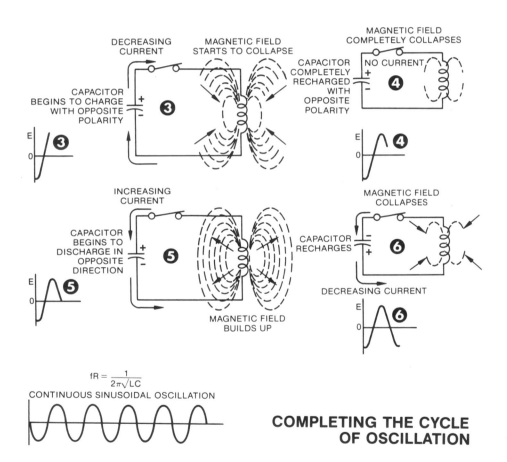

$$fR = \frac{1}{2\pi\sqrt{LC}}$$

CONTINUOUS SINUSOIDAL OSCILLATION

COMPLETING THE CYCLE OF OSCILLATION

Electronic Circuit Oscillation (continued)

The process now continues in reverse. Current starts to flow from the lower plate, through the coil, to produce a magnetic field opposite from the original. The magnetic field generates a counter-emf that resists current flow. When the capacitor is completely discharged at point 5, the current is zero. Now the magnetic field begins to collapse and generates a counter-emf that forces current flow to continue in the same direction to restore the original charge condition on the capacitor, just after point 6 in the diagram. This process repeats itself indefinitely if there are no losses. Thus, the voltage across the LC circuit is a sinusoidal oscillation with a frequency $(f_R) = \dfrac{1}{2\pi\sqrt{LC}}$.

If there were no resistance in the circuit or leakage in the capacitor, oscillation would continue indefinitely. However, resistance and leakage cannot be eliminated completely, and the damped oscillation seen earlier is produced. Oscillation can be made to continue indefinitely by replacing the lost energy to the LC circuit (called a *tank circuit* because it can store energy as described). This energy must be put back into the circuit at just the right moment, so that it will give the required extra *push* at the time when it can act to continue the current flow.

ELEMENTS OF AN ELECTRONIC OSCILLATOR

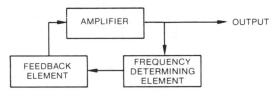

One way of supplying this electrical push would be to use a switch to connect a source of voltage across the capacitor just at the moment that the capacitor reaches its full charge. In this manner, the original charge will be exactly restored, and the push will be at exactly the time required to continue the existing oscillation. Obviously, it is impractical to use a switch to aid the required energy for oscillations taking place at thousands and even millions of times per second. The way to do it is to connect the LC tank circuit to an amplifier and *positive-feedback loop*, as shown in the diagram. This arrangement provides continuous oscillation.

Frequency stability is usually important in oscillators. In all cases, it's effects on the tank circuit or frequency-determining element that lead to frequency change.

Oscillator Configurations

There are many common configurations of oscillators. Generally they are named on the basis of the tuned-circuit and feedback configuration. In all cases, the oscillator consists of an amplifier, a frequency-determining element, and a feedback path. The sketches below show some typical oscillator configurations.

The tapped winding *Hartley oscillator* is one of the most popular LC oscillators; however, the *Clapp oscillator* is often found where highly stable operation is necessary. The *crystal oscillator* will be described later in this section in more detail, but is shown here to illustrate how all oscillators are *similar* in nature and operation, and the choice of which to use is mainly involved with how easily tuned or ganged the oscillator tuned circuit might be or how conveniently the output can be taken for a given application.

OSCILLATOR CONFIGURATIONS

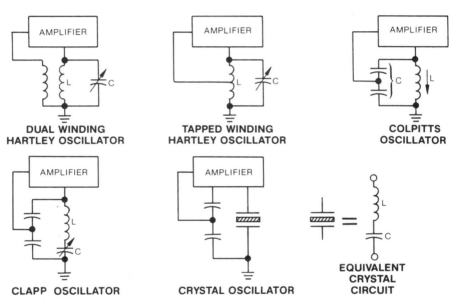

Generally, oscillators are operated Class A, at very low power levels, for improved stability. Buffer stages (RF amplifiers that isolate the oscillator from load) are used to increase the power level when necessary.

Oscillators start typically because of a small disturbance (turn-on transient or noise) that is regenerated through the positive-feedback loop and the tuned circuit. This signal builds up at the resonant frequency of the tuned circuit. Stable operation occurs when further buildup of the signal is prevented by limiting (clipping) in the amplifier.

Oscillators will be discussed further in subsequent volumes in terms of their role in operating transmitters and receivers. There, also, we will discuss frequency synthesizers that generate operating frequency *indirectly*. These are characterized by being flexible and stable in operation.

Hartley Oscillator

As a typical example of an oscillator, the Hartley oscillator will be described in some detail. You will find that the tapped-coil version of the Hartley oscillator is very popular and commonly used because it is simple and stable and uses few parts.

In the circuit shown, Q1 operates as an amplifier. The base-emitter circuit and the tuned-circuit LC provide the elements necessary for building an oscillator. The tuned circuit consists of L1 and C2. Notice that L1 is tapped (usually about one-third the way up from the bottom). L1 acts as an autotransformer; thus, it provides a voltage gain from the tap to the upper end. The transistor provides current gain (as a common-collector configuration) so that the signal from the emitter drives the autotransformer, providing a voltage gain to the top of the transformer. Since the common-collector configuration has 0-degree phase shift from base to emitter, and the coil provides 0-degree phase shift, the feedback phase is positive.

The capacitor C1 is a dc blocking capacitor that has low impedance at the operating frequency. Base bias is provided by resistor R1. In some cases, an emitter bias is added via R_E bypassed by C_E to improve stability of operation.

One of the advantages of the Hartley oscillator is that the signal can be taken from the collector or the collector can be connected to RF ground (bypassed by a capacitor) and the output taken from the emitter. For the collector output shown, the collector is fed dc via an inductor called a *radio-frequency choke (RFC)*. Capacitor C3 provides an ac bypass for any ac that might appear at the cold end of the RFC. The output is taken through coupling capacitor C4 to avoid interfering with the dc operating conditions of Q1.

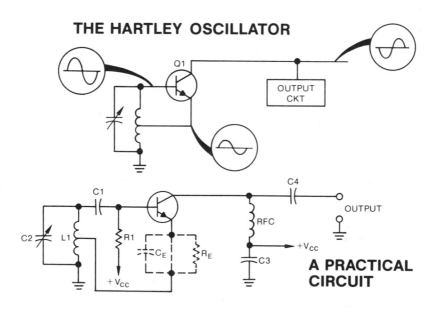

THE HARTLEY OSCILLATOR

A PRACTICAL CIRCUIT

Colpitts Oscillator

The Colpitts oscillator shown here is adaptable to a wide range of frequencies and can have better frequency stability than the Hartley oscillator. The distinguishing feature of the Colpitts is that feedback is obtained by means of a tap between two capacitors connected in series. This arrangement lends itself to the use of *permeability tuning* (shown by the dotted core and arrow on L1) and fixed capacitors, although a tuning capacitor (C6) can also be used across L1, as shown in the diagram.

The tank circuit consists of L1 together with C1 and C2 (and C6 if used). Comparison with the diagram of the Hartley oscillator on the previous page shows that the two circuits are very similar. As the tank circuit oscillates, its two ends are at equal and opposite voltage, and this voltage is divided across the two capacitors. The signal voltage across C2 is connected to the base, which is part of the signal from the collector. The collector signal is applied across C3 as a feedback signal whose energy is coupled into the tank circuit to compensate for losses. This feedback signal is in phase with the base signal, because the common-emitter stage introduces a 180-degree phase shift (base to collector), and the signal at the bottom of the coil is 180 degrees out of phase with the signal at the top (base end) of the coil.

Resistor R2 in the emitter circuit coupled with resistor R1 in the base circuit provide for proper biasing of Q1. Capacitor C1 provides a dc block for the base bias, while allowing the ac signal to pass to the tank circuit. The output is usually taken from the collector. The RFC functions as before to provide a high impedance to ac while allowing free flow of dc into the collector circuit.

BASIC COLPITTS OSCILLATOR

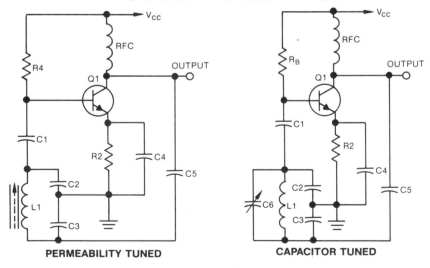

PERMEABILITY TUNED   CAPACITOR TUNED

Clapp Oscillator

There are many variations of the basic LC oscillator, and many of them have different names. All of them operate by using an LC resonant circuit to establish a tank circuit that oscillates at the resonant frequency. Then a transistor amplifier accepts a portion of the tank signal, amplifies it, and feeds it back into the tank circuit in phase with the existing tank signal. This feedback compensates for losses in the tank circuit and permits oscillation to continue. If a circuit has these features, it is an LC oscillator.

Many pages could be spent in describing additional oscillators. Little benefit would be gained because they all operate according to the above principles. An important LC oscillator is the *Clapp oscillator* because it has great frequency stability.

The Clapp oscillator is a simple variation of the Colpitts oscillator. The total tank capacitance is the series combination of C1 and C2. The effective inductance of L is varied by changing the net reactance by adding and subtracting capacitive reactance via C3 from inductive reactance from L1. Usually C1 and C2 are much larger than C3, but L1 and C3 are series resonant at the desired operating frequency. C1 and C2 determine the feedback and are so large compared to C3 that tuning has almost no effect on feedback. Signal output is usually taken from the collector. The Clapp oscillator configuration is very stable because of the series-tuned circuit with the circuit stray capacitances swamped by C1 and C2 so that the frequency is almost entirely determined by L1 and C3, even as the external circuit varies with time. The output from the Clapp oscillator can be taken from the emitter or the collector. In any event, it is best to lightly load the oscillator (take minimum power out) so that the frequency stability is least affected. Clapp oscillators are commonly used as stable variable-frequency oscillators (VFO) in applications where fixed-frequency operation is undesirable.

CLAPP OSCILLATOR

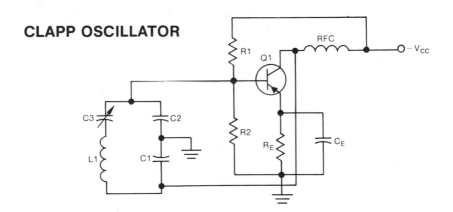

Crystal Oscillator

The crystal oscillator is an important oscillator that is almost invariably used as a fixed-frequency source because it provides the extreme frequency stability required in many applications by FCC regulations and/or the circuit design.

In your study of *Basic Electricity* you learned that the compression of certain types of crystalline materials produces an electric charge on their surfaces; this is known as the *piezoelectric effect*. Conversely, when a charge is put across these crystals, they will expand and contract.

This effect is used to stabilize the frequency of an oscillator. In such applications, an appropriate slice of quartz or other suitable crystalline material is placed between metal plates (or has an electrode plated on its surfaces) and mounted in a holder. The frequency of operation is controlled by the dimensions of the crystal and the way it is cut with respect to the growth of the original crystal; it must be cut to accurate dimensions and ground to the exact required thickness. Most crystals nowadays are supplied in sealed holders.

The crystal has a very high Q because of its natural frequency of mechanical vibration. Thus, the crystal acts as a very-high-Q series-resonant circuit. Crystal oscillators can be operated in either a parallel or series resonant mode. The frequency of vibration is determined by the crystal thickness; the thinner the crystal, the higher the frequency. Thus, for most applications, the crystal oscillator is operated at low power because added power in the crystal circuit produces greater vibration amplitudes; it is possible to fracture a crystal when excessive power is used. Also, because the crystal may become excessively thin at frequencies much above 20–30 MHz, frequency multipliers are commonly used to obtain higher frequency operation. You will learn about these schemes in later volumes.

It is the very high Q of the crystal as a resonant element that makes it a stable oscillator. Even quartz, which is very stable, changes its dimensions somewhat with temperature. To minimize even these changes, for some very stringent applications (broadcast-transmitter frequency control) an *oven* is used to hold the crystal at a constant temperature.

THE CRYSTAL

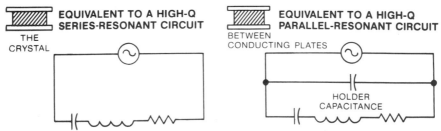

EQUIVALENT TO A HIGH-Q SERIES-RESONANT CIRCUIT

THE CRYSTAL

EQUIVALENT TO A HIGH-Q PARALLEL-RESONANT CIRCUIT

BETWEEN CONDUCTING PLATES

HOLDER CAPACITANCE

Colpitts Crystal Oscillator

The Colpitts crystal oscillator shown is extremely stable and adaptable for use over a wide range of frequencies. The fact that it is a Colpitts circuit is immediately recognizable from the tap between two capacitors connected across the tank circuit to supply feedback.

In this circuit the crystal is the complete tank circuit, both the inductor and the capacitor. As you know, the crystal resonant frequency is determined by the thickness of the crystal, and the resonant current is determined by this. The crystal inherently has a very high Q. Therefore, oscillation is obtained at a single frequency. The Colpitts crystal oscillator uses the crystal in the parallel-resonant mode. The purpose of C1 and C2 in this circuit is to provide a means for adjusting feedback for most efficient operation. These two capacitors also have a slight effect on the frequency of oscillation. Oscillation frequency can be adjusted very slightly (typically $\pm$ 0.005%) by adjusting the capacitance across the crystal. Any larger frequency change can be made only by using a different crystal made for use at the new fequencey.

When a good sine-wave output waveform is desired, the circuit is operated Class A. When maximum efficiency is the goal or when a high harmonic content is desired for frequency multiplication, the circuit can be biased for Class AB.

In the illustrated typical circuit, the voltage divider formed by R1 and R2 supplies base bias for Q1. Capacitor C3 bypasses the base of Q1. The transistor signal output voltage is divided across C2 and C1, both or either of which can be made adjustable to permit control of the amount of feedback to the emitter. In a common-emitter circuit such as this, the emitter and collector signals are in phase, and consequently the feedback has the positive relationship required to produce oscillation. The crystal operates in the series-resonant mode and looks like a low resistance at its resonant frequency.

If a crystal oscillator fails to operate and transistor bias seems correct, damage to the crystal can be suspected. The quickest check is to remove the suspected crystal and to substitute one *known* to be good.

COLPITTS CRYSTAL OSCILLATOR

Other Crystal Oscillator Circuits

Almost any oscillator configuration can be used with a crystal to form a crystal oscillator. The *Pierce oscillator* shown here is almost identical to the Colpitts circuit. The feedback path is from collector to base. The values of C1 and C2 are selected to form a capacitive divider that supplies the correct amount of feedback. As shown, one or both of these capacitors can be made variable to optimize feedback.

The second Pierce oscillator configuration is often used because it is simple and provides a convenient low impedance point for signal takeoff. The RFC provides emitter isolation while providing a dc path for the emitter current.

The *Butler oscillator* uses the series-resonance of the crystal. Thus, Q1 and Q2 are coupled together at the series-resonant frequency via their emitters. Q1 functions as an emitter follower that couples the signal from the collector of Q2 into the emitter of Q2 via series resonant crystal XTAL 1. In this way, a positive-feedback loop is formed. Resistors R1–R2 and R3–R4 are base-biasing resistors. R5 and R6 are emitter resistors. Capacitor C1 bypasses the collector of emitter follower (common collector) Q1. Feedback from the collector of Q2 is supplied to the base of Q1 via C2. Since Q2 is operated as a common-base stage, its base is bypassed by C4. Capacitor C3 and L1 form a tank circuit, tuned to the frequency of oscillation. The advantage of this circuit is that it will operate over very wide frequency ranges and is commonly used for low-frequency oscillators. Frequency multiplication can be accomplished by using an *overtone crystal* (designed for high harmonic content) and tuning the tank to the desired harmonic frequency.

CRYSTAL OSCILLATOR CIRCUITS—PIERCE CRYSTAL OSCILLATORS

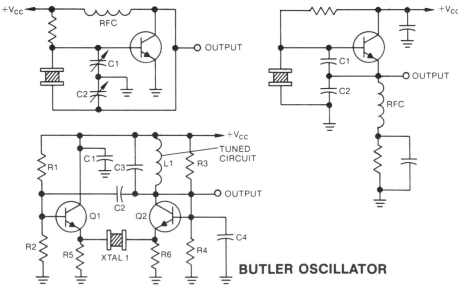

BUTLER OSCILLATOR

Crystal Oscillators in Synthesizers

Many types of oscillator circuits have been developed for use in special applications and in the higher frequency bands. Equipment designed to operate very stably on many frequencies on a selectable basis would require many crystals. For example, a 40-channel CB would require 40 crystal pairs (80 crystals) to cover all channels. The crystal pairs are needed to set the frequency of the transmitter and also to set the frequency of the local oscillation in the receiver. To avoid this problem, such equipment uses *frequency synthesizers* and *phase-locked loops* that actually use one or more oscillators working together to produce the necessary selection of very stable frequencies with only a very few crystals. You will learn about these in detail in the volumes on receivers and transmitters. Right now, it is important to get an initial familiarity with oscillator-circuit diagrams as they appear in schematic diagrams prepared by equipment manufacturers.

As an example, examine the circuit diagram of one of the crystal oscillators used in the CB radio learning system, the Messenger Model 4170. In this radio, the output of this oscillator is used in a synthesizer to produce the required selectable CB frequencies. For the present, consider only the example of this particular crystal oscillator.

The purpose of this oscillator is to produce a table output frequency for the synthesizer. The frequency of oscillation is 21.855 MHz as delivered by the crystal Y202. The circuit is a modified Pierce oscillator with the output taken from the emitter. Capacitor C202 provides feedback from emitter to base. The second capacitor (emitter to ground) normally seen in this configuration is supplied by the stray circuit capacity and the output. L202 (6.8 μH) is an RFC to isolate the RF signal while providing a dc connection to the emitter-bias circuit. As shown, the collector is at ac ground because of bypass capacitor C225. As you can see, this oscillator is very similar to the crystal oscillator described previously. As is commonly done, the output is taken from the emitter as a convenient low impedance point. In operation, the output from this oscillator is mixed with the synthesizer output to produce the transmit frequency. A similar oscillator is used to produce the local-oscillator signal, as you will see when you study the complete synthesizer in Volume 4 of this series.

DETAILS OF CRYSTAL OSCILLATOR

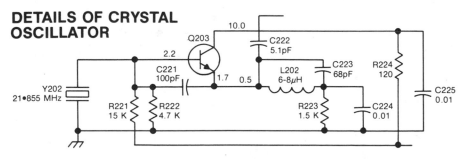

RC Oscillators

An RC oscillator consists of one or more transistor amplifier stages with provisions for feeding back a portion of the output signal in phase with the signal at the input via an RC network. These RC oscillators are almost invariably used as audio oscillators, because LC oscillators would become highly unwieldy at audio frequencies. The frequency of oscillation is determined by an RC filter at the point where the phase of the loop is exactly 360 degrees (in phase), typically 180 degrees in the amplifier and 180 degrees in the feedback loop.

The illustration shown here summarizes the concept of a *phase-shift oscillator*. Detail A shows the circuit of a stabilized common-emitter amplifier stage. Its operating point and current gain for small signals have been stabilized sufficiently to permit it to operate reliably over a collector voltage range of 2 to 24 volts.

Detail B shows a phase-shift network consisting of three 0.047-μF capacitors and three 4.7-K resistors. In such a network, each capacitor and resistor pair produces a phase shift on the input signal. At a particular frequency determined by the values of R and C, the signal is shifted 60· degrees by each RC pair and produces a total phase shift of 180 degrees for that frequency.

In detail C, the phase-shift network is used to feed back signal from the output of the amplifier to the input. The common-emitter stage produces a 180-degree phase shift from input to output. In addition, the feedback network produces a 180-degree phase shift from input to output. The result is that the signal fed back to the input is 360 degrees out of phase, or exactly in phase, with the signal on that base. The requirement for oscillation has been met and takes place at the frequency for which the phase shift is 360 degrees.

With the network illustrated at A and B, the frequency of oscillation is about 400 Hz. The addition of the 5.0-K variable resistor makes the circuit adjustable from about 200 to 400 Hz. Much wider frequency adjustment ranges are available by using ganged potentiometers with one section in each leg. Switches are also commonly used to provide for changing the network RC values to provide for wide frequency coverage.

RC PHASE-SHIFT OSCILLATOR

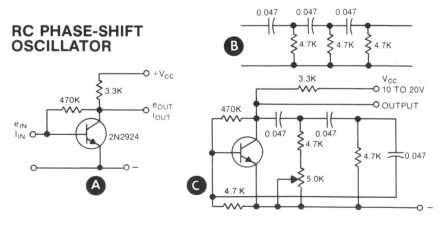

RC Oscillators (continued)

The oscillator described here is a phase-shift oscillator that has high-frequency stability. It uses an emitter-follower (noninverting) stage to drive the base of the common-emitter (inverting) amplifier stage.

The bridged-T network that determines the operating frequency operates as described below. This network in the collector of the amplifier stage consists of two RC filters connected in parallel as shown. The upper RC filter consists of the two 0.022-µF capacitors in series, with their junction connected to ground through the 4.7-K fixed resistor and the 5-K variable resistor in a T-configuration. Because the reactance of a capacitor decreases as frequency increases, this branch of the network acts as a *high-pass filter*. The lower T network consists of two 47-K resistors connected in series, and their junction is connected to ground through a 0.047-µF capacitor. Because capacitive reactance decreases as frequency increases, this branch of the network acts as a *low-pass filter*.

These two T-filters produce opposite phase shifts as a function of frequency. This, in turn, produces a very narrow band filter at the frequency where these effects cancel out. At this frequency, the attenuation is minimal, and the phase shift is 180 degrees (see figure). With respect to the attenuation (*not* phase shift), the circuit behaves much like a series-resonant circuit. Thus, the filter passes only the narrow band of frequencies in this bandpass with the required 180-degree phase shift needed for positive feedback. Oscillation can thus take place only at this frequency.

HIGH-STABILITY BRIDGED-T OSCILLATOR

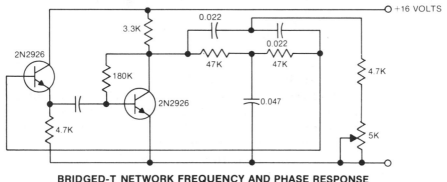

BRIDGED-T NETWORK FREQUENCY AND PHASE RESPONSE

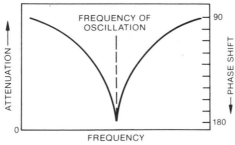

Experiment/Application—Colpitts Crystal Oscillator

Before we move onto the review of video/IF/RF/amplifiers and oscillators let us perform our last Experiment/Application of this volume. As you know, a circuit that exhibits a very high frequency stability is the crystal-controlled oscillator. You can check out the circuits for simple crystal oscillation by using a readily available 100-kHz crystal in a Colpitts oscillator configuration. Suppose you connected the circuit as shown below.

OSCILLATOR

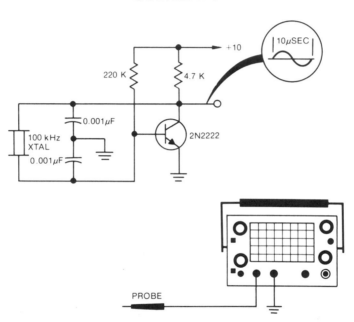

If you examine this circuit using the oscilloscope probe, you will see that the output is at 100 kHz and does not vary appreciably. You will also note that the signals on the base and collector are in phase as a necessary condition for oscillation to take place. You can vary the feedback by changing one of the 0.001-µF capacitors. Try ones rated at 0.01 and 0.0001 µF. You observe that oscillation stops with the 0.0001-µF capacitor, and may stop with the 0.01-µF one, oscillate at a slightly different frequency, or *squeeg*. This occurs when there is too much feedback, and consists of intermittent periods of oscillation such that the oscillation builds up to a high value and then cuts off and starts again.

Review of Video/IF/RF Amplifiers and Oscillators

VIDEO AMPLIFIERS PROCESS THESE KINDS OF SIGNALS

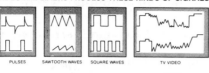

PULSES   SAWTOOTH WAVES   SQUARE WAVES   TV VIDEO

1. VIDEO AMPLIFIERS are essentially like audio amplifiers except for having a wider passband, typically from 30 Hz to 10 MHz.

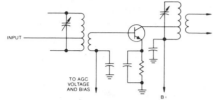

INPUT

TO AGC
VOLTAGE
AND BIAS

B

2. IF AMPLIFIERS are fixed-tuned RF amplifiers designed to amplify a specified band of frequencies on a highly selective basis.

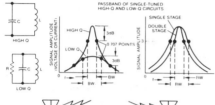

3. SELECTIVITY is a measure of a circuit's ability to reject undesired signals. The bandwidth of a tuned circuit determines its selectivity, and this in turn is determined by the circuit Q.

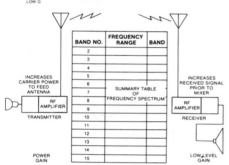

4. RF AMPLIFIERS are like IF amplifiers except that they are tunable, so they can preselect the incoming signal. RF amplifiers also preamplify the signal before the mixer so that the receiver is more sensitive, because RF amplifiers can have lower internal noise than mixers.

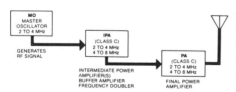

5. RF POWER AMPLIFIERS are designed to provide power. In principle, they are like RF amplifiers except for the fact that power output at one frequency is usually the most important consideration.

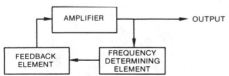

6. OSCILLATORS are amplified with positive feedback to a tuned circuit to cause a sustained oscillation at a frequency determined by the tuned circuit.

Self-Test—Review Questions

1. How does a video amplifier differ from an audio amplifier? Where are video amplifiers used?
2. What is the basic difference between an IF and RF amplifier? Why are IF amplifiers used?
3. Define center frequency and passband in IF and RF amplifiers.
4. Define selectivity and describe the filters used to obtain selectivity. Are such filters practical in RF amplifiers? Why are the base and collector connected to taps on IF and RF tuned circuits?
5. Describe some of the tuning methods for RF amplifiers. What are some of the special requirements on these tuning devices?
6. Discuss briefly the relationship between input noise, gain, and the need for RF amplifiers.
7. Define master oscillator (MO), intermediate-power amplifiers, and final-power amplifiers. Why are they necessary?
8. What are oscillators? Define LC, RC, crystal, and synthesizer types of oscillator circuits. Where are oscillators used?
9. What are the fundamental components of an oscillator circuit? Draw the basic configuration showing these elements. Describe how each works.
10. Show the elements of an RC oscillator. Identify the fundamental components. How does it work?

Learning Objectives—Next Section

Troubleshooting Amplifier Circuits

Troubleshooting Procedure Before Trouble Develops

Troubleshooting is the logical and systematic procedure for restoring normal operation to equipment that is operating poorly or not at all. The words *logical* and *systematic* are the key to success in troubleshooting—haphazard efforts waste time and can sometimes lead to even worse malfunctions. You may have seen an electronic technician make a repair after a very fast checkout. Such rapid troubleshooting is based on extensive experience in using logical and systematic procedures.

Because amplifiers are the most frequently used circuits in radio transmitters and receivers, hi-fi systems, and TV receivers, you now have much of the basic knowledge required to repair them. When you combine this with your knowledge of power supplies, you are familiar with enough basic principles to learn more about troubleshooting. That is the purpose of this section—to start you on the learning path to effective troubleshooting. What you learn here will not be limited to amplifiers but will apply to electronic equipment in general, and will be expanded in later volumes to cover additional circuits, functions, and equipment.

Troubleshooting has two fundamental parts. The first involves an understanding of equipment as it *should* function normally; the second involves procedures to be followed after a malfunction has occurred.

What to do *before trouble develops* is often ignored—it is to learn to use your test instruments to observe the *normal* operating characteristics of the equipment you will be troubleshooting. To do this:

1. Use the equipment manufacturer's instruction manual to learn the *correct operating procedure*. Perform each procedure as described.
2. Follow the equipment manufacturer's instructions; make all external adjustments called for, and observe their effects on operation.
3. Use the test instrument instruction manuals supplied and learn how to use the volt-ohmmeter, the AF and RF signal generator, and the laboratory oscilloscope.
4. Use the equipment instruction manual and the test instruments to observe the internal operation of the equipment. Learn how to apply correct input test signals and how to observe waveforms and operating voltages at all test points.
5. Learn to make all internal adjustments from the equipment instruction manual.

BEFORE TROUBLE DEVELOPS, LEARN TO USE

YOUR TEST INSTRUMENTS TO OBSERVE SYMPTOMS OF NORMAL EQUIPMENT OPERATION

Instruction Manuals

Although instruction manuals vary widely in quality and extent of content, they must be considered the most authoritative source of information. The only better source is an instructor/supervisor who is thoroughly familiar with the equipment and test instruments involved.

What can you expect to find in a well-prepared instruction manual for an equipment or test instrument?

1. A description of the equipment. This includes details of its purpose and outstanding features, indications of how it is different from other equipment of the same general type, and performance specifications.

2. A description of the operating controls and front-panel scales, dials, and other readouts.

3. An identification and location of connectors for equipment inputs and outputs.

4. A set of instructions for preparing the equipment for use. Included here are instructions for connecting antennas, microphones, external loudspeakers, and other input or output devices.

5. A step-by-step procedure for placing the equipment in operation, including a description of the characteristics of normal operation.

6. A step-by-step procedure for making adjustments to controls on the back, sides, or bottom that are not used in normal operation but are occasionally required for special adjustment of operating characteristics.

7. A review of routine maintenance procedures.

8. A description of the equipment's operating principles, including block diagrams of signal flow and any special processes taking place.

9. Instructions for servicing.

Manuals supplied for equipment intended for use by the general public often do little more than supply a list of service centers and/or a schematic diagram. Most manufacturers supply well-prepared *Service Manuals* for use by technicians, and these contain detailed component layout diagrams, block and schematic diagrams, and recommended test procedures. Service booklets of this type are usually available through electronic parts suppliers or from the manufacturer.

THE EQUIPMENT INSTRUCTION MANUAL TELLS YOU WHAT YOU MUST KNOW:

DESCRIPTION, OPERATING CONTROLS,
CONNECTORS, PROCEDURES
FOR: OPERATION, ADJUSTMENT,
SERVICING, AND SERVICING DIAGRAMS

Troubleshooting Procedure

Remember that *logical* and *systematic* are the key words in troubleshooting. When trouble develops, a logical and systematic approach along with good use of your head will be successful in troubleshooting. Here is a logical procedure:

1. Discuss the problem with the operator. What is the complaint? Did the malfunction develop gradually or all at once? Was the equipment dropped or subjected to unusual stress? Was any attempt made to adjust the special controls on the back, side, etc.? Did this or any other malfunction require *previous* servicing? The answers to these questions may simplify the steps that follow.

2. Examine the equipment for external signs of damage or exposure to excessive heat or moisture. If such signs are found, you may not have a simple malfunction but extensive internal damage.

3. If the equipment is even in partial operation, you may learn much by checking what happens as you adjust all the normal operating controls. It is possible that improper adjustment of the operator controls or other external adjustments is the cause of the trouble. Your knowledge of what these controls do in normally operating equipment may permit an immediate "fix" by control adjustment or provide a valuable clue concerning the location of an internal fault.

4. Check the operation of any input or output devices such as antennas, loudspeakers, microphones, etc. It is possible that a malfunction or faulty connection in one of these could be the cause of the complaint.

5. If equipment shows no signs of operation, check the connection to the power source, check to see that the power source is supplying required operating voltage, check the power cord for damage, and check to see what fuses in the equipment's external fuse holders might be blown. If you replace a fuse and that action restores normal operation, do not leave until you have operated the equipment for some time. It is possible that the fuse blew because of some internal trouble, not because of temporary overload. Check the marking on the fuse holder or the instruction manual to make sure that the correct size fuse is being used.

If there are no obvious external difficulties, it is usually necessary to remove the equipment from its cabinet or other protective enclosure in order to proceed with troubleshooting:

6. Follow the manufacturer's recommended procedure to remove the equipment from its protective enclosure. Not doing it their way may result in damage. Always disconnect the power cord from the power source before doing this; turning off the switch does not protect you or the equipment from serious injury.

7. Place the equipment on a workbench or other suitable surface. With the power off, make a close inspection for signs of overheating, broken connections, corrosion, or anything that appears at all unusual. If the equipment has plug-in circuit boards or other internal connectors of any kind, make sure that good electrical contact is being made and that there is no corrosion that may be interfering with good contact. Carefully examine the printed circuit boards for bad connections.

Troubleshooting Procedure (continued)

8. Use the *signal-tracing* or *signal-injection* method of trouble localization to find the immediate area of the malfunction or maladjustment. These methods are described later.

9. If the trouble was a maladjustment, such as an apparently misaligned IF transformer, think twice before making a readjustment. Unless equipment was used in a location of severe vibration, it is *unlikely* that such a misalignment can take place. It is more likely that a component has aged and changed performance. After replacing a component associated with a tuned circuit, it is often necessary to retune the circuit.

10. If the trouble is a malfunction or intermittent operation, use *both* voltage and resistance checks supplemented with signal tracing or signal injection to narrow the possible cause to a single component. Voltage and resistance data are often included with the schematic diagram. Intermittents are often the result of cracked conductors on a printed-circuit board or a poorly soldered component.

TROUBLESHOOTING

11. Replace the suspected component with an *identical* or suitable substitute part. Make sure that the connecting leads are arranged exactly as in the original circuit. Use proper precautions to protect parts from excessive heat from the soldering iron.

12. Place the equipment back in operation. If operation is restored, check the alignment of any necessary internal adjustments. Then replace the equipment in its cabinet, place it back in operation, check out all the external adjustments, and the job is done. If operation was not restored by the component replacement, go back and use signal tracing and injection to relocalize the difficulty.

Signal Tracing and Signal Injection

Signal tracing is the process of tracing the signal through the equipment. The place where the signal *fails to appear* at an expected point or *appears distorted* or *insufficiently amplified* is the immediate area of the trouble.

The process is best described by an example. Suppose you are troubleshooting the radio receiver of the transmitter-receiver learning system. Initially, you can connect a signal generator (an oscillator whose frequency and output level are adjustable) to the antenna terminals. The frequency of the signal generator is set to an RF frequency somewhere in the passband of the receiver's tuning range; tune the receiver to the *same* frequency. If you set the signal-generator controls to apply audio modulation to the carrier signal being applied to the antenna terminals, and you adjust the signal output for 100 mV, the signal being fed into the antenna terminals is a close simulation of a very nearby radio station.

SETUP FOR SIGNAL TRACING

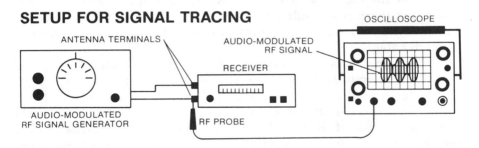

You can now use an oscilloscope to trace this signal from the antenna terminals to the loudspeaker. The stage where the signal fails to appear in the output, or is weaker or distorted, is the stage that is the *immediate area* of the trouble. The oscilloscope that you use must be sufficiently sensitive to detect the input signal, and it must have a so-called *low-capacity probe* that enables it to be applied to an RF stage without loading it down so much that it will not operate. At higher frequencies, a *detector probe* is used since the oscilloscope will not pass the RF carrier. The detector probe shows an output that is the modulated tone of the test oscillator.

A schematic diagram and parts-layout diagram are important. You must be able to locate any point in the signal path to see if the desired signal is there.

Start by applying the oscilloscope probe to the ungrounded antenna terminal. The oscilloscope controls are then adjusted until you see a wide band with the applied audio modulation on the screen. The RF variations are so fast that they appear as a blur on most scopes, and it is not necessary for troubleshooting to have an oscilloscope that can show the individual waveforms. What you see here is the signal that you are going to trace all the way through the receiver. A very strong signal is needed to see the waveforms in the input circuit. As you proceed through the receiver, you must reduce the input to prevent overloading.

Signal Tracing and Signal Injection (continued)

You can start by placing the probe on the collector of the RF amplifier and tuning the receiver to obtain maximum amplitude of the modulated RF signal on the screen. If no signal appears at this point, you know that there is a fault in the circuit between the antenna and the collector of the RF amplifier. What you can do then is go back to the antenna terminal and check for the presence of RF signal at each connection in the signal path between the antenna terminal and the collector of the RF amplifier. The circuit between the point where the signal last appeared and the next point where it failed to appear (or appeared badly distorted or low in amplitude) is the *immediate location* of the trouble.

The region between those two points must be checked for an open or short circuit or a faulty component. When you find the trouble, make the repair and recheck the circuit for proper operation. If the signal appears at the collector of the first RF stage properly amplified, you know that there is no fault up to that point and can then look for the signal at the next significant point, which is the input to the mixer or converter stage.

The signal at that point is the output of the RF amplifier stage and should be an amplified version of the original input. Lower the oscilloscope gain to keep the same amplitude signal on the screen. If the signal is as expected, the RF amplifier is working, and you can go to the next check point. If the signal is missing, distorted, or lacking in gain, the trouble is in the RF amplifier. Trace the signal back toward test point 2, looking for open or short circuits and faulty components. Be careful and make sure of the difficulty before you start to make adjustments or change components, because a voltage or signal level can be incorrect for many reasons.

FIRST STEPS IN SIGNAL TRACING
THROUGH THE RECEIVER

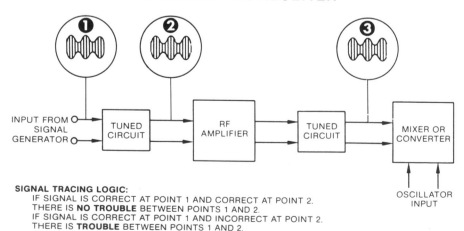

SIGNAL TRACING LOGIC:
IF SIGNAL IS CORRECT AT POINT 1 AND CORRECT AT POINT 2.
THERE IS **NO TROUBLE** BETWEEN POINTS 1 AND 2.
IF SIGNAL IS CORRECT AT POINT 1 AND INCORRECT AT POINT 2.
THERE IS **TROUBLE** BETWEEN POINTS 1 AND 2.
THIS LOGIC IS APPLICABLE BETWEEN **ANY** TWO POINTS.

Signal Tracing and Signal Injection (continued)

While you are at the mixer or converter check for local oscillator operation. Placing the probe in the vicinity of the oscillator tuned circuit should give enough pickup to show an *unmodulated* signal on the oscilloscope. If no unmodulated signal can be found even with the scope gain set at high, check the oscillator stage in detail with voltage and resistance checks. With no local oscillator signal, no IF signal will be present.

You continue to make checks in this manner until you reach the loudspeaker. The various checkpoints of interest are shown in the diagram. Even though you have not yet learned about how mixers or detectors operate, you know *what kind of signal* should be found, and you know how to check for the presence of that signal. When you have made the repair that is the cause of the trouble, the tone that the signal generator modulates onto its output signal should sound loud and clear from the loudspeaker. The input signal necessary to produce full output should be noted to be certain that the sensitivity is correct.

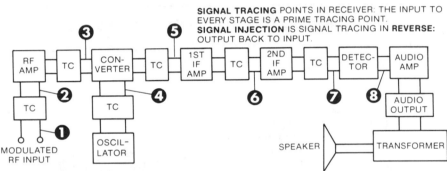

SIGNAL TRACING POINTS IN RECEIVER: THE INPUT TO EVERY STAGE IS A PRIME TRACING POINT.
SIGNAL INJECTION IS SIGNAL TRACING IN **REVERSE**: OUTPUT BACK TO INPUT.

Signal injection or *signal substitution* is really the same procedure as signal tracing, except that you start at the output and work *backward* toward the input. You start by connecting the oscilloscope to the receiver loudspeaker input terminals. Then, inject an *audio* signal into the driver amplifier by connecting an audio signal generator across the appropriate point. You should be able to *hear* the signal on the loudspeaker as well as *see* it on the oscilloscope.

Now you leave the oscilloscope connected where it is and you connect the audio signal generator to the input of the audio amplifier preceding the audio stage. If you can see and hear the signal, you know that the audio amplifier is operating.

Now you proceed toward the input, injecting audio, *IF* and *RF* signals at the same points that you observed for the presence of those signals during signal tracing.

The logic of the procedure is the *same* as in signal tracing. If the signal fails to appear at the scope and loudspeaker, the trouble is between that point and the previous point where it produced the expected output. Signal injection is more appropriate for receivers operating at frequencies well above the AM broadcast band, since even low-capacity oscilloscope probes will heavily load the RF and IF circuits.

Self-Test—Review Questions

1. Why do you think it is worthwhile to understand the operation of equipment before it becomes necessary to service it?
2. What information would you expect to find in the equipment instruction manual?
3. Describe a step-by-step troubleshooting procedure that you might follow to repair a piece of equipment.
4. Describe signal-tracing and signal-injection methods for localizing trouble.
5. Suppose you had a receiver that did not operate and you were troubleshooting by signal tracing. The correct signal appeared at the output of the mixer, and the first IF amplifier, but there was no signal at the output of the second IF amplifier. Where would you look for the trouble? Why?

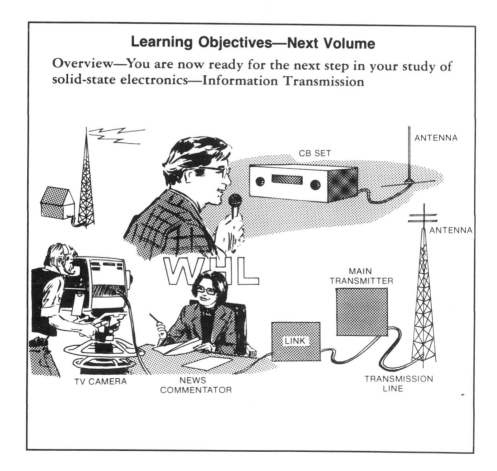

Learning Objectives—Next Volume

Overview—You are now ready for the next step in your study of solid-state electronics—Information Transmission

The dB Measurement Scale—Calculation

As presented on page 2–12, the definition of the decibel is given as

$$dB = 10 \log_{10} \left(\frac{P1}{P2} \right)$$

where P1 is the signal power at some terminal (usually the output) and P2 is the signal power at another terminal (usually the input) or the signal power at the reference level. A $\log_{10}$ (log to the base 10) is simply the power to which the base (10) must be raised to equal the quantity (in this case P1/P2). For example, if P1/P2 = 10, then $\log_{10}$ = 1(10^1 = 10); if P1/P2 = 100, then $\log_{10}$ (P1/P2) = 2 (100 = 10^2); if (P1/P2) = 4, then $\log_{10}$ (P1/P2) = 0.6 ($10^{0.6}$ = 4); etc. Losses are expressed as negative numbers of decibels. For example, a 10-dB loss would be

$$dB = \log_{10}\frac{1}{10} = -10 \text{ dB}$$

Logs have a number of other useful properties. The log of 1 = 0 (since 10^0 = 1), and therefore a device with a gain of 1 has 0 dB gain. Since logs are exponents, and the mathematical rules for exponents are to add them when multiplying and subtract them when dividing; we can do the same with decibels. Thus, two amplifiers with a power gain of 10 dB (gain of 10) in cascade would have a gain of 20 dB (10 dB + 10 dB). As you know, the total gain will be 100, which is the same as 20 dB. By the same token, if we were dealing with a loss, the loss would be −20 dB. This also follows from the property of logarithms, that log A/B is the same as log A − log B.

Examples of dB Power Calculations

$$dB = 10 \log \frac{3}{1} = 10 \log 3 = 10 \times 0.48 = 4.8 \text{ dB}$$

$$dB = 10 \log \frac{10}{1} = 10 \log 10 = 10 \times 1 = 10 \text{ dB}$$

$$dB_1 = 10 \log \frac{3}{1} = 4.8 \text{ dB}$$

$$dB_2 = 10 \log \frac{9}{3} = 4.8 \text{ dB}$$

Total gain = 4.8 + 4.8 = 9.6 dB

$$dB = 10 \log \frac{0.01}{1} = 10 \log 0.1 = -10 \text{ dB}$$

Practice logs until you understand them.

The dB Measurement Scale—Calculation (continued)

You can use logs for voltage gain as well as by using the relationship that you know:

$$P = \frac{V^2}{R}$$

If we substitute in

$$dB = 10 \log \frac{P1}{P2} = 10 \log \frac{V1^2/R}{V2^2/R}$$

and if the Rs are the same (rarely true, but we use decibels anyhow), then

$$dB_v = 10 \log \left(\frac{V1}{V2}\right)^2 = 20 \log \frac{V1}{V2}$$

Examples of dB Voltage Calculations

$$20 \log \frac{10}{1} = 20 \log 100 = 20 \times 2 = 40 \text{ dB gain}$$

$$20 \log \frac{14}{2} = 20 \log 7 = 20 \times 0.85 = 16.9 \text{ dB gain}$$

$$20 \log \frac{140}{20} = 20 \log 7 = 20 \times 0.85 = 16.9 \text{ dB gain}$$

$$20 \log \frac{20}{140} = 20 (\log 20 - \log 140) = 20 (1.30 - 2.15)$$

$$= -17 \text{ dB (loss)}$$

$$20 \log \frac{10}{0.01} + 20 \log \frac{100}{10} = 20 (2) + 20 (1) = 60 \text{ dB}$$

Work these calculations well, because you'll be using them.

Obviously, you can reverse the procedure by using a log table. For example, if you have an amplifier with 30-dB gain, what is the numerical gain?

$$30 = 20 \log_{10} x$$

or

$$\log x = 1.5$$

But the log is the exponent of 10; therefore, the numerical gain is

$$10^{1.5} = 31.6$$

Therefore, a 30-dB gain amplifier (voltage) has a voltage gain of 31.6.

The dB Measurement Scale—Drill

Try these problems, the answers are on page 2–190.

Input	Output	Power gain (dB)	Input	Output	Voltage gain (dB)
0.1 watts	50 watts	?	1 volts	10 volts	?
100 mW	2 watts	?	1 mV	5 volts	?
1 mW	100 mW	?	100 mV	50 volts	?
4 mW	1.8 watts	?	3 mV	3.8 volts	?
1,000 watts	150 kW	?	1.5 mV	100 volts	?
1,000 mW	150 watts	?	12 mV	16 volts	?

The following are the answers to the dB scale drill on page 2–189:

Input	Output	Power gain (dB) =	Input	Output	Voltage gain (dB) =
0.1 watts	50 watts	27.0	1 volts	10 volts	20.0
100 mW	2 watts	13.0	1 mV	5 volts	74.0
1 mW	1000 mW	30.0	100 mV	50 volts	54.0
4 mW	1.8 watts	26.5	3 mV	3.8 volts	62.0
1,000 watts	150 kW	21.8	1.5 mV	100 volts	96.5
1,000 mW	150 watts	21.8	12 mV	16 volts	62.5

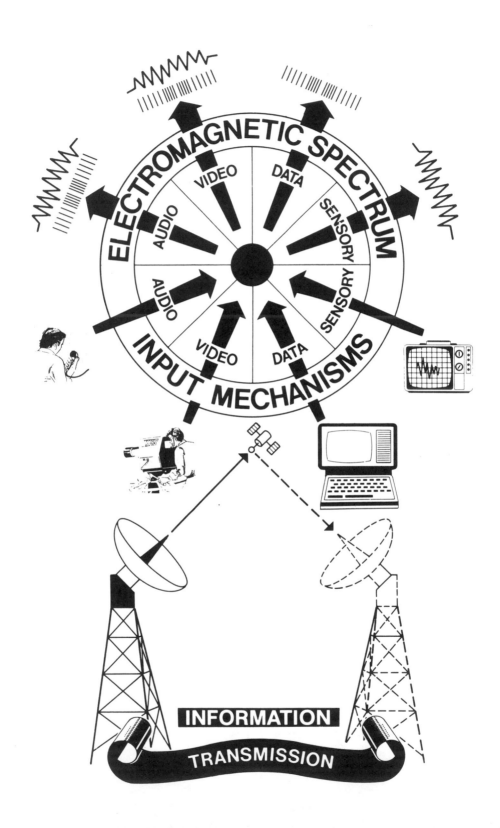

Basic Solid-State Electronics

COMMON·CORE

The Configuration and Management of Information Systems

INFORMATION TRANSMISSION
VOL. 3

RADIO WAVE PROPAGATION/TRANSMITTERS
TRANSMISSION LINES/ANTENNAS
AMPLITUDE MODULATION (AM)
SINGLE-SIDEBAND AM TRANSMISSION
FREQUENCY MODULATION (FM)
PULSE MODULATION/TV TRANSMISSION

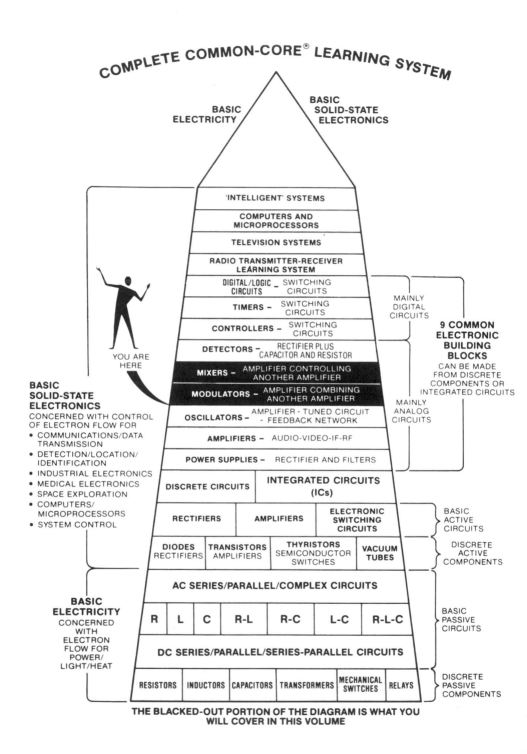

COMPLETE COMMON-CORE® LEARNING SYSTEM

BASIC ELECTRICITY

BASIC SOLID-STATE ELECTRONICS

'INTELLIGENT' SYSTEMS

COMPUTERS AND MICROPROCESSORS

TELEVISION SYSTEMS

RADIO TRANSMITTER-RECEIVER LEARNING SYSTEM

DIGITAL/LOGIC CIRCUITS — SWITCHING CIRCUITS

TIMERS — SWITCHING CIRCUITS

CONTROLLERS — SWITCHING CIRCUITS

DETECTORS — RECTIFIER PLUS CAPACITOR AND RESISTOR

MIXERS — AMPLIFIER CONTROLLING ANOTHER AMPLIFIER

MODULATORS — AMPLIFIER COMBINING ANOTHER AMPLIFIER

OSCILLATORS — AMPLIFIER · TUNED CIRCUIT · FEEDBACK NETWORK

AMPLIFIERS — AUDIO-VIDEO-IF-RF

POWER SUPPLIES — RECTIFIER AND FILTERS

DISCRETE CIRCUITS

INTEGRATED CIRCUITS (ICs)

RECTIFIERS

AMPLIFIERS

ELECTRONIC SWITCHING CIRCUITS

DIODES RECTIFIERS

TRANSISTORS AMPLIFIERS

THYRISTORS SEMICONDUCTOR SWITCHES

VACUUM TUBES

AC SERIES/PARALLEL/COMPLEX CIRCUITS

R | L | C | R-L | R-C | L-C | R-L-C

DC SERIES/PARALLEL/SERIES-PARALLEL CIRCUITS

RESISTORS | INDUCTORS | CAPACITORS | TRANSFORMERS | MECHANICAL SWITCHES | RELAYS

YOU ARE HERE

MAINLY DIGITAL CIRCUITS

9 COMMON ELECTRONIC BUILDING BLOCKS

CAN BE MADE FROM DISCRETE COMPONENTS OR INTEGRATED CIRCUITS

MAINLY ANALOG CIRCUITS

BASIC SOLID-STATE ELECTRONICS
CONCERNED WITH CONTROL OF ELECTRON FLOW FOR
• COMMUNICATIONS/DATA TRANSMISSION
• DETECTION/LOCATION/ IDENTIFICATION
• INDUSTRIAL ELECTRONICS
• MEDICAL ELECTRONICS
• SPACE EXPLORATION
• COMPUTERS/ MICROPROCESSORS
• SYSTEM CONTROL

BASIC ACTIVE CIRCUITS

DISCRETE ACTIVE COMPONENTS

BASIC PASSIVE CIRCUITS

DISCRETE PASSIVE COMPONENTS

BASIC ELECTRICITY
CONCERNED WITH ELECTRON FLOW FOR POWER/ LIGHT/HEAT

THE BLACKED-OUT PORTION OF THE DIAGRAM IS WHAT YOU WILL COVER IN THIS VOLUME

The Transfer of Information

In this and the next volume of *Basic Solid-State Electronics* you will learn about the mechanisms of *information transfer*—usually called *transmission* and *reception*. While it is convenient to separate this information-transfer system into transmission and reception, it is important to understand at the outset that they are *both part of the same process*; one without the other is of no value. The study of information transmission is important, even though your interest may be primarily in reception; only by studying transmission systems can you fully understand the attributes of the information signal to be received and, hence, appreciate the nature of the receiving mechanism!

The purpose of a transmission/reception system is to allow communication between two entities, either human or machine, in some agreed upon format. In some cases, the communication channel is unidirectional—for example, a radio or television broadcast. In other cases, the channel is bidirectional—as in a telephone or two-way radio. In any case, the essential elements of the information-transfer system are unchanged, and these are shown in the diagram.

INFORMATION TRANSFER

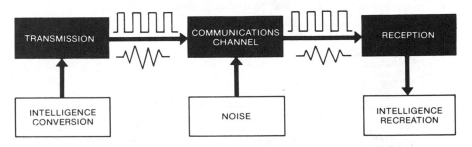

The essential step involves the *conversion* of the input signal information (or intelligence) into some *electrical form* (if it is not already available in that form). The intelligence (with some agreed-to format) is then *conditioned* so that it can *modulate* a suitable *carrier* medium. Modulation is the process of *impressing* the signal intelligence onto the carrier medium (electric current or space waves) in some agreed-to way.

At the receiving end, the receiver accepts the transmitted signal (electric current or space waves) and *demodulates* it to recover the original electrical signal, which, in turn, when applied to an appropriate *transducer* (for example, a loudspeaker), *recreates* the original information-signal input to the transmission/reception system.

As you can see from the chart on the following facing pages, the information to be transferred can take almost any form. The methods and processes of transmission and reception used are based on the most convenient and economical way to permit the transfer of information with *acceptable levels* of distortion and noise.

INFORMATION

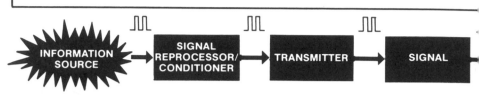

| INFORMATION SOURCE | SIGNAL REPROCESSOR/ CONDITIONER | TRANSMITTER | SIGNAL |

FROM MAN
SPEECH • VISION • DATA

SPEECH
MIND AND CENTRAL
NERVOUS SYSTEM
IDEAS/INTELLIGENCE

WORDS INTO CONTINUOUSLY VARYING
SOUND PRESSURES TRANSDUCED TO
ELECTRICAL SIGNALS

VISION
MIND AND CENTRAL
NERVOUS SYSTEM
IDEAS/INTELLIGENCE

PERCEIVED OPTICAL IMAGERY TRANSDUCED
INTO CONTINUOUSLY VARYING
ELECTRICAL SIGNALS

DATA
STORED INFORMATION-
PARALLEL SOURCE
IDEAS/INTELLIGENCE

INFORMATION IN ELECTRONIC MEMORY
ENCODED INTO DIGITAL PULSES

TO MACHINE
TO COMPENSATE FOR DISTANCE

INFORMATION TRANSFER SYSTEMS

TELEPHONE
AUDIO
VOICE-WORDS,
SENTENCES

VARYING SOUND PRESSURE INTO
CONTINUOUSLY VARYING ELECTRIC CURRENT

**TELEGRAPH/
TELETYPE/
FACSIMILE/
WORD/DATA
PROCESSING
DATA**
DISCRETE LETTERS,
WORDS, NUMBERS,
IMAGERY

SEQUENCES ENCODED INTO
SEQUENCES OF DISCRETE
INTERRUPTED ELECTRIC CURRENT

RADIO
AUDIO/DATA
VOICE, MUSIC,
DATA INPUT

AUDIO/DATA CONVERTED INTO
LINEAR VARYING ELECTRIC CURRENT

TELEVISION
AUDIO/VIDEO
VOICE/MUSIC/
PICTURES/
VISUAL DATA PATTERNS

AUDIO/VIDEO CONVERTED INTO
CONTINUOUSLY VARYING MULTI-
DIMENSIONAL ELECTRIC CURRENT SIGNAL

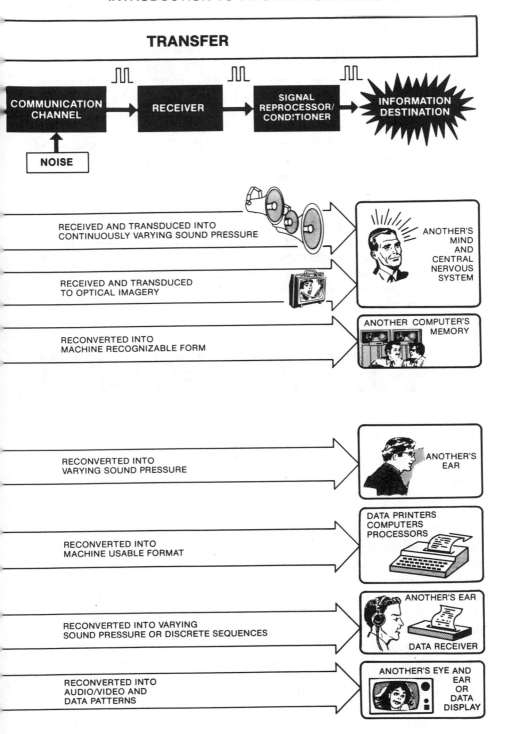

TRANSFER

COMMUNICATION CHANNEL → RECEIVER → SIGNAL REPROCESSOR/ CONDITIONER → INFORMATION DESTINATION

NOISE

RECEIVED AND TRANSDUCED INTO
CONTINUOUSLY VARYING SOUND PRESSURE

ANOTHER'S MIND AND CENTRAL NERVOUS SYSTEM

RECEIVED AND TRANSDUCED
TO OPTICAL IMAGERY

RECONVERTED INTO
MACHINE RECOGNIZABLE FORM

ANOTHER COMPUTER'S MEMORY

RECONVERTED INTO
VARYING SOUND PRESSURE

ANOTHER'S EAR

RECONVERTED INTO
MACHINE USABLE FORMAT

DATA PRINTERS
COMPUTERS
PROCESSORS

RECONVERTED INTO VARYING
SOUND PRESSURE OR DISCRETE SEQUENCES

ANOTHER'S EAR

DATA RECEIVER

RECONVERTED INTO
AUDIO/VIDEO AND
DATA PATTERNS

ANOTHER'S EYE AND EAR OR DATA DISPLAY

Information Theory

The objective of a communication system is to provide the *destination* (or receiver output) with information that is *as faithful a replica as necessary* of the message originally transmitted by the source (or transmitter input). This communication process is hampered, or distorted, by *noise*, which can be either *physical channel noise* or *mental semantic noise*. Noise places a *limit* on the performance of any communication system.

Information theory—first formulated by Dr. Claude E. Shannon, of the Bell Telephone Laboratories, back in 1948—suggests that if information is first properly *coded* and then sent through a channel, it will arrive at the destination much less distorted than if not properly coded. In the *coding* (or *encoding*), process information is converted from one scheme of symbols into another (for example, to a modulated carrier). The information passes through the channels in coded form and then is *decoded* (demodulated carrier) upon arrival at its destination. The *decoding* element of a communication system transforms the encoded information back into its original form so that the transducer at the destination can make proper use of it.

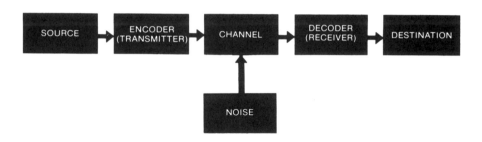

All communication systems are concerned with the flow of information, and according to Information Theory each system consists of five elements: a source, an encoder, a channel or channels, a decoder, and a destination. Information from the source is encoded for transmission. The medium through which this selection of symbols is conveyed is called a *channel*. Channels (which can take the form of air, wires, records, etc.) vary in their ability to transmit information to the receiver, or final destination. (See the above diagram.) Channels have different *loading capacities*. There is an upper limit on the amount of information that a channel can transmit, called *channel capacity*. Noise is added to the channel, and it can corrupt the encoded information in many different ways. The decoder (receiver) is designed to *minimize* the effects of the noise on the recovered information that is delivered to the destination for use.

The Transmission of Information

You have arrived at the step in the COMMON-CORE learning ladder where you can put together elements of a transmission system—for the transfer of information. In the process of studying transmission systems, you will need to learn about two more of the common building blocks: modulators and mixers; these will be covered in your study of this volume. You have already learned about power supplies, amplifiers, oscillators, and some transducers, and you are now ready to start to dissect and understand the anatomy of transmission technology.

When next to each other people transmitted information by

PERSON TO PERSON— ADJACENT COMMUNICATION

AUDIO SIGNALS

VOICE SOUNDS/WORDS

VISUAL SIGNALS

PICTORIAL SIGNS/
DIGITAL WORDS/
SIGN LANGUAGE

When *not* next to each other people transmitted information by

PERSON TO PERSON— DISTANT COMMUNICATION

AUDIO SIGNALS
HORNS/CYMBALS/DRUMS

VISUAL SIGNALS
SMOKE/FLAGS/PENNANTS/
LIGHT SIGNALS (MIRRORS)

And when not only individuals but *groups* of people became involved, there had to be developed a set of *rules* whereby the information transmitted was received and efficiently *understood* by all. This meant the development of *language*. By dictionary definition, "language is the means, vocal or otherwise, of expressing or communicating thought or feeling."

When words—oral or sign—came to be written—by means of picture writing, marks, pictographs, hieroglyphics; and then letters and words, etc.—for the storage and/or transmission-reception of information, there had to be developed *sets of rules* for various groups of transmitters and for various groups of receivers. This led to the development of the nine chief language families of the world for the spoken and written word.

The Transmission of Information (continued)

Now with the development of modern electronic communication systems with computer/microprocessor control, we need *still more languages—machine languages*—to govern the processing and storage of the information to be transmitted and received. You will learn more about machine languages in Volume 5. Also, the information to be transmitted can be the equivalent of *all* forms of human thought and feelings—audio signals, video signals, words, physiological and scientific data, electrical and magnetic sensory pulses, etc.

Moreover, these information signals now can be transmitted, via analog or digital techniques, next door or around the world by *direct* (or *guided*) means or by *indirect* (or *propagated*— the "wireless") means.

DIRECT OR INDIRECT TRANSMISSION

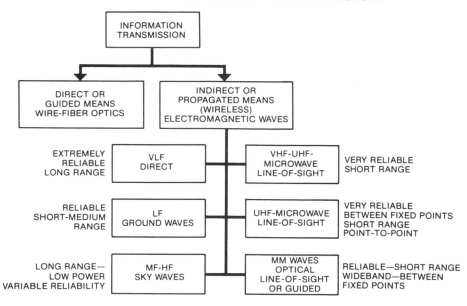

As you may know, information in the form of electrical signals can easily be transmitted by wire (guided), and you are probably most familiar with this in the telephone (and telegraph). Modern direct (guided) systems use not only direct means like the telephone line, but also can use a carrier for specific purposes. *Light beams* can function as the carrier of information by using a *fiber optic* conductor to replace the metal wire.

Here we will concentrate on indirect (propagated) transmission by electromagnetic waves from very low frequencies (VLF) to optical frequencies, although some of the techniques to be described can be used with transmission line (guided) techniques. You should remember that transmission and reception are used to transfer information from point to point, and we will study the various ways this can be done.

The Transmission of Information (continued)

As you will learn, the amount of information that can be transferred depends directly on the *bandwidth* of the *information-transfer link*. Thus, a 1-MHz channel can transfer a thousand times as much information per second as a 1-kHz channel. Or alternately, a given amount of information can be transferred in a thousandth of the time. As the channel bandwidths have increased, it has been necessary to go to higher and higher frequencies, because, as you shall see, the carrier frequency must be many times the signal bandwidth for effective transmission. This has pushed data links, multichannel telephone links, and other data-transfer systems first into the VHF/UHF range, then into microwaves, and ultimately into millimeter (mm) waves and optical wavelengths as the need for faster data transmission became necessary. Nowadays, optical links can carry the entire contents of a set of the encyclopedia in a few seconds. (These millimeter waves are waves having wavelengths of 1 mm or less, but below the infrared range of about 0.01 mm.)

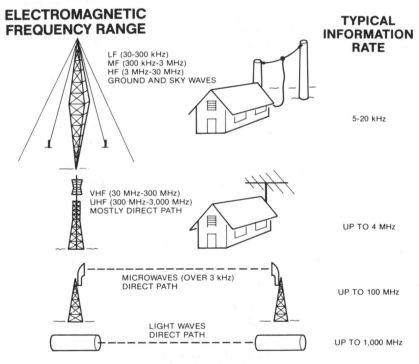

At present, optical information-transfer links are becoming popular. You probably are familiar with optical fibers and their use to carry thousands of telephone calls at one time on one thin fiber. Recently the use of unguided optical links (transmission in air) has begun to become important for use over limited distances. These provide security from eavesdropping and are reliable broad-band communication links.

The Radio Frequency Spectrum

You should understand that *radio frequency* (RF) does *not* mean the specific portion of the electromagnetic spectrum used *only for radio broadcasting*, but rather has a *much broader definition* that includes *all* of the electromagnetic spectrum used to *transmit information*. Thus, the *RF spectrum* encompasses the entire electromagnetic spectrum from low-frequency navigation aids to radio broadcasting, TV, communication, radar.

You know that the information signal or message is *superimposed* on an RF carrier wave for radio transmission. You also know that the RF spectrum covers a wide range from less than 10 kHz, to higher than optical frequencies, to x-rays. Usually, in radio communication and data transmission we are concerned with the region between 100 kHz and 30,000 MHz, although higher and lower frequencies are also used. You may wonder *how* RF frequencies for transmission are chosen, and *why* one frequency is chosen over another? In practice different frequency ranges propagate somewhat differently, and so different frequency ranges are appropriate for different applications. A rough division of the RF spectrum showing how radio waves are propagated is shown on the next page. We will discuss these modes of propagation in the next few pages. Generally speaking, radio wave propagation is divided into three broad ranges: (1) *ground wave* or *direct*, where the signal propagates *along* the earth's surface (or *through* it); (2) *sky wave*, where the signal propagates via *reflections* from the ionosphere (a layer of charged particles in the upper atmosphere and beyond); and (3) *line-of-sight*, where the wave propagates like optical or light signals in *straight* lines and hence requires that two stations be approximately within sight of each other.

RADIO FREQUENCY SPECTRUM PROPAGATION

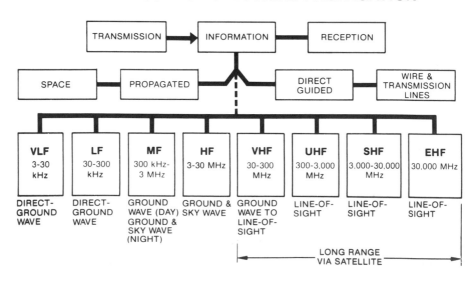

ELECTROMAGNETIC FREQUENCY SPECTRUM WITH ASSOCIATED EQUIPMENT SYSTEMS

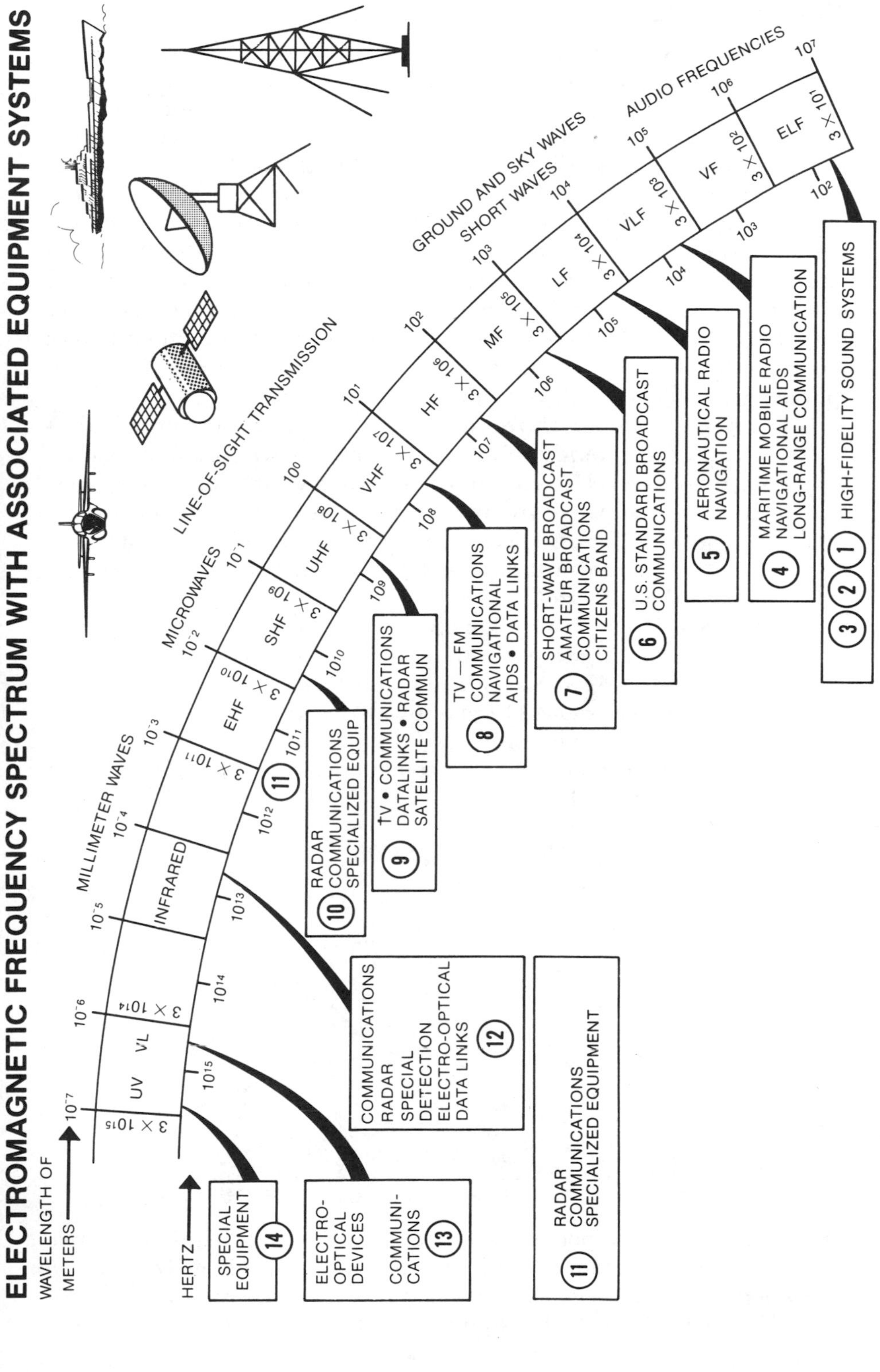

The Ionosphere

At frequencies above about 500 kHz and below 30 MHz (although occasionally at higher and lower frequencies), transmission occurs by *both* sky and ground waves. In this frequency range, most long-distance single- or multiple-hop communication occurs via the sky wave. The sky wave is a signal that is reflected back to the earth via the *ionosphere*. The ionosphere is a region high above the earth where the rarified air is ionized mainly by the sun's ultraviolet rays. This ionized layer reflects or absorbs radio waves permanently in the band between 3 and 25 MHz; under some circumstances, however, these characteristics can extend down to 500 kHz and up into the VHF range. Although long and very long waves also are affected, we use these mostly in the ground wave mode, so the effects of the ionosphere are not so important. The ionosphere consists of the following principal regions:

D layer—50–90 km above the earth's surface; usually only in daylight; reflects VLF and LF, absorbs MF, and weakens HF.

E layer—About 110 km above the earth's surface, usually most intense in daylight, important for HF propagation in daytime for distances greater than 1,000 miles. Important at night for MF sky-wave propagation at distances greater than 100 miles.

F_1 layer—About 175–250 km above the earth's surface; exists only in daylight; not usually important in sky-wave propagation; MF signals mainly absorbed.

F_2 layer—About 250–400 km above the earth's surface; the principal reflector for sky waves in the HF range; important in most long-distance communications; at night it merges with the F_1 layer.

EARTH SKY WAVES

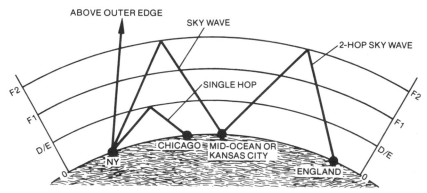

The effects of the ionosphere diminish and are generally negligible above 50 MHz, except for some absorption between 50 and 200 mHz. Thus, *all communication* that takes place *into space* generally uses frequencies above *400 MHz* since there is little or no absorption or reflection.

Wave Propagation Modes

You know that the function of an antenna in a transmission system is to radiate electromagnetic energy into space. Once this energy is released from the antenna, it travels through space where some of its energy may be picked up by a receiving antenna, or dissipated until it is too weak to be received, or is reflected from an object (as is the case with radar transmission).

It is important to know what happens to a radiated wave (namely, what its path is, if it is absorbed by the earth, if it is reflected by the ionosphere, etc.) in order to tell how far the wave will travel before it becomes too weak to be picked up. In addition, it is important to know how the radiated wave is reflected, so the location where the reflected wave strikes the earth again can be determined.

When a radiated wave at lower frequencies leaves the antenna, part of the energy travels through the earth, following the curvature of the earth, and it is called the *ground wave* or *surface wave*. The rest of the energy is radiated in all directions into space. The waves that lie in the line of sight between the transmitter and the horizon are called *space waves* or *direct waves*. Waves leaving the antenna at an angle greater than that between the antenna and the horizon are *sky waves*. Sky waves that travel outward at only a small angle with respect to the surface are reflected back to the surface by the ionosphere. Those that travel upward at greater than a certain critical angle penetrate the ionosphere and are not directed back down.

GROUND WAVE   **SKY WAVE**

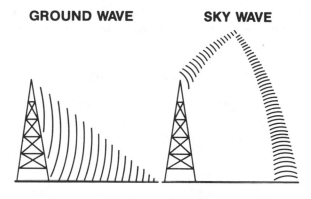

LINE-OF-SIGHT

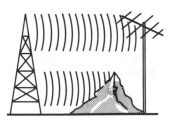

Frequency-Range Relationships

The following is an outline of the components of radiated waves used for transmission at various frequencies:

From 10 to 500 kHz, *VLF* and *LF* (very-low-frequency and low-frequency bands)—the ground wave is largely used for medium- and long-range communication since its stability is not affected by seasonal and weather changes. For communication over very long distances, extremely high power is used.

From 500 to 3 MHz, *MF* (medium-frequency band)—the range of the ground wave varies from 15 to 200 miles. Sky-wave transmission is excellent at night for ranges up to several thousand miles. In the daytime, however, sky-wave transmission becomes erratic, especially at the high end of the band.

From 3 to 30 MHz, *HF* (high-frequency band)—the range of the ground wave decreases rapidly and sky-wave transmission is somewhat unpredictable, depending on the seasonal factors that affect the ionosphere. This is the band used for long-distance communication (day and night). By proper choice of operating frequency, relatively consistent performance is obtainable over great distances with little power.

From 30 to 300 MHz, *VHF* (very-high-frequency band)—neither ground waves nor sky waves are very usable, and space-wave transmission (direct or line-of-sight) finds major application.

From 300 to 30,000 MHz and above, *UHF–SHF–EHF* (ultra-high-frequency, super-high-frequency, and extremely-high-frequency bands)—line-of-sight transmission is used almost exclusively. As the frequency increases above 30 MHz, the need for line-of-sight transmission becomes more critical, and above about 500 MHz becomes absolutely necessary for any reasonable performance.

TRANSMITTING FREQUENCIES...

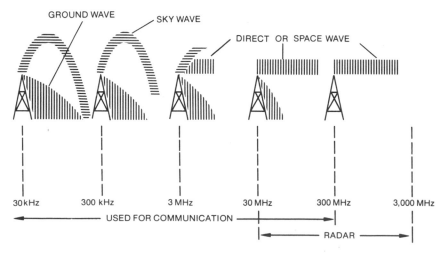

Field Strength

To estimate whether a signal will be picked up with enough strength to be usable, we must know its *field strength*, as well as the antenna characteristics and the receiver sensitivity. To calculate field strength, we use a "fictional" antenna called an *isotropic radiator*, which radiates uniformly in *all* directions. For this antenna, the free-space propagation characteristics are:

$$P_R = \frac{P_T}{4\pi D^2}$$

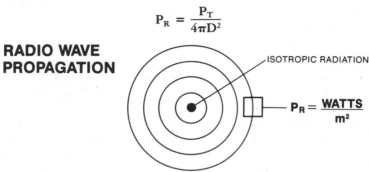

RADIO WAVE PROPAGATION

ISOTROPIC RADIATION

$$P_R = \frac{\text{WATTS}}{\text{m}^2}$$

where P_R is the power density in watts per square meter if the transmitted power (P_T) is in watts and the distance (D) is in meters. Thus, at a distance D, an antenna with a cross-section of 1 square meter will pick up an amount of signal equal to P_R. If the transmitting antenna has gain, we multiply P_T by this gain to get the total field strength as you will learn later in this volume when you learn about antennas.

Often the field of strength is desired in volts/meter (for a wire antenna). The intrinsic impedance of free space has been found to be equal to 120π or about 377 ohms. Therefore, we can use the power formula to convert the power equation to a voltage equation

$$P_R = \frac{V_R^2}{R} = \frac{V_T^2}{120\pi} = \frac{P_T}{4\pi D^2}$$

Thus, $V_R^2 = 120\pi \times P_T/4\pi D^2$ and $V_R(\text{volts/meter}) = \sqrt{30P_T/D}$. You can calculate line of sight over the earth by the equation $D = \sqrt{2h1} + \sqrt{2h2}$, where D is the line-of-sight distance in statute miles when h1 and h2 are the transmitting and receiving antenna heights in feet.

This relationship works very well for free space; ground wave and sky wave propagation, however, are another matter. Here, the conductivity of the ground, its dielectric constant, the nature of the antenna, the properties of the ionosphere, and terrain characteristics are predominant; therefore, it is not possible to use these simple relationships to calculate field strength, and we must make use of complicated tables and graphs found in radio engineering handbooks. These problems also occur when antennas are not in free space even for frequencies above HF into the VHF range; antennas are not considered to be in free space, because their height above the ground is limited.

Regulation of the RF Spectrum

The U.S. Federal Communications Commission (FCC) was established in 1934 to allocate portions of the RF spectrum to different functions, licensing users, assigning maximum operating power, and monitoring the performance of broadcast stations.

The FCC has entered into international agreements to allocate specific uses for various portions of the RF spectrum. For example, domestic U.S. radio broadcasting has been given the spectrum from 535 kHz to 1,605 kHz to use for amplitude-modulated (AM) broadcasts. This range is divided into 107 channels, each having a 10-kHz bandwidth. Frequency and power assignments are made to enable a city or community to have a number of stations appropriate to its area, population, and interests. These frequency and power assignments are selected so that nearby stations will not interfere with each other.

Frequency-modulated (FM) broadcasting was introduced to offer improved transmission and freedom from interference from electrical disturbances. For these purposes, the FCC assigned the spectrum from 88 MHz to 108 MHz for commercial and educational FM broadcasting. This region is divided into 100 channels, each with a 200-kHz bandwidth.

RF SPECTRUM

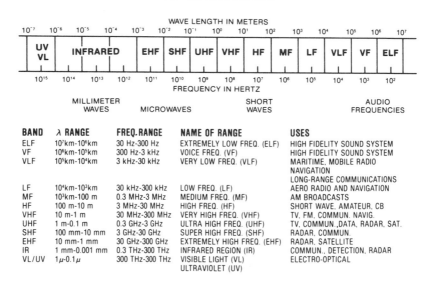

BAND	λ RANGE	FREQ. RANGE	NAME OF RANGE	USES
ELF	10^7km–10^6km	30 Hz–300 Hz	EXTREMELY LOW FREQ. (ELF)	HIGH FIDELITY SOUND SYSTEM
VF	10^6km–10^5km	300 Hz–3 kHz	VOICE FREQ. (VF)	HIGH FIDELITY SOUND SYSTEM
VLF	10^5km–10^4km	3 kHz–30 kHz	VERY LOW FREQ. (VLF)	MARITIME, MOBILE RADIO NAVIGATION LONG-RANGE COMMUNICATIONS
LF	10^4km–10^3km	30 kHz–300 kHz	LOW FREQ. (LF)	AERO RADIO AND NAVIGATION
MF	10^3km–100 m	0.3 MHz–3 MHz	MEDIUM FREQ. (MF)	AM BROADCASTS
HF	100 m–10 m	3 MHz–30 MHz	HIGH FREQ. (HF)	SHORT WAVE, AMATEUR, CB
VHF	10 m–1 m	30 MHz–300 MHz	VERY HIGH FREQ. (VHF)	TV, FM, COMMUN. NAVIG.
UHF	1 m–0.1 m	0.3 GHz–3 GHz	ULTRA HIGH FREQ. (UHF)	TV, COMMUN.,DATA, RADAR, SAT.
SHF	100 mm–10 mm	3 GHz–30 GHz	SUPER HIGH FREQ. (SHF)	RADAR, COMMUN.
EHF	10 mm–1 mm	30 GHz–300 GHz	EXTREMELY HIGH FREQ. (EHF)	RADAR, SATELLITE
IR	1 mm–0.001 mm	0.3 THz–300 THz	INFRARED REGION (IR)	COMMUN., DETECTION, RADAR
VL/UV	1μ–0.1μ	300 THz–300 THz	VISIBLE LIGHT (VL) ULTRAVIOLET (UV)	ELECTRO-OPTICAL

With the growing interest in radio broadcasting, the FCC established the frequency band from 26.965 MHz to 27.405 MHz as the Citizens Band (CB). This band contains 40 channels, each with a bandwidth of 10 kHz and a maximum allowable transmitter power output of 4 watts. Amateur radio operation is permitted with a variety of transmission methods in many bands spread across the spectrum.

Types of Emissions

The FCC has also designated certain types of modulation or emission for use in various frequency bands in the United States under different conditions. You will learn about most of these in this volume. Most of the usual types of signals are listed below. A table of standard bands is also included (see the figure on page 3–8 for more details about various portions of the spectrum and associated equipment).

TYPE OF MODULATION	TYPE OF TRANSMISSION	SYMBOL
AMPLITUDE MODULATION (AM)	NO MODULATION—CW ONLY	A0
	TELEGRAPHY—ON-OFF KEYING	A1
	TELEGRAPHY—ON-OFF KEYING OF AUDIO TONE (MCW)	A2
	TELEPHONY—DOUBLE SIDEBAND—FULL-AMPLITUDE CARRIER (RADIO BROADCASTING AND COMMUNICATION)	A3
	TWO INDEPENDENT SIDEBANDS—WITH REDUCED CARRIER	A3B
	SINGLE-SIDEBAND SUPPRESSED CARRIER	A3J
	FACSIMILE	A4
	TELEVISION (PICTURE)—VESTIGIAL SIDEBAND	A5
FREQUENCY MODULATION (FM)	NO MODULATION	F0
	TELEGRAPHY—FREQUENCY-SHIFTKEYING—(TWO FREQUENCIES)	F1
	TELEGRAPHY—ON-OFF TONE MODULATION (FM)	F2
	TELEPHONY (RADIO BROADCASTING AND COMMUNICATION, INCLUDING STEREO)	F3
	FACSIMILE (BY DIRECT FM)	F4
	TELEVISION (PICTURE)	F5
PULSE MODULATION	SIMPLE PULSED CARRIER WITHOUT OTHER MODULATION	P0
	TELEPHONY (DATA TRANSMISSION—MULTIPLEXED OR SIMPLE VOICE—ETC.)	
	(A) AMPLITUDE-MODULATED PULSES	P3D
	(B) WIDTH-MODULATED PULSES	P3E
	(C) POSITION MODULATED PULSES	P3F
	(D) DIGITAL CODED GROUPS	P3G

STANDARD BANDS

WAVELENGTH RANGE	BAND NO.	FREQUENCY RANGE	BAND
$10^7 - 10^6$ m	2	30 TO 300 Hz	ELF (EXTREMELY LOW FREQUENCY)
10^6 m $-$ 100 km	3	300 TO 3,000 Hz	VF (VOICE FREQUENCY)
100 km $-$ 10 km	4	3 TO 30 kHz	VLF (VERY LOW FREQUENCY)
10 km $-$ 1 km	5	30 TO 300 kHz	LF (LOW FREQUENCY)
1 km $-$ 100 m	6	300 TO 3,000 kHz	MF (MEDIUM FREQUENCY)
100 m $-$ 10 m	7	3 TO 30 MHz	HF (HIGH FREQUENCY)
10 m $-$ 1 m	8	30 TO 300 MHz	VHF (VERY HIGH FREQUENCY)
1 m $-$ 10 cm	9	300 TO 3,000 MHz	UHF (ULTRAHIGH FREQUENCY)
10 cm $-$ 1 cm	10	3 TO 30 GHz	SHF (SUPERHIGH FREQUENCY)
10 mm $-$ 1 mm	11	30 TO 300 GHz	EHF (EXTREMELY HIGH FREQUENCY)
1 mm $-$ 100 μm	12	300 TO 3,000 GHz	LONG-WAVE INFRARED
100 μm $-$ 10 μm	13	0.3×10^{13} TO 3×10^{13} Hz	FAR INFRARED
10 μm $-$ 1 μm	14	0.3×10^{14} TO 3×10^{14} Hz	NEAR INFRARED
1 μm $-$ 100 nm	15	0.3×10^{15} TO 3×10^{15} Hz	ULTRAVIOLET AND VISIBLE LIGHT

Radio Transmitter-Receiver Learning System

In Volume 1 you were introduced to the block diagram of a complete radio transmitter-receiver system and learned that this would be the *learning vehicle* which would take you from basic concepts and circuits to a working knowledge of complete radio transmitter-receiver systems. This block diagram is presented below.

The arrangement of this learning system is typical of that used when a radio station transmits a program to a radio receiver in your home. It also applies to the arrangement used when one amateur radio station transmits a message to another or when a Citizens Band (CB) operator in one automobile talks to another operator several miles (or kilometers) away. Although these various systems differ in size, cost, operating range, and sound quality, they all employ the basic arrangement shown in the figure.

In previous volumes you learned about power supplies; audio, video, IF, and RF amplifiers; and oscillators—all the basic electronic building blocks that are used to make up complete radio transmitters and receivers except modulators and mixers. Now that you are about to enter the world of transmitters, it is worthwhile to review the operation of the transmitter-receiver system, to firmly establish a frame of reference upon which to base the details of this volume.

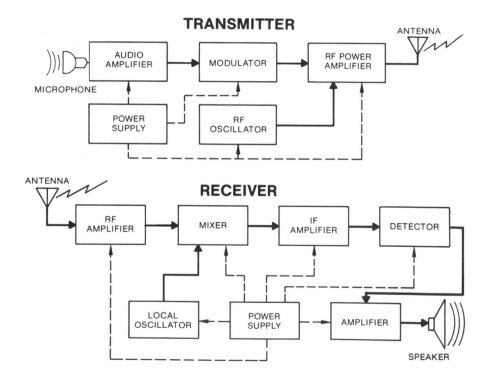

Summary of AM Transmitter-Receiver Operation

Although the electrical signals that represent sound (audio) waves travel very easily through a conductor, they are *incapable* of being *radiated* (*propagated*) through air and space to a distant point. For electrical signals to radiate into space without difficulty, their *frequency* must be *considerably higher*. For example, the lowest frequencies used in commercial broadcasting in the United States are about 540 kHz, and frequencies of hundreds of megahertz or more are commonly used in communications.

Because of this natural physical phenomenon it is necessary to generate a signal at a frequency *high enough* to be radiated to a distant point and to *superimpose* information on this signal. The *high-frequency signal* is known as the *carrier*, and the process of modifying the carrier to superimpose the message (the information) is known as *modulation*.

There are two types of modulation used today. These are *amplitude modulation* (AM), where the *carrier amplitude* is varied, and *frequency modulation* (FM), where the *carrier frequency* is varied. We will study AM systems first and FM systems later. TV systems use both AM and FM.

In the transmitter, a carrier signal is generated and modulated by the voice (audio) signal. The receiver picks up this signal and *demodulates* it by removing the carrier signal so that only the *electrical equivalent* of the *original voice* (audio) *signal* is left.

As shown in the block diagram, in both the transmitter and receiver a *power supply* converts power from a source to voltages suitable for use by the various circuits. In the transmitter, a *radio-frequency (RF) carrier* is generated by the *oscillator* followed by RF amplifiers. A *microphone* is used to convert the sound pressure of the original voice message to an audio signal, and an *audio amplifier* steps up this signal to a usable voltage. For an AM transmitter, *both* the *voice* and the *constant-level RF carrier* signals are fed into the *modulated* amplifier, which raises and lowers the amplitude of the carrier so that the resulting peaks and valleys *duplicate* the *variations* in the original voice signal. This *modulated RF carrier* can be fed either directly to the *antenna* or into a *power amplifier*, which steps it up to a higher power level before it is fed to the *antenna*.

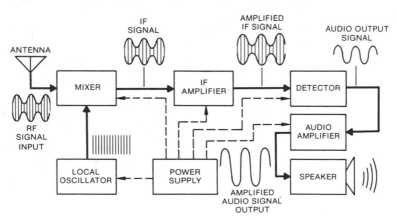

Summary of AM Transmitter-Receiver Operation (continued)

At the receiver, it is necessary to *remove* the carrier signal from the received amplitude-modulated RF signal and leave *only* the electrical equivalent of the original audio signal. Usually, the received signal is *very weak*, often only a few millionths of a volt. The weak signal picked up by the receiver *antenna* is first *stepped up* by an *RF amplifier* to raise it to a level suitable for *processing* to begin. Because amplification at this frequency is *inefficient*, the modulated carrier is *reduced* to a considerably lower *intermediate frequency* (IF), at which it is much easier to amplify. This is accomplished by the *mixer*, which takes in the *modulated RF carrier* and a *constant-level carrier* at an offset frequency produced by the receiver *local oscillator*. The output from the mixer is essentially a signal that duplicates the amplitude-modulated RF carrier, but the new carrier is at the intermediate frequency (IF) equal to the difference between (or sum of) the incoming-signal RF and local oscillator frequencies. Considerable amplification can now be accomplished by the *IF amplifier* at a fixed frequency. The *AM detector* (or *demodulator*) is a rectifier that removes the intermediate-frequency carrier component so that only the original audio signal remains. Now, an *audio amplifier* boosts this signal to a level sufficient to drive a *speaker* to duplicate the original voice message.

The pages that follow will review the outstanding features of radio transmitters in much greater detail. From your work in Volume 2 you are already familiar with the basic audio RF and IF amplifier stages and RF oscillators that are used in transmitters. The emphasis now will be on *how* these circuits *work together*, and you will study examples from transmitters in use today.

AM RECEIVER OPERATION

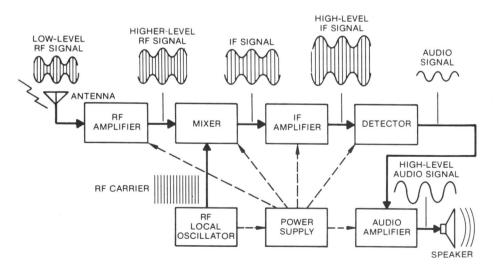

A Simple Transmitter

The simplest transmitter consists of an *oscillator*, which generates a high-frequency carrier signal. The oscillator—and the type of oscillator really doesn't matter—could be connected to an *antenna* to make up a complete transmitter. The antenna in such a case would radiate a signal that is constant in amplitude and at the same frequency as the oscillator.

If your home radio set picked up the constant-amplitude signal from such a transmitter, you would hear *nothing* because the detector output would be only a dc signal representing the carrier alone! Such a signal contains almost *no intelligence* (information)—except whether the carrier is *present* or *not*. To *add intelligence* (information) to the signal, the carrier *must be modulated*. The simplest thing would be to turn the carrier on and off with a key to produce the dots and dashes of the Morse code (a language).

A modulated signal of this type contains intelligence since information can be obtained from it. If the transmitted carrier were modulated by a short burst of carrier, a longer burst of carrier, and another short burst (dot-dash-dot), the radio would produce a sound like *dit-dah-dit*, which the radio operator understands in Morse code language as the letter *R*. This *on-off approach* was the method first used to electrically convey information over long distances. Other methods of adding intelligence to the oscillator signal will be considered in this volume.

A SIMPLE TRANSMITTER

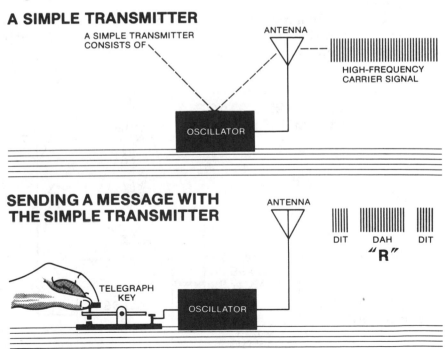

A SIMPLE TRANSMITTER CONSISTS OF

ANTENNA

HIGH-FREQUENCY CARRIER SIGNAL

OSCILLATOR

SENDING A MESSAGE WITH THE SIMPLE TRANSMITTER

ANTENNA

DIT   DAH   DIT

"R"

TELEGRAPH KEY

OSCILLATOR

A Simple Transmitter (continued)

By itself, an oscillator is a poor transmitter—first, because it usually has low power output, and second, because modulating it may interfere with its stability. The basic three-stage transmitter to be reviewed here is far superior and is typical of transmitter configurations. A block diagram of the basic three-stage transmitter is shown. The solid-state amplifier stages are operated as Class B for high RF efficiency. The *master oscillator* (MO) generates the RF carrier signal, which is almost invariably crystal controlled, although in amateur radio use a self-excited, tunable oscillator may be used. In general, transmitters are designed for fixed-frequency use. For some applications, however, several interchangeable or switchable crystals may be used for different operating frequencies. As you will see, for multichannel operation, a *frequency synthesizer* is used to provide the stability of crystal control with tunability.

The *intermediate power amplifier* (IPA) (an RF amplifier) amplifies the RF signal and isolates (or buffers) the master oscillator from the final power amplifier. The IPA is therefore called a *buffer amplifier*. The IPA may also act as a frequency multiplier to multiply the oscillator frequency to a higher one. Thus, the final output frequency need not be at the oscillator frequency. The operation of a frequency multiplier will be explained later. The *final power amplifier* (PA) (an RF amplifier) amplifies the output from the buffer and delivers it through a transmission line to the antenna.

The basic configuration shown here is the complete RF portion of a transmitter. Provisions for modulating the RF carrier to add information are considered in later sections of this volume.

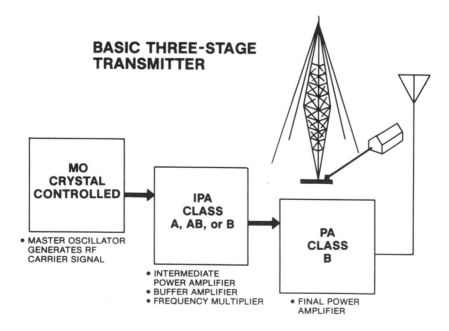

BASIC THREE-STAGE TRANSMITTER

MO
CRYSTAL
CONTROLLED

- MASTER OSCILLATOR
  GENERATES RF
  CARRIER SIGNAL

IPA
CLASS
A, AB, or B

- INTERMEDIATE
  POWER AMPLIFIER
- BUFFER AMPLIFIER
- FREQUENCY MULTIPLIER

PA
CLASS
B

- FINAL POWER
  AMPLIFIER

Example of a Practical CB Transmitter

The basic three-stage transmitter shows the fundamental parts of the RF section of a transmitter system. A practical system may have more stages to accomplish the necessary *signal processing*. It is more realistic to consider these three parts as *basic sections*, each of which may contain *several stages*. We will consider initially a CB (Citizens Band) AM transmitter system as an example of a solid-state, highly stable transmitter system.

An example of a very popular 40-channel CB transmitter-receiver is the one shown below. The frequency range is from 26.965 MHz to 27.405 MHz, which makes provision for 40 channels with a 10-kHz bandwidth. Over twenty million CB transceivers built by this and other manufacturers and based on the same general principles are in use today. The transmitter portions of this unit will be described in this volume as an example of a practical transmitter circuit, and the receiver portions will be described in Volume 4 as an example of a practical receiver.

As shown here in the block diagram, this transmitter is composed of three basic sections—master oscillator, intermediate power amplifier, and power amplifier, each of which contains several stages. The frequency-stability requirements of this system (0.01%) are such that *crystal control* is required. To avoid using 40 separate crystals for the transmitter, a technique called *frequency synthesis* is used to give 40-channel operation with just three crystals. Each of the three basic transmitter sections will be considered on the pages that follow.

BLOCK DIAGRAM OF TRANSMITTER SECTION OF JOHNSON MODEL 4170 CB TRANSCEIVER

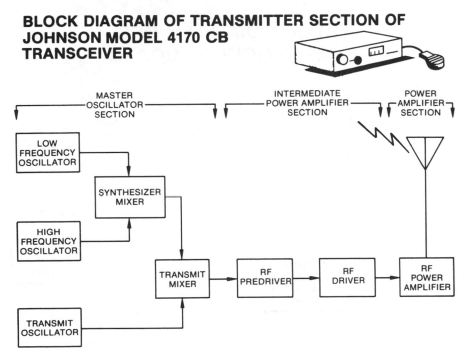

40-Channel CB Frequency Synthesizer—Phase-Locked-Loop Operation

Multichannel communications equipment requires high-frequency stability for both the transmitter and the receiver. To accomplish this stability without a great number of crystals, frequency synthesizers are used. These synthesizers use just a few crystals, with mixing and frequency division to synthesize the desired frequencies for both the transmit signal and for the receiver local oscillator. Many of the digital circuits necessary to form a frequency synthesizer are packaged together into an integrated circuit (IC). Thus, the bulk of the necessary circuitry comes in compact form. Most digital synthesizers make use of a phase-locked loop (PLL) and voltage-controlled oscillator (VCO) to accomplish the frequency synthesis. Therefore, before you can understand frequency synthesizer operation, you will have to learn something about these circuits. An understanding of frequency synthesizers is also necessary to understand digitally tuned TV, FM, and communication receivers.

The PLL consists of a *voltage-controlled oscillator* (an oscillator whose frequency is controlled by dc voltage input), a phase detector, and a low-pass filter. A buffer amplifier is sometimes added so that an output can be taken from the VCO without affecting circuit operation. For some uses, the entire loop is packaged as an integrated circuit by a number of different manufacturers, and it has a wide range of applications. Here, we will consider its use as a 40-channel frequency synthesizer for CB radio.

BLOCK DIAGRAM OF PHASE-LOCKED-LOOP IC

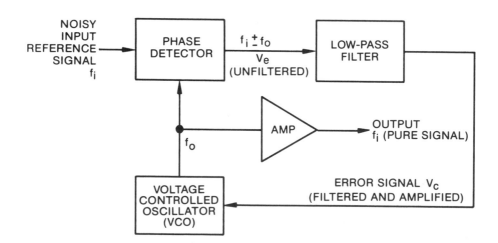

40-Channel CB Frequency Synthesizer—Phase-Locked-Loop Operation (continued)

During operation without an input signal, the VCO oscillates at a particular free-running frequency f_o. If an input signal is applied at frequency f_i, the phase detector compares the phase of this input with the frequency of the VCO. The output of the phase detector is a voltage proportional to the phase difference (or error) between f_o and f_i and is consequently known as the *error voltage* V_e. The error voltage is filtered to remove high-frequency components resulting from the combination of the two frequencies and noise. Then it is amplified and fed to the VCO input with a polarity to reduce the frequency difference between the VCO and the input frequency to zero.

Receiving an error-voltage input causes the VCO to change its frequency f_o so as to approach and become equal to that of the incoming signal. When this takes place, the difference between f_o and f_i approaches zero and the VCO is locked to the input phase and frequency. At that time, the error voltage out of the phase detector, and hence out of the low-pass filter and amplifier, also approaches zero.

You may ask why a VCO is used since its output frequency is the same as its input signal. This is because the VCO output is a pure signal, while the input may be corrupted with noise and other frequency components. The VCO, then, functions as a high-quality filter to remove the unwanted harmonics, noise, etc., and thus deliver a clear signal.

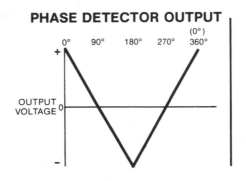

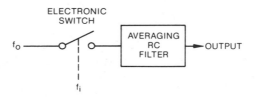

40-Channel CB Frequency Synthesizer—Phase-Detector Operation

Phase detector operation is clearly an essential part of phase-locked-loop (PLL) operation. The phase detector can be considered a *switch* operated by one signal after the other, followed by an RC integrator. Although the circuit configuration may vary, the essential elements always consist of the switch and filter.

As shown in the diagrams, the input frequency f_i is used to operate the electronic switch so that the switch is closed for the positive half cycle of f_i. (In practice, less than a half cycle of switch closure is used.) The portion of the half cycle of f_o gated through depends on its phase with respect to f_i. This portion of f_o is a pulsating dc voltage. This pulsating voltage is smoothed in the RC-averaging filter to produce a dc voltage output from the phase detector whose polarity and magnitude depends on the phase between the two signals, as shown.

PHASE-DETECTOR OPERATION

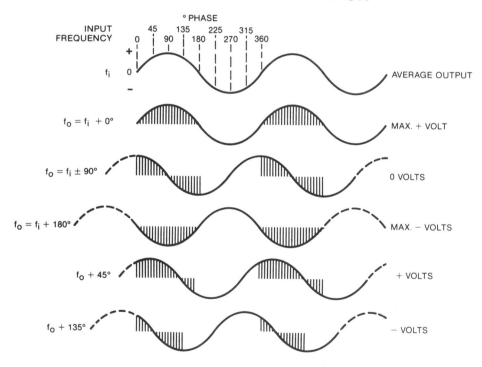

You can see from the figure showing phase detector output that positive or negative voltages are generated, depending on the relative phase, and this constitutes the error signal. Note that the null point of the phase detector output characteristic is zero at 90 degrees in phase; therefore, the output from the VCO loop is at 90 degrees in phase with respect to the input. This does not matter, since at null ($V_e = 0$) the phase shift is constant and thus has no effect on system operation.

40-Channel CB Frequency Synthesizer Operation—Mixer Operation

To understand the 40-channel CB frequency synthesizer operation, you must also understand *mixer operation*. As you see, mixer operation is also essential for receiver operation. A mixer accepts two different frequency inputs (f1 and f2) and produces sum (f_s) and difference (f_d) frequencies (as well as the original signal) at output. The output is tuned to the desired output frequency—invariably either the sum or difference frequency. This tuned circuit rejects the undesired frequencies. Mixers can have many circuit configurations, but their basic function is the same.

MIXER CONFIGURATIONS

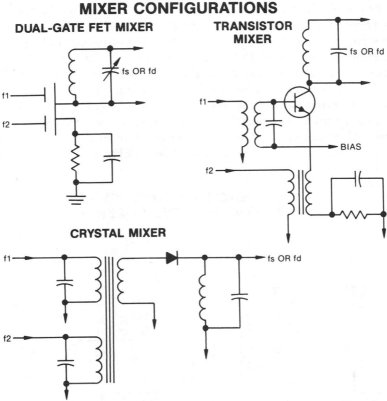

As you can see, the essential elements of a mixer consist of a way to couple the two input signals into the active element, which can be, for example, a FET, a transistor, or a diode. The essential property of a mixer is that one signal is sufficiently strong to drive the active element into nonlinear operation, while the second signal changes the nonlinearity slightly in such a way as to produce the sum and difference outputs. It will be shown mathematically in Volume 4 how this happens; here, however, you will have to accept it as fact. The output signals, f_d or f_s, can be larger or smaller than the input signals, depending on the conversion gain. The conversion gain is a measure of the output signal at f_d or f_s to the input signal at f1 or f2.

40-Channel CB Frequency Synthesizer Operation

The block diagram of the 40-channel frequency synthesizer shown here uses a phase-locked loop (PLL) and voltage-controlled oscillator (VCO), described earlier. The channel-selector switching system uses a binary-coded decimal (BCD) code to set up the *divide-by* number in the programmable divider of the synthesizing circuit. Light-emitting diodes (LEDs) are used to form the display of the selected channel number in the same manner that they are formed in a pocket or desk calculator. Because an understanding of these circuits requires a knowledge of computer basics, the description of this feature will be postponed until Volume 5.

As the block diagram shows, the output of the synthesizer is a signal at the frequency of the selected channel less 4.3 MHz. This output is fed to the transmit (Tx) or receive (Rx) mixers. This output signal is produced by mixing the output of the voltage-controlled oscillator (Q201) with the output of a fixed-frequency oscillator that generates a signal at 21.855 MHz. The VCO output and the fixed-frequency oscillator output are mixed in the synthesizer mixer (Q204). The resulting feedback frequency is divided by the programmable divider in U201, which operates under the control of the channel-selector switch. The programmable divider output has a nominal output frequency of 10 kHz, which is fed into the phase detector.

FREQUENCIES IN SYNTHESIZER
CIRCUIT FOR CHANNEL 1 OPERATION

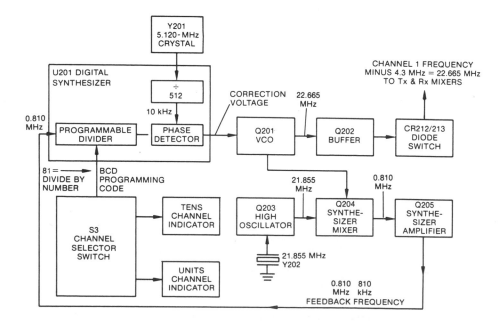

40-Channel CB Frequency Synthesizer Operation (continued)

Also fed into the phase detector is a 10-kHz reference signal. The reference signal is generated from crystal Y201, which controls the oscillator in digital frequency synthesizer IC U201. This crystal oscillator generates a 5.12-MHz output signal. This 5.12-MHz signal is divided by 512 to generate a precise 10-kHz reference signal for the phase detector. The output from U201 is a dc correction voltage that adjusts the frequency of the VCO as required to make the programmable divider output to the phase detector exactly equal to 10 kHz. When this condition exists, the output of the VCO is exactly equal to the frequency of the selected channel minus 4.3 MHz.

Assume that you wish to transmit on channel 1. To do this, the output to the transmit mixer must be the selected channel frequency of 26.965 MHz minus 4.3 MHz, or 22.665 MHz. This means that the output of the VCO must be at 22.665 MHz. If the VCO output frequency is correct, when mixed with the 21.855-MHz output from the fixed-frequency oscillator, the resulting feedback frequency is equal to the difference between these signals, or 0.810 MHz (810 kHz).

BLOCK DIAGRAM OF 40-CHANNEL
CB FREQUENCY SYNTHESIZER

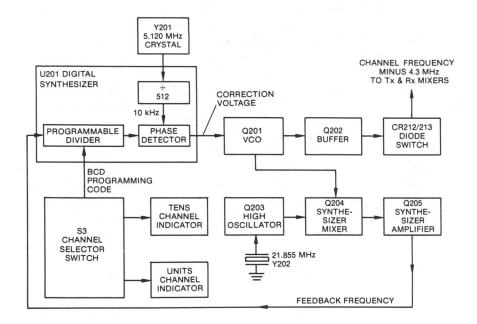

40-Channel CB Frequency Synthesizer Operation (continued)

This 810-kHz signal is fed into the programmable divider. When the channel-selector switch is set to channel 1, the channel-selector switch issues a binary digital programming code that sets the programmable divider to divide the frequency of its input signal by 81. Dividing 810 kHz by 81 produces 10 kHz, which is fed into the phase detector. The other input to the phase detector is the 10-kHz reference signal.

Thus, if the VCO output signal is higher or lower than 22.665 MHz, the output from the programmable divider will be higher or lower than 10 kHz. As a result, the phase detector will produce a correction voltage that adjusts the VCO output frequency to 22.665 MHz. The *divide-by* number produced by the channel-selector switch is always the number that is required for dividing the feedback frequency to produce a 10-kHz output to the phase detector when the VCO frequency output is correct. The figure shows the key circuit frequencies for each channel.

The schematic diagram on the following page shows the details of the synthesizer that has been described on the previous pages. Digital frequency synthesizer U201 is an IC. Crystal Y201 (5.12 MHz) is connected to the other components for the oscillator circuit which are located within the IC. Also located within the IC is the frequency divider that divides the 5.12-MHz oscillator signal by 512 to generate the 10-kHz reference signal, the phase detector, and the programmable divider.

Voltage-controlled oscillator Q201 is a FET that uses a variation of a Colpitts oscillator circuit. The VCO frequency is set to approximately 22 MHz with no control input voltage. As you know, the control circuit adjusts this VCO output frequency so that it is equal to the frequency of the selected channel minus 4.3 MHz. This is accomplished by feeding a sample of the VCO output signal from Q201 to the base of the synthesizer mixer Q204. The 21.855-MHz output signal from crystal oscillator Q203 is also fed to Q204. The resulting difference frequency is coupled through C227 to the base of Q205, the common-emitter synthesizer amplifier. Amplified feedback signal from the Q205 collector is connected to the programmable divider of U201.

The output of U201 (terminal 8) is a correction voltage that is fed to varactor diode CR211 connected across T201 to form part of the tuned tank circuit for the VCO. The capacitance of varactor diode changes as the applied voltage varies (as described in the section on FM transmitters in this volume), and thereby corrects the VCO frequency as required to produce the frequency of the desired channel minus 4.3 MHz.

DIGITAL SYNTHESIZER SCHEMATIC

40-Channel CB Frequency Synthesizer Operation (continued)

Buffer amplifier Q202 uses a two-gate MOSFET. A two-gate MOS-FET requires that both gates be biased positively for it to produce an output. Gate 1 operates as the equivalent of a common-emitter circuit and amplifies the output of mixer Q204; gate 2 is connected to terminal 7 of U201. When the system is *phase-locked*, a positive voltage is produced at terminal 7 of U201. This biases gate 2 positively and enables the MOSFET Q202 to conduct the VCO output to the diode switches. If phase-lock does not exist, no positive voltage is produced on terminal 7, and Q202 does not conduct the VCO output signal to the diode switch, thereby preventing an incorrect frequency from being transmitted.

The diode switch contains two diodes that are forward-biased by the transmit-receive switch. CR213 is biased to conduct the VCO output signal to the transmit mixer when the push-to-talk transmit button is pressed. CR212 conducts the VCO output signal to the receive mixer, thus enabling the receiver to operate when the transmit button is re-leased.

What the 40-channel phase-locked-loop frequency synthesizer has accomplished is to produce any selected one of a series of 40 RF signals. Each of these has a frequency 4.3 MHz below the frequency assigned to a CB channel. The selected assigned channel frequency is obtained by mixing the selected output of the PLL frequency synthesizer with the output of a 4.3-MHz fixed-frequency oscillator. The sum frequency from this mixing is the desired assigned channel output. The synthesizer output frequency is used directly in the receiver. This will be described in Volume 4.

The arrangement of the transmit oscillator and mixer in the 40-channel CB set is quite similar to most digitally tuned synthesizers. In some, however, more of the circuitry (particularly the VCO) is made up of IC elements.

As shown in the diagram on the following page, the fixed-frequency 4.3-MHz oscillator (Q20) is a modified Colpitts circuit using a parallel-resonant crystal (Y7) to produce a stable signal at 4.3 MHz. The output of this oscillator is taken from the emitter, for greater stability, and fed to gate 1 of dual-gate MOSFET transmit mixer Q21.

The output of the PLL frequency synthesizer is fed to gate 2 of Q21. In Q21, the two signals are mixed to produce sum and difference frequencies as well as the two original frequencies. However, the tuned circuits at the mixer output (T15 and C106, and T16 and C112) are tuned to pass only the sum signal from mixer Q21. This sum is equal to the assigned frequency of the selected channel.

TRANSMIT OSCILLATOR AND TRANSMIT MIXER

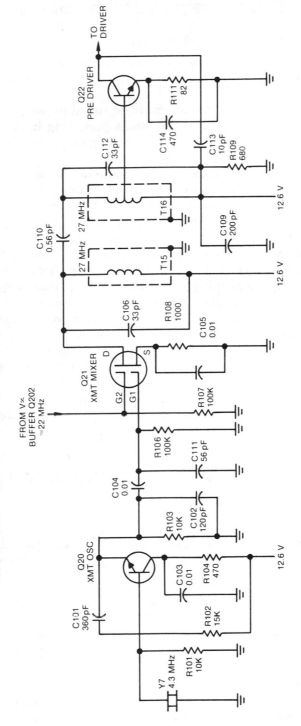

Buffers and Frequency Multipliers

Two basic circuits used in most transmitters, whether simple or complex, are the *buffer amplifier* and the *combined buffer amplifier and frequency multiplier*.

The buffer actually is an amplifier tuned to the same frequency as the master oscillator or amplifier stage preceding it. As the name suggests, it buffers or isolates the previous stage from later amplifier stages while also providing amplification.

Buffering can be obtained while also doubling, tripling, or otherwise multiplying the frequency of the oscillator. This enables the high-stability features of a low-frequency oscillator to be used, while producing at little extra cost a much higher operating frequency. The phase-locked-loop frequency synthesizer discussed earlier requires buffer amplifiers because the presently available IC units cannot produce the power levels required in the intermediate power amplifiers.

The frequency-multiplier circuit produces an output-signal frequency that is a multiple of the input frequency. This is accomplished by tuning the stage output to the desired harmonic (integer multiple of the input frequency). However, there are limits to the multiplication obtainable per stage because each harmonic signal level is reduced in amplitude by about $\frac{1}{2N}$, where N is the harmonic number (2 or greater). The most practical multipliers are doublers and triplers. Therefore, to increase the frequency by a factor of 6, a doubler and tripler are used in series; to increase frequency by a factor of 9, two triplers are used in series; etc.

FREQUENCY DOUBLING AND TRIPLING

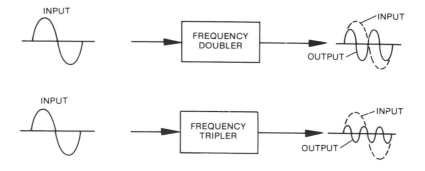

On the next page we will examine a typical doubler circuit—that is, one in which the output frequency is twice the input frequency—and see how it works.

Frequency Multipliers—The Doubler

The circuit of a frequency doubler appears to be the same as that of an RF amplifier which operates at the input frequency. The only differences are that the collector circuit will be tuned to twice the input frequency.

The doubler circuit is operated as Class C, with the collector output resonant to twice the input signal frequency. The pulses of current at the same frequency as the input signal flow from the emitter to the collector, energizing the collector-output circuit. These current pulses are rich in harmonics, and the amplifier has maximum collector impedance at the selected harmonic. Thus, the gain is greatest at the selected harmonic frequency. For this case, maximum output occurs at twice the input signal frequency.

The tuned circuit provides the missing RF cycle and since the pulses of current always arrive at the same time during alternate cycles of the doubled frequency, the output circuit is energized at the right time. As you remember from your study of resonance, the current is stored alternately in the inductor and capacitor. This tuned circuit *oscillation* continues even when no signal is present, but decreases depending on the circuit Q. The loss over a single cycle is small, so the power is taken from the stored energy during the missing half cycle. As you can see, the only significant difference between an RF amplifier and a frequency multiplier lies in the resonant frequency of the output circuit.

DETAILS OF FREQUENCY DOUBLING AND TRIPLING

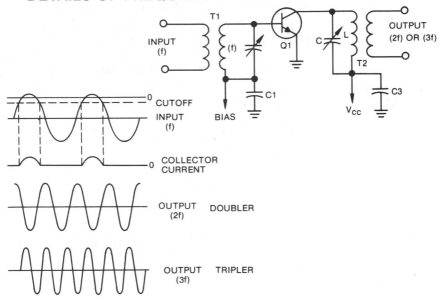

Neutralization

When the input and output circuits of an RF (or IF) amplifier are tuned to the same frequency, any positive feedback between input and output can result in oscillation. Such oscillation produces distortion and spurious radiation that significantly interferes with normal operation. Positive feedback can take place through the collector-to-base capacitance or through inductive or capacitive coupling between the input and output circuits. Modern amplifiers do not have this problem because of careful control of coupling and use of semiconductors with low internal capacitance. However, such feedback can be a problem with some older designs or semiconductors.

A process called *neutralization* is used to prevent such oscillation. This process involves including in the RF amplifier a circuit that precisely counterbalances the positive feedback with an exactly adjusted equal amount of negative feedback.

Collector neutralization is so named because it involves installing a negative-feedback circuit between the collector and the input circuit. The method is shown in the diagram. The coil in the output tuned circuit has a center-tap located at RF ground by means of capacitor C4. Because points A and B are at opposite ends of coil L1–L2, their RF potentials are 180 degrees out of phase with each other.

Now the neutralizing capacitor C_n is connected between point B and the base. Remember that the collector-to-base capacitance acts as if the capacitor C_f were connected as shown by the dotted line between point A and the base. The result is that the phase of the voltage fed from the collector to the base (through L1–L2 and point B) is opposite in phase to the voltage fed through C_f. Capacitor C_n is adjusted so that the negative feedback voltage fed back through it can be made exactly equal to the positive feedback coming through C_f. The negative feedback thus exactly cancels the positive feedback, and no oscillation takes place. Other neutralization schemes can be used, but all operate on the principle of balancing the positive feedback with the negative-feedback signal.

COLLECTOR NEUTRALIZATION

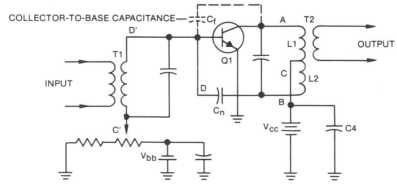

40-Channel CB IPA

As an example of an intermediate power amplifier that employs no frequency multiplication, consider the IPA arrangement in the 40-channel CB transceiver used as an example in earlier discussion. In this system, the output of the transmit oscillator and transmit mixer considered earlier require intermediate power amplification in order to drive the final power amplifier to its full authorized 4-watt output. In the equipment used as an example, these stages are known as the *RF predriver* and *RF driver*.

As shown in the schematic diagram, the output of the transmit mixer is coupled to the RF predriver (Q22) and the RF driver (Q23). These are straightforward RF power amplifier stages. They employ common-emitter circuits and npn transistors and have tuned inputs and outputs as shown. As mentioned earlier, the output frequency of the transmit mixer is in the frequency band between 26.965 MHz and 27.405 MHz for channels 1 through 40. This same band of frequencies is amplified by Q22 and Q23 without frequency multiplication. The final power amplifier output from Q24 is tuned and matched to the antenna via a coupling network in its output.

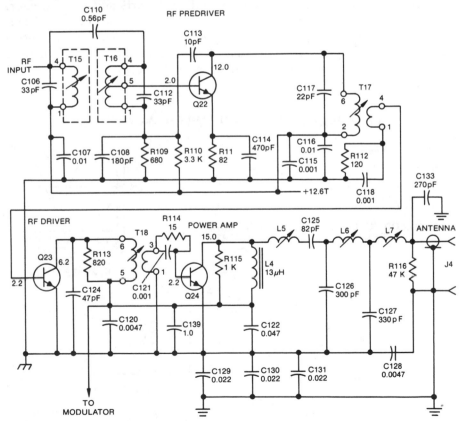

Final Power Amplifiers

The final stage of the transmitter is the *power amplifier*. A single-ended, low-power transistor-type power amplifier is shown here.

In addition to other considerations, the final amplifier stages are usually the *hottest* stages in terms of temperature. Because of the heat generated by the high currents in these stages, the components are usually cooled externally.

All of the early vacuum-tube amplifiers radiated heat due to the hot filament as well as the result of plate dissipation. Transmitters as small as a few watts required ventilation or a blower fan to keep cool. Modern low-power solid-state transmitters have convection cooling fins and heat sinks for conduction cooling. Transmitters up to about 10 watts can be operated continually with no special cooling requirements. However, the final amplifier stages of large transmitters use forced circulation of air, circulating water, circulating oil, or other devices to carry away the heat generated. Some older transmitters use solid-state devices for low-level stages and vacuum tubes for high-power stages. This is because solid-state devices of sufficient power-handling capability were initially not available. Nowadays, new transmitter designs feature solid-state amplifiers for power levels up to several hundred watts and even greater. However, *very-high-power* transmitters *still* use a vacuum-tube final amplifier. Also, many existing installations use vacuum tubes for intermediate power stages. The power amplifiers described here can be used as drivers for a higher power amplifier or as an output stage.

FINAL POWER AMPLIFIER WITH WAVEFORMS

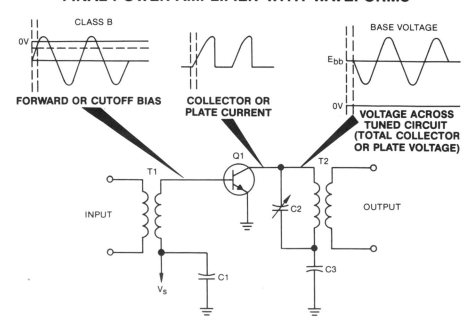

A Complete 40-Watt CW Transmitter

The circuit shown is a complete 50-MHz 40-watt CW (continuous-wave) transmitter that can be used by licensed radio amateurs to transmit from one point to another as long as they are near direct line of sight because of the way 50-MHz signals behave.

The crystal oscillator consisting of transistor Q1 and crystal Xtal; the tuned output C3–L3 is transformer-coupled to the base of Q3 which is gain-controlled by the forward arm of the VSWR bridge. The low-level amplifier output is tuned circuit coupled to the driver amplifier Q5, and its output is coupled to the final power amplifier Q2. Inductors L7, L11, and L5 keep the RF signal from coupling through the power source. The overtone crystal generates a 50-MHz signal, so no frequency multiplication is used, and the low-level amplifier serves as a buffer.

The output transistor is an RCA 40341 overlay transistor that develops 40 watts of power at 50 MHz. The filter networks at the output of the power amplifier reject harmonic and spurious frequency signals and match the transistor output to 50 ohms. The output signal is coupled through the SWR bridge 50-ohm coaxial cable.

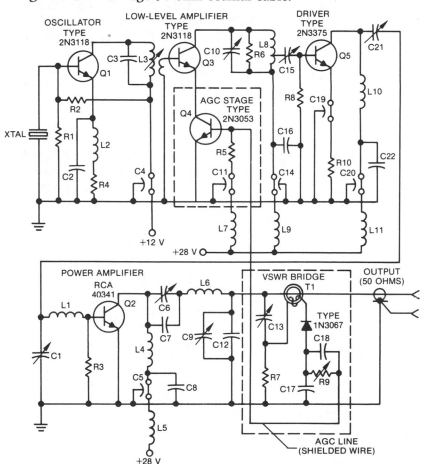

Review of Propagation and Transmitter RF Systems

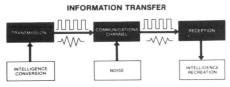

1. **INFORMATION TRANSFER** involves transmission and reception. Input intelligence in electrical form modulates a carrier and is recovered at the receiving end.

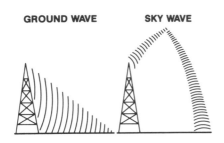

2. **THE ELECTROMAGNETIC SPECTRUM** includes all frequencies used to transmit information. Waves travel along the ground, in space, or reflect from the *ionosphere*. Wave propagation mode differs with frequency.

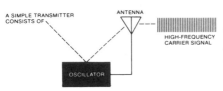

3. **SIMPLEST TRANSMITTERS** emit a steady carrier wave, interrupted by keying according to Morse code. Others superimpose an intelligence signal on the carrier by *modulation*.

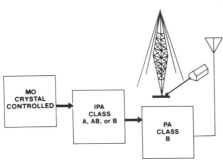

4. **THE BASIC CB TRANSMITTER** has three sections: a master oscillator, intermediate power amplifier, and power amplifier.

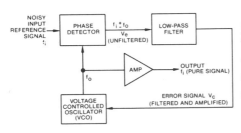

5. **THE CB MASTER OSCILLATOR** uses *frequency synthesis* to provide signals at 40 frequencies with only 3 crystals. It usually uses a *phase-locked loop* (PLL) to accomplish this.

6. **BUFFERS** are transmitter RF amplifiers. Their output can be at the input frequency or at a higher multiple, when they are called *multipliers*.

Self-Test—Review Questions

1. Discuss how radio transmission fulfills its role in the transfer of information.
2. Discuss briefly the relationship between the modes of propagation for different wavelengths.
3. Briefly discuss the difference between direct ground-wave and sky-wave transmission. What and where is the ionosphere?
4. What is the free-space field strength (power and voltage) of a signal at 10 km when the transmitted power is 1 kW with an antenna gain of 0 dB?
5. Briefly describe the function and operation of a phase detector and a mixer.
6. Why is crystal control or frequency synthesis almost always used in transmitter frequency control?
7. Draw a block diagram for a phase-locked loop. Describe the function of each element in the loop. Describe how it operates and why it is used.
8. Draw a block diagram of a frequency synthesizer and describe how it works.
9. Why are IPAs and power amplifiers used in transmitters? What is the fundamental difference between buffers, IPAs, and frequency multipliers?
10. Why are vacuum tubes still used in high-power transmitters?

Learning Objectives—Next Section

Transmission Lines—Antennas

Overview—Now that you know something about propagation, you're ready to study transmission lines and antennas. This study will cover transmission line losses, antenna patterns and impedance, and more.

ANTENNA PATTERNS

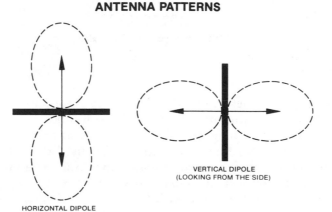

HORIZONTAL DIPOLE
(LOOKING FROM ABOVE)

VERTICAL DIPOLE
(LOOKING FROM THE SIDE)

Transfer of Energy to Antenna

The output of the transmitter is connected to a transmitting antenna for the radiation of RF energy into space so that this energy can be picked up by remote receiving antennas. This energy is propagated by ground waves, sky waves, or directly. To transfer the amplified RF from the output circuit of the final power amplifier into space, additional circuits are required. These include coupling circuits, transmission lines, and antennas. Just as a loudspeaker in audio systems transfers electrical audio energy from electronic circuits into acoustic energy in the air, so the antenna is the *means* of transferring *RF energy* from the transmitter circuits *into space*. The transmission line carries energy from the transmitter to the antenna, and the coupling circuit couples energy from the final power amplifier tank circuit to the transmission line. In some cases, a coupling circuit is also used between the transmission line and the antenna.

While it would seem to be logical to study coupling circuits first, we will study transmission lines first and then see how they interact with coupling circuits and antennas.

CB TRANSMISSION SYSTEM

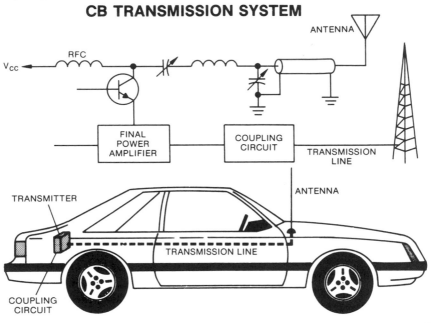

Transmission lines are used to carry the information signal from the final amplifier coupling circuit of the transmitter to the antenna. The transmission line may be a short line such as that used in mobile and CB units or a very long line such as those used to feed high-power commercial antennas. Since transmission lines and wave guides are not without loss, it is important to keep their length to the *minimum* necessary. Generally, losses increase with frequency.

Transfer of Energy to Antenna (continued)

Some simple transmission lines use two wires spaced with insulators or a dielectric. *TV twin lead*, with which you are probably familiar, is this type of line. Another type, called *coaxial cable*, consists of a central conductor with a concentric outer shield isolated by a dielectric. Coaxial cables can use a solid insulating (dielectric) medium or air insulation with periodically spaced insulating disks or a spiral wound dielectric to provide an air space. At microwave frequencies and above, specially shaped *wave guides* (usually in the form of round or rectangular tubing) are used to carry the signal from the transmitter to the antenna.

Transmission lines have a property known as *characteristic impedance*, which will be discussed on the next page. For maximum power transfer, the transmission line should be "matched" or terminated by its characteristic impedance. This "matching" is done by the coupling network to be described later.

The characteristic impedance of a parallel line is determined by the ratio of the diameter of the conductors and the spacing between the conductors as ell as the dielectric constant of the material between the conductors. In coaxial cables, it is the ratio of the diameter of the outer conductor to the inner conductor and the dielectric constant of the intervening medium. Variation of these parameters provides for lines of different characteristic impedance.

EXAMPLES OF TRANSMISSION LINES

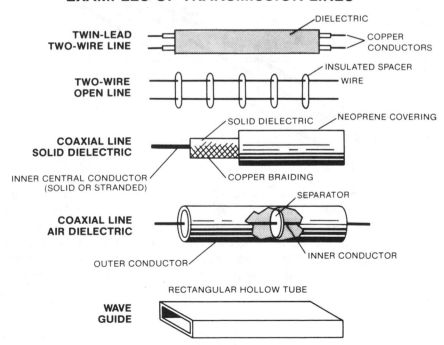

Transmission Line Losses

Transmission lines can be considered as a confined form of spatial transmission. As you know, space has a characteristic impedance of 377 ohms. Transmission lines have characteristic impedances that depend on the dielectric material and spacing. Nowadays, for almost every application you will encounter, coaxial cable is used as the transmission line for transmitters and most receivers. For TV receiver work, a two-wire parallel line, *twin lead*, is also used. The most popular twin lead has a characteristic impedance of 300 ohms. Coaxial cables most popularly used have a characteristic impedance of 50 or 75 ohms. The graph on this page gives the loss in various transmission lines. The numbers refer to the numerical designation for the specific line, which has been standardized for most manufacturers. For coaxial cables, each type of cable has a connector, specific for the type of cable, which should be used, with mating connectors at the transmitter and antenna ends.

ATTENUATION OF A–N CABLES VERSUS FREQUENCY

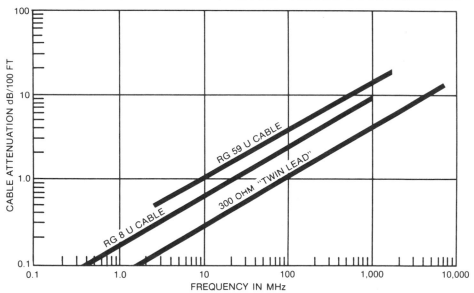

As you can see, losses increase with frequency and length, and therefore line length should be as short as possible, particularly as the frequency increases. While almost any small coaxial cable can be used for low-power applications, larger cables (including those with an air dielectric having diameters up to several inches with a rigid inner and outer conductor) are needed for high powers (up to 50 kW or more).

Standing Wave Ratio—Match

For maximum power transfer, the line impedance (Z_o) should be matched to the antenna (load) impedance (Z_{ant}). If a line is not terminated in its characteristic impedance, it is said to be *mismatched*, and all of the RF energy traveling down the line is not absorbed by the load. Energy not absorbed and radiated by the antenna (load) is *reflected back* on the transmission line. If the transmitter end of the line is matched, this reflected energy will be absorbed there; if there is a mismatch, some energy will be reflected; etc. The forward and reflected waves add to each other and form a resultant wave called a *standing wave*. The energy in this standing wave is energy not being radiated into space and so represents a loss. The ratio between the forward wave and the reflected wave is known as the *standing wave ratio* (SWR). The ratio between the sum of forward voltage and the reflected voltage divided by the difference is the *voltage standing wave ratio* (VSWR).

An example of getting maximum power transfer from a transmission line to a load is the case of a line feeding an antenna. If a type of CB antenna called a *whip antenna* is used, the impedance at its feed point is usually adjusted to 50 ohms. Therefore, in order to get maximum power transfer from the coaxial transmission line to the antenna, the characteristic impedance of the line should be 50 ohms or close to it. When this is the case, the line is said to be *matched* to the antenna—that is, the reflected signal is zero (or close to it).

MATCHING LINE IMPEDANCE TO ANTENNA IMPEDANCE

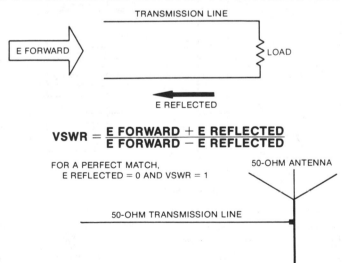

TRANSMISSION LINE

E FORWARD

LOAD

E REFLECTED

$$VSWR = \frac{E\ FORWARD + E\ REFLECTED}{E\ FORWARD - E\ REFLECTED}$$

FOR A PERFECT MATCH,
E REFLECTED = 0 AND VSWR = 1

50-OHM ANTENNA

50-OHM TRANSMISSION LINE

Standing Wave Ratio—Match (continued)

You may ask what having a match means. Strictly speaking, a match exists when the terminating impedance of a line is the *complex conjugate* of the line impedance. The complex conjugate is the same as the impedance vector (R + X) except with the sign of the reactance reversed—that is, vector (R − X). Usually, you try to make the load impedance resistive so that the transmission line is resistively matched and there is no reactance involved, since reactance can make the line sensitive to length (which is undesirable).

As you remember, + X is inductive and − X is capacitive; therefore, if the transmission line looks inductive, then the load should look capacitive by an equal amount. Of course, the line could look capacitive, in which case, the load would be inductive. The losses due to mismatch can be calculated readily by using the data on losses for the transmission line under matched conditions and with the VSWR, from the graph below. For example, if the VSWR is 4 and the line loss is 6 dB, the VSWR loss will be an additional 1.9 dB, for a total loss of 7.9 dB.

TRANSMISSION LINE LOSSES DUE TO MISMATCH

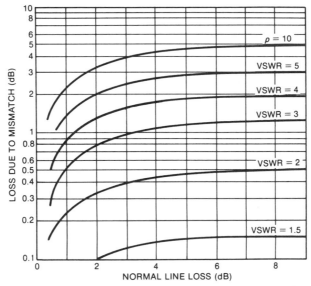

In solid-state transmitters, it is usually important that the VSWR be kept reasonably low since *solid-state amplifiers can be destroyed by a high VSWR*. For this reason, they should *not* be operated *without a load* or *with a badly mismatched load*. For this reason, and to minimize interference, transmitters are often tuned up using a *dummy load* which has the characteristic impedance of the transmission line or antenna and which can dissipate the transmitter power.

Experiment/Application—SWR Measurement

As you have learned, it is important to have a good match (low SWR) for best power transfer. The device for making these measurements is known as an *SWR meter*. Basically, it measures the ratio of forward power to reflected power and has a scale calibrated to indicate the SWR directly, depending on this ratio. As you know, the VSWR is simply

$$\text{VSWR} = \frac{V_f + V_r}{V_f - V_r}$$

where V_f is forward voltage and V_r is reflected voltage. A typical VSWR meter can be made using the bridge principle to measure the forward and reflected power. To make meaningful measurements, some power is required. The signal source should generate at least 50–100 mW with an impedance of 50 ohms.

SWR METER CIRCUIT

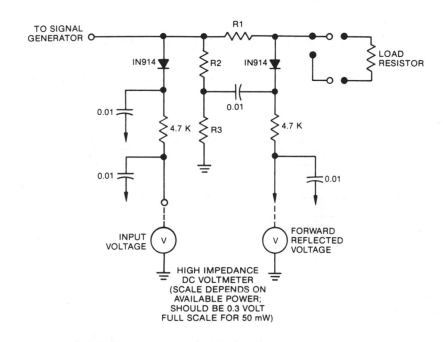

Note: R1 = R2 = R3, which should equal transmission line nominal impedance, and should also equal signal generator output impedance. (Scale depends also on available power; it should be about 0.3 volt at full scale for 50 mW.)

Experiment/Application—SWR Measurement (continued)

If you set up the circuit shown on the previous page, you can measure the effects of mismatch. If you set up the system so that the input voltage is kept constant, the V_r and V_f can be read directly from the meter on the output.

To make the measurements, the meter is calibrated by making R_L = ∞ (open circuit). The voltmeter is set to give a full-scale reading (V_f). If you try various values of R_L (including short and open), you will find that the V_r will vary. The SWR can be calculated using the formula above. If R_L = 50 ohms, the system should be in balance and V_r = 0. You can also calculate the SWR obtained by using the formula

$$SWR = \frac{R_L}{R_{line}}$$

where R_L is the load resistance and R_{line} is the coaxial line impedance. If you plot the result, you will obtain the following curve, which shows VSWR vs. error in load impedance.

VSWR VERSUS ERROR IN LOAD IMPEDANCE

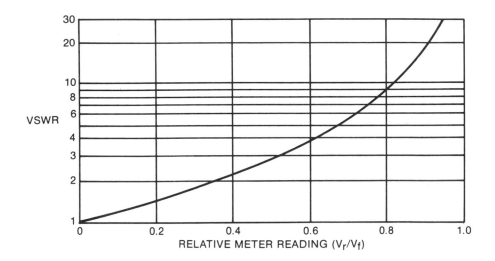

Now that we have completed our first Experiment/Application of this volume we will move onto the study of *matching transmission lines*. This section will focus on lines without loss ending in both open and short circuits.

Matching Transmission Lines

In a transmission line without loss ending in an open circuit (VSWR = ∞), the standing waves produce an effective minimum current (*maximum voltage*) at the open end and at each half wavelength from the end. At odd multipliers of λ/4 from the end, the current is maximum and the voltage is minimum.

In a transmission line without loss ending in a short circuit (VSWR = ∞), the *opposite* is true. The shorted end and each half wavelength from the end have minimum voltage and maximum current. At odd multipliers of λ/4 from the end, the voltage is maximum and current minimum.

At the points between the λ/2 and λ/4 points from the end, the effective impedance is either inductive or capacitive (depending on the relative phase of the resistive and reactive components). With the open circuit, the line is capacitive between the open end and λ/4 and inductive between λ/4 and λ/2 from the end. The opposite is true with a shorted line, where the line appears inductive from the shorted end to λ/4 from the end and capacitive at all points from λ/4 to λ/2 from the end.

For an improperly terminated line, the load is not resistive at the characteristic impedance of the line. Therefore, a properly chosen line length can be used to improve the match. This is rarely done, because the line must be of a specific length and may have standing waves on it. Usually, it is important to have the *best match* with *minimum VSWR* to have the most efficient system for any line length.

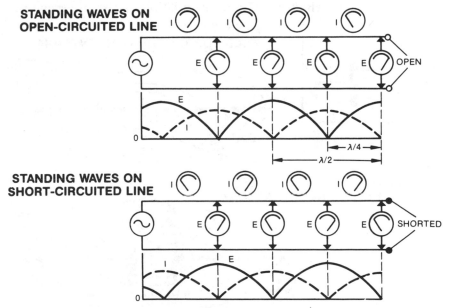

How an Antenna Works

If the wires of an open-ended transmission line are bent back a quarter wavelength from the open end, at right angles to the line, a simple antenna is formed called a *half-wave dipole*, a *doublet*, or a *Hertz antenna*. The voltage and current distribution on the antenna are the same as on the original transmission line except that the ends of the line are open—representing a voltage maximum and a current minimum.

Whenever there is a difference of voltage between two points, an electric field is set up between these points. You learned in *Basic Electricity* that when a capacitor is charged, one plate will be positive and the other negative. As a result, an electric field having a direction toward the positively charged plate is built up between the capacitor plates, as shown. Similarly, the voltage difference between the two wires of an antenna also generates an electric field having a pattern and direction, as shown.

Besides this electric field, there is also a *magnetic field* generated by the antenna current. The plane of this magnetic field is at *right angles* to the direction of current flow and therefore is at right angles to the antenna line, as shown. The electric and magnetic fields must therefore be at right angles to each other.

These electric and magnetic fields alternate about the antenna, building up, reaching a peak, collapsing, and building up again in the opposite direction, at the same frequency as the antenna current. In the process of building up and collapsing, a *portion* of these fields *escapes* from the antenna, becoming the *electromagnetic waves that radiate* through space and thereby convey the transmitted information to distant receivers.

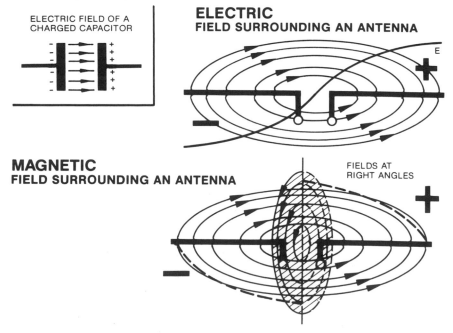

ELECTRIC FIELD OF A
CHARGED CAPACITOR

ELECTRIC
FIELD SURROUNDING AN ANTENNA

E

MAGNETIC
FIELD SURROUNDING AN ANTENNA

FIELDS AT
RIGHT ANGLES

Antenna Patterns

The direction of the electric field lies in the plane of the antenna and by convention determines the *polarization* of the antenna. Thus, a horizontal dipole is called *horizontally polarized* and a vertical dipole is called *vertically polarized*. Because receiving and transmitting antennas are really the *same* in operation, it is important for best results to have *both* antennas with the *same* polarization. However, since lower-frequency transmitted signals usually interact with the ground (earth) for a resultant depolarization of the wave, this is less important at frequencies below VHF.

As you can see from the figures on the preceding page, the electric field is largest broadside to the dipole. This results in maximum radiation in that direction and minimum radiation along the axis of the antenna. Thus, *maximum* signal is radiated (or received) in the *broadside* direction.

ANTENNA PATTERNS

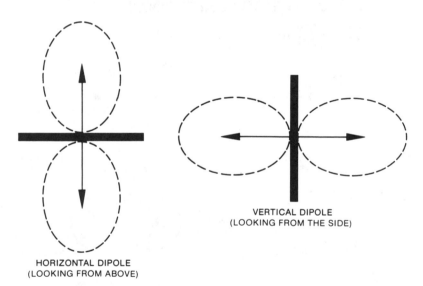

HORIZONTAL DIPOLE
(LOOKING FROM ABOVE)

VERTICAL DIPOLE
(LOOKING FROM THE SIDE)

As you can see, a *horizontal dipole* will radiate well in *two directions*. For example, an east-west of the dipole conductors gives best results in a north-south direction. A *vertical dipole*, on the other hand, will radiate well in *all horizontal directions*, but will have *little radiation vertically* (up or down).

You will find many many variations of antenna shapes and configuration. Typically, however, they are defined in terms of their characteristic input impedance, and an appropriate line is used to match this impedance either directly or via a coupling (impedance-matching) network.

Directive Antennas

In some cases, transmission (or reception) is of interest from only a given direction. In this case, *multielement antennas* are used. These elements are adjusted in phase to reinforce in the desired direction. They provide directivity and increased signal strength in the desired direction at the expense of other directions. The commonly used TV receiving antenna is an example of this. The multiple towers of AM broadcast stations and multielement antennas at TV stations also are examples of directive antennas.

Directive antennas are used in AM broadcasting to minimize interference between stations. At higher frequencies, they are used in FM and TV to increase the transmitted power in desired directions. Similarly, the additional gain of a directive antenna is useful to improve received signals. While in many cases directive antennas are fixed, some (particularly TV receiving antennas) are mounted on a motor driven rotator so that the direction can be varied for best results.

DIRECTIONAL ANTENNAS

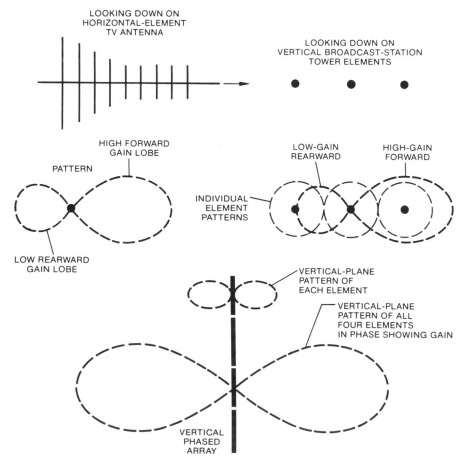

Basic Antennas

The *half-wave dipole* or *Hertz antenna* is one type of basic antenna which finds wide application in many types of transmitting and receiving equipment. The impedance at the feed point is about 75 ohms, and hence is matched with a 75 ohm open line or coaxial cable.

Another basic antenna is a *vertical quarter-wave* antenna, sometimes called a *Marconi antenna*. If one of the elements of a half-wave dipole antenna is removed and the wire that originally went to that element is grounded, the result is a Marconi antenna. The earth actually acts as a *mirror* that produces an image of the other quarter-wave element, so that the earth and the real quarter-wave element form an effective half-wave dipole. The current maximum and voltage minimum points are at the base of the antenna, as shown. The Marconi antenna is used in situations where a full half-wave would be inconvenient because of height (for example, AM broadcast).

When a Marconi antenna is used, the earth directly beneath the antenna must be a good electrical conductor. This is why AM broadcast station antennas are located in marshy or wet areas, because these areas have good electrical ground conductivity. Sometimes copper rods are driven into the ground or are buried radially at the base of the antenna to improve the ground conductivity. For some applications (at higher frequencies like HF), a vertical quarter-wave antenna may be mounted at some distance above the ground. A simulated ground plane is provided by using grounded metal rods at least a quarter wavelength long and placing them at the base of the antenna.

BASIC ANTENNAS

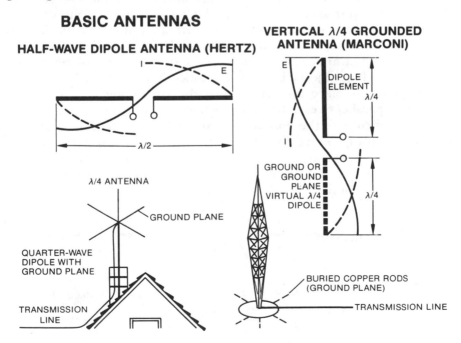

HALF-WAVE DIPOLE ANTENNA (HERTZ)

VERTICAL λ/4 GROUNDED ANTENNA (MARCONI)

Experiment/Application—Polarization and Directivity Effects

Suppose you used a VHF signal generator as a transmitter and an FM receiver to measure signal strength of the received signal. You can use these to measure the effects of polarization mismatch on the received signal if you set them up as shown.

ANTENNAS-POLARIZATION AND DIRECTIVITY EFFECTS

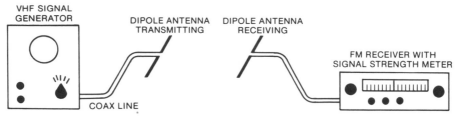

If you set up the signal generator in a clear area and adjust its frequency for the FM broadcast band (88–108 MHz), it will act as a signal source. If the output is connected to a dipole antenna via a coax line (or twin lead if a folded dipole is used), the antenna will radiate a signal whose polarization is the same as the plane of a dipole. If you now set up an FM receiver with some sort of signal-strength meter, and a similar dipole, you can observe the effects of polarization mismatch and the directivity of these antennas.

Set the signal output from the signal generator so that the FM receiver is unsaturated—that is, the signal-strength meter is slightly below full scale when the two antennas are in the same plane (as shown). Rotate the receiver antenna slowly to the vertical position and you will note that the signal strength decreases, with a minimum when the two antennas are perpendicular. At this point, the two antennas are cross-polarized. The minimum may not be very deep because of reflections from surrounding objects, but it will be clearly observable.

If you now place the antennas in the same plane (for maximum reception), parallel to each other, and rotate the receiver antenna so that it is perpendicular to the transmitting antenna (the end of the receiving antenna pointing toward the transmitting antenna) you will find that there is a minimum signal when the receiving antenna is perpendicular. This is what you would expect since you know that this is the minimum response direction for a dipole antenna.

Radiation Resistance

In a half-wave dipole antenna, the voltage at the center is minimum (practically zero) whereas the current is maximum. If you will recall the characteristics of a series-resonant circuit, you will remember that the voltage across it is also minimum and the current through it is maximum. At its center, a half-wave dipole is equivalent to a series-resonant circuit when operated at its resonant frequency. Its resonant frequency is approximately given by $f_r = 468/L$, where L is length in feet. A generator that supplies power to a series-resonant circuit works into a low value of pure resistance since X_L and X_C cancel each other—the resistance being mainly the resistance of the wire in the inductor.

Similarly, a transmission line works into a pure resistance when a resonant half-wave dipole is connected to it. This resistance consists of both the resistance of the wire and a resistance called the *radiation resistance*. The resistance of the wire is usually negligible, so only the radiation resistance is considered.

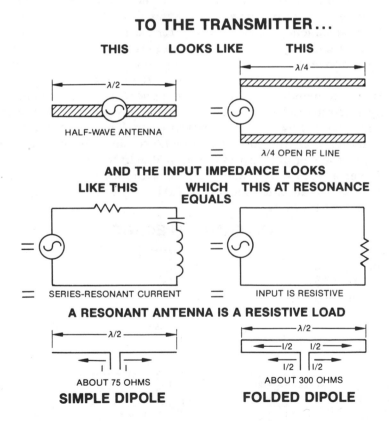

TO THE TRANSMITTER...

The radiation resistance is *not* a real resistance. It is an equivalent resistance which, if connected in place of the antenna, would dissipate the same amount of power as the antenna radiates into space.

Antenna Impedance

For a resonant half-wave dipole, the radiation resistance is about 73 ohms, measured at the center of the antenna (see illustration on previous page). Additional elements can be added in parallel to a half-wave dipole to raise its impedance. The popular *folded dipole* is an example of this, as shown. It has a characteristic impedance of about 300 ohms at resonance. The 4:1 change comes about from the fact that the current in each parallel conductor is half the value in a conventional dipole. The spacing of the conductors is kept much less than a wavelength so that the currents in the conductors are in phase.

Since a half-wave dipole acts like a series-resonant circuit, it will show either inductive or capacitive reactance as the frequency of the signal applied to it varies from resonance.

When the frequency of the RF is right, the dipole is exactly a half wavelength long and is series-resonant, with its impedance resistive and equal to the radiation resistance. In transmitting, it is always desirable that the antenna present a *resistive load* to the transmission line so that a maximum amount of power will be radiated by the antenna.

If the frequency of the transmitter goes *up*, the antenna will be *longer* than a half wavelength. The series circuit is then operating at a frequency above its resonant frequency, and since inductive reactance is larger than capacitive reactance, the antenna appears *inductive*.

If the frequency of the transmitter goes *down*, the antenna will be slightly *shorter* than a half wavelength. The series circuit is then operating at a frequency below its resonant frequency. The capacitive reactance is thus larger than the inductive reactance, and the antenna appears *capacitive* to the transmitter. As a practical matter, the frequency can be varied over a small range (about ±5%) without seriously affecting the match. Thus, an antenna cut for 30 MHz will usually work well from about 28.5 to 31.5 MHz.

ANTENNA IMPEDANCE

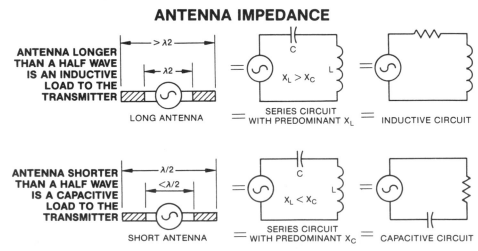

Tuning the Antenna

You have seen that as the frequency of the transmitter is varied, the electrical length of the antenna varies, as does the impedance at its input. Since it is desirable to have the antenna impedance *resistive* for all transmitter frequencies (for maximum radiated power), the antenna can be resonated by adding inductors or capacitors to effectively increase or shorten its electrical length.

For example, if a vertical quarter-wave grounded antenna is *less* than a quarter wavelength long, its input impedance at its base will be resistive and capacitive. The antenna can be *electrically lengthened (resonated)* by adding the right size inductor to cancel the capacity, thus leaving the antenna resistive. The inductor must be placed in series with the antenna at its base, as shown.

If a vertical quarter-wave grounded antenna is *longer* than a quarter wavelength, the input impedance at its base is resistive and inductive. The antenna can be *electrically shortened* by adding the right size capacitor to cancel the inductance, thus leaving the antenna resistive.

In some cases, the antenna is deliberately cut short because an antenna of proper length would be unwieldy. For example, a resonant λ/4 antenna in the Citizens Band would be about 17 ft long. To reduce this to manageable size, the antenna is electrically lengthened by adding an inductor (called a *loading coil*) either at the base of the antenna or at some point along its length. Of course, there are severe ohmic losses since the coil is located at the high current point. Thus, antenna efficiency suffers appreciably unless the coil is very carefully designed.

In some cases, the antenna is artifically lengthened by adding a top loading capacitor. This consists of a flat plate or other structure at the top of the antenna to act as a capacitor.

METHODS OF CORRECTING THE ELECTRICAL LENGTH OF A GROUNDED ANTENNA

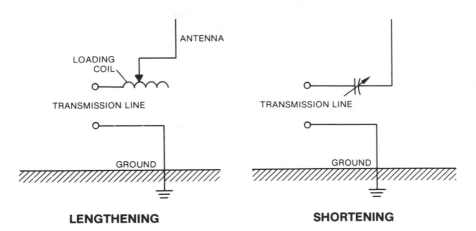

LENGTHENING   SHORTENING

Antenna-Coupling Circuits

Solid-state RF power amplifiers are quite unlike their vacuum-tube counterparts. Vacuum-tube power amplifiers usually have a *high output impedance* (high voltage, low current) and have *relatively little output capacitance*. Thus a conventional parallel-tuned tank circuit will have relatively high Q, and the coupling to the antenna is via a step-down transformer (line or inductive coupling). Solid-state power amplifiers are characterized by having a *low output impedance* (low voltage, high current) with *high output capacitance* and hence present a low-impedance, reactive load to the output. Because of the low output impedance, a conventional parallel-tuned circuit will be extremely inefficient (low Q). Thus, the coupling circuit for a solid-state amplifier generally consists of a network that transforms the vector (R − X) to a new vector R that is equal to the transmission line impedance (Z_o).

COUPLING CIRCUIT
COUPLES RF FROM POWER AMPLIFIER TO TRANSMISSION LINE

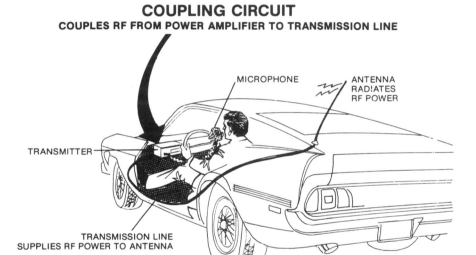

The output resistance of a solid-state amplifier can be determined from

$$R_o = \frac{\left(\dfrac{V_{cc}}{2}\right)^2}{P_o}$$

where R_o is the output resistance in ohms, V_{cc} is the supply voltage, and P_o is the power output. For a typical CB system (4 watts = P_o) and a supply voltage of 13 volts, the output resistance is

$$R_o = \frac{\left(\dfrac{V_{cc}}{2}\right)^2}{P_o} = \frac{\left(\dfrac{13}{2}\right)^2}{4} = 10.6 \text{ ohms}$$

The capacitance may be in the range of 20–50 pF. Thus, the network must tune out the circuit capacitance and transform 10.6 ohms to 50 or 75 ohms.

Antenna-Coupling Circuits—Coupling Networks

In your work, you will encounter many different coupling networks for solid-state amplifiers. It is important to realize that their purpose is the *same*, and it is usually a matter of convenience (or efficiency) that results in a given choice. Usually, to minimize losses, only the capacitive elements are tuned, although in lower-frequency, low-power applications, the inductors may be the tuned elements. This is the case in the CB transceiver learning system you have been studying. The schematic diagram of the output coupling is shown below.

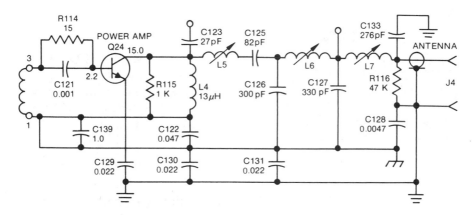

The matching section shown consists essentially of the impedance transformer L5, C125, and C126 in a series-resonant configuration followed by a low-pass filter (to reduce harmonic output) consisting of L6, C127, L7, and C133. Thus, the output is converted to 50 ohms by the impedance transformer before being filtered and delivered to the 50-ohm transmission line.

TYPICAL ANTENNA COUPLING CIRCUITS— COUPLING NETWORK

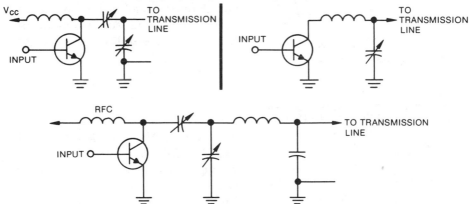

The schematic diagrams above show some of the typical coupling configurations you may encounter in your work.

Review of Antenna Systems and Amplitude Modulation

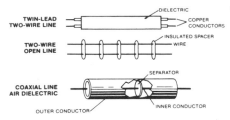

1. TRANSMISSION LINES have a basic property called *characteristic impedance*, whose value relates to cross-section geometry. Most coaxial lines have a characteristic impedance of 50 or 75 ohms, open-wire lines 300–600 ohms.

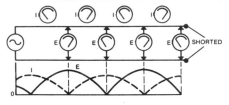

2. A LINE IS MATCHED when it is terminated in its characteristic impedance. When it is not matched, standing waves develop. SWR is the ratio of E_{max} to E_{min}.

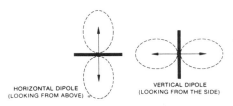

3. A SIMPLE ANTENNA is the *half-wave dipole*. It has directivity perpendicular to its length. Two or more antennas can be combined to form an *array* with greater gain.

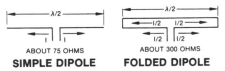

4. RADIATION RESISTANCE simulates the load offered by an antenna coupled to space. Antennas also have reactance unless they are resonant.

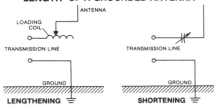

5. ANTENNAS MUST BE MATCHED to transmission lines for maximum efficiency and power transfer. Variable capacitors and/or inductors can be used for this matching.

Self-Test—Review Questions

1. What is the role of the antenna in transmission?
2. List some of the types of transmission lines used to carry energy from the transmitter or the antenna.
3. Why is it important that the transmission line be kept as short as practical?
4. What is SWR? What is VSWR? What is mismatch, and how does it affect performance?
5. What are the conditions required for match? Why is it important that the VSWR be kept to low values in solid-state amplifiers?
6. How does an antenna work to radiate energy?
7. Show the radiation patterns for horizontal and vertical dipole antennas. Define the horizontal and vertical polarization.
8. Why is the impedance of a folded dipole equal to 300 ohm, while a single dipole is about 75 ohm?
9. What are the impedance characteristics of antennas that are shorter than half wavelength? Longer?
10. What are antenna-coupling networks? What do they do? Why are they usually necessary?

Learning Objectives—Next Section

Overview—You now have a background in propagation, generation, and radiation of electromagnetic energy—all the tools needed to learn about the AM transmission of information that you will study next.

HIGH-LEVEL AMPLITUDE MODULATION

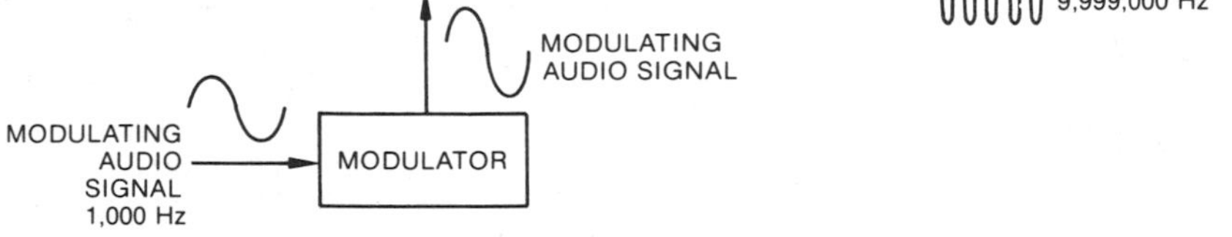

RF CARRIER 10,000,000 Hz → FINAL POWER AMPLIFIER → MODULATED RF OUTPUT

UPPER S.B. 10,001,000 Hz

CARRIER 10,000,000 Hz

LOWER S.B. 9,999,000 Hz

MODULATING AUDIO SIGNAL

MODULATING AUDIO SIGNAL 1,000 Hz → MODULATOR

What Modulation Is

Modulation is the technique used to *impress* information onto a *continuous-wave carrier signal*. There are four basic methods (with many variations) by which this is accomplished in a radio transmitter.

In the simplest method, *continuous-wave* (CW) transmission, the carrier signal is interrupted, or turned on and off, with a hand key or other device. The result is the familiar *dot-dash* or *dit-dah* signal. CW transmission was the method used by Guglielmo Marconi in 1895 when he sent the first radio message for a distance of over a mile. This method is now used primarily for long-distance communication by amateur radio and special communication systems. The signals are not readily audible, so receivers that detect this modulation have an oscillator (beat-frequency oscillator) tuned near the intermediate frequency to beat with the incoming signal to produce an audible tone.

In *modulated-continuous-wave* (MCW) transmission, a fixed-frequency audio signal of constant amplitude is used to amplitude modulate the carrier with constant tone. A key is used to turn tone on and off, as in CW transmission. MCW is used mainly for emergency transmission, and audio can be picked up by any AM receiver *without* the special oscillator.

Modern, high-speed communication systems use *frequency-shift keying* (FSK), where the space (0) is represented by one frequency and the mark (dot, dash, or 1) is designated by a frequency that is slightly different. Modern data transmission systems use very complicated digital multiplexed transmission that you will learn about later.

COMPARING CW, MCW, AND FSK TRANSMISSION

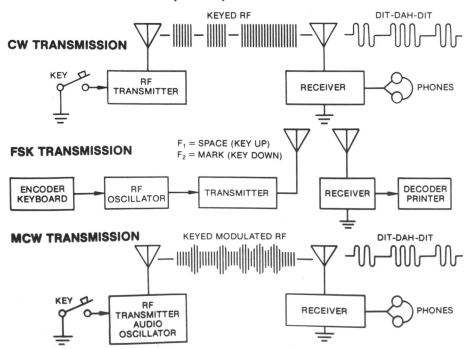

What Modulation Is (continued)

In *amplitude-modulated* (AM) transmission, the *amplitude* of the carrier is varied to *correspond* with the audio or *digital* signal generated by means of a microphone or other input transducer. This is the method of transmission used by Reginald A. Fessenden when he transmitted the first voice radio message in 1900. In 1906, he transmitted music by the same method of modulation. This is the type of transmission picked up today by a standard AM radio receiver.

Frequency modulation (FM) is another method of tansmission used to transmit voice and other signals. The *frequency* of the RF carrier signal is *shifted* or *deviated*, and the *rate* of deviation is *equal* to the *frequency* of the *input* signal. The first FM broadcast station was constructed by Major A. Armstrong in 1933.

Pulse modulation (PM) can be by either AM or FM. In AM pulse modulation, the carrier is usually switched on or off to represent a pulse or no pulse, respectively. This, as you can see, is a special case of AM. In FM pulse modulation, the carrier is shifted between two frequencies (f1 and f2) where f1, for example, is a pulse on and f2 is pulse off. Usually, FM pulse modulation is called FSK ((*frequency-shift keying*).

Other more complex methods of modulation are used to transmit pictures (video) by wire or radio, to transmit both pictures (video) and sound (audio) by means of TV, to transmit the left and right sound signals by stereophonic broadcasting, and to transmit the readouts (analog and digital) of scientific instruments in laboratories, aircraft, and spacecraft. All these methods are, however, based on the three fundamental methods of modulation that have been mentioned here.

COMPARING AM AND FM TRANSMISSION

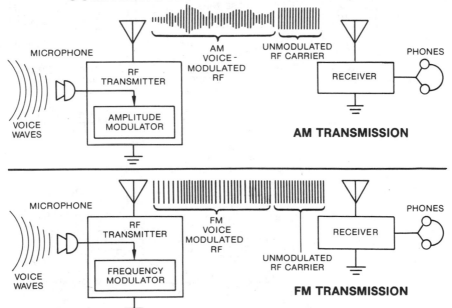

The Effects of Modulation

Here we discuss the amplitude modulation *combining* of *the information signal* with the RF *carrier signal*. The output will contain four different frequencies:

1. The *information signal*, which, for purposes of discussion here, we will consider to be an audio tone at 1,000 Hz.
2. The *unmodulated carrier RF output*, which we will consider here to be a 10-MHz signal.
3. The *sum* of frequencies 1 and 2 (10,001,000 Hz), which is known as the *upper sideband*.
4. The *difference* between frequencies 1 and 2 (9,999,000 Hz), which is known as the *lower sideband*.

The modulated signal consists of a carrier plus two sidebands that are 1,000 Hz away from the carrier on each side. The same information is in both sidebands, so that only one is really necessary. This principle is used in *single-sideband (SSB) systems*, which you will study later. Furthermore, the carrier is needed *only in the detection process* in the receiver, and, if its frequency is known, it can be reintroduced at the receiver. Therefore, if we transmit *only one* sideband (present only when the signal is modulated), we have communicated *all* of the *necessary information*.

THE AM CARRIER IS THE SUM OF THE 3 RF SIGNALS PRODUCED BY MIXING

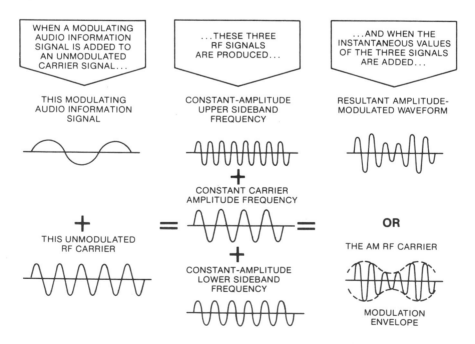

WHEN A MODULATING AUDIO INFORMATION SIGNAL IS ADDED TO AN UNMODULATED CARRIER SIGNAL...

...THESE THREE RF SIGNALS ARE PRODUCED...

...AND WHEN THE INSTANTANEOUS VALUES OF THE THREE SIGNALS ARE ADDED...

THIS MODULATING AUDIO INFORMATION SIGNAL

CONSTANT-AMPLITUDE UPPER SIDEBAND FREQUENCY

RESULTANT AMPLITUDE-MODULATED WAVEFORM

+

CONSTANT CARRIER AMPLITUDE FREQUENCY

+

THIS UNMODULATED RF CARRIER

= = OR

+

CONSTANT-AMPLITUDE LOWER SIDEBAND FREQUENCY

THE AM RF CARRIER

MODULATION ENVELOPE

More about Sidebands

If a 500-kHz carrier is modulated with a 2,000-Hz audio tone, the signal will contain, in addition to the carrier frequency, the *sum* (502 kHz) and *difference* (498 kHz) *frequencies.* The range of frequencies transmitted from the lower sideband to the upper sideband is known as the *bandwidth* of the transmitted signal. In the above example, the bandwidth is 4 kHz (498 kHz to 502 kHz). If the modulating audio signal frequency is reduced from 2,000 to 1,000 Hz, the bandwidth will be only 2 kHz (499 kHz to 501 kHz). It is the *sidebands* and *not* the carrier frequency that contain the information in the signal. If, for example, a receiver were to pick up only the carrier and *exclude* the sidebands, the only information available would be whether the carrier were on or off.

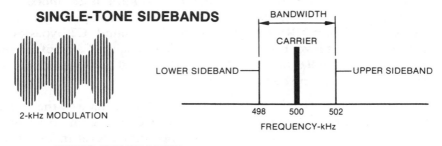

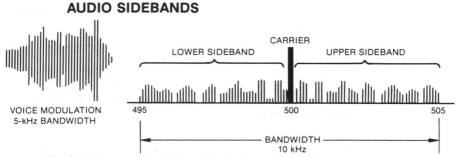

In broadcast transmission, the modulating signal may contain frequencies as high as 20 kHz. As a result, broadcast transmissions contain many complex sidebands. Thus, the transmission bandwidth may cover a range of frequencies as high as ±20 kHz around the carrier unless restricted. Generally, for voice or music transmission, a bandwidth as narrow as ±5 kHz or even less is adequate. For *high fidelity*, a *greater* bandwidth is desirable. There is a common misconception that AM transmission inherently is *incapable* of high fidelity. This is *not* so; it is only the *available bandwidth* of transmission that *limits* fidelity.

As you have probably observed, the upper and lower sidebands contain the same information and are essentially mirror images of each other. Therefore, only the upper or lower sideband is necessary to convey information. Therefore, the same information can be transmitted in half the band. This is called *single-sideband transmission,* and you will learn about this later in this volume.

Collector Modulation

The final RF power transistor amplifier can be modulated by applying the audio information signal to the base, emitter, or collector. Collector modulation is the most efficient and is the only one that can be used to obtain either low or high percentages of modulation. Collector modulation will be described here in detail, with a schematic diagram and applied signal levels shown on this page and the collector current and output signals emphasized on the following page.

An unmodulated RF signal is applied to the base of the transistor through transformer T1 or another suitable coupling. Fixed forward bias is applied to this base through the voltage divider formed by R1 and R2. The lower terminal of the T2 secondary winding is grounded to RF by capacitor C1. During operation, dc stabilization is provided by the bias developed across R3 from the emitter to ground. C2 bypasses the emitter to RF to prevent negative feedback. Capacitor C6 is sufficiently large to bypass the RF signal at this point but small enough not to have a significant effect on the audio signal.

Because of the fixed negative bias applied to the base by the voltage divider, the amplifier operates as a Class A or Class AB amplifier when a small input signal is applied. As the input-signal level increases, the amplifier operates more nearly as a Class B. As stated earlier, Class C operation of transistor power amplifiers is not usually practical.

Note that the audio transformer T3 is in series with the RF transformer T2. Thus, the instantaneous amplitude of the audio signal adds to and subtracts from the instantaneous voltage at the collector due to the RF signal. This *changing value* of collector voltage is the basis for achieving amplitude modulation.

COLLECTOR-MODULATED TRANSISTOR AMPLIFIER

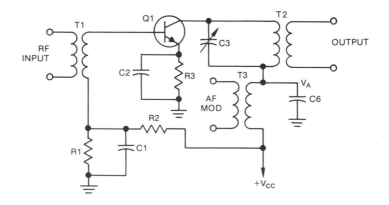

Collector Modulation (continued)

If no audio information signal is applied to T3, the RF carrier in the collector flows as a result of the RF signal applied to the base through T1. During this time, an amplified, constant-level RF signal appears at the output of T2.

The situation changes when an audio signal is applied to the input of T3. The audio signal voltage across the secondary winding of T3 is now in series with the effective collector voltage. Then, depending on the polarity of the audio voltage, the instantaneous value of the audio voltage will add to or subtract from the effective collector voltage.

Thus, when the upper end of T3 becomes negative, this voltage subtracts from the positive voltage of V_{cc}, and the result is that the effective collector voltage becomes more negative. In contrast, when the upper end of T3 becomes positive, this voltage adds the effective collector voltage, and the collector becomes more positive than V_{cc}. As the effective collector voltage varies, so does the power output from the transistor. The result is that the level of the RF power output signal rises and falls at the same rate as the audio signal, duplicating the changes in the level of the audio signal; in other words, it is *amplitude modulated*. It is apparent that power from the audio signal is necessary to achieve AM modulation. The audio signal power necessary is equal to half the RF power output. Thus, for a 10-watt transmitter, a 5-watt audio modulation signal is necessary. This will be discussed later when modulation percentage is considered.

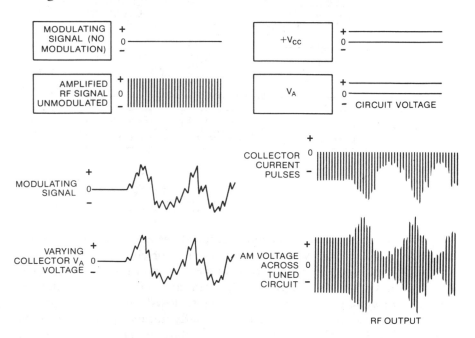

Methods of Amplitude Modulation

In the examples given earlier, the carrier was amplitude modulated by varying the voltage to the collector of the final RF amplifier in accordance with the modulating signal. If the output of the final amplifier is directly (and hence linearly) proportional to the collector voltage, this method works fine; however, this requires a modulator that has an audio power output equal to half of the power input to the final amplifier to achieve complete (100%) modulation. You will learn more about this in the next few pages. In some cases, as you will see, the final power amplifier and the preceding driver are modulated simultaneously to achieve the highest percentage of modulation.

It is also possible to do the modulation at low levels and follow the low-level RF amplifier by linear (Class B or Class A) amplifier stages since these amplifiers have a power output that is linearly proportional to the drive power. A Class C amplifier (not commonly used in solid-state amplifiers in any event) does not have a linear input-output relationship and hence *cannot* be used subsequent to the modulated stage.

HIGH-LEVEL AMPLITUDE MODULATION

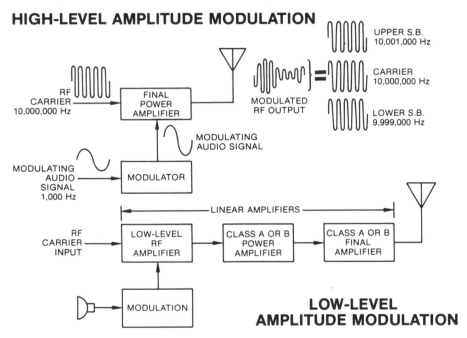

As you can see, low-level modulation saves audio power since the modulator need only drive the low-level collector circuit. This arrangement is, however, less efficient (from an RF standpoint) than a high-level modulation. As you will learn a little later in this volume, single-sideband transmission systems almost invariably use low-level modulation followed by linear amplifiers because it is usually necessary to generate the single-sideband signal at low levels using relatively complex circuitry.

AM Transmitter

An example of an amplitude-modulated transmitter is shown here in block diagram form. It is the *transmit* portion (mode) of a CB transceiver, which is a combination transmitter and receiver. Some audio circuits are used in both the *transmit* and *receive* sections and some are switched out and not used in one mode or the other. This transceiver will be recognized as the AM transmitter described earlier, but now the emphasis will be on the modulator and modulation.

As described earlier, the transmitter oscillator (at 4.3 MHz) is mixed with the local oscillator signals developed by the synthesizer, producing all the carrier frequencies required for the CB transceiver. The *mixer* output (at the transmitter output frequency) is coupled to the RF predriver that acts as a buffer stage and RF amplifier. The predriver output is coupled to an RF driver where the signal is amplified to the proper level to drive the final power amplifier.

The audio information signal used to modulate the carrier is controlled by the microphone push-to-talk (PTT) switch. The PTT switch connects the audio amplifier to the transmitter when in the transmit condition, and connects the audio amplifier to the loudspeaker in the receive position. In transmit, the audio signal developed by the microphone is coupled to an audio preamplifier. The preamplifier output drives the audio amplifier to the proper level to modulate the RF carrier.

The audio signal modulates the output of the RF driver and the final power amplifier. A coaxial relay, controlled by the push-to-talk switch, switches the antenna from the receiver input to the transmitter output. The modulated RF signal is then coupled to the antenna through the coaxial relay when transmitting.

BASIC BLOCK DIAGRAM OF AM CB TRANSMITTER

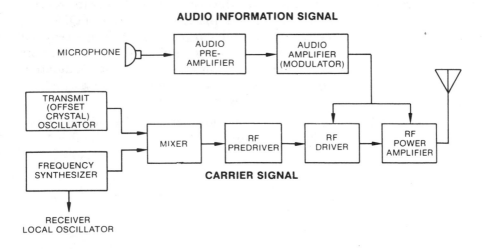

AM Transmitter—Audio Circuits

When the microphone push-to-talk (PTT) switch is held, the operating supply voltage (+13.8 volts) is switched from the receive circuits to the transmit circuits. This activates the oscillator and other circuits used for transmit only and removes power from circuits that are used for receive only. The PTT switch also switches the audio output from the speaker to the modulation transformer (T14) and couples the microphone output to the audio preamplifier (Q15) input.

The audio signal from the microphone is coupled to the base of Q15, the audio preamplifier. EP3 is a ferrite bead that forms an RF choke to keep the RF from entering Q15. The output signal from the preamplifier is RC coupled through C81 to the input of IC audio power amplifier U3. U3 is a high-gain amplifier that can produce up to 4 watts of audio output and is used to modulate both the RF driver and final RF amplifiers.

The amplified audio output from U3 (pin 1) is coupled through capacitor C89 and through the PTT switch to the modulation transformer T14. Frequency compensation on U3 and capacitor C78 (collector of Q15) limit the high-frequency response of the audio amplifier. The low value of input-coupling capacitors C79 and C81 restrict the low-frequency response. To achieve a high percentage of modulation without exceeding 100% modulation (to be described very shortly) on peak signals, part of the audio output from U3 is used to control the gain of Q15.

A part of the audio signal from U3 is taken from the secondary of the modulation transformer T14 through capacitor C93 to diode CR19. Diode CR19 rectifies the audio signal, and the resulting positive voltage is applied to the emitter of Q15 as reverse bias after being filtered by R88, R89, C92, and C77.

Therefore, the stronger the audio level at T14, the larger the reverse bias. This increases the current through the emitter resistor of Q15, and, since the base is held constant, the collector current of Q15 is reduced and hence the gain is decreased. This tends to keep the signal level almost constant with large variations in microphone output. The modulation signal is coupled through the modulation transformer T14 to the collectors of the RF driver and final power amplifier.

On the following page we present the schematic of the audio preamplifier and audio amplifier modulator.

AUDIO PREAMPLIFIER AND AUDIO AMPLIFIER MODULATOR

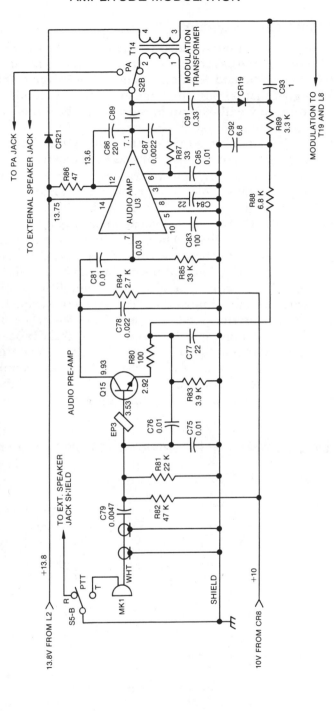

AM Transmitter—RF Section

Information content is added to the RF carrier by the modulator. The audio and RF signals are combined in the driver and final amplifiers to produce the modulated RF signal. The output is the amplitude-modulated RF signal that is coupled to the antenna. The following text and the schematic below will explain in more detail how the RF driver and power amplifier are modulated.

The unmodulated carrier signal from the synthesizer-mixer output transformer is applied to the base of the RF predriver transistor Q18. The output of the predriver collector is tuned to the transmit frequency range by coupling transformer T18 primary. The secondary of T18 couples the RF carrier signal to the base of the RF driver Q19.

The output of Q19 is coupled via another tuned transformer circuit to the final or power amplifier. The audio signal from modulation transformer T14 is applied to the collectors of both the RF driver Q19 and the power amplifier Q20. The audio signal varies both collector voltages simultaneously. Thus, the output of the power amplifier is the modulated RF carrier with the audio information contained in the modulation. The maximum average carrier power output of the power amplifier Q20 is 4 watts. This is the limit imposed by the FCC for CB operation. The modulated carrier is coupled to the antenna via the coupling network described earlier. Both the RF driver and the power amplifier (Q19–Q20) are modulated to achieve a higher percentage modulation than could be obtained from modulating only the final power amplifier.

MODULATING THE RF DRIVER AND POWER AMPLIFIER

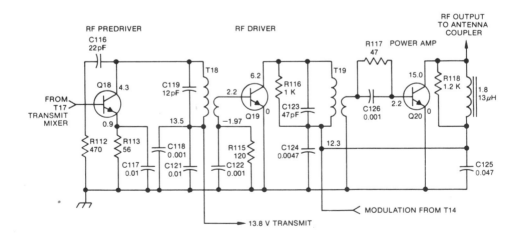

Percentage Modulation

The term *percentage modulation* is a measure of the extent to which the carrier is modulated. If the carrier is modulated 100%, the maximum amplitude of the modulated carrier is twice that of the unmodulated wave and the minimum amplitude is zero. In communication, the goal is the highest modulation percentage because the RF signal is then transmitted at maximum power. Increasing carrier level above the unmodulated value is often called *upward* or *positive modulation*, while decreasing carrier level is called *downward* or *negative modulation*.

If the maximum amplitude of the modulated wave is more than twice that of the unmodulated wave and the minimum amplitude is zero for any part of the cycle, the carrier is *overmodulated*, or the percentage modulation is more than 100%. In general, overmodulation is *undesirable*, and negative modulation that leads to carrier cutoff is *illegal*. This is because the transients generated when the carrier is suddenly cut off or turned on, as it is with excessive negative modulation, will generate additional sideband signals that can interfere with other transmissions on nearby frequencies. On the other hand, positive modulations of up to 125% are permitted. This asymmetrical modulation—accomplished by asymmetrical gain characteristics in the audio portion of the transmitter—leads to relatively little overall distortion. Such *supermodulation* gives a transmitter additional range of coverage for a given nominal carrier power and is quite useful. Many AM stations use this technique to increase their coverage area.

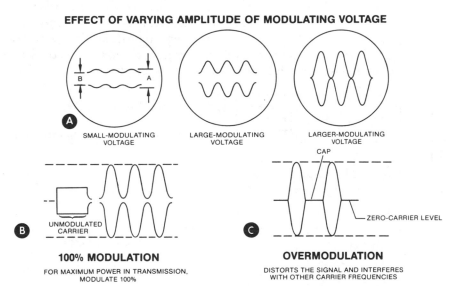

EFFECT OF VARYING AMPLITUDE OF MODULATING VOLTAGE

SMALL-MODULATING VOLTAGE

LARGE-MODULATING VOLTAGE

LARGER-MODULATING VOLTAGE

CAP

ZERO-CARRIER LEVEL

UNMODULATED CARRIER

100% MODULATION

FOR MAXIMUM POWER IN TRANSMISSION, MODULATE 100%

OVERMODULATION

DISTORTS THE SIGNAL AND INTERFERES WITH OTHER CARRIER FREQUENCIES

Percentage Modulation (continued)

Usually it is necessary to know the percentage of modulation rather exactly. If the maximum amplitude of the modulated wave is less than twice that of the unmodulated wave and the minimum amplitude is more than zero, the percentage modulation is less than 100%. This is the most common condition. To avoid overmodulation, typically most AM transmitters operate within an average percentage of modulation between 85% and 95%. For AM broadcast stations, the FCC requires an average of 85% modulation (minimum).

The exact percentage modulation can be calculated using the following formula:

$$\% \text{ modulation} = \frac{H_{max} - H_{min}}{H_{max} + H_{min}} \times 100$$

where H_{max} is the maximum height of the modulated wave and H_{min} is the minimum height. These values can be measured directly from oscilloscope pictures.

In the figures shown, H_{max} is 8 boxes and H_{min} is 2 boxes. The percentage modulation is therefore 60%. If H_{max} is 9 boxes and H_{min} is 1 box, the percentage modulation is 80%.

Since modulation percentage control is so important, it must be continuously monitored at commercial AM broadcast stations. Here they use an instrument—the *modulation monitor*—which employs the principle outlined above to allow for monitoring upward, downward, and total modulation percentage. The modulation monitor measures the average carrier amplitude as well as positive and negative peaks, and performs the calculation which in turn is displayed on a meter. An indicator lamp flashes when excessive modulation is present to alert the operator.

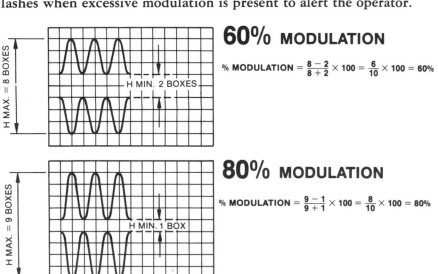

60% MODULATION

% MODULATION $= \frac{8-2}{8+2} \times 100 = \frac{6}{10} \times 100 = 60\%$

80% MODULATION

% MODULATION $= \frac{9-1}{9+1} \times 100 = \frac{8}{10} \times 100 = 80\%$

Using the Oscilloscope to Determine Modulation Percentage

The oscilloscope can be used to indicate the percentage to which the output of a transmitter is modulated. This method is often used to check equipment such as our CB learning system where modulation percentage is measured only on occasion. It can also point out distortion existing in the modulation. If a pickup loop, which is connected to the scope input terminals, is brought close to the output circuit of a modulated transmitter, the scope will show the modulation pattern if it has sufficient bandwidth or if the connection is made directly to the deflection plates.

If the modulating voltage is a sine wave and the sweep (called the *time base*) inside the oscilloscope is used, the patterns shown below are obtained. The first pattern is useful in determining the presence of distortion.

The second pattern would indicate that the positive peaks of the modulating voltage are not causing corresponding peaks in output current. This may be due to improper bias, saturation, or insufficient excitation of the power amplifier stage by the preceding stage.

If the transmitter output shows breaks in the modulation pattern, as shown in the third pattern, the transmitter is *overmodulated*. This is usually due to an excessive modulating signal but may also be due to insufficient carrier voltage on the RF power amplifier input.

The last pattern shows a signal with about 100% modulation. In this case, the peaks are at twice the unmodulated value, and the minima are just at the baseline.

TIME-BASE MODULATION PATTERN

NORMAL SINE WAVE
MODULATION

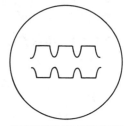

IMPROPER ADJUSTMENT
CAUSES FLAT POSITIVE
PEAKS

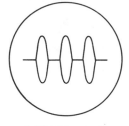

OVERMODULATION

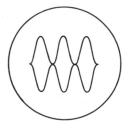

100% MODULATION

Using the Oscilloscope to Determine Modulation Percentage (continued)

The trapezoid figure is another type of oscilloscope pattern which is often used to determine the percentage of modulation of an AM signal. The advantage of using trapezoid figures over time-base modulation patterns to analyze percentage modulation is that they are easier to interpret. To produce the trapezoid figure, the audio modulating signal is used as an external horizontal-sweep signal instead of the internal sweep of the oscilloscope. The vertical-deflection circuits are connected or coupled to the modulated RF output of the transmitter.

A typical setup for showing trapezoid figures is illustrated here. The vertical input of the scope is coupled to the output coil of the power amplifier, and the horizontal input is coupled to the audio output of the modulator. Usually attenuators are necessary in the line to prevent overload. Often, to avoid bandwidth problems in the oscilloscope, the vertical-deflection circuit is bypassed and the RF signal is fed directly into the vertical-deflection plates.

Shown are the two oscilloscope presentations for various modulation percentages. Either of these configurations enables you to measure the maximum amplitude (peak) and minimum amplitude (valley) of the RF—and thus modulation percentage.

TO RF OUTPUT — VERT.   HORIZ. — TO AVOID MODULATING SIGNAL

EFFECT OF VARYING MODULATION PERCENTAGE

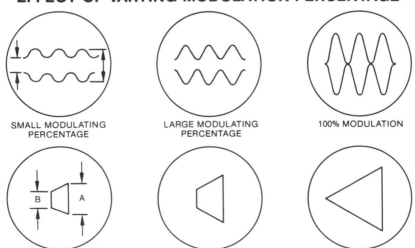

SMALL MODULATING PERCENTAGE

LARGE MODULATING PERCENTAGE

100% MODULATION

Amplitude Compression

To solve problems of occasional high-amplitude peaks in modulation, either a *speech clipper* or a *volume compressor* is often used. In operation, the preamplifier gain is adjusted so that the peaks are clipped (limited) in the clipper. Thus, these peaks are removed from the signal. The filter is necessary to eliminate the high-frequency transients produced by the clipper. Since this clipping introduces distortion for music, a *volume compressor* is often used. This device has a variable gain amplifier, with the gain controlled inversely to the signal level. Thus, when the audio signal is weak, the gain is high; and when the audio signal is strong, the gain is low. In this way, the fluctuations in signal level are reduced.

SPEECH CLIPPER TO IMPROVE MODULATION PERCENTAGE

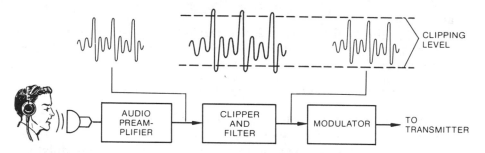

VOLUME COMPRESSION TO IMPROVE MODULATION PERCENTAGE

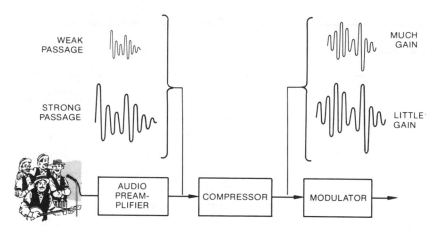

In the compressor, the audio input signal is rectified and used to control the gain of an amplifier through which the signal is passed. This is similar to the circuit that you observed a few pages earlier for the CB learning system.

AM Single-Sideband Power Relationships

In the conventional double-sideband (DSB) AM signal, the information component in each sideband is in phase. For a sine-wave modulating signal, the power in each sideband is equal to 25% of the carrier power at 100% modulation. For both sidebands, therefore, the power is 50% of the carrier power. This power must come from somewhere, and indeed it represents the audio power supplied (in addition to the dc power input) to the final RF amplifiers by a high-level modulator. This is the reason why an audio power of 50% of the RF amplifier dc power input is required for 100% modulation. A reduction in modulation percentage reduces the sideband energy content (power). Since the sidebands have 50% as much power as the carrier, they represent only one-third (33.3%) of the total power because

$$P_T = P_c + P_s$$

where P_T is the total power, P_c is the carrier power, and P_s is the total sideband power. $P_T = 1.5$ times the carrier power for 100% modulation, and therefore

$$\text{Sideband power} = \frac{P_s}{P_T} = \frac{0.5}{1.5} = \frac{1}{3} \text{ (100\% modulation)}$$

Thus, each sideband has half of the total sideband power or $\frac{1}{6}$ of the total power (16.7%). Since speech and music have a peak-to-average-value ratio much greater than for a sine wave, these average modulation percentages cannot be made as high as for a sine wave unless compression or limiting are used, as described earlier.

TOTAL POWER IN CARRIER AND SIDEBANDS

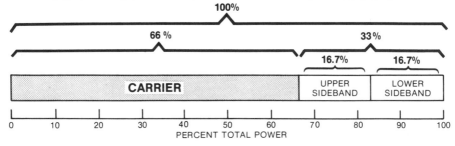

There are several points to be noted in a double-sideband AM signal: (1) The information-bearing sidebands represent only one-third ($\frac{1}{3}$) of the total radiated power. (2) The upper and lower sidebands contain the *same* information. (3) The radiated signal bandwidth is *twice* the bandwidth of the highest modulating frequency. It is apparent that double-sideband AM transmission is *relatively inefficient*; however, it leads to *very simple receiver designs*, as you will learn in Volume 4. It is also apparent that if only a *single* sideband were transmitted, *all* the information necessary would be transmitted. This approach, called *single-sideband transmission*, will be described next.

Experiment/Application—AM Percentage Modulation

You can use a very simple setup to see how modulation percentage can be measured with an oscilloscope. What you need to do is really build a small transmitter. This is most easily done using a signal generator and a simple modulated power amplifier driven by an audio oscillator. Suppose you wired up the circuit shown below. The RF signal comes from a signal generator tuned to 455 kHz, and the audio signal for modulation is a tone from an audio generator. The modulated stage uses a simple collector-modulation scheme.

MEASURING MODULATION PERCENTAGE

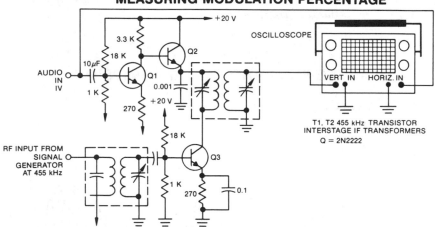

With no audio input, and the horizontal sweep on internal, the scope picture should appear as in part A, since the carrier level does not change. If an audio input is added, the carrier will be amplitude modulated, and the scope picture will appear as in part B. If the horizontal input is coupled to the audio oscillator, you will see the familiar trapezoidal picture shown in part C.

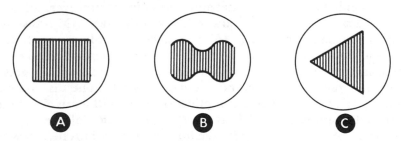

You can vary the percentage modulation by changing the audio signal generator output, and you will observe that the depth of modulation changes. You can overmodulate and see the effects of carrier cutoff by putting a large audio input into the modulator.

Single-Sideband Transmission

As stated on the previous page, *all* of the information to be transmitted is in *each* sideband; therefore only one sideband needs to be transmitted. If we had a system that *suppressed* one sideband, we could transmit the same information in half the bandwidth, thus conserving frequency spectrum (twice as many stations over the same frequency range). Such transmission is called *single-sideband transmission*. However, we still would be transmitting the carrier, and this represents an inefficiency since the carrier contains no information. We can carry the process one step further and *suppress* the *carrier* as well as one sideband. This is called *single-sideband suppressed-carrier transmission* and is the technique used today in what we call simply *single-sideband transmission* or SSB.

SINGLE-SIDEBAND AM TRANSMISSION

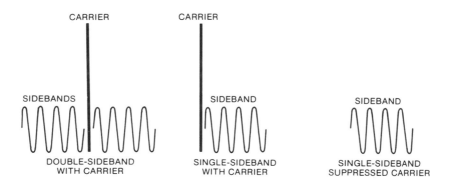

There are important advantages for SSB transmission. As you know, for a given transmitter power output, only 16.7% of the energy appears in each sideband. For a single-sideband transmitter, however, all of the transmitted signal is sideband power. Thus, for a 1-kW AM transmitter, the single-sideband content is 167 watts. For an SSB transmitter, the single-sideband power will be 1 kW. This greatly increases the range of the transmitter. In addition, with no modulation, no signal is present. This greatly reduces interference between signals. The disadvantages are mainly in greater receiver complication. You will learn about interference reduction and special receiver problems in Volume 4. At the present time, SSB transmission is used for communication only. However, it has important potential for AM broadcast use. As you know, stereo requires two channels. One way to accomplish this is to use the sidebands separately for each channel.

SSB Transmitter Block Diagram

The basic single-sideband signal is obtained as shown in the block diagram. Usually, the SSB signal is developed in two steps. A balanced modulator is used to generate a dual-sideband suppressed-carrier signal. This signal is then filtered to pass the desired sideband and reject the other. The result is a single-sideband suppressed-carrier signal.

In an SSB transmitter the RF carrier is applied to a balanced modulator along with the audio signal. This results in a double-sideband suppressed-carrier signal at the output. By any one of several techniques to be described, the upper or lower sideband is suppressed, leaving the SSB signal. This signal is amplified in a linear RF amplifier (Class A, AB, or B). The output is then coupled to the antenna. Often, the SSB signal is generated at a lower frequency than is to be transmitted since the modulator-filter is more easily built at lower frequencies. In this case, the signal is converted to proper frequency by mixing it with an appropriate carrier signal to raise it to the desired frequency for transmission.

An important aspect of SSB transmitters involves the extent to which the carrier and the unwanted sideband are suppressed relative to the desired signal, since the rejection of these unwanted signals is not infinite. Typically, at least 30 dB (1,000:1) rejection is desired. To achieve this performance, careful adjustment of the equipment is necessary. It is also important that the RF amplifiers be linear; otherwise nonlinearity can cause a reduction in the sideband rejection.

BLOCK DIAGRAM-SSB TRANSMITTER

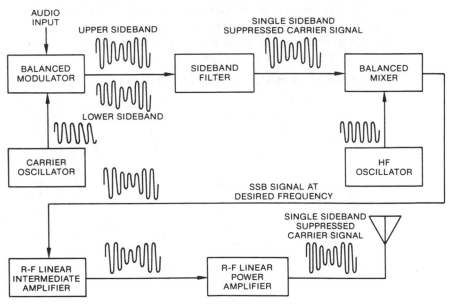

Transistor Balanced Modulator

The first step in single-sideband transmission is to modulate the carrier so that two sidebands are produced and the carrier is suppressed. This is typically accomplished by a circuit called the *balanced modulator*. As you can see, this circuit appears similar to a push-pull amplifier.

As shown in the diagram, the audio modulating signal is applied to the bases of Q1 and Q2 through center-tapped transformer T1 and consequently appears out of phase at the bases of Q1 and Q2. Also, the RF carrier is applied to the bases of Q1 and Q2 through balancing potentiometer R3, and consequently is in phase at bases of Q1 and Q2.

With only an RF signal applied, the arm of balancing potentiometer R3 is adjusted so that equal and opposite RF currents are produced in the primary of output transformer T2. The balancing potentiometer is used to compensate for differences in Q1 and Q2. Because the output currents of Q1 and Q2 are out of phase, no RF signal voltage is developed across the T2 primary winding. Thus, there is no RF output from the T2 secondary winding. When the audio signal is applied, the currents through the transistors are unbalanced, since the audio signal is out of phase at the bases of Q1 and Q2. Therefore, an RF signal voltage appears across T2 which is proportional to the unbalance (modulating audio signal). Thus when an audio signal is present, an RF output is produced at the secondary of T2.

The transistors are operated in a linear mode, and the RF output is linearly proportional to the unbalance induced by the audio signal. Capacitors C1 and C6 are RF and audio bypass capacitors; therefore they are relatively large. Capacitors C2, C3, and C4 are chosen to have a low impedance to the RF frequency but a high impedance to the audio signal. Transformer T2 is an RF transformer, and the audio component is not passed on to the next stage. In some cases, this transformer is tuned to the RF frequency. In other cases, an untuned transformer is used.

TRANSISTOR BALANCED MODULATOR

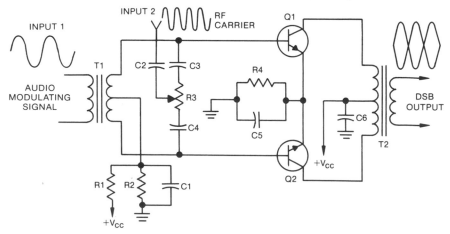

An IC Balanced Modulator

The circuit shown here can be used as an integrated-circuit balanced modulator that functions in essentially the same way as the transistor circuit described on the previous page. Since it is an IC, the balance of the transistors will be excellent, and no adjustment is necessary. The basic circuit is a balanced differential amplifier (Q1–Q2) driven from a constant-current source (Q3). The modulating signal is applied to the base of the constant-current source that drives the emitters of transistors Q1 and Q2. The RF signal is therefore in phase at the collectors of Q1 and Q2, and hence no RF output is produced across T2.

The audio signal is supplied to the base of one of the amplifiers (Q1). However, as you will note, Q1 and Q2 are driven by constant-current source Q3. Thus, as the current in Q1 is changed by the variation of the input audio signal, the current in Q2 must also vary, but in the *opposite* direction. Thus, the conditions in output transformer T2 are the same as for the previous circuit (see the previous page). A carrier suppression of greater than 25 dB (about 300:1) is claimed for this circuit with a carrier of 10 mV and a modulating signal load of 31.5 mV. As you can see, the signal levels used are low to preserve linearity. Nowadays, there are many ICs that act as balanced modulators for SSB signal generation.

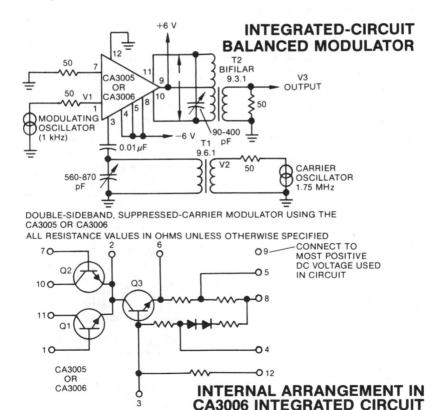

INTEGRATED-CIRCUIT BALANCED MODULATOR

DOUBLE-SIDEBAND, SUPPRESSED-CARRIER MODULATOR USING THE CA3005 OR CA3006
ALL RESISTANCE VALUES IN OHMS UNLESS OTHERWISE SPECIFIED

INTERNAL ARRANGEMENT IN CA3006 INTEGRATED CIRCUIT

Balanced Diode-Ring Modulator

The balanced diode circuit will also produce a suppressed-carrier signal. As you will see, the operating principle is the same as before; the RF carrier current in the output transformer flows in a direction opposite from the center tap, while the sidebands unbalance this condition to produce an output.

During the first half of the RF cycle (see the left diagram) CR2 and CR4 are forward biased, and CR1 and CR3 are reversed biased. Because T1 and T3 are center-tapped transformers, the RF current divides equally along the paths marked by the arrows. Consequently, the RF current flow through the two halves of the primary of T3 are equal and opposite, and no net RF current is developed across the ends of T3.

During the next half cycle (see the right diagram), the diode bias conditions are reversed. Although the resulting currents are in the opposite direction, in the two halves of the primary winding of T3 the currents are still equal and opposite, and no RF carrier voltage is developed across the ends of T3.

When the audio signal is applied via T1 (typically at 0.1 of the amplitude of the carrier), the actual bias (sum of audio and RF) applied to each diode is varied, leading to an unbalance in the currents through the two primary halves of T3.

Thus, with no audio signal, the RF carrier signal components are equal and produce no RF carrier output voltage. Because of the difference in diode-biasing produced by the audio signal, the sideband-current components are not equal in the primary of T3, and RF upper- and lower-sideband signals are produced. Because T3 is an RF transformer, the audio component is not transmitted to the next stage. IC diode packages containing all four diodes carefully matched are available and are often used when this configuration of balanced modulator is used.

FIRST HALF CYCLE
CR2 AND CR4 CONDUCT

SECOND HALF CYCLE
CR1 AND CR3 CONDUCT

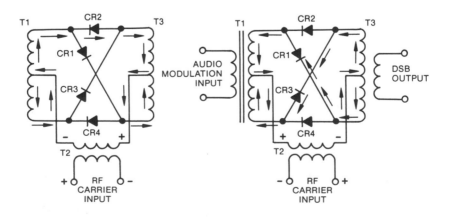

Experiment/Application—DSB Balanced Modulator

You have seen how a balanced modulator suppresses the carrier signal to produce a double-sideband suppressed-carrier output signal. You can perform a simple experiment to illustrate this point. Suppose you hooked up the circuit shown below using an RF signal generator as the carrier source and an audio signal generator as the modulating signal.

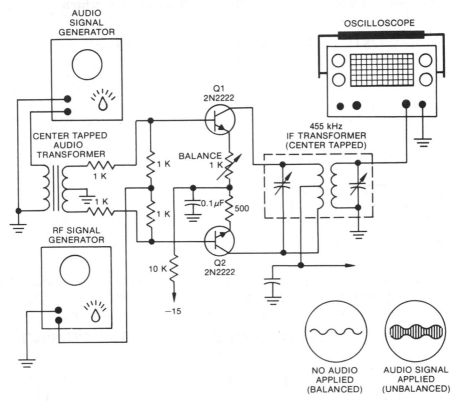

NO AUDIO APPLIED (BALANCED)

AUDIO SIGNAL APPLIED (UNBALANCED)

The exact operating frequency is not important, except that the IF transformer should be properly tuned. With no audio input, the balance adjustment is used to balance the circuit so that no (or little) RF appears on the oscilloscope. This occurs because the RF signals are in phase at the collector and hence cancel each other in the push-pull output circuit. This phasing occurs because the RF signal to the bases of Q1 and Q2 is the same.

If you now apply an audio signal to the input as shown, the push-pull audio signal (from the center-tapped audio transformer) unbalances Q1 and Q2 and produces an output proportional to this unbalance, as shown. The major points to note are the lack of carrier with no modulation and its presence only when a modulating signal is applied. Note that the amplitude of the sideband signal is proportional to the amplitude of the modulating signal.

Methods of Sideband Suppression—Filtering

Subsequent to carrier suppression, the simplest way to eliminate the undesired sideband is by filtering it out by means of a suitable filter. While this is easy to do in principle, it is difficult to do in practice because the upper and lower sidebands can be very close when a low modulation frequency is used. For example, a 50-Hz component of the modulating signal will produce sidebands ±50Hz from the carrier, which as a practical matter makes them difficult to separate.

FILTER METHOD OF SSB GENERATION

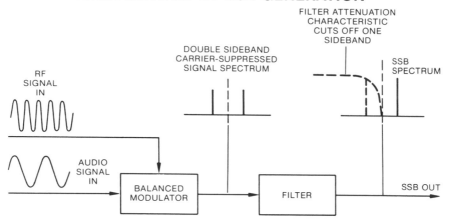

A filter with very sharp frequency cutoff characteristics is necessary. In addition, f_o must be carefully controlled since it would affect the suppression of the unwanted sidebands if it varied by even a few hertz.

EFFECT OF FREQUENCY SHIFT ON SSB FILTERING

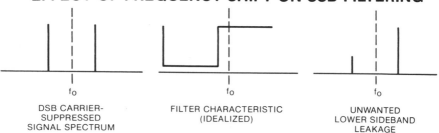

If f_o shifts to f_o', the filter is not able to suppress all of the unwanted sideband. Some comes through, particularly at low audio frequencies. If f_o' shifts the other way, some of the desired sideband signal is lost. The same problem will arise if the passband of the filter shifts as a result of component aging or temperature changing of component values, etc.

In spite of these problems, the filter method is used in commercial equipment designed for voice communication where low-frequency response is not terribly important and thus eases the filtering and frequency-stability problem. The special filters for this application are usually packaged as a single unit using quartz or ceramic elements in the filter.

Methods of Sideband Suppression—Phasing

A once popular method for sideband suppression to generate an SSB signal from a DSB signal is known as the *phasing method*. This method was originally put forth by R.V.L. Hartley, who also developed the Hartley oscillator. In this method, one of the two sidebands is reversed in phase and added to the original DSB signal, resulting in cancellation of one sideband and enhancement of the other. The phasing method is used mainly because it is easier to implement in some respects and because it does not have the critical frequency-control requirements of the filter method.

In this method, two balanced modulators are used. Although they are identical, as shown below, one (B) is fed with modulating and RF carrier signals that are shifted 90 degrees in phase with respect to the other (A).

PHASING METHOD OF SIDEBAND SUPPRESSION

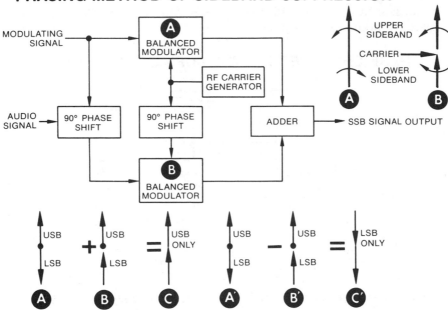

As shown in the vector diagram, balanced modulator A produces two sidebands that have a 180-degree phase relationship to each other—that is, the same sideband components on each side of the carrier are 180 degrees out of phase instantaneously. On the other hand, the phase shifting of the RF carrier combined with the phase-shifted modulating signal causes both sidebands to be in phase at the output of balanced modulator B. When the sideband signals from A and B are added vectorially, as shown, the result is the upper sidebands (USB) add together and the lower sidebands (LSB) cancel. The lower sideband can be obtained by reversing the phase of the network or by placing the RF phase shifter in the line to balanced modulator A.

Methods of Sideband Suppression—Phasing (continued)

An easier way is to change the adder to a subtractor. Thus, the upper sideband is suppressed and the lower sideband is enhanced, as shown.

Expressed arithmetically, using + for in phase and − for out of phase with respect to the carrier, the relationships are

$$\text{for USB:}\quad \overbrace{(\text{USB} - \text{LSB})}^{A} + \overbrace{(\text{USB} + \text{LSB})}^{B} = 2\ \text{USB}$$

$$\text{and for LSB:}\quad \overbrace{(\text{USB} - \text{LSB})}^{A} - \overbrace{(\text{USB} + \text{LSB})}^{B} = -2\ \text{LSB}$$

While 90-degree phase shifters are easy to build for the single-frequency RF carrier, they are very difficult to build to cover the multi-octave frequency range of the modulating signal. It turns out that only a 90-degree relative phase shift is required, so therefore the audio phase shifter is split into two with partial phase shifts in each leg feeding the balanced modulators, as shown.

PHASE-SHIFT METHOD OF SIDEBAND SUPPRESSION

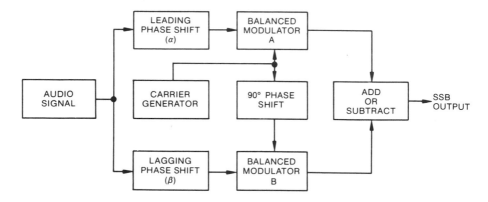

As shown, two phase shifters for the modulating signal are used, and now the only requirement on the modulating phase shifter is that $\alpha + \beta = 90$ degrees at all times. This is much easier to manage since to maintain a relative phase shift of 90 degrees over the modulating band accomplishes the same result and is considerably easier. For any system, the SSB signal is generated at a relatively low frequency and then reconverted to the appropriate frequency for amplification.

Single-sideband operation has become very popular for CB and other communication systems. Typically, these systems have a switch so that conventional AM or single sideband is available with a choice of the upper sideband (USB) or lower sideband (LSB). In the circuit shown on the next page, a balanced diode modulator is used to generate the double-sideband suppressed-carrier signal. This DSB signal is then filtered using a crystal filter to remove the unwanted sideband.

A CB Single-Sideband Transmitter

In the CB SSB transmitter shown below, the balanced modulator (CR604, CR605, CR606, and CR607) has the carrier (at 7.8 MHz) applied via balancing potentiometer R609. The audio input is applied to RF transformer T602. The DSB suppressed-carrier signal is applied to the sideband filter FL501, which feeds the Q501 RF amplifier. The output of Q501 is fed to one base of the FET mixer Q701. The other base is supplied with the frequency synthesizer output that is 7.8 MHz below the desired output frequency. The sum output from the mixer is amplified and provides the RF output signal for transmission. Because the crystal filter frequency is fixed, the synthesizer output frequency is shifted slightly (2.5 kHz) to center the desired sideband in the filter band.

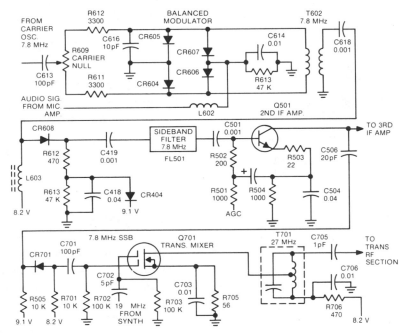

Since no carrier is normally present in an SSB transmitter, it is necessary to use some other basis for power measurement than is used for conventional AM. The maximum AM power output allowed for a CB AM transmitter is 4 watts. For single-sideband operation, the power is measured as *peak effective power* (PEP). This is the peak power output from the SSB signal when 100% modulated. This corresponds to a PEP of 12 watts as allowed by FCC regulations. Since the transmitter is only delivering power when modulated, and voice transmission has a high peak-to-average value, clipping circuits are used to improve this ratio so that the maximum energy content can be obtained in the signal. Since one-third of the power is in each sideband, a 100% modulated SSB signal has a peak level of 12 watts for a 4-watt-average power transmitter. This allows the same energy in the information-carrying portion (the sidebands) as was available for conventional AM modulation.

Review of Amplitude Modulation Systems

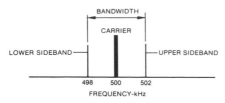

1. THE INFORMATION SIGNAL is impressed on the carrier by *modulation*. The frequency spectrum of an amplitude-modulated signal consists of a carrier and two *sidebands*.

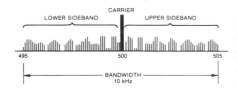

2. SIDEBAND COMPONENTS in amplitude modulation extend each side of the carrier by the modulation frequency they represent. With voice or music, sidebands contain many frequency components.

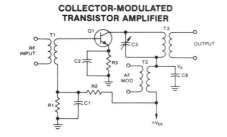

3. AMPLITUDE MODULATION is accomplished by varying current in a stage of the RF section in accordance with the AF signal. One common method is *collector modulation*.

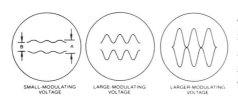

4. DEGREE OF MODULATION is indicated as a percentage, and for AM, is dependent upon AF power relative to RF power. Full modulation is at 100%.

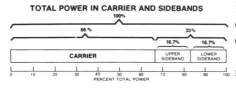

5. IN AN AM SIGNAL, two-thirds of the power is in the carrier and one-third in the two sidebands, for 100% modulation. AF power in the modulator is then half the RF power.

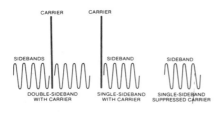

6. IN A SINGLE-SIDEBAND SIGNAL, the carrier and one sideband are suppressed. This increases efficiency because all power goes into the information portion of the signal.

Self-Test—Review Questions

1. Discuss modulation from an information-transmission standpoint.
2. What are sidebands? When are they generated? Where is the information in an AM-transmitted signal?
3. List with a brief description some methods for amplitude modulation. Show block diagrams.
4. Define percentage modulation. Why should it not exceed 100% for downward modulation? Is it possible to have more than 100% upward modulation? How?
5. Why are amplitude compressors used in voice communications? What do they do?
6. What are the power relationships between the carrier and the sidebands in an AM system?
7. Why is the bandwidth of an AM signal controlled? What is the difference between single- and double-sideband transmission?
8. What is single-sideband transmission? Why is it used? Are there advantages? How does an SSB transmitter differ from a conventional AM transmitter?
9. Draw the schematic diagram to show a transistor balanced modulator and describe how it works and its role in development of an SSB signal.
10. List some of the techniques for sideband suppression, and briefly describe how each works.

Learning Objectives—Next Section

FM Transmission

Overview—Now that you understand AM transmission, you are ready to learn about FM transmission and how it is similar to and different from AM.

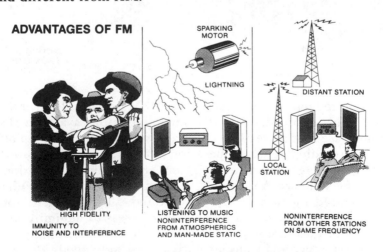

ADVANTAGES OF FM

SPARKING MOTOR

LIGHTNING

DISTANT STATION

LOCAL STATION

HIGH FIDELITY
IMMUNITY TO NOISE AND INTERFERENCE

LISTENING TO MUSIC NONINTERFERENCE FROM ATMOSPHERICS AND MAN-MADE STATIC

NONINTERFERENCE FROM OTHER STATIONS ON SAME FREQUENCY

Why FM?

Frequency modulation (FM) has become very popular for some applications such as high fidelity broadcasting, TV audio, and VHF-UHF communications. One of the major advantages of FM is that it can be relatively immune to interference from electrical noises like sparking motors, lightning, etc. This is essentially a property of the *receiver*, based on the method of transmission, and will be discussed in Volume 4. To transmit the same information as in an AM system, an FM system generally requires a *greater bandwidth*, particularly if its noise- and interference-reduction properties are to be maximized. This is so because as the bandwidth of an FM system narrows and approaches an AM system's bandwidth, the noise-reduction properties of the FM system approach that of the AM system. Another valuable property of an FM system (again a property of the *receiving system*) is that when two interfering stations are present, only the *stronger* one is heard if the signal strength difference is appreciable. This very effectively eliminates interference from other stations on the same frequency.

ADVANTAGES OF FM

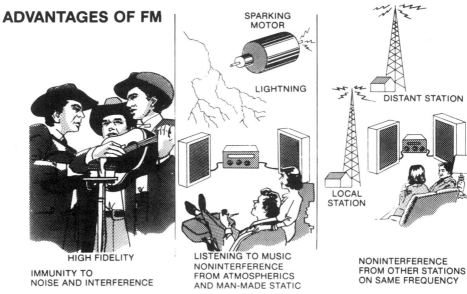

HIGH FIDELITY

IMMUNITY TO
NOISE AND INTERFERENCE

SPARKING MOTOR

LIGHTNING

LISTENING TO MUSIC
NONINTERFERENCE
FROM ATMOSPHERICS
AND MAN-MADE STATIC

DISTANT STATION

LOCAL STATION

NONINTERFERENCE
FROM OTHER STATIONS
ON SAME FREQUENCY

As pointed out earlier, both AM and FM are capable of producing the *same* fidelity of broadcast if sufficient bandwidth is available. Because spectrum space is at a premium at longer wavelengths (because of their utility in providing reliable operation over greater than line-of-sight distances), the longer-wavelength regions are crowded with stations. For example, the standard AM broadcast band (535 to 1,605 kHz) has 107 channels spaced by 10 kHz. By complicated calculations, stations are assigned operating frequencies in different areas to minimize interference. Many of these are so-called *clear-channel stations* that are protected against interference and transmit with surprisingly good fidelity.

Why FM? (continued)

FM transmission, because of its relatively great bandwidth requirements, is confined generally to the UHF-VHF regions where more spectrum space is available but transmission is limited to almost line of sight. For example, the FM broadcast band extends from 88 to 108 MHz, and consists of 100 channels, each 200 kHz wide. The availability of this channel space for FM allows for not only high fidelity of transmission for all stations, but also allows for easy transmission of multiple signals for stereo, etc. You will learn how this is done in this section.

TV audio uses FM transmission and nominally occupies a band that is part of the TV station channel-frequency allocation. FM is used for these applications primarily because of immunity to AM as well as for fidelity. As you will see later in Volume 4, the use of FM for TV audio allows for a very simple system of reception of TV sound, which must have high immunity to AM for operation. Narrow-band FM used for UHF and VHF communications typically has a channel width of 10 kHz.

In summary, FM has noise-immunity advantages and can (under the proper circumstances) provide higher fidelity. These are obtained at the expense of greater bandwidth than equivalent AM. Thus, FM operation is generally confined to the UHF-VHF region where line-of-sight conditions of propagation prevail, and more spectrum space is available.

DISADVANTAGES OF FM

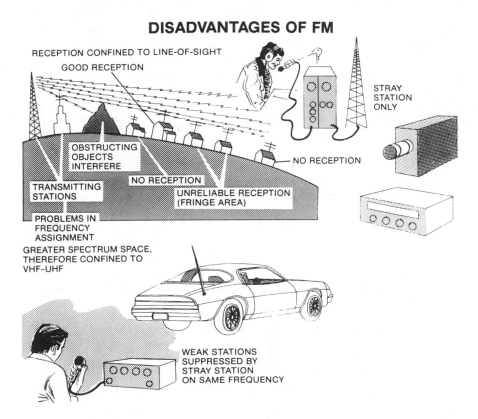

Unmodulated and Modulated Carriers

Understanding the nature of frequency modulation can be most easily accomplished by examining the simplest form of FM transmitter. Suppose you have an RF oscillator, such as a Hartley oscillator whose frequency is controlled by an LC circuit. Buffer amplifiers and power amplifiers usually following the oscillator are omitted in this discussion.

As you know, the RF signal generated by the oscillator is an RF sine wave of constant amplitude. With a fixed inductor and variable capacitor, the frequency of the generated signal is determined by the setting of the variable capacitor. The frequency increases as the capacitor is moved toward minimum capacitance, and the frequency decreases as the capacitor is moved toward maximum capacitance.

Assume, for example, that when the variable capacitor is set to the center of its range, the frequency generated is 1 MHz. Frequency modulation is obtained by varying the capacitance over some range from this center position. The frequency range covered is the frequency deviation.

If the capacitor shaft is mechanically oscillated over the range about its center position at a rate of 40 times per second, the frequency of the oscillator will also shift back and forth about the center frequency at a 40-Hz rate. If the capacitor shaft is moved at a rate of 400 or 4,000 times per second, the shift rate will increase to 400 or 4,000 Hz, respectively. If the capacitance varies over the same range regardless of the driving frequency, then you can see that the frequency change or frequency deviation would be the same in all cases. However, the rate of change of carrier frequency would be different in each case, since the frequency would have to shift 10 times faster at 400 Hz as compared to 40 Hz.

RATE OF RF CARRIER FREQUENCY SHIFT EQUALS RATE OF INPUT SIGNAL CHANGE

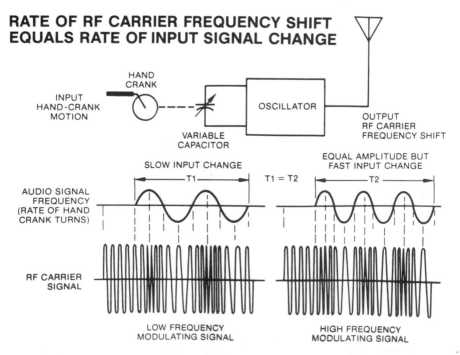

Unmodulated and Modulated Carriers (continued)

Actually, the two factors are not separable. If you were to increase the frequency shift by a factor of 10 at 40 Hz, the rate of change would be the same as the 400-Hz signal at one-tenth the frequency shift.

As you learned on the previous page, an FM signal has two properties: (1) the magnitude of the frequency shift or frequency deviation, and (2) the rate at which the carrier is shifted. The transmitter capacitor shaft may be varied by only a small amount on each side of its center position—for example, varying the frequency 1 kHz above and below the 1-MHz center frequency. On the other hand, the capacitor shaft may be varied at the same rate (same number of variations per second) but by a larger amount—for example, varying the frequency 10 kHz above and below the 1-MHz center frequency. As you can see, the number of variations is held constant although the frequency deviation is changed (the number of times the RF traverses the band is constant). On the other hand, the total band covered is proportional to the frequency deviation.

MAGNITUDE OF RF CARRIER FREQUENCY SHIFT IS DIRECTLY PROPORTIONAL TO AMPLITUDE OF INPUT SIGNAL CHANGE

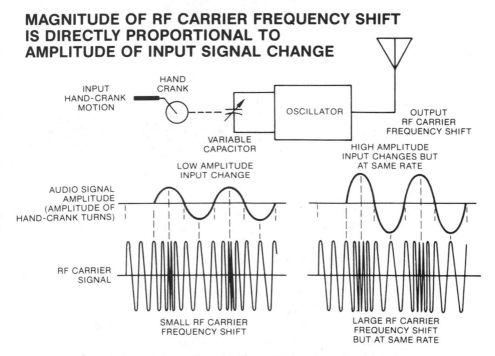

In both cases, the *same tone* would be heard in an FM receiver. However, the volume of the tone *would increase with the magnitude of the frequency deviation.* The tone produced by the ± 1-kHz shift will be lower in volume by 10:1 as the same tone produced by the ± 10-kHz shift.

The frequency of the modulating signal determines the rate at which the carrier frequency is shifted, and the amplitude determines the frequency deviation. The symbol f_m is used to designate the modulating signal frequency, Δf denotes the frequency deviation, and f_o usually denotes the carrier center frequency (no modulation frequency).

FM Sidebands

When a carrier signal at frequency f_o is *amplitude modulated* by a constant-level single audio frequency (f_m), the *upper* sideband is a constant-level signal at a frequency equal to the carrier frequency *plus* audio frequency ($f_o + f_m$). The *lower* sideband is a constant-level signal equal to the carrier frequency *minus* the audio frequency ($f_o - f_m$).

In the case of frequency modulation, sidebands are also developed, but the situation is *much more complex*. Because the frequency of the carrier is being changed during modulation, many sidebands are produced. The analysis that leads to how the sidebands are developed is very complex. Therefore, we will confine our study to the results of this analysis. It is sufficient to know something about how and where sidebands are produced and how this affects broadcasting. If a 1-MHz carrier at f_o is amplitude modulated by a constant-level 1-kHz audio signal (f_m), two constant-level sidebands will be produced at 1.001 MHz ($f_o + f_m$) and at 0.999 MHz ($f_o - f_m$). In an AM system only these sidebands occur.

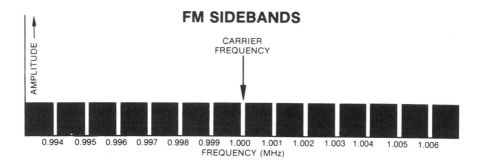

On the other hand, if a 1-MHz carrier (f_o) is frequency modulated by an audio signal at 1 kHz (f_m), a constant-level sideband will be produced at 1.001 MHz ($f_o + f_m$), and another will be produced at 0.999 MHz ($f_o - f_m$). However, unlike AM, additional sidebands will also be produced at 1.002 MHz ($f_o + 2f_m$) and 0.998 MHz ($f_o - 2f_m$), at 1.003 MHz ($f_o + 3f_m$) and 0.997 MHz ($f_o - 3f_m$), at 1.004 MHz ($f_o + 4f_m$) and 0.996 MHz ($f_o - 4f_m$), and so forth. In general, the magnitude of these sidebands becomes smaller the further from the carrier frequency. The magnitude and frequencies of these sidebands depend on the magnitude and frequency of the modulating signal. Thus, both the frequency and the amplitude of the modulating signal are important in determining the sideband content of an FM signal. It should also be noted that the carrier frequency may not be present at all under some FM conditions. This is in contrast to an AM signal where, for a single tone, *only one* pair of sidebands is produced, its amplitude proportional to the amplitude of the modulating signal, and the carrier is constant in amplitude.

Only some of the FM sidebands contain sufficient power to be important. The number of significant sidebands decreases as the frequency of the modulating signal increases and increases as the amplitude of the modulating signal increases.

FM Sidebands (continued)

In AM, the power in the carrier frequency is not affected by the development of the sidebands. In FM, the generation of sideband power is at the expense of the carrier in accordance with the frequency and amplitude of the modulating signal. Under some conditions it is possible for the power in the carrier to be zero.

A complexity of sidebands is constantly created and changed. However, the total power (carrier and sidebands) cannot exceed the unmodulated carrier power. If the major close-in sidebands are received, good fidelity is possible. Bandwidth limitations restrict the amount of frequency deviation and modulating frequencies that can be transmitted to ensure that important sidebands are included in the received signal band.

AMPLITUDES OF VARIOUS SIDEBANDS PRODUCED WHEN A CARRIER IS FREQUENCY MODULATED BY A 5,000-Hz TONE

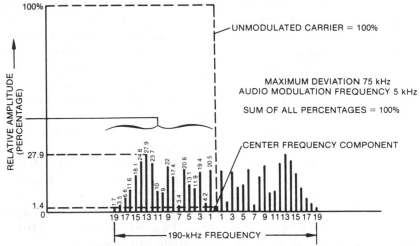

AMPLITUDES OF VARIOUS SIDEBANDS PRODUCED WHEN A CARRIER IS FREQUENCY MODULATED BY A 15,000-Hz TONE

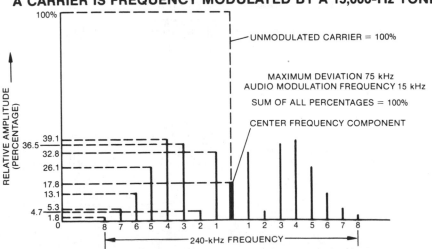

Determining Modulation Percentage

Modulation percentage is the ratio between the unmodulated carrier amplitude and the maximum (or minimum) carrier amplitude under modulation. When the peak-to-peak amplitude of the modulated carrier changes between zero and twice that of the unmodulated carrier, the modulation percentage is defined as 100% since the carrier cannot decrease below zero. While in principle there is no similar limit on FM, since we could, in theory, deviate the frequency as much as we wish, there are practical limits imposed. In FM systems, the practical limitation to carrier signal frequency deviation is the limit imposed by the channel bandwidth available. For example, a frequency deviation of ±75 kHz (150 kHz total) is the maximum allowed for commercial FM broadcasting in the United States. Thus, when the audio modulation produces a carrier frequency deviation of ±75 kHz, the degree of modulation is defined as 100%. The audio (aural) transmission for TV stations use FM with ±25 kHz defined as 100% modulation. Narrow-band FM communications systems have deviations as narrow as ±5 kHz.

In contrast to AM systems, modulations beyond 100% do not result in carrier cutoff distortion, since the carrier is always present. Distortion can be introduced if the FM deviation becomes nonlinear or if the receiver cannot handle the signal bandwidth. Since frequency modulated signals may have sidebands greatly in excess of the highest modulating frequency, and thus create interference, specifications on the maximum output at frequencies outside the band (at 100% modulation) are usually set for FM transmitters. For example, in the FM broadcast band it is required that emissions be down 25 dB between 120 and 240 kHz of the unmodulated carrier, 35 dB down between 240 and 600 kHz, and more than 80 dB down at greater than 600 kHz.

Many information signals can be transmitted simultaneously on a single FM carrier; however, the sum of the modulating signals for these must *not* exceed the modulation percentage limits specified.

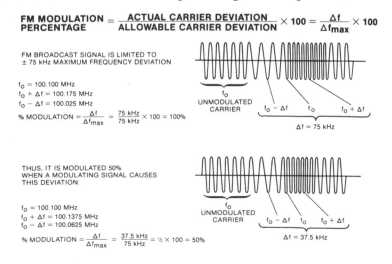

$$\text{FM MODULATION PERCENTAGE} = \frac{\text{ACTUAL CARRIER DEVIATION}}{\text{ALLOWABLE CARRIER DEVIATION}} \times 100 = \frac{\Delta f}{\Delta f_{max}} \times 100$$

FM BROADCAST SIGNAL IS LIMITED TO
± 75 kHz MAXIMUM FREQUENCY DEVIATION

$f_o = 100.100$ MHz
$f_o + \Delta f = 100.175$ MHz
$f_o - \Delta f = 100.025$ MHz

$\% \text{ MODULATION} = \frac{\Delta f}{\Delta f_{max}} = \frac{75 \text{ kHz}}{75 \text{ kHz}} \times 100 = 100\%$

f_o
UNMODULATED
CARRIER $f_o - \Delta f$ f_o $f_o + \Delta f$

$\Delta f = 75$ kHz

THUS, IT IS MODULATED 50%
WHEN A MODULATING SIGNAL CAUSES
THIS DEVIATION:

$f_o = 100.100$ MHz
$f_o + \Delta f = 100.1375$ MHz
$f_o - \Delta f = 100.0625$ MHz

$\% \text{ MODULATION} = \frac{\Delta f}{\Delta f_{max}} = \frac{37.5 \text{ kHz}}{75 \text{ kHz}} = \frac{1}{2} \times 100 = 50\%$

f_o
UNMODULATED
CARRIER $f_o - \Delta f$ f_o $f_o + \Delta f$

$\Delta f = 37.5$ kHz

Modulation Index

The relationship of the bandwidth of an FM signal, and the modulating signal frequency (f_m) and amplitude, and the frequency deviation and the modulating frequencies' amplitude is very complicated. As a practical matter, significant sidebands are defined as those that exceed 1% (-20 dB) of the unmodulated carrier level. The number of significant sidebands and the bandwidth of the signal depend more on the frequency of the modulating signal (f_m) than on the amplitude of that signal. If the modulating frequency is 1 kHz, as described earlier, the sidebands are 1 kHz apart and the significant sidebands would occupy a given band. However, if the modulating frequency were 10 kHz, the sidebands would be spaced 10 kHz apart. Now the bandwidth required to include the significant sidebands would be much wider than for the 1-kHz modulating signal even if the 10-kHz signal had a lower amplitude.

Although this results in a complex relationship, it is made even more complex by the fact that the *amplitude* of the modulating signal affects the maximum frequency deviation of the carrier signal. The wider this deviation, the larger the number of sidebands that can have significant amplitudes.

A relationship called *modulation index* is used to describe the number of significant sideband frequencies and the bandwidth. The modulation index (MI) is defined as the ratio of the carrier deviation (Δf) to the modulating signal frequency (f_m) or

$$MI = \frac{\Delta f}{f_m}$$

MODULATION INDEX DESCRIBES SIGNIFICANT SIDEBANDS AND SIGNAL BANDWIDTH

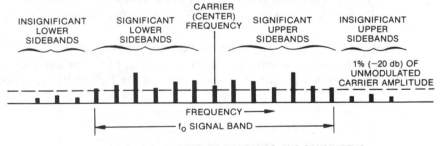

MODULATION INDEX, NUMBER OF SIDEBANDS, AND BANDWIDTHS

MODULATION INDEX	N = NUMBER OF SIDEBANDS ABOVE AND BELOW CARRIER FREQUENCY	TOTAL BANDWIDTH = $2 \times N \times f_m$
0.5	2	$4 \times f_m$
1	3	$6 \times f_m$
2	4	$8 \times f_m$
3	6	$12 \times f_m$
4	7	$14 \times f_m$
5	8	$16 \times f_m$
6	9	$18 \times f_m$
7	11	$22 \times f_m$
8	12	$24 \times f_m$

Basic FM Transmitter Design

Since frequency, not amplitude, carries the information signal, FM transmitters invariably use an FM oscillator, which you will learn about shortly, to drive an amplifier chain. Unlike an AM transmitter, *amplitude linearity* is *not* important—so Class A, AB, B, or C amplifiers can be used. Usually Class B or C stages in all but the lowest levels are used because they provide highest efficiency. Except for the need to be sure that the bandwidth is adequate to handle the FM signal, AM and FM RF amplifier stages are *very similar* in design.

BASIC FM TRANSMITTER

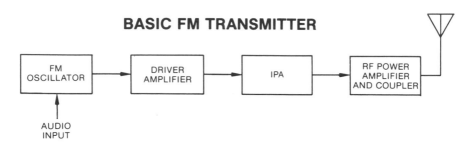

Because it is not necessary to accommodate a doubling in amplitude of the output (at 100% modulation) as in an AM system, FM transmitters generally run at a higher power output for a given final amplifier configuration. Also note that frequency multipliers are used only in very special systems since a frequency doubler will, for example, also double the frequency deviation. As you will see, in some cases this fact has been used to advantage for FM transmitter designs, where a small frequency deviation is used to provide good linear FM, and this is multiplied in frequency multipliers to provide the desired deviation. For example, a doubler will double the modulation index since Δf increases by a factor of 2 but f_m is unchanged.

Some care in FM transmitter design is necessary to avoid the introduction of incidental amplitude modulation. One of the primary causes of this is too narrow a passband (bandwidth) in the RF amplifier stages. If the passband is not flat, the output will change as a function of frequency, thus introducing an amplitude modulation on the RF signal.

The heart of an FM transmitter is the *FM oscillator*. While there are many schemes for generating an FM signal, all systems depend simply on varying the oscillator frequency in accordance with the modulating signal input. While almost any type of information can be transmitted by FM, we will in this section discuss *mainly audio* or voice transmission since this is one of the commonest uses of FM transmission. To accomplish voice modulation, a microphone is used in conjunction with an audio amplifier to cause deviations in the FM oscillator carrier frequency. The rate of the frequency shift must be proportional to the frequency of the signal going into the microphone, and the magnitude of the frequency shift must be proportional to the magnitude of the modulating signal input.

FM Oscillators—Varactors

One of the most popular methods for accomplishing FM modulation is based on the use of a *varactor* or *voltage variable capacitor* (VVC). The varactor is a special form of semiconductor diode that is operated in the reverse-bias mode. In the region of the junction, the space charge layers behave like capacitor plates that are displaced away from each other in proportion to the increase and decrease of the applied reverse voltage. The varactor is used as part of the tuned circuit of an LC oscillator. The resonant frequency of a tuned circuit is proportional to the square root of capacitance; therefore the FM oscillator will not deviate directly in proportion to the modulating signal. However, if the deviation is kept small the FM signal that results is quite linear. In the circuit shown, the carrier frequency is determined by the inductor (L) and the capacitor (C) plus the varactor capacitance with a dc bias. The audio signal and bias are applied to the varactor via an RFC (radio frequency choke), which allows the modulating signal to pass but presents a high impedance to the RF signal. When a modulating signal is applied, the varactor capacitance changes, causing the oscillator frequency to shift as well—resulting in FM. As you can see from the graph, as the varactor voltage decreases the capacitance increases, and vice versa. Also, the long-term frequency control of this FM oscillator configuration is poor because of component drift with temperature, aging of components, dc bias change, etc. Variations of this circuit to allow for better long-term frequency stability are commonly used in FM transmitters.

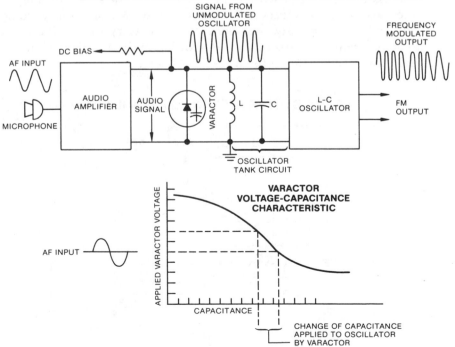

BLOCK DIAGRAM OF FM VARACTOR MODULATOR

FM Oscillators—Transistor Reactance Modulator

The circuit shown here accomplishes frequency modulation by making use of the fact that the *output capacitance* of a transistor (the collector-emitter capacitance) *decreases* as the collector voltage *increases* and *increases* when the collector voltage *decreases*.

In the circuit shown, transistor Q1 is the FM oscillator. The tank circuit of the oscillator is formed by capacitor C1 and the T1 winding between terminals 1 and 3. Feedback necessary for oscillator operation is provided by the winding between terminals 4 and 5. The frequency-modulated carrier signal output appears between terminals 6 and 7. Capacitors C2, C3, and C4 are RF bypass capacitors. R1 is the emitter-bias resistor, and R2 and R3 provide base bias for Q1.

Transistor Q2 is the reactance modulator. Its audio modulating signal input comes through coupling capacitor C6 and bias network R5–R6 provides bias (along with emitter R/C, R4, and C5) to the base of Q2. The collector-emitter capacitance is shown by the capacitor (C_e) in dotted lines from the Q2 collector to emitter.

When voltage at the Q2 base increases and decreases as determined by the applied audio signal, the collector voltage across RL increases and decreases in the same way. Thus, the output capacitance (C_e) varies in accordance with the applied modulating signal. The oscillator output is frequency modulated in accordance with the applied signal.

A decrease in C_e has the effect of decreasing the capacitance across the oscillator's resonant tank circuit, and the frequency of oscillation increases. Similarly, an increase in C_e effectively increases the capacitance across the oscillator's resonant circuit, and oscillator frequency decreases. As with the varactor FM oscillator shown previously, some method for long-term frequency control is necessary to provide for a stable carrier frequency. As for the varactor FM oscillator, the frequency excursion must be restricted for linear operation.

TRANSISTOR REACTANCE MODULATOR

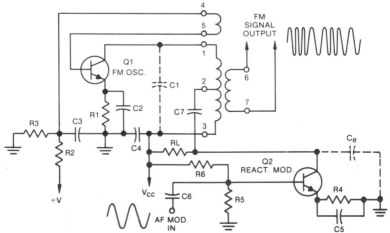

Drift-Stabilized FM Transmitter

The FM oscillators just described are subject to long-term *frequency drift*. This drift is caused by changes in the oscillator tank (LC) circuit due to temperature variations, aging, etc. In AM transmitters, the most reliable method of avoiding this drift is to use a crystal oscillator. However, this method cannot be used directly in an FM transmitter, since the frequency of a crystal oscillator cannot be shifted easily.

To obtain the required frequency deviation while obtaining frequency stability, it is necessary to make use of a crystal oscillator in an *indirect* way. One method is shown in the block diagram. The FM transmitter components previously described are used as the major part of the system. Added to this are a crystal oscillator, a frequency converter (mixer), and a discriminator.

The mixer receives signals from both the FM oscillator and the crystal oscillator. As you know, one of the outputs from the mixer is equal to the difference between the two input signals. Thus, the mixer output frequency increases and decreases as the FM oscillator frequency deviates from the crystal oscillator frequency. This difference signal is fed to a *frequency discriminator*. You will learn more about discriminators in Volume 4. However, the discriminator output is a dc voltage whose magnitude and polarity are related to the frequency error between the crystal oscillator and the FM oscillator. A low-pass RC filter in the discriminator output attenuates the FM signals that appear on the output of the discriminator, leaving only the long-term (dc) voltage due to a carrier frequency error to be applied to the FM oscillator. Thus, if the error is zero, the dc output from the discriminator is zero; and if it is not, the output is a positive or negative voltage proportional to the magnitude and direction of the frequency error. This voltage adjusts the bias on the varactor of the FM oscillator to shift its center frequency in such a way as to drive the discriminator output to zero, which means that the carrier frequency is correct.

DRIFT-STABILIZED
FM TRANSMITTER

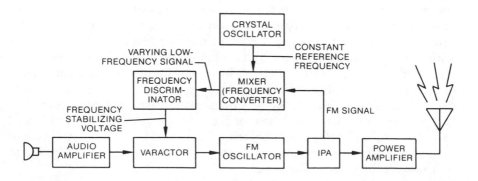

Crystal-Stabilized FM Transmitter

The drift-stabilized transmitter described on the previous page suffers from problems with the discriminator stability. The development of high-speed *IC counters* has made it possible to use alternative schemes for frequency control of FM transmitters to provide much greater stability. These schemes depend on the use of frequency dividers. As you will learn later (in Volume 5), a *digital binary counter* divides by 2, producing one output pulse for each two input pulses. Using these devices, a frequency can be divided by 2 (the *inverse* of frequency multiplication). As the frequency is *divided*, the modulation index *decreases*. If the RF frequency is divided enough times, the FM deviation becomes very small, but the long-term carrier shifts can still be observed as a phase change.

CRYSTAL-STABILIZED FM TRANSMITTER

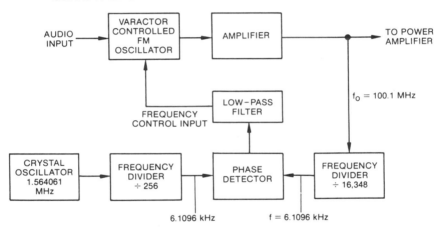

Suppose you had an FM transmitter with an output frequency of 100.1 MHz (channel 265 in the FM broadcast band). If the frequency is divided by 16,384 (2^{14} or 14 binary dividers), then if the modulation index was 5 originally, it is 0.0003 after division, and the center frequency is 6.1096 kHz. A crystal oscillator at 1.564061 MHz divided by 256 (2^{8} or 8 binary dividers) is also 6.1096 kHz. If these two signals are compared in a phase detector, frequency variations from the FM oscillator center-frequency shift produce an output from the phase detector proportional to the frequency error. If this is applied to the FM oscillator with the correct polarity, the system stabilizes when the FM oscillator frequency is exactly 100.1 MHz. A low-pass filter minimizes the effects of the modulating signal from deviation while responding to the slow drift of the FM oscillator. Since a properly constructed phase detector is almost free of drift, this system of frequency control is very stable and can easily hold the 100.1-MHz carrier frequency to less than ± 1 kHz of variation. Often, for best frequency stability the crystal oscillator is kept in a temperature-controlled oven.

Crystal-Controlled FM Transmitter

While, as you know, a crystal oscillator will provide for a very stable operating frequency, the exact frequency of oscillation depends not only on the crystal dimensions but to a small extent on the external circuit. For example, a capacitor in parallel with the crystal can be used to change its frequency very slightly—typically between 0.001% and 0.005%. This phenomenon is known as *pulling*, and while a crystal frequency can be pulled as much as 0.005%, linear pulling can only be obtained over a much smaller range (0.001%). Thus, a crystal oscillator can be frequency modulated over a small range by a parallel varactor. Generally, the available deviation is too small to be useful directly, but the deviation can be increased to the desired value by using frequency multipliers.

CRYSTAL-CONTROLLED FM TRANSMITTER

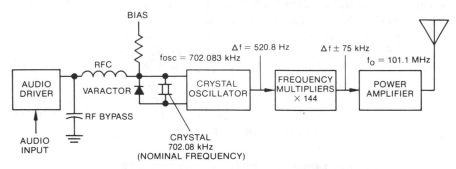

As shown above, the crystal oscillator is frequency modulated by a varactor in a way very similar to that for the basic FM oscillator you already learned about. Typically, the linear FM deviation available is only around ±0.0005% (0.000005). For an FM station on 101.1 MHz (±75-kHz nominal maximum deviation), the FM deviation is ±0.0742% (0.000742). Therefore, to achieve the appropriate deviation (Δf) at the output frequency of 101.1 MHz, the required frequency multiplication is 0.000742/0.000005 = 148. Thus, we would use a 144× multiplier and slightly increase the FM deviation required by the crystal oscillator. Since an output frequency of 101.1 MHz is desired, the crystal frequency will be 101.1/144 = 702.083 kHz, and the deviation is 520.8 Hz at the oscillator frequency (520.8 × 144 = 75 kHz). As you can see, the crystal oscillator at 702.08 kHz is frequency modulated by the varactor across the crystal to achieve a frequency deviation of ±520 Hz for the maximum audio input signal level since this represents ±75 kHz or 100% modulation at the output frequency of 101.1 MHz. The RFC (radio frequency choke) and RF bypass capacitor isolate the audio and RF circuits from each other. A convenient chain of multipliers to achieve 144× would be 2×3×4×3×2 (doubler, tripler, quadrupler, tripler, and doubler). Since this multiplication can be done at very low power levels, such a multiplier chain is quite practical. The multiplier output could be fed to a conventional power amplifier as for any FM transmitter.

Phase-Modulated Indirect-Method FM Transmitter

The FM transmitter configurations considered up to this point use the *direct* method of modulation. In these transmitters the RF oscillator is directly frequency modulated with the modulating signal. As you know, some of these systems use fairly complicated methods for carrier frequency control, and the possibility exists that the oscillator frequency can drift beyond the operating range of the control circuit. Direct crystal-controlled systems require extensive chains of frequency multipliers. To avoid this, other methods for generating FM signals can be used.

A basic FM scheme in widespread use for broadcast stations is known as the *indirect* method, because the frequency deviation is not introduced at the source of the RF carrier signal (the oscillator). Instead, the oscillator frequency is fixed, and the frequency deviation is introduced indirectly in one or more of the stages following the oscillator, as will be described. The basic advantage of this method is that the oscillator is crystal-controlled, and will thus maintain a stable center frequency without the need for separate stabilizing circuits. The deviation is produced after the signal is generated by the oscillator. In this scheme, a method known as *phase modulation* is used to provide FM. It should be clearly understood that the final output signal of an *indirect* FM transmitter is the *same* as that produced by a *direct* FM transmitter. The differences are confined to the methods used to obtain the FM signal, and *not* in the characteristics of the output signal itself.

PHASE-MODULATED INDIRECT-METHOD FM TRANSMITTER

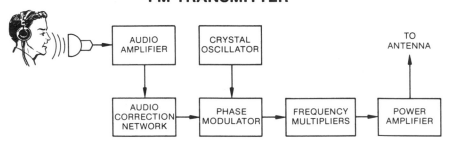

The phase-modulated transmitter is shown above. Since phase modulation produces very little frequency deviation, a frequency-multiplier chain is required in this approach. These multipliers are followed by the power amplifiers and antenna. The modulating signal input drives the audio amplifier, and the signal is amplified to a level suitable to operate the audio correction network which drives the phase modulator. All but these last two circuits are identical to those studied earlier.

Frequency can be defined as a measure of the rate of change of phase. A change in the rate of change of phase is a frequency shift; therefore if the rate of change of phase can be altered, a frequency shift occurs. This is the basis for phase-modulated indirect FM transmission.

Phase-Modulated Indirect-Method FM Transmitter (continued)

Frequency is defined as the *rate of change of phase*. For example, a 10-kHz signal will go through 10,000 360-degree phase shifts per second. If an advancing phase shift of 360 degrees is added to the 10-kHz signal in 1 second, then there would be 10,001 360-degree phase shifts per second—or, the frequency has been shifted by 1 Hz. By the same reasoning, if the phase shift is retarded by 360 degrees in 1 second, the result would be 9,999 360-degree phase shifts per second, or 9.999 kHz. Thus, as you can see, *phase modulation can result in frequency modulation*. As you might suspect, the amount of FM deviation available is quite small.

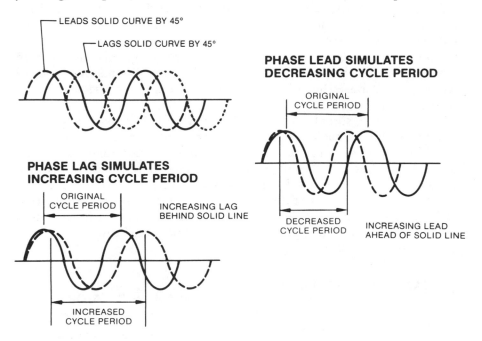

The top left of the above diagram shows a sine wave drawn in a solid line. The dotted curve shows a sine wave that lags 45 degrees behind the solid curve, and the dash-dot curve shows a sine wave that leads the solid curve by 45 degrees. These curves show that the phase of a sine wave can be advanced or retarded in time by means of phase shifting.

The second diagram shows the result of using a phase-shifting network to smoothly shift the phase of the sine wave while it is being generated. When the phase is smoothly shifted so that it lags the original wave, the peaks occur later; this is equivalent to increasing the cycle period or *lowering the frequency*. If the phase is smoothly shifted so that it leads the original wave (third diagram), the peaks occur sooner; this is equivalent to shortening the cycle period or *increasing the frequency*. Thus, smoothly shifting the phase of a signal is equivalent to smoothly shifting its frequency. Therefore, if the *phase* of an RF carrier is *shifted* by a modulating signal in the proper way, a *frequency-modulated signal* will result.

Basic Transistor Phase Modulator

As you know, a simple form of phase-shifting network consists of a capacitive or inductive reactance connected in series with a resistor. When a signal of constant frequency is connected across the series combination, the signal output across the resistor (or reactance) is shifted in phase with respect to the input signal. If the resistance is varied, the phase shift also varies. The constant-frequency RF signal from a crystal oscillator can be applied to this phase-shifting network, and its phase can be shifted by varying the resistance (or reactance).

By replacing the resistor with a transistor, the resistance in the phase-shift network can be varied in accordance with the amplitude changes in an audio signal of constant frequency. As the audio signal voltage varies, the transistor collector current also varies. Increases in collector current lower the collector resistance, and this causes an increasing shift in the phase of the input RF signal. Decreases in the collector current raise the collector resistance, and this causes decreasing shift in the phase of the input RF signal.

Thus an RF signal of constant frequency, such as that coming from a crystal oscillator, can be phase shifted in accordance with the amplitude of a modulating signal input. Also shown is a varactor phase shifter that uses the varactor as a variable reactance. Typically, because of linearity problems, the total phase excursion is limited to ±20–30 degrees. In practice, a single circuit of this type cannot produce a sufficient amount of linear phase shift, and several networks are connected in cascade to produce additional shift. For example, four networks can provide peak phase shifts of ±80–120 degrees.

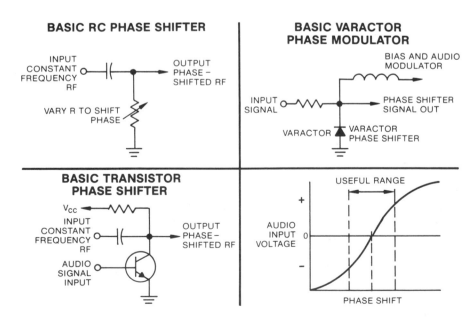

Available FM by Phase Modulation—Audio Correction Network

As you know, the amount of FM available by phase modulation depends on the signal amplitude and the rate of change of phase (modulating frequency). For example, if the total available phase shift is ±180 degrees, the total FM available is equal to the modulating frequency times the phase shift. Thus, a 40-Hz tone of a given amplitude with a ±180-degree phase shifter will give an FM deviation of ±20 Hz (±180 degrees × 40 = ±7,200 degrees = ±20Hz). On the other hand, the same amplitude tone of 15 kHz with the ±180-degree phase shifter will provide an FM deviation of ±7500 Hz (±180 degrees × 15,000 = ±2,700,000 degrees = 7,500 Hz). Since we would like the *deviation* to be the *same* for the *same amplitude input signal*, we must modify the input signal so that the same deviation is produced regardless of modulating frequency. This is done by the *audio correction network*. If you work it out, you will find that the audio signal should be attenuated by 2:1 (6 dB) for each octave. This is precisely what an RC network does when connected as shown in the diagram.

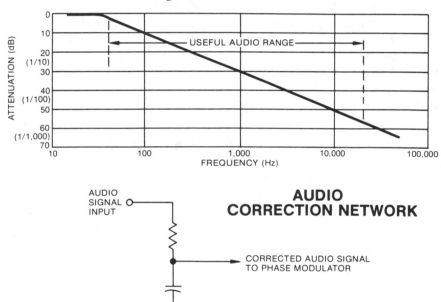

Therefore, an RC network in the audio line provides for the correction of the modulating signal so that a *uniform FM signal deviation* is obtained at all modulating frequencies.

If the resistance and capacitive reactance are equal (3-dB point) at some frequency below the lowest modulating frequency of interest, the RC network shown will have the property of uniformly attenuating signals at frequencies above the 3-dB point. This is all that is necessary to provide a constant deviation irrespective of the modulating frequency.

Available FM by Phase Modulation—Audio Correction Network (continued)

As you see by the foregoing calculations, the typically available FM deviation by phase modulation is small (± 20 Hz for the example). To achieve ± 75-kHz deviation for commercial broadcast requires a frequency multiplication of 3,750 (75,000/20). At a fundamental frequency of 100 kHz, the final frequency would be about 375 MHz, which is not the desired frequency for an FM broadcast station. To solve this problem, frequency multiplication is used, resulting in a final frequency of 375 MHz with the required deviation at the output of the multiplier. This 375-MHz signal is then down-converted via a mixer and second crystal oscillator to the desired frequency, since the mixing process does not affect the FM deviation. For a final transmitter frequency of 101.1 MHz, the frequencies selected would be as shown.

The choice of 476.1 MHz rather than 255.9 MHz for the fixed oscillator input, is to prevent the lower frequency from getting into the multiplier chain. This is done mainly as a convenience to prevent interference, since either 476.1 MHz or 255.9 MHz mixed with 375.0 MHz will provide an output at 101.1 MHz (476.1 − 375 = 101.1 or 255.9 − 375 = 101.1). Obviously the 476.1-MHz frequency is generated by a crystal oscillator-multiplier chain since a crystal oscillator operating in the fundamental mode is impractical. As for the direct FM system using multipliers, the multiplication is done at low level so that the circuits required are low power and single.

COLLECTOR-MODULATED TRANSISTOR AMPLIFIER

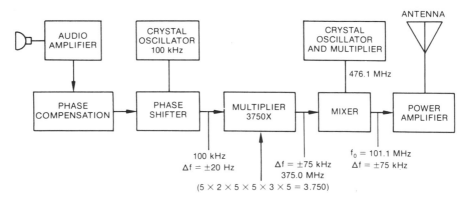

As the bandwidth of a modulating signal increases, the noise in the system will increase in direct proportion. For an audio signal, most of the signal power is in the frequency range below 4kHz, which means that the signal-to-noise ratio above about 4 kHz will be poorer because less signal power is available.

Pre-Emphasis

This situation could be improved if we increase the modulation percentage for these higher-frequency signals (thus improving their signal-to-noise ratio). A *pre-emphasis* network in an audio line increases the amplifier gain at high frequencies so that these frequencies represent increased modulation percentage. Obviously, this would result in strongly accented high frequencies in the received signal, so another network (called a *de-emphasis network*) is used to restore the response to normal. (You will learn about these when you study receivers in Volume 4.)

It has been found that the power in audio signals drops off typically at about 6 dB per octave at higher frequencies (2:1 for each doubling of the audio frequency). Thus, if the gain of the audio amplifier is increased by 6 dB per octave, the modulation percentage at all frequencies will tend to equalize.

PRE-EMPHASIS NETWORK

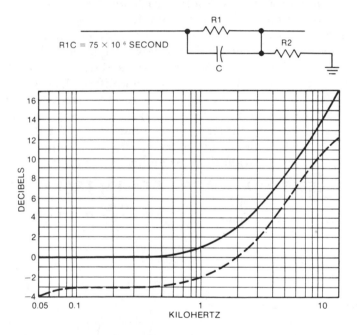

A typical pre-emphasis network is the RC configuration shown here. The FCC has set standards in the United States for FM broadcast and TV audio utilizing a network with a time constant of 75 microseconds (μseconds)—(the product of R and C equals 75×10^{-6} seconds). This results in the pre-emphasis curves shown above. For some applications, other pre-emphasis characteristics are used.

Fleet Communications FM Transmitter—Summary of Operation

Frequency modulation is used in military aircraft and private communication systems in addition to standard FM broadcast stations. The block diagram shown is for the transmit section of a UHF (450 MHz) two-way FM fleet communications transceiver. A complete block diagram is shown. As described previously, a transceiver is a combination transmitter and receiver that has some circuits for transmit only, others for receive only, and some (usually audio) shared by *both* the transmit and receive modes.

The audio section is fairly straightforward, with a four-stage IC amplifier that includes diode series limiters and a splatter filter (clipper). The series limiter limits the high-amplitude transient peaks in the audio signal, and the splatter filter (clipper) suppresses the wide-band frequency components produced by limiting from causing interference in adjacent channels. Feedback is used to increase the stability of the IC audio amplifiers. The output of the audio section is coupled to both oscillator/doublers where it is used to frequency modulate the crystal oscillator with a varactor, as described earlier. The desired channel is selected by using a channel-selector switch.

The modulated signal, doubled from the fundamental frequency of the crystal oscillators, is then fed to a pair of frequency doublers. This signal is then amplified by the RF predriver and driver amplifier stages before being coupled to the final power amplifier. The output of the final amplifier is then coupled through a varactor frequency tripler to obtain the proper transmit frequency. An RF filter suppresses the fundamental signal into the tripler. This is a narrow-band FM system that operates with ±5-kHz deviation. Since the highest audio frequency is about 5 kHz, the modulation index is about 1.

DETAILED BLOCK DIAGRAM OF FM TRANSMITTER

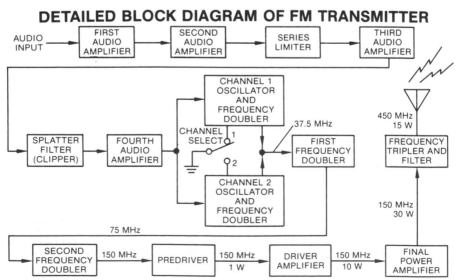

Fleet Communications FM Transmitter—IC Audio Section

The frequency mutiplication is 24 ($2 \times 2 \times 2 \times 3$). Thus, for 100% modulation (± 5kHz) at the output, the crystal frequency deviation is ± 5 kHz/24 = ± 208 Hz, which is easily achieved. Below we will study the IC audio section of the transmitter.

The audio section of the transmitter consists of a four-stage IC audio amplifier. The integrated circuit U301 contains four identical audio amplifiers, and it is used for all the speech processing for the transmit section only. All four stages use feedback to control gain and improve their stability.

The first audio amplifier (U301A) provides a high-impedance input and a low-impedance output to drive the second audio stage (U301B). U301A has unity voltage gain (R302/R301 = 1) and is an impedance-matching stage only. The second stage (U301B) has a gain of about 45 dB (R305/R304 = 174 = 20 log 174 = 45 dB). Both the positive and negative signals are limited by parallel opposing-polarity diodes CR301 and CR 302. When either of these diodes conducts (voltage across greater than about 0.6 volt), the feedback resistor R305 is shunted and the amplifier gain is reduced to a low value, thus limiting the audio output signal to about ±0.6 volt at peak. This occurs on the ac signals only, since capacitor C304 blocks dc from the diodes. The audio signal is fed to U301C via diodes CR303 and CR304.

The anodes of CR303 and CR304 are held at about + 0.4 volt of dc. If the positive peak audio signal level exceeds 0.4 volt, CR303 stops conducting, while if the peak negative signal exceeds 0.4 volt, diode CR304 stops conducting. In either case, modulating signals greater than 0.4 volt will not pass to amplifier U301C. U301C, the third audio stage, provides an additional 20-dB gain at low frequencies (R311/R310 = 10 = 20 log 10 = 20 dB). Capacitor C315 reduces the gain of U301C at higher frequencies to suppress higher frequencies (splatter) generated due to limiting. C315 has an impedance of about 100 K at about 3 kHz. Therefore, the gain of the amplifier is down by about 2:1 (3 dB) at 3 kHz since the feedback impedance is now the parallel combination of R311 and C315. Beyond 3 kHz, the frequency response drops off at 6 dB per octave. This filtering prevents adjacent channel interference (splatter) from high-frequency signals without reducing intelligibility of speech.

Amplifier U301D both amplifies and provides further splatter filtering. The feedback circuit acts as a low-pass filter using the network C308, R314, and R315. At low frequencies, C308 presents a high impedance and the gain is unity (R315/R313 = 1). Capacitor C309 functions to reduce the gain of U301D below unity at frequencies above 3 kHz as in stage U301C.

The output from the audio amplifier is used to directly frequency modulate the crystal oscillator/doubler stage of the RF section of the FM transmitter, which will be described next.

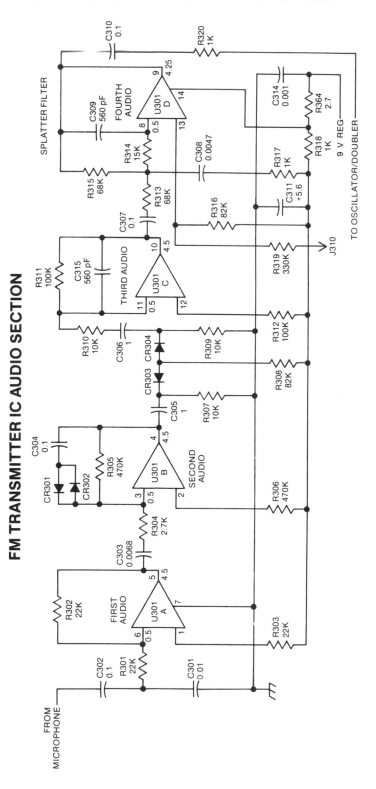

FM TRANSMITTER IC AUDIO SECTION

Fleet Communications FM—Transmitter Section

The audio signal—amplified, limited, and filtered by the audio amplifier—is used to frequency modulate the crystal oscillators. Two oscillators are provided so that one of two transmitting frequencies can be chosen. The oscillator output is then frequency multiplied prior to power amplification by the drivers, amplifiers, and frequency tripler to be described later. The desired oscillator is selected by grounding the emitter resistor of the selected oscillator.

The processed audio signal is coupled to potentiometers R321 and R324 to adjust the audio level into the FM oscillators. These potentiometers are used to adjust the frequency deviation produced by the audio signal. Frequency modulation of both oscillators is accomplished by audio modulating the varactors CR327 and CR322. The varactors provide frequency modulation, as described earlier.

The oscillators use a modified Colpitts circuit with the frequency set by either of the transmit crystals (Y301 or Y302). Since the frequency is multiplied by 24, the crystal frequency is 1/24 of the output frequency or

$$\text{Oscillator frequency} = \frac{\text{channel output frequency}}{24}$$

Therefore, for an operating frequency of 480 MHz, the crystal frequency is 480 MHz/24 = 20 MHz.

OSCILLATOR/DOUBLER—FLEET COMMUNICATIONS FM TRANSMITTER

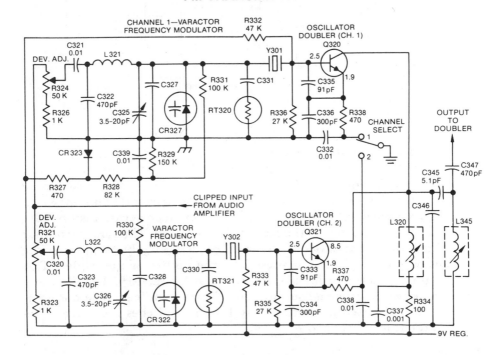

Fleet Communications FM—Transmitter Section (continued)

The output from each oscillator is tuned to the second harmonic of the crystal frequency. Therefore, the single-stage oscillator serves as an oscillator/frequency doubler/frequency modulator. Fine tuning of the center frequency is done via variable capacitors C325 and C326, which pull the crystal frequency slightly to make fine adjustments of crystal frequency. RT320 and RT321 are thermistors (temperature-sensitive resistors) that in conjunction with capacitors C330 and C331 provide improved stabilization over a temperature range.

The output from these oscillators and doublers is coupled to two more doublers of the type we have previously discussed. The output from the second additional doubler is at eight times the crystal frequency ($2 \times 2 \times 2 = 8$). This output is further amplified in three stages: a predriver, a driver, and a final power amplifier.

FLEETCOM II 558 SCHEMATIC DIAGRAM

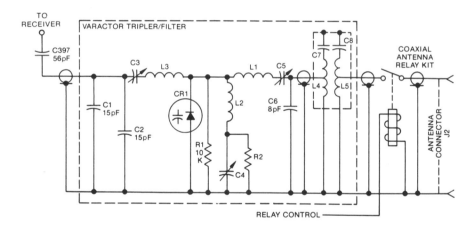

The output from the final amplifier is at the 30-watt level, but only at one-third the desired frequency. A high-power varactor is used as a frequency tripler. One of the useful properties of varactors is that they make very efficient multipliers. In this case, varactor CR1 takes in 30 watts at $f_o/3$ and delivers 15 watts at f_o (50% efficient). This approach was taken because efficient, reliable, inexpensive power transistors were not available at the time of the design of this system. The varactor output is fed to a filter that passes the desired frequency (f_o) but suppresses the strong output at $f_o/3$ that is present at the output of the tripler. This filter suppresses the undesired $f_o/3$ signal by 80 dB (10^8). The filter and output network antenna-coupling circuit are combined. The output is fed to the antenna via antenna relay K1 that switches the antenna between the transmitter and receiver.

FM Transmitting Antennas—Wave Propagation

VHF, UHF, and higher frequencies are basically limited to line-of-sight transmission. Thus, FM broadcast transmitting antennas (88–108 MHz) are always located as high as possible. The FCC specifies that FM broadcast stations radiate with *horizontal* polarization, but they allow stations to radiate an equal amount of power to that authorized with *vertical* polarization, too. This allows for good reception with either a rooftop antenna or with vertical whip antennas like auto antennas and portable radios. The same effect is obtained by using a circularly polarized (CP) antenna so that equal signal is received at any position of the receiving antenna. The most popular antennas for FM broadcast stations are the circular antenna and the circular CP antenna. In many cases, you find FM transmitting antennas arrayed in stacks. These arrays, as you know, increase the gain in the desired direction (horizontally). For other UHF-VHF communications service, vertical polarization is usually used, and whip antennas are most popular.

FM TRANSMITTING ANTENNAS

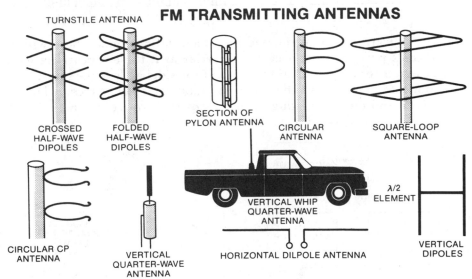

TURNSTILE ANTENNA

CROSSED HALF-WAVE DIPOLES

FOLDED HALF-WAVE DIPOLES

SECTION OF PYLON ANTENNA

CIRCULAR ANTENNA

SQUARE-LOOP ANTENNA

CIRCULAR CP ANTENNA

VERTICAL QUARTER-WAVE ANTENNA

VERTICAL WHIP QUARTER-WAVE ANTENNA

HORIZONTAL DILPOLE ANTENNA

$\lambda/2$ ELEMENT

VERTICAL DIPOLES

The line-of-sight range (R in miles) between two stations of height h1 and h2 in feet is approximately equal to

$$R = \sqrt{2h_1} + \sqrt{2h_2}$$

For a station transmitter at 1,000 ft and a receiver at 30 ft, the line-of-sight range R is

$$R = \sqrt{2h_1} + \sqrt{2h_2} = \sqrt{2 \times 1,000} + \sqrt{2 \times 30}$$
$$= \sqrt{2,000} + \sqrt{60} = 52.5 \text{ miles}$$

Intervening objects can interfere and reduce this line-of-sight range. On the other hand, some reception is obtained a short distance beyond this point. The above calculation assumes that the antennas are a reasonable distance above ground.

Multiple Signal Transmission

In some cases it is desirable or necessary to be able to transmit two or more differently modulated signals at the same time over the same channel. As you learned in AM transmission, one way this can be done is to transmit different modulating signals on the two sidebands. Another way is to modulate the carrier with subcarriers. Subcarriers are signals that will, in turn, be used to carry the modulation information on the subcarrier sidebands. Suppose you modulated a carrier with subcarriers at 10 kHz, 20 kHz, and 30 kHz. You would have the following spectrum.

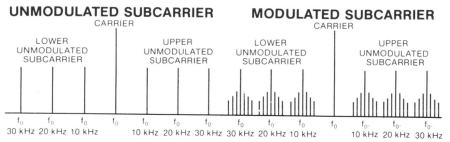

If the subcarriers are modulated by a modulating signal, there will be sidebands produced about each subcarrier as if these subcarriers were the carriers for the modulating signal. If the sidebands of the subcarriers do not extend into each other, then the subcarrier can be received in the receiver, and the modulating signals recovered from each one.

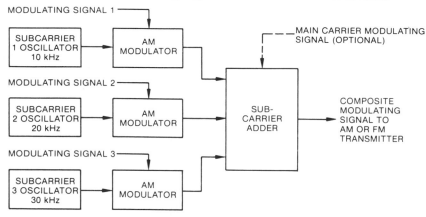

By using the subcarrier approach, multiple signals can be transmitted. Since you should not modulate over 100%, then each subcarrier signal modulates the main carrier by a fraction of the total, adding up to 100%. It is also possible to directly modulate the main carrier with a modulating signal so long as its sidebands do not interfere with the subcarrier sidebands. You can use balanced modulators to suppress the carrier and/or subcarriers without loss of information. The composite modulating signal can be applied to either FM or AM; however, these techniques are generally restricted to FM. These multiple modulation techniques are referred to as *frequency-division multiplexing*.

FM Stereo Broadcasting

As you learned in Volume 2, stereo recording has essentially re-placed monophonic recording both in tape and in disk recordings. Most FM broadcasting is in stereo to take advantage of this capability. The problem is how to transmit the two stereo signals (left and right channel) on one FM channel. Furthermore, it is required that monophonic FM receivers operate to produce reception of stereo broadcasts without mod-ification. As you remember from your study of high fidelity in Volume 2, stereo disk recordings are played on monophonic systems by summing the right and left channel signals into a single signal that has all the audio information in it. A similar idea is used to provide mono/stereo compatibility in FM stereo broadcasting.

RECORDING AND REPRODUCTION OF STEREOPHONIC SOUND

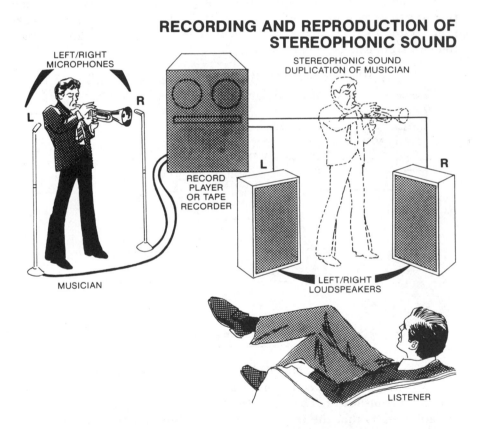

LEFT/RIGHT
MICROPHONES

L   R

RECORD
PLAYER
OR TAPE
RECORDER

MUSICIAN

STEREOPHONIC SOUND
DUPLICATION OF MUSICIAN

L   R

LEFT/RIGHT
LOUDSPEAKERS

LISTENER

As you might suspect, subcarrier modulation or multiplexing tech-niques are used to provide FM stereo broadcasting. These multiplexing techniques are also used to send other information on FM subcarriers, as you shall see on the next several pages of this volume. The basic idea consists of combining the left (L) and right (R) channel signals to form sum (L + R) and difference (L − R) signals and transmitting these to the receiver by multiplexing techniques. At the receiver, the L + R and L − R signals can be processed to recover the L and R signals.

FM Stereo Multiplexing

FM stereo broadcasting is accomplished by multiplexing several signals onto a single FM carrier. Both the left (L) and right (R) signals are passed through low-pass filters to restrict the frequency response to 15 kHz. These signals are then pre-emphasized and applied to the matrix circuit where L + R and L − R audio signals are formed. The L − R signal is fed to a balanced modulator also fed with a 38-kHz subcarrier. The output from the balanced modulator is a double-sideband suppressed-carrier L − R AM signal. The 38-kHz subcarrier is generated by doubling a 19-kHz CW carrier signal called the *pilot* signal. Thus, three signals are formed for transmission, an L + R audio signal, L − R signal consisting of the double-sideband suppressed-carrier centered at 38 kHz, and the 19-kHz pilot signal. The pilot signal is necessary at the receiver to demodulate the L − R signal. In some cases, the pilot signal is generated by frequency dividing from a 38-kHz oscillator. In any case, the two signals must be generated.

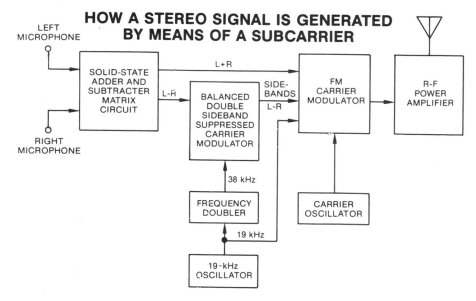

The L + R, L − R, and 19-kHz pilot are added together to form a composite signal. This composite signal is used to modulate the carrier of the FM transmitter. The amplitudes of the three signals (at peak level) are adjusted so that the L + R and L − R signals modulate the carrier by 45% each. The pilot signal modulates the carrier by 10% (total = 45 + 45 + 10 = 100%). It should be noted that the signal-to-noise ratio of stereo FM is somewhat poorer than for monophonic transmission. This is true because the equivalent maximum modulation percentage is less for stereo than for mono.

FM Stereo Multiplexing (continued)

To summarize, the L + R signal frequency modulates the carrier directly; the suppressed subcarrier L − R signal is an amplitude-modulated subcarrier signal centered 38 kHz away from the main carrier. The pilot signal (19 kHz) is a constant tone. These signals produce the spectrum shown below, which is fed into the FM transmitter.

FM STEREO MULTIPLEXING SIGNAL

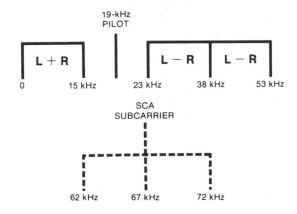

You may wonder why the simple approach of putting the two signals (left and right) on two subcarriers isn't used to avoid the complexity shown. The reason is that using two subcarriers would require a special stereo receiver to receive these broadcasts. By making the L + R signal (the monophonic signal) available as a direct FM signal, conventional monophonic FM receivers can decode just this portion automatically and thus receive either stereo or mono broadcasts. Thus, this approach is compatible with both mono and stereo receivers.

The FCC allows the use of other subcarriers from an FM broadcast station. These subcarriers, called *SCA* (subsidiary channel authorization), are narrow-band (5 kHz) and located above 53 kHz (for example, centered at 67 kHz). These are used to carry special information. For example, they can be used to key FM receivers off during commercials (advertising) so that FM receivers feeding public address systems in airports or supermarkets are free of such commercials. Special messages or advertisements for stores are also carried on these SCA channels. A special receiver equipped to receive these subcarriers is necessary; however, their presence does not interfere with reception as a conventional mono or stereo receiver. To provide for these channels, the modulation percentage on the stereo channels must be appropriately reduced. While several of these subcarriers are possible in a given FM channel, typically only one is used.

FM Stereo Multiplexing—Information Recovery

The operation of the receiver in picking up the FM RF carrier signal and demodulating it is covered in Volume 4. However, a short discussion is included here so that you will have some understanding of the *entire* process. For simplicity, let it be assumed that the output of the FM detector in the receiver is the spectrum of the stereo multiplex signal shown on the previous page. The left and right signals are reconstituted in a stereo decoder. A block diagram of decoder operation is shown below. The composite signal (0–53 kHz) is fed to a low-pass filter (0–15 kHz) that allows only the L + R audio signal to pass. Another filter passes the DSB L − R signal (23–53 kHz). The 19-kHz pilot signal is recovered in a special filtering circuit and is doubled to 38 kHz. This carrier is added to the L − R signal and AM detected to produce the L − R audio signal. These separated L+R and L − R signals are then combined in a matrix that adds and subtracts these signals appropriately. As you can see from the figure, the recovered L + R and L − R signals are converted into L and R signals by addition and subtraction. Thus, the capability is present for reception of both mono and stereo in a single receiver. The output is the separated L and R signals from the original input to the transmitter.

FM STEREO RECEIVER
DECODER OPERATION

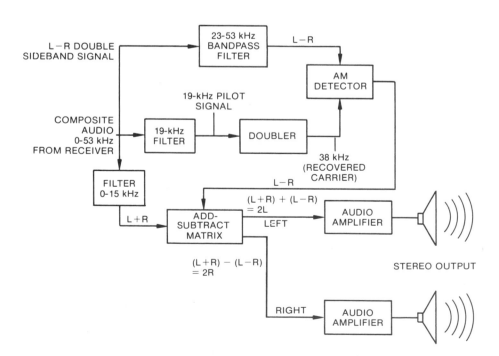

Review of Frequency Modulation Systems

LISTENING TO MUSIC
NONINTERFERENCE
FROM ATMOSPHERICS
AND MAN-MADE STATIC

1. **FREQUENCY MODULATION ADVANTAGES** are: less subject to atmospheric and man-made interference and, due to wideband operation at VHF, greater AF fidelity.

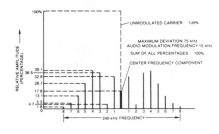

2. **FM SIDEBANDS** are infinite in number. However, only those in a limited bandwidth (± 75 kHz for FM broadcasts) are needed.

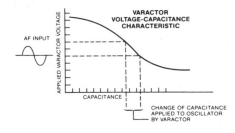

MODULATION INDEX, NUMBER OF SIDEBANDS, AND BANDWIDTHS

MODULATION INDEX	N = NUMBER OF SIDEBANDS ABOVE AND BELOW CARRIER FREQUENCY	TOTAL BANDWIDTH $2 \times N \cdot f_m$
0.5	2	$4 \cdot f_m$
1	3	$6 \times f_m$
2	4	$8 \times f_m$
3	6	$12 \times f_m$
4	7	$14 \times f_m$
5	8	$16 \times f_m$
6	9	$18 \times f_m$
7	11	$22 \times f_m$
8	12	$24 \times f_m$

3. **THE NUMBER OF SIGNIFICANT SIDEBANDS** in FM depends on ratio of the *deviation* to the modulating frequency, called the *modulation index*, which thus determines bandwidth.

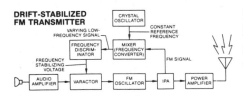

4. **FREQUENCY MODULATION CAN BE PRODUCED** by introducing a variable reactance into an oscillator circuit. Examples of this (called the direct method) are *varacators* and *reactance* modulators.

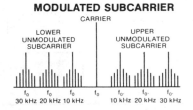

5. In **INDIRECT FM**, a crystal oscillator output goes through a *phase modulator*. An *audio correction network* is needed to change PM to FM.

MODULATED SUBCARRIER

6. In **MULTIPLEX TRANSMISSION**, two or more modulation signals can be sent on the same carrier, using subcarriers in the modulation, themselves modulated. FM stereo uses this principle.

Self-Test—Review Questions

1. What are some of the advantages and disadvantages of FM transmission vs. AM transmission?
2. Define modulation percentage in FM. What is the modulation index?
3. Describe briefly any basic differences between an FM and AM transmitter.
4. Draw a block diagram for and describe a crystal-stabilized FM transmitter. How does it differ from a crystal-controlled FM transmitter?
5. Why is pre-emphasis used in FM broadcast transmission?
6. What is a subcarrier? What is it used for?
7. Describe the way that the audio signals for FM stereo are derived.
8. How are the audio signals applied to the carrier in FM broadcasting? Describe the resultant FM stereo signal. Define multiplexing.
9. What is the pilot signal? What is it now for? How are the FM stereo signals recovered?
10. What is SCA? What is it used for?

Learning Objectives—Next Section

Overview—You now understand the AM and FM transmission of information and are now ready to consider pulse modulation and TV transmission, which are extensions of what you know.

THE TV TRANSMITTER

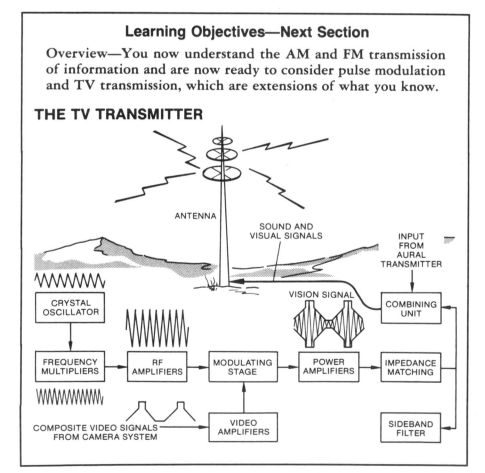

Introducing Pulse Modulation—Analog and Digital Transmission

Your study of pulse modulation in this volume will be brief and is primarily an introduction to TV or video-information transmission. You will study pulse systems in some detail in Volume 5 of this series. Some detail on pulse receivers will be covered in Volume 4.

Pulses are generally described as *relatively abrupt voltage* or *current changes of short duration*—generally from zero to a positive or negative value (or from a negative or positive value to zero). The term *short duration* is relative. Pulses are described in terms of their *shape, polarity, amplitude, duration* (width), and their *rate of occurrence*—repetitive rate or frequency (number of pulses per unit of time, usually per second).

A great deal of information is transmitted by analog systems where the voltage or current waveform has an amplitude and frequency analogy to the input information signal. The telephone, AM and FM radio, TV, and facsimile are examples of analog transmission. In addition, many sensors for temperature and pressure are analog in nature. As you will see in Volume 5, this *analog* information can be *converted* to *digital (or pulse)* form for transmission—and conversely, *digital* signals can be *converted* back to *analog* form. Why, then, bother with digital systems at all? The answer is that for some applications digital transmission results in *better fidelity* and *more accurate transfer* of information. In addition, *digital* or *pulse transmission* allows *digital devices* like computers, electronic office systems, computer terminals, etc. *to talk directly* to each other. It should be emphasized that digital transmission is *not restricted* to *radio* transmission methods but is equally useful over *wire lines* and *between units* of the *same system*.

Here we will be primarily concerned with how pulse (digital) information is placed on a carrier or other transmission media and some of the special requirements on transmitters for this application. As you will see, the definition of pulse or digital transmission is very broad.

THE BASIC CHARACTERISTICS OF PULSES

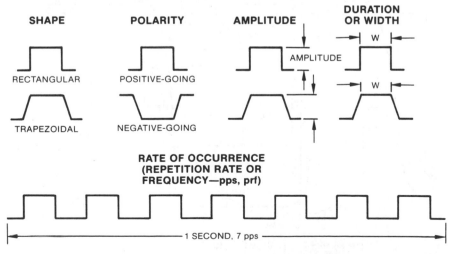

Pulse Terminology

The waveform shown is a rectangular, positive-going pulse to show some of the common terminology associated with pulses:

Period is the time that elapses from a point on one pulse to the same point on the next pulse in a repetitive pulse train. The point selected is usually the center of the leading or trailing edge where the waveform is steepest (usually almost vertical).

Rise time is the elapsed time for the waveform voltage to rise from the 10% to 90% of final amplitude. Sometimes the limits are different, but they are usually 10–90% unless specified otherwise.

Decay time is the time required for the pulse to fall from the 90% amplitude value to the 10% amplitude point. If any other limits are used, they are usually identified.

Pulse width is usually specified as the time between the 50% amplitude point on the leading edge to the 50% amplitude point on the trailing edge.

Amplitude is defined as peak or average peak pulse height, which, in a well-defined pulse, will be the same point. However, in a wave with overshoot or droop (see below), the peak amplitude is the maximum point and the average amplitude is the amplitude of the flat-topped portion of the wave.

Overshoot is the presence of short spikes on the leading and/or trailing edge of the pulse in excess of the nominal pulse height.

Droop is usually only associated with low-frequency response of long-duration pulses. It is the amount that the amplitude decreases from its maximum amplitude during the pulse interval.

PULSE TERMINOLOGY (FOR POSITIVE-GOING PULSES)

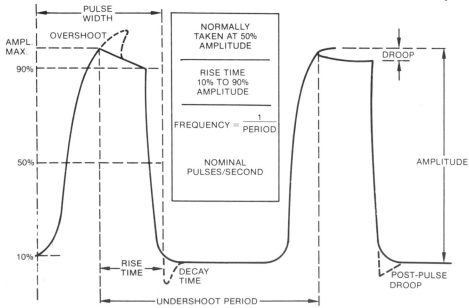

Types of Pulse Modulation

The simplest type of pulse transmission, and indeed the earliest form of radio, is the transmission of Morse code by a telegraph key. As you know, the Morse code consists of a series of dots and dashes (pulses or marks), where the dashes are about three times as long as a dot, with spaces within a letter equal to a dot and spaces between letters about equal to a dash—with a somewhat longer space between words.

For a simple Morse-type code, it is only necessary to turn the *carrier* on and off, and the *presence of* or *lack of* the carrier provides the *information*. Another way to transmit the information is by keeping the carrier on at all times and *changing the frequency* so that spaces are defined by the carrier frequency at f_o and a mark (dot or dash) is defined by being at frequency $f_o + \Delta f$. This technique, called *frequency-shift keying* (FSK), is widely used in digital transmission today. A very useful variation of FSK involves the use of modulated CW (MCW), where the modulating frequencies (tones) are changed for mark and space and the carrier frequency is kept constant.

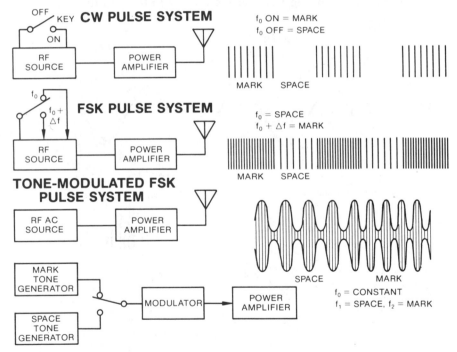

Obviously, these are only simple cases of pulse modulation, and there are many other methods for transmitting information by pulses. As shown above, a pulse train is usually involved—that is, the information is not in a *single* pulse, but resides in a *group* of pulses, which collectively convey one piece of information. For example, a Morse code A is a dot and dash (· —), and a B is a dash followed by three dots (— ···), etc. Other types of coded pulse signals or pulse groups can also easily be transmitted by these means representing digital data of almost any kind.

Types of Pulse Modulation (continued)

In the pulse modulation described earlier, the *information transmitted* consisted of the presence or lack of a *carrier* or *tone* or *particular frequency*. There are many other schemes for pulse modulation. One of the most straightforward is known as *pulse-amplitude modulation* (PAM).

The example of PAM shown in the diagram indicates how a sine wave can be transmitted by using it to modulate the amplitude of the pulses in a steady train of pulses. The signal can be considered to be a series of samples of the modulating waveform. You may ask, why not just transmit the sine wave directly since this would be easier to do? The answer is that other signals can be *multiplexed* into the intervals between samples. For example, another signal (signal 2) could be sampled during the spaces between the samples for signal 1. This process can be expanded to handle many signal channels that are sampled sequentially and transmitted over the same carrier channel. For example, this is the way multiple telephone signals can be transmitted on a single carrier. The basic restriction (to be taken on faith) is that the sampling frequency has to be at least twice the highest signal frequency to be transmitted.

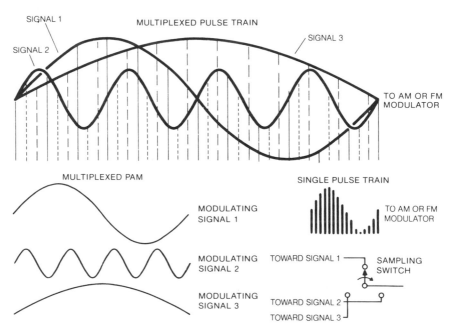

As you can see, for each channel, the pulse train has an amplitude related to the sampling time and has an outline that resembles the original signal. For the multiplexed case, the three signals generate three pulse trains that are time sequential. Obviously, many more signals could be multiplexed into a given time. At the receiving end, a similar sampling switch, synchronized to the transmit-end sampling switch, is used to separate the different channels.

Types of Pulse Modulation (continued)

There are many ways that pulse signals can be generated to carry almost any information desired. In addition to simple pulse modulation and PAM, there is, for example, *pulse-width modulation* (PWM), where the pulse width is proportional to the signal amplitude. (This is sometimes called *pulse-duration modulation* (PDM).) When the information is in the timing of pulses (as in Morse code or FSK) but the pulse amplitudes are constant, the modulation is known as *pulse-code modulation* (PCM).

PULSE-WIDTH MODULATION (PWM OR PDM)

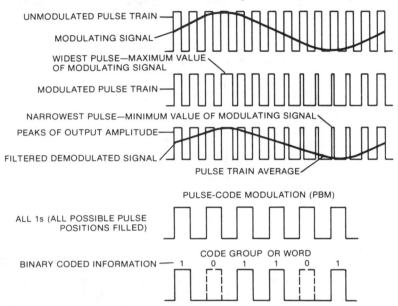

Pulse-code modulation is very important for transmission of digital data. The information in digital form consists of a group of pulses that make up a *code group* or *word*. Because the presence or absence of a pulse is easily detected, a system called *binary coding* is used where a pulse represents a one (1) and absence of a pulse represents a zero (0) in the pulse group. You will learn about binary coding in Volume 5 of this series. At this point, let's just say that such message formats are used to generate pulse trains.

TV information signals represent a combination of *pulse* and *pulse-like* signals and have properties like both pulsed and analog signals. You will study TV transmission in the next section of this volume.

The field of digital data transmission is a very large and rapidly growing one. We will discuss applications and detailed techniques mainly in Volume 5. Also in Volume 5 you will learn about the great number of digital ICs using MSI, LSI, VLSI and microprocessors that are currently in use today and allow for very complicated digital computation and control functions.

Bandwidth Requirements for Pulse Transmission

The bandwidths required for pulse transmission are generally quite different than for audio information signals. Many years ago, a mathematician named Fourier found that any wave or waveform could be made up from a series of sine and cosine waves (spectrum) of varying frequency amplitudes and phases. The *inverse* case is also true—that is, any wave is made up of a series of sine and cosine *terms* (spectrum) of varying frequencies, amplitudes, and phases. To faithfully reproduce a pulse, it is necessary to transmit these and sine and cosine terms faithfully.

For a perfect pulse (ideal flat top and zero rise time), the spectrum is infinite and has the characteristics shown.

SIN X/X DISTRIBUTION

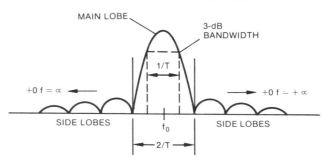

This is the well known *sin x/x distribution*, and it has the property that the width of the main lobe is equal to 2/T, where T is the pulse width. Thus, a 1-microsecond pulse has a main lobe width of 1 MHz, a 10-microsecond pulse has a main lobe width of 200 kHz, and a 0.1-microsecond pulse has a main lobe width of 10 MHz. It has been found that a system capable of transmitting and receiving a pulsed signal with a 3-dB bandwidth equal to half of the main lobe width will give a rather good reproduction of a pulse. Therefore, it is not necessary to have a system for pulse transmission and reception with *infinite bandwidth*, but as a practical matter a system with a bandwidth (3 dB) equal to 1/T is sufficient. While the internal character of the spectrum of a single pulse and a pulse train differ somewhat, the overall bandwidth is still about the same. Since most pulses aren't perfectly rectangular anyhow, their side lobes are even less important, and hence very little is lost by excluding them. You may ask why bother with pulse transmission because it requires so much bandwidth? The answer is that while you need greater bandwidth for pulse transmission, you can send data at a *much higher rate*, too. For example, a 1 = microsecond pulse system can transmit data at a 1-MHz rate. Thus, the bandwidth is consistent with the data rate, *nothing is lost* compared to an analog system that can send information with a 1-MHz signal content.

Bandwidth Requirements for Pulse Transmission (continued)

The *information content* and *bandwidth* are *closely related*. A system with a bandwidth of 5 kHz can transmit information at a 5-kHz rate. A system with a bandwidth of 5 MHz can transmit information at a 5-MHz rate. This is an extremely important point. *It is the channel bandwidth that limits information transfer.* While this may not have been apparent earlier, it is important that you understand it here. For example, to transmit high-fidelity sound requires a 15-kHz channel width, but to transmit a TV broadcast picture takes a 4-MHz channel width. As you probably realize, these are single-sideband or video/audio channel widths, but that is all that's necessary to transmit all of the information data.

THE CHANNEL WIDTH IS PROPORTIONAL TO THE INFORMATION DATA RATE

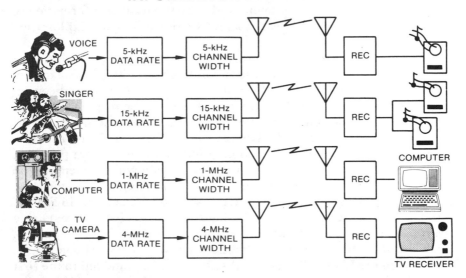

As you might suspect, the bandwidth of the transmission system involves not only the transducer, but also the transmitter itself if modulation is carried out at low level. Thus, the modulator, RF amplifier, and power amplifier stages must have the necessary bandwidth to support the desired transmission bandwidth. In addition, it doesn't really matter (from a data-rate standpoint) whether the transmission is via AM or FM. Practical considerations limit system bandwidths at VHF/UHF frequencies to about 1–5% of the carrier frequency. To achieve very wide bandwidths, *optical fiber* and *laser data transmission* are being used. Here, the wide-band signals are used to modulate a light beam that is used directly to transmit to a receiver in the line of sight, or the light signal is conducted by an optical fiber to a remote receiver. Today, systems with bandwidths of many tens of megahertz or even hundreds of megahertz are used for high-speed data transmission and the transmission of a multiplicity of narrower band signals on a multiplexed basis.

Introduction to Television—Brief History

Television is one of the great accomplishments of our technological civilization. Most of the concepts in television were put forth before World War I, but it was not until after World War II that black-and-white television as we know it came into being. The introduction of modern color television in 1953 and its subsequent explosive growth starting in 1962 has resulted in color TV in almost every home. Currently in the United States, there are about 1,000 TV stations, most capable of color transmission. It is estimated that 98% of the homes (about 73 million) have at least one TV set. Of these, about 80% have color TV receivers. Furthermore, about 20% are connected to cable TV systems (CATV). The development of low-cost TV systems has also led to the rapid growth of closed-circuit television (CCTV), and the ability to record TV on tape has greatly increased the utility and flexibility of TV systems, not only for entertainment and dissemination of information to the public, but also into all areas of scientific research, space exploration, remote sensing, etc.

The idea of transmitting pictures was originally suggested in 1873 when it was discovered that selenium would change its *resistance* in proportion to the *light intensity* falling on it. Since radio had not been invented, these schemes involved the use of wires. The original proposal to have a mosaic of cells connected by wires to electric lamps at the receiving end was suggested by George Carey of Boston. This very complicated scheme was replaced in about 1880 by the important principle of *scanning* each element in the picture *sequentially* while depending on *persistence* of *vision* to see the *complete* picture. This suggestion came originally from W. S. Sawyer in the United States and M. LeBlanc in France. In 1908, A. Campbell Swinton in England proposed a basic system using *scanning electron beams* at *both* the *transmitting* and *receiving end* of the system, which embodies the basic ideas in modern television. Experiments by C. F. Jenkins in the United States and J. L. Baird in England led to the first demonstration of TV in 1926. Early systems used a *mechanical scanning* device and transmitted very crude pictures. The image of the cartoon character "Felix the Cat" was very popular as a test image in these early systems.

FROM THIS...

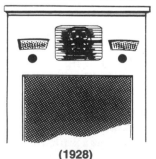

(1928)

TO THIS

(NOW)

Introduction to Television—Brief History (continued)

Vladamir Zworykin, in 1923, patented the *iconoscope*, an *electronically scanned* TV camera tube. The original cathode-ray tube (CRT) was invented by K. F. Braun in Germany in 1897. The iconoscope, perfected in 1932, and improvements in the CRT led to the demonstration of an *all electronic* TV system, based on A. Campbell Swinton's original idea. Parallel developments (mainly in England by I. Schoenberg and J. L. Baird) led to the world's first transmission of regular TV programs in 1936 by the British Broadcasting System.

As you can see, *the early development of TV was truly international in character* with major contributions from the United States, Great Britain, France, Germany, and Russia. Subsequent to World War II, TV developed rapidly in the United States. Questions of standards for TV became important, and the United States developed systems using a picture (frame) rate of 30 per second (half the power-line frequency of 60 Hz); in Europe (notably Great Britain) a frame rate of 25 per second (half the power-line frequency of 50 Hz) was established as standard. Standards of 525 scanning lines were adopted in the United States, while 405 (later 625) scanning lines were adopted in Europe. As you will learn, these standards have *important implications* for *picture quality, resolution,* and *channel bandwidth.*

By 1951, there were about 10 million TV receivers in use; by 1959, about 50 million receivers were in use in the United States alone and 25 million more throughout the rest of the world. Experiments with color TV began as early as 1928; however, a German patent in 1904 disclosed some of the basic principles. Under the leadership of P. Goldmark, the CBS Company developed a color TV system that was used in the United States for a brief period in the early 1950s.

Introduction to Television—Brief History (continued)

This system was not received with favor, since it was *not compatible* with existing black-and-white transmission. Meanwhile, the RCA Company was deeply involved in developing a system (based on a patent by G. Valensin of France in 1938) that *would be compatible* with black-and-white receivers.

A committee, the National Television Systems Committee (NTCS), formed in the United States by the FCC and industry to develop standards for color TV led to the standards for color TV that are in use today. In the United States, color broadcasting began in 1954. Modifications of this system for color transmission were standardized more recently in various parts of Europe. Sale of color receivers was slow initially because of limited programming; however, the aggressive efforts of RCA and its broadcast affiliate NBC succeeded in making color the present-day preference. The sales of color TV sets in the United States exceeded sales of black-and-white sets in 1971, with the difference increasing dramatically with the entry of solid-state devices.

Meanwhile, dramatic changes have taken place in TV camera design with the development of highly sensitive, miniature camera tubes and solid-state devices. Cameras that can be held in one hand and provide color outputs are currently available. In addition, TV recording systems have become smaller, lighter, and very inexpensive, allowing for *video recording* of almost anything desired. Newer technology (such as the *video disk*) is important not only for home entertainment, but also as devices for *mass information storage and retrieval*. In addition, dramatic reduction in component costs have made color TV systems, as well as black-and-white, readily available. At the present time, the United States, Canada, Mexico, and part of South and Central America, Japan, Korea, and others use the NTSC system. Our study of TV systems will consider the NTSC system since it is the only one you will encounter here.

VIDEO RECORDING

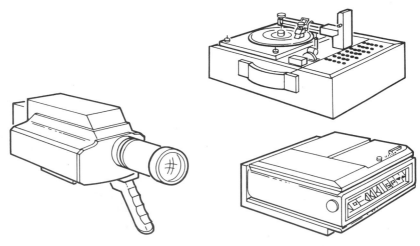

The Basic TV Transmission-Reception Process

The major problems a practical TV system has to solve can be stated simply as follows: A pretty girl is singing in a television studio; how is it possible to *transmit* to a television receiver some 50 miles away an *instantaneous, moving picture of her appearance and simultaneously a faithful reproduction of the sound*? The block diagram shows, in broadest outline, how this is done.

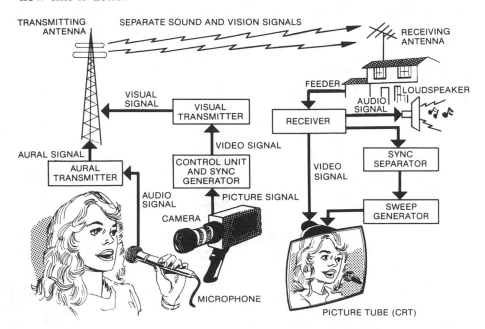

The sound waves generated by the singer's voice are converted by a microphone into electrical impulses, which are amplified and conveyed to an FM transmitter and then radiated from an antenna as in a normal radio broadcast. This is the aural (or audio) signal.

Simultaneously, the girl's appearance is *scanned* by a television camera which is capable of detecting changes in the intensity of the light reflected from her and of converting these changes, again, into electrical impulses. To these impulses (the picture or visual signal) are added *synchronizing and blanking pulses* by a control unit in the studio, and the resulting video signal is fed to a separate video transmitter, where it is imposed on an AM RF carrier wave. The resulting visual signal is taken to the *same* antenna as that used for the aural signal; and the two modulated carriers are *radiated together* on *different* frequencies.

In this volume you will learn about how the TV picture is formed and transmitted. In Volume 4 of this series, you will see how the two separate signals are collected by the receiving antenna and passed to the TV receiver. The receiver converts the two signals back into sound via a loudspeaker and into a picture displayed on a picture tube (cathode-ray tube).

Light and Color

As you learned earlier, the electromagnetic spectrum is a continuum that extends from dc to x-rays and beyond. Within this spectrum, well above the wavelengths used for radio, TV, radar, and communication, lies the so called *visible region* that encompasses the narrow range of wavelength from about 400 to about 700 nm (nanometers). A nanometer is equal to 10^{-9} meters (one-billionth of a meter). You can see we are dealing with *very short wavelengths*. While you feel radiation in the infrared region as heat, and ultraviolet radiation tans your skin, the sensation known as *vision* is stimulated only over the range of 400–750 nm.

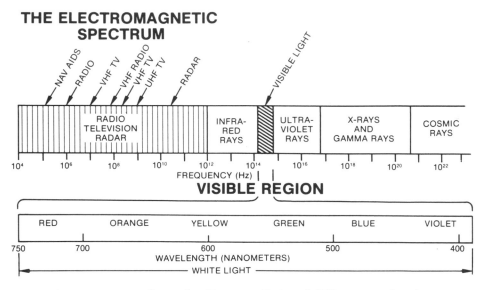

As you can see from the diagram, light of *different wavelengths* causes *different visual sensations* and we call these differences *color*. White light consists of *all* of the colors together in about equal proportions. Black, of course, is the *absence* of light. We perceive light from three basic sources: (1) *direct radiation*, from a source such as the sun or electric light; (2) *transmitted*, such as through sunglasses, a window, or a colored plastic filter; and (3) by *reflection* from a surface. For example, the sun shines through a window, and we use the light that falls on a newspaper to see it as reflected light.

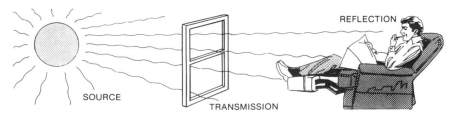

Any of these processes affect the *character* of the light perceived either by the eye, a photographic camera, or a *TV camera*.

Light and Color (continued)

In TV we are mainly interested in light *reflected* from objects. Obviously, this depends very strongly on the light source intensity and color as well as the reflecting object. But for our purposes here, we will assume that the *source* is *white* light. This reflected light can be described by two qualities—*brightness* (or *luminance*) and *hue* (or *chrominance*). The intensity of a light source is measured in units called *the foot-candle*, which is the intensity of the source relative to the intensity of a special candle at a distance of 1 ft. The sun typically has an intensity of about 10,000 foot-candles on a bright summer day. Full moonlight has an intensity of about 1 foot-candle. The brightness (or luminance) of reflected light is measured in a unit called the *foot-lambert*. One foot-lambert is about 0.32 foot-candle. Hue (or chrominance) refers to color of light reflected from an object, and this is determined not only by the *color* of the illuminating light, but also by how light is *absorbed* and reflected by the object. For example, a leaf looks green in white light because *all* the colors *except green* are *absorbed* and *only green* is *reflected*. If we were to observe a leaf in red light (or through a red filter), the leaf would look *black* because the *red light* is *absorbed* and *no red light* is *reflected*.

ONLY GREEN LIGHT REFLECTED   **NO REFLECTED LIGHT**

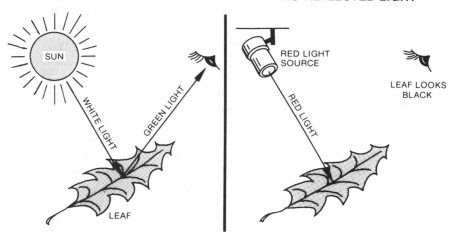

Thus, it is the *spectral character* of the light source as altered by the *reflecting object* that we perceive as color. The brightness, of course, is determined by how strong the light source is and how much of the light is reflected. Thus, we can say that *our eye is capable of estimating the wavelength of light.* While the subject of color vision is extremely complicated and not really well understood, these definitions will do for our study of TV. It is only necessary that you realize that *brightness* (luminance) and *hue* (chrominance) are the *important factors* not only in TV but also in vision. It is then the job of a TV system to reproduce to the viewer a pattern of light that has the same brightness and hue of the original image.

Light and Color (continued)

It has long been known that white light can be *split* into *three* so-called *primary colors* (red, blue, and green), and almost *any color* can be *synthesized* using the right combination of these colors. For example, red and blue make purple and red and green make brown. Therefore, you can take an object, illuminate it with white light, pass the reflected light through three primary color *filters*, and see how much of the reflected light is in each part of the spectrum, as shown.

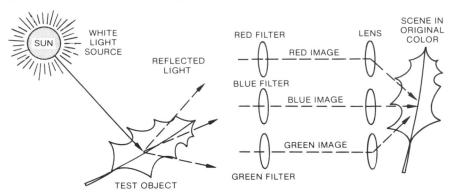

If we observed the light through each filter, we would see the amount of red, blue, and green light reflected from the test object. If the test object were a leaf, we would see it as a green object via the green filter, but it would appear black via the red and blue filters (no reflected light). For other objects, differing amounts of light would be seen through the filters, depending on the color of the object in different areas. If we put lenses in the system and focused the red, blue, and green images on a screen so that they *coincided*, we would see the scene or object just about as it would be if we received it *directly*. While this is a *highly simplified explanation*, basically, it is *true*. If we now consider just one point on the object, we can describe it in terms of its brightness in the three primary colors. That is, we can specify it in terms of the *amount of red light*, the *amount of blue light*, and the *amount of green light*. Carrying this one step further, if we know the *ratio* of these components to each other and also what the *total brightness* was, we have *all the information* needed to form an *exact image* of that point. *Now can you see what we're driving at for TV?* If we can *generate a brightness signal* for a scene and *specify* the *ratio* of the *three colors* in the brightness signal, we can *recreate* the scene in *color* because we have *all the information* we need!

In TV transmission, we transmit the brightness and color information separately. The *brightness-information signal* is the *black-and-white* TV picture. The *color-ratio information* is transmitted separately on *subcarriers*; at the receiver these are *recombined* and *displayed* as a *color image* in a color TV set; or, if the color information is *ignored*, we can receive a *black-and-white image*.

How Light Becomes an Electrical Information Signal

As a preliminary to your study of TV camera systems, you need to know how light can generate or control an electric current. One of the things that is poorly understood about light, and other electromagnetic radiation, is that while it has a wavelike nature, these waves occur in packets of energy called *quanta*. Thus, it can also be treated as though it were *particle-like* in nature. For light, these energy packets or particles are called *photons*. The photon is the smallest amount of light it is possible to have. Even the smallest amount of discernible light has many many photons in it. We will consider light as being made up of photons. There are three basic types of photo effects of interest to us: the *photoelectric effect*, the *photoconductive effect*, and the *photovoltaic effect*.

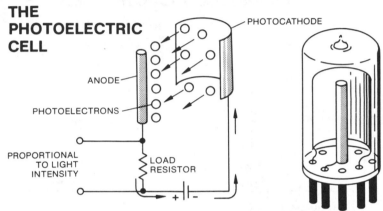

When photons fall on a surface coated with certain materials, the energy of the photons is used to dislodge electrons from the material. Many materials exhibit this property, such as germanium, silicon, selenium, zinc, sodium, potassium, and cesium. One of the most sensitive is a surface of silver oxide coated with a thin layer of cesium. The number of electrons produced depends on the number of photons (or light intensity). Therefore, if we could collect these electrons, we would have a *current proportional* to the *light intensity*. A simple *photoelectric cell* (sometimes called *photoemissive*) consists of a coated surface (the *cathode*) in a vacuum with another electrode (the *anode*) to collect the electrons. The anode has a positive voltage on it to attract the electrons so that they do not just fall back onto the surface of the cathode.

The photoconductive effect depends on the fact that certain materials such as selenium and cadmium sulfide *change* their *resistance* in *proportion* to the *light intensity* falling on them. This effect comes from the fact (as you know) that electron mobility affects the resistance of a material. High-electron-mobility materials are conductors, and low-mobility materials are insulators. In these photoconductive materials, electron mobility depends on light intensity. The *higher* the light intensity, the *greater* the electron mobility and hence the *lower* the resistance.

How Light Becomes an Electrical Information Signal (continued)

Thus, a photoconductive cell can *control* a current flow and *produce* an electrical signal *proportional* to light intensity.

The third common technique for obtaining an electrical signal proportional to light is known as *photovoltaic effect*, since a potential difference is generated across the cell. This depends on the fact that certain semiconductor materials such as silver oxide or selenium will show a charge separation across them in the presence of light. The amount of charge is dependent on the light intensity, and thus these effects can be used to produce electrical signals from light.

THE PHOTOCONDUCTIVE CELL

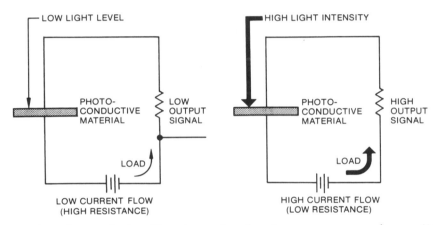

The photovoltaic effect is used today in some very modern solid-state TV cameras for space and similar applications. One of its major advantages is that it can be made as a compact rugged structure and lends itself to the use of digital technology to a greater extent than the other devices. At the present time, solid-state cameras of this type are expensive and do not provide adequate picture quality for TV broadcasting, but this will undoubtedly change in the near future.

THE PHOTOVOLTAIC CELL

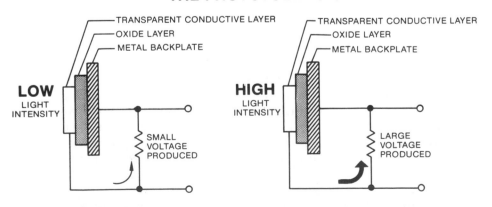

The Scanning Process—Introduction

One way to generate a set of electrical information signals corresponding to a scene to be televised would be to have an array of photocells on which the scene was focused by a lens. Each cell would correspond to the smallest element of the picture desired (*resolution element*). Obviously, it would be best if there were a very large number of elements because this would improve the *definition* (*resolution*) of the picture. As you will learn later, a typical TV picture contains over 250,000 elements. If information from each of these elements were transmitted separately, 250,000 *channels* or *subcarriers* would be needed! While this is possible to do, it is impractical. Therefore, the TV image is not sent out *simultaneously*, but is sent out *sequentially*—that is, the output from each of these photocells in the array is connected sequentially (one at a time), and the information on each resolution cell or element is sent out in a *serial string*. That is repeated over and over again. The process of sequential examination described above is known as *scanning* (or *sequential scanning*) and is the basis for modern TV transmission and reception.

Persistence of vision and of the viewing screen make the picture *appear* as if it were sent *all at the same time*. Obviously, if the pictures were scanned at too slow a rate, the persistence would be possibly insufficient to prevent fading of the image between scans leading to flicker of the image.

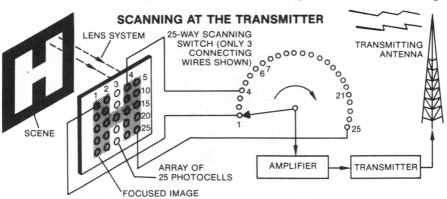

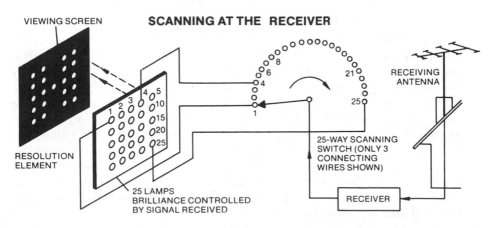

The Scanning Process—The Scanned Raster

In addition, too slow a scan will result in *blurring* of moving scenes. On the other hand, as you will see, the *bandwidth required increases as the scanning frequency increases.*

This sequential process is shown in very simple form on the previous page for a 25-element picture. As you can see, the photocells are scanned from left to right, line by line, until all 25 elements are scanned, and then the process repeats continuously. Thus, a sequential signal is produced with each element representing the *signal level* appropriate to that element. The rate at which each line is scanned is called the *line frequency* or *line rate*. The rate at which the entire picture is scanned is called the *frame rate*. At the receiver, a similar arrangement connects an array of lamps in the *same sequence* via an *identical switch* so that each lamp is excited to produce a light output related to the light input to the photocell at the transmitting end. The transmitter and receiver switches move precisely together so that when the transmit arm is at a given photocell, the receive switch arm is at the corresponding lamp. Thus, the two scanning elements (switch arms) move together and are said to be *synchronized*. In this way, the *pattern* of light and dark elements is *reformed* at the receive end.

THE SCANNING RASTER

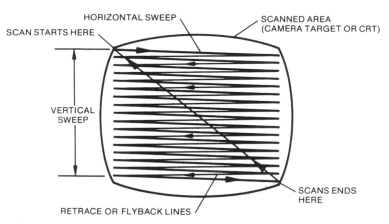

Obviously such a simple mechanical system as shown on the previous page will not provide a usable picture. As you will learn a little later, the TV system used in the United States and elsewhere transmits pictures with a total of 525 lines at a 30-Hz frame rate. To do this, electronic scanning methods using electron beams are used at both the transmit and receive end. In the scanning methods used in TV, the lines are scanned uniformly from left to right. At the end of the line, the beam flies back very quickly, and the scanning of the next line starts. As the beam scans lines horizontally, a slower *vertical scan* moves the beam from the *top* to the *bottom* of the scanned area. At the bottom, the beam flies back rapidly to the top left to commence the process again. The scanned area is called the *raster*.

The Scanning Process—The Scanned Raster (continued)

The area scanned by the electron beam is either a target in a TV camera tube or the viewing screen of a cathode ray tube (CRT) in the receiver. The inactive time of the beam, the *retrace* or *flyback* time, is *blanked* (beam turned off) so that the retrace lines are *not visible.* As you will learn shortly, the TV transmission information signal contains *synchronizing pulses* to start the *horizontal line sweep* and the *vertical sweep.* These pulses are used in conjunction with sweep circuits to control the electron beams in *both* the camera and the CRT at the receiver. As you will learn shortly, the scanning beam in the camera tube scans the target upon which the picture to be televised is focused by a lens system. This produces an output-current signal that is proportional sequentially to scene brightness point by point for each line. At the receiving end, this signal is used to modulate the intensity of the electron beam at the receiver CRT. This beam strikes a fluorescent screen at the face of the CRT to recreate the scene.

BEAM SCANNING

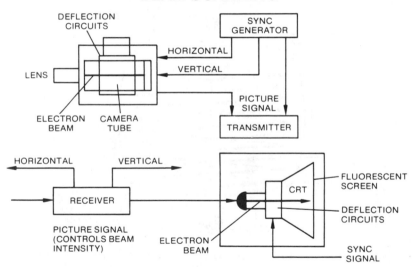

The circuits that cause the beam scanning are called *deflection circuits*, and these are present in *both* the receiver and the camera. The synchronizing pulses, generated at the transmitter, are used to generate waveforms that drive deflection coils (or other deflection mechanisms) in the TV camera tube. The waveform generated is such that the beam is scanned from left to right and from top to bottom. These synchronizing pulses also form part of the TV data transmitted and are separated from the TV picture video in special circuits of the receiver. These pulses are used to drive deflection circuits that are identical in function at the receiver. Thus, the TV signal contains information on the brightness of each point in the picture and the information necessary to locate that point at the receive end.

The Scanning Process—Interlacing

To avoid flicker, it has been determined that the pictures must be presented at a minimum rate of 48 Hz (48 times per second). For technical reasons, it is desirable to have the frame rate very close to a multiple of the power-line rate, since this reduces effects due to power-supply ripple. The United States standard is about 60 Hz, which ensures freedom from flicker. To transmit 525 lines at a 60-Hz rate requires a *very wide bandwidth transmission*. An ingenious way around this problem is called *interlaced scanning*. In interlaced scanning, the vertical scan is done in two steps. First the *odd* lines are scanned (1, 3, 5, 7, etc.), then the beam retraces vertically and the *even* lines are scanned (2, 4, 6, 8, etc.). These vertical scans occur at the 60-Hz rate, but the total scan period is now only 1/30 sec (1/60 second for odd lines and 1/60 second for even lines). Each of the vertical scans is called a *field*, and the *complete scan* (odd and even) is called a *frame*. Thus, the field rate is 60 Hz and the frame rate is 30 Hz.

INTERLACED SCANNING

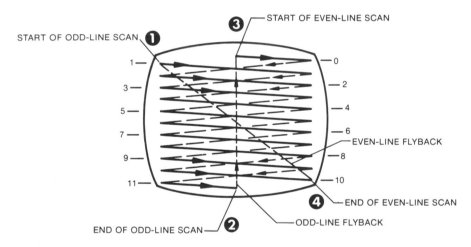

The vertical distance between lines is small, and the interlacing allows each area to be scanned in 1/60 second, thus minimizing flicker but allowing for a *reduction* of *2:1 in bandwidth* of the transmitted signal. In interlaced scanning, the interlacing is easily accomplished if the number of vertical lines is an odd number. As you can see from the figure, the odd vertical scan is terminated in the middle of the last odd horizontal line and the even lines begin on the half line at the top. For an odd number of lines (for example, 525), the number of lines per field is 525/2 or 262½; therefore the vertical sync takes place between the odd and even scan at the half-line point. This results automatically in accomplishment of interlaced scanning. Remember, the sync pulses occur every 1/60 second, but alternate sync pulses within the 1/30-second frame rate occur on the half line, and this results in an interlaced scan.

The Scanning Process—Beam Deflection

As you will see, an *electron beam* is used to do the scanning—that is, a finely focused stream of electrons is *deflected horizontally* and *vertically* to produce a scanning raster. An electron beam is like a conductor carrying current and *generates its own magnetic field*. In addition, the electrons are *charged particles*, and hence can be influenced by an *electrostatic field*. Thus, the electron stream can be deflected by either an electrostatic or magnetic field. We will study magnetic deflection since electrostatic deflection is used mainly for CRTs in oscilloscopes and special applications.

To achieve deflection of an electron beam, we would like to have the beam move smoothly and uniformly from one side of the screen to the other (horizontally or vertically). To do this, we must increase the magnetic field uniformly from a starting value to some maximum value, to carry the beam to the other edge of the screen, then quickly return and do this again. To do this, we use a *sawtooth waveform*. The sawtooth is started with a synchronizing pulse generated at the transmitter, and, as you know, this synchronizing signal is sent to the receiver as well. Thus, *both* the electron beam in the *camera tube* and the one in the *receiver move together*, starting at the *same time* for each line or frame scanned.

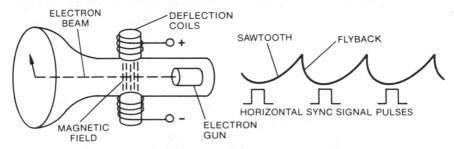

You know from the left-hand rule and Faraday's Law in *Basic Electricity* that when a magnetic field interacts with a conductor, a force is exerted on the conductor at right angles to the field. In this case, we have an electron beam that is free to move in the field.

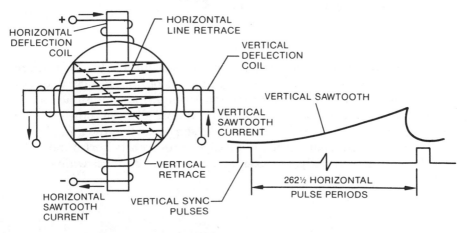

The Scanning Process—Scanning Frequencies

As you can see from the diagram on the previous page, many horizontal lines occur for one vertical sweep. Therefore, we obtain a raster scan where the beam sequentially covers the entire area to be scanned.

For monochrome or black-and-white commercial TV, the vertical scan rate is 1/30 second per frame (1/60 second per field) and 525 vertical lines. Thus, the horizontal scan rate can be calculated as follows:

H scan rate = no. lines × frame rate = 525 × 30 = 15,750 Hz.

Thus, horizontal sync occurs at a 15,750-Hz rate, or the time for one horizontal line is 1/15,750 second or 63.5 microseconds.

In color TV, the chrominance subcarrier must be very accurately related to the scanning frequencies to prevent interference between the sideband of the brightness signal and the chrominance signal. The chrominance subcarrier frequency is exactly 3.575954 MHz. The horizontal scanning frequency is set to exactly 2/455 times the subcarrier frequency or 15,734.264 Hz, and the vertical rate is 2/525 times the line frequency or 59.94 Hz. As you can see, these differences are not significant except for the fact that they must be *very accurately maintained* at the transmitter if the brightness and color signal sidebands are not to interfere with each other.

U.S. TV BROADCAST SCANNING STANDARDS

	MONOCHROME	COLOR
FRAME RATE	30 Hz	29.97 Hz
FIELD RATE	60 Hz	59.94 Hz
HORIZONTAL LINE FREQUENCY	15,750 Hz	15,734.264 ± 0.044 Hz
HORIZONTAL LINE TIME	63.5 μSECONDS	63.556 μSECONDS
CHROMINANCE SUBCARRIER FREQUENCY	—	3.575954 MHz
ASPECT RATIO	4/3	4/3

For many other applications like closed-circuit TV (CCTV) where better picture resolution is desired, the number of lines is increased. For example, many systems use 855 or 945 horizontal lines, and some special TV systems use more than 1,200 lines. In almost all cases, however, the vertical frame rate is 30-Hz interlaced. As shown above, the aspect ratio for TV is 4/3. This means that the picture is four units horizontally for each three units vertically. For example, a picture 12 inches wide should be 9 inches in height.

TV Signal Bandwidth and Resolution

The resolution and maximum frequency or signal bandwidth can be calculated on the basis of the fact that the horizontal and vertical resolution are the same. Since the smallest vertical picture element is a scanning line, there are N possible picture elements vertically for an N-line picture. Because of an aspect ratio of A, this leads to a horizontal resolution of A × N. For the simple checkerboard pattern shown below, the waveform is a square wave during scanning interval consisting of alternating light and dark intervals.

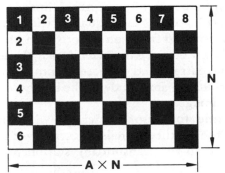

WITH THE ASPECT RATIO 4:3, TOTAL NUMBER OF ELEMENTS IN THE PICTURE IS

$$AN^2 = \frac{4}{3} \times 36 = 48$$

Since TV resolution is defined in terms of black-and-white element pairs— that is, one pair is equivalent to the resolution or the number of resolution cells is actually $AN^2/2$. Since this occurs at the frame rate (P), the maximum picture frequency can be defined as $PAN^2/2$ Hz.

For a 525-line, 30-Hz system, the maximum frequency can be calculated as

$$f_{max} = \frac{PAN^2}{2} = \frac{30 \times 4 \times 525 \times 525}{3 \times 2} = 5.5 \text{ MHz}$$

Actually, the signal bandwidth is somewhat less because retrace times make the actual number of available lines somewhat less (typically, about 455). This is partially offset by the fact that horizontal retrace takes some line time, too. Because these retrace times are a similar percentage (about 15%) for both horizontal and vertical retrace, the aspect ratio is not materially affected. If we can use this as a basis for calculation, we arrive at a bandwidth (maximum picture frequency) of about 4.2 MHz. This is the video-signal bandwidth and would require an 8.4-MHz RF band for double-sideband transmission.

By using a technique called *vestigial-sideband* transmission, which will be described later, it is possible to confine the signal into an RF band of 6 MHz, including the sound channel. The transmission of color information does not appreciably increase the required bandwidth because the color information is provided on a subcarrier in such a way that its sidebands are interlaced (between) the sidebands of the brightness signal.

The TV Camera—Introduction

The most important piece of equipment in the television studio is the TV camera; for it is there that the picture signal originates. This TV camera is continually watching every movement, every change of shape, and every tonal content of the scene, and converting what it sees into a stream of electrical information signals. Thus it is a transducer, just as a microphone is, except that its output-information signal is a *light* pattern rather than a *sound* pattern.

To perform this feat, the TV camera makes use of a special type of cathode-ray tube called a *camera tube*, which transforms the image to be televised into an equivalent picture composed of individual electric charges. This charge-image pattern is then scanned from top to bottom in a series of horizontal lines by a narrow beam of electrons which *reads* the electrical information contained in the pattern and converts it into a stream of electrical signals, each proportional in amplitude to the brightness of a particular point of the original image.

After amplification (and mixing, for reasons which you will shortly see, with synchronizing pulses), the picture signals, together with the separate sound signals, are carried to the transmitting station, where they are radiated.

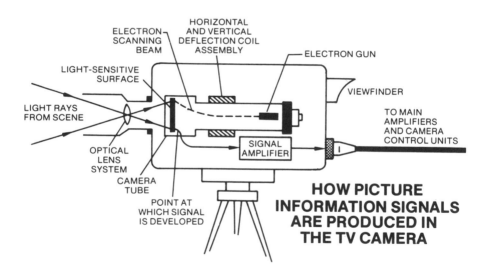

HOW PICTURE INFORMATION SIGNALS ARE PRODUCED IN THE TV CAMERA

Television cameras, like photographic cameras, contain an optical lens system that collects the light from the scene and focuses it onto the light-sensitive surface. In the photographic camera, this surface is usually a section of a spool of film. In the TV camera, it is a specially coated surface inside the camera tube which is known as a *photocathode*, or (in some types of camera tube) a *mosaic*. This photocathode itself forms part of a *target assembly*. While there are many different types of camera tubes, varying in operating principle internally, they all have the same function.

The TV Camera Tube

This function is to convert the optical image focused on their photocathode into a stream of electrical signals representing the image brightness in each resolution element.

The heart of the TV camera is the *camera tube*. Early TV cameras used a camera tube called the *iconoscope* that was mentioned earlier, but this is now obsolete. There are basically two types of TV camera tubes in use today. These are the *image orthicon* and the *vidicon*. The simplicity, small size, and low cost of the vidicon camera have made it the most popular for black-and-white and color in all but the most exacting TV studio use. Here, the image orthicon is still in use, but is gradually being replaced by vidicon cameras. One of the major advantages of the orthicon is its high light sensitivity. The image orthicon uses *photoemissivity* and *secondary emission* as the basis for its operation. The vidicon uses *photoconductivity*. Both use a scanning electron beam to gather the information from the target.

Before you learn about how these camera tubes work, you must learn something about secondary emission. Secondary emission occurs when electrons strike certain surfaces such as the silver oxide-cesium surface described in photoemissivity, except that in this case electrons are the energy source. When electrons strike these surfaces, a shower of electrons is released—that is, more electrons are typically released than strike the surface. This represents a *current gain* and advantage is taken of this to *increase the signal current* from an electron stream.

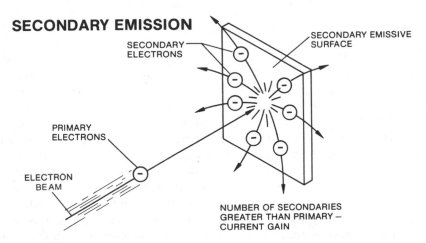

SECONDARY EMISSION

SECONDARY ELECTRONS

SECONDARY EMISSIVE SURFACE

PRIMARY ELECTRONS

ELECTRON BEAM

NUMBER OF SECONDARIES GREATER THAN PRIMARY — CURRENT GAIN

Secondary emission is illustrated. In secondary emission, incident primary electrons excite the emission of secondary electrons. The ratio of secondary electrons to primary electrons is called *the secondary emission ratio* and is typically between 2 and about 10. If these secondary electrons are accelerated in an electrostatic field and focused on another secondary emitting surface, electron current gain results. Several of these surfaces, properly configured, can give current gains of up to 10^8 times. These devices (used inside the image orthicon) are called *electron multipliers*.

The TV Camera Tube—The Orthicon

The orthicon camera tube is sensitive and used in high-quality studio work but is expensive, bulky, and rather tricky in operation.

The scene to be televised is focused onto a photocathode that is photoemissive. Thus, an electron image is formed that has an electron density in each element *proportional* to the number of photons (light intensity) impingent on it. This image is accelerated toward a target by a small positive voltage, and is kept focused by the arrangement of the electrodes in the image section and by a focus coil that surrounds the tube. It is focused onto the target, and the electrons strike the target with sufficient energy to liberate secondary electrons that are captured by the positively charged mesh directly in front of the target, leaving the target with a *positive charge distribution* that *corresponds* to the *original light image*. Because the semiconducting target is so thin the charge distribution also appears almost immediately on the *other side* of the target as well.

THE IMAGE ORTHICON CAMERA TUBE

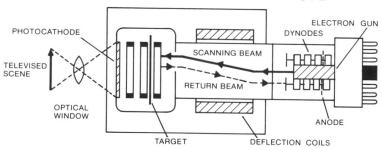

A fine electron beam, formed by the electron gun, is deflected by the deflection coils (magnetic) to scan the target to form the signal. A relatively high potential and high electron velocity is needed, but it is undesirable to have this strike the target and liberate secondary electrons, so the beam is slowed down, almost to a stop, by the decelerator grid. The decelerator field also is shaped to cause the beam to strike the target perpendicular to the target surface at all points. As the beam scans the target, some of these slow electrons are attracted to the positively charged areas of the target, neutralizing the charge. Those not used to neutralize charges are turned around and form the return signal. These return electrons represent the excess over what was necessary to recharge the target and thus are *proportional* to the *scene brightness*, with *fewest* electrons in *white* (bright) areas and *most* in *dark* areas. The return beam is focused on an electron multiplier that *amplifies* the returned electrons to form the output-signal current. This current flowing through a load resistance forms the image-signal voltage. Since the beam scans the target sequentially, the output signal is a *sequential* information signal that *corresponds* to the scene being televised. The electron beam neutralizes the target so that a new charge image is formed to be scanned again for the next frame.

The TV Camera Tube—The Vidicon

The vidicon camera tube has replaced the image orthicon in all but the most critical or special applications. Since its original development in 1952, it has been steadily improved and modified and is capable of truly remarkable performance. In addition to vidicons for use in TV cameras, *special vidicons* for many scientific and unique applications are available. The vidicon camera tube is simple, rugged, and inexpensive. It is now used in color TV cameras, portable TV applications, and CCTV.

The vidicon uses *photoconductivity* as the basis for operation. A thin conducting transparent film (the signal electrode) is applied to the inner surface of the faceplate with a thin (few thousandths of a millimeter) photoconductive surface deposited over this. Each element of the photoconductive surface can be considered as a leaky capacitor element with one side connected to the signal electrode and the other connected via the scanning electron beam.

THE VIDICON CAMERA TUBE

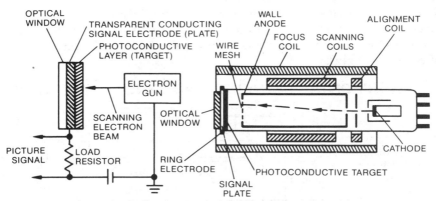

Initially, the electron-beam side of the photoconductive surface is uniformly charged by the beam. This leaves a charge on each point on the photoconductive layer. Because the surface is photoconductive, light at each point varies the conductivity, allowing these charges to be leaked off through the target to the signal electrode plate by an amount dependent on light intensity. Thus, a *charge pattern* is left on each photoconductive element that is *greatest* where *little* or no light fell and *least* in *bright* areas. The electron beam is decelerated and made perpendicular to the target during the scan in a manner analogous to that for the orthicon. Where there are positive areas on the photoconductive surface, the electron beam recharges these, and there is a current flow through the target capacitance to the signal plate. A resistance in series with the signal plate therefore contains the video-signal information. The photoconductive target material is critical in determining the performance of vidicon camera tubes. The most popular material for TV use is lead oxide. As with orthicons, vidicon camera tubes have a variety of trade names for similar units. Typically, vidicons for TV are about 1 inch in diameter and 4–7 inches in length.

Blanking and Synchronization

Regardless of the type of camera tube used, the video output information signal during the scan is *basically* the *same*. The only requirement is that the *proper polarity* of the modulating signal be preserved. The United States TV standard (as in most of the world) is that the *brightest signal* (reference white level) represents *least power* from the transmitter (about 12.5% of the peak output) and the *black signal* (reference black level) represents about 67.5% of the peak output from the transmitter. Thus, a *decrease* in *brightness* of a picture element causes an *increase* in *radiated power*. The establishment of the black level in the camera can be rather complex and is not necessary in our discussion of TV transmission.

To prevent the generation of spurious undesirable signals during flyback (vertical and horizontal), the beam in the camera tube is turned off (or otherwise prevented from striking the target) during the retrace time. To do this, *blanking pulses* are generated that *turn the beam off* in the camera tube and these are also *transmitted* so that the retraces of the electron beam in the *receiver are also suppressed*. These blanking pulses (blacker than black) are transmitted at about 75% of the peak carrier level. Blanking pulses begin at the end of the picture interval but slightly before flyback begins. The blanking is continued until the active portion of the scan begins.

HORIZONTAL BLANKING AND SYNCHRONIZATION

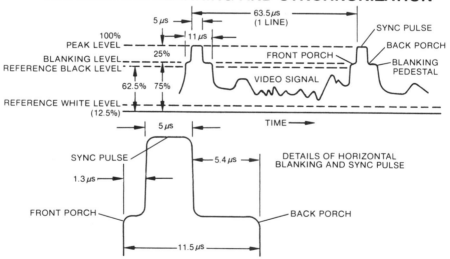

The illustration above shows the formation of the composite signals for the *horizontal* lines. The horizontal *sync pulses* are *superimposed* on the blanking pulses. The blanking commences shortly before the sync pulse (front porch). The sync pulse causes the sawtooth generator to reset (flyback) and the next line scan to start. The back porch keeps the beam blanked until the sawtooth signal is under way. The blanking interval represents about 18% of each horizontal line or about 11.5 microseconds (0.18 × 64).

Blanking and Synchronization (continued)

The *vertical* blanking and synchronization is more complicated. This is because the vertical retrace takes many horizontal line intervals to complete and because of the interlaced scan. The vertical blanking interval is set by standards at about 1,250 microseconds. If the horizontal-sync pulses were not transmitted during this interval, the horizontal sawtooth generator could drift and during the resynchronizing time, the top portion of the picture could be badly distorted. To prevent this, the vertical-sync pulse consists of a series of very wide horizontal-sync pulses, since only the leading edge of the horizontal-sync pulse is necessary to trigger the horizontal sawtooth generator.

VERTICAL BLANKING AND SYNCHRONIZATION

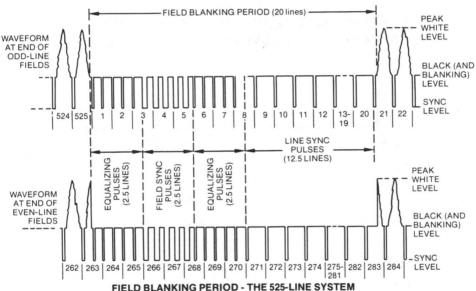

FIELD BLANKING PERIOD - THE 525-LINE SYSTEM

As you can see, at the beginning of the vertical blanking period there is a series of equalizing pulses that occur at the half-line period. Because of the nature of the horizontal sawtooth generator, it will only respond to sync pulses near the line period, so the alternate half-line pulses are ignored, and horizontal scanning continues uninterrupted. The vertical-sync pulses are a series of wide horizontal-sync pulses that are treated in a *special way in the receiver* to make the actual vertical-sync pulses. Following the vertical-sync pulses, additional half-line rate-equalizing pulses are transmitted. On the even-line fields, the line (horizontal) sync pulses commence a half line sooner than on the odd-line fields. This effectively delays the horizontal-sync pulses by a half line on the odd-line fields, as is necessary to achieve interlacing. The way this vertical-synchronizing system operates will be described in more detail in Volume 4 of this series. It is sufficient here to know that a *blanking and synchronizing signal* of this type is *generated* for *transmission*.

Blanking and Synchronization—Timing

As you can see, the blanking and synchronization occur at *specific related times*. As you realize, these blanking and synchronizing pulses are *timing pulses* that regulate when the beam is to be turned off and on, and when the next sweep (horizontal or vertical) is to be initiated. Since these signals control *both* the camera tube electron beam and the receiver CRT electron beam in the *same way*, the whole complex is *synchronized* by these timing pulses. As you will see in Volume 4, these pulses are stripped off (separated) in the receiver to control the sawtooth oscillators that in turn control the beam motions. As you might also realize, any interference with this timing or synchronization will cause severe disruption of the received image since it will disrupt the timing of the beginning of the sweep.

SYNC PULSES KEEP THE CAMERA AND CRT SCANNING BEAMS SYNCHRONIZED AND TIMED

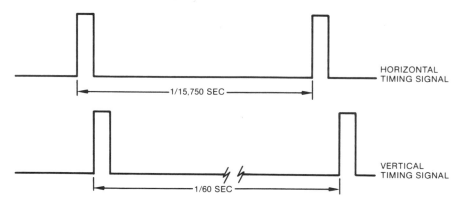

For TV, the synchronizing signals have some special timing requirements that you learned about earlier, so that horizontal synchronization is maintained during the vertical blanking and retrace interval and the vertical-sync pulses are designed to allow for interlace in the vertical circuits. However, these special features still do not obscure the basic function of these synchronizing signals, which is to cause the start of beam deflection *synchronously* in *both* the transmitter and receiver deflection systems.

This introduction to timing (or timers) by means of pulses—one of the last of the nine common electronic building blocks—will be discussed again in receiving systems in Volume 4 and in much more detail in your study of pulse and digital circuits in Volume 5. There you will learn how timing signals control sequences of events so that an entire system can be controlled in time. For example, the timing of radar signals, computer logic, digital data transmission, control system logic, and many more are controlled by timing pulses similar in many respects to TV timing signals.

The Complete Camera System

For commercial TV broadcast, the TV cameras are part of a complex system of synchronizing, switching, and monitoring that allows transfer from camera to camera, to video tape, to motion picture equipment, etc. These are combined into one or more studio complexes that provide as an output the composite TV video information signal that was described earlier. Generally, in these installations one master synchronizing unit is used to drive all cameras so that transition from one input device to another can take place smoothly. Modern TV cameras often have complex lens systems (lens turret and/or zoom lenses) as well as special dollies to carry the cameras to any area of the studio. In addition, most TV cameras use a small CRT as a monitor mounted so the operator can see the TV image being televised by the camera and hence provide the best picture possible from a focus and composition standpoint.

TYPICAL STUDIO CAMERA

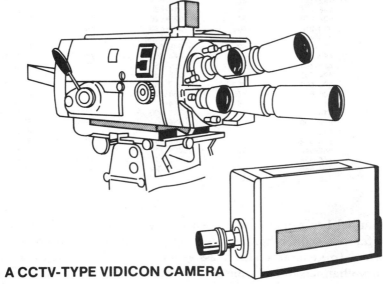

A CCTV-TYPE VIDICON CAMERA

For CCTV (closed-circuit TV) or portable use or for nonstudio applications, the camera system often includes all of the circuitry necessary to develop the composite signal output. When several CCTV cameras are used together, a common synchronizing unit is used to control all of the cameras. Remote control of focus, the lens system, and operating parameters are often incorporated into these systems so that the cameras can be installed at remote points, but can still be operated efficiently.

While for many CCTV systems, features such as interlacing are omitted for simplicity, the basic operating principles are nonetheless the same. The important thing from our standpoint is that there is a *composite output signal* from the camera system. For a CCTV system, this signal is sent to a remote point by coaxial cable, while for commercial TV, the signal is used to amplitude modulate a transmitter.

TV Channel Allocation

In the United States, there are 82 available commercial and educational TV channels. Channels 2 to 13 are in the VHF region, and channels 14 to 83 are in the UHF region. For reference, these are listed below.

NUMERICAL DESIGNATION OF U.S. TELEVISION CHANNELS

CHANNEL NUMBER	BAND (MHz)	CHANNEL NUMBER	BAND (MHz)	CHANNEL NUMBER	BAND (MHz)
2	54–60	29	560–566	57	728–734
3	60–66	30	566–572	58	734–740
4	66–72	31	572–578	59	740–746
5	76–82	32	578–584	60	746–752
6	82–88	33	584–590	61	752–758
7	174–180	34	590–596	62	758–764
8	180–186	35	596–602	63	764–770
9	186–192	36	602–608	64	770–776
10	192–198	37	608–614	65	776–782
11	198–204	38	614–620	66	782–788
12	204–210	39	620–626	67	788–794
13	210–216	40	626–632	68	794–800
14	470–476	41	632–638	69	800–806
15	476–482	42	638–644	70	806–812
16	482–488	43	644–650	71	812–818
17	488–494	44	650–656	72	818–824
18	494–500	45	656–662	73	824–830
19	500–506	46	662–668	74	830–836
20	506–512	47	668–674	75	836–842
21	512–518	48	674–680	76	842–848
22	518–524	49	680–686	77	848–854
23	524–530	50	686–692	78	854–860
24	530–536	51	692–698	79	860–866
25	536–542	52	698–704	80	866–872
26	542–548	53	704–710	81	872–878
27	548–554	54	710–716	82	878–884
28	554–560	55	716–722	83	884–890
		56	722–728		

Channel 1 was deleted in the early days of commercial TV because of interference with other services. As you can see, each channel is 6 MHz wide. In general, channel assignments are not made on adjacent frequency channels in the same area because of the possibility of mutual interference. Since TV transmissions are in the VHF and UHF range, reception is limited to line of sight, as discussed earlier in this volume. For this reason, TV transmitting antennas, which in many respects resemble FM transmitting antennas, are located at the *highest possible point* to accommodate the greatest possible audience.

As shown above, the total TV channel width is 6 MHz. You learned earlier that the TV video signal is actually about 4.2 MHz wide. The picture signal for TV uses amplitude modulation, and for conventional double-sideband AM transmission, the bandwidth required is 8.4 MHz. Obviously, something must be done to resolve this problem. As you might suspect, a form of *single-sideband transmission* is used.

The TV Channel—Visual and Aural

The figure below shows how the 6 MHz channel is used.

TV CHANNEL UTILIZATION

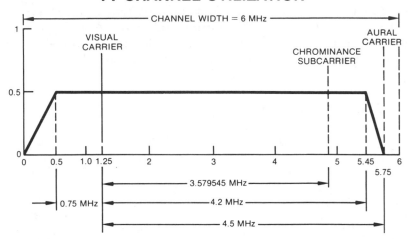

As you can see, the visual transmission is double sideband for signal components in the vicinity of the carrier (±0.75 MHz) and tapers to single sideband for signal components above 0.75 MHz. This method of transmission is known as *vestigial-sideband transmission*. Also note that the total video bandwidth (1.25 MHz to 5.45 MHz) is 4.2 MHz, which causes some loss of resolution in the transmitted signal, as described earlier. The signal at the band edge (0 MHz) and at the sound carrier (5.75 MHz) is reduced to a low level—greater than 100:1 (20 dB) reduction. The sound carrier is positioned exactly 4.5 MHz above the picture carrier. You will learn more about the chrominance subcarrier located exactly 3.579545 MHz above the picture carrier when you learn about color TV transmission a little later in this volume. Color transmission requires that the frame, line frequency, and the color subcarrier be very precisely related in frequency, so the color subcarrier sidebands interleave and do not interfere with the brightness signal.

The *aural* (sound) transmission is done using *FM*. A separate aural transmitter is used, and its RF signal may be multiplexed into the TV transmitting antenna or be transmitted on a separate antenna. Typically, its power output is between 10 and 20% of the peak visual transmitter power. A frequency deviation of ±25 kHz represents 100% aural FM modulation. Fidelity and pre-emphasis requirements are the *same* as for FM transmission, which you learned about earlier. At present, the aural transmission is monophonic.

The Visual Transmitter

The TV visual transmission is accomplished by amplitude modulation of the carrier, as described earlier, with the blackest signals (peaks of sync pulses) representing 100% peak modulation. Because of the need to carefully preserve the fidelity of the video signal, and because of its wide bandwidth, the modulation is invariably done at low levels with Class A and Class B linear RF amplifiers following. Also, as you might suspect, the carrier is crystal controlled by means of a lower-frequency crystal oscillator and frequency multipliers to obtain the final RF frequency.

THE TV TRANSMITTER

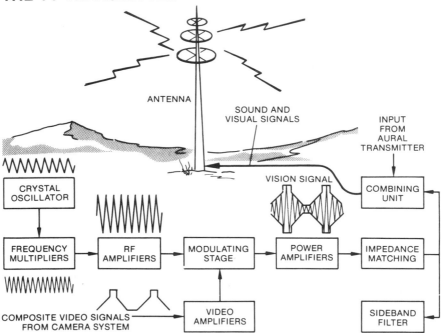

As shown, the TV transmitter output is multiplexed with the output of the aural transmitter and fed to a common antenna. The basic difference between the AM transmitters you studied earlier and the TV transmitter is the need for careful design to provide for good amplitude linearity and phase stability following modulation. Another basic requirement is that the amplifier stages subsequent to the modulator have a wide enough bandwidth to handle the wide-band TV signal. While some of the unwanted sideband energy is removed in low-level stages, the final shaping of the transmitter output is done in a sideband filter. This is a complicated RF filter that shapes the frequency response of the final transmitted signal to confine it in the way shown on the previous page to provide for vestigial-sideband transmission.

Color TV Transmission—The Color Camera

As you learned earlier, essentially any color can be made by proper combination of three primary colors. In color TV transmission, the colors used are red, blue, and green. You also know that the brightness (luminance) and color (hue or chrominance) information can be separated and transmitted separately. Also, a color subcarrier is used to transmit chrominance information. One thing that may have struck you was how carefully the line, frame, and color subcarrier frequencies are related. Without going into detail, these frequencies are related in such a way that the sidebands from the brightness signal and the color signals interleave, and do not fall upon one another. This allows them to be separated at the receiver without mutual interference.

The color camera is basically *three identical black-and-white cameras* used with *special filters* and *image splitters*. There are now available single-camera-tube color cameras, but for simplicity in this explanation, a three-camera tube system will be described first. Three identical vidicons with identical deflection circuits are arrayed with a lens, mirror, and filter system, as shown in the diagram.

THREE-COLOR VIDICON CAMERA

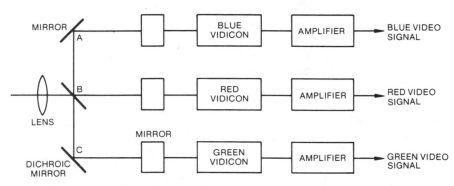

The scene to be televised is focused on the photocathodes of the three vidicons via the image-splitting (dichroic) mirror that splits the light into *three paths* (A, B, C), as shown. This mirror is partially silvered on both sides so that some of the light is reflected from the front surface (C), some of the light is reflected from the back surface (A), and some goes straight through (B). Mirrors in the light paths A and C reflect the light from the dichroic mirror surfaces into the vidicons. Blue, red, and green filters (transmit only these colors) are interposed in the paths as shown so that only the component of color passed by the filter excites the vidicons. The path lengths A, B, and C are carefully adjusted to be exactly the same so that the image is focused on all three photocathodes simultaneously. Thus, the three vidicons produce signal outputs that represent the brightness of the three component primary colors for the image being televised.

Color TV Transmission—The Color Camera (continued)

The scanning rasters for each vidicon are very carefully adjusted with respect to size, shape, and position so that all three scanning electron beams examine the *same point* on the televised scene at the *same time*. Thus, three output signals are produced that contain the red, blue, and green video information from this scene. As you might suspect, if these were simply added together, a monochrome image would result.

THREE VIDICONS SCANNED SIMULTANEOUSLY

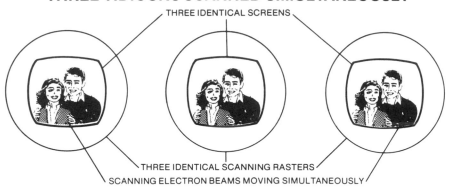

THREE IDENTICAL SCREENS

THREE IDENTICAL SCANNING RASTERS
SCANNING ELECTRON BEAMS MOVING SIMULTANEOUSLY

While a monochrome signal is obtained by adding the red, blue, and green camera outputs, some color TV cameras use a fourth vidicon with *no filters* to generate a monochrome output. Using a fourth vidicon allows for better control of the color tubes without concern for the monochrome image quality.

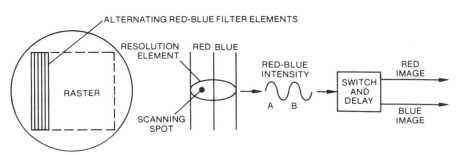

ALTERNATING RED-BLUE FILTER ELEMENTS

RESOLUTION ELEMENT   RED BLUE

RASTER

RED-BLUE INTENSITY

A   B

SWITCH AND DELAY

RED IMAGE

BLUE IMAGE

SCANNING SPOT

A new single-tube color TV camera has been developed that uses color filters integral with the vidicon faceplate within alternating vertical red, blue, and green strips. The photocathode image is scanned so that separate red, blue, and green outputs are available at each element in sequence. To *separate* the color images, a *switching matrix* is used to isolate each color, and the outputs are *aligned in time* by *delaying* one output with respect to the other. These single-tube color cameras are used in industrial TV and similar applications, including home color TV systems, because they are very simple, inexpensive, and compact. They do not, however, at the present time produce studio-quality images, both from a color fidelity and a resolution standpoint.

Experiment/Application—The Composite TV Signal

It is very complicated to build even a simple TV transmitter; however, you can see the composite TV signal (video plus sync and blanking) by observing the signal from a simple black-and-white TV receiver that is tuned to a local TV station. *Care* is necessary, because high voltages are present in a TV receiver; however, the composite signal output is readily obtained directly from the grid terminal of the CRT, which is usually at a relatively low voltage.

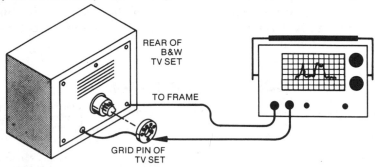

The signal from the grid of the CRT is applied to the vertical plates of the oscilloscope, and the internal oscilloscope horizontal sync and sweep are used. The grid pin can be located from reference to the receiver schematic diagram. If you set the horizontal sweep to 30 Hz, you will see the entire frame. If you set the horizontal sweep so that a single TV line is shown, you can see the horizontal blanking pedestal, sync pulse, and video line. If it is a color transmission, you should be able to see the 3.58-MHz sync burst on the back porch of the pedestal.

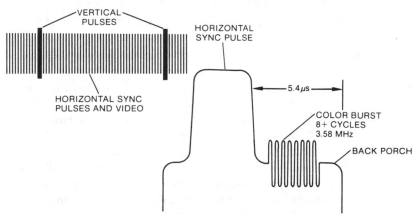

You will notice that while the pedestal and sync are constant, the video signals change as the scan changes. By manipulation of the scope controls, you will be able to see most of the detailed structure of the composite video signal. This signal looks the same as the video output from the TV camera in the studio.

The NTSC Color Signal

The NTSC color signal is transmitted as a pair of suppressed-sub-carrier double-sideband signals in quadrature (original carriers before suppression are 90 degrees apart in phase). These color subcarriers (chrominance signals) are transmitted at relatively low resolution compared to the brightness (luminance) signal. This can be done because it has been found that the color resolution can be reduced, if the brightness signal has high resolution, without significantly degrading the picture quality. You can see an example of this in the color comics (in news-papers), where the detail is provided by black lines and the colored areas are roughly filled in.

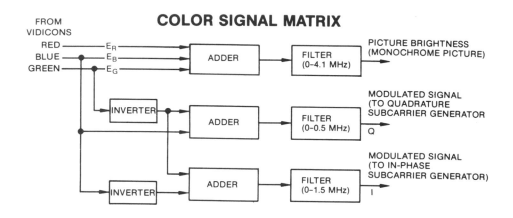

E_M = monochrome signal = $0.3\,E_R + 0.59\,E_G + 0.11\,E_B$
E_I = in-phase color signal = $0.6\,E_R - 0.28\,E_G - 0.32\,E_B$
E_Q = quadrature color signal = $0.21\,E_R - 0.52\,E_G - 0.31\,E_B$

The three outputs from the color camera are combined in a matrix, as shown. The equations give the proportion of each color signal (E_R, E_G, E_B) added to produce the monochrome signal (E_M), the in-phase signal (E_I), and the quadrature signal (E_Q). The inverters are used to allow for subtraction, as required. These three video signals constitute the basic video outputs. The E_I and E_Q signals are now ready for modulating the subcarrier prior to addition to the monochrome signal.

The monochrome signal is suited to provide for an excellent black-and-white image when used alone. It is the signal that provides the brightness information for the color signal as well. Thus, a monochrome receiver will use only the brightness signal.

The video signals E_I and E_Q are inputs to balanced modulators that you studied earlier. Since these are double-sideband signals, no further filtering is needed. A block diagram of the necessary circuits is shown on the next page.

The NTSC Color Signal (continued)

THE TOTAL NTSC COLOR SIGNAL

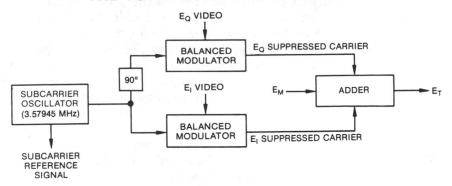

$$\text{Modulating video} = E_T = E_M + E_I' \cos(\omega t + 33 \text{ degrees})$$
$$+ E_Q' \sin(\omega t + 33 \text{ degrees})$$
$$E_I' = -0.27 (E_B - E_M) + 0.74 (E_R - E_M)$$
$$E_Q' = 0.41 (E_G - E_M) + 0.48 (E_R - E_M)$$
$$E_M = 0.3 E_R + 0.59 E_G + 0.11 E_B$$

If you work out the arithmetic, you will find that these equations agree with those on the previous page. The terms sin ωt and cos ωt represent the quadrature relationship between E_I and E_Q.

To *demodulate* the color signals at the *receiver*, the color subcarrier must be precisely recovered. Therefore, in the color receiver, a 3.579545-MHz signal must be generated that is exactly tuned in frequency and phase to the transmitter subcarrier. To do this, a *sample* of the *subcarrier reference* must be sent to the *receiver*. This is done by adding the subcarrier reference sample to the transmitted signal. This signal is added to the back porch of the horizontal blanking pulses.

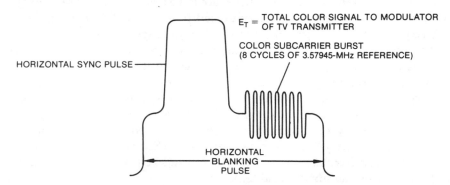

In the receiver, this color subcarrier burst is used to lock the subcarrier oscillator. It is also used to switch the receiver between color and monochrome.

Special Requirements on Color TV Transmitters

As you have seen, the NTSC color signal requires *good control* of the *phase* of the transmitted signal so that there is the proper relationship between the color burst and the color subcarriers. If this phase is not controlled, there will be undesired color shifts in the received signal that can lead to strangely colored people and objects as the picture changes. In simple monochrome systems, a loss of high-frequency bandwidth results only in some loss of resolution. In color systems, the results are *catastrophic* since the color information can be either lost or seriously distorted. Of course, the color signal can be received as a monochrome signal, even on a color receiver, if the color picture quality is inadequate.

At the present time, the NTSC system is used in most of North, Central, and South America and in Japan and South Korea as well as elsewhere; it provides excellent performance.

U.S., JAPAN, CANADA, MEXICO, PART OF SOUTH AMERICA, ETC.	BRITAIN AND FEDERAL REPUBLIC OF GERMANY	FRANCE AND SOVIET BLOC
NTSC	**PAL**	**SECAM**
60-Hz, 525-LINE, SUPPRESSED I AND Q AM SUBCARRIER, CONSTANT PHASE	50-Hz, 625-LINE, SUPPRESSES I AND Q AM SUBCARRIER, ALTERNATING PHASE	50-Hz, 625-LINE, FM SUBCARRIER WITH SPECIAL PHASING

To overcome some of the phase-control problems common in the early days of the NTSC system, the PAL (Phase-Alternation Line) and the SECAM (System Electronique Couleur Avec Memoire, or Sequential with Memory) systems have been developed. The PAL system reverses the phase of the subcarrier which is changed every line. A *switching signal* is transmitted as well as the color burst so that the correct phase can be used for demodulation. In other respects, it is like the NTSC system. In the SECAM system, the color subcarrier is FM rather than AM, and the color subcarrier is alternately modulated by color difference signals. It also requires special switching signals. For these systems, improved color performance is claimed under conditions where multipath signals are present. Multipath signals are signals that arrive at the receiving antenna after being reflected from a nearby object. Less sensitivity to the phase control problem is also claimed. At the present time, all three systems provide excellent color performance. Efforts are under way in areas not using the NTSC system to standardize on one system. The choice of 525- or 625-line systems has no important effect on the choice of NTSC, PAL, or SECAM but is related to the resolution desired in use in those areas and channel bandwidth available.

Satellite Systems

TV transmission, because of bandwidth requirements, is confined to the VHF region and above where line-of-sight conditions prevail. Transmission of TV by coaxial cable and similar means is well developed in North and Central America and in Europe, but transoceanic TV is not practical using wire transmission. Satellite stations are being used *increasingly* to handle domestic communication and TV relaying and international satellite communications also handle intercontinental TV transmission. These satellites are generally in stationary (synchronous) orbits. Special stations are needed on the ground to send and receive the signals from these satellites. Information (including TV) is sent up to the satellite typically on the band between 5.925 and 6.425 GHz, and relayed back in the band between 3.7 and 4.2 GHz, although new bands are being opened for use.

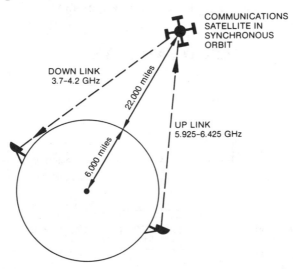

A *synchronous orbit* for a satellite is one where the satellite takes exactly the *same time* to circle the earth as it takes the earth to rotate through one revolution. Since both complete one revolution in the same time, the satellite *appears stationary* above a point on earth. This occurs at a height of about 22,000 miles. RCA, Western Union, AT&T, and Comsat General are some of the corporations that have communications satellites in orbit that are used for U.S. domestic communication. In these cases, a single satellite will cover the continental United States and adjoining areas. Up- and down-link ground stations are located at convenient points, and signals and data are carried locally on conventional land lines.

Satellite Systems (continued)

Intelsat is an international corporation that operates a set of three satellites in orbit over the North Atlantic, Indian, and Pacific Oceans. These are placed so that one of these is *visible* from *any point* on the earth. As a result, communication to essentially anywhere is possible.

The satellite acts *only* as a *relay*. That is, it receives the transmitted signal in one frequency band and translates it *directly* to another for retransmission to the ground station in another frequency band. Because it is a linear device, it can handle many signals at one time. Since it is wide band (500-MHz channel), it can easily handle many TV signals as well as thousands of voice or low-data-rate channels. The received signal is amplified and then down-converted to the transmit frequency using a local oscillator and mixer that shifts the received signal directly. This transmit signal is amplified and transmitted, as shown.

RELAY SATELLITE

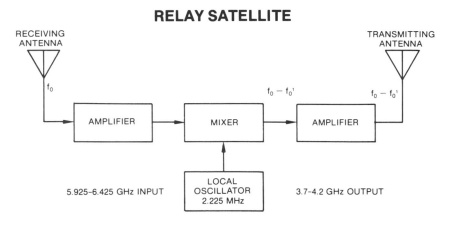

As you can see, the signal is not reduced to video and then used to remodulate a new carrier, so it is possible to relay *without distortion*. The frequency is shifted so that there will be *no interference* between TV transmission and reception. Also, many users can be on simultaneously without interference if they are on different frequencies. Transmission and reception by ground stations is usually done via FM or by pulse-code modulation.

There are other such satellites in operation, and many new ones are planned to provide flexible communications and TV relay throughout the world. It is possible (but expensive) to listen to the relay of TV *directly* from satellites to home receivers. Not only is standard TV available, but special CATV material is also available. This is particularly helpful in areas with poor or no ground communications networks or when the population is so dispersed that a ground-based TV station would serve too few people.

Troubleshooting Transmitters

As with any troubleshooting, common sense is important. The rules for troubleshooting transmitters vary, depending on the size of the transmitter. Obviously, one would approach the high-power final stage of a broadcast transmitter with vacuum tubes at high voltages in a different manner than one would a low-power CB set with solid-state components and a mere 13 volts as its maximum voltage level. Here, we will confine our attention to solid-state transmitters. This covers most transmitters in use today; even for large high-power transmitters using vacuum-tube final amplifiers, the exciter (driver) is usually solid state today.

Before attempting to service a transmitter, it is important to obtain from the owner (user) as much information regarding symptoms, external effects, etc. It is also important to determine immediately whether the system appears to operate at all. Most solid-state transmitters can be operated only when the output is loaded. Make sure that either an antenna or dummy load is properly connected before applying power.

The troubleshooting chart on the following page is an aid to the establishment of an orderly procedure for troubleshooting. Generally, it is important to isolate the problem to an RF or audio/video problem, as shown. When this has been determined (always make sure there is no power-supply problem first!), the conventional signal-tracing approach can be used to isolate the defective stage and then the component.

Servicing by means of signal tracing is simple. For example, if the audio modulator portion is believed to be defective, use an audio oscillator as a microphone input signal and then proceed stage-by-stage from input to output to determine where the signal is not present and/or where its level is insufficient or distorted, etc. In this way, the defective stage can be isolated and then the repair effected. Similar techniques can be used for the RF section. In any event, signal tracing will tell you where to concentrate your efforts. Be careful about what you see, and use your head! Don't be fooled by an apparent problem in one stage that is really caused by the defective next stage. Be *very slow* to change settings of components such as tuning capacitors, potentiometers, etc. They were set properly when the set was in operation. They should be all right once repair is made. Messing with adjustments prematurely can be a disaster! Logic, careful thinking, and relating cause and effect will enable you to solve almost any transmitter troubleshooting problem.

All manufacturers provide information on test equipment needs, alignment, adjustment, and troubleshooting—as well as schematic diagrams, parts references, layout drawings, and other information. In addition, explanatory material is usually included to provide you with information on how unusual circuits operate. Servicing modern solid-state hardware, particularly on printed-circuit boards with ICs, can be especially difficult if no servicing information is available. Often, such information is provided to the purchaser. Check with the owner about what information is available if you do not have complete servicing information.

TROUBLESHOOTING TRANSMITTERS

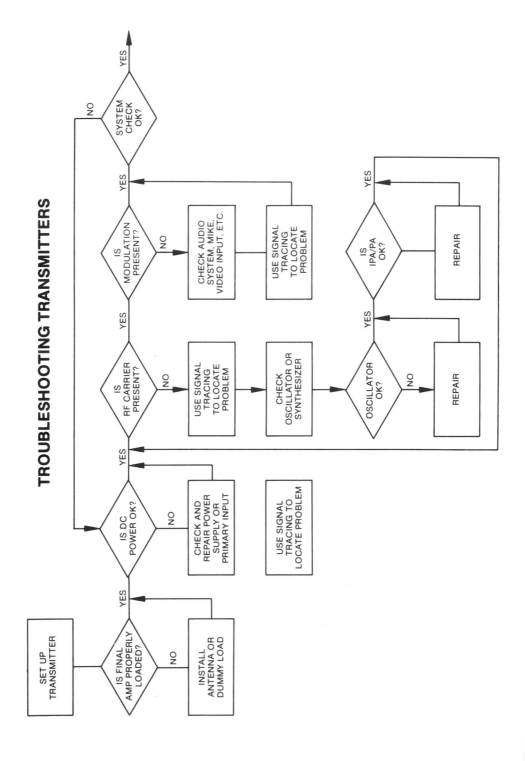

Review of Introduction to Television Systems

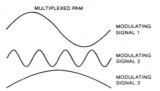

1. A PULSE is a relatively abrupt voltage or current change of short duration. Series' of pulses can be used to transmit information in what is called a *digital system.*

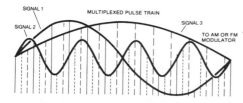

2. PULSE MODULATION can be *pulse amplitude modulation* (PAM), *pulse width modulation* (PWM), or *pulse code modulation* (PCM). As with other systems, information content determines bandwidth.

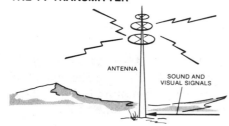

3. A TV SYSTEM converts an image and sound to electrical impulses and transmits them via radio waves to a receiver where they are reconverted to image and sound.

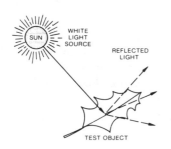

4. TV IMAGES are converted to electrical impulses by use of *photosensitive* materials that react electrically to changes in light intensity. Color is added by doing this for each of the three primary colors: *red, blue,* and *green.*

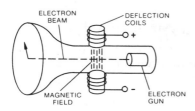

5. SCANNING converts an image to a *video signal.* A beam of electrons is swept by a magnetic field from *deflection coils* through tiny *elements* of the image, one at a time.

THE SCANNING RASTER

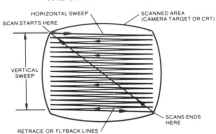

6. In the SCANNING PATTERN, the beam sweeps horizontally, then "flies back" to start a new line, moving down the screen. At the bottom, it flies back to the top to start again.

INTERLACED SCANNING

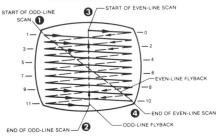

7. INTERLACE is the action by which every other line is scanned first (*first field*) and then the lines between (*second field*). Two fields make a *frame*.

THE IMAGE ORTHICON CAMERA TUBE

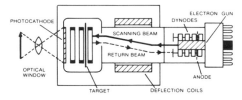

8. THE IMAGE ORTHICON CAMERA TUBE is sensitive and of high quality, but more bulky and expensive than the *vidicon* tube.

A CCTV-TYPE VIDICON CAMERA

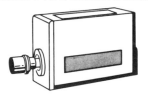

9. THE VIDICON CAMERA TUBE is now more common. It uses *photoconductivity*. For color, three vidicons can be used with *dichroic* mirrors to produce *color primary signals*.

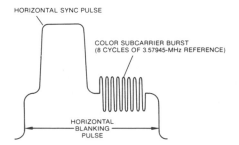

10. In THE NTSC TV SIGNAL, two *chrominance* signals modulate two separate phases of a subcarrier that are parts of the modulation of the video carrier.

Self-Test—Review Questions

1. Define pulse modulation. What are pulse width, overshoot, and droop. List the various types of pulse modulation.
2. What is the relationship between information-transfer rate and band-width? What compromise is usually made to keep the bandwidth to a reasonable value?
3. Outline the basic TV transmission-reception process.
4. What is the role of scanning in TV systems? Define the scanning raster. How is an interlaced scan obtained? What are the scanning frequency relationships necessary?
5. Describe the operation of vidicon and orthicon TV camera tubes.
6. Describe the operation of a complete TV camera system. Draw the output waveforms including blanking and synchronization pulses.
7. How are the various signals (visual and aural) arranged in a TV channel? What is the channel width in the United States? List the standards for TV transmission (color and black-and-white). What is vestigial-sideband transmission?
8. How does a color camera differ from a black-and-white camera? Define chrominance and luminance. How is the color information to be transmitted generated? How is it reduced to two signals? How is the color information transmitted?
9. What is the NTSC color TV signal? Describe in detail the specifications and the location within the TV channel of the various portions of the signal. How do PAL and SECAM differ from NTSC?
10. How is color subcarrier reference information transmitted? What is this information needed for? What does its presence or absence signify?
11. Why have satellite systems become important in modern communication? What is a synchronous orbit? Why are UHF and higher frequencies used for satellite systems?

Learning Objectives—Next Volume

Overview—Following the Epilogue to Information Transmission we will be ready to bring you to the next step in your study of solid-state electronics— the study of receivers in Vol. 4.

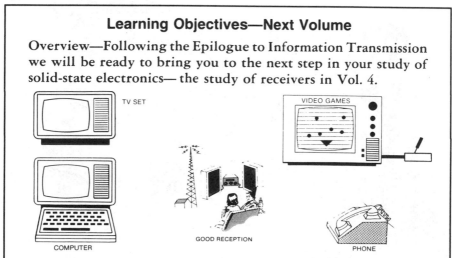

TV SET

VIDEO GAMES

COMPUTER

GOOD RECEPTION

PHONE

Epilogue to Information Transmission

You have now reached the point at which you understand how the transmitting portion of an information-transfer system operates. You know the nature of many of the signals and the information to be transferred via either direct or indirect means. You are now prepared to see how the other half of the information transfer system—the receiver—operates.

You should be aware that although we use audio and TV video as convenient means for the discussion of information transfer, it is not necessary that the systems you are learning about be *limited* to these. A TV system can be looked upon more generally as a device for the transfer of patterns of light and dark, and as such can transfer not only video based on a remote scene, but also alphanumerics derived from a disk, computer-generated pictures, or any information transfer that leads to the presentation of light patterns to an interpreter (operator). In the same sense, radio can be considered as the information-transfer system that positions a diaphragm at a remote point, in accordance with the position established at the input end. Clearly, audio is only one use of a system with this capability.

During your studies in Volumes 4 and 5, therefore, try to think of the transmission-reception system in much broader terms as an information-transfer system, where the concept of information transfer is as broad as your imagination will allow it to be.

EPILOGUE TO INFORMATION TRANSMISSION

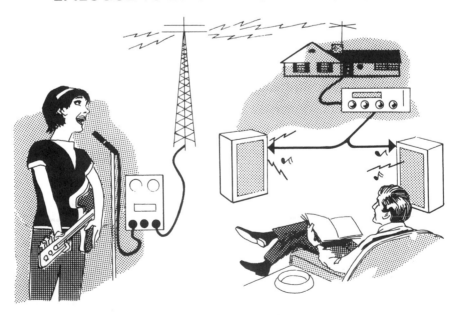

Basic
Solid-State
Electronics
VOL. 4

Basic
Solid-State
Electronics

COMMON CORE

The Configuration and Management of Information Systems

INFORMATION RECEPTION
VOL. 4

AUDIO/VIDEO/DATA/SENSORY RECEPTION
RECEIVING ANTENNAS
AM/FM/COMMUNICATION RECEIVERS
BLACK-AND-WHITE/COLOR TV RECEIVERS
VIDEO RECORDING/DISPLAY TERMINALS
TROUBLESHOOTING/ALIGNMENT

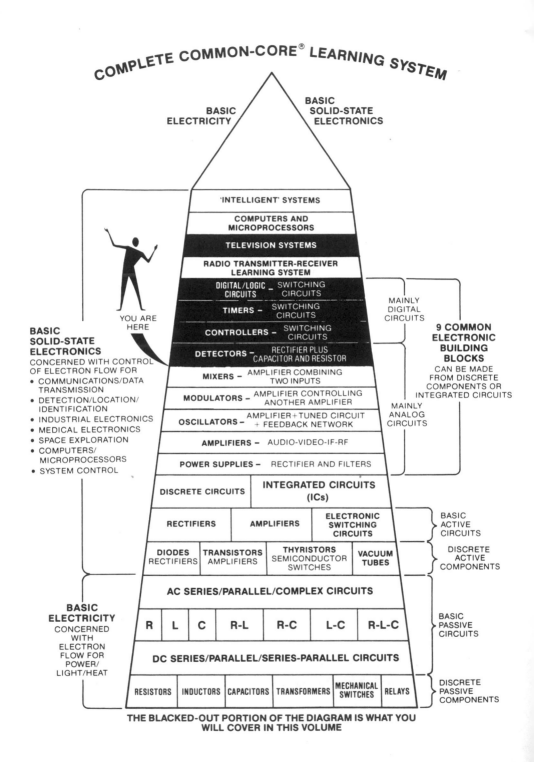

COMPLETE COMMON-CORE® LEARNING SYSTEM

BASIC ELECTRICITY

BASIC SOLID-STATE ELECTRONICS

'INTELLIGENT' SYSTEMS

COMPUTERS AND MICROPROCESSORS

TELEVISION SYSTEMS

RADIO TRANSMITTER-RECEIVER LEARNING SYSTEM

DIGITAL/LOGIC CIRCUITS – SWITCHING CIRCUITS

TIMERS – SWITCHING CIRCUITS

CONTROLLERS – SWITCHING CIRCUITS

DETECTORS – RECTIFIER PLUS CAPACITOR AND RESISTOR

MIXERS – AMPLIFIER COMBINING TWO INPUTS

MODULATORS – AMPLIFIER CONTROLLING ANOTHER AMPLIFIER

OSCILLATORS – AMPLIFIER + TUNED CIRCUIT + FEEDBACK NETWORK

AMPLIFIERS – AUDIO-VIDEO-IF-RF

POWER SUPPLIES – RECTIFIER AND FILTERS

DISCRETE CIRCUITS | INTEGRATED CIRCUITS (ICs)

RECTIFIERS | AMPLIFIERS | ELECTRONIC SWITCHING CIRCUITS

DIODES RECTIFIERS | TRANSISTORS AMPLIFIERS | THYRISTORS SEMICONDUCTOR SWITCHES | VACUUM TUBES

AC SERIES/PARALLEL/COMPLEX CIRCUITS

R | L | C | R-L | R-C | L-C | R-L-C

DC SERIES/PARALLEL/SERIES-PARALLEL CIRCUITS

RESISTORS | INDUCTORS | CAPACITORS | TRANSFORMERS | MECHANICAL SWITCHES | RELAYS

YOU ARE HERE

BASIC SOLID-STATE ELECTRONICS
CONCERNED WITH CONTROL OF ELECTRON FLOW FOR
• COMMUNICATIONS/DATA TRANSMISSION
• DETECTION/LOCATION/ IDENTIFICATION
• INDUSTRIAL ELECTRONICS
• MEDICAL ELECTRONICS
• SPACE EXPLORATION
• COMPUTERS/ MICROPROCESSORS
• SYSTEM CONTROL

BASIC ELECTRICITY
CONCERNED WITH ELECTRON FLOW FOR POWER/ LIGHT/HEAT

MAINLY DIGITAL CIRCUITS

9 COMMON ELECTRONIC BUILDING BLOCKS
CAN BE MADE FROM DISCRETE COMPONENTS OR INTEGRATED CIRCUITS

MAINLY ANALOG CIRCUITS

BASIC ACTIVE CIRCUITS

DISCRETE ACTIVE COMPONENTS

BASIC PASSIVE CIRCUITS

DISCRETE PASSIVE COMPONENTS

THE BLACKED-OUT PORTION OF THE DIAGRAM IS WHAT YOU WILL COVER IN THIS VOLUME

Information Reception

Information reception is the process whereby an original message—voice, music, picture, alphanumeric data, or sensory information—is recovered from a transmitted signal.

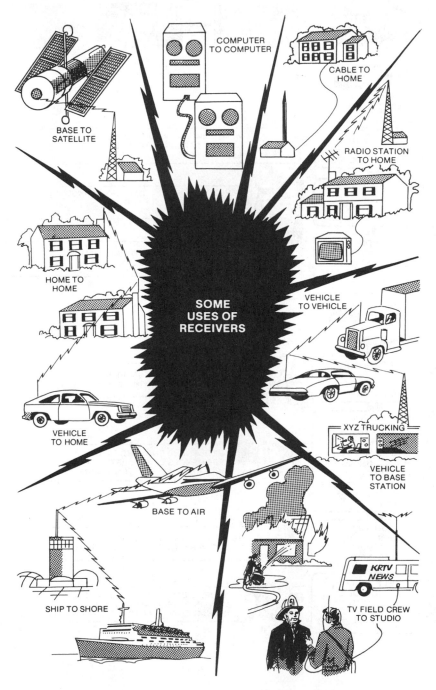

INFORMATION

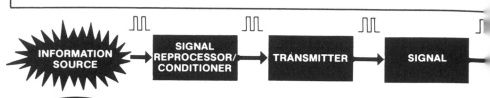

INFORMATION SOURCE → **SIGNAL REPROCESSOR/ CONDITIONER** → **TRANSMITTER** → **SIGNAL**

FROM MAN
SPEECH • VISION • DATA

SPEECH
MIND AND CENTRAL
NERVOUS SYSTEM
IDEAS/INTELLIGENCE

WORDS INTO CONTINUOUSLY VARYING
SOUND PRESSURES TRANSDUCED TO
ELECTRICAL SIGNALS

VISION
MIND AND CENTRAL
NERVOUS SYSTEM
IDEAS/INTELLIGENCE

PERCEIVED OPTICAL IMAGERY TRANSDUCED
INTO CONTINUOUSLY VARYING
ELECTRICAL SIGNALS

DATA
STORED INFORMATION-
PARALLEL SOURCE
IDEAS/INTELLIGENCE

INFORMATION IN ELECTRONIC MEMORY
ENCODED INTO DIGITAL PULSES

TO MACHINE
TO COMPENSATE FOR DISTANCE

INFORMATION TRANSFER SYSTEMS

TELEPHONE
AUDIO
VOICE-WORDS
SENTENCES

VARYING SOUND PRESSURE INTO
CONTINUOUSLY VARYING ELECTRIC CURRENT

**TELEGRAPH/
TELETYPE/
FACSIMILE/
WORD/DATA
PROCESSING
DATA**
DISCRETE LETTERS
WORDS, NUMBERS
IMAGERY

SEQUENCES ENCODED INTO
SEQUENCES OF DISCRETE
INTERRUPTED ELECTRIC CURRENT

RADIO
AUDIO/DATA
VOICE, MUSIC
DATA INPUT

AUDIO/DATA CONVERTED INTO
LINEAR VARYING ELECTRIC CURRENT

TELEVISION
AUDIO/VIDEO
VOICE/MUSIC/
PICTURES/
VISUAL DATA PATTERNS

AUDIO/VIDEO CONVERTED INTO
CONTINUOUSLY VARYING MULTI-
DIMENSIONAL ELECTRIC CURRENT SIGNAL

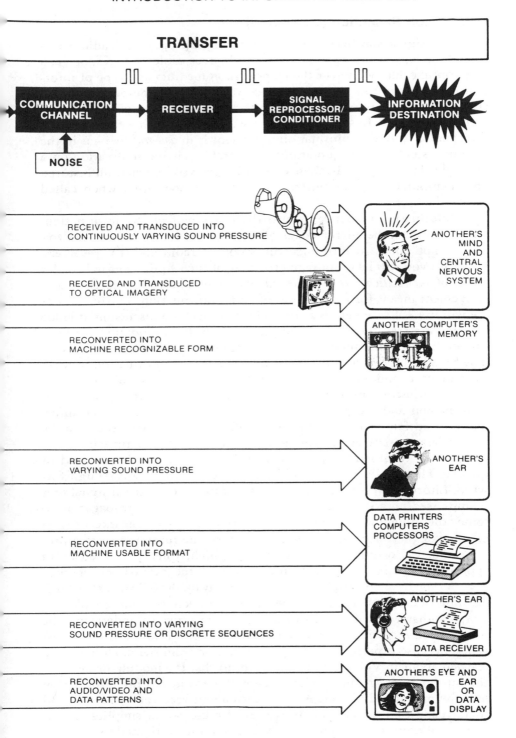

TRANSFER

COMMUNICATION CHANNEL → RECEIVER → SIGNAL REPROCESSOR/CONDITIONER → INFORMATION DESTINATION

NOISE

RECEIVED AND TRANSDUCED INTO CONTINUOUSLY VARYING SOUND PRESSURE

RECEIVED AND TRANSDUCED TO OPTICAL IMAGERY

ANOTHER'S MIND AND CENTRAL NERVOUS SYSTEM

RECONVERTED INTO MACHINE RECOGNIZABLE FORM

ANOTHER COMPUTER'S MEMORY

RECONVERTED INTO VARYING SOUND PRESSURE

ANOTHER'S EAR

RECONVERTED INTO MACHINE USABLE FORMAT

DATA PRINTERS COMPUTERS PROCESSORS

RECONVERTED INTO VARYING SOUND PRESSURE OR DISCRETE SEQUENCES

ANOTHER'S EAR

DATA RECEIVER

RECONVERTED INTO AUDIO/VIDEO AND DATA PATTERNS

ANOTHER'S EYE AND EAR OR DATA DISPLAY

Information Reception (continued)

The signal may arrive indirectly through propagated radio waves or light beams via an antenna, or directly over cable or wires, or via a fiber optic cable. Whatever the transmission medium and type of information, it is the function of the receiver to isolate the desired signal from any others present and to deliver a *true reproduction* of the original signal to the user. In some cases, the user may elect to have the receiver *filter* and/or store the information to present it in *non-real time*—as in the case of a stock-market quotation or a hotel or airline reservation on a video display terminal. In these cases, the received information is stored and manipulated by computer/microprocessor, then used when called for by the user.

Information reception is also the other half of the information-transfer system, the mirror image of information transmission. This volume is not going to introduce the study of radio and TV receivers, however; you will be encouraged to think of such equipment in broader terms of *information transfer*. We will use radio and radio receivers as a convenient means for a discussion of audio information transfer, just as we will use TV receivers as a convenient means for a discussion of video information transfer. These receivers will lay the foundation for your broader understanding of audio/video/data/sensory information transfer. The broader aspects of control and management of information transfer by computers and microprocessors will be addressed in Volume 5.

The illustrations on the previous two pages show the symmetry of transmission and reception. As you can see, the purpose of the information-transfer system (transmission and reception) is to relay audio, video, data, or sensory information from a source to a point where it is to be used, recorded, or processed with some acceptable standard of quality. The details may vary greatly, but the basic idea is straightforward. Information is transduced into some form of electrical signal that is impressed on a suitable carrier (radio or optical) for propagation to some other point (or many points). Reception starts with the interception of some of the information-bearing carrier. The receiving system processes out the original electrical signal, which, when presented to an appropriate transducer, results in recovery of the desired information.

As you have now progressed over halfway up the COMMON-CORE learning ladder, there is only one new common electronic building block—*detectors*—that you will need to learn about in order to understand the information-receiving system. Actually, a detector is also called a *demodulator*, because it performs the *opposite* function from the modulator you learned about in Volume 3. You will recall that the modulator *impressed* the information *onto* the carrier. A demodulator *recovers* the information *from* the modulated carrier. As you learn about receivers, you will find that, in addition to demodulators, they also use power supplies, amplifiers, oscillators, and mixers—just as these circuits were used to assemble the transmitter portion of the information-transmission system; it is only demodulators that are unique to receivers.

RF Spectrum

In your study of transmitters in Volume 3 you learned that frequencies from subaudio to visible and ultraviolet were used in information-transmission systems. You also learned that broadcasting in most of the spectrum is, of necessity, strictly regulated by United States and international governmental agreements. These regulations do not apply to the noncommercial interception and use of broadcast information. Anyone can build, or buy, the receiving system necessary to monitor radio broadcasts, TV transmissions, satellite communications, aeronautical radio transmissions, or even radar or sonar. The only limitation is that the information cannot be sold or rebroadcast. Also, the receiver must not accidentally radiate signals from its own antenna. In this volume you will study in detail systems ranging in frequency from the AM broadcast band to the UHF TV band. The important thing to remember is that the reception process is for the recovery of the information transmitted in the original signal with *as great fidelity* as is necessary for the intended use.

The electromagnetic spectrum, presented first in Volumes 2 and 3, is shown on the following page for your convenience. A table of standard bands, also presented first in Volumes 2 and 3, is shown below. You will find, in the interests of spectrum conservation, that the signal bandwidth is held to a minimum consistent with the information-transfer requirements. For example, the RF bandwidth of an FM broadcast station is 200 kHz and for TV transmission up to 6 MHz. These bandwidths are not available at lower frequencies, so these transmission/reception systems are confined to VHF and UHF. On the other hand AM broadcasting, short-wave communications, and similar applications require relatively narrow bandwidths (± 5 kHz) and can effectively use the lower short- and medium-wave bands. The properties of the ionosphere can strongly influence the range of operation of a system in some frequency ranges. This, in turn, has effects on frequency allocation. Thus, as you can see, the management of the RF spectrum—its use and allocation—is a very complicated matter.

STANDARD BANDS

WAVELENGTH RANGE	BAND NO.	FREQUENCY RANGE	BAND
$10^7 - 10^6$ m	2	30 TO 300 Hz	ELF (EXTREMELY LOW FREQUENCY)
10^6 m $-$ 100 km	3	300 TO 3,000 Hz	VF (VOICE FREQUENCY)
100 km $-$ 10 km	4	3 TO 30 kHz	VLF (VERY LOW FREQUENCY)
10 km $-$ 1 km	5	30 TO 300 kHz	LF (LOW FREQUENCY)
1 km $-$ 100 m	6	300 TO 3,000 kHz	MF (MEDIUM FREQUENCY)
100 m $-$ 10 m	7	3 TO 30 MHz	HF (HIGH FREQUENCY)
10 m $-$ 1 m	8	30 TO 300 MHz	VHF (VERY HIGH FREQUENCY)
1 m $-$ 10 cm	9	300 TO 3,000 MHz	UHF (ULTRAHIGH FREQUENCY)
10 cm $-$ 1 cm	10	3 TO 30 GHz	SHF (SUPERHIGH FREQUENCY)
10 mm $-$ 1 mm	11	30 TO 300 GHz	EHF (EXTREMELY HIGH FREQUENCY)
1 mm $-$ 100 μm	12	300 TO 3,000 GHz	LONG-WAVE INFRARED
100 μm $-$ 10 μm	13	0.3×10^{13} TO 3×10^{13} Hz	FAR INFRARED
10 μm $-$ 1 μm	14	0.3×10^{14} TO 3×10^{14} Hz	NEAR INFRARED
1 μm $-$ 100 nm	15	0.3×10^{15} TO 3×10^{15} Hz	ULTRAVIOLET AND VISIBLE LIGHT

ELECTROMAGNETIC FREQUENCY SPECTRUM WITH ASSOCIATED EQUIPMENT SYSTEMS

WAVELENGTH OF
METERS

HERTZ

AUDIO FREQUENCIES

GROUND AND SKY WAVES
SHORT WAVES

LINE-OF-SIGHT TRANSMISSION

MICROWAVES

MILLIMETER WAVES

ELF

VF

VLF

LF

MF

HF

VHF

UHF

SHF

EHF

INFRARED

VL

UV

(1) HIGH-FIDELITY SOUND SYSTEMS

(2)

(3)

(4) MARITIME MOBILE RADIO
NAVIGATIONAL AIDS
LONG-RANGE COMMUNICATION

(5) AERONAUTICAL RADIO
NAVIGATION

(6) U S STANDARD BROADCAST
COMMUNICATIONS

(7) SHORT-WAVE BROADCAST
AMATEUR BROADCAST
COMMUNICATIONS
CITIZENS BAND

(8) TV – FM
COMMUNICATIONS
NAVIGATIONAL
AIDS • DATA LINKS

(9) TV • COMMUNICATIONS
DATALINKS • RADAR
SATELLITE COMMUN

(10) RADAR
COMMUNICATIONS
SPECIALIZED EQUIP

(11) RADAR
COMMUNICATIONS
SPECIALIZED EQUIPMENT

(12) COMMUNICATIONS
RADAR
SPECIAL
DETECTION
ELECTRO-OPTICAL
DATA LINKS

(13) ELECTRO-OPTICAL
DEVICES
COMMUNI-
CATIONS

(14) SPECIAL
EQUIPMENT

Radio Transmitter-Receiver System

In Volume 1 of this series, you were introduced to the block diagram of a complete radio transmitter-receiver system and learned that this would be the learning *vehicle* which would take you from basic concepts and circuits to a working knowledge of complete radio transmitter-receiver systems.

The arrangement is typical of that used when a radio station transmits a program to a radio receiver in your home. It likewise applies to the arrangement used when one amateur radio station transmits a message to another or when a Citizens Band (CB) operator in one automobile talks to another operator several miles away. Although these various systems differ in size, cost, operating range, and signal quality, they all use the basic arrangement shown here.

In previous volumes you learned about power supplies; audio, video, IF, and RF amplifiers; modulators; oscillators; mixers; and transducers—all the basic electronic building blocks that are needed to make up complete radio transmitters and receivers. Now that you are about to enter the world of receivers it is appropriate to review the operation of the transmitter-receiver system, to firmly establish a frame of reference on which to base the details of this volume.

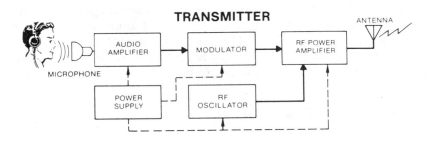

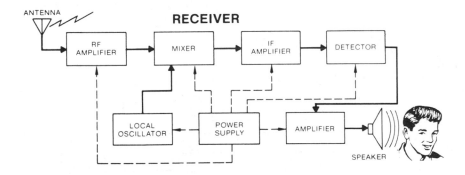

Summary of Transmitter-Receiver Operation

In your study of transmitters in Volume 3, you learned that although the electrical signals (analogs) that duplicate sound waves travel very easily through a conducting wire, they are incapable of radiating readily through air and space to a distant point. For electrical signals to radiate conveniently into space without a wire, their *frequency* must be *considerably higher*. Because of this, it is necessary to generate a high-frequency signal known as a *carrier*, and to superimpose the information on this signal by means of a process known as *modulation*.

In the AM transmitter, a fixed radio-frequency (RF) carrier wave is generated by the *oscillator*. A *microphone* is used to convert the acoustical pressure of the voice message to an electrical signal, and an *audio amplifier* steps up this signal to a usable voltage level. Both the voice and constant-level RF carrier signals are fed into the *modulated* RF amplifier. In an AM system, the signal changes the *amplitude* of the carrier in the modulated RF stage so that the resulting peaks and valleys duplicate the variations in the input signal. Generally this is done by modulating the final power amplifier; it can, however, be done at low level if the subsequent power amplifiers are linear in operation. The output signal is then fed to a suitable antenna system. The block diagram of the AM transmitter operation—presented first in Volume 1—is shown at the bottom of this page for your convenience.

In an FM system, the audio signal varies the carrier *frequency* of the oscillator by an amount equal to the audio signal amplitude, and at a rate equal to the signal frequency. This frequency-modulated RF carrier is fed into a *power amplifier*, which steps it up to a suitable power level. This output signal is fed to the *antenna* to be radiated.

AM TRANSMITTER OPERATION

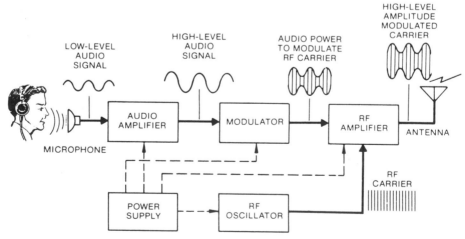

Summary of Transmitter-Receiver Operation (continued)

At the receiver it is necessary to remove the carrier signal from the received, modulated RF signal and leave only the electrical equivalent of the original information. The problem is that the received signal is quite weak, often only a few millionths of a volt. Furthermore, selective circuits are necessary so that the desired signal can be picked out from among all the various signals present from other transmitters on nearby frequencies. The weak signal, picked up by the receiver *antenna*, is first stepped up and preselected by an *RF amplifier* to raise it to a level suitable for signal processing to begin. Because amplification and signal selection at the carrier frequency are not always convenient, the modulated carrier is usually converted to a lower *intermediate frequency* (IF), which is independent of the incoming carrier frequency. This conversion is accomplished by the *mixer*, which takes in both the modulated RF carrier and a separate local oscillator signal. The oscillator frequency differs from the received carrier by a fixed *intermediate frequency* (IF). The variations in the mixer output signal duplicate those of the original received signal but at the lower frequency of the IF carrier. Considerable amplification can now be accomplished by the narrow-band fixed-tuned *IF amplifier*. The *detector* (or *demodulator*) removes the intermediate-frequency component from the output of the IF amplifier, and only an electrical duplicate of the original voice signal remains. Now an *audio amplifier* boosts this signal to a level sufficient for the *speaker* to convert it back to sound duplicating the original voice message.

The pages that follow will review the outstanding features of radio receivers in much greater detail. Stress will be placed on the organization of these systems in terms of the basic building blocks and how these circuits interconnect and work together.

AM RECEIVER OPERATION

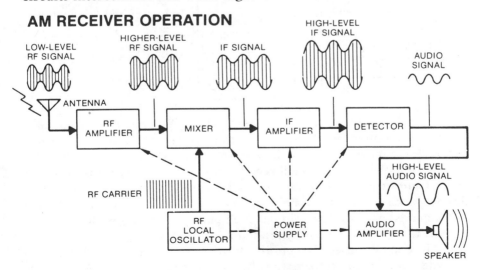

Modes of Reception—AM

The mode of receiver operation is determined by the method *originally used* to impress information onto the continuous-wave (CW) carrier signal. In Volume 3 you learned about the various types of modulation.

As you have learned, in the simplest method of AM, A1 (telegraphy), the carrier is interrupted, or simply turned on and off. The result is the familiar dot-dash or dit-dah signal. This method is used primarily by amateur radio and special long-distance communications systems. If the receiver is tuned to exactly the same frequency as the transmitter, the intelligence appears as a level shift in the output, with no audible signal present. If an audio output is desired, a beat-frequency oscillator (BFO) is used. The BFO is tuned to a frequency a few hundred hertz from the IF and mixed with it. This produces an audio tone at the receiver output when the transmitter is on.

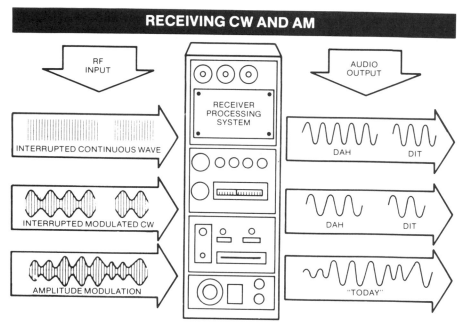

A2 telegraphy (interrupted modulated CW) uses fixed-frequency amplitude modulation of the carrier to provide an audible tone at any AM receiver output without the use of a BFO.

A3 telephony, as you will recall from your study of transmitters, is normal amplitude modulation in which the information is in the double sidebands of the RF carrier. For a reduced-carrier system, the modulation is designated as A3B; for a single-sideband transmission with a reduced carrier, the modulation is designated A3J.

Since AM causes the amplitude of the carrier signal to vary in accordance with the information being transmitted, and AM receiver must be configured to *detect* or *demodulate* these amplitude variations to recover the original information.

Modes of Reception—FM

In FM transmission, the information signal causes the frequency rather than the amplitude, of the RF carrier to change. F1 (telegraphy) uses frequency-shift keying (FSK) to cause the carrier to shift between two frequencies, with one frequency representing the carrier *on* and the other frequency representing the carrier *off* (analogous to A1 modulation). F2 uses a tone, as in A2 modulation, except that frequency—not amplitude—is varied by the tone. F3 telephony is used in FM broadcasting (including stereo) and in communication systems. In telephony, the information-signal frequency controls the *rate* at which the RF carrier frequency is changed, and the information-signal *amplitude* controls the amount by which the carrier frequency is changed—called its frequency *deviation*.

Reception of FM requires a method of detecting the frequency variation of the RF carrier. In your study of FM and TV receivers in this volume you will learn several different methods of detecting FM. The FM demodulator or detector is often called a *discriminator*. Below is a figure describing FM, TV, and FAX reception.

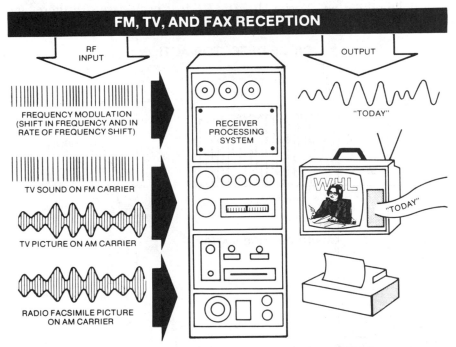

Regardless of the method of modulation used, the receiver has a very small signal at its input. This signal must be amplified and then demodulated to reproduce the information originally sent by the transmission system. This volume will describe the techniques used in modern receivers to enable them to retrieve the information originally sent, and will cover AM and FM reception, communication, and television (both color and black-and-white) systems.

The Diode Detector

Detection, or demodulation, is the process of recovering from an RF carrier the information that was originally impressed on it. For AM, detection means recovery, from the modulated wave, of a voltage that varies in accordance with the carrier amplitude. In most cases, this recovery is accomplished by rectification of the modulated wave, using a diode. The diode operates as a detector in the same way it operated as a rectifier, and you will note that the diode AM detector is almost identical to the half-wave rectifier/filter that you studied in Volume 1. For a steady carrier wave with no modulation, a dc voltage is produced. For an AM wave, the carrier wave *varies*, and hence the output is a *varying dc* or, in our case, the *envelope* of the carrier, which is the desired information.

Consider the simple diode rectifier with load R_L. When an unmodulated carrier signal is applied to the detector, the unfiltered output across R_L consists of half cycles of the input carrier and is thus pulsating dc. When an amplitude-modulated carrier is applied, the output consists of RF carrier pulses whose average (dc) amplitude varies in accordance with the modulation. Notice that the RF half cycles are still there, and the average value is well below peak value.

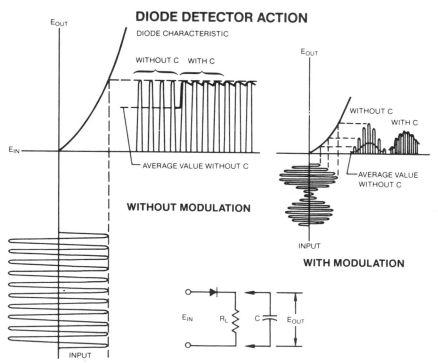

When a capacitor (C) is placed across R_L, the pulsations are smoothed out, just as in a capacitor input filter for a power supply. Now, output is steady dc for a steady carrier and a varying dc representing the modulation signal, but now the modulation signal is near peak value.

The Diode Detector (continued)

The capacitor value is chosen so that there is little discharge between the carrier pulses, but the capacitor can charge and discharge to follow the modulating signal.

In practice, the AM detector circuit usually contains additional RF carrier filtering (R1 and C1) to remove any components of the pulsating carrier wave produced by the diode. In addition, the output from the diode has an average dc value (taken over a long term of several tenths of a second) that is proportional to the average carrier amplitude. Thus, two signals are available, by suitable filtering, from the detector output—one representing the variation of the carrier due to the modulated carrier input and the other representing the average carrier level.

FILTER PASSES ONLY DC TO MODULATION FREQUENCIES

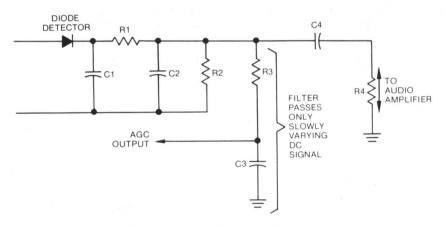

The diagram shows a practical diode detector for receiving AM. The diode detector produces the unidirectional carrier pulses, as described previously. Capacitors C1 and C2 and resistors R1 and R2 act as a low-pass filter to pass signals ranging from dc through the highest modulation frequency of interest. Because of the need for the filter to recover the modulation, the carrier frequency has to be many times the modulating frequency. This signal output is passed to the volume control R4 via the dc-blocking capacitor C4 so that only the desired audio information is sent to the audio amplifier.

Resistor R3 and capacitor C3 function as a low-pass, long-time-constant filter that removes the audio signal, leaving only a slowly varying dc signal that represents the average carrier level. This signal can be fed back to the IF amplifier and used to control the input for some applications to hold the carrier at the detector constant in level (to be discussed later). This is called *automatic gain control (AGC)* voltage.

As you can see, the diode detector functions to strip the modulation signal from the carrier to provide the audio signal. To do this, a filter is used at the detector output to filter the carrier but still allow the variations due to modulation to come through.

Crystal Sets

To aid you in understanding modern radio receivers and receiving circuits, it is helpful to review some of the circuits and receivers that were developed during the evolution from the earliest AM radio receivers up to the modern AM/FM radio sets used today. Most of these early circuits are still used in modern radios, with solid-state devices replacing their vacuum tube counterparts.

The first receivers were used in the early 1900s and were called *crystal sets*. In those days, AM only was used for broadcasting. In its simplest form, the crystal set consisted of an antenna, a tuned circuit, a crystal detector, and a pair of earphones. The crystal detector operated on the principle that certain materials (for example, the mineral galena) would act as a detector if a sharpened fine wire (called the catwhisker), was allowed to rest lightly against a sensitive spot on a chunk of the mineral. Actually, the junction between the galena and the catwhisker was a *semiconductor point-contact junction*, which you learned about in Volume 1, and it *functioned as a diode*. Thus, such a device is a *detector*.

The antenna picked up any signals in the air—in those days there were very few—and the crystal (a primitive diode rectifier) allowed the antenna currents to flow through the headphones on one half cycle of the RF, but blocked the other half cycle. These pulses of RF were filtered in the headphones to make them operate so as to produce an output that was proportional to the modulating wave. Crystal sets at best had one tuned circuit before the crystal, and therefore the selectivity was very poor. Because no amplifiers were used, sensitivity was so bad that crystal sets could be used only for receiving strong local stations. Today these sets have only limited practical application.

Tuned Radio-Frequency Receiver

When the vacuum tube became available, crystal sets were replaced by *tuned radio-frequency (TRF)* receivers that use vacuum tubes as RF amplifiers to provide better selectivity and sensitivity than the old crystal sets could. A vacuum-tube detector does the same thing as the crystal detector. After the detector, the audio signal is amplified in an audio amplifier. The output of the audio amplifier can be a powerful signal, which can be used to drive a loudspeaker or several pairs of earphones.

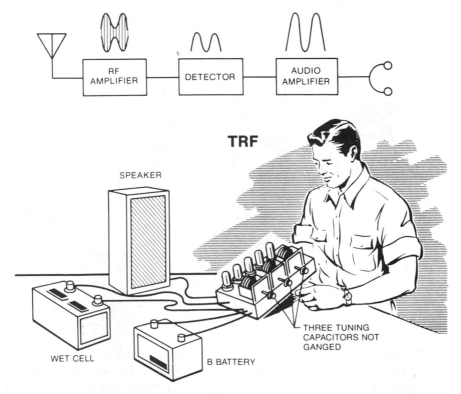

To increase sensitivity, the TRF receiver contains one or more stages of RF amplification *ahead* of the detector. The main purpose of these amplifiers is to provide *additional* selectivity and sensitivity. *Selectivity* indicates how well a receiver receives a desired signal and also rejects unwanted signals; *sensitivity* is a measure of the receiver's ability to pick up a weak signal. (Selectivity and sensitivity will be discussed in greater detail later in this volume.) In general, the more RF amplifier stages used, in TRF receivers, the greater the selectivity and sensitivity will be. On this and the following few pages you will review some of the outstanding points about RF amplifiers. Since the RF amplifier stage is designed primarily for voltage amplification, any transistor or IC suitable for voltage amplification may be used, provided the component has the necessary frequency-response capability to provide amplification at the desired frequency. The RF amplifiers discussed here are the same as those you learned about in Volume 2.

TRF Receivers/RF Amplifiers

TRF receivers have a minimum of two, and usually more, tuned circuits, one associated with each RF amplifier and one with the detector. In the early days of the TRF receiver, each variable capacitor was coupled to its own individual tuning knob. To tune your radio to a station, you had to adjust each knob individually until each tuned circuit was resonant to the frequency of the desired station.

The need for individual tuning knobs was eliminated by having the variable capacitors of all tuned circuits mounted on *one shaft*. This is called *ganged tuning*. Since all capacitors are varied together, then all tuned circuits should be resonant to the same frequency at the same time—making it easier to tune for maximum sensitivity and selectivity.

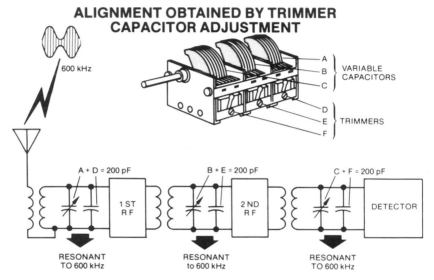

ALIGNMENT OBTAINED BY TRIMMER CAPACITOR ADJUSTMENT

Unfortunately, no two capacitors can be manufactured *exactly alike*; neither can two individual circuits be exactly alike. If nothing were done to compensate for these differences, the tuned circuits in a receiver would be resonant to slightly different frequencies—resulting in reduced receiver selectivity and sensitivity. Such a receiver is said to be *out of alignment*. The problem of misalignment is solved by adding small variable capacitors, called *trimmer capacitors*, in parallel with the main variable-tuning capacitors. Sometimes an adjustment is made in the coil of the tuned circuit rather than on the capacitors. Here, an iron-cored slug is moved in and out of the coil, causing inductance to vary (slug tuning).

In receivers covering only one band, trimmers are usually located on the ganged capacitors, one for each section. In receivers using band switching, additional trimmers for each range are usually mounted on, and in parallel with, the individual coils or the individual coils are adjustable. Trimming is usually done at the high end of the dial since this is usually where error in tuned circuits due to components will have the most serious effect on selectivity and sensitivity.

Experiment/Application—A Crystal Receiver

You can build a simple receiver that will allow detection of local AM radio stations. Only a detector, tuned circuit, and pair of headphones (transducer) are necessary. Suppose you hooked up the circuit shown.

SIMPLE CRYSTAL RECEIVER

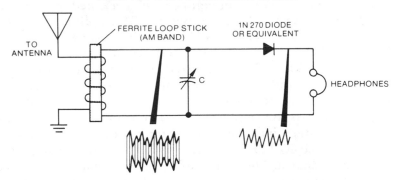

With this circuit, you can receive broadcasts in the AM band for local powerful stations. You must choose the capacitor and coil so that they tune over the AM band. These components are available from any radio-parts distributor. Once the receiver is built, you can observe the signal before and after detection by using an oscilloscope. If you do this, you will notice that the detector acts as you would expect—that is, it takes the intelligence-modulated carrier and strips off the information. You will need a good antenna to get reasonable results because there is no amplification in a simple crystal detector.

You can improve the performance markedly by adding an RF amplifier stage. You can build a simple RF amplifier with a 2N2222 transistor and power supply. Suppose you hooked up the circuit shown below.

RF AMPLIFIER

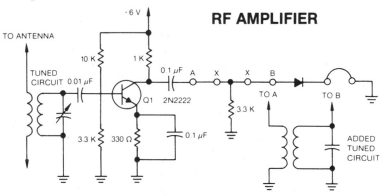

If you try this circuit, you will find that the sensitivity is much improved. You can make an even greater improvement by adding another tuned circuit at the detector to replace the 3.3-K resistor: break the circuit at points X and connect points A and B of the tuned circuit to points A and B of the circuit as shown. With two tuned circuits you will find the selectivity improved and that the detector works better.

Modern IC TRF Receiver

A modern single-TRF receiver can be constructed using a single IC to replace all of the RF amplifier, detector, and audio preamplifier functions previously performed by vacuum tube or discrete transistor circuits and numerous individual components.

The National LM172 is an eight-pin IC packaged in a round can (5/16 inch in diameter) and can be used for RF amplifier and detector functions at frequencies below 2 MHz. The circuit shown is used in the AM broadcast band (535–1,605 kHz) and has sufficient sensitivity for use in urban areas. The LM172 contains three wide-band amplifiers. One amplifier with the tuned circuit at its input is used as an RF amplifier. This tuned circuit also functions to replace the antenna (as will be discussed later). The first RF amplifier has a gain-control voltage input such that as the dc voltage increases on this input, the gain decreases. The output of the tuned-RF (TRF) amplifier is fed to a broadband (untuned) RF amplifier, as shown. The output of this broadband amplifier is then fed to the detector that demodulates the RF signal and also provides automatic-gain-control (AGC) voltage output. The detector output is fed to audio amplifier Q1, which drives the loudspeaker. The detected output is filtered by R4–C4 to provide automatic gain control for input signals greater than 50 MV at pin 2. As the carrier input tends to increase, the AGC voltage also tends to increase. This voltage, applied to the tuned-RF amplifier, tends to reduce its gain, which in turn tends to hold the carrier level constant.

NATIONAL LM172
INTEGRATED CIRCUIT

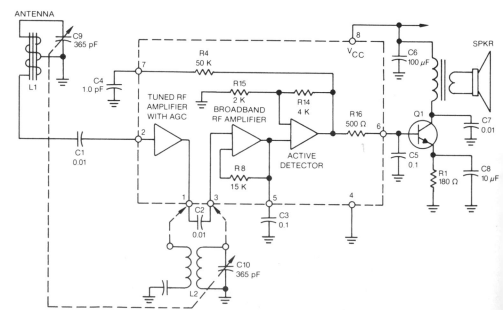

Modern IC TRF Receiver (continued)/The Limits of Receiver Sensitivity—Noise

With additional gain either preceding the module or between pins 1 and 3, it may be used for monitoring loran signals (1.8–2.0 MHz), or the numerous navigational and informational channels below 550 kHz. Since this particular circuit has only one tuned circuit (L1–C9), its selectivity is poor. The selectivity can be improved by adding a second tuned circuit between pins 1 and 3 of the IC, as shown in the previous figure by the dotted lines, instead of the coupling capacitor illustrated. The audio output at pin 6 is maintained between 2.1 and 2.4 volts by the AGC action and is used to bias the class-A audio amplifier Q1 directly for greatest circuit simplicity. Now let us move on to the next subject.

You may wonder why any number of amplifiers cannot be strung in cascade to provide any amount of gain desired—so that *any* signal would be detectable. The answer is that you can do this, but a point is soon reached beyond which no further improvement in sensitivity will occur. The reason for this is that it is not signal level *alone* that determines receiver sensitivity, but the *signal-to-noise ratio*. The signal-to-noise ratio (S/N) is the ratio of signal power to noise power at some point in the receiving system (typically, the output). Receiver noise comes from inside and outside the receiver. External noise comes from interfering signals, atmospheric effects (like lightning), operating electrical equipment, automobile ignition, etc. External interference or noise is sometimes collectively called *static*. Internal noise comes from irregularities—motion of electrons normally present in conductors or semiconductors—and is often called *thermal noise* or *hiss*.

SOURCES OF NOISE IN RECEIVERS

External noise predominates at frequencies below about 2 MHz, and becomes gradually less important, becoming negligible (except in extreme cases) above 30 MHz. Improvement in the receiver internal noise is of no value if it is below the external noise level. Most receivers currently in use below 2 MHz are limited by external noise almost entirely. You can hear this noise (static) on all AM broadcast receivers by tuning to a spot where no station is heard; the pops, hisses, and crackles of static are only too evident. At VHF and above, the internal receiver noise predominates, and hence keeping it at a minimum becomes important. This internal noise is due to the nonuniform flow of electrons in conductors and semiconductors and can be reduced—but never eliminated—by careful receiver design. You can hear this noise on an FM receiver tuned to where no station is heard. You'll hear a steady hiss due to internal noise.

The Limits of Receiver Sensitivity—Noise (continued)

Since both external and internal noise are *broadband*—that is, they are rather uniform over a broad range of frequencies—the signal-to-noise ratio can be improved by making the receiver as *selective* as possible, covering only the frequency bandwidth of the desired signal so that noise at other frequencies is excluded. For example, if a signal has a total bandwidth of 10 kHz, a receiver with a 20-kHz bandwidth will admit twice as much noise as is necessary and hence have a 2:1 poorer S/N ratio than a receiver with a bandwidth of 10 kHz. On the other hand, if the receiver bandwidth is narrowed to 2.5 kHz, there will be no improvement in the S/N ratio because signal is being excluded along with noise by about the same amount.

INTERNAL NOISE DEPENDS MAINLY ON THE FIRST STAGE NOISE CHARACTERISTICS

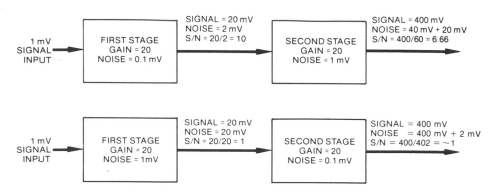

As you can see from the figure, it is the first-stage internal noise that is important in minimizing internal noise. Suppose you had two amplifier stages, each with a gain of 20. One produces the equivalent of 0.1 MV of noise at its input and the other produces 1 MV. As shown, if these stages are put in cascade to produce a gain of 400 (20 × 20), the S/N ratio at the output is radically different for the two configurations, depending on which of the stages comes first. Obviously, the first-stage noise level is most important. Since RF amplifiers can be made with a minimal noise contribution, their use is desirable to provide maximum sensitivity where thermal or hiss noise is predominant. Mixers in super-heterodyne receivers tend to have much more internal noise than RF amplifiers and therefore RF amplifiers are more commonly used on receivers, particularly those for use above a few megahertz. The internal receiver noise is often specified for a given receiver in terms of the input signal necessary to give a specific signal-to-noise ratio at the output. For sophisticated receivers, the internal noise is specified in terms of *noise figure*, which is a measure of how far the receiver under test departs from an ideal receiver.

The Superheterodyne Receiver

The TRF receivers used in the early days of radio broadcasting were marvels of their time and far superior to the crude crystal sets. However, since all of the gain prior to demodulation was done at RF, and since early vacuum tubes had limited RF-gain capability, tube-type TRF receivers had limited sensitivity. Furthermore, the need to have multiple variable-frequency tuned circuits tracking precisely over a 3:1 (535- to 1,605-kHz) tuning range limited the degree of selectivity that could be obtained. The superheterodyne receiver, developed during World War I and introduced for broadcast use in the 1920s, satisfied the then vital need for a highly selective, sensitive receiver. Since the late 1920s, essentially all receivers are of the superheterodyne type. The superheterodyne receiver and FM were both invented by Major Armstrong, one of the truly great pioneers of early radio and radio broadcasting.

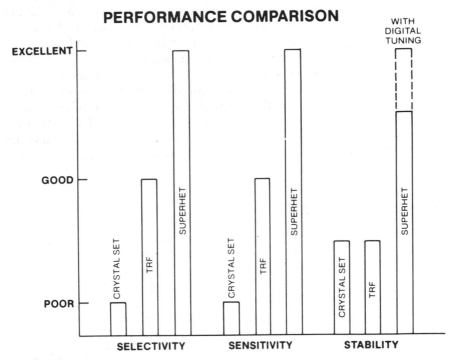

As you can see from the chart shown here, the superheterodyne receiver (often called a *superhet* for short) is greatly superior in all aspects of selectivity, sensitivity, and stability. The advantages of the superheterodyne receiver lie in the intermediate-frequency portion of the receiver. As you will learn, the RF amplifiers, detector, and audio circuits for a superhet receiver are no different than for a TRF receiver. The difference lies in the fact that in a superhet receiver, the incoming RF signal is converted to a fixed frequency (the IF), at which most of the amplification and selectivity are obtained. These can be carefully controlled because the IF amplifier needs to operate at only one frequency.

The Superheterodyne Receiver (continued)

As stated on the previous page, the superheterodyne receiver avoids the shortcomings of the TRF by converting the RF signal to a new, usually *lower frequency*, called the *intermediate frequency* (IF). This is accomplished by mixing the original RF signal with a second RF signal, generated in the receiver local oscillator (LO). The combination of the original RF signal with the LO signal produces the IF signal. As you remember from your study of transmitters in Volume 3, a mixer is a device that accepts two RF signal inputs (f1 and f2) and produces at its output two signals, the sum of the input frequency (f1 + f2) and the difference of the input frequency (f1 − f2). Usually the difference frequency is chosen as the intermediate frequency. This IF signal is identical to the original RF signal, except that it is displaced in frequency by the difference between the RF and LO signals (IF = RF − LO). The intermediate frequency is fixed because the difference between the signal input and LO frequencies is kept constant and does not vary, regardless of the frequency to which the receiver is tuned. This is done by having the LO frequency change with receiver tuning (ganged tuning capacitors) so that the constant difference is maintained regardless of the incoming signal frequency. Since the IF is fixed and lower than the lowest carrier frequency (455 kHz for standard AM broadcast), IF amplifiers can be designed to give rather large gains and high selectivity. The only parts of the superhet that differ from the TRF are the *local oscillator* and the *mixer* (the IF amplifier is actually a fixed-tuned lower-frequency RF amplifier).

SUPERHET

MIXER

RF AMPLIFIER

IF AMPLIFIER

DETECTOR

AF AMPLIFIER

LOCAL OSCILLATOR

Selectivity of the Superhet—Signal Response

When you tune a superheterodyne receiver with an IF of 455 kHz to a station with a frequency of 880 kHz, you are setting the tuned RF circuit to 880 kHz and at the same time you are automatically tuning the local oscillator to 1,335 kHz. Two signals—one of 880 kHz, the other of 1,335 kHz—are fed into the mixer stage. The output of the mixer is one signal at a frequency of 455 kHz, which is the *difference* or *beat frequency* between its two inputs, and another of 2,215 kHz, the *sum* of 1,335 and 880. The 2,215-kHz signal is so far from 455 kHz that it is not amplified by the IF amplifier.

If, at the same time, the antenna picks up another station at a frequency of 1,100 kHz, that signal, if strong enough, could get by the first tuned circuit and would then be mixed with the local oscillator output in the mixer stage. This undesired signal of 1,100 kHz would produce a beat frequency of 1,335 − 1,100 or 235 kHz, which would not be amplified by the IF amplifier tuned to 455 kHz.

ACHIEVING SELECTIVITY

DESIRED SIGNAL	RF FREQUENCY (kHz)	OSCILLATOR FREQUENCY (kHz)	MIXER OUTPUT FREQUENCY (kHz)	FREQUENCY SELECTED BY IF AMPLIFIER
→	880 1,100	1,335 1,335	455 235	←
→	880 1,100	1,555 1,555	675 455	←

The IF amplifier tuning does not vary. It is always tuned to 455 kHz, so you can see that only the *difference* signal produced by the desired station (880 kHz) will be amplified by the IF amplifier. Since the *sum* signal produced by an undesired signal of 1,100 kHz produces a beat signal whose frequency is different from the intermediate frequency, that beat signal is not amplified. Thus, the superhet has selected the proper input signal on the basis of the frequency of the difference signal produced in the mixer stage.

In order to hear the 1,100-kHz station, the receiver would have to be *retuned*. Turning the knob changes the frequency to which the RF amplifier is tuned and, at the same time, changes the local oscillator frequency. A two- or three-section ganged tuning capacitor does the trick. Tuning the receiver does not affect the IF stages. When the RF and mixer tuned circuits are set at 1,100 kHz, the ganged capacitor section that tunes the oscillator is at the right position to put out a signal of 1,555 kHz; the intermediate frequency remains at 455 kHz.

Now it is the 1,100-kHz input signal that produces the 455-kHz beat frequency. The beat produced by the 880-kHz signal would be the difference between its frequency and the 1,555-kHz local oscillator frequency, or 675 kHz, which will not be amplified by the IF stages.

Selectivity of the Superhet—Image Response

In order for the superhet to work properly, the local oscillator must be adjusted so that it will always tune to a frequency at a fixed difference from the desired RF signal. Thus, as the receiver—that is, the RF and mixer tuned circuits—are tuned from 550 to 1,600 kHz, the local oscillator should tune from 1,005 to 2,055 kHz. Then, any signal picked up at the frequency to which the receiver is tuned will produce an IF signal of 455 kHz (which is the standard intermediate frequency for AM broadcast receivers). Since all of the IF gain is implemented at a common fixed frequency, all of the stages can be made a fixed narrow bandwidth, and a high degree of selectivity is thereby obtained.

Perhaps you have already noted that there is a problem with the selectivity of the superhet receiver. On the previous page it was stated that the intermediate frequency is the difference between the signal frequency and the local oscillator (LO) frequency. For example, for a station at 880 kHz the local oscillator is set at 1,335 kHz to produce an IF of 455 kHz. Note, however, that an input signal at 1,790 kHz will produce a difference between the local oscillator frequency and the signal exactly equal to the IF (1,790 − 1,335 = 455 kHz). This undesired signal is called the *image*, and it is always located at a point twice the value of the IF away from the desired signal.

SELECTIVITY OF THE SUPERHET

DESIRED SIGNAL	IF	LOCAL OSC
880 kHz	455 kHz	1,335 kHz
90.3 MHz	10.7 MHz	101.0 MHz

IMAGE

1,790 kHz
111.7 MHz

Obviously if the LO is on the low side, the image is below the desired signal by twice the IF. The only discriminants against the image are in the RF circuits. For example, the mixer input tuning will provide some selectivity against the image. At lower signal frequencies, the selectivity of the mixer input circuit is adequate to keep the image responses down; at signal frequencies above the broadcast band, however, the additional selectivity of a tuned RF amplifier is important in reducing the image response. You will also notice later that as higher radio frequencies are used, the intermediate frequency chosen for the IF amplifier is increased to move the image further away from the desired response. For example, a broadcast receiver has an IF of 455 MHz, while a CB radio (at 29 MHz) may have an IF of 4.5 MHz, and an FM receiver (at 100 MHz) may have an IF of 10.7 MHz.

Selectivity of the Superhet—Double Conversion

Since the selectivity and gain of an IF amplifier depend on the frequency (both better at lower IFs)—a low IF is desirable for some applications. On the other hand, if the received signal is broadband, or if the input frequency is higher, a higher IF is desirable because broadband circuits are easier to design at these higher frequencies. For example, TV receivers (to be discussed later) require an IF bandwidth of almost 6 MHz; for these, the IF is between 40 and 50 MHz. Radar receivers commonly use 30 MHz, 60 MHz, or even higher IFs for the same reason. When a combination of high frequency (VHF and above) and high selectivity is desired, *double* conversion is often used. For such applications, there are two IF amplifiers—one with a high IF to provide best image rejection and a second one with a low IF to provide best selectivity.

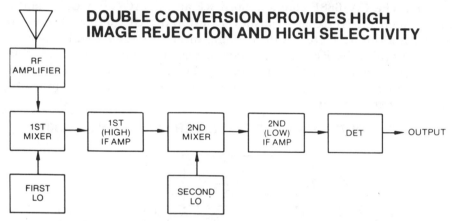

DOUBLE CONVERSION PROVIDES HIGH IMAGE REJECTION AND HIGH SELECTIVITY

The double conversion superhet has two mixers and local oscillators. The first mixer takes the high-frequency received signal and mixes it with the signal of a high-frequency (first) LO to produce a high-frequency (first) IF (to minimize image response). This is amplified (or in some cases just bandpass filtered) and applied to the second mixer, which mixes it with a lower-frequency (second) LO to produce a second (low-frequency) IF (for selectivity). In communication receivers, when tuning range is narrow, the first LO is crystal controlled for better stability at the higher frequency. The circuits between the first and second mixers must then be broadband, because the first IF must vary over the tuning range (because the first LO is fixed). The second LO is then made variable and used for tuning; the lower frequency allows greater stability for this. Tuning ranges over different frequencies are obtained by providing several selectable first-LO crystal frequencies.

When the tuning range is too great for this, the tuning must be done in the first LO, with the second LO fixed and usually crystal controlled. As you will learn, special crystal filters are also used in communication receivers to improve their selectivity.

Modern IC Superhet Receiver

The above can be constructed using a single IC to perform the oscillator, mixer, IF amplifier, detector, and audio preamplifier functions.

The RCA CA3088E is an IC suitable for AM receiver applications up to 30 MHz. Since many high-frequency receivers use IF amplification at the higher frequencies, the high-frequency characteristics can be very important. In the AM broadcast receiver shown, double-tuned transformer-coupled circuits are used in the IF. However, other forms of bandpass filters (such as ceramic filters), which you will learn about later, would be equally suitable. Because of the size (the CA3088E is a single IC chip), the IF and mixer tuned circuits are external to the chip, as are some of the larger audio-coupling components and the audio volume control. The CA3088E contains both RF and IF AGC functions, and a tuning-meter driver that furnishes a buffered AGC output voltage to use with a tuning meter. A dc voltage, internally regulated by a zener diode, supplies the RF amplifier, converter, and first IF amplifier. You can see how simple an IC of the proper design can make something as complicated as a complete superhet receiver very simple from an external-circuit standpoint.

TYPICAL AM BROADCAST RECEIVER USING THE CA3088E

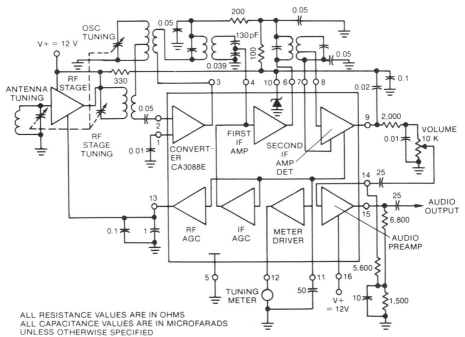

ALL RESISTANCE VALUES ARE IN OHMS
ALL CAPACITANCE VALUES ARE IN MICROFARADS
UNLESS OTHERWISE SPECIFIED

The IF amplifier output drives the detector, which strips off the modulation and also can provide for AGC. The detector output is fed to the external volume control, and the volume control output is fed to an audio preamplifier (included in the IC). The output of the audio preamplifier is delivered from the IC for input to an audio power amplifier.

Review of Radio Reception and Receiver Characteristics

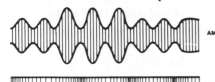

1. **THE TWO BASIC MODES OF RECEPTION** are amplitude modulation (AM) and frequency modulation (FM).

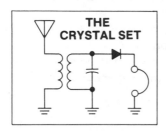

2. **A DIODE DETECTOR** rectifies the received signal to develop a pulsating dc signal output whose envelope is the waveform of the desired intelligence.

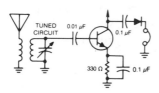

3. **THE SENSITIVITY AND OUTPUT** of a detector can be increased by preceding it by one or more RF amplifier stages, forming a tuned radio frequency (TRF) receiver.

4. **NOISE** generated in its circuits affects a receiver's ability to receive weak signals—especially in the circuits in the front end.

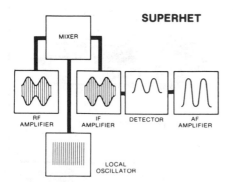

5. **IN A SUPERHETERODYNE** the received signal is mixed with a local oscillator signal to produce an IF, at which optimum amplification and selectivity are possible.

Self-Test—Review Questions

1. Briefly list the elements of a receiving system and describe their function.
2. List and identify the FCC-designated modes of transmission and reception.
3. Describe the operation of a simple diode detector.
4. How does a crystal set work?
5. Draw a block diagram of a TRF receiver. Describe its operation.
6. Draw a block diagram of a superhet receiver. Describe its operation.
7. Compare the TRF and superhet receivers. What are their advantages and disadvantages?
8. What causes an image in superhet receivers? How do RF selectivity and double conversion minimize image responses?
9. What factors limit receiver sensitivity? How can it be maximized?
10. Compare the role of the second detector in a superhet to the detector in a TRF receiver.

Learning Objectives—Next Section

Overview—In the next section you will study various types of antennas and then proceed to a detailed discussion of AM radio receivers, including their contents and characteristics.

VERTICAL ANTENNA

HORIZONTAL ANTENNA

Receiving Antenna Functions

Just as transmitting is the mirror image of receiving, so receiving antennas are the mirror of transmitting antennas.

The purpose of the receiving antenna is to intercept the electromagnetic waves radiated from the transmitting antennas. When these waves cut across the receiving antenna, they generate a small voltage in it. This voltage causes a weak current to flow in the antenna-ground system. This feeble current has the same frequency as the current in the transmitting antenna. If the original current in the transmitting antenna is modulated, the receiving antenna current will vary in exactly the same manner. This weak antenna current, flowing through the antenna coil, induces a corresponding RF signal in the input circuit of the receiver.

**RECEIVER ANTENNAS INTERCEPT THE RADIO WAVES
SENT OUT BY THE TRANSMITTER**

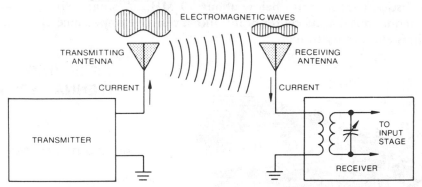

A receiving antenna should feed as much signal and as small an amount of undesired interference as possible to the receiver. It should be constructed so that the signal is not lost or dissipated before reaching the receiver. It should give maximum response for the frequency or band of frequencies to which the receiver is tuned. Thus, the condition of match and other parameters specified for transmitting antennas also apply for receiving antennas; however, the requirements are usually much less critical because receivers can be made comparatively sensitive and do not handle significant power. An antenna can also be *directional*, which means that it will give best response in the direction from which the operator wishes to receive.

The receiving-antenna problem is more easily solved when the receiver is operated in conjunction with a transmitter or with a transceiver, since the transmitting antenna is usually designed to incorporate the desirable features that have just been listed, and the *same* antenna is used for both transmitter and receiver. A switch or relay is used to connect the antenna to either the transmitter or receiver, depending on the operating function at a particular moment. However, when no transmitter antenna is available, it is necessary to use a *separate* receiving antenna, paying attention to the four considerations of noise, signal loss, operating-frequency response, and directivity.

Receiving Antenna Types

The simplest receiving antenna is merely a piece of wire of arbitrary length. In a field of a given strength (volts/meter), the longer the wire, the greater the voltage induced. Thus, one would be tempted to build an antenna of the greatest possible length. On the other hand, such an antenna could be very difficult to match so that the losses could outweigh the gains. A *resonant* antenna is desirable because it looks resistive and hence is easiest to match, particularly over a band of frequencies. A resonant antenna is one that has a length that is a multiple of a half wavelength. The basic antenna is the *half-wave dipole*—that is, a wire that is a half wavelength long. In theory, a half wavelength (in feet) is about equal to 500/f where f the frequency, is expressed in megahertz. As a practical matter, end effects and other phenomena cause the antenna to appear longer than it actually is physically, so a factor of 470/f is more realistic at frequencies below about 50 MHz. A single-wire half-wave antenna looks, at its center point, like a resistive impedance of about 70 ohms and is easy to match.

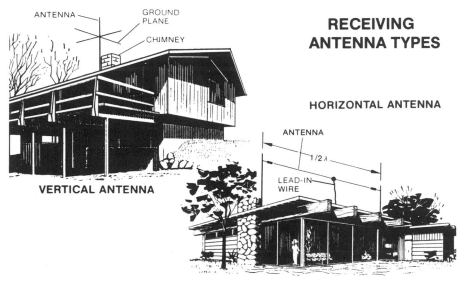

RECEIVING ANTENNA TYPES

Usually, a half-wave antenna is *horizontally polarized*—that is, the main axis of the antenna is horizontal. For some applications, as you will see, *vertically polarized (vertical)* antennas are used, as you will recall from your study of transmitting antennas. In any case, for a resonant antenna the voltage is minimum at the center and maximum at the ends, and the current is maximum at the center and minimum at the ends. If the antenna is fed at the center it is called a *current-fed antenna*. If it is fed at the end, it is called *voltage-fed*.

Usually, horizontal receiving antennas are used for short-wave applications and for medium-wave applications, as well as for TV and FM broadcasting. Vertical antennas are used for receiving in portable and in communications equipment.

Receiving Antenna Types—Broadcast/Short-Wave Receivers

The wavelength (λ) used for AM broadcast ranges from 200 to 600 meters. Antennas in the order of $\lambda/2$ are thus not practical. For home radio use, the configuration is almost always some variation of a loop antenna, as shown in the diagram. Loops operate somewhat differently from conventional wire antennas. Since the loop circumference is small compared to the wavelength, uniform in-phase currents exist around the loop conductors and the reception properties are relatively independent of loop geometry, and the antenna pattern for a small loop is constant in the plane of the loop. In the plane perpendicular to the loop, the voltage at the antenna terminals varies from a maximum off the sides of the loop to a sharp minimum or "null" in the direction perpendicular to the loop plane. The pattern is shown in (D). This directional property is quite noticeable when receiving weak stations; with strong stations, however, receiver AGC action compensates for the effect of θ variation in received signal amplitude, except in the nulls.

LOOP ANTENNAS

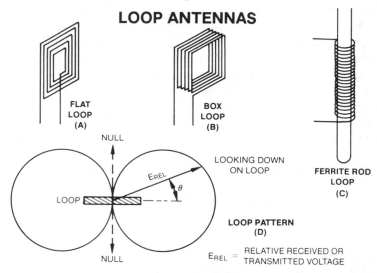

FLAT LOOP (A)

BOX LOOP (B)

FERRITE ROD LOOP (C)

LOOKING DOWN ON LOOP

LOOP PATTERN (D)

$E_{REL} = $ RELATIVE RECEIVED OR TRANSMITTED VOLTAGE

The ferrite *loopstick* antenna (C) is used in most AM radios and consists of a ferrite rod or slab wound with many turns of wire. The ferrite core intercepts the radio-wave magnetic field to generate an electromagnetic field in the core, and the variations of this core field, at the radio-frequency rate, convert the field to a variable current in the antenna coil surrounding it. This produces an RF voltage at the input of the receiver first stage. Also, the antenna circuit is usually tuned.

Because of the directional properties of loop antennas, they are used in *direction-finding equipment* to locate the direction from which a broadcast is coming. This is done by rotating the loop until the signal is *weakest* (or at null), and the direction then is perpendicular to the plane of the loop. The reason that the null is used is because it is much sharper and easier to find than the peak. Direction-finding receivers have the antenna mounted on a compass scale, so that direction can be read directly.

Receiving Antenna Types—Mobile Systems

Mobile operation in automobiles, trucks, and boats have special problems for operation at wavelengths at 30 MHz and below because of the physical length of the antenna required to achieve a half wavelength. Since an automobile is shielded on the inside, an external antenna is always required. Usually, mobile antennas take the form of a vertical rod or *whip* with a length much less than half a wavelength for frequencies below 30 MHz. Generally, antennas for mobile operation are very short for operation in the AM broadcast band, but approach a quarter wavelength at 30 MHz (Citizens Band). As you learned in your study of transmitting antennas, an antenna less than a half wavelength long looks capacitive (as well as resistive). For quarter-wave antennas with a ground plane, the same is true. To match a short antenna, an inductive reactance from the source is required. Alternatively, the antenna can be electrically lengthened by adding inductance in series with its output. Since these antennas are fed at a high-current point, a low-resistance (loss) coil is important if efficiency is to be maximized.

At VHF and UHF, it is possible to use antennas that are a quarter wavelength long with a ground plane that gives the equivalent of a half-wave antenna. Thus, VHF-UHF antennas do not generally need reactance in the line to achieve a matched system. As you remember from your study of antennas in Volume 3, the ground plane acts as a mirror that allows for a quarter-wave antenna to function as a half-wave antenna. To be effective, the ground plane must extend more than a quarter wavelength. On automobiles, the roof can act as a ground plane for some VHF and UHF applications. You often see CB antennas mounted on chimneys or towers with ground plane radials, as shown in the figure.

MOBILE ANTENNAS

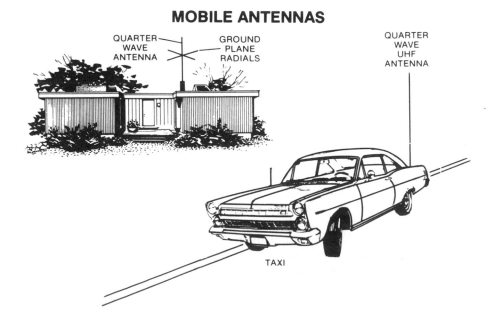

QUARTER WAVE ANTENNA

GROUND PLANE RADIALS

QUARTER WAVE UHF ANTENNA

TAXI

Receiving Antenna Types—FM and TV

FCC regulations in the United States require FM and TV stations to radiate horizontal polarization, but circular or elliptical polarization may be employed if desired. Most FM and TV stations transmit both horizontal and vertical polarizations, since this allows good reception with either horizontal rooftop antennas or with vertical whips and loop antennas. Communication stations usually use vertically polarized whip antennas for both transmission and reception.

Since VHF and UHF transmission is limited to line-of-sight, superior performance, especially in fringe areas, can be obtained with an elevated or rooftop antenna. Antennas can be either omnidirectional, for receiving signals from many directions, or directional with high gain, for receiving signals primarily from one direction.

Omnidirectional horizontally polarized VHF-UHF antennas usually take the form of a pair of horizontal crossed-folded dipoles. As you learned in your study of transmitting antennas, each individual dipole has a figure-8 pattern, and the sum of the crossed pair has nearly uniform gain in the horizontal direction. A single folded dipole bent into an S-shape can also provide fair approximation of omnidirectional coverage.

FM/TV RECEIVING ANTENNAS

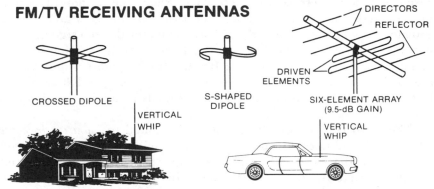

CROSSED DIPOLE

S-SHAPED DIPOLE

VERTICAL WHIP

DIRECTORS

REFLECTOR

DRIVEN ELEMENTS

SIX-ELEMENT ARRAY (9.5-dB GAIN)

VERTICAL WHIP

A directional receiving antenna provides for improved reception in one direction at the expense of other directions. If all (or most) of the transmitting stations are located in the same direction—the direction of the city or urban area—then the high gain associated with high antenna directivity can make the difference between good or poor reception. Most directional antennas are used for TV reception and consist of an array of dipoles, with one or more folded dipoles being used as the driven elements, and additional directors and reflectors added to increase the gain and directivity. These additional elements are referred to as *parasitic* elements, and they are coupled to the *driver* (folded dipole) antenna by the mutual space impedance between them. The current induced in the parasitic antenna will produce a radiation field which combines with that of the driver elements to yield the overall pattern. By choosing the number, spacing, and length of the parasitic elements, it is possible to control the gain, directivity, and bandwidth of the array. These directional arrays can be used horizontally or vertically.

RF Amplifier Stage

In a superhet receiver, the signal from the antenna may be connected directly to the mixer stage input, without using an RF amplifier stage. Except at low frequencies, however, you will find receivers usually contain stages of RF amplification preceding the mixer. The RF amplifier in these cases is performing three functions that improve receiver performance.

The major function of the RF amplifier is to *improve the signal-to-noise ratio*. The mixer stage usually produces much more internal noise than an RF stage of amplification. The signal, plus the mixer noise, is amplified by the following IF amplifier stage. As you know, a low-noise amplifier preceding a noisy amplifier provides a better overall signal-to-noise ratio. Thus, an RF amplifier will improve the overall signal-to-noise ratio in a superhet receiver. In some cases, two RF amplifier stages are used; usually, however, a single stage is sufficient to provide for good performance.

The second function of the RF amplifier stage is concerned with *selectivity*. You will recall that in the TRF receiver, the RF amplifier stages enabled the operator to select the desired signal from a group of signals whose frequencies were very close to each other. The RF amplifier in a superhet serves to prevent interference from the image signal whose frequency is twice the value of the IF away from the desired signal. Since the basic selectivity to signals at nearby frequencies is obtained in the IF amplifier, the selectivity of the RF amplifier is most important in terms of how well it rejects the image. In addition, the RF amplifier selectivity helps in keeping strong local signals at nearby frequencies from overloading the receiver.

The third function of the RF stage is related to *radiation from the oscillator*. It should not be forgotten that this oscillator is a low-powered transmitter. If there is no RF amplifier stage, the oscillator is connected through the mixer stage to the antenna. This antenna can radiate some energy from the oscillator and cause interference with reception in other nearby receivers at the local-oscillator frequency. This radiation may be reduced or prevented by using one or more stages of RF amplification, and by carefully shielding the oscillator stage.

BASIC FUNCTIONS OF RF AMPLIFIERS

- IMPROVES SIGNAL-TO-NOISE RATIO
- IMPROVES SELECTIVITY AND REJECTS IMAGE
- REDUCES LOCAL OSCILLATOR RADIATION

Local Oscillator

As you have already learned, the function of the local oscillator (LO) in a superhet receiver is to provide an unmodulated reference signal that *tracks* the desired RF signal at a fixed, offset frequency—typically, 455 kHz for the AM band. Local oscillators have circuits that are the same as the oscillators you studied previously, and these oscillators are very low in power output. Below are examples of typical operating frequencies.

f_0 FREQUENCY OF RF CARRIER	f_1 LOCAL OSCILLATOR FREQUENCY	INTERMEDIATE FREQUENCY
550 kHz	1,005 OR 95 kHz	455 kHz
1,000 kHz	1,455 OR 545 kHz	455 kHz
1,600 kHz	2,055 OR 1,145 kHz	455 kHz

The LO used in AM broadcast receivers and other low- or medium-frequency receivers is usually 455 kHz above the RF carrier, or 995 to 2,055 kHz for the case shown. Thus, while the RF frequency is changing by a factor of about 3:1 (540–1,600 kHz), the LO must change by about 2:1 (995–2,055 kHz) to maintain perfect tracking. This is done by controlling the relative size and shape of the plates of the two sections of the ganged tuning capacitor where capacitive tuning is used, and by relative tuning slug motion where variable-inductance tuning is used.

As you learned previously, tracking between tuned circuits can be maintained with appropriate care and the addition of trimming capacitors, etc. Similarly, the ganged-tuning capacitor used in a superhet receiver can be designed to accomplish the same objective. Usually, the LO capacitor is smaller and designed to tune an inductor over the desired LO range, while the RF tuning capacitors tune simultaneously over the desired RF signal input range.

LOCAL OSCILLATOR TRACKING

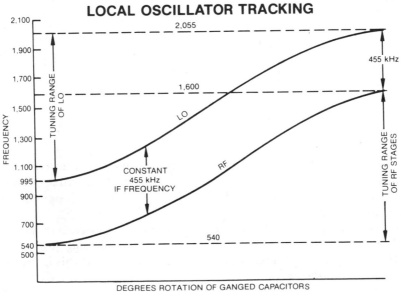

The Mixer—What a Mixer Does

The mixer is used in a superhet receiver to convert the incoming RF signal to the intermediate frequency. It must perform this frequency translation without distorting the incoming signal. As you know from your earlier study of mixers in Volume 3, the mixer accepts two input frequencies—the received signal and the local oscillator signal. The mixer *output* contains four different frequencies: (1) the RF signal, (2) the LO signal, (3) the sum of 1 and 2, and (4) the difference between 1 and 2. The difference signal is usually the desired signal in receiver mixer application. If the mixer is linear—that is, the output difference signal is directly proportional to the incoming RF signal—then the difference signal has all the information contained in it. Tuned circuits allow the desired signal (4) while discriminating against 1, 2, and 3 above.

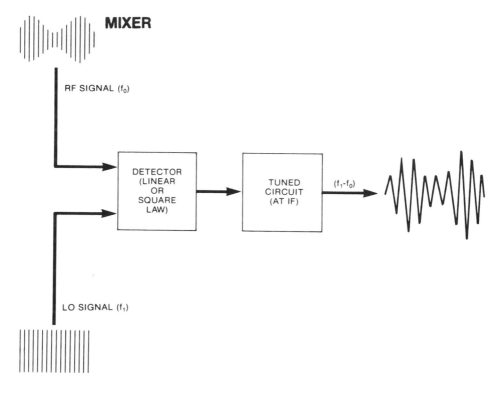

The desired IF signal is filtered by the primary winding of the first IF transformer included in the collector circuit of the mixer. The secondary winding of this IF transformer is coupled to the base of the first IF amplifier stage.

Mixers are sometimes called *frequency converters*, *frequency translators*, and *heterodyne detectors* or *first detectors*. The mixing or *beating* of two signals together in a mixer is sometimes called *heterodyning*, and this is the origin of the word *superheterodyne*.

Experiment/Application—Mixer Operation

You can see how mixers work by using a very simple test setup. You will need two signal generators, with at least one capable of tone modulation. These will be used as the signal and LO inputs. Suppose you connected the circuit shown below:

MIXER OPERATION

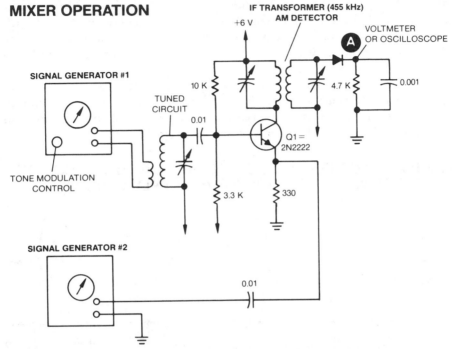

Set the output from signal generator 2 to maximum output. Set the output from signal generator 1 at least 10 times lower. Tune the system up to 1 MHz by setting signal generator 1 to 1,000 kHz and signal generator 2 to 1,456 kHz. Tune up the tuned circuit for maximum response at the detector output (maximum dc level). Then tune the IF transformer for maximum response at 455 kHz. If you tone modulate signal generator 1, you will find an ac voltage developed at the detector output. You can hear the tone if you replace the RC at the output of the detector by a pair of headphones or connect an oscilloscope at point A, as shown.

Without changing the signal generator outputs, tune signal generator 1 to 544 kHz. As you know, this is the image frequency. Compare the response at the correct and image frequencies. This will give you an idea of how much image rejection you get from the single tuned circuit at the input.

You can check the linearity by plotting the change in detected output level as the percentage tone modulation is varied. As you will see, the signal produced at the IF is a faithful (linear) reproduction of the tone-modulating signal.

IF Amplifier Operation

The intermediate-frequency amplifier is fixed tuned to the constant difference in frequency between the incoming RF signal and the local oscillator. As you know, this difference signal comes from the mixer. The tuning of the IF amplifier stage is usually accomplished by means of tuned IF transformers. The one associated with the amplifier input circuit is called the *input* IF transformer, while the one associated with the amplifier output circuit is called the *output* IF transformer. If more than one IF stage is used, *interstage* IF transformers are also employed between IF stages.

Since this amplifier is designed to operate at only one fixed frequency, the IF circuits may be adjusted for high selectivity and maximum amplification. It is in the IF stage that *practically all* of the *selectivity* and *voltage amplification* of the superhet are developed.

THREE-STAGE IF AMPLIFIER TUNED CIRCUITS PROVIDE SELECTIVITY AND AMPLIFIERS SUPPLY SENSITIVITY

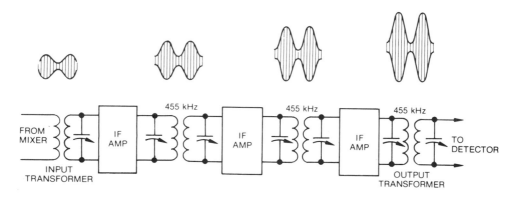

The intermediate frequency used in most AM broadcast superhet receivers is 455 kHz, although other intermediate frequencies are also used, as described previously. Using a low intermediate frequency, such as 175 kHz, results in high selectivity and voltage gain, but also increases the possibility of image-frequency interference. A high intermediate frequency reduces the possibility of image interference, but requires higher Q IF transformers and more stages to achieve the desired selectivity. In many cases, you will find the IF transformers tapped. This is done to minimize loading of the tuned circuit by the low transistor input impedance that can reduce the Q and hence the selectivity of the tuned circuit. The output of the last IF stage is fed to a detector that you studied earlier to recover the modulating signal.

In many high-performance superhet receivers used for communications applications and FM reception, crystal filters are used to provide very good control of the IF passband response. These will be discussed later in this volume.

Automatic Gain Control

As you know, the detector output generates a dc signal whose amplitude is proportional to the carrier signal level. This dc signal can be used to provide *AGC*, or *automatic gain control*.

AGC is usually applied to the first IF amplifier in a superhet receiver, and its function is twofold. First, AGC tends to keep the *receiver output constant* as the signal fades—due to either atmospheric conditions or changing tuning to a weaker station. And second, it *prevents amplifier overload* on a strong signal. If an RF amplifier is used, the AGC is also applied to the RF stage. When RF AGC is employed, it is usually in the form of *delayed AGC*. Delayed AGC is used to avoid reduction of RF amplifier gain on weak signals, which can cause a loss of sensitivity because reduced gain can degrade the signal-to-noise ratio, as you learned previously. Delay is accomplished by requiring the detector output to rise above a fixed threshold level before the RF amplifier gain is reduced.

The block diagram shows a typical AGC system with delayed AGC for the RF amplifier. Typically, the AGC is applied to the base or emitter of the transistor to reduce the quiescent current through the transistor and hence the gain. As you will learn, this approach to gain control is not practical for broadband amplifiers, because bias changes cause large changes in transistor input capacitance and hence can cause detuning.

The RCA CA3088 IC provides both IF and delayed RF AGC. In the performance curve, the IF gain is continuously reduced as signal level increases, but the RF AGC voltage is virtually constant for IF input levels of less than 2 mV (delayed) and then increases rapidly for high-level signals.

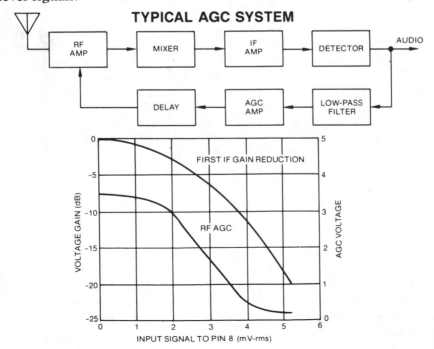

TYPICAL AGC SYSTEM

Automatic Gain Control (continued)

The AGC function can also be implemented by using a transistor as a variable resistance—as it is in the LM172 IC you studied when learning about the TRF receiver. This approach avoids the detuning effects of simple changes in transistor bias. In the LM172, an emitter-coupled transistor pair (Q2–Q3) is used as a series-shunt attenuator. The base of Q1 is held at $+2.1$ volts (dc) by the drop across diodes D1 through D3, and the base of Q2 is held at 1.4 volts (2.1 volts $- V_{BE2}$). Since the emitter of Q2 (and Q3) is held at $+0.7$ volt, the AGC voltage must be equal to or greater than 2.1 volts ($V_{BE3} + V_{BE4} + 0.7$ volt) in order for Q3 to conduct. For AGC voltages less than 2.1 volts, Q3 is cut off and Q2 functions as a simple emitter follower. When V_{AGC} equals 2.1 volts ($3 V_{BE}$), Q2 and Q3 conduct equally and share the current through *source* resistor R2. As the AGC voltage tries to increase beyond $3 V_{BE}$, Q3 turns on harder—conducts more—and Q2 starts to turn off—conducts less. Thus, the effective emitter resistance of Q2 (R_{E2}) increases, and the emitter resistance of Q3 (R_{E3}) decreases as the AGC voltage tries to rise above 2.1 volts. The sketch considers the emitter-coupled Q2–Q3 as a shunt attenuator adjusted by the signal to keep RF output constant.

AGC SYSTEM

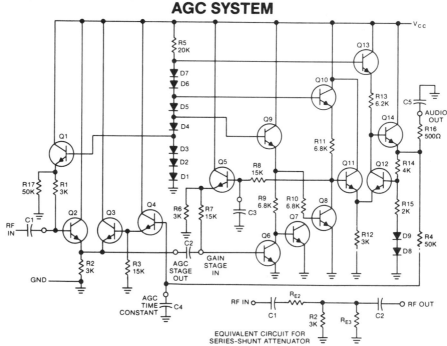

EQUIVALENT CIRCUIT FOR
SERIES-SHUNT ATTENUATOR

In any AGC system, care is necessary to keep the AGC free of any audio-output components. If the AGC filter allows low-frequency audio signals through, the receiver gain will be varied at that audio rate and the modulation will be erased, resulting in loss of low-frequency response. On the other hand, if the AGC frequency response is too slow, it will not respond rapidly enough to fading or station changes.

Experiment/Application—AGC Operation

Essentially all AM broadcast receivers use AGC. It would be some-what difficult to build such a receiver from parts, but it could be done. However, AM broadcast receivers are very common, and you can use a commercial AM receiver to advantage to study AGC action. To do this, you must be able to get at the circuits. Assuming that you can do so, obtain an AM receiver and locate the AGC line. *Do NOT use an ac/dc radio for this experiment.* You can find it most easily by locating the second detector and tracing the associated circuitry to find the AGC bus. When you have, make the following setup.

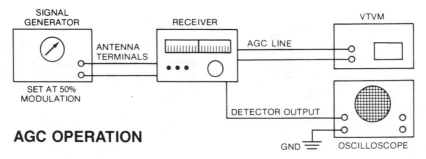

AGC OPERATION

You can observe the audio output with the oscilloscope and the AGC voltage with the voltmeter. You can vary the input-signal level from the signal generator to simulate differing signal strengths. Tune the signal generator to the receiver frequency (or vice versa). Set the signal generator to a low-level output (slightly above noise). Then, note the audio level on the scope and the AGC voltage. Record these as well as the input-signal level. Gradually increase the input-signal level and take readings until the receiver overloads. If you make a plot of the input-signal level (ordinate) versus AGC voltage and audio output (ab-scissa), you will get a curve that resembles the one shown below.

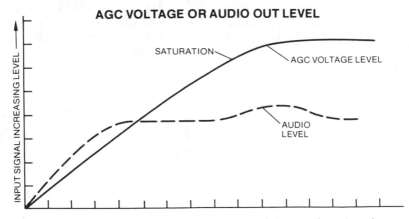

As you can see, the AGC voltage increases with increasing signal strength, but the audio output is relatively constant. You can see that the AGC action reduces the receiver gain to hold the audio output constant.

Review of Receiving Antennas and AM Receivers

1. **THE BASIC RESONANT AN-TENNA** is a half wavelength long. AM broadcast receivers use loops; mobile antennas are often quarter-wave verticals.

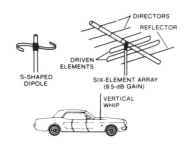

2. For FM, TV, and VHF/UHF in general, crossed dipoles, "S" dipoles, arrays, or mobile vertical whips are used.

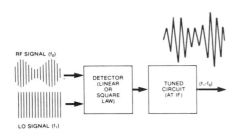

3. **RF AMPLIFIERS** precede mixers to help minimize image response. RF/mixer tuning must track with LO tuning to keep the IF signal frequency constant.

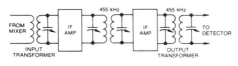

4. **THE IF AMPLIFIER** is fixed tuned to the difference between the LO and received-signal frequencies and develops practically all the selectivity and voltage amplification of the receiver.

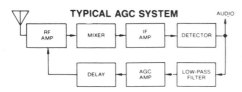

5. **AUTOMATIC GAIN CONTROL** (AGC) tends to keep the receiver output constant during changes of input signal strength. A dc voltage from the detector provides controlling bias.

Self-Test—Review Questions

1. Describe some typical receiving antennas.
2. Which do you believe make the best antennas? Why?
3. Draw a block diagram of a typical AM superhet receiver.
4. Describe the function of each block in question 3 and describe how it works.
5. Draw a diagram of an RF amplifier stage for an AM receiver.
6. Draw a diagram of a mixer/local oscillator.
7. Discuss the operation of mixers in some detail.
8. Draw a diagram of a typical IF amplifier/detector. Describe the function of the circuit elements.
9. How does an AGC work? What is it used for? What would a receiver do without AGC? Why is delayed AGC used on RF amplifiers?
10. What happens if the AGC filter has too short a time constant? Why?

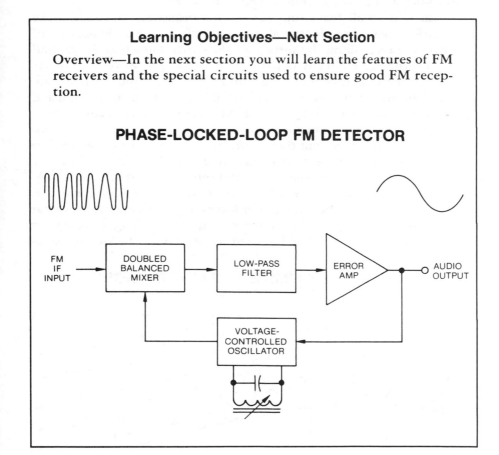

Learning Objectives—Next Section

Overview—In the next section you will learn the features of FM receivers and the special circuits used to ensure good FM reception.

PHASE-LOCKED-LOOP FM DETECTOR

FM
IF
INPUT

DOUBLED
BALANCED
MIXER

LOW-PASS
FILTER

ERROR
AMP

AUDIO
OUTPUT

VOLTAGE-
CONTROLLED
OSCILLATOR

Introduction to FM Receivers

FM modulation is used in information-transmission and reception systems where *high fidelity* and *noise immunity* are desired as in FM broadcasting. These two properties are, however, obtained at the expense of *greater bandwidth* than an equivalent AM system. Because of this, FM operation is used mainly in the UHF-VHF region. FM is also used in communications systems where multipath propagation, or other interference, may cause problems. Multipath propagation takes place when the signal comes not only directly to the receiving antenna but also as the result of reflections from surrounding objects. Since multipath problems are often associated with VHF-UHF propagation, FM is used primarily in these bands. The FM broadcast band extends from 88 to 108 MHz and consists of 100 channels, each 200 kHz wide, while an FM communications band is located at 450 MHz and at other frequencies. It should be understood, however, that AM is also extensively used for point-to-point and air-to-ground communications.

You will recall from your study of transmitters and radio-wave propagation that when the signal is radiated from the antenna, part of the energy travels along the surface of the earth (known as the ground wave) and the remainder is radiated into space (sky waves). At VHF-UHF frequencies and above, the ground waves are rapidly attenuated. In addition, the sky waves are *not* reflected back toward the ground and the receiving antenna as they are at lower frequencies. Because of this, *reception* is limited approximately to *direct line of sight* between the transmitter and the receiver, and the *effective range* is limited by the *curvature of the earth*. If either or both antennas are elevated, the effective range can be extended. In cases where hills and buildings obstruct the line of sight, a severe decrease in the quality and range of reception can occur. Actually, there is some bending of the waves so that reception somewhat beyond line of sight is possible. For reasonable antenna heights, reception is typically limited to ranges of less than about 50 miles (80 km) for noise-free operation.

RECEPTION LIMITED TO LINE OF SIGHT

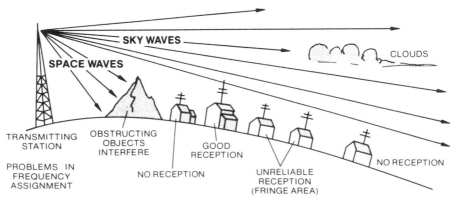

Introduction to FM Receivers (continued)

FM receivers invariably use a superheterodyne configuration, and the basic block diagram is the same as that for an AM receiver. However, the antenna, RF amplifier, and local oscillator usually operate in the VHF-UHF frequency range and require special consideration in design and construction in order to provide stable operation at these frequencies. The IF frequency for most FM sets is usually 4.5 or 10.7 MHz, and the IF amplifiers differ from their AM counterparts in a number of significant aspects that you will learn about shortly.

In both AM and FM receivers, the purpose of the stage following the IF amplifier is to extract the information from the carrier signal. Since the information was originally impressed on the carrier by modulating its frequency, the FM detector must sense *frequency*—not amplitude—variations. Thus, the basic differences between an FM and AM receiver lie in the *detector circuit*. Also, since FM signals may occupy a wider band for a given modulating signal, the *RF and IF bandwidths are typically greater*. Since FM is usually confined to higher frequencies, this wider bandwidth requirement is more tolerable.

The audio portion of an FM receiver is identical to that for an AM receiver, except that the audio portion of an FM broadcast receiver is usually capable of amplifying without distortion a wider frequency range than an AM broadcast receiver—typically 15 kHz versus 5 kHz.

In the sections that follow, you will learn about the performance characteristics of each of the circuits that are of importance to FM reception. Special emphasis and detailed study will be devoted to the IF amplifier and detector sections—the heart of the FM receiver and the only circuits that may be significantly different from an AM receiver. While the principles described here are applicable to any FM receiver, emphasis will be on FM *broadcast* receivers as a mechanism for studying FM receivers.

FM RECEIVER

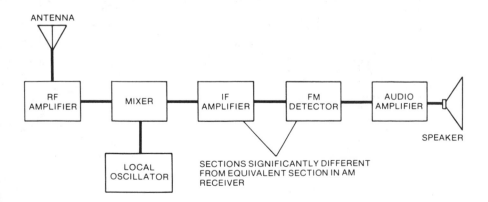

Introduction to FM Receiver Circuits

As an introduction to FM receivers, a complete FM broadcast receiver will be described on a stage-by-stage basis. First, a stage or group of stages will be described in terms of discrete, transistorized elements. Then, when appropriate, you will learn how these same stages have been designed in more advanced solid-state elements such as MOSFETs and ICs. The FM section of an AM/FM receiver will be used as the learning vehicle for the discrete transistor discussion. You will notice that the block diagram of this receiver looks very much like an AM broadcast receiver, and the only new elements are the limiter (Q8), and the discriminator (CR19–CR20). Thus, as you can see, the only basic difference between an FM and AM receiver involves the detector. In one case, the detector produces an output signal proportional to the *carrier amplitude*; in the other case, the output signal is proportional to the *frequency deviation* of the received signal.

As you have already learned, the FM broadcast band in the United States consists of 100 channels (channel numbers 201 to 301) covering the frequency range from 88 to 108 MHz. This is the frequency over which the RF amplifier's and mixer's input must tune. The first channel (201) is *centered* at 88.1 MHz and the last channel (301) is *centered* at 107.9 MHz. Since the IF is 10.7 MHz, the oscillator (usually tuned to the high side) must tune from 98.8 to 118.6 MHz.

FM SECTION: TYPICAL FM RECEIVER

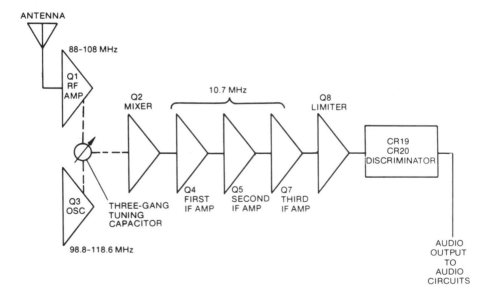

RF Amplifier, Mixer, and Local Oscillator

The FM RF circuits are tuned with a three-gang tuning capacitor (C1), which is coupled to the tuning dial. A fixed capacitor and a small trimmer capacitor are connected in parallel with each of the three sections to provide for proper alignment (see next page for schematic).

The FM signal at the antenna terminals, normally in the microvolt range, is coupled to the base of Q1 by input transformer T1. The input signal is amplified in RF amplifier Q1 and then coupled through C11 to the base of the mixer (Q2). A tuned circuit comprised of C1 (B) and inductor L2 provides for tuning of the RF amplifier mixer interstage. The local-oscillator signal is also coupled to the base of the mixer through C12. The oscillator stage, Q3, is tuned by one section of C1 and inductor L4. The oscillator coil has a movable core to permit oscillator alignment with the tuning dial.

The mixer stage is biased near cutoff by R8 and R9, and driven into conduction during the negative peaks of the oscillator signal. The RF signal, also present at the base of Q2, alternately adds to and subtracts from the oscillator signal in the mixer stage. The output of the mixer contains, in addition to the original RF and oscillator signals, the sum and difference frequencies. The desired difference frequency of 10.7 MHz is coupled through fixed-tuned transformer L5 to the first IF amplifier. The 100-MHz (nominal) input signals and the 200-MHz sum signal are shunted to ground through the transformer primary network and C19. As you can see, except for component values that reflect the operation in the VHF range, the circuits are essentially identical to those for AM. C14 in the LO tuned circuit has a temperature characteristic chosen so that mistuning does not occur as the temperature changes.

The development of MOSFET devices has made it possible to make considerable improvements in FM receiver performance. MOSFET devices are used in RF stages in preference to bipolar transistors because of their greater dynamic range, lower circuit noise, and higher gain. The primary improvements are their better performance due to less noise and ability to handle a greater range of input-signal levels. Conventional bipolar transistors show relatively larger change in input capacitance as a function of signal level, which can cause circuit detuning at higher frequencies; these effects are absent in MOSFETs. There is little loading of the input signal or drastic change in input capacitance even with extremely large input signals. The dynamic range of the MOSFET is about 25 times greater than bipolar transistors, but the differences in circuit source impedances reduce this to a practical value of about five times. The linearity of MOSFET RF amplifiers therefore ensures good performance at high signal levels.

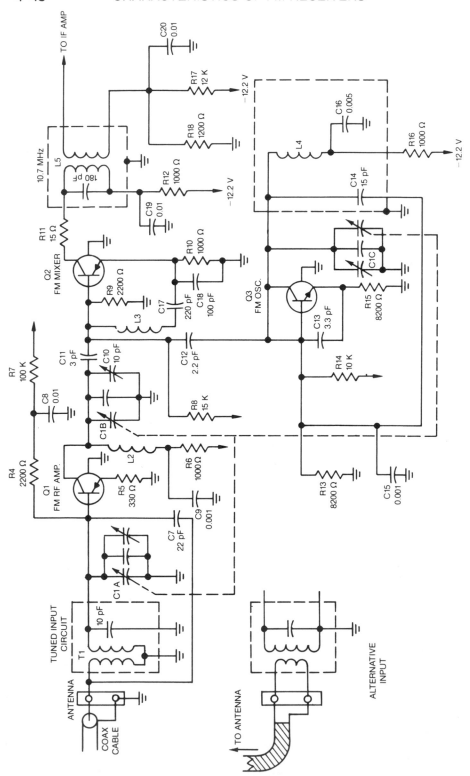

FM IF Amplifiers

In an AM broadcast receiver, the IF amplifiers are synchronously tuned (all tuned to the same frequency), with the result that the IF passband has a single peak and the tuned-circuit rolloff provides good selectivity. Since the maximum AM audio frequency is typically about 5 kHz, the double-sideband AM spectrum occupies 10 kHz. If the bandpass is just 10 kHz wide, the high-frequency (5-kHz) audio components are passed with 3 dB less gain than low-frequency components. While this results in some loss of fidelity, little harmonic distortion results.

The situation with FM IF amplifiers is quite different. An FM broadcast station is permitted a peak deviation of 75 kHz (150 kHz peak-to-peak). For FM reception, a more nearly rectangular passband than for AM is desired for two reasons. The first is harmonic distortion and phase shift of the relative frequency components; the second is channel assignment and selectivity.

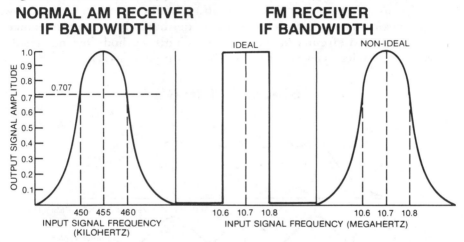

You will recall that FM deviation is a function of the amplitude or loudness of an audio signal and *not* its frequency. Therefore, if a loud tone, whether high or low frequency, is passed through an IF amplifier with limited frequency response, the result is a clipping or rounding of the peak of the audio waveform, causing harmonic distortion. The second factor, requiring a flat IF frequency response with sharp out-of-band rolloff, is the need for good adjacent channel rejection in FM receivers. Because of the line-of-sight limitation in the FM broadcast band, it is very possible to have large differences in signal strength which are *not* functions of the distance from the transmitting station but of relative antenna altitudes. Therefore, while the FCC limits channel assignments to minimize interference, it is entirely possible for a distant station to have a stronger signal at the user's location than that of a local station being listened to. If selectivity is not sufficient to attenuate this out-of-band station, its signal will affect the action of the limiter and can reduce, or even completely suppress, the signal from the local station. FM receivers therefore are designed to approximate the ideal response.

FM IF Amplifiers (continued)

Three methods are generally used to provide the desired FM IF response curve. The first method, known as *stagger tuning,* gives an excellent frequency-response curve but reduced gain. Three IF amplifier stages are tuned, for example, to 10.6, 10.7, and 10.8 MHz, respectively. The overall frequency response of this combination adds to produce the close-to-ideal response curve shown in the diagram below. The overall gain of the three stages is not much greater than that of a single stage. In addition, special alignment techniques are required. As a result, this method is not used as frequently as other methods for FM, but is frequently used for other wide-band IF applications such as TV and pulse receivers.

The second method of approximating the desired broad frequency response is to use three IF amplifiers, all tuned to the same center frequency, synchronously tuned. A broader response is obtained in each stage by using less-than-critical coupling in the transformers and by using transformer coils with low Q or by deliberately adding resistance across the tuned circuits to reduce the Q. This method does not result in very good selectivity (see the top figure on page 4–51).

STAGGER TUNING

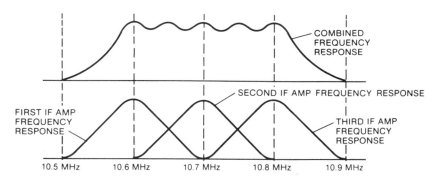

The third method also makes use of two or three IF tuned circuits tuned to the same center frequency. The first and third stages are designed and tuned as in the second method noted above. However, the transformer of the second stage is overcoupled, resulting in the familiar double-peaked frequency-response curve. When the individual response curves of the three stages are combined, the double peaks of the second stage have a significant effect in broadening the frequency response and steepening the skirts, producing a good approximation of the desired response curve. The good response characteristics obtained make this method a popular one (see the bottom figure on page 4–51).

THREE IFs WITH SAME CENTER FREQUENCY

ALTERNATE SINGLE-AND DOUBLE-PEAKED STAGES

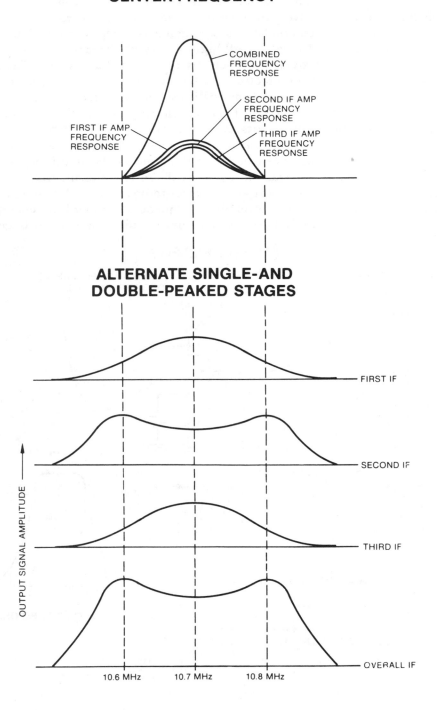

Ceramic Filters

Although electromechanical and crystal filters have been employed for many years, the advent of microelectronic circuits has resulted in a proliferation of the equipment and technology needed to produce low-cost ceramic filters. Nowadays, all high-quality FM broadcast receivers use these filters to provide good selectivity and passband shape. These filters are based on the piezoelectric effect in the ceramic material, and have the inherently very high Q normally associated with quartz crystals.

Ceramic filters can range from simple single-element devices to complex multiple-element units. However, even multiple-element units are small in size at 10.7 MHz. If a single-element, two-terminal filter is placed across the emitter resistor of an IF amplifier the amplifier has a gain about equal to the transistor β at f_o, the series-resonant frequency of the crystal element, because the series-resonant crystal essentially grounds the emitter at the intermediate frequency. At other frequencies, the gain is approximately R_L/R_E, which can be quite low if R_E is large.

CERAMIC FILTERS

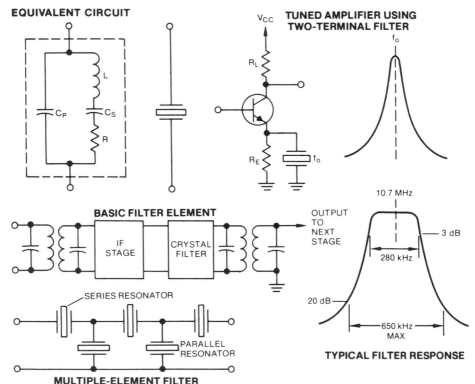

Ceramic Filters (continued)/IF Amplifier Circuit

When two or more elements are connected in a four-terminal arrangement and used between IF stages, substantial increases in bandwidth and selectivity can be obtained. Also, the input and output impedances can be controlled, and made unequal if necessary for interstage matching purposes. By controlling the coefficient of electromechanical coupling between filter elements, the bandwidth can be very accurately shaped, both in width and flatness. Ceramic filters are widely used in communication equipment and in FM receivers to provide high selectivity to avoid adjacent channel interference.

Now let us move on to our next subject—the study of the IF amplifier circuit.

The example of the IF amplifier shown on page 4–54 is part of the same receiver previously described with transistor-type RF amplifier, oscillator, and mixer. This receiver has three IF stages consisting of Q4, Q5, and Q6. The input and output of each stage is transformer-coupled with a standard FM tuned IF transformer at 10.7 MHz. Tapped transformer windings are used to provide better impedance matching, and coupling is adjusted to provide an optimal passband shape.

The signal from the mixer is coupled to the base of Q4 via an input transformer, and Q4 output is coupled through R20 to the second IF transformer L6. Both the primary and secondary windings of L6 are inductively tuned to the FM intermediate frequency of 10.7 MHz. The signal in the secondary winding of L6 is applied from the winding tap to the base of the second IF amplifier (Q5), where it is further amplified and coupled through R27 to another double inductively tuned IF transformer (L11), which is identical to L6.

From the secondary of L11, the amplified signal is supplied directly to the base of the third FM IF amplifier (Q6). This stage is the last amplifier stage for the IF signal. After amplification, the signal is again coupled through resistor R32 to a double inductively tuned transformer L13. The IF signal is then applied to the amplitude limiter stage Q8, to be described later.

The bases of the three IF amplifiers are biased at about − 1 volt by resistive dividers and decoupling networks similar to those shown for the RF amplifier and mixer stages. Resistors R20, R27, and R32 are used to provide suppression of undesired oscillation in the IF amplifier stages because of inadvertent feedback. The IF stages shown are neutralized to minimize instability and to minimize tuning changes as a result of change in input-signal levels.

After you review the schematic on the next page you will be introduced to the three basic FM detectors and their properties.

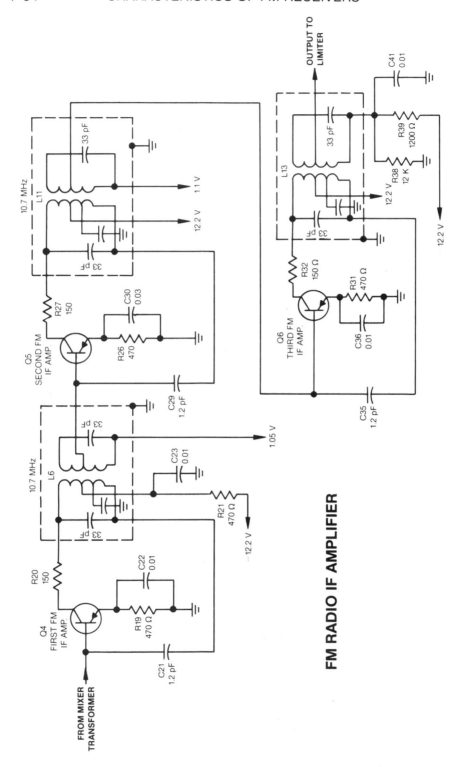

FM RADIO IF AMPLIFIER

FM Detectors—Limiting

In your earlier study of the use of frequency modulation for the transmission of information, you learned that the carrier-signal amplitude did not carry any information and that all intelligence was contained in the frequency variations of the RF carrier. Because of these factors, FM detectors are quite different from their AM counterparts and require completely new circuits.

These basic detector types are commonly used—the *slope-detector*, *phase-detector*, and *frequency-following* (or *phase-locked-loop*) detectors. Some of these detectors are sensitive to amplitude variations and, as you will learn shortly, some are not. If the FM receiver uses an amplitude-sensitive FM detector, then the detector stage must be preceded by a *limiter* to remove from the signal any amplitude variations that occur from interference signals caused by lightning flashes, neon signs, automobile ignition, microwave ovens, small appliances, and various other spark-producing electrical equipment. The limiter removes the signal-amplitude variations caused by these disturbances but leaves the original FM intelligence unaltered, since the frequency is not changed by limiting. It is the *limiter*—either as a separate stage or as the absence of amplitude sensitivity in the detector—that gives the FM receiver its well-known freedom from noise and immunity to interference from other signals. This immunity occurs because the weaker signal added to the stronger signal is stripped off by the limiter and does not appear at the limiter output.

LIMITER AND DISCRIMINATOR FUNCTIONS

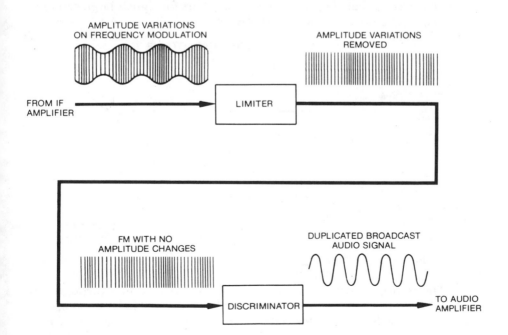

AMPLITUDE VARIATIONS
ON FREQUENCY MODULATION

AMPLITUDE VARIATIONS
REMOVED

FROM IF
AMPLIFIER

LIMITER

FM WITH NO
AMPLITUDE CHANGES

DUPLICATED BROADCAST
AUDIO SIGNAL

DISCRIMINATOR

TO AUDIO
AMPLIFIER

FM Detectors—Limiting (continued)

Limiters use saturation and cutoff to limit the positive and negative signal peaks. As shown below, operation is linear for small signals; but as the signal increases in amplitude at the input, the output does not increase because of limiting action. The limiter responds to the amplitude variations for very small signals (1); however, for higher level signals (2), the signal output remains constant.

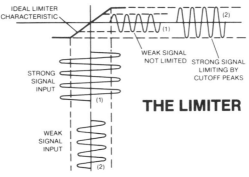

THE LIMITER

There are many circuit configurations for limiters. Basically, these depend on cutoff and/or saturation of diodes or transistors. The simplest limiter is a pair of back-to-back diodes. As you remember, there is a minimum voltage below which the forward bias is insufficient to narrow the depletion layer gap width, so that the diode does not conduct. If a pair of diodes is connected as shown, the input signal is unaffected by the diodes, but the diodes conduct (CR1 for positive peaks and CR2 for negative peaks) so that clipping or limiting occurs for signals large enough to overcome the offset voltage of the diode.

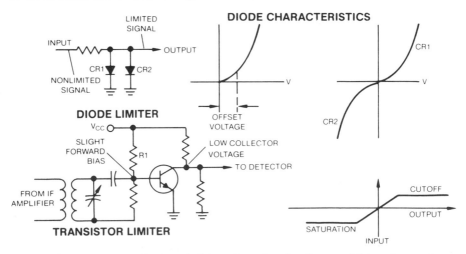

The transistor limiter is biased on by the forward bias through R1 and is driven to cutoff by relatively small negative signals. By keeping the collector voltage very low (1.5–2 volts), saturation occurs with relatively small positive signals. In this way, limiting action takes place.

FM Detectors—Slope Detectors

The simplest FM detector is the slope detector that uses one resonant circuit tuned to one side of the carrier, converting frequency deviation to an amplitude variation because of the unbalanced frequency response. This amplitude variation can then be detected by a simple AM (diode) detector. Because of its sensitivity to carrier detuning/drift, this simple detector has little practical application, and the *balanced slope detector* (or *diode discriminator*) is used instead. This configuration uses two slope detectors, one tuned to either side of the carrier frequency.

In the circuit shown on the next page, the final IF transformer has a center-tapped secondary. The primary is tuned to the IF frequency (10.7 MHz for FM broadcast) and the two halves of the secondary (L_x–C_x and L_y–C_y) are tuned, for example, to 10.6 and 10.8 MHz, respectively. If no modulation exists on the FM carrier, the IF frequency is at 10.7 HMz, and the detected currents through R_x and R_y are equal and opposite, so the voltage across R_L (equal and opposite currents) is zero. If no modulation exists but the local oscillator is detuned, the IF is not at 10.7 MHz, and a steady-state dc error voltage is produced across R_L because I_x does not equal I_y. This error voltage can be used to correct the oscillator frequency (automatic frequency control, or AFC), as will be described later. Unfortunately, the slopes of the two tuned circuits are neither linear nor constant, so although the circuit shown will work, it is rarely used in a practical FM receiver. It is important to understand how the slope detector works, since this helps in understanding more widely used FM detectors.

OPERATION OF BASIC DIODE SLOPE DETECTOR

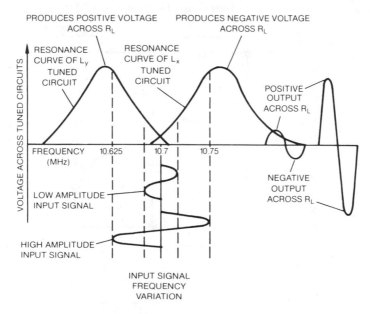

FM Detectors—Slope Detectors (continued)

Let us assume that the RF carrier has been frequency modulated by a low-deviation (low-amplitude) 1,000-Hz tone to produce a total carrier deviation of ± 15 kHz. The IF signal will then deviate 15 kHz to each side of the IF frequency at a 1,000-Hz rate. When the IF signal is 15 kHz higher than 10.7 MHz, more current will flow through R_x and a smaller current through R_y. The net result is an imbalance of current through R_L and a positive voltage at the output. Now, as the IF signal swings back across 10.7 MHz—following the instantaneous envelope of the 1,000-Hz modulation—an increasingly larger current flows through R_y and a decrease occurs in the current through R_x. Thus, the current through R_L, which is the sum of the currents, will produce a negative voltage at the output. Thus, the output follows the variations in the IF frequency. Since the IF frequency was changed at a 1,000-Hz rate, a 1,000-Hz signal appears at the audio output.

If the frequency deviation increases, as it would with a louder 1,000-Hz tone, larger difference current will flow through R_L and a large 1,000-Hz voltage appears at the audio output.

You will note that when no modulation is applied the voltage across R_L is zero, and therefore this detector is not sensitive to variations in carrier amplitude. A fixed-frequency offset will produce a dc voltage (+ or −), depending on how far the frequency is offset. This dc voltage can be used to provide automatic frequency control. However, when modulation is applied, the output is proportional to the amplitude of the input signal as well as its frequency. Therefore, an IF limiter is required when a slope detector is employed. Sometimes this detector is called a *high/low FM detector*. It is not usually used in FM broadcast receivers because it is not very linear, and hence distortion results. It is mainly used in special broadband applications in radar receivers.

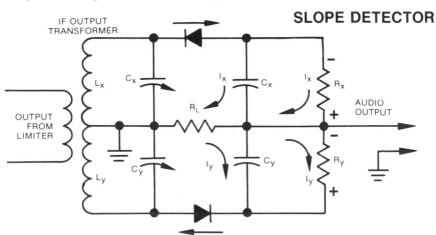

SLOPE DETECTOR

IF OUTPUT TRANSFORMER

OUTPUT FROM LIMITER

AUDIO OUTPUT

FM Detectors—Quadrature Detectors

An important group of PM detectors is based on the combining of *quadrature* voltages (signals that are 90 degrees or $\pi/2$ radians out of phase). You learned during your study of transmitters how quadrature mixers were used to generate both SSB and FM signals.

Assume two signals (E_a and E_b) are 90 degrees out of phase. The 90-degree phase shift in E_b is produced either by capacitor C or inductor L. If these two signals are added vectorially, the resultant is as shown in diagram A. Since the secondary of the transformer is tuned, the output at E_a shifts in phase relative to E_b above and below resonance. As you remember, above resonance X_C is less than X_L and the parallel circuit looks capacitive; below resonance X_C is greater than X_L and the parallel circuit looks inductive. Therefore, as the FM signal varies in frequency, the phase of E_a shifts relative to E_b, as shown in diagrams B and C. This phase shift (ϕ) is proportional to the FM deviation, as shown. If these two signals are added, properly detected, and then filtered to remove all carrier-frequency components and higher harmonics, the output is a signal proportional to $E_a E_b \cos \phi$. Therefore, FM detection can be accomplished by providing two signals which are in quadrature when the IF (or RF) signal is at its center frequency of 10.7 MHz and which deviates from this phase relationship as the frequency varies. This principle forms the basis for two of the most common FM detectors, the *Foster-Seeley discriminator* and the *ratio detector*.

QUADRATURE DETECTOR

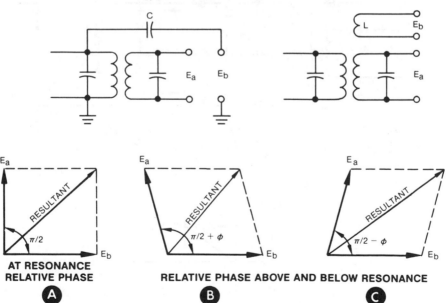

AT RESONANCE
RELATIVE PHASE

RELATIVE PHASE ABOVE AND BELOW RESONANCE

A   B   C

FM Detectors—Foster-Seeley Discriminator

In the Foster-Seeley discriminator bypass capacitors C1 and C2 maintain the cathodes CR1 and CR2 and the end of RF choke L at ground, but are too small to bypass the audio signals. Voltage E_p appears across the choke, and the secondary voltage appears equally across each half, shown here as E_A and E_B (see A).

Voltages E_A and E_B are equal but 180 degrees out of phase with respect to the secondary center tap. The primary voltage E_p is coupled through capacitor C and appears 90 degrees out of phase with the total secondary voltage across L. Voltages E_p and E_A add to produce E_1, and voltages E_p and E_B add to produce E_2. The rectified currents I_1 and I_2 flow in opposite directions and the voltages across R1 and R2 are *equal and opposite* at resonance (see A). Thus, the audio output voltage is zero when $R_1 = R_2$ and $E_1 = E_2$. This reduces noise and AM. When E_1 exceeds E_2 (above resonance), the output voltage is positive (B). When the carrier swings below resonance, E_2 exceeds E_1 and the audio output is negative (C). The audio output circuit is usually taken from the top end of R1, as shown. Since any offset in frequency appears at the junction of R1–R2 as a dc voltage, AFC voltage is also available at this point. This discriminator is sensitive to AM signals off center frequency and therefore requires a limiter.

FOSTER-SEELEY DISCRIMINATOR

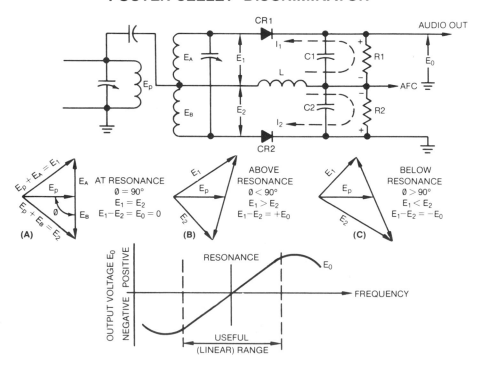

FM Detectors—Ratio Detector

The ratio detector is a quadrature detector that is relatively insensitive to short-term amplitude variations in the carrier. Therefore, a separate limiter is *not* needed, and this circuit is very popular in FM receivers. You will note that the ratio detector circuit is very similar to the Foster-Seeley discriminator with the principal change being the reversal of one of the diodes and the addition of C3. Because of the way the diodes are connected, the total voltage (sum) across C1 and C2 is constant for a given carrier level. However, the difference (at the junction of C1 and C2) is dependent on the instantaneous frequency—or audio signal. Thus, an audio output can be taken from this point. The time constant for R1 + R3 and C3 is long compared to the lowest audio frequency of interest (typically 0.1 second), so that the total voltage across R1 + R2 can be changed only slowly.

If the input-carrier level should shift slowly, the voltage across C3 will readjust to the new carrier level. However, if the voltage shifts abruptly, the voltage across C3 cannot shift abruptly, since current is required to charge (or discharge) C3. Thus, with rapid carrier shifts due to noise, part of the current must be supplied to C3, and this current is diverted from across R1–R2 into C3. Conversely, if the carrier abruptly shifts downward, less current goes into C3 and more goes through R1–R2. Thus, the total voltage across R1 and R2 is held constant. Long-term changes in carrier level—as between stations, or for slow fading, allow the average voltage across C3 to vary slowly, and hence the peak-to-peak audio output voltage from a ratio detector is not only proportional to the FM deviation but also the average carrier level.

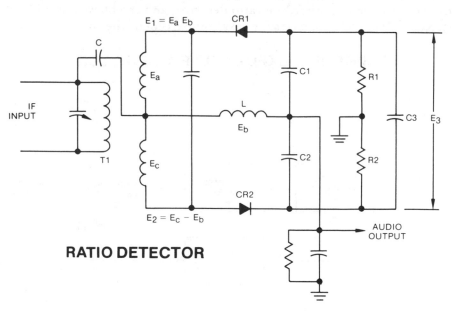

RATIO DETECTOR

FM Detectors—Phase-Locked Loop

The PLL (phase-locked-loop) detector is a variation of the PLL you learned about in your study of transmitters. Because of its freedom from tuning elements and the ease with which it can be implemented in IC form, the PLL is very popular for FM receivers. Most PLL ICs use a double-balanced mixer as the phase detector since the double-balanced mixer suppresses level changes. Thus, the PLL FM detector needs no limiter. The FM input signal (10.7 MHz) is mixed with the output of the voltage-controlled oscillator (VCO), which is at a nominal frequency of 10.7 MHz also. The mixer output contains both the sum and difference frequencies—21.4 MHz and the VCO tuning error (difference signal). The sum signal is removed by the low-pass filter, and the difference (error) signal is amplified and drives the VCO into frequency and phase coincidence with IF input.

The input IF signal changes frequency in accordance with the information content and the PLL amplifier develops a signal equal and opposite to the original modulation that keeps the VCO tuned to the instantaneous IF frequency. Thus, the output of the error amplifier is a signal that duplicates the original frequency modulation impressed on the carrier back at the FM transmitter.

Since the PLL frequency discriminator or tracker contains no tuning elements, the total circuitry is readily implemented in IC form and has found broad application in modern FM systems. Furthermore, since the VCO error voltage (from which the audio output is derived) is proportional to frequency only, no limiter is needed.

On the next page we will evaluate and compare the Foster-Seeley discriminator and limiter, the ratio detector, and the phase-locked-loop detector. Included is a chart for easy comparison of the three detectors.

PHASE-LOCKED-LOOP FM DETECTOR

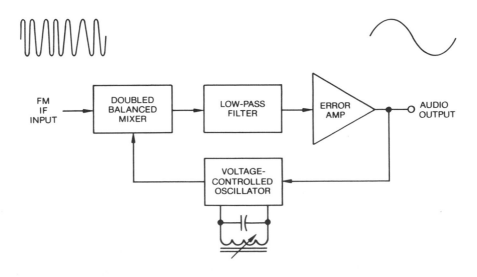

Evaluation of FM Detectors

Both the limiter-discriminator and ratio detector arrangements have advantages and disadvantages when compared with each other. Because of the nature of the relative advantages and disadvantages, there can be no clear-cut statement of which is better. Each type is preferred by some designers, and it is essentially a matter of economics as to which is used.

The important advantage of the limiter-discriminator arrangement is that it is relatively a simple matter to balance the two sides of the discriminator and obtain excellent reproduction of the audio-frequency signal. One disadvantage of this arrangement is that the limiter does not operate unless the incoming signal has sufficient amplitude to cause the limiting action to take place. When limiting action does not take place, the amplitude variations in the signal result in interfering noise and signal distortion. This means that high-gain RF and IF stages must be used to boost the signal amplitude into the limiter. In a number of FM receivers, two limiters are used in a cascade arrangement to assure that adequate limiting action will take place. Even under these conditions, signals of very low amplitude will result in amplitude modulation into the discriminator, and hence interference and noise.

COMPARING FM DETECTORS

TYPE	FOSTER-SEELEY (DISCRIMINATOR + LIMITER)	RATIO DETECTOR	PHASE-LOCKED LOOP
LIMITATIONS	HIGH GAIN RF AND IF STAGES REQUIRED TO ASSURE LIMITING ACTION	REQUIRES BROAD IF RESPONSE TO ACHIEVE LOW DISTORTION; DIFFICULT TO BALANCE	IMPRACTICAL EXCEPT AS IC PACKAGE
ADVANTAGES	EXCELLENT AUDIO REPRODUCTION; EASY TO BALANCE; AUDIO OUTPUT LEVEL DEPENDS ONLY ON FM DEVIATION	NO LIMITER STAGE REQUIRED; WEAK SIGNALS NOT SUBJECT TO AM INTERFERENCE; AUDIO OUTPUT LEVEL DEPENDS ON BOTH AVERAGE CARRIER LEVEL AND ON FM DEVIATION	EXCELLENT AUDIO REPRODUCTION; EASY TO BALANCE; LOW COST; HIGH RELIABILITY

The important advantage of the ratio detector is that it is not sensitive to short-term amplitude variations in the incoming signal, as was shown in the description of that circuit. Therefore, the ratio detector circuit eliminates the need for the limiter stage, or stages, and does not depend on the use of RF and IF stages of very high gain. The noise level rises with weak signals, but distortion does not take place until the signal amplitude is too low for reception. One disadvantage of the ratio detector is that special care must be taken to balance the two sides of the detector; otherwise some of the insensitivity to amplitude modulation will be lost.

The phase-locked-loop FM demodulator has the outstanding features of reliability, simplicity, and low cost.

De-Emphasis

In your study of FM transmitters, you learned that the signal-to-noise ratio at the high-frequency end of the audio band was improved by passing the modulating signal through a pre-emphasis network. This is possible because there is little signal energy in the high-frequency components, and these can be increased without fear of overmodulation. In this way, noise introduced after pre-emphasis and before de-emphasis will be reduced by the pre-emphasis. The U.S. standards for FM broadcast call for *pre-emphasis* of the higher frequencies in accordance with a 75-μsec time constant. The same *de-emphasis* must be applied at the output of the FM detector to obtain a linear audio output. A series RC network with a 75-μsec time constant serves this purpose. If there is high-frequency falloff in any of the audio stages following the detector, then the *net* rolloff must be in accordance with the equivalent 75-μsec time constant. The de-emphasis circuit is normally placed immediately after the detector since it can also filter IF signals from the audio and prevent the IF signals from feeding through to the audio stages.

DE-EMPHASIS NETWORK

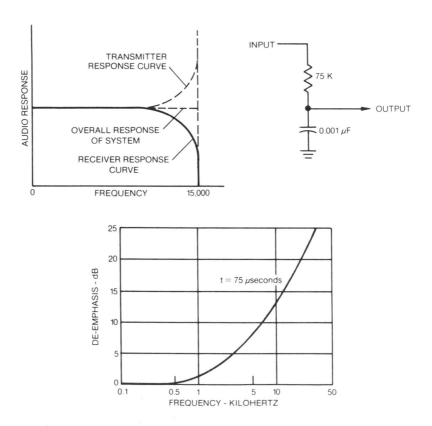

Limiter/Discriminator Circuit

The example of FM limiter and discriminator FM receiver circuit to be discussed here is part of the same receiver previously described with solid-state RF amplifier, oscillator, mixer, and IF stages. The output of the last IF amplifier stage (Q6) is fed to the limiter Q8 and a frequency-discriminator composed of CR19 and CR20. The limiter Q8 is a transistor IF amplifier stage that limits the signal amplitude to a fixed level so that variations in RF signal amplitude are not detected and produced in the audio output. The limiter is needed because the FM detector is of the Foster-Seeley type—that is, amplitude sensitive. The amplitude is limited by operating the stage between cutoff and saturation, which ensures clipping of both positive and negative peaks at all usable signal levels. A low collector voltage on Q8 ensures limiting with low-voltage inputs.

Conversion of the FM signal to an audio signal is accomplished by the Foster-Seeley-type frequency discriminator shown below. The limiter output is coupled from the collector of Q8 to a tuned IF transformer (10.7 MHz). However, this transformer has a center-tapped secondary winding to provide for operation of the Foster-Seeley discriminator. As shown, the quadrature phase shift is obtained from an inductor connected by the center tap, as described previously. Both the primary and secondary windings are tuned to the IF frequency. The audio output is taken from the tap of RC network (K1), and the IF signal leakage is bypassed to ground through the two 220 pF capacitors. The circuit diagram and operation of this discriminator are exactly as described previously. Also, the audio output can be fed to any audio amplifier.

FM RECEIVER LIMITER AND DISCRIMINATOR STAGES

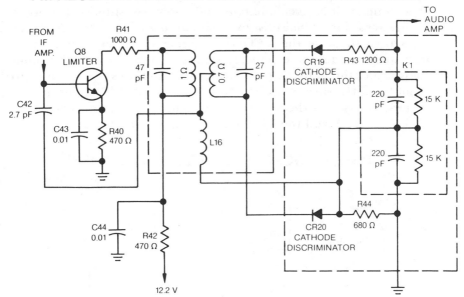

Digitally Tuned FM Receiver

The phase-locked-loop (PLL) system illustrated here in block diagram form provides a method for synchronizing the frequency of an oscillator with the frequency of any suitable reference signal. This system is packaged as an IC and makes it practical and economical to obtain *automatic electronic tuning* in radio and TV receivers. As you will see, it is almost identical with the corresponding part of the digitally tuned CB transmitter that you studied in Volume 3.

The most important part of the phase-locked-loop system is the voltage-controlled oscillator (VCO), which has the capability of being tunable over the entire range of frequencies required by a voltage input. In this respect, it is like the varactor-tuned FM oscillator you learned about when you studied transmitters. For FM automatic tuning, the oscillator that functions as the local oscillator must cover the entire range of LO frequencies (98.8 to 108.6 MHz). The reference-frequency source operates at a lower convenient frequency; consequently, the VCO output is divided down by a counting circuit to the reference frequency or a harmonic of the reference frequency, which is usually produced by a stable crystal oscillator. You will learn how counter circuits work in Volume 5 when you study digital circuits.

A phase detector compares the phase of the VCO signal to the phase of the reference signal and produces a + or − dc error signal when there is a phase (or frequency) difference. For example, if the VCO output were at a higher frequency than the reference frequency, its signal would lead in phase. The phase-comparator output would be a negative tuning voltage. Similarly if the VCO frequency were lower in value than the reference frequency, a positive tuning voltage would be produced.

The tuning voltage is coupled through a low-pass filter that removes rapid variations due to noise, signal, and transient effects. The filter output is used to drive the VCO to a higher or lower frequency as required to keep the phase-detector ouput near zero. The dc tuning voltage is proportional to the phase difference and thus proportional to the frequency difference. When the frequency-divided VCO output frequency exactly matches the reference frequency, no tuning (correction) voltage is produced, and the VCO maintains the correct frequency.

PHASE-LOCKED-LOOP DIGITAL TUNING

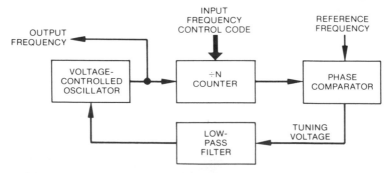

Digitally Tuned FM Receiver (continued)

Using frequency synthesis with a VCO in a phase-locked loop, it is possible to preselect several stations or tune each station in the entire range of stations (100) by properly choosing the proper value for N in the frequency divider. The tuning method consists of simply selecting the number for N in the ÷ N counter and the loop will tune the VCO to the proper local oscillator frequency. In all other respects the FM broadcast receiver, TV, or CB is identical to a more conventionally tuned receiver, except that the tuning is always exact for the channel selected.

A 2.56-MHz crystal-controlled oscillator provides the stable reference signal. This reference signal is divided by 256 (dividing by 2 eight times). Therefore, the crystal-controlled reference signal being fed into the phase comparator is a stable 10-kHz signal. The other signal into the phase comparator is also a 10-kHz signal that is frequency divided down from the voltage-controlled oscillator output signal by N. The VCO is designed to operate as the local oscillator in the 98.8- to 118.6-MHz range, and the oscillator will be tuned above the frequency of the FM station.

PHASE-LOCKED-LOOP AUTOMATIC TUNING

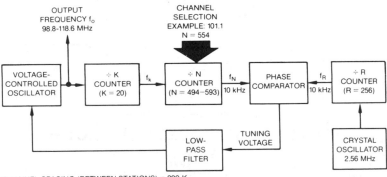

CHANNEL SPACING (BETWEEN STATIONS) = 200 K

Initially, the VCO output is divided by 20 by a fixed counter. Thus, the frequency range of signals representing the VCO frequency is now 4.94 to 5.93 MHz. To select a particular station or channel, a number is now programmed into the variable (channel-select) counter. For the FM band and the system shown, this number is any number (N) from 494 to 593. A value of N equal to 494 produces a 98.8-MHz VCO output for the first station on the dial (88.1 MHz); a value of N equal to 593 produces a 118.6-MHz VCO output for the last station on the dial (107.9 MHz).

For example, to tune to a station at 100.1 MHz (100 on the dial), the value of the number N would be calculated as follows: 100.1 + 10.7 MHz = 110.8 MHz: 110.8/20 MHz = 5.54 MHz and N = 554; 5.54MHz/554 = 10 kHz.

When the selected value of N is programmed into the ÷ N counter, the counter output will be exactly 10 kHz if the VCO is on frequency. Comparison of the divided VCO frequency with the reference signal at exactly 10 kHz produces a tuning error voltage to drive the VCO to the proper frequency and hold it exactly at that frequency.

FM Stereo

Most high-fidelity home audio systems use stereophonic reproduction, and stereo recording—both tape and disk—has almost entirely replaced monaural recording. You learned about stereo audio systems in Volume 2 and how the two channels required for stereo (left and right) are prepared for transmission in Volume 3. The principle of FM stereo transmission will be reviewed here.

The format for stereophonic transmission is set by the FCC, and requires the use of a *main* channel and a *subchannel*. Furthermore, the information transmitted on the main channel must be transmitted so that it can be received, without degradation, by a monophonic FM receiver. Monaural or monophonic reproduction of a stereo signal is done by simply adding the left and right outputs to form a single channel (L + R).

FORMATION OF FM STEREO SIGNAL

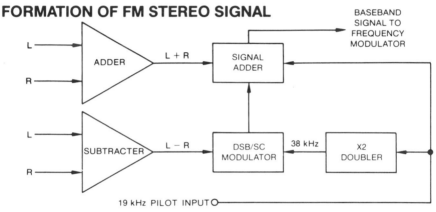

STEREOPHONIC BASEBAND SPECTRUM

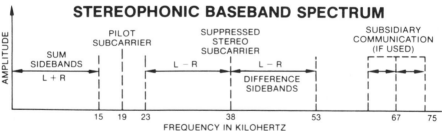

At the transmitting station, the L (left) and R (right) signals are combined to form the sum (L + R) and difference (L − R) signals. The sum signal represents the total monophonic information. The L − R signal is used to amplitude modulate a 38-kHz subcarrier in a double-sideband (DSB) suppressed-carrier modulator. The 38-kHz signal is derived from a 19-kHz ± 2-Hz *pilot* subcarrier. The L + R signal (50 Hz– 15 kHz), the (L − R) sideband signal (23 kHz–53 kHz), and the 19-kHz pilot are added together and used to frequency modulate the carrier. The composite baseband stereo signal spectrum and a block diagram showing the formation of the FM stereo signal are shown in the figure above. An extra channel for subsidiary communciation, also sometimes added to the modulation, will be described later.

FM Stereo Matrix Decoder

Receiving stereo programs requires recovery of both the L + R and L − R signals and recombining them to produce the original left and right channels of information. The output from the detector of an FM broadcast receiver contains the entire baseband spectrum from 50 Hz to 53 kHz. The sum channel (L + R) is readily recovered by passing this signal through a 15-kHz lowpass filter. Recovery of the L − R signal requires the *reinsertion* of the 38-kHz suppressed carrier in the *correct phase relative to the sidebands*, and the envelope detection of the 38-kHz amplitude-modulated signal. The 38-kHz signal is obtained by using the 19-kHz pilot signal to phase lock the locally generated 38-kHz source or by doubling the 19-kHz pilot signal recovered from the discriminator output by a narrow-band filter.

The *FM stereo matrix decoder* is so named because it uses a *resistive matrix* to recombine the L + R and L − R signals to recover the original left and right signals. This is the form of most of the early FM stereo decoders, but it has been largely replaced by phase-locked-loop circuits.

FM STEREO MATRIX DECODER

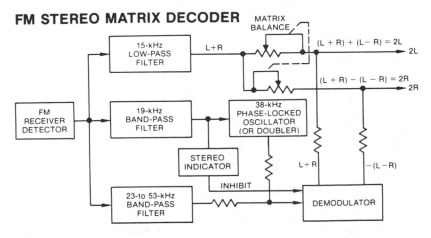

The composite baseband signal (containing the L + R, pilot, and DSB L − R signals) is fed to three tuned filters as shown. The L + R signal is recovered by the 15-kHz low-pass filter. The output of the 19-kHz pilot filter is fed to a 38-kHz oscillator or frequency doubler to generate a 38-kHz reference signal that demodulates the L − R signal.

The output of the 23- to 53-kHz filter (the upper and lower sidebands of the stereo subcarrier) is added to the 38-kHz reconstituted carrier and demodulated (detected) to provide the original L − R signal. The L + R and L − R signals are added algebraically, as shown, to produce the left- and right-channel signals.

In most FM receivers, the presence of the 19-kHz pilot signal is detected, and the detected signal is used to light a stereo-indicator light. When no pilot signal is present (indicating monophonic transmission), the L − R channel is inhibited and the L + R signal is supplied to both audio channels of the receiver automatically.

FM Stereo Time-Multiplex Adapter/Decoder

The time-multiplex adapter/decoder is an envelope-detection method wherein the composite L + R and L − R stereo signals are sampled at the phase-synchronous rate of 38 kHz. This approach to decoding has an advantage over the conventional matrix method in that neither a 15-kHz low-pass or a 23- to 53-kHz bandpass filter is required, and the balance problem is eliminated. This can result in significant cost reduction and improved system performance.

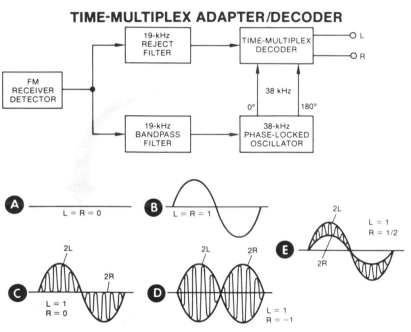

TIME-MULTIPLEX ADAPTER/DECODER

WAVEFORMS AT FM RECEIVER DETECTOR OUTPUT WITH PILOT SIGNAL REMOVED

Consider first the composite signal that exists at the input to the decoder for various combinations of L and R signals. For convenience, the 19-kHz pilot signal is not shown. If L and R signals are both zero, than L + R and L − R are also zero (A). Since the subcarrier is suppressed, the detector output (after the 19-kHz filter) is also zero (A). If L and R are equal and in phase (B), L − R equals zero, and sampling will yield identical outputs for both L and R. When L and R are not identical (C through E), the composite signal contains a 38-kHz component that has *envelopes* of 2L and 2R. In the time-multiplex system, the composite signal is sampled at a phase-coherent 38-kHz rate, and the envelope signals are thereby recovered—yielding the L and R channel outputs directly. The time-multiplex decoder or sampler samples the positive peaks at the 0-degree point (in phase) and the negative peaks at 180 degrees (out of phase) at the 38-kHz reference rate. As you can see from the illustration, this directly results in separation of the left and right audio signals at the output of the time-multiplex decoder without the need of the filters used in the matrix decoder described previously.

FM Stereo Time-Multiplex Adapter/Decoder (continued)

The operation of an FM stereo time-multiplex adapter/decoder can be illustrated by referring to the transformer-driver, diode-detector configuration shown. The composite multiplex signal is added to the in-phase (0-degree) 38-kHz reference in transformer T1 and fed to diodes CR1 and CR2. Similarly, the composite signal is added to the out-of-phase (180 degree) 38-kHz reference signal and fed to diodes CR3 and CR4. These two composite waveforms are shown in diagrams A and B.

The composite signals at points A and B are rectified by diodes CR1–CR2 and CR3–CR4, respectively, and produce the output signals shown by waveforms 1 through 4. The outputs from diodes 1 and 4 (waveforms 2 and 3) are summed to produce the left-channel output, and the outputs from diodes 2 and 3 (waveforms 1 and 4) are combined to provide the right-channel output. The two 150-K resistors (effectively in parallel) and the 0.001-μF capacitor constitute the 75-μsec de-emphasis network (75 K × 0.001 = 75 μsec).

TIME-MULTIPLEX ADAPTER/DECODER

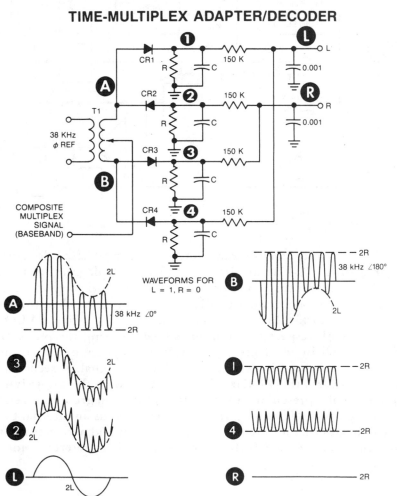

Modern IC FM Stereo Demodulators

The improvements in phase-locked-loop technology led to the development of IC stereo multiplex decoders that do not require the use of *any* resonant circuits, provide automatic stereo/monaural switching, and include a driver to operate the stereo-lamp indicator. Thus, the entire stereo decoding is done in a single IC. The heart of the IC stereo demodulator is a 76-kHz VCO whose output is counted down by three frequency dividers to provide a 38-kHz signal and two 19-kHz signals in phase quadrature.

MODERN IC FM STEREO DEMODULATOR

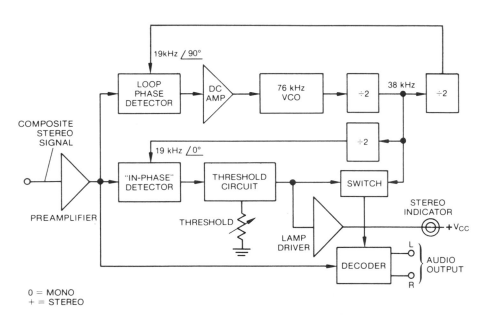

The composite stereo signal containing the 19-kHz pilot signal is compared to the 19-kHz quadrature signal (at 90 degrees) in the loop phase detector, and the resultant error signal is used to phase lock the 76-kHz VCO (RC-controlled oscillator). As you remember from your study of phase-locked loops in Volume 3, the output from a VCO is at 90 degrees with respect to the input signal. Shifting the phase of the feedback signal by 90 degrees produces an in-phase output from the VCO needed to properly decode the FM stereo signal. A second phase detector compares the 19-kHz in-phase signal with the pilot signal. If the pilot signal is present it will, by nature of the 0-degree phase relation, produce a unipolar dc output. If this output exceeds an internal threshold, it indicates that a stereo program is present and activates the indicator lamp and the switch that allows the 38-kHz reference into the decoder.

Modern IC FM Stereo Demodulators (continued)/Subsidiary Communication Authorization (SCA)

The phase-locked 38-kHz signal, switched into the decoder when a stereo program is present, is combined with the composite stereo input to provide the left- and right-channel signals, as described previously. The decoder can be either the matrix-decoder or time-multiplex configuration. De-emphasis for the L and R channels is normally applied external to the IC demodulator. Modern stereo FM receivers almost invariably use a single IC for FM stereo demodulation, although not all are specifically of the type described on the previous page. Some still require externally tuned components for operation.

Now let us move on to our next subject—the study of subsidiary communication authorization (SCA).

FM broadcast stations are often authorized to transmit a second set or more of DSB signals to carry special information. For example, special commercial messages, directed to specific places, or tones to cut out advertising can be transmitted via SCA (subsidiary-communication authorization) channels. These channels have a bandwidth of 10 kHz (± 5 kHz) and are formed just like the L $-$ R channel except that the carrier frequency is higher—for example, 67 kHz. This frequency is chosen so that the SCA channel falls between the L $-$ R channel upper-sideband limit (53 kHz) and the fourth harmonic (76 kHz) of the 19-kHz pilot signal, as shown in the diagram. As you can see, the SCA channel is only another sideband signal on the carrier which can be recovered by the filtering and demodulation techniques described earlier.

The SCA channel can have subcarriers on it which provide up to five separate information signals—for example, tones of various frequencies to allow for decoding at the receiver using appropriate filters. Thus, either voice transmissions and/or tones can be used to provide control information for use at the receiver. In many cases, music in public places like airports, restaurants, shopping malls, and department stores is obtained from FM stereo receivers. The SCA channel is used to provide signals that inhibit the audio output during commercials or news broadcasts so that only the music is heard.

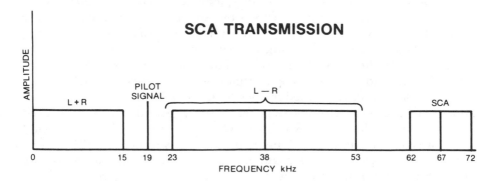

SCA TRANSMISSION

Automatic Frequency Control (AFC)

Because FM receivers operate in the VHF/UHF range, it is difficult to design the local oscillator so that it will be stable under different temperatures and line-voltage conditions, etc. Thus, many FM receivers are equipped with automatic frequency control (AFC). Of course, modern receivers using a frequency synthesizer do not need this feature because the local oscillator signal is stable and accurate. The AFC signal is derived from the discriminator output and is fed to the LO, which has an additional voltage-controllable tuning element such as a varactor to provide for limited tuning range (usually a few hundred kilohertz).

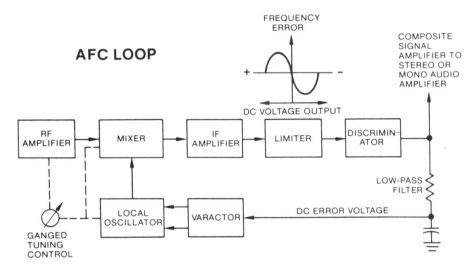

As you will recall, the output from some FM discriminators, such as the Foster-Seeley discriminator, is zero at the carrier intermediate frequency. With no modulation, if the carrier is mistuned, a dc offset (error) voltage is produced, and is proportional to the mistuning. This voltage can be used in a feedback loop to correct the tuning error, as shown in the diagram. If the receiver is not quite properly tuned or the LO drifts slightly, a dc voltage appears at the discriminator output in addition to the audio signals. A low-pass filter (typically, with a time constant of 0.1 to 0.5 second) filters out the audio signal, leaving only the dc frequency error component. If this is applied in the proper polarity to a voltage-controllable tuning element such as a varactor in the LO tuned circuit, the LO will be retuned until the dc voltage drops to zero (or very close to it), which means that the LO is properly tuned. Thus, the LO is kept properly tuned for best reception. Since the ratio detector does not have zero output at the correct tuning point, it cannot be used in a simple AFC loop as shown above. Receivers with an AFC loop usually have a switch to disable it. This is because it is sometimes difficult to listen to a weak station on a channel adjacent to a strong station because the AFC will pull the receiver tuning to the strong station.

Review of FM Receivers

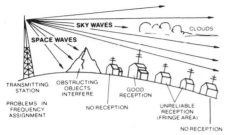

RECEPTION LIMITED TO LINE OF SIGHT

1. **FREQUENCY MODULATION** (FM) is used where high fidelity and noise immunity are important. For full benefit, it requires large bandwidths and is broadcast on VHF, where transmission is largely limited by *line of sight*.

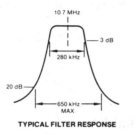

TYPICAL FILTER RESPONSE

2. **FM RF, LO AND MIXER STAGES**, except for frequency, are like those in other superhets. The *IF amplifier* must be carefully designed for the right *selectivity* curve.

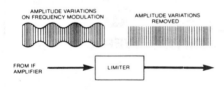

LIMITER AND DISCRIMINATOR FUNCTIONS

3. **FM DETECTORS** must produce a voltage that changes linearly with carrier frequency. Most popular are the *Foster-Seeley*, *ratio detector*, and *phase-locked loop* types. Some are sensitive to AM and must be preceded by a *limiter*.

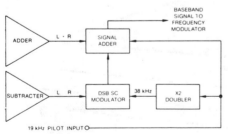

4. **STEREO REPRODUCTION** requires separate left and right channels. Transmission uses a *main channel* and a *subchannel*. Reception requires a *stereo multiplex demodulator*.

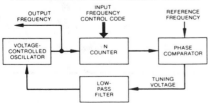

PHASE-LOCKED-LOOP DIGITAL TUNING

5. **AUTOMATIC FREQUENCY CONTROL** (AFC) uses an error signal from the discriminator to keep the LO on frequency. Some receivers use *PLL synthesizers*, which provide their own AFC.

Self-Test—Review Questions

1. Describe the basic difference between frequency and amplitude modulation.
2. Why is FM used?
3. What are the basic differences between an FM and an AM receiver?
4. Describe the operation of FM detectors. Why is limiting required? Discuss limiters. Compare various FM detectors.
5. Why is de-emphasis used? How does it function?
6. Describe how FM stereo is received and decoded.
7. What is SCA?
8. What is AFC? How does it work? Why is it needed?
9. Describe briefly how a digitally tuned FM receiver works. Do you need AFC on a digitally tuned FM receiver?
10. Why is FM confined to higher frequencies? Are FM broadcast antennas different from AM broadcast antennas?

Learning Objectives—Next Section

Overview—In the section that follows you will study communication receivers, including those for CB, fleet, amateur, and satellite communication.

TYPICAL FLEET
COMMUNICATION FM RECEIVER

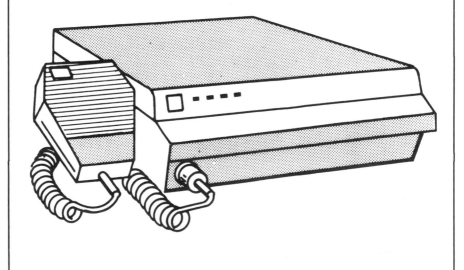

Introduction

This section will be devoted to a discussion of communication receivers. These receivers constitute the largest group of receivers except for the standard broadcast AM and FM radio receivers. In most cases, the transmitter and receiver are combined into a single transceiver to be used for *both* transmission and reception. This is done because costs can be reduced by using common circuits for both applications. The major categories and frequency assignments of public and private communication systems are shown in the table on the following page. Since, as indicated by its name, a communications receiver is primarily for reception of information, intelligibility is more important than fidelity and tone quality. Also, it must often operate at reduced input signal levels (compared to those in broadcast reception) and cope with high levels of interference.

The more than 10 million CB operators in the United States represent the largest group of communication receiver users. In order to minimize interference, CB transmitter power is limited to 4 watts, and there are 40 different frequencies of operation identified as channels 1 through 40. The CB frequency band is at approximately 27 MHz, with 10-kHz separation between channels.

Fleet communication systems represent the next largest group of operators. These systems are reserved for commercial trucking, taxi, and other basic service organizations, and their power output can be up to 50 watts to provide greater distance coverage. In this service, cochannel and adjacent channel interference is normally less severe than for amateur and CB operation, because frequency assignments are more restrictive.

Other large users of radio communication are public systems such as police, fire, medical emergency, and other community-service groups. These are mostly limited power systems, because their vehicle operations are usually only short distances from the home stations.

Many radio amateurs (over 250,000) use special bands for communications, as shown on the following page. Amateur radio operation includes almost all forms of communication, from telegraphy to TV. Amateur equipment includes some of the most sophisticated receiving systems, since much operation is at minimum signal level and at high density of frequency-band population.

Another special category includes the microwave relay stations used by the telephone, TV, and data-processing networks. Such systems also use satellite relay to speed messages around the world.

The receivers to be considered on the pages that follow operate on the same principles as those described previously. Only brief descriptions will be given of the individual stages, with the emphasis on special features and the interconnection of the stages to make up a complete system.

COMMUNICATION RECEIVERS

FREQUENCY CHART

Amateur

1,800	–2,000	kHz
3.500–	4.000 MHz	
7.000–	7.300	
14.00 –	14.35	
21.00 –	21.45	
28.00 –	29.70	
50.00 –	54.00	
144.0 –	148.0	
220.0 –	225.0	
420.0 –	450.0	
1,215	–1,300	
2,300	–2,450	
3.300–	3.500 GHz	
5.650–	5.925	
10.00 –	10.50	
24.00 –	24.25	
48.00 –	50.00	
71.00 –	84.00	
152.0 –	170.0	
200.0 –	220.00	
240.0 –	250.00	

Above 275.0

Citizens Radio (Personal radio services)

26.96 – 27.23	MHz	
462.5375–462.7375		
467.5375–467.7375		

Land Mobile (Communication on land between base stations and mobile stations or between mobile stations)

Land Transportation (Taxis, trucks, buses, railroads)

30.56 –	32.00	MHz
33.00 –	34.00	
43.68 –	44.61	
150.8 –	150.98	
152.255 –	152.465	
157.45 –	157.725	
159.48 –	161.575	
451.0 –	454.0	
456.0 –	459.0	
460.0 –	462.5375	
462.7375–	467.5375	
467.7375–	512.0	
1,427	–1,435	

Broadcast Remote Pickup

1,605	–1,715	kHz
26.10 –	26.48	MHz
161.625 –	161.775	
166.25		
170.15		
450.0 –	451.0	
455.0 –	456.0	

Land Mobile (Communication on land between base stations and mobile stations or between mobile stations)

Public Safety (Police, fire, highway, forestry, and emergency services)

1,605	–1,750	kHz
2,107	–2,170	
2,194	–2,495	
2,505	–2,850	
3.155 –	3.400	MHz
30.56 –	32.00	
33.01 –	33.11	
33.41 –	34.00	
35.19 –	35.69	
37.01 –	37.43	
37.89 –	38.00	
39.00 –	40.00	
42.00 –	42.95	
43.19 –	43.69	
44.61 –	46.60	
47.00 –	47.69	
150.98 –	151.4825	
153.7325–	154.46	
154.6375–	156.25	
158.715 –	159.48	
162.0125–	173.2	
451.0 –	454.0	
456.0 –	459.0	
460.0 –	462.5375	
462.7375–	467.5375	
467.7375–	512.0	
1,427	–1,435	

Industrial (Power, petroleum, pipeline, forest products, factories, builders, ranchers, motion picture, press relay, etc.)

1,605	–1,750	kHz
2,107	–2,170	
2,194	–2,495	
2,505	–2,850	
3.155 –	3.400	MHz
4.438 –	4.650	
25.01 –	25.33	
27.28 –	27.54	
29.70 –	29.80	
30.56 –	32.00	
33.11 –	33.41	
35.00 –	35.19	
35.69 –	36.00	
37.00 –	37.01	
37.43 –	37.89	
42.95 –	43.19	
47.43 –	49.60	
151.4975–	152.0	
152.465 –	152.495	
152.855 –	153.7325	
154.46 –	154.6375	
157.725 –	157.755	
158.115 –	158.475	
173.2 –	173.4	
216.0 –	220.0	
451.0 –	454.0	
456.0 –	459.0	
460.0 –	462.5375	
462.7375–	465.5375	
467.7375–	512.0	

Citizens Band (CB) Radio

The purpose of the CB radio is to provide citizens with a means of communicating meaningful information such as highway and weather conditions while traveling, reporting breakdowns or sickness, and allowing the traveler to communicate with the outside world. CB is not restricted to mobile uses, and it can be used between fixed stations as well.

There are over 10 million CB sets licensed for use in the United States today. Most are transceivers (transmitter/receiver combinations). Manufacturers of CB transceivers provide information to the buyer of the restrictions and licensing requirements for each transceiver. Actually, it is only the transmitter portion that requires licensing. Most manufacturers supply a license application form with the transceiver. No special skill is required for CB operation, but a knowledge of rules and regulations is required. CB operates at about 27 MHz, and CB transceivers use AM for information transfer.

With a mobile installation and an average antenna, the signal doesn't usually travel more than 5–10 miles (8–16 km) and often much less. Currently, there are 40 channels available in the United States, and maximum output power of 4 watts for either fixed or mobile stations.

U.S. Citizens Band Frequencies (40-Channel)

CHANNEL	FREQUENCY (MHz)	CHANNEL	FREQUENCY (MHz)
1	26.965	21	27.215
2	26.975	22	27.225
3	26.985	24··	27.235
4	27.005	25··	27.245
5	27.015	23··	27.255
6	27.025	26	27.265
7	27.035	27	27.275
8	27.055	28	27.285
9	27.065	29	27.295
10	27.075	30	27.305
11	27.085	31	27.315
12	27.105	32	27.325
13	27.115	33	27.335
14	27.125	34	27.345
15	27.135	35	27.355
16	27.155	36	27.365
17	27.165	37	27.375
18	27.175	38	27.385
19	27.185	39	27.395
20	27.205	40	27.405

··NOTE VARIATION IN SEQUENCE.

The FCC, which is the U.S. regulating agency, has some sophisticated equipment for monitoring CB operation, and their prime concern is to encourage operators to follow the rules so that everyone can use CB for the purpose it was designed—citizens' communications.

Citizen Band (CB) Radio (continued)

Prior to expansion of the band to 40 channels in 1977, CB receivers and transmitters were limited to 23 channels. These ranged in frequency from 26.965 to 27.255 MHz, or a total spread of less than 300 kHz. Currently, 40 channels are available covering the bands from 26.965 MHz to 27.405 MHz in 10-kHz steps, as shown on the previous page. The receiver to be described here is the receiver section of the 40-channel CB transceiver that you studied in Volume 3. There, you learned how the frequency synthesizer operated to provide for 40-channel operation with just three crystals. You also learned that the IF was 4.3 MHz, and that the synthesizer basically produced the LO signal; the transmit signal was generated by mixing this LO signal with the output from a 4.3-MHz crystal oscillator. The sum signal from this mixer (LO + 4.3 MHz) provides the transmit signal.

SIMPLIFIED BLOCK DIAGRAM OF RECEIVER IN CB TRANSCEIVER

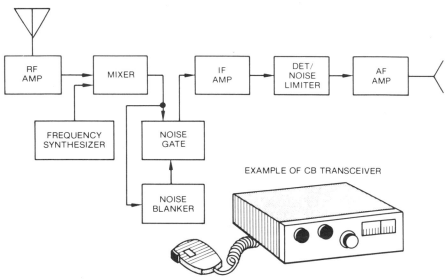

EXAMPLE OF CB TRANSCEIVER

This receiver contains signal-processing techniques and special features widely employed in modern AM communications receivers and is, therefore, a good medium for introducing this subject. A feature of the receiver is the synthesizer that allows for operating on any of the 40 channels by merely setting a selector switch for the desired channel.

Other features are a noise-blanking circuit to eliminate noise when no signal is being received, audio noise limiting, and the use of crystal filters in the IF section to provide sharply tuned IF amplification (which prevents interference from adjacent channels). Otherwise, the circuits and their arrangement are essentially the same as those described for a conventional AM receiver.

CB Radio RF Amplifier and Mixer

Input signals to the receiver are obtained when the transmitter is not keyed on. In the receive mode, the antenna is automatically connected to the receive input circuit. The antenna signal is coupled through the tuned transformer T1 to the base of RF amplifier Q1. Automatic gain control is provided from the Q21 collector to the emitter of the RF amplifier to compensate for input-signal level variations. The ouput of the RF amplifier is tuned-transformer coupled to the base of the receiver mixer Q2. At the same time, the proper synthesized local oscillator signal is also coupled from the cathode of CR201 to the base of Q2. Automatic gain control is also provided to the mixer from the Q22 collector. Thus, the mixer receives the incoming signal, and the switching of the local oscillator determines the channel selection. The output of the mixer Q2 is applied to both T11 of the noise-blanking circuit and to IF transformer T3. The IF transformer is tuned to 4.3 MHz, which is the difference between the channel operating frequency and the output frequency of the synthesizer.

CB RECEIVER RF AMPLIFIER AND MIXER STAGES

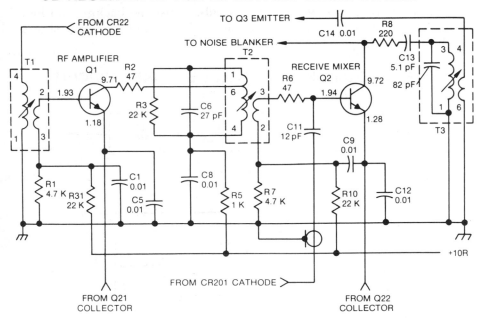

While the tuned circuits in the front end of the CB receive portion are adjustable, they are not adjusted as the channel is changed. Instead, the RF tuned circuits are made sufficiently broadband so that no tuning is required over the range of input frequencies (approximately 27 to 27.5 kHz). Thus, it is the setting of the LO frequency in conjunction with the high selectivity of the IF circuits that provides the frequency selection and selectivity of the receiver.

CB Radio Noise Blanker

The noise-blanking circuit in a communication receiver is used when receiving weak signals in the presence of noise pulses or impulse noise such as automobile ignition interference. It effectively gates the receiver off during the period of the noise pulse. If the impulses are short, the gated intervals are also short, and the impulse noise is deleted without loss of intelligibility. As with most noise blankers, the system described here uses a separate relatively broadband IF amplifier to operate the noise-gate generator, which in turn gates the IF amplifier input. The noise-gate IF is relatively wide-band to allow the circuit to react before the narrow-band IF amplifier. The separate IF is necessary also to be able to detect the duration of the noise pulse to establish the gate time in the signal IF. When the blanker is switched out, the signal from the mixer is coupled directly to the first IF stage. Setting the switch to the noise-blanking position couples the signal from the mixer through two noise-blanker IF stages, a noise detector, pulse amplifier, and noise gate.

Noise-blanking circuits are only useful when the noise is *impulsive*— that is, when it consists of sharp spikes of noise like ignition noise or electrical equipment noise. It is *not* useful against the background hiss noise generated by the receiver. The block diagram on this page shows the basic elements in the noise blanker.

BLOCK DIAGRAM OF NOISE BLANKER CIRCUIT

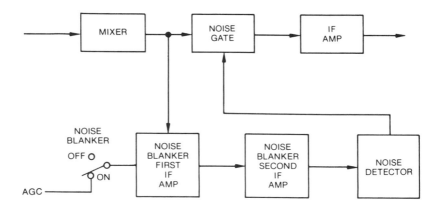

The mixer output is fed to a noise gate that is between the mixer and the first signal IF. The mixer also feeds the signal to the noise-blanker IF amplifier, which in turn feeds a second noise IF amplifier. These differ from the signal IF in having a wider bandwidth. The first noise-blanker IF is decreased by the AGC voltage when strong signals are present, since noise blanking is less necessary in this case. The IF output from the second noise IF amplifier is fed to the noise detector, which detects pulses exceeding the threshold. The gate pulses generated are fed to the noise gate to blank the IF input to the signal IF amplifier.

CB Radio Noise Blanker (continued)

With the noise-blanker circuit switched in, AGC voltage turns on the first noise-blanker IF amplifier Q10. The mixer output is coupled through the tuned IF transformer T11 and amplified in two noise IF stages, Q10 and Q11. The amplified signal is detected by the noise detector CR9. CR9 conducts only on the positive portions of the IF signal, charging C62 to a positive voltage that corresponds to the peak amplitude of the noise pulses. The audio portion of the signal is removed by the audio filter consisting of R54, C62, C63, and R57. The detected positive voltage forward-biases the pulse amplifier Q12. When Q12 conducts, the collector voltage decreases, and this negative-going pulse is coupled through pulse-shaper network C65, L5, and C15 to base of noise gate Q3.

REARRANGEMENT OF NOISE BLANKER CIRCUIT

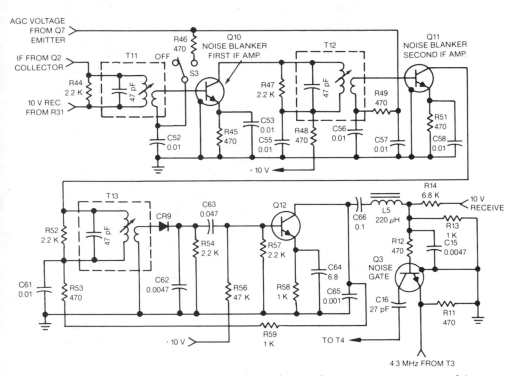

A negative-going voltage on the base of Q3 acts as a reverse bias to cause Q3 to cut off. Since the negative bias on Q3 corresponds to IF noise pulses present on the Q3 emitter, the noise gate is cut off only when the noise pulse or pulses are present.

The AGC voltage is connected to Q10 and Q11, the noise-blanker first and second IF stages, to vary the gain of the noise-blanker amplifiers. The negative AGC voltage acts as a reverse bias to decrease the gating-pulse amplitude when a large amplitude signal is received. Such a large amplitude signal overrides the noise, and the noise-blanker circuitry is not necessary while receiving high-level signals.

CB Receiver IF Amplifiers

Bandwidth and selectivity of the receiver are controlled by two 4.5-MHz crystal filters and the three IF amplifiers that have coupling transformers also tuned to 4.5-MHz.

When the noise-blanking circuit is not gating the signal off, the mixer output is coupled through the noise gate Q3 to the crystal input filter through matching transformer T4. The output of the filter is also impedance-matched to the first IF amplifier by tuned transformer T5.

CB RECEIVER IF AMPLIFIER

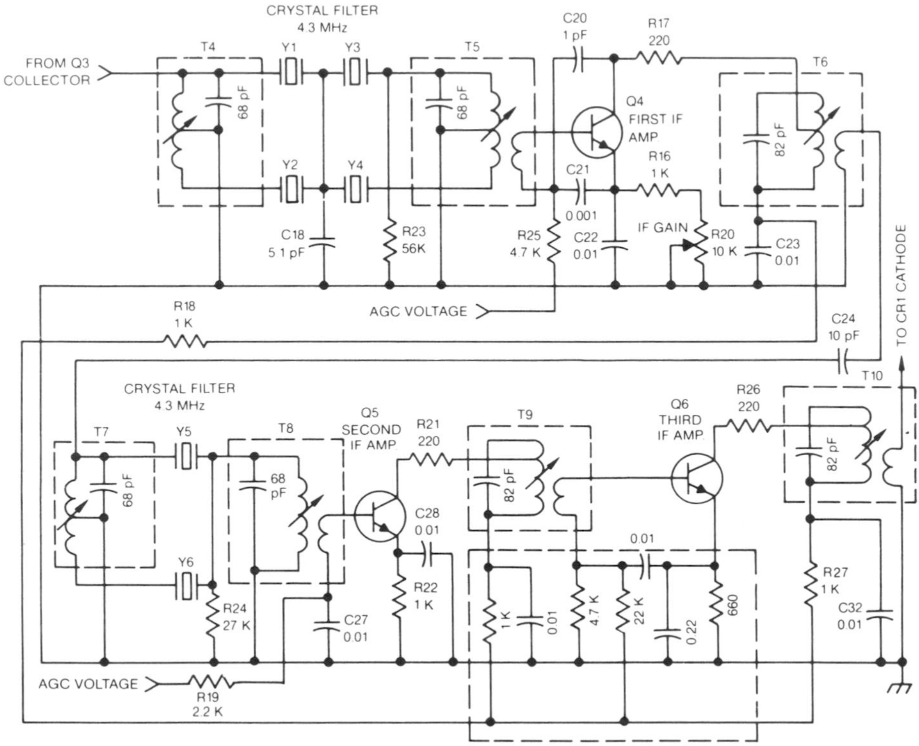

The first IF amplifier amplifies the IF signal to a sufficient level to be filtered again by the second crystal filter. Tuned transformers T7 and T8 match the input and output impedances to provide maximum signal transfer. The crystal filters together with transformers T4 through T10 provide a flat-top, steep-side IF response that establishes the high selectivity necessary to avoid interference from adjacent channels only 10-kHz away. The choice of the 4.3-MHz IF provides good image rejection because of the selectivity of the RF amplifier. While double conversion could have been used to achieve selectivity, the inexpensive crystal filter in the IF provides the same benefits without the additional complication. Provisions are made to apply an AGC voltage to the IF stages to hold the signal level constant over a wide range of input-signal strength.

CB Detector, Noise Limiter, and Audio Switch

The amplified signal from the third IF amplifier Q6 is coupled through IF transformer T10 to CR1, the audio detector shown below. With the IF signal applied, current flows through CR1 only during the negative portion of the IF signal. This causes the 0.047-μF capacitor in the temperature-compensated RC network, U2, to charge to the peak value of the rectified voltage on each negative half cycle. This capacitor and C35 (0.001 μF) shunt the 4.3-MHz component of the IF to ground and effectively filter all IF signals. The time constant of the RC network in CR1 output allows the audio modulation to pass unhindered.

CB RECEIVER DETECTOR, AUDIO NOISE LIMITER, AND AUDIO SWITCH

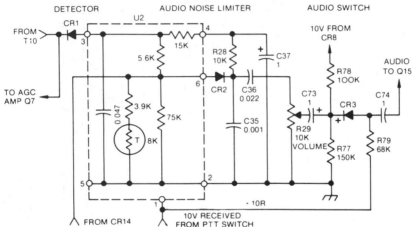

Diode CR2 provides noise limiting. It is forward biased by the 10-volt receive line through the resistive divider in the RC network U2. Bias is set so that CR2 conducts and passes the audio signal developed by CR1 through the 5.6-K resistor in U2, through CR2 and C36, to the manual volume control R29. The RC circuit associated with C37 has a very long time constant, so a dc voltage proportional to the peak carrier level is developed across C37.

A portion of the audio signal discharges C37 and develops a current through R28 to keep CR2 conducting. On noise spikes that exceed the nominal peak carrier (modulation) level, the voltage on C37 does not change, but since the noise spike is negative, CR2 will be reverse biased and stop conducting momentarily on all signals (for example, noise spikes that exceed the peak carrier level). Peak clipping occurs on modulation and noise peaks. This prevents peaks from interfering with audio reception.

After being coupled through C36 to the manual volume control, the signal is coupled through the audio switch diode CR3. This diode is biased on by the 10-volt receive line through R79. Capacitor C74 (1 μF) couples the audio signal to the preamplifier. The audio switch is biased off during transmission by applying the 10-volt transmit-on signal to the cathode of CR3 through R78.

CB Receiver Audio Section

The audio signal from the detector is RC coupled to the base of the audio preamplifier Q15, where it is amplified and fed to the power amplifier U3. C78 and R84 in the collector circuit of Q15 attenuate frequencies above 3 kHz. C81 and R85 form a high-pass filter and attenuate signals below 500 Hz.

EP3 is a ferrite bead, which, when slipped over a wire, increases the inductance of the wire lead. EP3 and C76 act as a filter to prevent any RF signals that may be picked up by the wiring during transmit from being detected by the emitter-base junction of Q15 and appearing as an audio output.

CB RECEIVER AUDIO PREAMPLIFIER

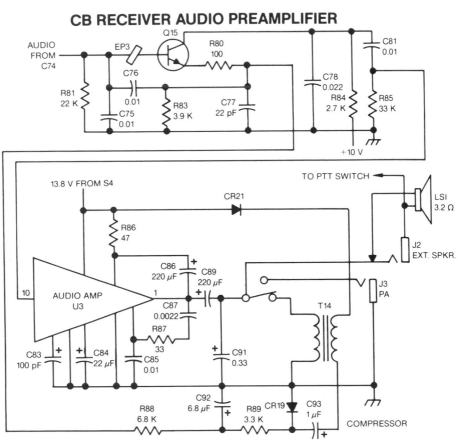

The audio amplifier U3 is an IC linear operational amplifier that produces 4 watts of output power at pin 1 of U3. This output is coupled through the external speaker jack J2 to the loudspeaker. Note that the speaker has an adjacent jack J2 for connecting an external speaker. By using J3, an external power amplifier can be connected to U3. As you learned in Volume 3, this audio power amplifier is also used as the high-level modulator for transmission. Transformer T14 drives the collector circuits of the final power amplifier in the transmit mode.

CB Automatic Gain Control and Squelch

Automatic Gain Control (AGC) is developed by coupling a portion of the signal from the IF output T10 to the AGC detector CR4 and amplifiers Q7, Q21, and Q22. AGC is applied to Q1, Q2, and Q5 in the receiver and Q10 and Q11 in the noise-blanker circuit.

A portion of the amplified AGC voltage is applied to the squelch amplifier Q13 and squelch gate Q14. The squelch gate disables the audio preamplifier by reverse biasing Q15 when there is no *on-frequency* signal at the receiver. The squelch level R73 is adjusted by the operator so as to cut off any signals that are considered to be too weak to be of interest.

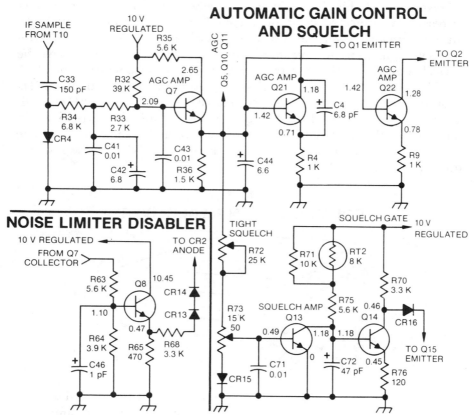

When a sufficiently strong signal is received, AGC voltage is developed on the emitter of Q7, and it is applied to Q13 at the level selected by R73. This causes Q13 collector voltage to increase and forward biases squelch Q14. When Q14 conducts, Q15 conducts and signals pass.

The audio-noise limiter-disabling circuit Q15 places a positive voltage on the anode of the audio-noise limiting diode CR2 discussed previously. This turns on CR2 to prevent clipping and distortion. When a strong signal is received, positive voltage is developed at the collector of AGC amplifier Q7. This turns on Q8, causing CR13 and CR14 to conduct, thus keeping CR2 in conduction—and therefore disabling its limiting function when strong signals are received.

Fleet Communication FM Receivers

Fleet communication receivers serve trucking and taxi fleets and service organizations. They operate in the 150-MHz range and the 450- to 512-MHz region, but other ranges are used as well. In the 450- to 512-MHz range, narrow-band frequency modulation (± 5 kHz) is usually used. These sets generally consist of a base station, which can operate at levels of up to 50 watts, and mobile units operating at power-output levels of up to 15 watts. More than one base station can be used at the same frequency or in the same band, and almost all units are transceivers.

The receiver presented here as an example is similar to the FM receiver described earlier, except for several special circuits. It receives either of two selected crystal-controlled channels at about 450 MHz, although many newer sets use frequency synthesizers. Oscillator frequency-tripling circuits, often used in transmitters, are used to raise the crystal-controlled oscillator frequency to the UHF operating band. A double-conversion superhet receiver is used with a first IF of 10.7 MHz (standard for FM receivers) and a second IF frequency of 455 kHz (standard for AM receivers). This dual-conversion technique is often used in receivers in the VHF–UHF range and above and combines the image rejection of the high-frequency IF with the selectivity of the low-frequency IF.

TYPICAL FLEET
COMMUNICATION FM RECEIVER

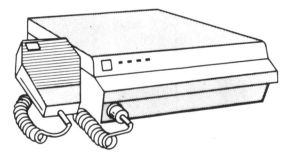

Other special circuits are the squelch filter, squelch gate, and squelch tail eliminator. The squelch circuits keep the receiver audio amplifier turned off until a carrier is received. This avoids the necessity of listening to receiver noise when no signal is present. Another special feature of this system is the *call guard*. When no communication is being received, the receiver audio output is turned off. A special call-guard tone is transmitted with the message intended for a specific receiver in a network or net of receivers. When this tone is received, the receiver audio output is turned *on*, and the message can be heard. With the squelch and call guard operating, only messages intended for a particular unit or group of units will be received.

Fleet Communication FM Receivers (continued)

The figure below is a block diagram of the receiver portion of the E. F. Johnson Fleetcom® FM transceiver. You will recall that you studied the transmitting portion of this system in Volume 3. As shown, the signal input from the antenna is filtered in a bandpass filter to reject undesired signals and the image signal. An RF amplifier is omitted in this receiver to reduce cost. The output of the bandpass filter feeds the first mixer directly. The first local-oscillator signal is crystal controlled and produces a first IF signal at 10.7 MHz to the high-frequency IF amplifier. This provides good image rejection. The amplified 10.7-MHz signal is then converted to 455 kHz via the second mixer and second local oscillator. This 455-kHz signal is then amplified to provide high selectivity, (better than that readily obtainable at 10.7 MHz). The output signal from the second IF is applied to a limiter/discriminator to produce an audio output.

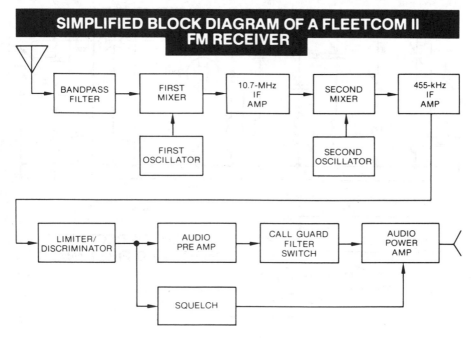

SIMPLIFIED BLOCK DIAGRAM OF A FLEETCOM II FM RECEIVER

The audio signal from the discriminator is preamplified and fed to the audio amplifier. The preamplifier output is also applied to the call-guard filter. The audio output is restricted if the call-guard tone is absent, even though a signal may be present. A call-guard filter in the audio line suppresses the tone during reception. The squelch line operates from the discriminator output. When the signal level exceeds the manually set squelch threshold, the audio amplifier is enabled to allow the signal through. Both the squelch and call guard can be disabled by the operator when desired.

Fleet Communication RF Filter/First Mixer/First Oscillator/ Triplers

As shown below, the antenna relay couples the antenna to the receiver input when in the receive mode. The incoming RF signal at about 450 MHz is coupled through antenna relay K1 and a bandpass filter, tuned by C101, C102, C103, C104, and C105, to the first mixer. The bandpass filter is wide enough in response to receive either channel without retuning but narrow enough to avoid the image signal 21.4 MHz away.

The first mixer, Q103, mixes the RF input signal with the output of the third local-oscillator tripler (Q101). The first mixer uses a MOSFET described earlier. The first oscillator frequency is chosen to produce an IF signal at 10.7 MHz. The mixer output is then coupled to the 10.7-MHz IF amplifier, Q201, through a conventional IF input transformer tuned to 10.7 MHz.

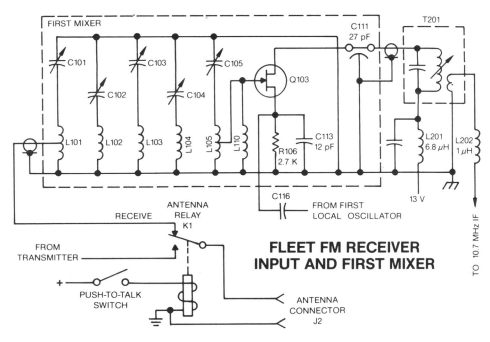

FLEET FM RECEIVER INPUT AND FIRST MIXER

The figure on the following page is a diagram of the first oscillator. Channel selector switch S2 grounds the emitter resistor of the appropriate channel oscillator—Q211 for channel 1 or Q212 for channel 2. Both oscillators are crystal controlled using a modified Colpitts circuit. Their outputs are tuned to the third harmonic of the crystal, and they function as frequency triplers. Transformers T208 and T209 provide the output tuning. A passive temperature-compensating technique is used with RT201 to control the effective capacitance of C291. The frequency is adjusted by C269 and can be extended by changing the value of capacitor C271.

FLEET FM RECEIVER
FIRST OSCILLATOR AND FREQUENCY TRIPLERS

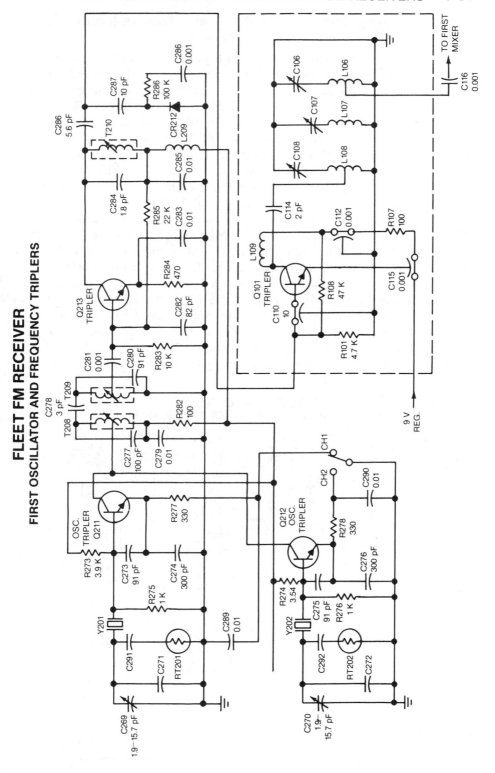

Fleet Communication First Oscillator and Triplers (continued)/ First IF Amplifier and Second Mixer

On the previous page, the tripled output of the oscillator is coupled through transformers T208 and T209 to the base of the second tripler Q213 through capacitor C281. The second tripler output, tuned to nine times the oscillator frequency, is coupled through transformer T210 and capacitor C286 to the base of the third tripler Q101. The signal from the third tripler (27 times the crystal frequency) is filtered by C106, L106, C107, L107, C108, and L108 to remove the undesired harmonics. The filtered signal is coupled to the source of the MOSFET first mixer Q103 through C116. The crystals have fundamental frequencies (in megahertz) equal to the channel frequency ± 10.7 MHz/27.

As shown below, transistor Q201 is a FET amplifier tuned to 10.7 MHz. It amplifies the signal at 10.7 MHz prior to conversion to the second IF. The output is tuned by L204 and coupled to a crystal filter Z201 by capacitor C204. Z201 has a center frequency of 10.7 MHz and a bandwidth of 13 kHz. Matching to the second mixer is adjusted by L205. The second mixer Q202 accepts the 10.7-MHz signal output from the crystal filter and the output of the second oscillator. The second local oscillator is tuned to 11.155 MHz, and the mixer produces a difference frequency of 455 kHz. The mixer output is coupled to the low-frequency IF amplifier Q203 by transformer T202 tuned to 455 kHz.

FLEET FM RECEIVER
FIRST IF AMPLIFIER, CRYSTAL FILTER, AND MIXER

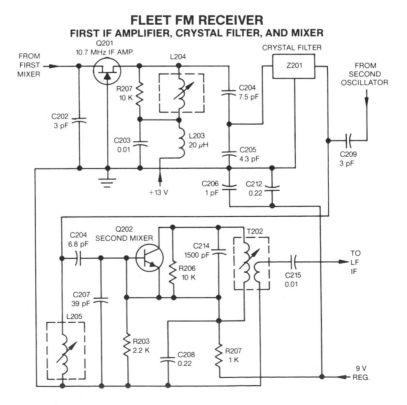

Fleet Communication Second Oscillator and IF Amplifiers

As shown in the schematic diagram below, the second oscillator Q214 operates as a parallel-mode Colpitts oscillator with the frequency controlled by crystal Y203 and positive feedback controlled by C263 and C264. The crystal operates as a parallel-resonant element and oscillates at 11.155 MHz (or 10.245 MHz as an optional *low-side* injection frequency). The signal from the second oscillator Q214 is coupled to the base of the second mixer Q202 through capacitor C209.

The second mixer output, containing the original frequencies plus the sum and difference of these frequencies, is tuned to the difference frequency, 455 kHz, by the tuned IF transformer T202. The signal is amplified by the two common-emitter IF amplifiers Q203 and Q204. The output of the second IF amplifier is coupled to the limiter/discriminator IC U201 by IF transformer T204.

FLEET FM RECEIVER SECOND OSCILLATOR AND 455kHz IF AMPLIFIER

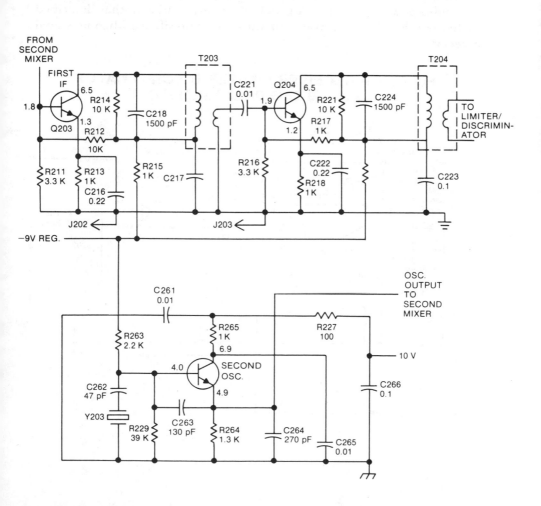

Fleet Communication Limiter/Discriminator and Audio Preamp

As shown in the schematic diagram below, U201 is a monolithic IC that includes a three-stage FM IF amplifier/limiter, a doubly balanced quadrature FM detector, and an audio preamplifier. The limited IF signal is coupled from pin 8, through C232, to the 455-kHz tuned network (C239/T205). This tuned circuit is external to the IC because of its size. A signal is fed into pin 9, 90 degrees out of phase with the original IF signal, which is combined with the normal (in-phase) IF signal in a double-balanced mixer in the U201 FM discriminator. The audio modulation (recovered as you learned in your study of FM detectors) is amplified in U201 and then fed to the de-emphasis network C230, C237, C255, R232, R233, and R234—which provides about 12 dB per octave noise reduction above 3 kHz.

The de-emphasis network plus feedback network C247 and R231 across amplifier U202A provide approximately 6 dB per octave de-emphasis from 700 to 3,000 Hz and compensate for pre-emphasis provided in the transmitter. The amplifier U202A provides an audio output for the audio power amplifier. A squelch circuit, similar to that described earlier for the CB transceiver, provides for audio silence when no signal is received.

FLEET FM RECEIVER LIMITER-DISCRIMINATOR AND AUDIO PREAMPLIFIER

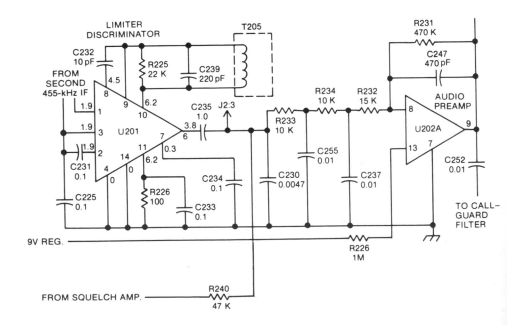

Fleet Communication Audio Section

The *call-guard* feature of the transceiver uses low-frequency audio tones to activate this feature. These are removed from the audio signal by the call-guard filter. The *call-guard* filter U202B functions as a high-pass filter with a low-frequency cutoff of approximately 340 hertz to remove the *call-guard* tone from the receiver audio output. The output of this stage is coupled to the manual volume control.

The audio signal passed by the call-guard filter is fed through the volume control R241, amplified by audio amplifier Q206 and audio driver Q207, and then coupled by transformer T206 to the audio output stage.

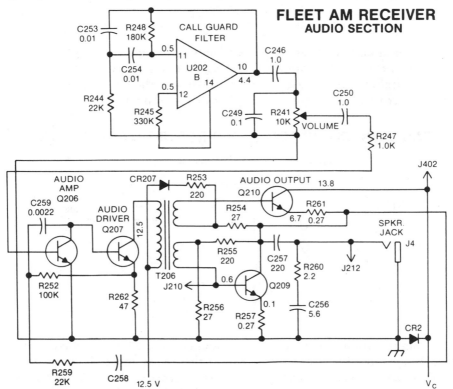

FLEET AM RECEIVER
AUDIO SECTION

Transformer T206 provides the split phase required to drive push-pull output amplifiers Q209 and Q210. Operating bias is developed by R254 and R256 for Q210 and Q209, respectively. A dc bias stability is provided by R257 and R261.

Negative-feedback network C258 and R259 provides for better linearity and frequency response of the audio amplifier, which results in better sound quality. Capacitor C258 couples the amplified audio signal to the speaker jack. Series network C256 and R260 prevents oscillation caused by the inductive effects of the speaker voice coil.

Amateur Radio/Short-Wave Receivers

Amateur radio is very popular in many countries and offers a way for individuals to communicate by radio and to experiment with transmission and reception. Radio amateurs have made many important contributions to radio and communications technology. The amateur-radio frequency bands are distributed over the spectrum, with the more important ones lying between 1,800 kHz and the UHF region. The table below shows the lower frequency bands that are most popular:

1.8 - 2.0 MHz	28.0 - 29.7 MHz
3.5 - 4.0 MHz	50 - 54 MHz*
7.0 - 7.3 MHz	144 - 148 MHz*
14.0 - 14.35 MHz	220 - 225 MHz*
21.0 - 21.45 MHz	*USUALLY INDIVIDUAL RECEIVERS.

To accommodate this range of frequency coverage in a single receiver requires a method for rapidly changing from one band to another. The band-switching receivers use a selector switch in the RF section to change the inductors to allow for band switching, as shown.

BANDSWITCHING FOR WIDE FREQUENCY COVERAGE

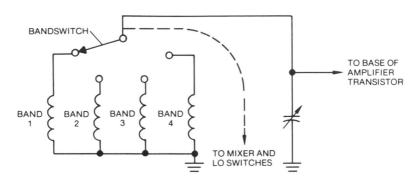

Similar inductor/switch arrangements are used at the mixer input and in local-oscillator circuit to switch these appropriately at the same time. Usually, these switches are ganged together so that a single control is used to switch rapidly and conveniently from band to band. In some cases, the band coverage is continuous; that is, the receiver bands are arranged so that the next band begins where the previous band left off. In this way, the entire short-wave range from the broadcast band through the Citizens Band and higher can be covered in a single receiver. These receivers are used to listen to short-wave broadcasts and communications. Usually, special filter circuits in the IF allow for adjustment of the IF bandwidth to match the signal bandwidth so that interference can be minimized. A BFO (beat-frequency oscillator) tuned a few hundred hertz from the IF can be introduced at the second detector. This allows CW signals to be converted to tones by the mixing action at this second detector, which produces a beat-frequency output between the IF and the BFO.

Receivers for Public Service/Navigation/Communications/ Weather

There are many specialized navigation systems that use the low-frequency ranges below the standard broadcast band because of their stable propagation characteristics. In addition, there are ship emergency frequencies (for example, 500 kHz) used internationally for distress calling. The VHF–UHF range is used for any number of specialized public services. For example, the VHF region slightly above the FM broadcast band is used for air-to-air and air-to-ground communication and navigation. The VHF frequency range between 148 and 174 MHz is used extensively for police and fire department communication, as well as the UHF bands. These bands are also used by boats for intercommunication and for communication with the U.S. Coast Guard in coastal waters.

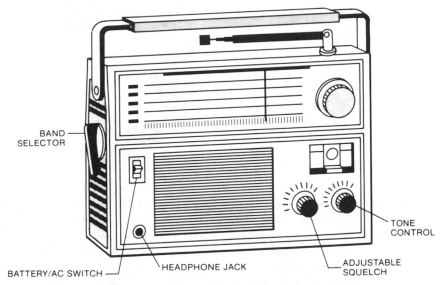

BAND SELECTOR

TONE CONTROL

BATTERY/AC SWITCH

HEADPHONE JACK

ADJUSTABLE SQUELCH

Of particular interest is the weather broadcasts in the 170-MHz range provided by the U.S. Weather Service. These broadcasts are continuous and provide instant access to local weather as well as weather forecasting. The weather information is provided on a 24-hour basis so that all within range of these stations can get weather information as needed. While these are very useful to individuals who own a suitable receiver, they are very important to ships and pleasure boats in coastal waters since they provide information on storms and other marine conditions important to the boat operators.

The National Bureau of Standards radio stations WWV provide transmissions at 5, 10, 15, and 20 MHz. These stations not only provide very accurate radio transmission for equipment calibration, but also provide time signals and special information via propagation.

All of these systems use receivers that do not differ materially from the receivers you have already studied. The only differences lie in the *antenna system* and the *tuning range* of the RF portion of the receivers.

Satellite Systems

As you know, radio-frequency propagation at VHF and above is pretty much limited to line of sight. On the other hand, the available spectrum space to carry the increasing load of information is severely limited at longer wavelengths. Furthermore, except at low frequencies, operation in the high-frequency range is often unreliable. Satellite relays solve the line-of-sight problem very neatly and allow the use of the spectrum space available at VHF and above. Nowadays, most communication satellites are in synchronous orbits—that is, they stay in one spot relative to a point on the earth. The altitude for a synchronous orbit is about 22,000 miles. While it may seem that this is doing things the hard way, it is important to realize that *only three satellites* can provide for communication to *any point on the earth*, including ships at sea and aircraft!

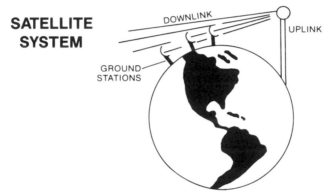

SATELLITE SYSTEM

Modern communication satellites are really complex relay stations. They receive transmitted signals from appropriate ground transmitting stations, convert these to signals at a different frequency, amplify the signals, and rebroadcast these to appropriate ground receiving stations. Both AM and FM transmission are used. Since highly directive, high-gain antennas can be used both for transmitting and receiving at the ground station end, the transmitter power required at the ground transmitter and in the satellite is not great.

A wide variety of information is sent via satellites, including not only communications such as telephone calls but also TV programs, radio programs, the facsimile editions of entire newspapers, digital data, or any other information that is desired. The use of satellite relays is increasing, and nowadays they are replacing other means of relay for long distances.

Satellites are owned either by governments or by corporations. Because of the great expense of building and orbiting a satellite, many corporations owning satellite facilities are actually formed by several companies to share the cost and the facilities. These special-communication corporations then rent use of these facilities just as land communication facilities (wire and coax cable) are rented for use by private corporations.

Single-Sideband Suppressed-Carrier Receivers

Single-sideband (SSB) transmission is widely used for communication, as you learned in Volume 3 on transmission systems. The basic differences between a single-sideband suppressed-carrier receiver and a conventional receiver are (1) increased selectivity because only one sideband is to be received and (2) the carrier must be reinserted at the detector for the detection process to take place. The E. F. Johnson Company Viking 4740 is a CB transceiver that has a single-sideband transmit and receive mode. Refer to Volume 3 for the details of the SSB transmitter portion. As you may remember, in this transceiver a crystal filter is used to suppress the unwanted sideband, while the carrier is suppressed by a balanced modulator. The same filter is used on receive to filter out unwanted interfering signals on the unused sideband. Thus, the filter FL501 is multiplexed for both transmit and receive, but is used in the same way for both. You will recall that the transceiver uses a frequency synthesizer to produce a local-oscillator output signal 7.8 MHz below the transmitted frequency.

FILTER AND DETECTOR SYSTEM FOR SSB RECEPTION

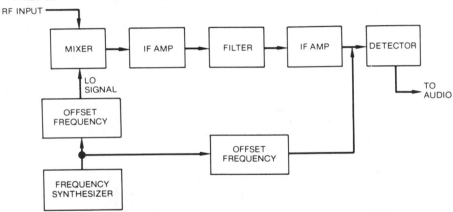

The figure shown is a partial block diagram of the filter and the detector system for SSB reception. Rather than use two filters [one for the upper sideband (USB) and one for the lower sideband (LSB)], a single filter is used, and the local-oscillator frequency is shifted slightly to allow for reception of either sideband. When this is done, the injected detector carrier must also be shifted; otherwise the sideband data for one case will also be inverted. With these exceptions, the transceiver circuits are essentially identical to the transceiver receiving circuits you studied earlier. The table shown below indicates the frequencies used for DSB, USB, and LSB transmission and reception.

MODE	CARRIER FREQ	SYNTHESIZER OUTPUT FREQ (LO)	IF SIGNAL BAND	DETECTOR CARRIER
DSB	f_o	$(f_o) - 7.8$	$7.7975 - 7.8025$	
LSB	f_o	$(f_o - 0.0025) - 7.8$	$7.7975 - 7.8025$	7.8025
USB	f_o	$(f_o + 0.0025) - 7.8$	$7.7975 - 7.8025$	7.7975

Single-Sideband Suppressed-Carrier Receivers (continued)

You can see how DSB, USB, and LSB reception can take place using a single filter. For example, if a filter centered at 7.8 MHz covering the band from 7.7975 to 7.8025 MHz (±2.5 kHz) is used, it can provide for selectivity for DSB reception by centering the carrier at 7.8 MHz, as shown, so that both sidebands are passed by the filter. To provide lower sideband only, we offset the carrier by 2.5 kHz to 7.8025 MHz. The lower sideband is now located in the filter passband. For the case shown, the carrier is reinserted at 7.8025 MHz to retain the correct relationship between the sideband components. Similarly, the upper sideband can be selected by shifting the local oscillator (or transmit signal) down to 7.7975 MHz so that the sideband lies in the filter passband.

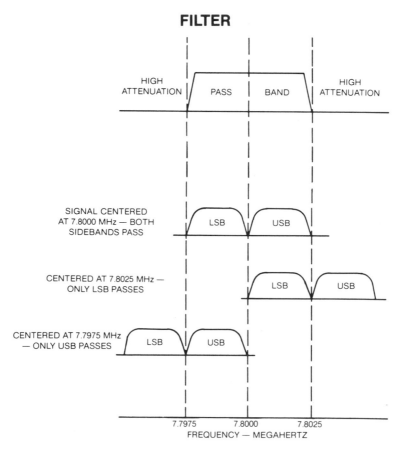

FILTER

FREQUENCY — MEGAHERTZ

As you can see from the figure, the choice of upper or lower sideband reception is accomplished with a single filter by simply shifting (offsetting) the IF carrier as shown. Note that the proper frequency for detector injection must be used so that the sideband does not become inverted.

Digital Communications Receivers

The principles of receiver operation that you learned about previously are directly applicable to receivers designed to receive digital signals. Thus, a conventional AM or FM superhet receiver that you learned about earlier can be used for reception of digital or pulsed signals. The only difference lies in the *bandwidth* of the receiver circuits. You will learn about the nature of the digital signal and its information content when you study digital circuits in Volume 5. You will also remember from Volume 3, the study of transmission systems, that the bandwidth required for a pulsed signal is about equal to the reciprocal of the pulse width. Thus, 1-μsec pulses require an overall receiver bandwidth of about 1-MHz in the RF/IF portion. A bandwidth of half this is sufficient for the detector/video amplifier portion. Since bandwidth is related to tuned circuit Q, the Q must be reduced if it is too high.

WIDE-BAND AMPLIFIER

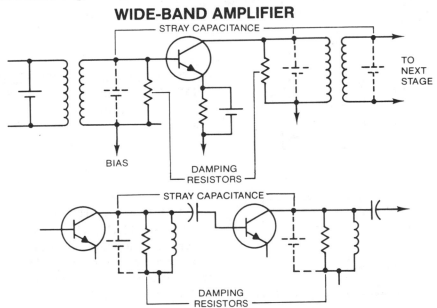

The usual configuration for wide-band amplifiers uses capacitance coupling and single tuned circuits, as shown. Since $Q = \omega L/R$, the damping resistor serves to reduce the Q of the circuit. As you may remember, Q is also defined as $\Delta f/f_o$, where Δf is the bandwidth. Therefore reduced Q means greater bandwidth. In the following study of TV receivers, you will see that damping resistors are used on the tuned circuits to reduce the Q by the desired amount. In fact, TV receivers are good examples of pulse or digital signal receivers, since they must deal with signals that are pulse-like in their characteristics.

As you can see, the only difference between a digital or pulse receiver and a conventional receiver lies in the bandwidth of the RF/IF system and bandwidth of the amplifiers (video) following the detector.

Review of Communication Receivers

EXAMPLE OF CB TRANSCEIVER

1. COMMUNICATION RECEIVERS are for receiving information as opposed to entertainment. They are most often parts of *transceivers*.

DIAGRAM OF NOISE BLANKER CIRCUIT

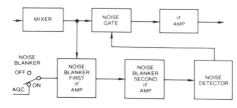

2. CB RECEIVERS feature special techniques such as synthesizer frequency selection and *noise limiting*.

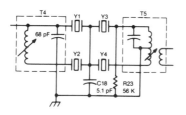

3. IF RESPONSE AND SELECTIVITY are other features of CB receivers. To minimize interchannel interference, *ceramic crystal filters* are most often used.

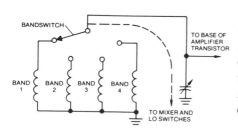

4. FLEET COMMUNICATION AND AMATEUR RECEIVERS are other communication types. Amateur receivers must cover a wide range of frequencies and do it by use of *band switching*.

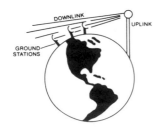

5. SHIP EMERGENCY, WEATHER, AIR-TO-GROUND, SATELLITE COMMUNICATION, and accurate frequency and timing broadcasts are some of the other transmissions for which receivers are used.

Self-Test—Review Questions

1. What is a transceiver?
2. What modes of transmission/reception are permitted for communication systems?
3. Draw a block diagram of a CB-receiver/digital-frequency synthesizer. Describe how it operates.
4. Why are synthesizers desirable?
5. Compare single- and double-sideband reception.
6. What are some of the essential differences between a broadcast receiver and a communication receiver?
7. What do most short-wave receivers use to cover a wide range of frequencies?
8. What are some of the special broadcast services available from WWV operated by the U.S. federal government?
9. Why would anyone want to use a satellite communication system?
10. What is the difference between a pulse or digital receiving system and a conventional voice system?

Learning Objectives—Next Section

Overview—You will next proceed to a study of the television system as a whole. This will review what you studied about TV in Volume 3 and prepare you for TV receiver details that follow.

CONVERTING THE VIDEO SIGNAL BACK TO A PICTURE

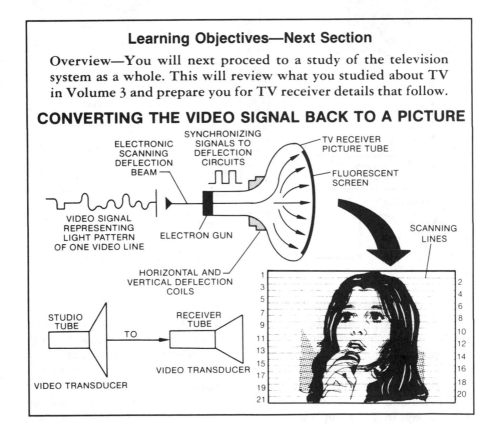

The Television System

Television (*distant vision*) is an electronic system that conveys both *visual* (video) and *aural* (audio) *information* from one location to another. When these locations are *far apart*, the process includes the impression of the video and audio information on an RF carrier and radiation through space. When the transmission and reception locations are *nearby*, the information can be carried, at baseband, over wires from the camera/microphone to the receiver/monitor. This is called *closed circuit TV (CCTV)*. *Cable TV (CATV)* is also very popular. In CATV, a master receiver system ·is used to pick up and amplify transmitted signals, and these are distributed to conventional TV receivers via coaxial cable.

THE TV TRANSMISSION AND RECEPTION PROCESS

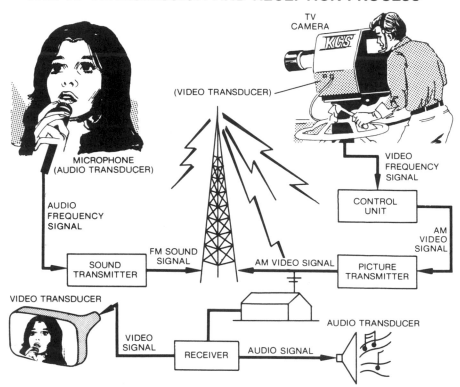

In the original scene, the sound information is converted to audio and used to frequency modulate the aural RF carrier. The TV camera scans the scene and converts the visual information to a video electrical signal. The video is combined with blanking and synchronizing signals and used to amplitude modulate an RF carrier that is 4.5 MHz lower in frequency than the sound carrier. The transmitted AM and FM signals are received using AM and FM receiving techniques with which you are now familiar. What is *new* in TV reception is *how the AM video signal is processed to recover the synchronizing/blanking signals and the visual information and uses them to provide a duplication of the original scene on the receiver picture tube.*

Types of TV Systems—CCTV and CATV

As you learned in Volume 3, the video and audio (aural) information signals are impressed on RF carriers for transmission through space. If the signals are used directly *without* going to RF, the system is called *closed circuit TV* (CCTV) and the connection is made via coaxial cable. In this case no RF is involved, and the video and audio signals are sent *directly* on the cable. When this is done, *no transmitter* is involved and the RF, IF, and detection portions of the receiver are not needed. The *TV monitors* used in CCTV are essentially TV receivers containing *only the video and CRT circuits* so that the input signals at baseband (video and audio) are used *directly*, as will be described later. TV systems of this type are widely used for security systems and monitoring of industrial processes. In some cases, the video and audio signals are converted to low-power RF signals (in a converter that is actually a miniature transmitter) so that a conventional TV receiver can be used at the receiving end. This is done because conventional TV receivers are readily available and, strangely enough, less expensive than TV monitors because conventional TV receivers are produced in such great quantities.

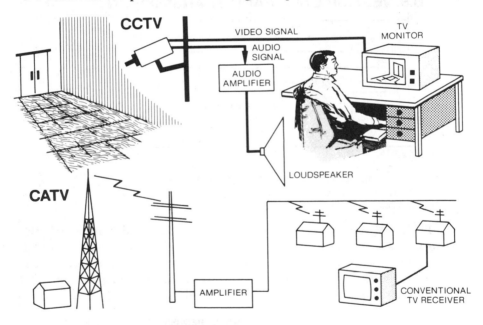

CATV or cable TV was originally designed to provide TV to homes in areas where reception was poor. A large, complicated, and expensive antenna system was erected, and the received signals were amplified and distributed at RF to homes via a coaxial cable. Nowadays, these systems are in use almost everywhere, even in good signal areas, because they have been expanded not only to provide excellent reception of local stations but also distant stations and *special programming* of movies, sports events, and local community events. These special programs are available only to CATV subscribers.

TV Frequency Allocations

The audio (sound) portion of a TV program is transmitted monaurally by frequency modulating a separate RF carrier specifically for sound transmission. For this purpose, an RF bandwidth of 50 kHz (± 25-kHz deviation) is adequate. In contrast, the video signal required for good picture definition has frequency components of up to 4 MHz and higher. As you know, to broadcast this signal by conventional double-sideband methods would require a double-sideband bandwidth of 8 MHz plus additional unused bandwidth to separate it from its own sound channel and from adjacent TV channels. To allocate such a bandwidth to a large number of TV channels would use up too much of the radio spectrum. The method finally established in the United States as standard is to use a combination of double- and single-sideband transmission known as *vestigial-* (or part-) *sideband transmission.* A 4.0-MHz bandwidth is assigned to the upper sideband, but only a 1.25-MHz bandwidth is given to the lower sideband. This provides good picture quality and saves bandwidth. It requires, however, a special design of the receiver IF stages.

U.S. VESTIGIAL SIDEBAND TRANSMISSION

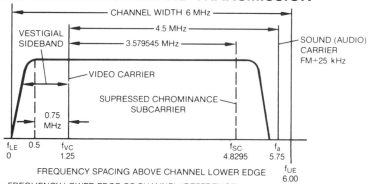

f_{LE} = FREQUENCY LOWER EDGE OF CHANNEL (REFERENCE)
f_{UE} = FREQUENCY UPPER EDGE OF CHANNEL

Each TV channel is assigned a 6-MHz bandwidth in which the sound RF carrier is at a frequency 4.5 MHz higher than the picture RF carrier. The location of these carriers within the channel assignment is shown in the diagram. Frequency allocations have been assigned to 82 TV channels. Details of assignments for typical channels are below.

US TELEVISION CHANNEL FREQUENCIES

CHANNEL	FREQUENCY RANGE	PICTURE CARRIER (MHz)	SOUND CARRIER (MHz)
2	54–60	55.25	59.75
3	60–66	61.25	65.75
4	66–72	67.25	71.75
5	76–82	77.25	81.75
6	82–88	83.25	87.75
7	174–180	175.25	179.75
8	180–186	181.25	185.75
9	186–192	187.25	191.75
10	192–198	193.25	197.75
11	198–204	199.25	203.75
12	204–210	205.25	209.75
13	210–216	211.25	215.75
14–83 UHF	470–890		

VHF (brace spanning channels 7–13)

From Picture to Video Signal to Picture

The video signal carries information from the scene in the transmitter studio to the face of the TV screen in the receiver. The manner in which this is done will be reviewed very briefly here.

The basic system uses a TV camera tube in the transmitter studio as the transducer to convert light-signal inputs to electrical signals and the picture tube in the TV receiver as the transducer to convert electrical signals back into light output. In the TV camera, a lens system focuses an image of the scene on the face of the camera tube. This face has a light-sensitive surface made up of minute areas that have photoelectric properties. The magnitude of the charge on each of these minute areas is proportional to the amount of light at that part of the image.

HOW PICTURE SIGNALS ARE PRODUCED BY THE TV CAMERA

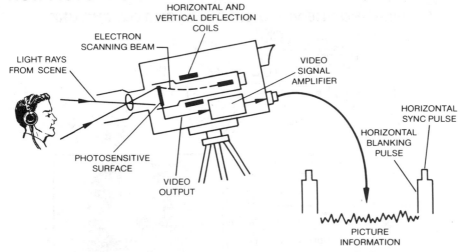

An electron gun in the camera tube focuses a beam of electrons directed toward the light-sensitive surface. Electric currents flow through coils of wire (deflection coils) placed around the neck of the camera tube and form magnetic fields that deflect the electron beam during its travel to the light-sensitive surface. The beam is scanned sequentially—vertically and horizontally—in a series of 525 horizontal lines, one above the other, to cover all parts of the image on the light-sensitive surface. The *horizontal* scan is at uniform speed from left to right with a much faster return (*retrace* or *flyback*) to the left. The *vertical* scan is at uniform speed from top to bottom with a much faster retrace to the top of the screen.

The light-sensitive element produces an electrical output signal when the electron beam strikes it, as described in Volume 3, and this signal is connected to the input of a video amplifier. The magnitude of the voltage produced is proportional to the light intensity at that point on the photosensitive surface. This video information signal is mixed with synchronization and blanking information, and is used to amplitude modulate the RF picture carrier from transmission to the receiver.

From Picture to Video Signal to Picture (continued)

At the receiver, the amplitude-modulated RF carrier is converted back to a video signal. This signal is used to drive the grid in an electron gun in the receiver picture tube (*cathode ray tube*, or *CRT*). The electron gun generates a fine beam of electrons that is focused onto a fluorescent screen at the face of the tube. The beam causes the spot on the screen to flow with a brightness that is proportional to the beam intensity. When a *peak* in the video signal, corresponding to a *bright* spot on the face of the camera tube, is applied to the grid of the electron gun, *more* electrons flow into the beam, causing a brighter spot to appear. When a *lower* point in the video signal, corresponding to a *darker* spot on the face of the camera tube, is applied to the grid of the electron gun, *fewer* electrons enter the beam, causing the spot to be less bright.

CONVERTING THE VIDEO SIGNAL BACK TO A PICTURE

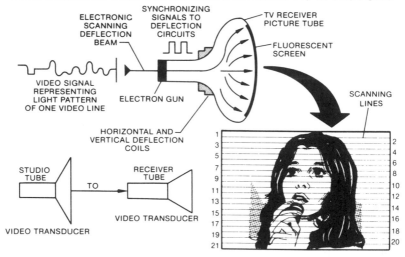

While the *beam intensity* is *varied* in accordance with the video signal, the beam is also *scanned* across the face of the CRT in a *pattern* duplicating the motion of the beam scanning the face of the TV camera tube. The two beams are kept in step (synchronized) by *synchronizing signals*, which are separated from the video signal to *lock* the scanning in the CRT beam to that of the camera tube. The actual scanning is usually accomplished by magnetic fields generated by vertical- and horizontal-deflection coils around the neck of the picture tube through which electric currents flow, duplicating those flowing through coils in the camera tube.

Thus, the *intensity* of the electron beam in the CRT *duplicates* the intensity of the current at the output electrode of the camera tube; also, the *relative position* of the moving spot on the CRT screen *matches* the position of the spot in the camera tube. This *duplication of spot position and intensity* produces an image on the picture-tube screen which duplicates the image produced on the face of the camera tube. The *image* formed on the CRT consists of a *pattern of varying-intensity light* laid down *sequentially*.

Interlaced Scanning

Since the image is formed sequentially, some means must be found to keep the *beginning* of the image *in view* while the rest of the image is being formed. The fluorescent screen and the human eye work together by *retaining* the spot images long enough so that a *complete image* is seen. The process is repeated fast enough so that *no flicker* is observed. Black-and-white picture tubes use a P4 phosphor as the fluorescent material. This is actually a mixture of phosphors to produce white light when struck by the electron beam. The P4 phosphor has a medium persistence so that the image from one scan begins to fade by the time the next scan begins.

If the scanning process that produces the TV image is slowed down, the image appears to flicker. What happens is that there is a *noticeable* delay between the formation of one image and the next. Not only is there a noticeable blank time between images, but the top of the image could fade before the moving spot reaches the bottom of the screen. To avoid this effect in modern TV, 30 complete 525-line pictures are scanned every second. Even though this is a fast rate, flicker can occur, so it is reduced even more by *interlaced scanning.*

INTERLACED SCANNING

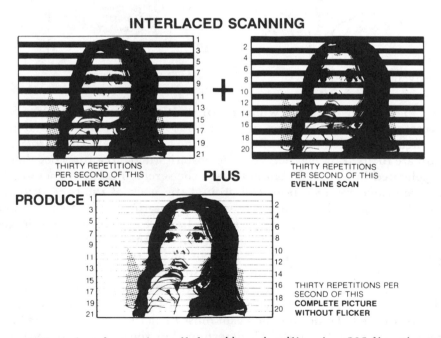

THIRTY REPETITIONS PER SECOND OF THIS **ODD-LINE SCAN**

PLUS

THIRTY REPETITIONS PER SECOND OF THIS **EVEN-LINE SCAN**

PRODUCE

THIRTY REPETITIONS PER SECOND OF THIS **COMPLETE PICTURE WITHOUT FLICKER**

In interlaced scanning, all the *odd-numbered* lines in a 525-line picture (lines 1, 3, 5, etc.) are scanned in 1/60 second. Thus, only 1/60 second elapses between the formation of the top and bottom lines, and there is no noticeable flicker in that time. Then the process repeats for the *even-numbered* lines (lines 2, 4, 6, etc.) in the next 1/60 second. The complete 525-line image formed every 1/30 second is known as a *picture* or *frame*. The images formed every 1/60 second by all the odd-line or even-line scans are known as *fields*.

The Picture Tube (CRT)

The TV picture tube (CRT) is a glass envelope or bulb having the approximate shape shown in the diagram. At the viewing end, the tube is approximately rectangular and the width of the rectangle is 4/3 of the height. These proportions (called the *aspect ratio*), correspond to those of the image scanned at the TV camera in the studio. Early CRTs were circular rather than rectangular. However, the 3/4 rectangular aspect ratio was still used, which resulted in wasted CRT faceplate area. Circular CRTs are still used for many non-TV applications. As its opposite end, the tube narrows down to a cylindrical neck usually between 1 and 2 inches (2.54 and 5.08 cm) in diameter. The neck terminates in a base with a number of pins to which electrical connections are made to the various elements of the electron gun inside. Connection to the high-voltage final anode is made to a terminal on the glass envelope.

THE TWENTY-THREE INCH PICTURE TUBE

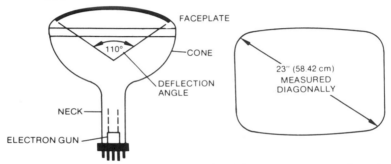

The inside of the viewing end of the CRT is coated with the fluorescent material. The beam of electrons is formed by the electron gun located in the neck of the tube close to the base. The electrons from the gun are focused by magnetic or electrostatic fields to a fine spot that is accelerated by a high potential toward the screen. Magnetic focusing is accomplished by a separate coil around the neck of the tube through which an adjustable dc current flows. Electrostatic focus, which is almost universally used in newer sets, is accomplished by adjusting the voltages of one of the elements (focusing electrode) in the gun assembly. Thus, the beam travels straight down the axis of the tube toward the center of the screen and a bright, fine spot is formed at the screen.

Magnetic coils (or sometimes electrostatic deflection electrodes) deflect the electron beam across the face of the CRT, so that an image can be formed. To do this the beam must be deflected from its straight path through a deflection angle determined by the geometry of the tube. Early tubes were much longer than the one shown and had much smaller deflection angles because efficient deflection circuits did not exist. Public demands for large screens and sets with short front-to-back dimensions have led to present-day tubes with deflection angles of 110 degrees or greater, resulting in much shorter tubes for a given screen size. Tube *size* is always measured diagonally. Popular sizes are available in about 1-inch (2.54-cm) increments between 5 and 23 inches (12.7 and 58.42 cm).

The Picture Tube (CRT) (continued)

Operation of the picture tube is based on vacuum-tube principles. In a vacuum, heated oxide surfaces emit large quantities of electrons that accumulate in a cloud. If an electrode near the oxide surface is made positive with respect to the oxide surface, the electrons travel toward it. The surface that emits the electrons is known as the *cathode*, and the positive electrode is known as the *anode*.

In a picture tube (CRT) there is an electron gun assembly similar to that shown in the diagram. A cylindrical cathode with an oxide-coated end emits electrons when heated. Inside the cathode cylinder is a heater element. When a current is passed through the heater (or *filament*) it raises the temperature of the cathode, and electrons are emitted.

THE ELECTRON GUN ASSEMBLY   THE FINAL ANODE

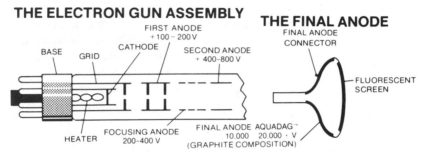

Picture tubes typically contain three anodes, as shown here. Electrostatically focused tubes may have an additional focusing electrode between anodes 1 and 2. The first anode is charged with a positive voltage of about 100 to 200 volts with respect to the cathode and is shaped like a drum with a small round hole punched in the center of each end. The second anode has a positive voltage of about 400 to 800 volts and is shaped like a hollow cylinder. The final anode is a metallic layer plated over the conical portion of the inside surface of the CRT envelope and has a positive voltage from about 10,000 to 20,000 volts (or more) applied. Connection to this anode is usually made via a button on the wall of the CRT. The glass envelope usually has a graphite-composition conducting layer that is grounded coating the outside of the conical portion. The capacitor formed by this coating and the final anode is used to provide filtering of the high voltage for the final anode.

When the electrons are attracted by the positive charge on the first anode, they pass through the small holes because of the much greater attraction of the fields from the second and final anodes. Passing through those small holes forms the electrons into a narrow beam, and the cylindrical shape of the second anode produces an electric field that maintains that beam shape. For electrostatically focused tubes a special additional anode, typically located between anodes 1 and 2, is used as a focusing electrode, and the voltage is adjustable to provide electrostatic focus. In other cases, beam focusing is magnetic. In this case, the beam is confined (or focused) by a concentric, steady magnetic field. Under the influence of the powerful electric field of the final anode, the beam accelerates to the screen and forms a bright spot of light at the center.

The Picture Tube (CRT) (continued)

The grid (control electrode) is a metal disk with a small central hole. It is located between the cathode and the first anode, but relatively close to the cathode. The grid has a negative voltage with respect to the cathode, and, because it is close to the cathode, this negative voltage has a *greater effect* on the electron flow than the anodes. Since these voltages are relative, the grid can be made relatively negative by applying a negative voltage to the grid with the cathode grounded or a positive voltage to the cathode with the grid grounded.

If the grid is made 50–100 volts more negative than the cathode, all the electrons are repelled back to the cathode, and no electrons can enter the beam. Thus, no spot is produced on the screen. If the negative voltage is reduced slightly, the strong positive field of the first anode attracts a few electrons through the grid opening. A beam containing only a very few electrons is formed, and this beam is accelerated toward the screen by the high positive voltages on the anodes. When this beam strikes the screen, a weak spot is formed on the screen.

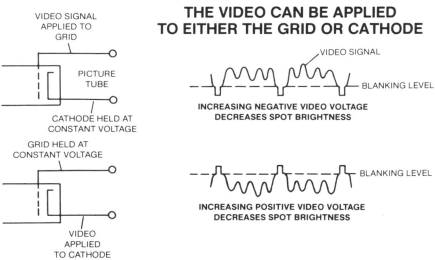

THE VIDEO CAN BE APPLIED TO EITHER THE GRID OR CATHODE

VIDEO SIGNAL APPLIED TO GRID

PICTURE TUBE

CATHODE HELD AT CONSTANT VOLTAGE

GRID HELD AT CONSTANT VOLTAGE

VIDEO APPLIED TO CATHODE

VIDEO SIGNAL — BLANKING LEVEL

INCREASING NEGATIVE VIDEO VOLTAGE DECREASES SPOT BRIGHTNESS

BLANKING LEVEL

INCREASING POSITIVE VIDEO VOLTAGE DECREASES SPOT BRIGHTNESS

As the negative voltage on the grid is reduced more and more, increasingly larger numbers of electrons enter the beam, and the spot becomes brighter and brighter. For a typical black-and-white TV picture tube, a voltage change of only about 30 volts is sufficient to change the spot intensity from total black to maximum useful brightness.

In the TV receiver, the video signal is applied between the cathode and grid with the polarity such that an increasing signal level—toward the blanking level—results in decreasing brightness. Thus, a black-level signal is sufficient to cut off the flow of electrons completely. The amplitude variations in the video signal cause corresponding brightness variations in the spot on the screen via the grid. When the spot is moved across the screen in horizontal and vertical lines (scanned), these brightness variations produce the picture on the CRT screen. As you can see from the illustration, the video signal can be applied to either the grid or the cathode, as long as the correct polarity is used.

Scanning the Electron Beam

A TV frame consists of 525 horizontal lines scanned sequentially from top to bottom of the frame in $\frac{1}{30}$ second. Each frame consists of two interlaced fields, each containing 262½ horizontal lines scanned in $\frac{1}{60}$ second. Thus, to form each such frame requires that the electron beam be scanned 262½ times in a horizontal direction for each scan in a vertical direction.

In *Basic Electricity*, you learned that a circular magnetic field is formed around a wire when an electric current flows through the wire. The direction of the magnetic field is determined by the direction of current flow, according to the left-hand rule. You also learned that if a current-carrying wire is placed between the poles of a magnet, a force will act on the wire. The direction of motion will be to the left or right, according to the current direction and magnetic polarity, as described in the three-finger rule. This motion is the basis of electric motor operation.

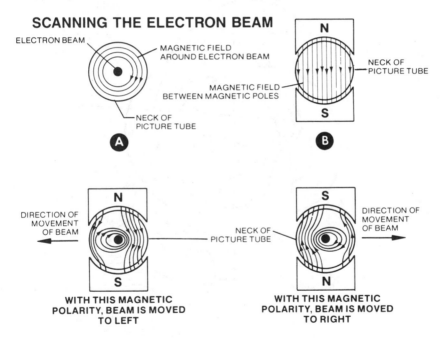

SCANNING THE ELECTRON BEAM

ELECTRON BEAM

MAGNETIC FIELD AROUND ELECTRON BEAM

MAGNETIC FIELD BETWEEN MAGNETIC POLES

NECK OF PICTURE TUBE

A

N

NECK OF PICTURE TUBE

S

B

N

DIRECTION OF MOVEMENT OF BEAM

NECK OF PICTURE TUBE

S

WITH THIS MAGNETIC POLARITY, BEAM IS MOVED TO LEFT

S

DIRECTION OF MOVEMENT OF BEAM

N

WITH THIS MAGNETIC POLARITY, BEAM IS MOVED TO RIGHT

This motion is also the basis for electron-beam deflection in a picture tube. The electron beam in its path from the gun to the screen has the same properties as current or electron flow in a wire—each consists of electrons in motion or an electron stream. Even though this stream is not confined by a wire, in a CRT it forms a magnetic field around itself in the same manner as if the wire were there.

Now, if magnetic poles are placed above and below the beam, the beam is deflected by the force generated by the interaction of the two magnetic fields in the same manner as the current-carrying wire. Since the beam has very little mass, it can be deflected very rapidly. The direction of the deflection is reversed by reversing the polarity of the magnetic poles.

Scanning the Electron Beam (continued)

In a television receiver, the magnetic poles are formed by coils of wire—the deflection coils—close to the neck of the tube, as shown in the diagram. Now the strength and polarity of the magnetic field can be changed as desired by changing the magnitude and direction of the current flow through the coils.

A pair of coils above and below the electron beam (H1 and H2 in the diagram) produces the desired horizontal deflection. Another pair of coils at the left and right of the beam (V1 and V2 in the diagram) produces the desired vertical deflection. The composite set of coils is usually called the *deflection yoke*. These coils are fitted around the neck of the CRT at the point where the neck flares into the cone.

DEFLECTION COILS

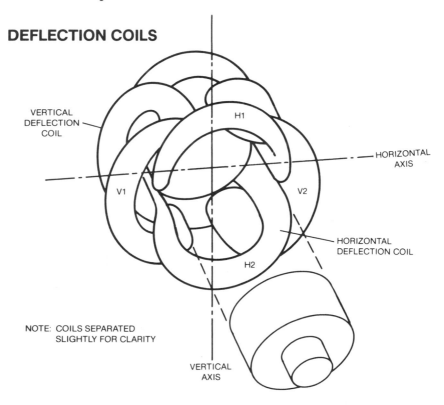

VERTICAL DEFLECTION COIL

H1

HORIZONTAL AXIS

V1

V2

HORIZONTAL DEFLECTION COIL

H2

NOTE: COILS SEPARATED SLIGHTLY FOR CLARITY

VERTICAL AXIS

To produce one of the interlaced frames of a complete picture requires simultaneous vertical and horizontal motion of the electron beam by the deflection coils. To sweep the beam uniformly from the extreme left to the extreme right with a very fast retrace to the left requires a sawtooth magnetic field that makes 262½ cycles in 1/60 second (or 262½ × 60 = 15,750/second). During that same period, another sawtooth magnetic field produces a uniform deflection from top to bottom with a very fast retrace to the top every 1/60 second or 60/second. Since the scan is interlaced, two 1/60 second fields make up one frame at a 30-Hz rate.

Scanning the Electron Beam (continued)

To obtain a sawtooth of current through an inductance requires a constant applied voltage since, as you may remember, $\Delta_I/\Delta_T = E/L$. A sawtooth is defined as having a constant rate of rise, or Δ_X/Δ_T is a constant where Δ_X is any quantity. Thus for a sawtooth magnetic field, the current must vary linearly, or Δ_I/Δ_T is a constant. Therefore E/L is a constant, which means that E is a constant since L (the inductance) does not change. Therefore, rectangular waves of voltage across the coil produce a sawtooth of current. During retrace, the current is abruptly interrupted, as shown. Thus, the magnetic field collapses, bringing the beam rapidly to the beginning of the scanned line or field. Because the deflection yoke coils have resistance, a small sawtooth component is added to produce a linear magnetic flux, as shown. The sawtooth component compensates for the voltage drop across the resistance as the current increases. Thus, the voltage across the inductance remains constant.

WAVEFORMS

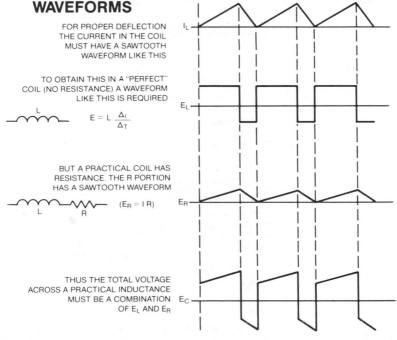

FOR PROPER DEFLECTION THE CURRENT IN THE COIL MUST HAVE A SAWTOOTH WAVEFORM LIKE THIS

I_L

TO OBTAIN THIS IN A "PERFECT" COIL (NO RESISTANCE) A WAVEFORM LIKE THIS IS REQUIRED

$E = L \dfrac{\Delta_I}{\Delta_T}$

E_L

BUT A PRACTICAL COIL HAS RESISTANCE. THE R PORTION HAS A SAWTOOTH WAVEFORM

$(E_R = I\,R)$

E_R

THUS THE TOTAL VOLTAGE ACROSS A PRACTICAL INDUCTANCE MUST BE A COMBINATION OF E_L AND E_R

E_C

The simultaneous horizontal and vertical scanning produces the *raster* of horizontal lines, one above the other. When you switch to a channel that is not in use, you can see the array of horizontal lines, all of uniform density, that make up the raster. You know that the electron beam is deflected from left to right by a sawtooth wave. Starting at the extreme left, the spot on the screen is moved at uniform speed to the extreme right with a very fast retrace or *flyback*. If you look closely, you may see the retrace on the TV set if it is tuned to an unused channel. The TV video signal has a blanking pulse as part of the synchronization that applies a high negative voltage to the grid of the CRT during the retrace period to keep retrace lines from appearing on the screen.

Scanning the Electron Beam (continued)

You know that the scan is interlaced and that an odd number of horizontal scan lines (525) makes up a complete frame. As described earlier, interlaced scanning is used to prevent flicker. This is done by scanning a field consisting of all the odd horizontal lines in a picture (⅟₆₀ second), and then scanning a field consisting of all the even horizontal lines (⅟₆₀ second) to make up the complete field (⅟₃₀ second). Because a picture is divided into two fields, each field consists of exactly 262½ horizontal lines. The scan for line 1 starts at the extreme upper left. The last odd-line scan thus ends with a half-line scan, and the next even-line frame must start with only half a horizontal scan line. The odd/even line scans are synchronized by the synchronizing signals present during the blanking period. As you learned in Volume 3, the odd/even line scans are automatically produced by the synchronizing signal for the vertical sweep so that an interlaced scan is produced from the transmitted signal, not as a result of any special circuits in the receiver.

INTERLACED SCANNING

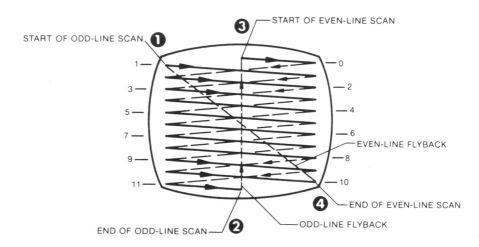

As you know, horizontal synchronization is maintained during the vertical blanking period. The vertical-sync pulse is serrated by narrow notches that occur at the horizontal-sweep rate to maintain horizontal sync. The region of the vertical-sync pulse preceding and immediately following the vertical-sync pulse contains equalizing pulses that occur at twice the horizontal-line rate. Since the horizontal-sweep circuits are sensitive to only the pulses that occur in the vicinity of the natural period of the horizontal-line oscillator, horizontal sync continues as before during the equalizing-pulse period; on alternate sweeps, however, the vertical sync starts earlier by one equalizing pulse (half line). To keep the sweep lines constant, the equalizing pulses after the vertical sync contain one additional pulse. Thus, effectively, the vertical-sync pulse is moved half a horizontal line every other sweep to provide for interlacing.

The Video Signal

Now what you need to learn more about at this time is *how* the picture on the receiver screen is *kept stable* by synchronizing the scanning in the CRT with the scanning in the camera tube. These signals were discussed in detail in Volume 3. In the figure below, details are shown only for five even-numbered scans over lines 200 through 208. The white level is shown as the lowest part of the scan and the black level as the highest. It is important to note that a black (blacker than black) blanking pulse is inserted at the end of each scan. Its purpose is to blank the retrace of the electron beam by turning the beam off.

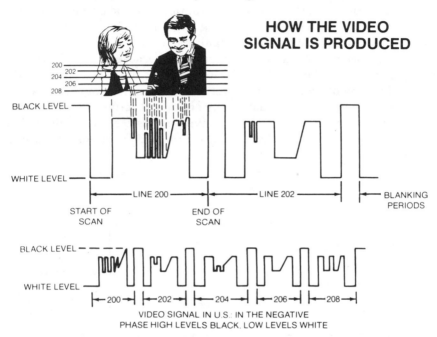

HOW THE VIDEO SIGNAL IS PRODUCED

VIDEO SIGNAL IN U.S. IN THE NEGATIVE
PHASE HIGH LEVELS BLACK, LOW LEVELS WHITE

As an example, suppose you trace out the details of scan number 200. At the extreme left is the bright background of the scene, which produces a corresponding width of maximum white signal (minimum carrier level). Then the scan passes to the very dark hair of the scene and produces a matching width of maximum black signal (high carrier level). The left side of the cheek again produces a maximum white signal, but the level then varies rapidly over gray levels as the scan passes over the varying shades of gray of the nostrils and right cheek. Maximum black is then produced by the hair at the right side of the image except for three small highlights. At the extreme right of the image is the maximum white of the background. At the end of each line is a slightly blacker-than-black-level signal (the *blanking pulse*) that is introduced into the video to turn the CRT beam *off* during retrace. Thus *horizontal-blanking pulses* are added to the video signal, starting shortly before retrace starts and ending slightly after the new line scan begins. A similar *vertical-blanking pulse* is inserted for the vertical retrace.

Sync Pulses

The horizontal-sync pulses occur during the horizontal-blanking period and represent 100% (maximum) modulation of the picture carrier. A horizontal-sync pulse occurs on each horizontal-blanking pulse and is used to initiate the horizontal-sweep retrace and beginning of a new sweep line, as shown.

During the vertical retrace, it is necessary to have a *blanking pulse* so that the vertical retrace will not appear on the screen. It is also necessary to have *vertical-sychronizing pulses* to start the vertical sweep at precisely the right moment. The operation for vertical blanking and synchronizing is similar to that for the horizontal, with the exceptions noted below. The illustration on the following page shows the structure of the transmitted signal during vertical blanking.

Vertical retrace is slower than horizontal because the vertical sweep is slower; therefore the blanking period must be longer. Typically, the vertical blanking of each field requires the same amount of time as about 18 horizontal lines, so that actually only about 490 active horizontal lines are seen in a complete frame. Vertical-sync pulses are much wider than horizontal-sync pulses, and they are distinguished from horizontal-sync pulses on this basis, as you will learn later.

SYNC PULSES

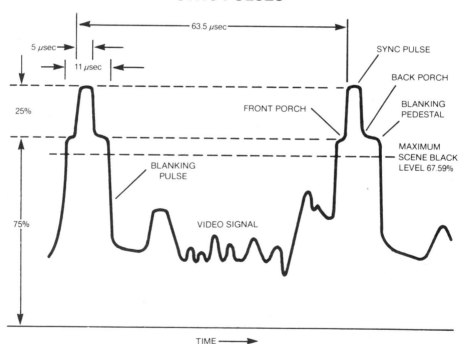

Sync Pulses (continued)

Provision must be made to continue supplying horizontal-sync pulses during the vertical-sync period so that the horizontal scan will not be out of sync when it must begin to scan the top line of the next field. Finally, because a half frame is made up of 262½ horizontal scans, provision must be made for synchronizing the half of a horizontal line that occurs alternately at the top and bottom of every other field. All these requirements are met by having a vertical-blanking and synchronizing pulse such as that shown here.

When vertical blanking occurs, a series of 5 narrow equalizing pulses are generated at twice the line rate (5 pulses in 2½ horizontal line intervals during field 1 and 6 equalizing pulses for field 2). Every other one of the equalizing pulses is used to keep the horizontal oscillator in sync. The vertical-sync pulse consists of 6 or more pulses at twice the horizontal rate, except that these pulses are very wide, as shown. The vertical-sync pulse is followed by 6 more equalizing pulses for field 1, but only 5 (2½ lines) equalizing pulses for field 2. Thus, the second field starts half a line later, yielding an interlaced scan between field 1 and 2.

In the receiver, the sync separator distinguishes between the narrow horizontal-sync pulses and the wide series of 6 pulses that make up the vertical-sync pulse.

FCC STANDARD TELEVISION
VERTICAL SYNCHRONIZING SIGNAL

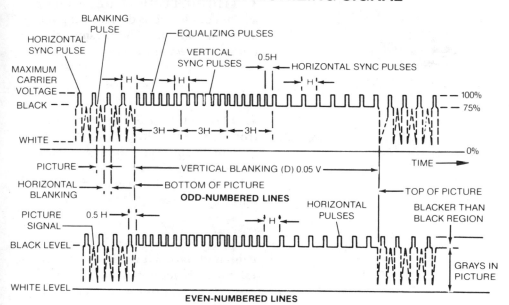

Review of Television Reception

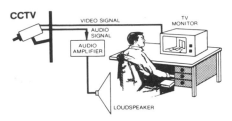

1. **TV CONVEYS VISUAL AND AURAL INFORMATION** either through the air via radio waves, or through conductors in *closed circuit TV* (**CCTV**) or *cable TV* (**CATV**).

2. **INTERLACED SCANNING** covers the image, both horizontally and vertically, in two steps. Each step is called an *interlaced field*.

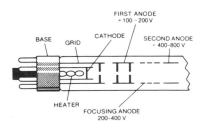

3. **THE PICTURE TUBE** generates a beam of electrons directed at a fluorescent screen. The beam originates in an *electron gun*.

4. **THE VIDEO SIGNAL** results from different voltage outputs as the camera beam scans light and dark elements. In the receiver, the picture tube beam intensity is varied to build the image.

5. **SYNC PULSES** reach a 100 percent modulation level. Each horizontal pulse is mounted on a *blanking* pulse. Vertical pulses are formed of groups of wider horizontal pulses.

Self-Test—Review Questions

1. What are CATV, CCTV? How are they different from commercial TV?
2. Describe briefly the essential elements of a TV system.
3. What is interlaced scanning?
4. Why is interlaced scanning used?
5. Draw a sketch of a CRT. Show how it operates to produce a TV image.
6. How is the electron beam in CRT scanned?
7. Draw a sketch of a TV signal.
8. What are the frame field and line rates for commercial TV in the United States?
9. Sketch and describe the black (blanking level) and the horizontal- and vertical-sync pulse characteristics. What do they do?

Learning Objectives—Next Section

Overview—Now that we have reviewed the fundamentals of the TV system, we next proceed to a study of the details of TV receivers, starting with the black and white type.

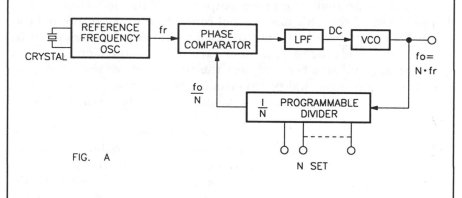

FIG. A

Typical TV Receiver—Block Diagram

Now that you are familiar with the signals used for transmission of television information, you are ready to learn the details of how black-and-white TV receivers operate.

The following page is an overall block diagram of a typical black-and-white TV receiver. The VHF and UHF tuners are fed from their respective antennas and include the RF amplifiers, mixers, and local oscillators. The UHF tuner output is fed to the VHF tuner so that the RF amplifier and mixer can be used as a preamplifier to provide additional gain and to provide double conversion to improve the image rejection in the UHF range. The IF amplifier usually consists of several stages with the necessary tuned circuits to provide selectivity and to provide the special shaping of the passband required to receive the vestigial-sideband transmission. The IF signal is demodulated to video by the video detector. This detector also produces as the output a 4.5-MHz sound carrier that is fed to the FM sound circuits via the video amplifier. These sound circuits include amplifiers, the FM demodulator, and the audio amplifiers to drive the loudspeaker.

The video signal from the detector is amplified for application to the grid/cathode of the CRT to control the beam intensity in accordance with the video data. An AGC circuit is provided via the ouput video to control the gain of the IF amplifier and the tuner RF amplifier.

An output from the video amplifier is also used to drive the sync separator, which develops horizontal- and vertical-sync pulses for the horizontal- and vertical-sweep generators. These signals provide the information necessary to keep the sweeps in the receiver synchronized to the transmitted signal. The sweep outputs drive the deflection yoke so that the beam of the CRT moves synchronously with the scanning beam in the TV camera, while the video signal modulates the beam intensity in accordance with the brightness of each point on the picture.

The high voltage necessary for the final anode of the CRT is obtained from the horizontal-sweep circuit during retrace. A low-voltage power supply provides the low voltages necessary for operation of the receiver circuits.

As before, we will use an actual system, a portable black-and-white TV receiver, as a learning system not only to help understand the individual circuit functions, but also to understand how the TV receiver functions as a system. Here, the learning system will be the model 12HB1 portable television receiver manufactured by the Zenith Radio Corporation. The 12HB1 has 26 transistors, 27 diodes, and a single IC. It covers VHF Channels 2 through 13 with memory fine tuning and UHF Channels 14 through 83. Some modern black-and-white TV sets use more ICs, but this set has been chosen because it clearly illustrates the necessary principles with easily understood circuits.

BLOCK DIAGRAM OF A TYPICAL BLACK-AND-WHITE TV RECEIVER

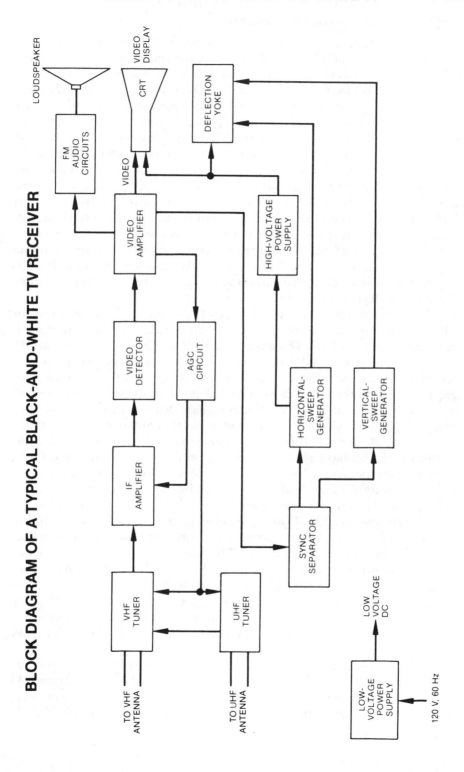

Receiver Section—Block Diagram

The block diagram of the receiver circuits that produce the video, sound, and synchronizing-signal complex for the other sections of the receiver is shown on the following page. Note that the blocks are marked with the circuit-identification numbers (Q1, Q2, etc.) to relate this diagram to the detailed schematics to be described shortly. The receiver must operate on either VHF or UHF channels, and separate tuners are used to process these signals most efficiently. The VHF tuner accepts the input from its own antenna and contains an RF amplifier stage (Q1), mixer (Q2), and local oscillator (Q3). The UHF tuner is similar, containing an RF amplifier (Q5) and local oscillator (Q6); the mixer, however, is a crystal diode (CR14). The UHF mixer output from CR14 is at VHF and is fed to the VHF tuner to be converted to the IF of the receiver. Thus, in UHF operation, double conversion is used.

The output of the VHF tuner mixer stage (Q2) is an IF signal containing the video IF carrier at 45.75 MHz and a sound IF carrier at 41.25 MHz. These combined signals are amplified by a three-stage video IF amplifier (Q101, Q102, Q103).

Output from the final (third) video IF is fed into the video detector CR101. The output from this stage consists of the video signals at baseband and the sound FM IF signal at 4.5 MHz. This 4.5-MHz sound IF signal is generated by CR101 acting as a mixer between the video and sound carriers, producing a 4.5-MHz sound IF signal. These video signals are amplified by the video driver Q801 that simultaneously amplifies the video, sync, and sound IF (4.5-MHz) signals. Video outputs to the sync separator and 4.5-MHz sound IF are taken from video amplifier Q801.

The AGC for TV receivers is somewhat different from the AGC systems that you have studied. The reason for this is that the average carrier level changes with scene brightness; therefore the average carrier level cannot be used for AGC. The sync pulses are of constant amplitude (100% modulation) and can be used for AGC. To accomplish this, the sync pulses are gated by Q403, the AGC gate. The AGC gating pulses are derived from the horizontal-sweep circuits so that only the sync pulses are fed to the AGC output (Q402). These pulses are filtered to produce a dc level that is proportional to the peak carrier level (sync level). This signal is applied to the first IF amplifier (Q101). An AGC voltage is also applied to the RF amplifiers (Q1 and Q5) via AGC delay (Q401). The AGC delay prevents the application of AGC voltage to the RF amplifiers except on very strong signals. Thus, full sensitivity is available for weak signals.

The remaining circuits are described next. The output from Q801 drives the final video amplifier Q803 which drives the CRT cathode. As you can see, at least in block-diagram form the receiver section of a TV receiver has the same elements as the receivers you studied earlier.

RECEIVER SECTION BLOCK DIAGRAM

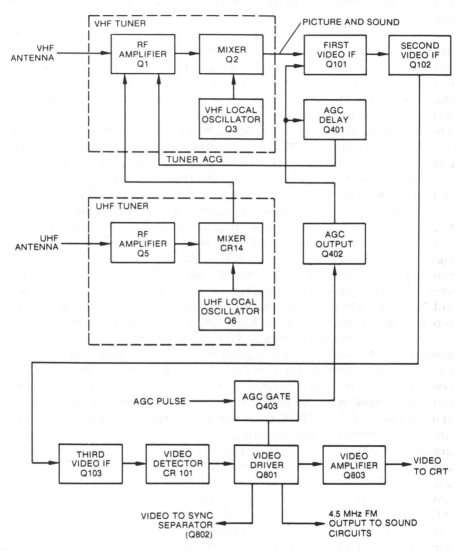

Sweep- and Sound-Circuit Section—Block Diagram

The 4.5-MHz sound IF output from video amplifier Q801 is applied to IC1001. This IC contains an IF amplifier, FM demodulator (ratio detector), and an audio preamplifier. The output from IC1001 (audio) is applied to audio driver Q1001, which drives the complementary-symmetry audio-output amplifiers Q1002 and Q1003. These drive the loudspeaker. The sound IF audio system is similar to the one for FM receivers.

The video output from video amplifier Q801 is also applied to sync separator Q802. The sync-separator (Q802) output consists of sync pulses stripped from the incoming video signal. These sync pulses (horizontal and combined vertical) are further separated (by RC networks to be described later) into horizontal-sync pulses and vertical-sync pulses to control the horizontal and vertical sweeps.

The horizontal-sync pulses are applied to the horizontal APC (automatic phase control) Q501 that controls the frequency of the horizontal oscillator Q503. The horizontal-sync pulses are compared in phase with the retrace pulses from the high-voltage transformer driven by the horizontal output Q504. An error voltage is developed in Q501 that drives the frequency of the horizontal oscillator so that the sync pulse and retrace pulse are locked in phase, thus synchronizing the horizontal oscillator to the horizonal-sync input. The output from the horizontal oscillator is amplified and shaped in horizontal driver Q502 that drives the horizontal output Q504. The output from Q504 drives the deflection yoke and the high-voltage transformer TX503. The current necessary for horizontal retrace is much greater than for the sweep because the retrace time is short. Thus, a high-voltage pulse is generated during retrace of the horizontal sweep. The pulse appears also at the output transformer TX503. This output pulse from TX503 is rectified by CR504 to provide the 12 kV dc high voltage to the CRT final anode. The high-voltage filter capacitor is formed by the final anode inside coating and the bulb outside coating, as described earlier. Because of the high frequency involved (15,750 Hz) and the low current requirements of the final anode, this filtering is sufficient. The AGC pulse is derived (as is the horizontal-retrace pulse for the horizontal APC) from a separate winding on the high-voltage transformer.

The vertical-sync pulses are applied to the vertical oscillator Q601–Q602 to lock it directly to the sync. The output from the oscillator is shaped and amplified by vertical amplifiers Q603 and Q604 that drive the vertical driver Q605. The vertical driver Q605 drives the complementary-symmetry pair of transistors (Q606–Q607) that drives the vertical-deflection coil.

A low-voltage power supply (not shown) is of conventional design, providing an unregulated 24-volt dc signal to the audio circuits with a regulated +22-volt output for all video circuits and a regulated +11 volts for tuner operation. Input is 120-volt ac at 60 Hz.

SWEEP-AND SOUND-CIRCUIT SECTION BLOCK DIAGRAM

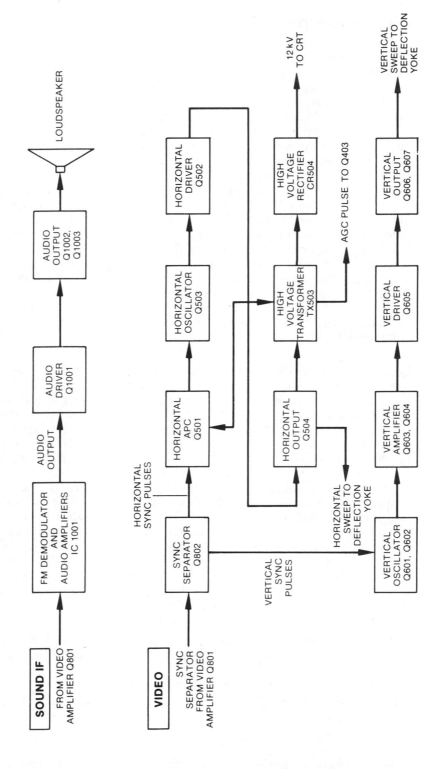

ELECTRONIC TUNERS

Electronic tuning systems are equipped with a Phase Lock Loop (PLL) frequency synthesizer tuner which features stable reception without drift. To increase the ease of operation, a microprocessor is used in the control circuit. Channel selection can be accomplished in two ways: direct channel selection where the desired channel number is selected by a command of double figures or sequential channel selection where the desired channel is selected by channel up/down scan operation. If there is a difference between the frequencies of a receiving carrier and the normal frequency, fine tuning will be done by turning on the AFC or AFT switch.

Fig. A shows a basic diagram of the PLL frequency synthesizer.

The Reference Frequency Oscillator generates the reference frequency (fr) by means of its crystal oscillator. The Phase Comparator compares the phase difference between fr and fo/N, as well as their frequencies and tunes out the difference. The Low Pass Filter (LPF) functions not only to eliminate high frequency components generated by the phase comparator, but also to determine the PLL's synchronization and response characteristics. The Voltage Control Oscillator's (VCO) frequency varies according to the control voltage. The I/N Programmable Divider changes the dividing ratio according to the divisor (N) set beforehand. If "fr" is the reference frequency and "fo" is the output frequency, the following formula will be achieved when the PLL is completely locked.

$$fr = fo/N \longrightarrow fo = N \bullet fr$$

Because N is an optional integer in the above formula, "fo" changes according to the step of "fr". The above is a brief explanation of the basic block. However, Fig. A, an illustration of the above formula, shows a principle and basic block diagram and characteristics of its components. Among various systems in the field, the pre-scaler system is the most popular. Fig. B illustrates a block diagram of a PLL frequency synthesizer of a pre-scaler system. In a basic pre-scaler tuning system, an ultra high speed fixed divider, also referred to as a pre-scaler, is inserted between the VCO and the programmable divider. The divider is able to function at a lower frequency, thus providing faster access to a given frequency. With this type of configuration, the comparison frequency can drop at the ratio of the pre-scaler's dividing ratio. If the S/N ratio of the PLL is considered, this drop becomes a problem. Some models resolve the problem by using the pulse-swallow system. In the pulse-swallow system, a high-speed device is used for the high-speed components, increasing the operation frequency without affecting the programmable divider's steps or a dropping the comparison frequency. A basic block diagram is shown in Fig. A, and when PLL is completely locked, the following formula is achieved.

$$fo = N \bullet fr$$

If a certain constant P (P = plus integer) is introduced, Np is the quotient of N divided by P, and the remainder is A. N can be shown by the following formula.

$$N = P \bullet Np = A$$

When the above formula is arranged by adding [AP-AP] to its right side, the formula becomes:

$$N = P \bullet Np + A + AP\text{-}AP$$
$$= A(P+1) + P(Np\text{-}A)$$

The above formula represents the operation of the pulse-swallow system. The details are:

P and P + 1:   Pre-scaler's dividing ratios. The pre-scaler is required to provide two dividing ratios. It is easier and more economical, however, to use a pre-scaler than high-speed programmable dividers.

A :   Swallow counter's counting value
Pre-scaler's dividing ratio is changed according to this value. This also represents the lower figure of the whole dividing ratio.

Np :   Programmable counter's counting value
This represents the upper figure of the whole dividing ratio.

The following is an explanation of the operation of the circuit referred to in the formula.

First, preset A into the swallow and Np into the programmable counter. The resulting dividing ratio of the pre-scaler is $1 \div (P + 1)$, which counts input pulses.

When the swallow counter counts input pulses of A (P = 1), the counter becomes "0" and the dividing ratio of the pre-scaler is changed to $1 \div P$. After that, the swallow counter remains in the "0" condition until it receives a Preset Enable Input (PE IN) signal.

The programmable counter which has counted by the point of A counts the remaining pulses of (Np-A). As the dividing ratio of the pre-scaler is $1 \div P$, the PE signal will be turned off when the counter becomes "0" after counting input pulses of P (Np-A). By repetition of the above steps, the following dividing ratio is realized.

$$N = (P+1)A + P(Np\text{-}A) = P \bullet Np + A$$

The pulse-swallow counter consists of a 5-bit swallow counter and a 12-bit programmable counter. The pre-scaler's dividing ratio can be changed to 1/128 and 1/126. The reference frequency (fr) is 5 kHz.

The channel selection system is composed of a Tuner/Timer/Display Control (T/T/D CTL) Microprocessor, a PLL Controller, and a Tuner Unit. The remote information receiver in this system receives information signals from the infrared remote control circuit and transfers its data to T/T/D CLT microprocessor.

Channel selection can be accomplished by using the main panel or by using the infrared remote control. When using the main panel, one of preset channels can be selected by pushing the CH UP/DOWN key. When the infrared remote control is used, a desired channel can be selected at random by using the 10 numbered keys (0 to 9) or the
CH UP/DOWN key.

When using either the main control panel or the infrared remote control, as a channel number is chosen, the information is sent to the PLL controller by the function of the T/T/D CTL microprocessor. When the PLL controller receives the information of a channel number, it selects a band corresponding to the channel and sends it to the tuner unit as 5-bit data. Then, the PLL controller presets the dividing ratio of the PLL programmable divider. After the above steps are completed and the PLL is completely locked, the TV maintains stable reception of the selected channel.

TYPICAL TUNERS

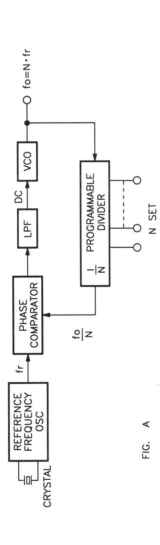

FIG. A

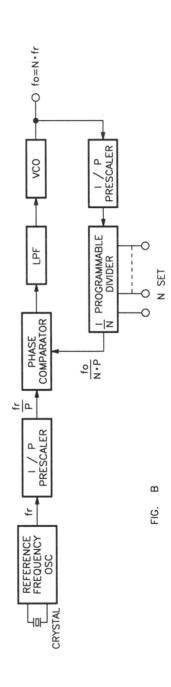

FIG. B

Video IF Amplifier

Most TV tuners have local oscillators tuned to the high side of the desired signals. The nominal IF (picture carrier) for the present-day TV is at 45.75 MHz, and the sound carrier is at 41.25 MHz. The IF amplifier must pass all signals in the band between the picture and sound carriers plus an additional 0.75 MHz for the lower vestigial sideband of the TV picture. Thus, the IF amplifier must be broadband to amplify signals between 41.25 and 46.5 MHz. Because of the vestigial sideband, the lower frequencies in the picture signal (± 0.75 MHz) are transmitted double sideband while the higher frequencies (above 0.75 MHz) are transmitted single sideband. This, of course, does not apply to the aural carrier.

Since the sideband power at low video frequencies is twice that of the higher frequencies, special shaping of the IF passband in this region is necessary to equalize the sideband power at all frequencies. This is accomplished by shaping of the IF passband as shown in the illustration. As you can see, the IF amplifier gain is slowly reduced, starting at a point 0.75 MHz below the picture carrier, and is zero at a point 0.75 MHz above the picture carrier. At any given frequency in this region (± 0.75 HMz) the sum of the responses on both sides of the carrier is equal to the response of the rest of the lower IF sideband.

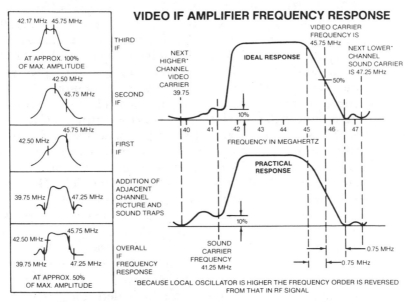

Obtaining an IF amplifier with such a frequency response is not as difficult as you might believe. The shape of the response of any individual stage can be controlled to some degree by adjusting the amount of coupling between primary and secondary windings of the interstage transformer. Damping resistors across inductors reduce the circuit Q and broaden the amplifier passband. The shape can be modified even more by using several IF stages with the interstages tuned to slightly different frequencies in the IF band. Such a tuning configuration is called *staggered tuning*.

IF Amplifier—Video Detector

The IF amplifier has three stages (Q101, Q102, and Q103). Stages Q101 and Q103 are common-emitter stages, while Q102 is in a common-collector configuration. Overcoupling of transformers, stagger tuning, and traps are used to shape the IF passband. As shown in the schematic diagram on the following page, the input from the tuner is applied to the base of the last IF stage (Q101). In addition to the input tuning (L101), several wave traps are present to reduce adjacent channel interference and to provide passband shaping. The series-tuned trap L107–C125 is tuned to the lower adjacent channel video (39.75 MHz) to suppress this signal. In like manner, L105, C104, and C105 are tuned to the higher adjacent channel to suppress the adjacent channel audio carrier. Series-tuned traps L108–C122 and L106–C123, provide for additional shaping of the IF passband. As shown, Q101 is neutralized by negative-feedback capacitor C129 to minimize tuning changes with varying AGC voltage. AGC is applied to the base of Q101 through R101 as shown. As the AGC voltage becomes increasingly positive (increasing signal level), Q101 draws more current, increasing the voltage drop across R105 (1.8 K). This lowers the collector voltage on Q101, reducing its gain proportionately as the AGC voltage increases. This AGC scheme is the same as that used for the RF amplifier in the UHF tuner, except that it is not delayed. You will study the AGC circuits later in this section. Q101 is coupled to Q102 via tuned interstage L114 in parallel with C110 and the circuit stray capacitance. Resistances R106 in parallel with R110 reduce the Q of this interstage, providing broadbanding. Q102 provides impedance matching between Q101 and Q103, having a high input impedance and a low output impedance (common collector). The Q103 output is a double-tuned circuit (L103–L104) that couples the output of Q103 to the detector CR101, as well as providing some passband shaping. L103 and L104 are capacitor-coupled to form an overcoupled transformer that provides for a wide passband with good selectivity. The output of the video detector diode CR101 is negative-going video. The IF signal is filtered out by the LC network shown, at the output of CR101. The video-detector filter network uses relatively short time constants, as you can see. This is necessary because the detector must pass video frequencies that extend up to 4 MHz and in fact must also pass the 4.5-MHz sound carrier that will be FM demodulated. The negative-going video output from the detector contains the AM video information. In addition to demodulating the AM video signal, the diode detector acts as a mixer that heterodynes the picture carrier against the sound carrier. This produces a difference frequency (4.5 MHz), which is the sound IF carrier. Note that this sound IF carrier is always at 4.5 MHz regardless of the local oscillator tuning. The 4.5-MHz sound carrier is amplitude modulated by the TV picture carrier. This AM is stripped off the FM sound carrier by limiting before FM demodulation. Sometimes, the limiting is not good enough, and you can hear a sound buzz that corresponds to the 60-Hz field rate in the sound channel.

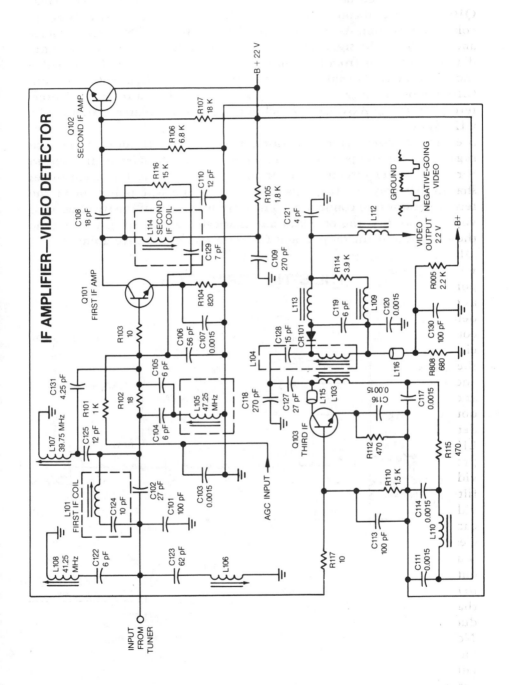

Video Driver—Sync Separator

The detected negative-going video is applied to the video driver Q801. This signal contains the 4.5-MHz sound IF carrier as well as the video information. As shown in the diagram below, Q801 is an emitter follower. Thus the signal at the emitter of Q801 is also negative going. The video output from the emitter of Q801 is used to drive the video output (see page 4–136) and to drive the sound IF circuits. In addition, a sample of the negative-going video attenuated by a resistor network is applied to the AGC gate, which will be discussed later as part of the AGC system.

The sync-separator transistor is normally cut off but is driven into conduction by the positive-going video signal. The current flowing in the base/emitter circuit causes it to act like a diode that back-biases R/C so that only the very tips of the video (the sync pulses) drive the sync separator into conduction. The output consists of the stripped sync pulses. Note that the base bias automatically adjusts so that only sync pulses appear at the output, and these pulses are negative going.

THE SYNC SEPARATOR

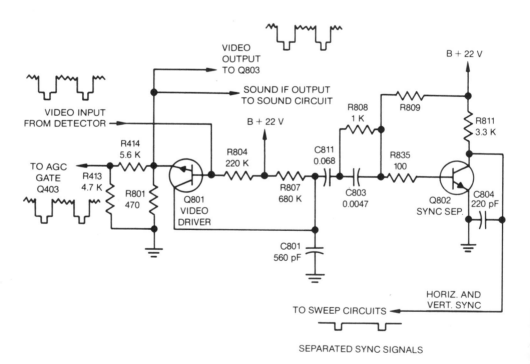

Video Driver—Sync Separator (continued)/Video Output—CRT Circuits

For the circuit shown on the previous page, a small resistance (R807) in the collector of video driver Q801 provides an inverted (positive-going) video signal that is capacitively coupled to the base of sync separator Q802. Thus, these positive pulses applied to the base of Q802 cause current to flow in the base-emitter junction of Q802. This back-biases the base of Q802 so that it is cut off except for the most positive part of the video waveform, the sync pulses. Thus, the collector output from Q802 consists of the sync pulses only, separated from the video signal, and these are used to provide synchronization of the horizontal and vertical sweeps.

Now let us move onto the study of video output/CRT circuits.

The video output/CRT circuits are shown on the next page. The negative-going video from the video driver (Q801) is applied to the base of the video output amplifier Q803 through contrast control R813. R813 is a voltage divider that adjusts the video level into the Q803 base. Since R813 controls the video level, it controls the peak-to-peak amplitude of the signal applied to the CRT cathode. Therefore, it controls the ratio of minimum to maximum picture contrast. Q803 is a common-emitter amplifier stage; therefore its output at the collector is positive going. Since these positive-going signals must reduce the beam current, the video is applied to the CRT cathode. The collector supply for Q803 is +180 volts, obtained from the high-voltage power supply to allow for generating the large video voltage (50–100 volts) necessary to fully modulate the CRT beam. The video signal is capacitance-coupled via C807 to the CRT cathode. The brightness control R829 adjusts bias on the CRT cathode to set the average intensity of the beam.

The CRT uses a fixed-focus voltage on the focus electrode provided by voltage divider R824–R832. The grid and second anode are operated at a fixed voltage provided from the 180-volt supply. The capacitor/resistance network momentarily holds the 180 volts of dc when the set is turned off to prevent a bright spot from appearing when the sweeps collapse, thus preventing the fluorescent screen from being burned from excessive beam current.

A parallel-tuned trap (at 4.5 MHz) in the emitter of Q803 provides a high impedance at the sound IF to reduce the gain of Q803 at 4.5 MHz to prevent sound signals from appearing at the video output. To ensure positive blanking during retrace, positive vertical- and horizontal-retrace blanking pulses from the sweep circuits are applied to the emitter of Q803. These form large positive-going blanking pulses at the collector of Q803, to ensure adequate blanking at all times.

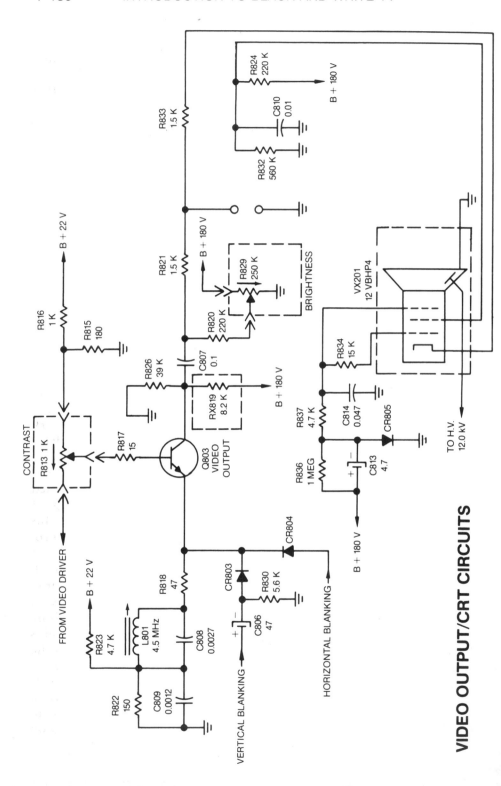

VIDEO OUTPUT/CRT CIRCUITS

AGC Circuits

TV AGC circuits are a little more complicated than a simple AGC for various reasons (see 4–39). The video signal during the blanking-pulse interval can be used for determining AGC. To do this, the video signal must be gated during blanking into a suitable peak detector that holds the peak blanking level so that a control voltage proportional to signal strength is generated for AGC—that is, proportional to the video amplitude during the blanking/sync-pulse interval (see figure).

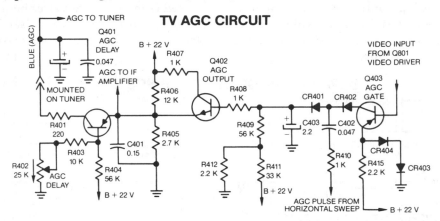

TV AGC CIRCUIT

The AGC gate (Q403) allows the video from the video driver to be gated through during the retrace time of the horizontal sweep. As shown, Q403 has its emitter clamped to 1.3 volts dc by the forward-biased diodes CR403 and CR404. The base of Q403 is biased from the output network of Q801 (video driver) so that the negative video signals tend to drive Q403 into conduction. The collector voltage for Q403 is derived from the negative-going AGC pulse that occurs during retrace. Thus, the only time Q403 is active is during retrace. The video signal thus is sampled during the blanking/sync-pulse interval. Since the video input to Q403 is negative going, the video sample at its collector is positive going. This causes diode CR401 to conduct, storing a charge on capacitor C403 proportional to the peak video level. After many samples are stored, C403 represents the average signal level during the blanking/sync interval. The bias network in the base of the AGC ouput Q402 allows this charge to leak off slowly if no video signal is present. Q402 functions as an emitter follower with the dc voltage stored on C403 on its base. This dc voltage becomes increasingly positive as the signal level increases and is fed to the base of the first IF amplifier (Q101) to reduce its gain as the signal level tends to increase. The output of Q402 is also applied to the emitter of Q401 (common-base amplifier). Potentiometer R402 adjusts the base bias on Q401 so that it is normally cut off, and it is only when the emitter of Q401 becomes sufficiently positive that Q401 begins to conduct and provide AGC voltage to the RF amplifier. Thus, AGC action for the RF amplifier is delayed and is only applied when dealing with strong signals. In this way, full sensitivity is maintained.

Separating the Horizontal and Vertical Sync

As you learned, the sync-pulse train consists of a series of pulses at the horizontal-line rate of 15,750 Hz. These are narrow pulses about 4 μsec wide during all intervals except during the vertical-sync interval when these pulses are made to be almost the full horizontal period in width (about 60 μsec). However, the leading edges of the horizontal-sync pulses are always spaced precisely by the line interval, even though their width may vary. The variable width provides the basis on which the horizontal and vertical sync are separated from each other after being separated from the video signal. The separation of the sync pulses is accomplished by using RC circuits called *differentiators* and *integrator circuits*, as shown.

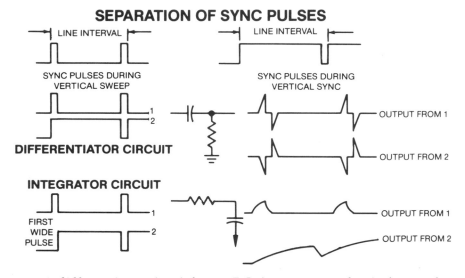

SEPARATION OF SYNC PULSES

A differentiator circuit has an RC time constant that is shorter than the pulse length. Thus, the output, as you know from your study of amplifiers, consists of the leading and trailing edges of the pulses, as shown. You will note that the leading edges of a uniform pulse train like the horizontal-sync pulses are the same regardless of the length of the intervening pulse. By passing the separated sync signal through a differentiator, a series of spikes representing the horizontal sync is obtained; these spikes are continuous, even during the vertical-sync interval. To recover the vertical sync, an integrator circuit is used. Here the time constant for the RC shown is long compared to the normal horizontal-sync pulse time. The capacitor of the RC cannot charge much during the horizontal-sync pulses, and discharges between them result in no buildup. On the other hand, during the vertical-sync pulse interval the horizontal pulses are very wide, and these do build up in the integrator to produce the vertical-sync pulse output, as shown.

In this way, the differentiator (short time constant) or high-pass filter separates the horizontal-sync pulses, while an integrator (long time constant) or low-pass filter separates the vertical-sync pulses.

Vertical-Sweep Circuits

The sync pulses separated by Q802 (sync separator) are negative going at the collector of Q802. These sync pulses are applied directly to the integrator circuit of the vertical sweep circuits as shown on the next page. The integrator consists of R625, C608, R624, and C607. The diode network CR601–CR604 acts as a clipper to prevent large noise pulses from triggering the vertical oscillator improperly. Diode CR601 is forward biased to a 0.5-volt level and acts as the return path for the emitter of Q601. Diode CR604 clips the negative peaks in excess of −0.5 volt.

The vertical oscillator is a collector-coupled multivibrator (Q601–Q602), shown in simplified schematic form in the diagram.

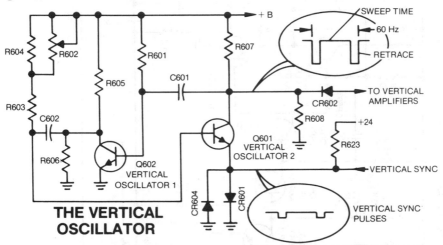

THE VERTICAL OSCILLATOR

Assume that Q602 has just become conducting via base current from R601. The negative-pulse output from the collector of Q602 is coupled to the base of Q601, cutting Q601 off. C602 recharges through network R604, R602, and R603, and when the voltage rises sufficiently, Q601 conducts and its collector voltage decreases. This negative-going waveform is coupled via C601 to the base of Q602, which is cut off until C601 recharges through R601. The time constant C601–R601 is very short compared to the time constant of R604, R602, R603, and C602; therefore, Q601 off time is much longer than its on time. This results in the waveform shown at the collector of Q601. While the on time of Q601 is determined by R601–C601, the off time is set by R602, R603, R604, and C602. Since the time constant of this network is variable (R602), the frequency of oscillation is adjustable by R602.

Negative-going sync pulses from the integrator cut off diode CR601. This causes Q601 to become nonconducting, thus forcing the cycle to be synchronized to the incoming vertical-sync pulses. As you can see, a high measure of mistriggering immunity is achieved because during the active time of the sweep (Q601 cut off), noise pulses cannot affect Q601 since it is already cut off. The vertical-sweep gate generated by the multivibrator is coupled through diode CR602 to the next stage of the vertical-sweep circuit, shown in the complete schematic.

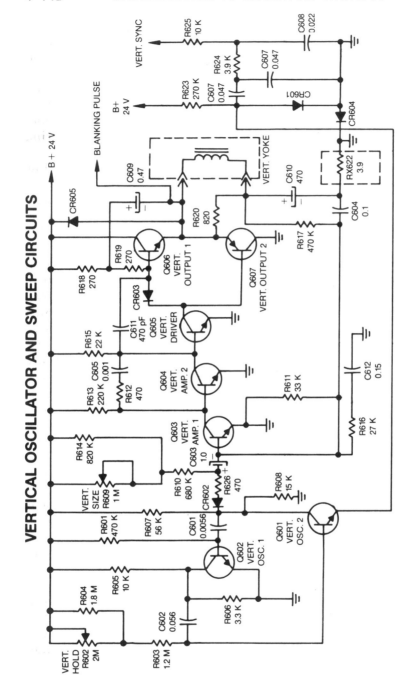

VERTICAL OSCILLATOR AND SWEEP CIRCUITS

Vertical-Sweep Circuits (continued)

The complete schematic of the vertical-sweep circuits is shown on the previous page. The vertical-synchronized sweep gate from the collector of Q602 (the vertical oscillator) is applied to the Q603 base via CR602 and C603. When Q601 collector is positive, CR602 is nonconducting and C603 charges linearly through resistor network R609, R614, and R610, forming the vertical-sweep sawtooth. When Q601 becomes conducting, diode CR602 conducts, discharging capacitor C603 during the vertical retrace. Thus, a vertical sawtooth of voltage is formed. As you will remember, for a linear sweep, a sawtooth of current must flow through the deflection yoke. The current through the vertical deflection yoke is sampled by RX822. Thus, a voltage is generated across RX822 which represents the current flow through the vertical deflection yoke. This voltage is compared to the sawtooth generated at the input of Q603. The resultant is an error voltage applied to the vertical-amplifier circuits such that a sawtooth current is developed through the vertical-deflection yoke which will be linear and the same as the input sawtooth at Q603. The feedback amplifier consists of Q603, Q604, Q605, Q606, and Q607. Q603 amplifies and inverts the error voltage (sawtooth) waveform, which is then dc coupled into amplifiers Q604 and Q605. The negative-going waveform at the collector of Q605 is applied to the bases of the complementary-symmetry transistor pair Q606 and Q607. During the more positive portion of the waveform at the Q605 output, Q606 conducts with Q607 cut off, while during the more negative portion of the input waveform, Q607 conducts while Q606 is cut off. Diode CR603 smooths the switching action. As you can see, the output transistor pair is quite similar to the complementary-symmetry audio amplifiers that you studied earlier. The emitters of Q606 and Q607 are connected together and drive the vertical-deflection coils directly to provide the vertical sweep of the CRT trace.

As you can see, the basic vertical sync from the sync separator is used to determine the frequency of oscillation of the vertical-sweep oscillator, which in turn develops the vertical sweep. The vertical oscillator is free running so that the raster appears even where no signal is present. For synchronization to occur, it is necessary to have the free-running frequency of the oscillator slightly lower than the frequency of the sync signal. Otherwise the vertical oscillator will have started the next sweep before the sync pulse arrives.

Now that we have completed an in-depth study of vertical-sweep circuits we are ready to tackle our next subject—the study of horizontal-sweep circuits, the horizontal oscillator, and horizontal APC.

Horizontal-Sweep Circuits—Horizontal Oscillator—APC

The sync pulses from the sync separator are applied to the horizontal-sweep circuits via the differentiator, as shown in the partial schematic diagram on the next page. The diagram shows the horizontal oscillator and the horizontal APC (*automatic phase control*) that keeps the horizontal-oscillator frequency locked to the incoming sync pulses. The APC approach to horizontal-frequency control is used because it provides an extremely high degree of immunity to noise that can cause very annoying horizontal jitter. The noise immunity comes from the fact that the horizontal-oscillator frequency is controlled by averaging many samples so that oscillator frequency changes can only occur slowly. Because of limited pull-in range, the horizontal oscillator must be very stable in frequency, with only small corrections provided by the APC.

The horizontal oscillator Q502 is a conventional Hartley oscillator (base-emitter coupled) with its free-running frequency determined by inductor T501 and parallel capacitor C508. A portion of the oscillator voltage is coupled through C507 to the collector of Q501. Diode CR503 prevents the collector from swinging negative. Thus, the collector of Q501 consists of positive pulses. C507 can be considered to be in parallel with the horizontal-oscillator tuning capacitor C508. As the base voltage on Q501 varies, the amount of apparent capacitance from C507 varies, providing a fine adjustment of the frequency of the horizontal oscillator.

The input to Q501 base is obtained from the phase detector that compares the sync pulses to an APC pulse obtained from the horizontal flyback. The APC pulses are shaped into a sawtooth waveform by network RX506, C520, C502, and C503. The negative-going horizontal-sync pulses are applied to the diodes CR501 and CR502, causing them to conduct. These clamp a portion of the sawtooth generated from the APC pulse to ground. The point where this occurs depends on the phase relationship between the sawtooth and the sync pulse, producing a voltage at the input of R504 that depends on the point where the sawtooth is clamped to ground. Thus, varying voltage is filtered by R504, C506, R505, and C505 and applied to the base of Q501 to vary its collector impedance, which adjusts the frequency of Q502 to the value that leads to a voltage from the phase detector that locks the horizontal oscillator in phase with the sync pulse.

The schematic diagram of the horizontal-sweep circuits (less those portions previously discussed) is shown on page 4–145. As shown in this schematic, the output from the horizontal oscillator is applied to the base of horizontal driver Q502. This horizontal driver plus the horizontal output (Q504) will be discussed in detail on page 4–144. On the following page we have the schematic diagram for the phase detector and APC.

PHASE DETECTOR AND APC

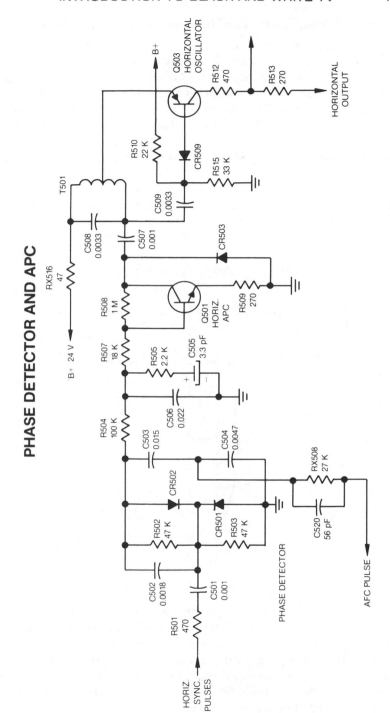

Horizontal-Sweep Circuits—Horizontal Driver and Output

The voltage swing at the base of Q502 is sufficient to drive it between cutoff and saturation. Thus the collector voltage of Q502 consists of a square-wave signal at the horizontal frequency. This signal is applied to the base of the horizontal output Q504 via transformer T502. Q504 is normally cut off (base and emitter at ground potential). The signal from T502 drives Q504 into conduction (saturation), which causes transformer T503 to become saturated, causing increasing current through the transformer primary and the deflection yoke, deflecting the beam from the center of the screen to the right-hand edge. When Q504 is cut off, the magnetic field in the yoke collapses, causing the beam to retrace to the left-hand edge of the screen. The magnetic field in the yoke begins to build up as the transformer unsaturates, deflecting the beam toward the center of the screen, as shown by the waveforms below.

HORIZONTAL DRIVER AND OUTPUT

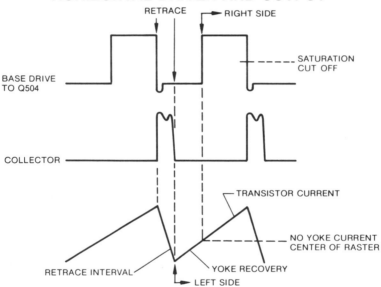

The inductive circuit formed by the deflection yoke and transformer T503 primary in parallel provides an inductive kick when the horizontal-output transistor Q504 is abruptly cut off, yielding the waveform at the Q504 collector shown above. When Q504 is conducting, current is drawn through the deflection yoke to provide the sweep. When Q504 cuts off, a 100-volt pulse appears at its collector. This is caused by the collapse of the magnetic field of the yoke, resulting in retrace of the horizontal sweep. This pulse is also used by T503 to provide high voltage, as described on the next page. A sample of the horizontal-retrace pulse is taken from the collector of Q504 via RX518 to the emitter of the video output transistor to provide horizontal blanking, as described earlier.

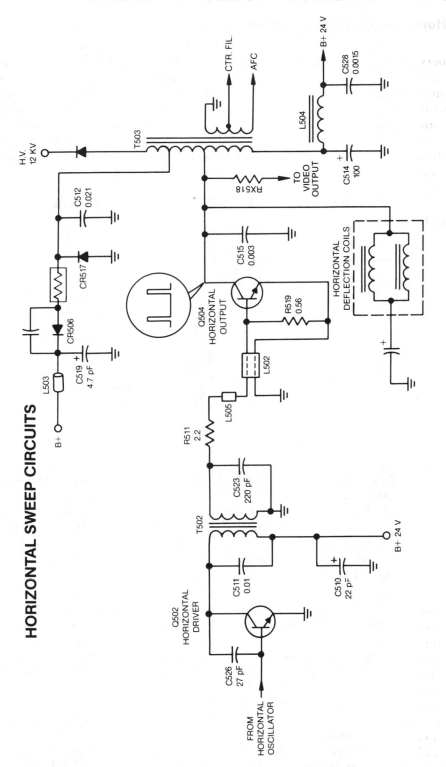

HORIZONTAL SWEEP CIRCUITS

High- and Low-Voltage Power Supplies

The high voltage for the CRT second anode (12 kV) and +180 volts for operation of some of the video circuits and vertical-sweep circuits as well as CRT bias are obtained from T503. As shown in the diagram (see previous page for component value), the 100-volt retrace pulses at the collector of Q504, the horizontal output, are stepped up in the autotransformer primary of T503. The +180-volt output is obtained from an appropriate tap on T503 and rectified and filtered by CX506 and C519. Diode CRX517 clamps the bottom of the pulse waveform to ground so that the full pulse voltage available at the tap is obtained across C519.

HIGH- AND LOW-VOLTAGE POWER SUPPLIES

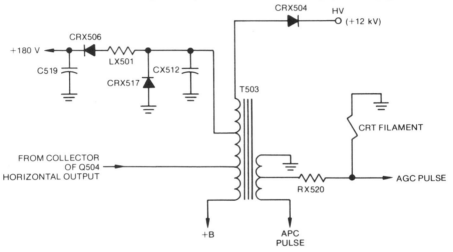

High-voltage dc (+12 kV) is obtained from the stepup of the collector pulse by autotransformer T503. This is rectified by diode CRX504. As described earlier, the high-voltage filter capacitor is formed by the final anode and an external conductive coating on the CRT envelope. Voltage for the CRT filament (a low-power filament) as well as the APC pulse and AGC gating pulse are obtained from a separate secondary on transformer T503. This method for obtaining filament voltage may seem unusual, but it is simple and convenient; in many TV sets, however, the filament voltage is taken from the low-voltage power supply transformer.

As shown by the schematic on the following page, the low-voltage power supply is of conventional design, providing an unregulated +24 volts of dc to the audio and sweep circuits. A value of +22 volts and a regulated +11 volts are provided for the video portion and the tuner, respectively.

For the circuit shown, a loss of sweep causes a loss of bias, but the bias voltage is held, as discussed earlier, long enough for the filament to cool so that no spot is formed. Otherwise, a disadvantage of the approach shown here would be that a failure of the sweep would also cause loss of CRT bias, leading to a bright spot that can burn the CRT screen.

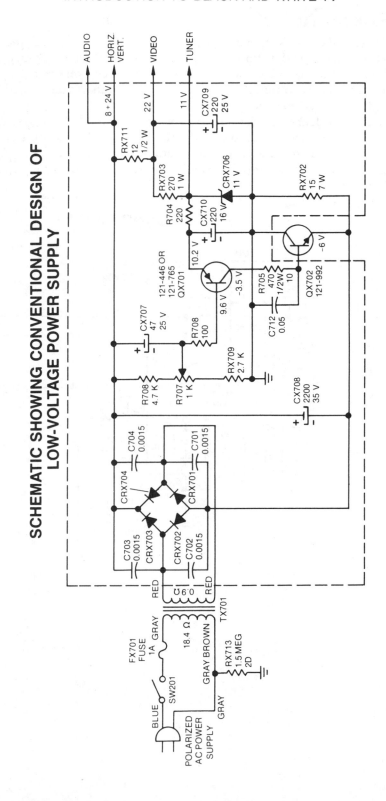

SCHEMATIC SHOWING CONVENTIONAL DESIGN OF
LOW-VOLTAGE POWER SUPPLY

FM Sound Circuits

The sound circuits shown here and on the next page accept the 4.5-MHz IF carrier signal from the video driver Q801 and amplify and demodulate it in the same way as the FM radio receivers you learned about earlier in this volume. The resulting audio signal is amplified and used to drive a loudspeaker.

In the TV learning system, the sound IF demodulator system is contained in a single integrated circuit (IC1001), which also contains the ratio detector and first stages of audio amplification. The 4.5-MHz IF transformer (T1002) at the input to the IC and the quadrature coil (T1001) are external to the IC.

FM SOUND CIRCUIT

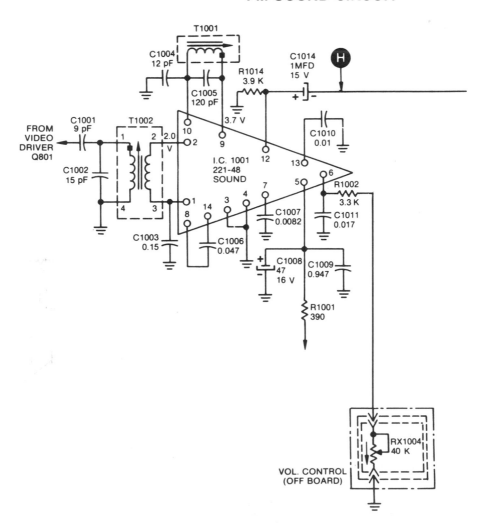

FM Sound Circuits (continued)

The input to IC1001 is the tuned transformer (T1002) coupled lightly via C1001 (8 pF) from the video driver Q801. The output of IC1001 is an audio signal that is fed to the audio amplifier. Volume is controlled by RX1004, which is connected to pin 6 of the IC. The audio output of IC1001 is fed to the complementary-symmetry amplifier (Q1001, Q1002, and Q1003) that you studied in Volume 2 of this series.

The audio output signal of the IC is developed across emitter resistor R1014, as shown. The audio power amplifier that receives the output signal from the IC consists of driver Q1001 and output transistors Q1002 and Q1003 connected in complementary symmetry. Q1003 is a pnp transistor, while Q1002 is an npn. They are connected as an emitter-follower combination that is biased to operate as a Class B amplifier. Use of this emitter-follower circuit permits connecting the output loudspeaker without the use of an output transformer.

FM SOUND CIRCUIT

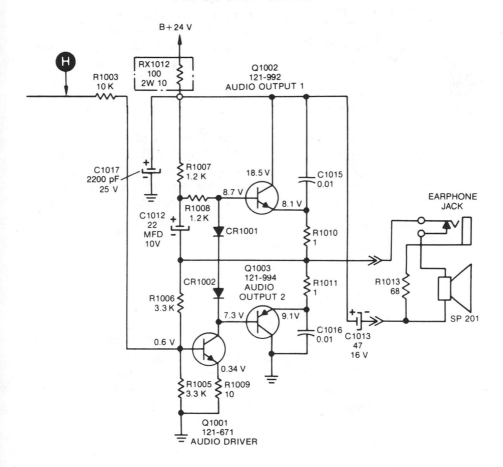

Review of Black-and-White TV Receivers

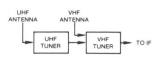

1. BOTH VHF AND UHF TU-NERS are included in TV receivers. The output of the UHF tuner feeds through the VHF tuner.

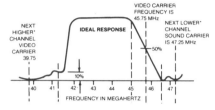

2. VIDEO IF AMPLIFIERS typically have three transistor stages and feed a diode video detector. They must have a response appropriate for proper video and sound reception.

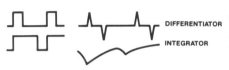

3. GATED AGC is provided by opening the AGC circuit only during horizontal sync pulses. Horizontal and vertical sync pulses are separated by a *differentiator* and an *integrator*.

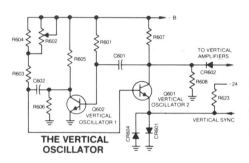

4. HORIZONTAL AND VERTICAL SWEEP CIRCUITS generate the voltages to deflect the CRT beam. Each has an oscillator and amplifiers. The horizontal sweep uses APC for noise immunity.

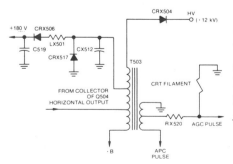

5. LOW VOLTAGE AND HIGH VOLTAGE POWER SUPPLIES are used. High voltage (and usually some low voltage) is obtained from the pulses in the *horizontal output transformer*.

Self-Test—Review Questions

1. Draw a complete block diagram of a black-and-white TV receiver.
2. Describe the function of each block from question 1.
3. Draw a typical TV channel showing location of video and sound carriers. What is the overall response shape of a TV video IF passband? How is this shaping accomplished? Why is it made this way?
4. Why is the sound carrier at 4.5 MHz regardless of where the receiver local oscillator is tuned? Sketch and describe the elements of a TV sound system.
5. What are some of the basic differences between a radio receiver RF/IF-detector system and a TV RF/IF-detector system? Explain. How are TV sets tuned?
6. How does a TV AGC differ from a radio receiver AGC? Why? Why is the tuner AGC delayed?
7. How are the sync signals recovered from the video signal? How are they separated?
8. Draw a block diagram of a horizontal-sweep circuit, including the horizontal oscillator and horizontal AFC. Describe how each element operates.
9. Draw a block diagram of the vertical-sweep circuits, including the sync circuits. Describe how each element operates.
10. How is high voltage for the CRT final anode obtained?

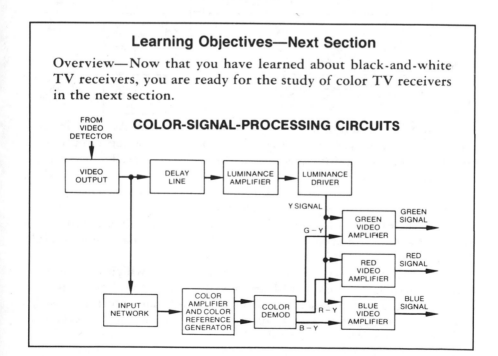

Learning Objectives—Next Section

Overview—Now that you have learned about black-and-white TV receivers, you are ready for the study of color TV receivers in the next section.

COLOR-SIGNAL-PROCESSING CIRCUITS

Introduction to Color Reception

You learned how a color TV signal was generated for transmission in your study of information transmission in Volume 3. Here, we will briefly review the form of the color signal. As you know, essentially any color can be made up from the three primary colors of red, green, and blue in the proper combinations. You also will recall that a given point in a TV scene can be described in terms of its *brightness* or *luminance* and its *hue* or *chrominance* (color). In black-and-white TV we used only the brightness or luminance signal. In color TV, a *second set* of signals is generated; these describe the *chrominance* and are transmitted on a *sub-carrier*. The only real difference between a color signal and a black-and-white signal lies in the *additional information* contained in the *color subcar-rier*. Thus, a color TV receiver is like a black-and-white receiver except for the special circuits to demodulate and regenerate the chrominance signal and the special CRT system to use this chrominance information to provide a color image.

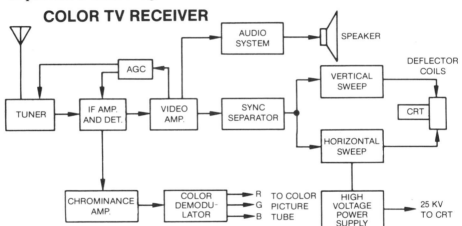

A block diagram of a color TV receiver is shown. As you can see, the RF, IF, audio, sync, and horizontal- and vertical-sweep circuits are essentially the same as for a black-and-white TV receiver. These circuits are similar to those you have already studied, except that more care is often used in the design of the IF circuits for color TV to ensure that all of the signals up to 4.2 MHz are received with good fidelity. In addition, the high voltage required for a color picture tube is generally higher (26–30 kV) and must be relatively stable, so that more careful design is required in this area as well.

The major difference lies in the video circuits, which include not only the necessary video amplifiers but also the circuits necessary to demodulate the chrominance signal and the additional circuits necessary to combine the luminance and chrominance information to recreate a color image on a color picture tube. Before you begin your study of the special circuits added to form a color TV receiver, the nature of the color TV transmission will be reviewed and the operation of the color picture tube or color CRT will be described.

The Color TV Signal

You will recall from your study of TV transmission that the color signals are generated at the studio by generating three signals using either three cameras with separate red-green-blue filters or by using a single tube with a special filter faceplate. These signals (E_R, E_G, and E_B for the red, green, and blue image signals, respectively) are then combined in a matrix, as shown, to form the luminance (or monochrome brightness) signal called E_Y. The E_R and E_B signals are also combined with the E_Y signals to form the E_I and E_Q video signals. E_I and E_Q are the chrominance signals, since they carry the color information. The E_I and E_Q signals are used to produce a suppressed-carrier double-sideband color-subcarrier signal centered at 3.579545 MHz, chosen at this frequency to ensure no interaction between the luminance and chrominance signals. The bandwidth of the color signals is carefully controlled, and the subcarrier is also carefully related to line and frame frequency.

COLOR TV SIGNAL

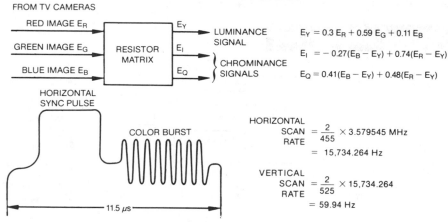

FROM TV CAMERAS

RED IMAGE E_R — RESISTOR MATRIX — E_Y → LUMINANCE SIGNAL

GREEN IMAGE E_G — E_I

BLUE IMAGE E_B — E_Q → CHROMINANCE SIGNALS

$$E_Y = 0.3\,E_R + 0.59\,E_G + 0.11\,E_B$$

$$E_I = -0.27(E_B - E_Y) + 0.74(E_R - E_Y)$$

$$E_Q = 0.41(E_B - E_Y) + 0.48(E_R - E_Y)$$

HORIZONTAL SYNC PULSE

COLOR BURST

$$\text{HORIZONTAL SCAN RATE} = \frac{2}{455} \times 3.579545 \text{ MHz}$$

$$= 15{,}734.264 \text{ Hz}$$

$$\text{VERTICAL SCAN RATE} = \frac{2}{525} \times 15{,}734.264$$

$$= 59.94 \text{ Hz}$$

11.5 µs

Since the chrominance information need not have very high resolution, because it does not carry the fine detail of the luminance information, the bandwidth of the E_I is limited to about 1.3 MHz and the E_Q to about 450 kHz. The band-limited E_I and E_Q signals are used to modulate the color subcarrier in quadrature, as will be described shortly.

To demodulate the suppressed-carrier chrominance signal, the carrier must be regenerated. Since phase is important, this carrier must be synchronized in phase with the subcarrier oscillator at the transmitter. To do this, a sample of the color subcarrier is transmitted on the back portion (back porch) of the horizontal-blanking pulses, as shown. This burst is recovered and is used to phase lock the receiver subcarrier oscillator. You will note that the horizontal- and vertical-scan rates are slightly different for color than for monochrome transmission. These differences are necessary to minimize any interference between the chrominance and luminance signals. These differences are so slight that the receiver can easily stay in sync when switching from monochrome to color or back. In fact, the old monochrome values are no longer used.

The Color TV Signal (continued)

The color information in E_I and E_Q must be combined so that they can be transmitted on the subcarriers, but in such a way that they can be recovered at the receiver. To understand how this is done, you will have to remember something about vectors and phase detectors that you also studied earlier. As shown in the diagram, the 3.58-MHz (3.58 MHz is often used as a shorthand way to express 3.579545 MHz) carrier is generated at the transmitter. It is used directly in the balanced modulator for the E_I signal, but it is phase shifted by 90 degrees for the balanced modulator for the E_Q signal. When the outputs from these balanced modulators are combined, the resultant is a single vector whose phase and amplitude depend on the relative magnitudes of E_I and E_Q. This single vector, containing both E_I and E_Q information, is the color signal that along with the E_Y and sync signals is used to modulate the video carrier. Obviously, if we can recover the E_Y, E_I, and E_Q signals at the receiver, we can put them into an appropriate matrix (add or subtract, as necessary) to reform E_R, E_G, E_B, and E_Y.

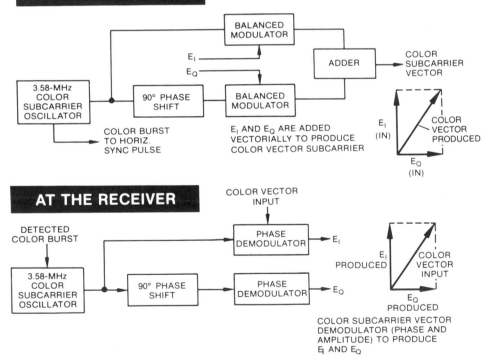

At the receiver, the color subcarrier is generated by locking a 3.58-MHz oscillator from the color bursts on the horizontal-sync pulses. This carrier signal is used directly in a demodulator (phase detector). The output from the phase detector is *only* the *in-phase* component of the color vector, which is E_I.

The Color TV Signal (continued)

As shown in the diagram on the previous page, an identical demodulator operated with a 3.58-MHz signal shifted 90 degrees produces only the E_Q (quadrature) component of the color vector. Thus, both E_I and E_Q are recovered from the color-subcarrier vector. The E_Y signal is recovered, as you have already learned, since it is the monochrome signal that is the same as a black-and-white video signal. Thus, the E_Y video, the E_I video, and the E_Q video are regenerated in the color TV receiver to produce the E_R, E_G, E_B signals necessary to operate the color TV picture tube.

As described on the previous page, the E_I, E_Q, and E_Y signals are recovered from the chrominance and luminance signals, respectively, to recover the red, blue, and green modulating signals. To do this it is now necessary to combine the E_I, E_Q, and E_Y signals in the right ratios and polarity. This is essentially what is done in a matrix similar to that used to generate the E_Y, E_I, and E_Q signals, which you learned about earlier. In most modern TV sets, this matrix operation is done in an IC containing the necessary resistors and inverters to reform the red, green, and blue signals (E_R, E_G, and E_B, respectively).

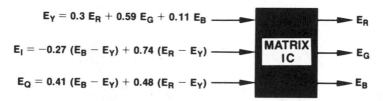

$$E_Y = 0.3\ E_R + 0.59\ E_G + 0.11\ E_B$$

$$E_I = -0.27\ (E_B - E_Y) + 0.74\ (E_R - E_Y)$$

$$E_Q = 0.41\ (E_B - E_Y) + 0.48\ (E_R - E_Y)$$

MATRIX IC → E_R, E_G, E_B

As you will see, the color picture tube has three electron guns that are designed to strike phosphors corresponding to a particular color— that is, a red gun that causes the screen to glow red, a green gun that causes the screen to glow green, and a blue gun that causes the screen to glow blue. You will note that at full intensity, the E_Y signal (monochrome) is equal to 1 ($0.3 + 0.59 + .11 = 1$). If this value is used to find E_I and E_Q, you will find that the sum of E_R, E_G, and E_B is equal to 1 also. Since we are dealing with difference signals, the difference $E_Y - (E_I + E_Q) = 0$, or all beams on full, giving a white picture if the CRT is properly adjusted. You will also find that each color can be recovered by substitution of the appropriate values in these equations. For example, if the picture is all blue, then $E_R = E_G = 0$. Substitution of these will show that only E_B remains. Similarly if $E_B = E_R = 0$, then only E_G remains; if $E_B = E_G = 0$, only E_R remains. You will see more clearly how these signals are used after you learn about the color picture tube.

It is important to realize that the red, blue, and green images are processed at the transmitter to form an E_Y, E_I, and E_Q signal, where E_Y is the monochrome signal and E_I and E_Q contain the information necessary, with E_Y, to form the red, blue, and green video signals. The recovery of E_Y, E_I, and E_Q at the receiver then likewise allows the recovery of E_R, E_B, and E_G at the receiver.

The Color Picture Tube

As you know, most colors can be generated by the appropriate combination of red, green, and blue (the three primary colors). If you look through a magnifying glass at a color TV set in operation, you will see that the color image is actually formed by clusters of tiny dots or strips of phosphor that glow red, green, and blue. These phosphor dots or strips are so small that they cannot be seen individually without a magnifier. A group of three spots (red, green, and blue) is called a *triad* and is somewhat smaller than an individual resolution element in the picture. If you watch these triads as the scene color changes, you will see that the relative brightness of the elements of the triad varies as the color changes from scene to scene. Since the phosphor dots or strips are so small, we see them collectively as a single spot whose color is determined by the *relative brightness* of the *individual elements*. The separate illumination of each of these spots is accomplished in a color TV picture tube using the three color signals E_R, E_G, and E_B described earlier.

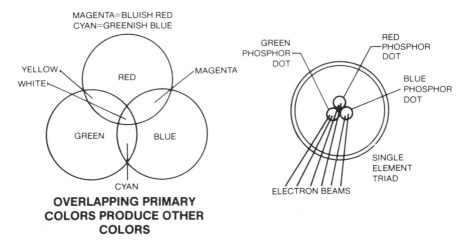

OVERLAPPING PRIMARY
COLORS PRODUCE OTHER
COLORS

Inside a typical color TV picture tube there are three electron guns, each producing a separate beam. These beams scan together, just as a single beam does in a monochrome picture tube. We can adjust the beam from each gun so that it strikes only a single kind of color phosphor dot—that is, the blue beam strikes only blue phosphor dots, the green beam strikes only green phosphor dots, and the red beam strikes only red phosphor dots. Then, if we adjust the *intensity* of each beam, we can produce in each triad *any color* we choose from white to any color to black (no beams on). We could receive a monochrome luminance signal and apply it to all three guns simultaneously after adjusting the ratio of the beam intensities to produce white light and have a satisfactory black-and-white image. This is what happens when we receive a monochrome picture on a color receiver. If we also individually modulate each beam at each point in accordance with the color information received, we would produce a color image, and this is exactly how color image is formed by driving each beam with its corresponding color video signal.

The Color Picture Tube (continued)

The basic problem in a color picture tube is making sure that the right beam strikes only the correct phosphor dots. An element known as the *shadow mask* is located behind the phosphor dot screen. Its purpose is to assure that the beam from the red gun passes through the shadow mask at an angle that will allow it to strike only the red-light-emitting phosphor. Similarly, the blue- and green-emitting phosphors can be struck only by the beams from the corresponding blue and green guns. This is accomplished basically by setting the guns and electron beams so that they converge through the holes in the shadow mask at an appropriate angle to give each beam the geometry necessary to permit it to strike only the appropriate color phosphor dots. The very fine registration necessary to achieve the purity of color by having each beam strike only its specific phosphor dots requires very careful control during the manufacture of the tube. Furthermore, most color TV sets use carefully regulated high voltage to maintain the registration and focus. Because some of the beam energy is lost by the shielding from the shadow mask, color CRTs usually use much higher final anode voltages to obtain a bright picture. In addition, since the beams cannot each occupy the same position in the deflection field, a set of adjustments to align each beam properly during deflection is necessary to maintain convergence of the beams to the proper color dots. These controls are called *convergence controls* and require resetting whenever the picture tube or its geometry is disturbed. The convergence controls in a conventional shadow mask color CRT are complex and difficult to adjust properly. Therefore, several of the newer color picture tube systems use somewhat different approaches to solve some of these problems. The phosphor dot screen is called a *P22 phosphor*. All modern color CRTs are rectangular with a 4/3 aspect ratio, as for black-and-white CRTs. Electrostatic focus is universally used.

OPERATION OF SHADOW MASK

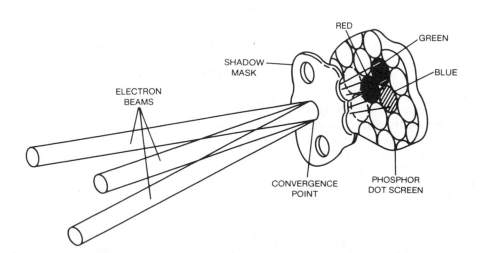

The Color Picture Tube (continued)

Several innovations have been used in color CRTs to reduce the critical aspect of the convergence controls and to simplify adjustment. One of these is the in-line gun structure developed by the Zenith Radio Corporation called the *Chromocolor®* picture tube. In this tube, the three guns are arrayed in a horizontal line. The phosphor is arranged in vertical strips and the shadow mask is also a vertical slit, so that the beams diverge through the slit to strike the color strips, as shown.

CONSTRUCTION OF COLOR PICTURE TUBE

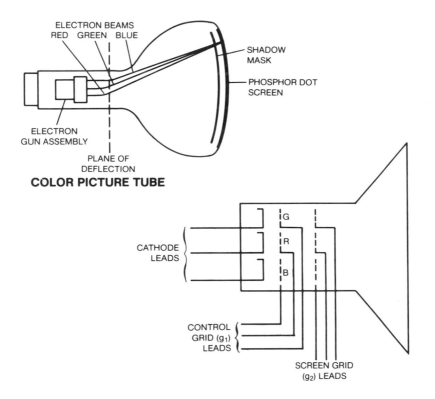

COLOR PICTURE TUBE

COLOR PICTURE TUBE SCHEMATIC

Since the convergence problems are limited to the horizontal direction only, a great simplification in adjustment is possible. Furthermore, the shadow-mask blocking area is reduced so that a brighter picture is possible for the same beam current.

The Color Picture Tube (continued)/Color Circuit Functions

Another innovation is the single-gun color picture tube. This tube uses a vertical strip phosphor system, just like the Chromocolor® tube, but the single beam is allowed to scan across all strips. Beam switching, as in the color single-gun camera tube, is used to separate the illumination of each vertical strip for each color.

In all color picture tubes, particularly in the larger sizes, the high voltage is sufficient to produce *soft x-rays*. While these are generally confined within the TV set to limits set by U.S. federal regulation, it is still a good idea to keep a *safe distance* (greater than 6 ft) when viewing a color TV receiver—particularly if the CRT is over 17 inches in size.

As you learned earlier, when the color TV signal is received, the processing is the same as for the black-and-white system, except for the *color-signal-processing circuits* not included in the black-and-white TV. The remainder of this section is devoted to describing those circuits.

The color-processing circuits in most modern TV receivers use ICs for many of the functions, so that the circuit functions are generally not obvious from the schematic diagrams. Therefore, in this section, while we will use a learning system approach as before, using the Zenith Radio Corporation model 19HC55, we will use a mixture of block diagrams and schematic diagrams to discuss the circuits unique to a color TV receiver.

COLOR CIRCUIT FUNCTIONS

1. EXTRACT COLOR INFORMATION FROM VIDEO
2. ADJUST INTENSITY OF BEAMS IN CRT
3. INDEPENDENTLY DRIVE THE R, G, B BEAMS
4. SUPPRESS COLOR INFORMATION DURING BLACK AND WHITE RECEPTION
5. PROPERLY CONVERGE CRT BEAMS

The color-processing circuitry of a TV receiver has *five functions*. First the *color information* must be *extracted* from the received signal, and the red, green, and blue *signals recovered*. Second, the *correct intensity* of each of the three electron beams must be *adjusted* so that the final color produced by the phosphor dots will *accurately duplicate* the colors in the original scene from the TV studio. Third, the red, green, and blue signals must be used to *drive* the three electron beams *independently*. Fourth, the colors must be *turned off* when only a black-and-white program is being received. Fifth, the three electron beams must always be made to *strike* the *correct dots* of the triad on the screen.

The pages that follow will describe how each of these objectives is achieved.

Color-Signal-Processing Circuits

The receiver color-signal-processing circuits extract the red (R), green (G), and blue (B) signals from the received signal. These are used to independently drive the three electron guns in the CRT. The block diagram shown here summarizes how this is accomplished in a typical color TV receiver.

The output of the video amplifier is the demodulated video signal, which is essentially a duplicate of the transmitted video. This signal contains the combined luminance (Y) and chrominance information, together with the blanking and sync signals and the color-sync burst.

COLOR-SIGNAL-PROCESSING CIRCUITS

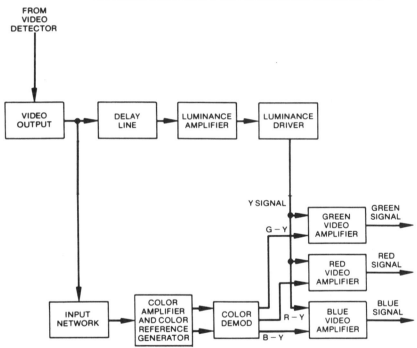

A 3.58-MHz wave trap at the output of the video amplifier is usually used to prevent the color subcarrier from entering the signal path to the luminance amplifier. A delay line at the input of the luminance amplifier delays the luminance signal in time so that it arrives at the output amplifiers *coincident* with the arrival of the corresponding chrominance signals, which are slightly delayed because of their narrower bandwidth and additional processing. After amplification in video amplifiers, the luminance signal is fed to the red-, green-, and blue-output amplifiers to be combined with R–Y, G–Y, and B–Y signals from the color or chroma sections.

Color-Signal-Processing Circuits (continued)/Video Output to CRT

The path of the chrominance signal is shown in the lower portion of the block diagram. The input signal to this path is the demodulated video signal, and it is first necessary to suppress the low-frequency components of the luminance signal. This is done by passing the video signal through a bandpass transformer that passes only signals in the 2.5- to 4.5-MHz range. The 3.58-MHz reference signal is generated by locking a local reference to the 3.58-MHz burst in the received signal. After amplification by the color amplifier, the signal enters the color demodulator where the R–Y, G–Y, and B–Y signals are recovered. The demodulator output is sent to individual color-output amplifiers, where they are each combined with the luminance signal for delivery to the three guns of the color CRT. As you can see, the Y signal is added to the R–Y, G–Y, and B–Y signals from the color section, yielding an R,G,B signal for modulating the color CRT. For black-and-white reception, the R–Y, G–Y, and B–Y signals are absent, leaving only the luminance signal Y to drive the three CRT grids.

In the discussion to follow here and on subsequent pages, we will start with the video ouput/CRT driver and work back to the video detector output at the IF. In this way, you will be able to see what signals are required at each step and how they are derived. You know that three signals (E_R, E_G, and E_B) are needed to drive the CRT guns. You know that the signals applied to these guns are simply proportions of the luminance (Y) signal for black-and-white reception. The schematic on the following page shows the video output circuits for the Zenith model 19HC55. The basic inputs, as shown, are the Y signal and the R–Y, G–Y, and B–Y signals.

As shown, the Y (luminance) signal is applied to the base of Q1201, the luminance driver. A resistance-divider network contains the precision resistors of the correct value to establish the magnitude of the Y signal into Q1201. All components that do not have identifying symbol numbers are in a separate IC. The output from the emitter of Q1201 is coupled to the emitter of the color-output amplifiers Q1205, Q1206, and Q1207 through resistor networks in U1202. Thus, Q1205, Q1206, and Q1207 act as common-base amplifiers for the Y signal. The outputs of these amplifiers are coupled to the cathodes of the CRT. The Y signal input is positive going, and since common-base amplifiers are noninverting, positive video is applied to the CRT cathodes. The R–Y, G–Y, and B–Y signals obtained from the color demodulators, to be described subsequently, are applied to the bases of the output amplifiers via resistive networks and emitter followers Q1202, Q1203, and Q1204. These negative-going signals at the bases of the output amplifiers are then added in the output of the output amplifiers. Thus, (B − Y) + Y yields B, etc. In this way, the three color signals for the CRT are derived from the color video and the Y (luminance signal).

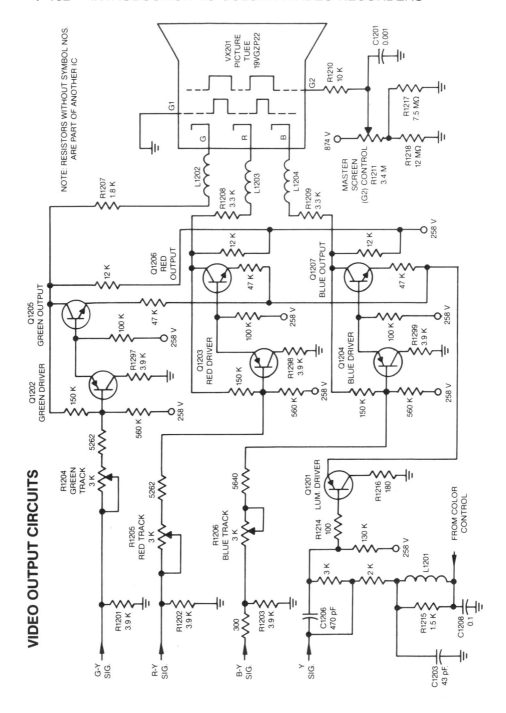

Low-Level Luminance Circuits

The low-level luminance circuits provide for video amplification of the luminance signal and the establishment of the necessary black-level control and chrominance-level control. All of these functions occur in a single IC in the Zenith model 19HC55, so this function will be discussed primarily in block diagram form, although a schematic diagram is included for reference.

As shown in the schematic diagram on page 4–164, the detected video is applied to the low-level luminance circuits via a delay line, described previously. The video to the chrominance circuits is taken from the video detector before the delay line. IC901, shown in the schematic, basically provides for video amplification of the luminance signal. The video signals are dc coupled, so it is necessary to establish a fixed, black level for the video. This is accomplished by clamping the back porch of the video horizontal-blanking pulse to a known level during each blanking period. In addition, the chroma signal (color reference) is brought into IC901 for adjustment of its gain consistent with the video gain (contrast) adjustment to ensure good tracking of the luminance and chrominance signals.

As also shown in the schematic diagram, the low-level luminance circuit is basically a single active IC (IC901) and an associated network of resistors packaged together as IC902 (resistors with values only). As illustrated, the video input is taken from the video detector and applied to the delay line (L208) (about 1.3 μsec) and also to the color demodulator. The delay line output is coupled to the IC via CR901, which clips negative-going noise pulses from the signal. C906 and L901 form a 3.58-MHz trap to recover the color subcarrier signal from the luminance signal.

The low-level luminance circuits are shown in block diagram form on page 4–165. IC901 contains the dual variable-gain video amplifiers operated by the contrast control, the gated clamp, and the output video amplifier. Not shown is a circuit that detects the vertical sawtooth. If the vertical sweep should fail, the CRT is biased off by a dc level applied to the output video amplifier that is dc coupled through the video output to the CRT cathodes. Since loss of horizontal sweep also causes loss of high voltage to the CRT, automatic protection against horizontal sweep failure is provided.

As shown, the video is applied to a variable-gain-controlled amplifier to provide video gain control (contrast control). The same type of variable-gain amplifier is also used to amplify the chrominance signal so that its gain is also controlled simultaneously. This is necessary because the chrominance data are taken before this gain control.

LOW-LEVEL LUMINANCE CIRCUITS

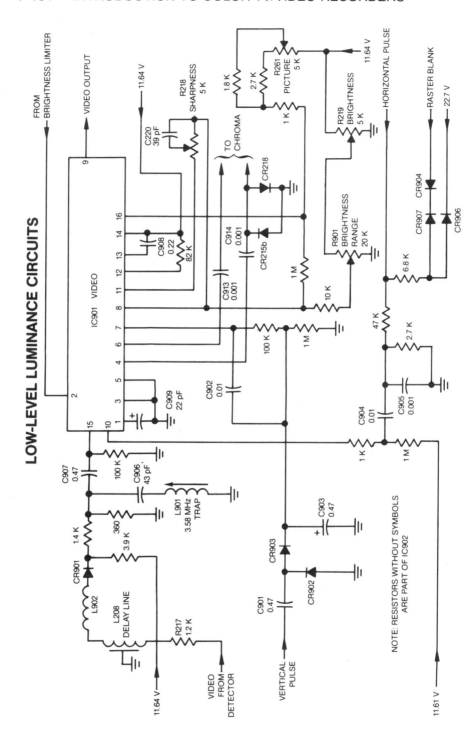

Low-Level Luminance Circuits (continued)

In many TV sets, a single control is used before the chrominance pickoff point. The output of the video amplifier has a dc component of the signal that depends on scene brightness. This variable dc would cause variation in the average CRT bias. To prevent this, the video signal is clamped to a fixed voltage (dc) by the gated clamp that establishes a fixed reference level. In this case, the clamp is driven by the retrace pulse from the horizontal sweep. This horizontal pulse is shaped and delayed in an RC network and used to drive an electronic diode switch that closes for the back porch interval of the horizontal-blanking interval. Since this level is fixed at the transmitter to represent the black level, all the video is referenced to this fixed level, as shown. The clamp output is fed to the video amplifier, as shown. Brightness is controlled by varying the dc level of the video output stage.

We will now proceed to a study of chrominance-processing circuits.

BLOCK DIAGRAM OF IC901

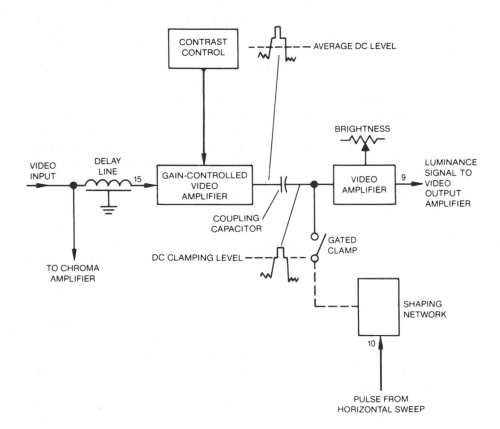

Chrominance-Processing Circuits

The last major element required to provide color TV reception is the recovery of the chrominance information. To do this requires that the 3.58-MHz color burst be separated and used to phase lock a local 3.58-MHz CW reference, and that this reference be used in conjunction with the chrominance video signal to obtain E_I and E_Q from which the color-difference signals R–Y, G–Y, and B–Y can be obtained. Since most modern color TV receivers use ICs, the overall IC circuits will be shown, followed by a block diagram description. It should be noted that many configurations of these circuits are used, however, the basic elements and functions described above are always present.

As shown schematically on the following two pages, the undelayed video input is applied to IC1001, which contains the color demodulators for recovery of the B–Y, R–Y, and G–Y signals. The video is coupled into IC1001 by L1002 and associated capacitances to eliminate low-frequency luminance signals. The phase reference is regenerated in IC1002. The horizontal blanking pulse is applied to IC1002 to gate the 3.58-MHz reference burst into the 3.58-MHz oscillator control. In the circuit shown, a 3.58-MHz crystal oscillator (CR1001) is used with IC1002 to provide a reference signal that is phase locked to the color burst. The phase-locked 3.58-MHz signal from IC1002 is then used in the phase detector/demodulator and matrix of IC1001 to provide the chrominance output signals. The E_I and E_Q signals recovered from the phase detector/demodulator are, in conjunction with the appropriate matrix, all contained in IC1001, so the E_I and E_Q signals are not a direct output.

The chrominance-processing circuits are shown in block diagram form on page 4–169. The video input is filtered to remove the low-frequency luminance components and applied to two phase detectors. The video input is also brought to a color-burst gate operated by the horizontal-blanking signal to gate the 8 cycles of 3.58 MHz through. The gated color burst is applied to a phase detector along with the output of the 3.58-MHz oscillator. The output of the phase detector is fed back to the 3.58-MHz reference oscillator circuit in such a way as to lock the phase of the 3.58-MHz reference oscillator to the color burst. The synchronized color-reference signal is then used to gate the video phase detectors. As shown, a 90-degree phase shift is included in one line to provide for demodulation of E_Q. The E_I output comes from the other phase detector, in which the 3.58 MHz signal has no phase shift.

The recovered E_I and E_Q signals are applied to the matrix shown to provide the R–Y, B–Y, and G–Y outputs for the video-output amplifier. The color-killer circuit shown is simply a detector for the 3.58-MHz color burst. When the color burst is present, indicating that a color program is being transmitted, the color killer allows the color demodulator to function normally. When no burst is detected, indicating that a black-and-white program is being transmitted, the color killer inhibits the demodulator circuits, and the R–Y, G–Y, and B–Y outputs are zero.

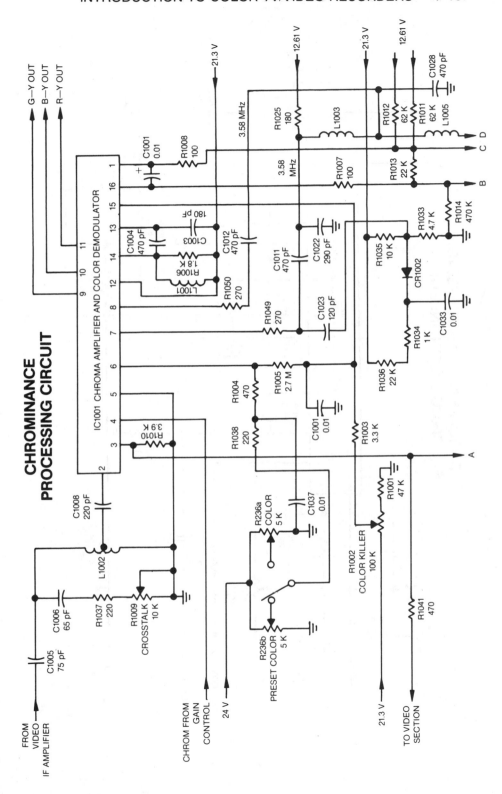

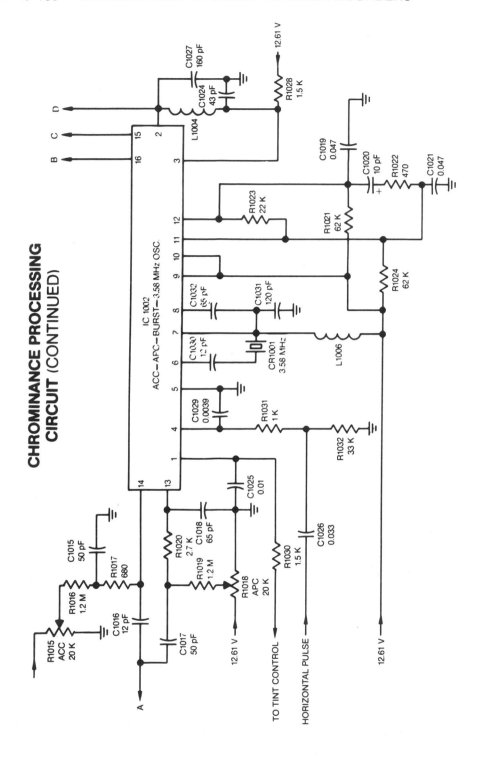

CHROMINANCE PROCESSING CIRCUIT (CONTINUED)

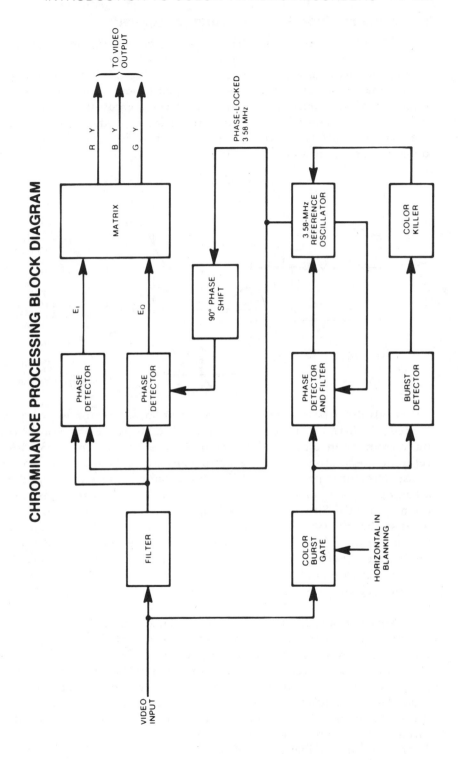

CHROMINANCE PROCESSING BLOCK DIAGRAM

Chrominance-Processing Circuits (continued)

The phase detector is at the heart of the chrominance-signal processing. You will recall from your study of phase detectors in Volume 3, that the phase detector provides an output signal that is proportional to the phase difference between two signal inputs. If the signals are in phase, maximum output is produced. If the signals are exactly out of phase, maximum output in the other direction is produced. If the signals are in quadrature (90-degree phase), no output is produced. The phase detector can be considered a switch operated by the reference signal.

PHASE DETECTOR

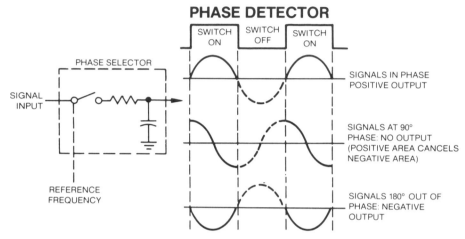

Consider first the operation of the phase detector in the phase-locked 3.58-MHz reference-oscillator circuit. Here, the phase of the color burst from the incoming video is compared to the phase of the 3.58-MHz reference oscillator. If a phase error exists, a voltage output from the phase detector is produced. This voltage can be filtered and applied to a tuning element such as a varactor across the crystal to adjust its frequency slightly. Since a feedback loop is involved, the circuit will adjust itself so that the two signals are in a fixed relative phase—actually, they will be in quadrature (90 degrees), since this is the null point of the phase-detector characteristic. Thus, the output of the reference oscillator is in quadrature and can be used directly for the E_Q demodulator. In a practical circuit, the 90-degree shift would be in the E_I phase-detector line.

Now that a stable phase-locked reference is available, it is a simple matter to apply it to two-phase detectors to recover the I and Q signals. The tint control on a TV receiver is actually a phase trimmer that shifts the phase of the reference (or one of the components of the reference) to compensate for any undesired phase shift in the color circuits. Thus, a phase-detected output is obtained which is aligned with the transmitted chrominance signal. The color (or hue) control adjusts the level of the chrominance-difference signals (R–Y, B–Y, G–Y). When turned completely down, only the luminance signal drives the output (black and white). As the level is increased, the intensity of the color image is increased.

Beam Convergence Control—Color Purity

As stated earlier, it is necessary to assure that the red, green, and blue beams strike only their corresponding color phosphor dots. If the illuminated dots on the picture tube screen do not *exactly match* those being scanned *at that instant* in the transmitter camera tube, the sharpness of original image will be lost, and distortion in color will take place.

The cause of the convergence problem is the fact that the distance between the electron guns and the screen increases as the beams are scanned away from the center of the screen. In addition, the beams do not occupy the same position in the deflection field. Because the distance increases in a regular predictable manner and the relative position of the beams in the deflection field is known and fixed, a correction can be made. All color TV receivers contain special *convergence-control circuits* that aim the electron beams at the same triad of dots and make corrections for the curvature of the screen and deflection.

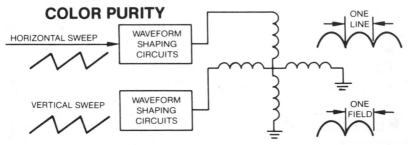

The key to making this correction is the fact that the amount of deflection of the electron beams away from the center is controlled by the magnitude of the electric current through the vertical- and horizontal-deflection coils. Since these currents determine exactly the angle that the beams will deflect, they can be used to make the precise correction required to make all the beams converge on the same triad at all times by driving them with the proper waveform.

The correction is made by placing a horseshoe-shaped electromagnet on the neck of the CRT, in back of the deflection coil. Vertical and horizontal coils are mounted on each magnet. Applied to each coil is a variable-signal waveform that is in step with the vertical- or horizontal-sweep signal. The magnitude of the magnetic fields produced by these coils can be made proportional to beam deflection, so that adjustment of the waveform and the current through these coils can be made to have the three beams always *converge* on the *same* triad of dots.

The horizontal and vertical sweeps are shaped by inductors, capacitors, and resistors to produce the desired waveforms for application to the convergence coils. Since the distances to all parts of the screen from the centers of deflection of the beam are not constant, some defocusing will result at the edges of the picture, if the center is focused. This defocusing will cause parts of a beam to leak through adjacent openings in the shadow mask, which is undesirable. To compensate for this, a correction voltage is often added to the focus voltage.

Features of Modern Color TV Receivers

The color TV receiver described earlier is typical and will allow you to understand how any color TV receiver operates. You will be able to understand circuit function, even though the circuit details may vary. There are, however, many variations of remote-control features and tuners. Most remote controls are *ultrasonic devices* that actuate an ultrasonic microphone in the receiver. Differing tones (or differing codes) determine what function is to be performed. For example, audio volume, contrast, or tint are often simply motor-driven controls that are actuated by the ultrasonic signals, so that the longer the appropriate button is depressed, the greater (or lower) the volume or other function becomes.

COLOR TV RECEIVERS

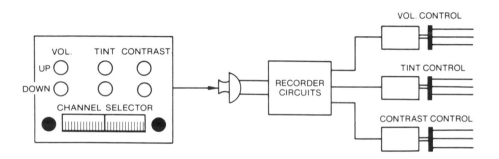

In simplest form, remote tuning is accomplished by having a motor, actuated by the remote control, that sequences the tuner through the channels, with the motor stopped when the remote control signal is removed by the viewer when the desired channel is reached. Usually no provision is made for fine tuning, with the assumption that the memory tuner has been preset earlier. In general, simple remote tuners do not provide access to the UHF channels.

Some TV sets provide for access of 10–18 or more channels that can be preset. Thus, the VHF and UHF channels available in a given area can be pretuned. It is then a simple matter to access these by a remote control. In some cases, this preset is mechanical, but in others it is done by presetting a dc voltage for each channel that operates varactor tuners, and then the only thing necessary to do is to select the proper voltage for a given channel and apply this to the tuning varactors. This approach provides a very flexible approach to remote tuning of both VHF and UHF channels. It is also possible to trim this applied voltage remotely, providing for simplified remote fine tuning. Because of difficulty with seeing the channel number on a TV set from a distance, some TV sets are designed to flash on the TV screen the channel number to which the set is tuned.

Features of Modern Color TV Receivers (continued)

The availability of low-cost microprocessors has greatly enhanced the flexibility of TV set control, both locally and remotely. For example, a microprocessor is used by the Zenith Radio Corporation in some of their deluxe-model TV receivers. This microprocessor is used in conjunction with a phase-locked-loop frequency synthesizer to provide for flexible remote electronic tuning of all TV channels as well as other special channels for CATV. While understanding of these digital circuits must await your study of digital systems in Volume 5, some insight can be provided here on system operation, since you already understand how phase-locked loops operate.

REMOTE ELECTRONIC TUNING

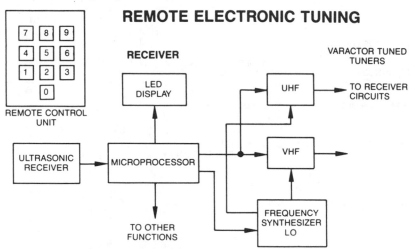

In operation, the input to the microprocessor is supplied as digital signals from the remote control. These are picked up by the ultrasonic receiver and fed to the microprocessor. The microprocessor interprets the received signal and generates voltages that tune the TV receiver tuners (RF, mixer, LO) to the proper channel. A frequency synthesizer and phase-locked loop are used to automatically fine tune the LO to the exact frequency for the channel selected. In addition, a digital signal is produced to operate a LED (light-emitting-diode) display (such as on a digital watch or clock radio) to indicate the channel selected. The channel-selector keyboard on the remote unit (or local-control unit) looks very much like a calculator keyboard. To select, for example, channel 21 requires only that the 2 and 1 pushbuttons be depressed, followed by ENTER. The receiver will automatically tune to channel 21, lock to the correct frequency for exact tuning, and display 21 on the LED channel display. The microprocessor is also used to route the other remote-control functions, such as volume, to the proper actuating circuits. It should be stressed that while some convenience feature like remote control or automatic fine tuning may vary widely from receiver to receiver, *the fundamental ideas are the same.*

Experiment/Application—TV Receivers

A lot can be learned by exploring a TV receiver with an oscilloscope and probe. If you set up a receiver and tune it to a channel that is functional, then all of the signals found in a TV receiver will be present. Thus, it will be easy to see TV signals in action. *CAUTION: Do NOT touch the high voltage portion of the receiver.*

EXPLORING A TV RECEIVER WITH OSCILLOSCOPE AND PROBE

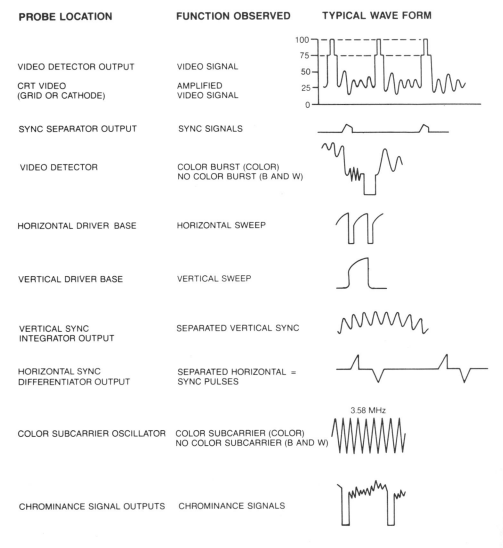

PROBE LOCATION	FUNCTION OBSERVED	TYPICAL WAVE FORM
VIDEO DETECTOR OUTPUT	VIDEO SIGNAL	
CRT VIDEO (GRID OR CATHODE)	AMPLIFIED VIDEO SIGNAL	
SYNC SEPARATOR OUTPUT	SYNC SIGNALS	
VIDEO DETECTOR	COLOR BURST (COLOR) NO COLOR BURST (B AND W)	
HORIZONTAL DRIVER BASE	HORIZONTAL SWEEP	
VERTICAL DRIVER BASE	VERTICAL SWEEP	
VERTICAL SYNC INTEGRATOR OUTPUT	SEPARATED VERTICAL SYNC	
HORIZONTAL SYNC DIFFERENTIATOR OUTPUT	SEPARATED HORIZONTAL = SYNC PULSES	
COLOR SUBCARRIER OSCILLATOR	COLOR SUBCARRIER (COLOR) NO COLOR SUBCARRIER (B AND W)	3.58 MHz
CHROMINANCE SIGNAL OUTPUTS	CHROMINANCE SIGNALS	

You can explore other portions of the video/audio section of the TV receiver, and it is of interest to examine the above waveforms with the TV set in sync and out of sync.

Video Tape Recording—Broadcast Type

In the early days of TV, recording of programs was done on film—either black and white or in color. Video recording on tape was impractical then, because of the signal bandwidth requirements. Improvements in recording technology have made it possible to record video *directly on magnetic tape*, and now such devices are commonplace and relatively inexpensive. To make such a recording on tape by the same techniques as for audio recording would require high tape speeds —a great deal of tape for a short program. The breakthrough that made TV recording practical was the development of the *rotating recording head.*

In these systems, the tape moves horizontally, as for a conventional recorder, but the head rotates, moving the head across the tape at high speed and laying down a strip of recorded data with each pass of the head over the tape. This strip of recorded data is *perpendicular* to the direction of motion of the tape. Typically, four heads are mounted on the rotating head assembly so that as one head leaves the tape, another moves onto the tape to continue recording.

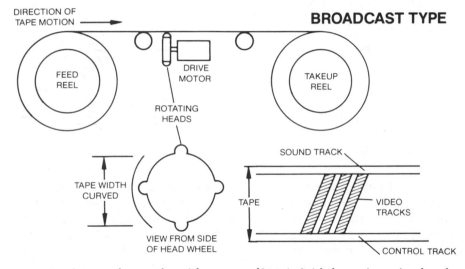

As shown above, the video recording is laid down in strips by the high-speed rotating head. The tape is curved into quarter circles so that the rotating head is in close contact with the tape during its traverse. A sound track and a control track are laid down in conventional fashion. The sound track is used to record the accompanying audio. The control track is used to carefully regulate and control the tape motion and speed. Typically, the tape speed is 15 inches (38.1 cm) per second. The head rotates at 14,400 rpm, and the width of each recorded strip is about 0.01 inches (0.025 cm). A 2-inch (5.08-cm) diameter head rotating at 14,400 rpm produces an equivalent tape speed of 1,500 inches (3,810 cm) per second. This allows for the recording of the high video frequencies necessary for high-fidelity TV tape recording. To achieve the best fidelity, the TV signal is used to frequency modulate a carrier.

Video Tape Recording—Portable Type

The portable and home type video tape recorders use a type of scan called a *helical scan*. In these recorders, the tape is wrapped partially around a fixed capstan that has the rotating recording head placed at an angle. The heads scan the tape at an angle. Because of this, each individual scan of the tape head can be made much longer, as shown below.

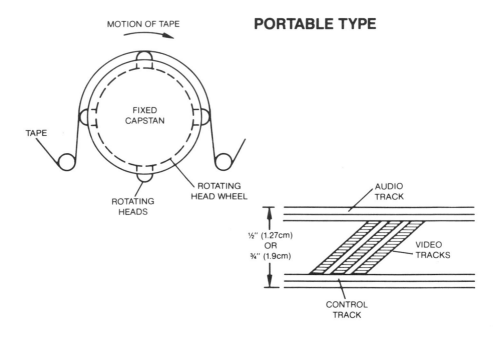

Depending on the capstan and rotating head wheel design, either two or four heads are used. In addition, the tape may be wrapped halfway around the capstan, as shown, or almost entirely around the capstan. The heads are actually rotated in a slot in the fixed capstan so that they can make contact with the tape for recording.

In other respects, the home or portable tape recorder is similar to the larger video tape recorders described on the previous page. Unlike the broadcast-type video tape recorder, there are no real standards and several types of machines are available with different characteristics; unfortunately, they are incompatible. In addition to ½-inch (1.27-cm) and ¾-inch (1.9-cm) tape models, open-reel and cassette designs are available, and, in addition, the recording technique varies somewhat. In a modern home tape recorder/player, up to 6 hours of recording can be obtained with a single tape cassette, with surprisingly good quality. Since an entire frame is recorded on one traverse of the head (half or full circle), it is possible to stop motion and look at a single frame by stopping the tape motion while the head continues to rotate.

Video Disk Players

Video disk recording for home entertainment and other purposes is becoming increasingly common. The reason for this development is primarily that phonograph disks continue to be the most popular for audio recording. They are easy to handle and change, and they are easy to mass produce—leading to low cost. Because of the complexity of recording and manufacture, video disk systems are used for playback only. Two basic systems have been devised. One system uses a laser beam to read the data from the disk. The other uses an extremely fine diamond stylus that acts as a capacitor plate to read the data from the disk. Video disks look like an ordinary phonograph record. Programs of all kinds are available in both color or black and white. Both systems produce a video signal fully compatible with current TV standards.

The laser beam player was developed by the North American Phillips Company and is marketed under the trade name of *Magnavision*®. In this system, the disk is rotated at a speed of 1,800 rpm. As with video tape, the signals are recorded and played back using FM. The sound channels (stereo) are subcarriers. The FM carrier is at 8.1 MHz, modulated by the video signal, including color signals. The sound subcarriers are at 2.1 and 2.3 MHz. The data are recorded as a series of microscopic pits of variable length, with a width of 0.4 µm (µm = micrometer = one-millionth of a meter). The grooves are spaced about 1.6 µm apart.

VIDEO DISK SYSTEM

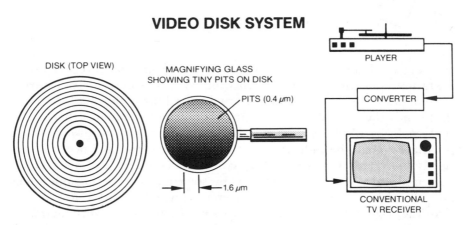

DISK (TOP VIEW)

MAGNIFYING GLASS SHOWING TINY PITS ON DISK

PITS (0.4 µm)

1.6 µm

PLAYER

CONVERTER

CONVENTIONAL TV RECEIVER

The disk is read with a laser beam that is split into three beams. The two outer beams are used to make the center laser beam track the recorded data. The output signal is converted to baseband video and audio and can be used directly or to modulate carriers, just as in a conventional TV transmitter, except for the very low power output (a few milliwatts). The carrier is set for an unused TV channel on a conventional TV receiver and is received just like any TV broadcast, so the player can be used with any TV receiver. The optical system used to read the disk is described on the next page.

Video Disk Players (continued)

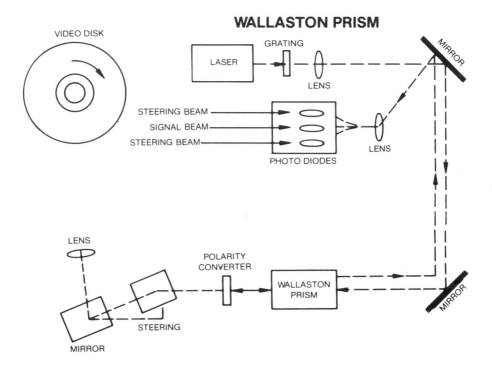

For the laser disk system, a laser beam is split into three beams by a grating, and these beams are focused by a lens onto a fixed mirror system, as shown. The light beams are directed through a *Wallaston prism* that bends the light differently for different polarizations—that is, *vertically* polarized light is bent in a direction *opposite* to that of *horizontally* polarized light. The light from the laser coming from the prism is vertically polarized, and the polarity converter converts it to a circularly polarized light that is directed via the steering mirrors through a lens onto the disk. The light reflected from the pits in the disk is less intense than the light reflected from the area between the pits, producing an intensely modulated reflected beam. This return beam is passed back through the steering mirrors to the polarity converter and prism. Because the beam is going in the opposite direction, it is deflected at the output of the prism in the opposite direction from the incoming laser beam; thus the laser and return beams are separated. The return beam is reflected by the fixed mirrors onto a set of photo diodes that produce the necessary output signals from the center beam and the beam-alignment data from the two outer beams. The video and audio output signals are generated by demodulation of the FM signal.

Video Disk Players (continued)

The RCA video disk system operates differently. The disk rotates at 450 rpm and is read by a diamond stylus. The stylus follows the groove just as with a conventional phonograph record, so no special circuitry is necessary to hold the stylus aligned. The information is coded as a series of transverse slots of varying width in the bottom of the groove. The diamond stylus has a thin metal electrode plated on its rear surface. The record itself is made from a conductive plastic or a thin plating of conductive material. These act as the plates of a capacitor, and the stylus reads the stored data by the *variations* of this *capacitance* as the record rotates.

DISK/STYLUS OPERATION

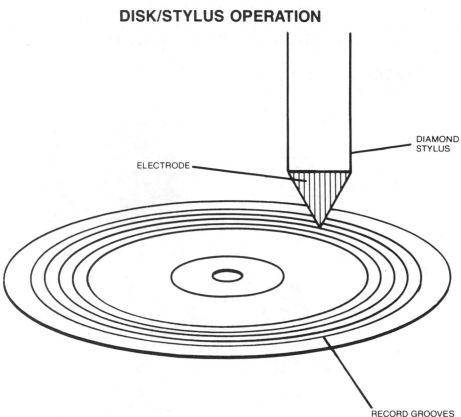

The recording technique, as noted previously, basically uses FM carriers. Separate carriers are used for the video and audio information; however, special subcarrier and encoding techniques allow for the use of a lower carrier frequency, which can lead to longer playing time per side than for optical techniques. While the lower rotating speed is an advantage that leads to a relatively low-cost system, special features such as stop motion and slow motion appear to be difficult to achieve.

TV Receiver as a Video Display Terminal

Now that you have learned all the aspects of the TV receiver when used as a TV receiver, we would like you to take another look but from a *different point of view*—that is, viewing the TV receiver as a *video display terminal* when used with a computerized information system.

As you will learn in Volume 5, the CRT terminals used with computers/microprocessors consist of the deflection and sweep circuits, the video portion, and the necessary power supplies. Additional hardware is used to provide the video and sync signals for the basic CRT display. These terminals have sweep characteristics very similar to those of a TV receiver—typically, a 60-Hz vertical field rate and about 500 lines. You also know that an interface is available which will take incoming video and convert it to an RF signal that is set for one of the TV channels. Thus, an incoming video signal is displayed on the TV receiver CRT.

TV RECEIVER AS A VIDEO DISPLAY TERMINAL

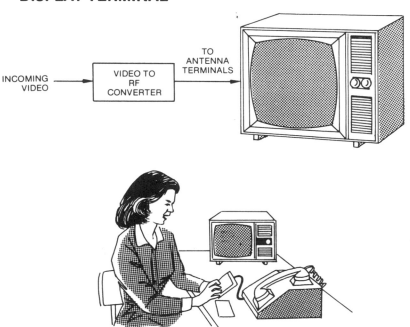

Digital data, a series of on-off signals that represent alphabetic or numeric characters, lines, or bars, etc., are really no different from TV information. Such digital data can easily be *formatted* (arranged) with the proper digital circuitry to provide a display of information on a regular TV receiver.

TV Receiver as a Video Display Terminal—Teletext

TV receivers can be (and are) widely used as *display terminals*. The necessary information from any source for display is stored in a memory device, and this memory device is scanned at the TV scanning rate to provide the necessary display. You will learn about these storage and interface devices when you study digital circuitry in Volume 5 of this series.

One of the new uses of TV (video) display terminals is in a *Teletext* mode. Teletext is data, usually alphanumeric, inserted into the TV transmission during the vertical (or horizontal) retrace interval. The data are displayed on the TV screen either alone or as a caption or separate information unrelated directly to the program material. Aside from providing simple captions, headlines, traffic information, supermarket specials, etc., it has obvious uses in providing subtitles for those with impaired hearing or providing translation of foreign languages. In addition, so-called *electronic home information services* are being developed which will use cable TV systems as a two-way link into textual and graphic material upon demand, and computer/microprocessor linkup as desired.

TELETEXT SYSTEM

The Teletext approach is the basic step in the use of TV for *home electronic retrieval of information*. A more elaborate system called *Viewtron* provides a much broader basis for *interactive operation* involving not only information retrieval and computer hookup but also access to a broadly developed *data base of information*. Future uses such as comparative shopping at home from supermarkets or other stores (using their existing price structure as stored in their computer systems for specific items) will be readily possible once these systems are in operation.

Now that we have completed our study of the TV receiver as a video display terminal we are ready to tackle the subject of troubleshooting receivers. It is important here to keep in mind the principles of troubleshooting that we have learned in the first three volumes.

Troubleshooting Receivers

Receiver troubleshooting is somewhat different than transmitter troubleshooting in that, as you might suspect, the signal-tracing procedure starts from the output end, as shown on the following page. As always, the first thing to make sure of is that there is primary power and that the power supplies are working properly. If this is so, it is then important to check the audio system. This is most easily done by injecting a low-level audio tone into the input of the first audio stage and tracing forward for the presence of the proper audio level if no signal or a weak or distorted signal appears at the output.

If the audio section is operating properly, the IF/detector system can be easily checked by loosely coupling an IF signal into the mixer collector. Most RF–IF signal generators have provisions for tone modulation, which is convenient. If the tone is heard at the receiver output and appears generally normal, then the local oscillator (LO) should be checked. No output from the LO will, of course, make the receiver totally inoperative. LO operation can be difficult to check unless a detector probe is available. Such a probe can be made as shown on the following page. A dc output indicates that the LO is functional, which means that the RF/mixer needs to be examined. This is most easily done by injecting a signal from an audio-tone modulated RF signal generator into the mixer input first and then, if this is working, directly into the RF input.

As with transmitters, it is extremely important *not to touch adjustments* until the receiver is functioning. Otherwise you could repair the defect but still not have a functioning receiver because of misalignment. This can lead you to believe that the receiver is still defective.

As with any electronic equipment, repair or troubleshooting is difficult without help in the form of schematic diagrams, voltage and signal levels at pertinent points, and layout diagrams. In addition, most equipment includes some relatively specialized alignment techniques. Thus, the use of manufacturers' data and alignment and troubleshooting procedures makes the job of troubleshooting much much easier than if you try it blind.

Generally, after repair it is wise to check the alignment of a receiver since misalignment can cause serious loss of sensitivity and selectivity. While this procedure differs for different receivers, the general approach will be described next; it can be used as a guide if no other data are available.

Now, study the following chart *carefully*, refer back to it if in doubt, and, above all, make sure that you have not left anything out in each step of the process.

The next section on receiver alignment will also help in the troubleshooting procedure.

TROUBLESHOOTING RECEIVERS

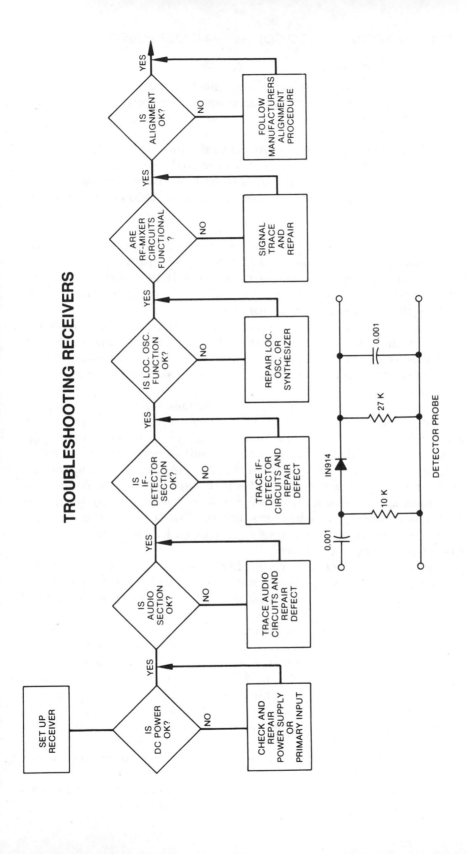

Receiver Alignment

Optimum performance of a superheterodyne receiver cannot be obtained unless it is properly tuned, or *aligned*. The process of alignment involves three things:

1. The IF amplifiers must be tuned to the intermediate frequency.
2. The tuned RF circuit(s) must be tuned to the frequency on the dial.
3. The local oscillator must be adjusted to give an output at each setting of the tuning dial that is above (or below) the RF by an amount equal to the IF.

The alignment procedure will be described by referring to an AM slug-tuned receiver, although an FM receiver is aligned in a similar way. The first step in the alignment is to tune the IF transformer to the IF. This is done by loosely coupling a signal from a signal generator set to the IF into the collector circuit for the mixer.

The antenna can be disconnected and a test oscillator coupled to the input through a small capacitor (27 pF). With the dial set at any convenient frequency, the test oscillator is adjusted to provide a signal at IF. The tuning-slug adjustments in the IF are then adjusted to provide a maximum output.

It should be noted that not all IF sections are aligned with every stage at the same frequency. Some are *stagger tuned*. As you have previously learned, this means that some IF stages are to be peaked at different frequencies, in order to broaden the overall response curve. The procedure then is to use a modulated signal from the generator and connect the probe to the detector output. Each IF stage is then peaked up at its own designated frequency (called for in service information). In the case of an FM receiver, with a limiter stage, the lowest possible signal input should be used, to avoid limiting action, or an RF probe can be used at a point just before the limiter.

The RF amplifier, mixer, and LO tuning are adjusted first at the high end of the band where the trimmer capacitors have maximum effect. Both the tuning dial and test oscillator are set at near the high end of the band, and then the trimmers are adjusted for maximum signal level at the output. Care must be taken to use a low-capacitance probe when measuring the RF signal in order to avoid detuning the RF amplifier output. Next, the signal generator and tuning dial are set to the low end of the band and the tuning-slug position is adjusted to provide maximum signal level. This tuning can change the high-frequency alignment. Therefore, the dial and generator are both tuned back to the high end, and the trimmers are readjusted for maximum output.

Alignment of the oscillator section is similar to the RF amplifier procedure with the trimmer being tuned first at the high end and the coil then tuned at the low end. Again, there will be an interaction between the effects of tuning, and the procedure should be repeated several times while tuning for a maximum at the IF output.

AM BROADCAST RECEIVER TUNING ELEMENTS

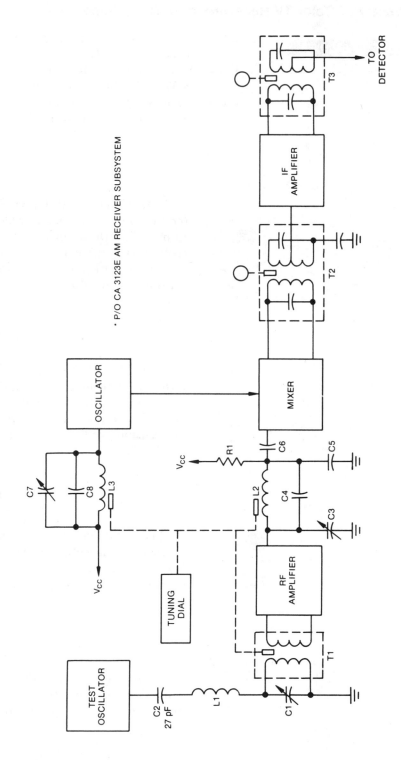

* P/O CA 3123E AM RECEIVER SUBSYSTEM

Review of Color TV Receivers and Video Recorders

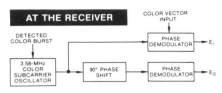

1. COLOR TV CHROMINANCE INFORMATION is transmitted on a subcarrier whose frequency is 3.579545 MHz. It is recovered at the receiver by developing two signals of that frequency in quadrature.

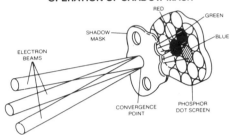

2. THE COLOR PICTURE TUBE (common version) has three electron guns. The three beams are controlled so that each goes through a hole in a mask and strikes its own phosphor color dot.

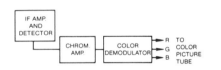

3. THE COLOR PROCESSING CIRCUITS receive the chrominance sidebands and burst signal from the video detector, demodulate in proper phases, and produce color signals for the CRT.

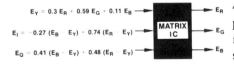

4. THE COLOR SIGNALS for the picture tube are obtained by matrixing the demodulated E_I, E_Q, and Y signals in proper proportion.

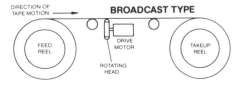

5. VIDEO TAPE RECORDERS became more practical as a result of the development of *helical scan* and *rotating heads.*

Self-Test—Review Questions

1. What is the difference between a black-and-white and color TV signal? Give details.
2. Define the terms luminance and chrominance.
3. Describe how a color image is formed.
4. What are the basic differences between a black-and-white TV picture tube and a color TV picture tube? Why are these differences significant?
5. Are there any significant differences between the RF/IF video-sync portions of a black-and-white and color TV receiver? Explain.
6. What is the color subcarrier reference? Where is it found? How is it recovered? What is it needed for?
7. Are there significant differences in the horizontal- and vertical-sweep circuits in black-and-white and color TV receivers? Specify and explain.
8. Draw a block diagram of the elements of the color portion of a TV receiver. Explain the function of each element. Describe how the chrominance signals are recovered.
9. Why are elaborate convergence and similar beam-control adjustments necessary in a color TV receiver?
10. Explain the advantages and disadvantages of video tape recording compared to video disk recording. For home use, what capability do tape machines have that disks do not have?

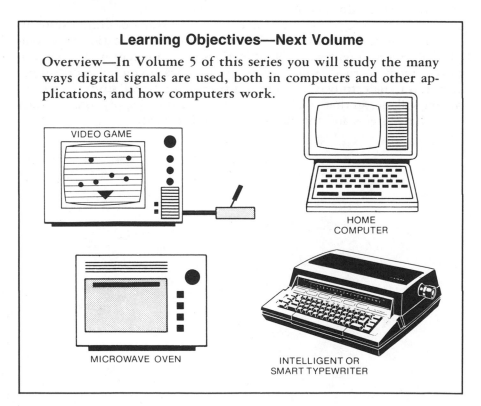

Learning Objectives—Next Volume

Overview—In Volume 5 of this series you will study the many ways digital signals are used, both in computers and other applications, and how computers work.

VIDEO GAME

HOME COMPUTER

MICROWAVE OVEN

INTELLIGENT OR SMART TYPEWRITER

Epilogue

At this point in your studies, you have learned *how information is transformed* into electrical signals by appropriate transducers, used to modulate a suitable carrier, and radiated through space or conducted by wire to another point. The receiver at the remote point converts the received signal back to a reproduction of the original electrical signal that will *reproduce the original information* when supplied to an appropriate transducer. As you learned, the range of information that can be transferred to a remote point is almost limitless, ranging from simple on-off control signals to audio/video/digital data.

The important point is that the *basic elements* of transmission-reception systems are *always the same*. While the nature of the input and output transducers may vary, and the bandwidth of the signal depends on the information content of the signal, the system elements are the same. If you keep this concept in mind, you will be able to understand any system, no matter how complex it may seem at first glance. You will be able to separate it into its component parts, and these will be the system elements you have already studied. Until now, the information transferred was primarily in analog form rather than digital; it should be stressed, however, that this is *not* of fundamental importance and does not significantly change the system, because digital signals are not materially different from any other information.

In Volume 5, you will concentrate on the study of digital systems. As you will see, although the manipulation of information is accomplished in a different way, the objectives are consistent with what you have already learned. Since digital systems are becoming increasingly important, their study is an important part of our overall study. However, as you will see, all that you have already learned will apply to your study of digital systems and the management of information.

Basic
Solid-State
Electronics
VOL. 5

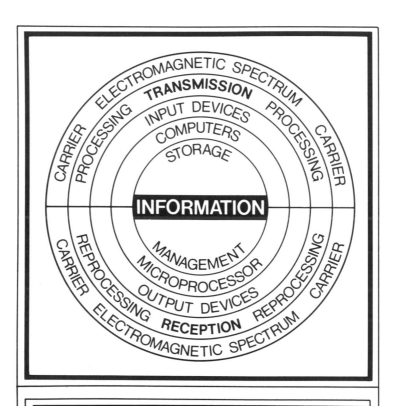

Basic Solid-State Electronics

COMMON CORE

The Configuration and Management of Information Systems

INFORMATION MANAGEMENT
VOL. 5

DIGITAL SYSTEMS/SYSTEM ELEMENTS
DIGITAL ARITHMETIC/TIMING AND COUNTING
COMPUTER/MICROPROCESSOR APPLICATIONS
INPUT/OUTPUT DEVICES
ALPHANUMERIC DISPLAY
DIGITAL CIRCUITS/TROUBLESHOOTING

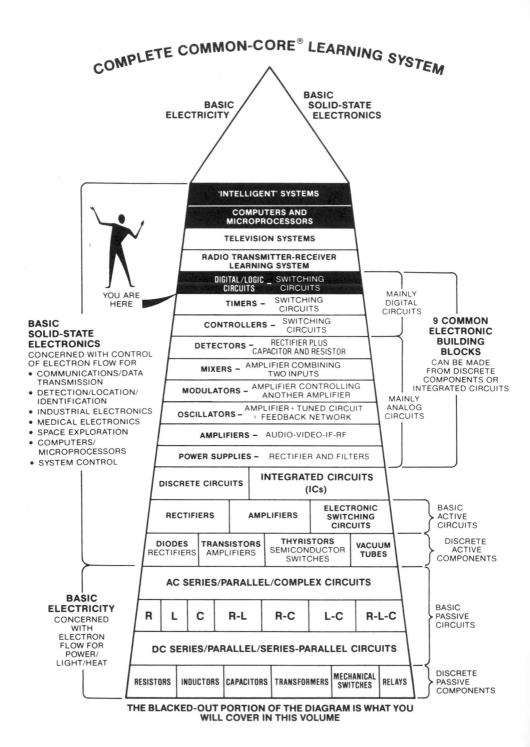

COMPLETE COMMON-CORE® LEARNING SYSTEM

BASIC ELECTRICITY

BASIC SOLID-STATE ELECTRONICS

'INTELLIGENT' SYSTEMS

COMPUTERS AND MICROPROCESSORS

TELEVISION SYSTEMS

RADIO TRANSMITTER-RECEIVER LEARNING SYSTEM

DIGITAL/LOGIC CIRCUITS – SWITCHING CIRCUITS

TIMERS – SWITCHING CIRCUITS

CONTROLLERS – SWITCHING CIRCUITS

DETECTORS – RECTIFIER PLUS CAPACITOR AND RESISTOR

MIXERS – AMPLIFIER COMBINING TWO INPUTS

MODULATORS – AMPLIFIER CONTROLLING ANOTHER AMPLIFIER

OSCILLATORS – AMPLIFIER + TUNED CIRCUIT + FEEDBACK NETWORK

AMPLIFIERS – AUDIO-VIDEO-IF-RF

POWER SUPPLIES – RECTIFIER AND FILTERS

DISCRETE CIRCUITS | INTEGRATED CIRCUITS (ICs)

RECTIFIERS | AMPLIFIERS | ELECTRONIC SWITCHING CIRCUITS

DIODES RECTIFIERS | TRANSISTORS AMPLIFIERS | THYRISTORS SEMICONDUCTOR SWITCHES | VACUUM TUBES

AC SERIES/PARALLEL/COMPLEX CIRCUITS

R | L | C | R-L | R-C | L-C | R-L-C

DC SERIES/PARALLEL/SERIES-PARALLEL CIRCUITS

RESISTORS | INDUCTORS | CAPACITORS | TRANSFORMERS | MECHANICAL SWITCHES | RELAYS

YOU ARE HERE

BASIC SOLID-STATE ELECTRONICS
CONCERNED WITH CONTROL OF ELECTRON FLOW FOR
- COMMUNICATIONS/DATA TRANSMISSION
- DETECTION/LOCATION/ IDENTIFICATION
- INDUSTRIAL ELECTRONICS
- MEDICAL ELECTRONICS
- SPACE EXPLORATION
- COMPUTERS/ MICROPROCESSORS
- SYSTEM CONTROL

BASIC ELECTRICITY
CONCERNED WITH ELECTRON FLOW FOR POWER/ LIGHT/HEAT

MAINLY DIGITAL CIRCUITS

9 COMMON ELECTRONIC BUILDING BLOCKS
CAN BE MADE FROM DISCRETE COMPONENTS OR INTEGRATED CIRCUITS

MAINLY ANALOG CIRCUITS

BASIC ACTIVE CIRCUITS

DISCRETE ACTIVE COMPONENTS

BASIC PASSIVE CIRCUITS

DISCRETE PASSIVE COMPONENTS

THE BLACKED-OUT PORTION OF THE DIAGRAM IS WHAT YOU WILL COVER IN THIS VOLUME

Man-Machine Systems vs. Internally Controlled Electronic Machine Systems

You are now about to study Digital Systems—mainly digital computers, data management systems, and control systems. These are electronic machine systems consisting of *digital logic circuits* that have been configured to perform the *human management functions* of *memory, calculation, decision making,* and *control.*

In Information-Transmission systems in Volume 3 and Information-Reception systems in Volume 4, you were primarily concerned with electronic man-machine information-transfer systems using analog circuits with some of the control aspect designed into the hardware but with the major portion of control supplied *externally* by *human intelligence* from a human operator.

CAN A MACHINE THINK?
Is There Artificial Intelligence?

No, a machine cannot think. Electronic machines—computers and microprocessors—can only follow instructions—so numerous and detailed that they consist of thousands upon thousands of steps in an inflexible sequence—and in doing so are more like the human mind than any machine has ever been. Still they are a long way from having the creativity of the human mind. So, when we refer to artificial intelligence, intelligent terminals, smart machines, etc., you'll know what we mean—only following instructions.

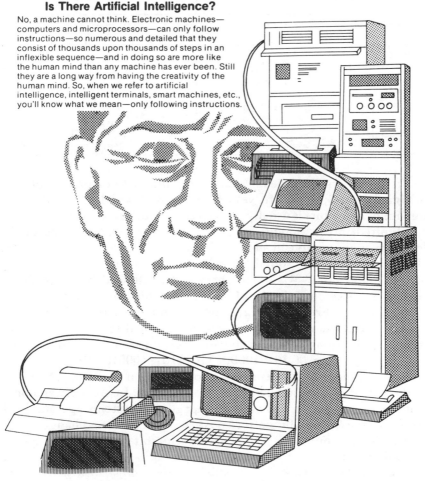

Man-Machine Systems vs. Internally Controlled Electronic Machine Systems (continued)

Now you are about to study an entirely new area of electronics—the world of digital logic circuits and digital systems—in which the control aspect (information management) of an electronic machine system *without* a human operator is *internally* supplied by a stored program of instructions—*artificial intelligence*—from an electronic memory.

DIGITAL SYSTEMS OPERATE BY
MANIPULATION OF BIT PATTERNS

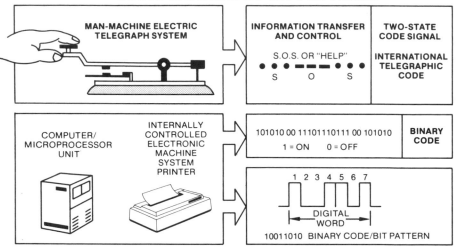

In digital systems, the *information* is contained in a pattern of a *two-state* signal (1 or 0; on or off; yes or no), each of which is called a *bit*. A sequence of many bits is usually used to convey information since a single bit is limited in the amount of information carried. The code used in telegraphy is an example of *bit patterns* that *carry intelligence*. In telegraphic code, for example, the signal for *help*, or *SOS*, is represented by a dot-dash bit pattern (. . . – – – . . .) of three dots, followed by three dashes, followed by three dots. (In the Morse telegraphic code, a dot is the basic bit, and a dash is three dots long; a space is equal to a single dot in length if it is between the elements of a letter and equal to a dash if between letters.) On this basis, we could conceive of a binary code with the equivalent—a bit pattern of 101010000111011101110001010100. As you can see from this binary code representation of *SOS*, it is the *manipulation* of *bit patterns* that results in the *transfer* of *information*. In precisely analogous fashion, you will see in this volume that digital systems operate by the *controlled manipulation* of bit patterns. As these bit patterns change, so does the information conveyed. And remembering bit patterns—by means of electronic memory—is a necessary part of digital system designs. So in summary, *digital systems consist essentially of the manipulation of bit patterns and the management and use of memory and associated logic.*

A Brief History of Digital Concepts and Computers

Originally, the *extension of counting* beyond fingers and toes (digits) and the *need* to *remember* a calculated or measured quantity probably involved the use of a pile of pebbles, with each pebble representing one item. As larger quantities had to be dealt with, the simple *one-to-one correspondence* between an item and the pile of pebbles became too great, so methods were devised to give certain stones higher value to simplify calculation. One way to do this was to draw a set of lines and spaces on the ground, with each given an ascending value.

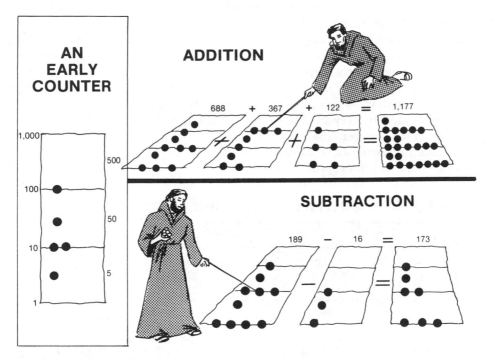

For example, here the *lines have values* of 1, 10, 100, and 1,000, while the *spaces have values* of 5, 50, and 500. As you can see, numbers can be added merely by placing the stones in the *correct positions* to *represent* each *number*; at the end, the *total* is represented by the *sum* of the stones *at each position times the value of the position*. As you can see, only one set of lines is needed with stones for each number added placed directly on the grid. Furthermore, by knowing that five 1s = 5, two 5s = 10, five 10s = 50, etc., pebbles can be removed and a single pebble placed in the next higher category. When this is done (for example, two 5s removed and one 10 substituted), the operation is known as *a carry*.

As you can see, if more pebbles need to be removed at a particular level than are present, the number at any level can be increased by *borrowing* from the next higher level. Thus, if four pebbles were required to be subtracted in the 1s column and only one was available, we could borrow five from the 5s column (replace a 5 with five 1s).

A Brief History of Digital Concepts and Computers (continued)

As you might suspect, the *abacus*, the world's most widely used nonelectronic calculator, is an extension of the pebbles-and-grid idea, but the *carry* is done by *hand*, as with the stones. The *mechanical carry* was introduced by a young French mathematical genius named Blaise Pascal in 1642. He marked the digits 0 to 9 around the edges of circular dials so that one revolution of a dial *would advance* (*carry*) the next dial by one-tenth of a revolution. This same idea is in use today in many mechanical devices—for example, the automobile odometer. Since *Pascal's machine* could *rotate* in *either direction*, it could therefore *subtract*. The German mathematician Baron von Leibnitz extended the ideas of Pascal to include *multiplication (successive addition)* and *division (successive subtraction)*. Unfortunately, these machines were quite unreliable (as well as expensive) and therefore were not widely used.

Even today, we use words and phrases associated with these early calculating devices or machines which imply the *moving* or *manipulation of pebbles*, such as carry and borrow. It is the Latin word for pebble, *calculus*, from which we get *calculate, calculator, calculation*, etc. The *basic ideas* of *counting* presented here *haven't really changed*, even in our most modern computers and microprocessors.

PASCAL'S ARITHMETIC MACHINE

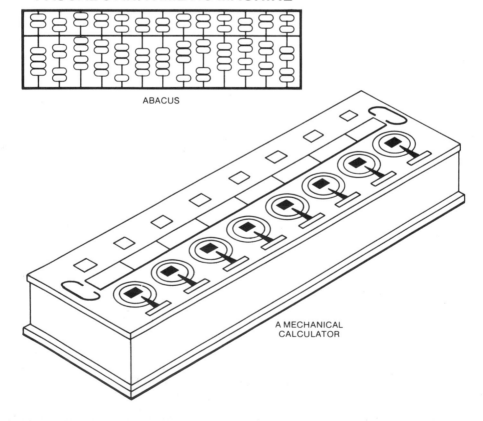

ABACUS

A MECHANICAL
CALCULATOR

A Brief History of Digital Concepts and Computers (continued)

The next major contribution in calculators was the work of Charles Babbage, a British subject who, starting in 1812, devoted his life to trying to build a *computing machine*. His main goal was to build a machine to do repetitive calculations for the creation of tables of powers, roots, and logarithms. Although he was not successful—because existing construction technology could not produce the intricate parts needed—Babbage's machine (which he called the *analytic engine*) was designed to hold a thousand 50-digit numbers in its *memory*, called the *store*. In the part of the machine called the *mill*, *arithmetic operations* (or *computations*) were to take place at the rate of 60 additions or subtractions of the 50-digit numbers in 1 minute or a multiplication of two 50-digit numbers in 1 minute. The truly unique feature of this machine, however, was its ability to *accept and execute an internal instruction set (program)*. The basic concepts embodied in Babbage's analytical engine are really the foundation of modern computing technology. In Babbage's times, electronics was nonexistent and all systems were *mechanical*. Thus, the basic ideas for digital technology were laid over 150 years ago; it is only since the development of electronics, however, that engineers have been able to put these ideas to use.

**CHARLES BABBAGE'S
ANALYTICAL ENGINE**

A Brief History of Digital Concepts and Computers (continued)

A major push into digital systems was brought about by the Constitution of the United States, which requires that a census be taken every 10 years. It took 7 years to tabulate the results of the 1880 census, and it was obvious that soon the census results would take longer than 10 years to compile. This problem was turned over to Dr. Herman Hollerith, a statistician and inventor from Buffalo, N.Y. He developed a system whereby *census data* was *punched onto cards* and *sorted* and *counted* by a machine he had invented. With this system, the 1890 census was completed in about 3 years. According to one story, when Dr. Hollerith was asked how large to make the original card, he threw a U.S. paper dollar bill on the table and said the cards should be the same size. Punch cards have remained that size to this day (although the size of the dollar bill was made smaller in 1929). The original card had 45 columns of *numerical information* punched with round holes. In 1928, IBM introduced rectangular holes and increased the punch capacity to 80 rows.

IBM PUNCH CARD
(INFORMATION STORAGE/RETRIEVAL)

The *foundation* of *modern digital logic* was set down by George Boole in 1854 when he described the system of logical algebra called *Boolean algebra*. Using Boolean algebra, Claude Shannon of Massachusetts Institute of Technology and later Bell Telephone Laboratories during the late 1930s and 1940s showed the *relationship* between *electronic switching circuits* and *Boolean logic*. During the late 1930s and the 1940s, computer technology developed rapidly from electromechanical relay-operated machines to electronic vacuum-tube systems with more than 18,000 vacuum tubes. These machines were very large, generated a lot of heat, and were unreliable because they used so many vacuum tubes. The first commercially available *electronic calculator* was introduced by IBM in 1949 and was called the *Card Programmable Calculator* (CPC).

A Brief History of Digital Concepts and Computers (continued)

The invention of the transistor at Bell Laboratories and its introduction in the early 1950s started the revolution in digital systems and computers. The high reliability, small size, and low power consumption of transistors made possible the development of *smaller* and *smaller systems* with *more* and *more capability*. The introduction of the *integrated circuit (IC)* in the late 1950s provided a major impetus to the development of *digital systems*, since it permitted logic system designers to purchase ready-made logic packages and not have to spend design time on contents.

COMPUTER-ON-A-CHIP

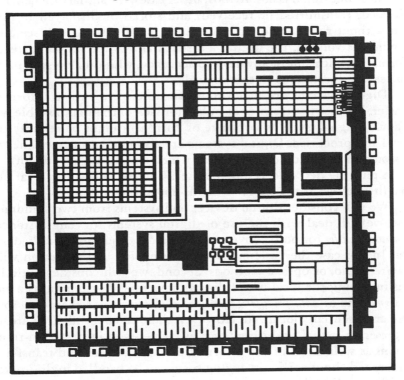

The original idea for a *stored-program computer* was probably conceived by a group at the Mathematical Laboratory of the University of Cambridge in England. This important concept has been expanded and developed in modern digital computers and systems, and *essentially all* digital computers are of the *stored-program type*. Large-scale integration (LSI), the proliferation of memory and digital logic ICs, and increased speed of operation have resulted in extraordinary application of digital systems in almost all aspects of living. In this volume, we will deal almost entirely with IC logic circuits and how they are used, alone or in conjunction with LSI computer chips or circuits, to perform truly marvelous feats of computation and/or control at low cost and with high efficiency.

Digital Information Management

In the earlier volumes of this series you learned about solid-state devices and systems and how they work in information transmission and reception. With the exception of a short introduction to pulse systems in TV, the systems you learned about were *analog* in operation. In this volume we will concentrate on *digital electronic* devices and systems. In some cases, an analog signal is converted to digital form (analog-to-digital conversion) or a digital signal is converted to analog form (digital-to-analog conversion). This is done so that the intervening signal processing, transmission, and reception can be in digital form even though the basic input signal is not. In most other cases, the signals are generated, processed, transmitted or received, and stored or displayed in digital form without reference to analog signals at all. As you will see a little later in this volume, the digital approach allows for simple transfer of information with almost any desired accuracy, while analog systems have some severe limitations in this regard.

Digital systems are much more flexible in regard to information transfer, because the transfer of pulses (or lack of them) is possible where an analog signal (such as a voltage or current) cannot be reliably transferred and processed. Digital systems are also more accurate. The transmission and reception of a signal in which simply an "on" or "off" condition must be detected is much easier. Current and voltage variations have little effect if the signal remains above the noise.

We will consider digital devices and systems from two standpoints. First, we will deal with the use of digital systems for long-distance *information transfer* for both converted analog data and for digitally derived data. In this case the basic mode of transmission and reception will be by wire, radio, or optical methods. Second, we will consider digital systems involving the local use of digital devices and systems as the means for *signal* and/or *data processing* and *control (information management)* so that these are essentially localized. However, these can be, and often are, used for *processing, control,* and *computation* as elements in information-transfer systems as well. Most of the information management and transfer that you learned about earlier in analog form can be handled in digital form. In addition, digital systems provide a unique environment for the *control* and *manipulation* of *information* that has no counterpart in analog systems. For example, computers, microprocessors, and many controllers would be essentially impossible without a digital approach.

One additional point must be made before we start our study of digital devices and systems—that is, that functions are packaged together in IC form almost exclusively, and therefore complete functional elements are available in IC chip form. Thus, we will be mainly concerned with how these devices operate together to form systems, without too much regard to their internal organization. So, as with our study of other solid-state devices, it is their *behavioral (input/output) characteristics* with which we will be concerned.

Digital Devices, Computers, Microprocessors, Calculators

You have probably heard a great deal about digital devices (or systems), computers, microprocessors, and calculators and may wonder what the differences are. Basically, the circuits or digital logic elements— the last of the nine common electronic building blocks—for all of these are the *same*. The differences lie in *how they are assembled and controlled*. While there are no hard-and-fast rules, there is some sense in the division of digital systems into these groups. For example, a digital device (or system) that does not incorporate a computer or microprocessor has its digital logic elements *fixed-wired* together to perform a *specific task*. Examples of this type of system include a digital watch, digitally tuned TV, and CB radio set.

HARDWARE-CONTROLLED
HARD-WIRED DIGITAL DEVICES
FIXED-PROGRAM CONFIGURATION

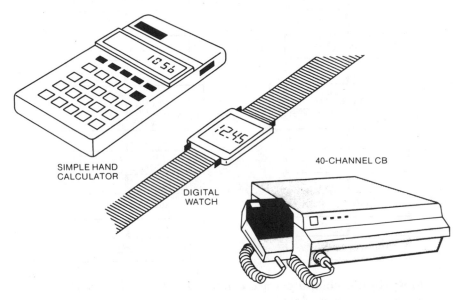

SIMPLE HAND
CALCULATOR

DIGITAL
WATCH

40-CHANNEL CB

Thus, when you set a digital watch to the correct time, digital logic and timing circuits in the watch will advance the initial time at fixed intervals so that when you read the watch later it will have the correct time and may also calculate month, day, time to go, etc. However, this digital system is designed to do *only* this specific thing and *nothing else*. Such digital systems are called *hard-wired* or *hardware-controlled* because the logic elements are in a *fixed, unchanging configuration*. This does not mean that there is no logical decision making in the system. It means that the logical *rules* of operation *cannot be changed* without *rewiring* the elements. A simple hand calculator (nonprogrammable) and a fixed-function timer are additional examples of hard-wired systems.

Digital Devices, Computers, Microprocessors, Calculators (continued)

Computers, microprocessors, and programmable calculators are more flexible because the things they can do are *controllable* by *software*. *Software* is a *set* of *logical instructions* written and supplied to the computers, microprocessors, or programmable calculators to enable them to perform some desired function *without* rewiring or the change of connections. The things that are done, therefore, are *controlled internally* by the *program software* and hence can be easily changed by merely *changing* the *software* (instructions) in the *memory* of the system. The *same* computer, microprocessor, or programmable calculator *can be instructed* to perform *many different tasks.*

INTERNALLY CONTROLLED DIGITAL SYSTEMS

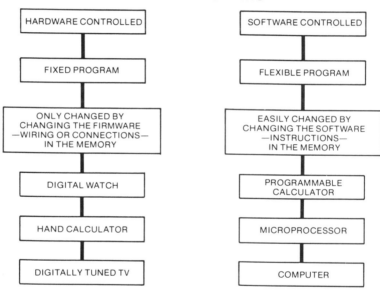

Most digital computers are called *general-purpose computers* because they can be programmed via software to perform many many tasks from scientific calculations to business and personnel record keeping, etc. Large general-purpose computers are characterized by a large memory capacity, high-speed operation, and the capability of servicing multiple users with differing program requirements. *Minicomputers* are less-expensive, smaller versions of general-purpose computers and are more limited in capability, but basically fill the same functional role as the larger general-purpose machines. Modern minicomputer systems, however, are being built with more and more capability, so that in many cases they are equal to or better than some of the older large computer systems. Both, however, retain the property of being *flexibly programmable* to do any job. Obviously, these systems are much more expensive than a simple digital logic device hard-wired to do a specific task.

Digital Devices, Computers, Microprocessors, Calculators (continued)

Microprocessors are also general-purpose computers with more limited capability. These usually have a restricted memory capability and are often assigned to do specific tasks in a system—although the *same* microprocessor, *differently* programmed, may do a *completely different* job in another system. When the microprocessor is assigned a specific task, such as controlling a microwave oven in your kitchen, the programming is generally fixed (or semi-fixed) in special memory devices. This type of system is said to be *programmed with firmware*—that is, a program that is *fixed* in memory but can be changed by making a *change* in the *memory device.* One of the most common examples of changing firmware is in the microprocessor-controlled video games, where the nature of the game is changed completely by replacing a firmware programming module.

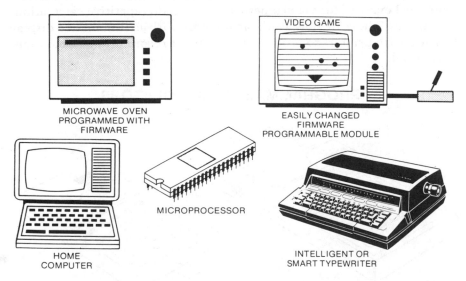

MICROWAVE OVEN
PROGRAMMED WITH
FIRMWARE

EASILY CHANGED
FIRMWARE
PROGRAMMABLE MODULE

MICROPROCESSOR

HOME
COMPUTER

INTELLIGENT OR
SMART TYPEWRITER

Microprocessors were originally developed in response to a problem that arose with the development of LSI (large-scale integration). LSI integrated circuits provide the capability of packaging 5,000–10,000 or more active semiconductor devices on a single chip! The original concept was to develop specific chips for a particular application. However, this was found to be extremely expensive because of the development costs. To avoid this problem, the Intel Corporation developed the microprocessor—a small general-purpose computer element that with appropriate peripheral hardware (external circuits) could be programmed with software or firmware to do any desired job. The development of microprocessors has proceeded very rapidly since their introduction in 1970, and there are now many types with a wide range of capability. Their capability has become so great and they are so inexpensive that they are used today for an enormously wide range of functions—some as simple as home appliance timers and controls, others as complex as the central element in rather sophisticated computers and terminals.

Digital Devices, Computers, Microprocessors, Calculators (continued)

Microprocessors have become so inexpensive (compared to comparable circuits) and can be programmed to perform so many tasks that they are rapidly replacing hard-wired digital devices for most applications. Microprocessors, using firmware programming, are now preferred for most control and timing applications because, as you might realize, the *same* device can be used for many different applications by merely changing its programming. In the past few years, their use has expanded enormously so that they are found in most control applications, from home appliances to sophisticated industrial and office information-processing systems.

The programmable hand calculator is an inexpensive flexible device that can be directed by simple software programming—inserted directly from the keyboard (or via magnetic cards)—to do repetitive calculations or routines. These are now available not only with an illuminated display but also with printers to produce a permanent record of the calculations performed.

PROGRAMMABLE CALCULATORS

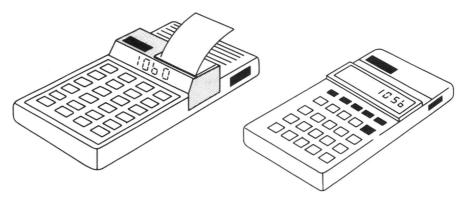

INCIDENTALLY, THE SIZE OF HAND CALCULATORS
IS DETERMINED NOT BY THE ELECTRONICS, BUT BY THE
SIZE OF THE BUTTONS NEEDED TO FIT YOUR FINGERS!

As you will see, the use of digital electronics is commonplace in all aspects of business, industry, government, scientific investigation, control, signal processing, and in many other areas. The basic circuit elements are relatively simple and common to all digital electronic systems. It is how they are used together to do a job that is important. Initially, we will study these basic circuit elements and then see how they are used in systems to do a variety of jobs.

Review of Digital Systems—Background Information

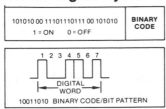

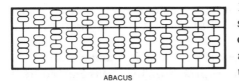

1. DIGITAL SYSTEMS operate using only one type of signal: a "bit" that is either on or off. Information is transmitted in bit patterns.

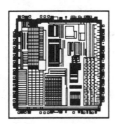

ABACUS

2. COMPUTING MACHINES started when pebbles or other small objects were moved to indicate count and later, Pascal's machine used interlinked dials for digits.

3. TODAY'S DIGITAL COMPUTERS have been greatly improved by the invention of the transistor and development of integrated circuits (ICs).

HOME COMPUTER

4. GENERAL PURPOSE COMPUTERS are those having flexible programming, so that they can be programmed for any of a wide variety of tasks.

MICROPROCESSOR

5. A MICROPROCESSOR (computer-on-a-chip) is a computer of more limited capability. It may use hardwired or less flexible programming.

Self-Test—Review Questions

1. Can machines think? Explain your answer.
2. What are the basic functional areas in information management?
3. What are the essential properties of a digital system? Where is the information in a digital system?
4. Using a grid and counters, show how the numbers 327 and 122 are added and subtracted.
5. What has been a major factor in the development of modern digital computer technology?
6. How are computers and microprocessors controlled?
7. Describe the essential differences between a hardware- (hard-wired), firmware-, and software-programmed digital computer or microprocessor.
8. Describe the essential differences between a calculator and a computer.
9. What is the major advantage of a microprocessor for electronic control applications?
10. List some applications that you know of for computers and microprocessors.

Learning Objectives—Next Section

Overview—In the next section, you will be introduced to the basic digital system elements that are the fundamental parts of all digital systems.

DIGITAL LOGIC
PACKAGE FLIP-FLOPS

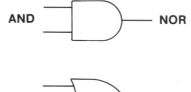

AND   NOR

OR   NOT

NAND   EXCLUSIVE OR

What a Digital Signal Is

One of the easiest things to generate electronically is a *signal* that is *on* or *off*. With modern electronic circuits, this can be done in a *fraction* of a *microsecond*. We can easily represent these on-off signals by a *positive* (or *negative*) *voltage* and *ground* (or *open* circuit). We define these positive (or negative) voltages and ground voltages as *states*. Usually, the positive (or negative) voltage is called a *1* (one) or *high state* and the ground (or near ground) state is called a *0* (zero) or *low state*.

A *single digital signal* can convey only a *very limited amount* of information. For example, such a single digital signal can convey the information that a lamp is off or on and that the lamp switch is either open or closed, as shown in the left-hand illustration. To overcome this limitation on the amount of information conveyed, *several signals* can be combined to provide *more information* (see right-hand illustration).

DIGITAL SIGNAL STATES

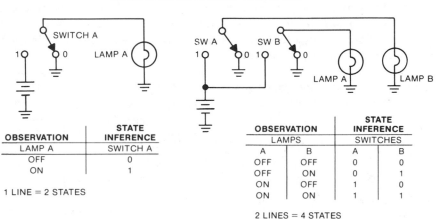

OBSERVATION	STATE INFERENCE
LAMP A	SWITCH A
OFF	0
ON	1

1 LINE = 2 STATES

OBSERVATION		STATE INFERENCE	
LAMPS		SWITCHES	
A	B	A	B
OFF	OFF	0	0
OFF	ON	0	1
ON	OFF	1	0
ON	ON	1	1

2 LINES = 4 STATES

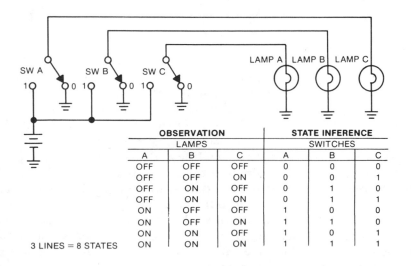

OBSERVATION			STATE INFERENCE		
LAMPS			SWITCHES		
A	B	C	A	B	C
OFF	OFF	OFF	0	0	0
OFF	OFF	ON	0	0	1
OFF	ON	OFF	0	1	0
OFF	ON	ON	0	1	1
ON	OFF	OFF	1	0	0
ON	OFF	ON	1	1	0
ON	ON	OFF	1	0	1
ON	ON	ON	1	1	1

3 LINES = 8 STATES

What a Digital Signal Is (continued)

As you can see from the figure on the previous page, the more lines, the more states that can be interpreted and the more information conveyed. In fact, if you examine the table you will see that for a number of lines N, there are 2^N possible combinations. For example, if 10 lines were available, the number of pieces of information that could be interpreted would be 2^{10} or 1,024. We can further define each line as able to carry a *bit* of information—with each bit having two possible states, high (or 1) and low (or 0). Thus, a pattern of *bits* (1s and 0s) makes up a *digital signal*. The circuit configuration that *generates* the bit pattern and the circuit configuration that *interprets* the bit pattern determines the *intelligence* (information) that is *transferred*.

The information thus transferred is in *parallel* form—that is, the signals are present on the three lines at the same time. While parallel digital-data transfer is used in some digital systems, in most cases it would be cumbersome to provide the *many lines* required for parallel data transfer. Another way to accomplish the same control task would be to use a *single line* with a *multibit digital word*. Suppose you could put a 3-bit word *sequentially* or *serially* on a line by using a sampling switch to sample each bit switch data output (lower left).

SERIAL DATA TRANSFER
3 SEQUENTIAL SAMPLES FROM SWITCHES A, B, C

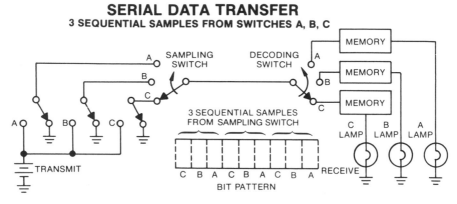

If the sampling switch is then rotated rapidly, it samples the data and produces the serial bit pattern at its output—repeated over and over again. Then, suppose you put a switch at the receiving end which also rotated, in sync with the sampling switch. You can see that the pulse sequence generated by the sampling switch is now separated into 3 lines by the decoding switch. If the sampling switch output has a memory device to remember the state of the line at the instant of sampling, then this signal can be used to operate the lamp. A table can be formed to show the relationship between the data switches (A,B,C), the bit pattern on the line, and the memory output to the lamps, as shown at the top of the next page. If you compare this table with the one on the previous page, you will see that the *results are the same*. Thus, both parallel and serial data transfer the same information. It is the *number* of *bits* that determines *how much information* is transferred.

What a Digital Signal Is (continued)

As you have learned, a digital word is a parallel or serial pattern of high and low (or 1s and 0s) states, with each element in the pattern called a bit. A digital word can be any desired length (or number of bits) depending on the system design and the amount of data to be transferred. A convention commonly used in digital systems is to transfer information in 8-bit sequences called *bytes*. A complete digital word may contain many bytes.

SERIAL DATA TRANSFER

DATA SWITCH			BIT PATTERN (REPEATING)			MEMORY OUTPUT			LAMP CONDITION		
A	B	C				A	B	C	A	B	C
0	0	0	0	0	0	0	0	0	OFF	OFF	OFF
0	0	1	0	0	1	0	0	1	OFF	OFF	ON
0	1	0	0	1	0	0	1	0	OFF	ON	OFF
0	1	1	0	1	1	0	1	1	OFF	ON	ON
1	0	0	1	0	0	1	0	0	ON	OFF	OFF
1	0	1	1	0	1	1	0	1	ON	OFF	ON
1	1	0	1	1	0	1	1	0	ON	ON	OFF
1	1	1	1	1	1	1	1	1	ON	ON	ON

BINARY BIT PATTERN REPRESENTATIONS

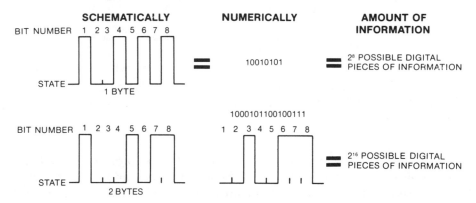

As you can see above, bit patterns can be represented easily by a bit sequence of 1s (high state) and 0s (low state), such as 10010101. This is the usual way to show bit patterns, rather than schematically, since it is much easier to interpret. Since only two states are involved, these digital signals are called *binary sequences* or *binary bit patterns*. You will see how these are interpreted and used as you continue your study of digital systems. The important thing to remember is that in all cases we are dealing with *the generation of bit patterns (information), transfer of these to another point, the manipulation of the bit patterns, and interpretation of the result for the specific thing to be done.* Since in digital systems we deal in relatively well-defined high–low states and nothing in between, and since we can make the bit pattern as long as we wish, *digital systems are well suited to the faithful transfer of information.*

Digital Logic—Boolean Algebra

The basic concepts concerning how digital systems can make logical decisions were established by George Boole (1815–1864), an English mathematician and logician. He pointed out the close analogy between the symbols of algebra and the symbols that could represent logical concepts restricted to the quantities 0 and 1. He showed that these logical symbols—applied to objects and concepts—can be added, subtracted, multiplied, and divided according to the same primary laws of combination as algebraic symbols. He expanded these concepts to show that symbolic treatment concerning premises of logic could be used to find any conclusions logically contained in those premises.

When digital systems came along, they required a systematic approach to design that was simple and logically correct. Since digital systems use only two states (0 or 1), digital designers have found Boolean algebra ideal for their use.

BOOLEAN ADDITION, MULTIPLICATION, AND COMPLEMENTATION

OPERATION	MEANING
A • B	A AND B
A + B	A OR B
$\overline{A}$	A NOT or NOT A

BOOLEAN LOGIC SYMBOLS

OPERATIONS	MEANING
(A + B)(C)	A OR B, AND C
AB + C	A AND B, OR C
$\overline{A}$ • B	A NOT, AND B
A + $\overline{B}$	A OR B NOT

COMBINATIONS OF LOGIC SYMBOLS

ADDITION	MULTIPLICATION	COMPLEMENTATION
0 + 0 = 0	0 • 0 = 0	$\overline{0}$ = 1
0 + 1 = 1	0 • 1 = 0	$\overline{1}$ = 0
1 + 0 = 1	1 • 0 = 0	
1 + 1 = 10	1 • 1 = 1	

In Boolean algebra the symbols *0* and *1* (low and high) respectively represent the concepts *off* and *on*, *false* and *true*, *none* and *all*, or the *non-conduct* and *conduct* states. In Boolean algebra only two conditions are possible. If a variable x is not equal to 1, it *must* be equal to 0. If x is not 0, it *must* be equal to 1.

The basic concepts of Boolean algebra make use of three unique symbols: A *plus (+)* symbol that is bolder (blacker) than the conventional mathematical + sign represents the Boolean concept of *OR*, a *star (*) or dot (·)* represents the Boolean concept of *AND*, and the mathematical *overhead bar (‾)* or *prime symbol (')* represents the Boolean concept of *NOT*. The NOT concept is also called the *complement* (or *inverse*). Thus, the complement of 1 is 0 and the complement of 0 is 1. The basic rules for using these symbols are shown in the illustration above.

Digital Logic Element—The AND Circuit

The Boolean concept of *multiplication* (i.e. A·B or AB) can be transformed into a practical and very useful circuit for digital decision making. This circuit is the AND circuit. An AND circuit can have two, three, or more inputs. AND circuit logic demands a *unanimous positive* set of *inputs* before it will produce a *positive output*. Only when *all* of its inputs are 1 (high) will it produce a 1 (high) output. When an AND circuit produces a 1 output, you know that all of its inputs are also in the 1 state.

In its most basic form, an AND circuit can be made from a pair of *switches connected in series*. The open or closed position of the switch represents the state of the input. If input A is *1*, this means that switch A is *closed*. If input A is *0*, this means that switch A is *open*. The same conditions exist for switch B. For there to be an output to the load (a 1 output) *both* switches must be closed. The presence of an output means that both inputs A AND B (or A·B) are high.

THE AND CIRCUIT (or AND GATE)

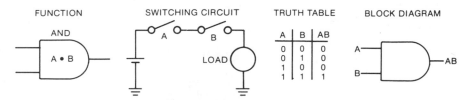

There are several possible combinations of inputs to an AND circuit. A *truth table* can be used to reveal all these possibilities and their effect on the output. The number of possible combinations is equal to 2^N where N equals the number of inputs. If there are two inputs (2^2), the truth table has four possible input combinations, as shown in the diagram.

As illustrated, when the input condition of both switches is 0 (open), the resulting A·B output is low. Each successive row shows other combinations of input conditions. Only the fourth combination, in which both inputs are high, results in a high A·B output.

For an AND circuit with three inputs, there are 2^3 or eight combinations of input conditions. Only in the last of these are all three inputs equal to 1, thus producing an A·B·C output of 1.

As you can see, the AND circuit can *function as a gate (switch)*. If, for example, the digital signal is at A, the B input can be used to allow or restrict the passage of the signal. When used this way, and AND circuit is called an *AND gate*.

Digital Logic Element—The OR Circuit

The Boolean concept of *addition* can also be used in digital circuits to provide *additional ability to make decisions*. This device is the *OR circuit or gate*. Like the AND circuit, it can have two, three, or more inputs. OR circuit logic requires that *at least one* of its inputs must be *high* before it will produce a *high output*. In the two-input circuit shown, either A *or* B must be high before a high output will be produced. In a three-input circuit, either A or B or C must be high for a high output to be produced. *More* than one of the inputs *can* be high, but at least one must be high for a high output to result. In other words, when an OR circuit produces a high output, you know that at least one of its inputs must be high.

THE OR CIRCUIT or OR GATE

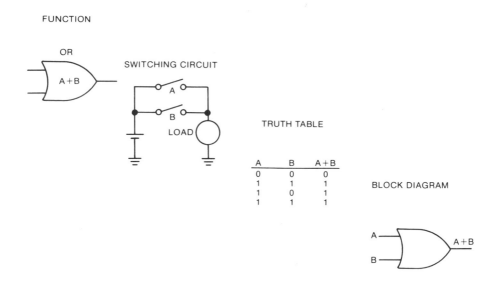

FUNCTION

OR

A + B

SWITCHING CIRCUIT

LOAD

TRUTH TABLE

A	B	A + B
0	0	0
1	1	1
1	0	1
1	1	1

BLOCK DIAGRAM

A
B
A + B

In its most basic form, an OR circuit can be made from a pair of *switches connected in parallel*. The open position of either switch is represented by a 0 (low), and the closed position is represented by a 1 (high). For there to be an output to the load (a 1 output), either one or both switches must be closed. The presence of an A or B output (shown in the block diagram as A + B) means that the stated conditions have been met.

The truth table for the OR circuit works in the same manner as the one for the AND circuit. The number of possible input-signal combinations is equal to 2^N where N equals the number of inputs. In the table for two inputs (four combinations), note that the only arrangement that does not produce a high output is the one in which both switches are open. In a circuit with three inputs there are eight combinations, and only one of these (all three switches open) does not produce an output.

Digital Logic Element—The NOT Circuit

The NOT circuit is the last of the basic digital logic circuits. This circuit produces the complement of the input at its output. It is also known as an *inverter*, which may be more meaningful to you. This circuit produces the *inverse* of what you supply to its input. If you apply a 1 to the input, the output is 0. If you apply a 0 to the input, the output is 1.

An example of how such a circuit could be made of basic components is shown in the circuit diagram here. If the input switch is open (a 0 input), the relay coil is not energized, the relay's switch contact applies battery voltage across the load, and the output is 1. If the input switch is closed (a 1 input), the relay coil attracts its switch contact, there is an open circuit between the load and its battery, and the output is 0.

THE NOT CIRCUIT (or INVERTER)

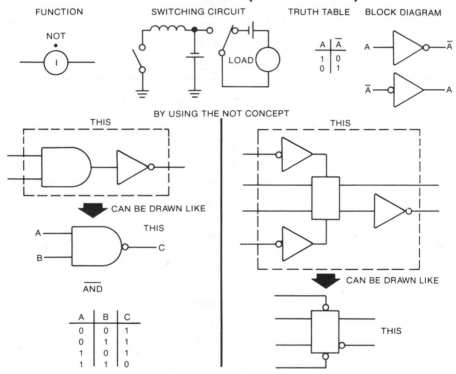

The NOT circuit is drawn as an amplifier block with a small circle to indicate output. A simple truth table summarizes the operation. The *bar* over an input or output symbol expresses the concept of NOT; thus $\overline{A}$ means not A. In many cases, the NOT function is included with other digital circuit elements to produce a complement output. When this is done, a small circle is used at output to indicate complementation. Thus, with the AND·NOT circuit shown, two high inputs will produce a low output while a high output is obtained if either input is low. Later, you will be introduced to devices that have both the normal and complement outputs available or can use either a normal or complement input.

Digital Logic Elements—NOR, NAND, Exclusive OR Circuits

Shown are some other combinations of the basic AND, OR, and NOT circuits. The *NOR* concept is a combination of OR and NOT operations. The basic circuit shows that load receives voltage through a normally closed circuit. When either input switch is closed, the relay coil is energized and attracts the contact to the open position. The truth table compares the outputs of the OR (A + B) and NOR ($\overline{A + B}$) circuits. The *NAND* circuit is a similar combination of AND and NOT operations.

THE NOR, NAND, AND EXCLUSIVE OR CIRCUIT DIGITAL LOGIC

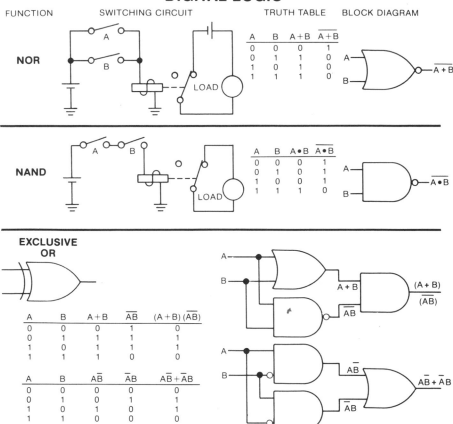

The *Exclusive OR* is a little more complicated. In the OR circuit, an output is produced when either A or B inputs are high. In an Exclusive OR circuit, a high output results when *different* inputs are applied at both A and B. It doesn't matter whether A or B is high or low; as long as they are *not* the same, a high output results. Two possible arrangements to produce this function are shown. Trace them through for yourself to see how they operate. Notice that the truth table for the Exclusive OR circuit is the same as the one for binary addition. The Boolean Algebra symbol for Exclusive OR is a + inside a circle ($\oplus$).

Variations in Drawing Digital Logic Circuits

Occasionally a logic element may be drawn differently to make its use in a circuit more obvious.

Some of the alternate forms in which the basic gates described previously can be drawn are shown in the diagram here.

ALTERNATE FORMS FOR BASIC GATES

COMMON NAME — TRUTH TABLE — THIS MAY BE SHOWN AS THIS

AND

A	B	OUT
0	0	0
0	1	0
1	0	0
1	1	1

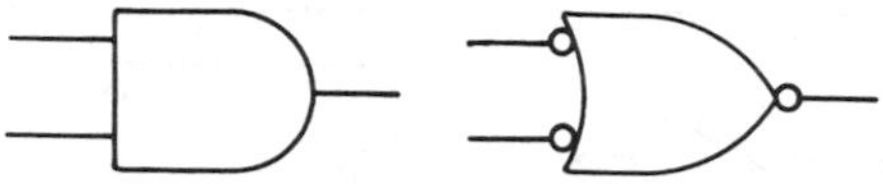

OR

A	B	OUT
0	0	0
0	1	1
1	0	1
1	1	1

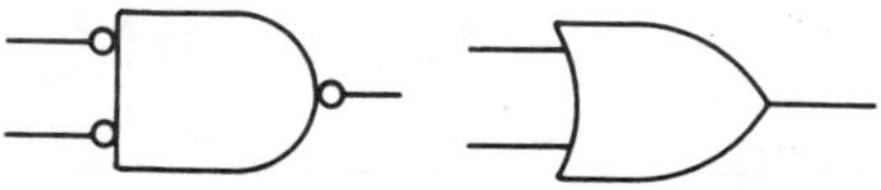

NAND

A	B	OUT
0	0	1
0	1	1
1	0	1
1	1	0

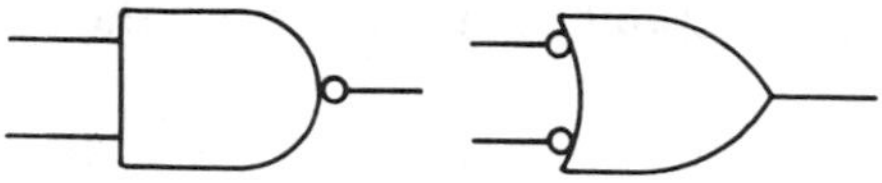

NOR

A	B	OUT
0	0	1
0	1	0
1	0	0
1	1	0

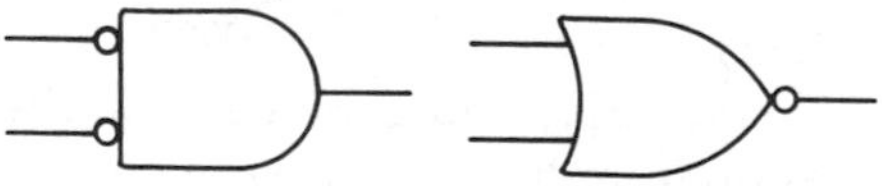

EXCLUSIVE OR

A	B	OUT
0	0	0
0	1	1
1	0	1
1	1	0

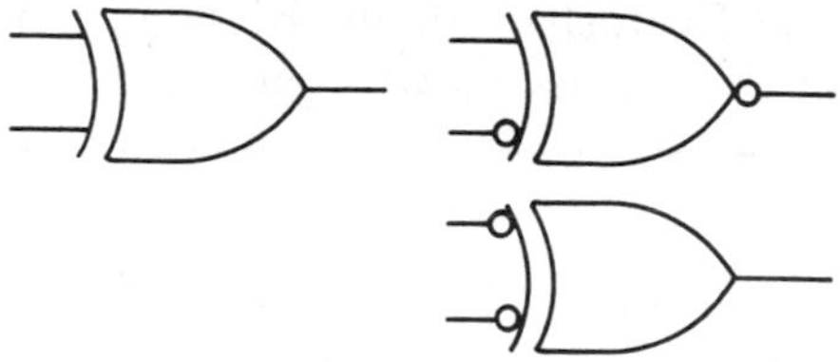

EXCLUSIVE NOR

A	B	OUT
0	0	1
0	1	0
1	0	0
1	1	1

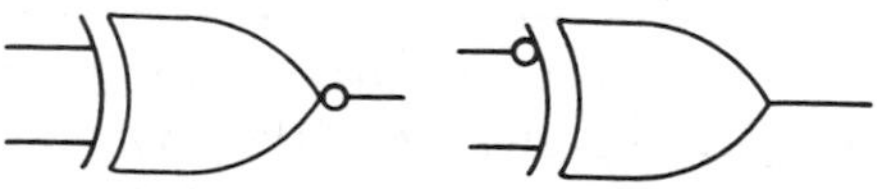

INVERTER (NOT)

A	OUT
0	1
1	0

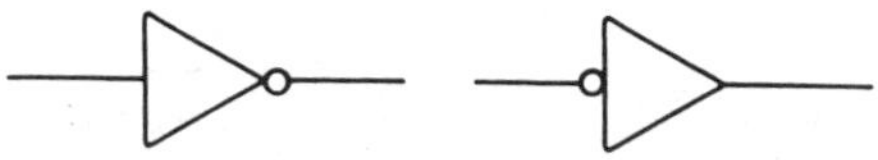

Experiment/Application—Digital Logic Elements

Since digital logic elements are invariably packaged as ICs, you can easily study their external properties to see for yourself how they behave.

Suppose you connected up a test circuit as shown below using an SN7408 quad two-input AND gate. Although there are four (quad) gates in this single IC chip, we will use only one. (Pins 1 and 2 are inputs, and pin 3 is the output.) If you leave S1 and/or S2 open, the output is at about ground potential (less than 0.5 volt); closing the switches will produce the result shown in the table below.

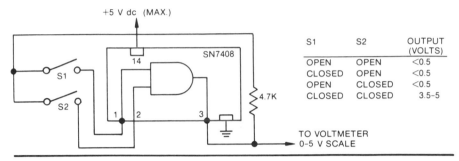

S1	S2	OUTPUT (VOLTS)
OPEN	OPEN	<0.5
CLOSED	OPEN	<0.5
OPEN	CLOSED	<0.5
CLOSED	CLOSED	3.5–5

If you substitute an SN7432 (quad OR) for the SN7408 (same pin connections), and check the output for various combinations of S1 and S2, you should get the results shown below.

	OR	
S1	S2	OUTPUT (VOLTS)
OPEN	OPEN	<0.5
CLOSED	OPEN	3.5–5
OPEN	CLOSED	3.5–5
CLOSED	CLOSED	3.5–5

Similarly, you can see how the NOR, NAND, and Exclusive OR gates operate by substituting suitable chips in the test setup. If you use an SN7402 quad NOR gate, an SN7400 quad NAND gate, and an SN7486 quad Exclusive OR in the same setup (same pin connections), you will get the results shown below for various switch settings.

	NOR			NAND			EXCLUSIVE OR	
S1	S2	OUTPUT (VOLTS)	S1	S2	OUTPUT (VOLTS)	S1	S2	OUTPUT (VOLTS)
OPEN	OPEN	3.5–5	OPEN	OPEN	3.5–5	OPEN	OPEN	<0.5
OPEN	CLOSED	<0.5	OPEN	CLOSED	3.5–5	OPEN	CLOSED	3.5–5
CLOSED	OPEN	<0.5	CLOSED	OPEN	3.5–5	CLOSED	OPEN	3.5–5
CLOSED	CLOSED	<0.5	CLOSED	CLOSED	<0.5	CLOSED	CLOSED	<0.5

The NOT circuit is simply an inverter. (Often six of these are packaged together as a "hex inverter".) The SN7407 is a simple example of this. To test this circuit, you connect the switch input to pin 1 and the voltmeter/resistor output to pin 2; pin 3 is open. The results are:

S1	OUTPUT (VOLTS)
OPEN	3.5–5
CLOSED	<0.5

Examples of Digital Logic Circuits—Combination Circuits

Many arrangements of components can be assembled to produce the logic circuits that have been described. Resistors and voltage sources can be combined to produce AND and OR gates; these are known as *resistor logic* (RL) circuits. Diodes can also be used to form these two basic circuits; these are known as *diode logic* (DL) gates.

Transistors offer much more positive current control, and their saturated (on) and cutoff (off) modes are like the operation of a switch. In these digital logic circuits any of the common-base, common-emitter, or common-collector arrangements can be used, although the common-emitter circuit is the one most widely used.

Transistors can be used in such digital logic circuits in combination with resistors, diodes, and other transistors. Resistor–transistor logic gates are known as *RTL circuits*, diode–transistor logic circuits are known as *DTL gates,* and transistor–transistor logic circuits are known as *TTL gates.* When the transistor is operated in the current mode, the arrangement is known as a *CML (current-mode logic)* or as an *ECL (emitter-coupled logic) gate.* TTL and ECL are the most commonly used configurations today. A few of these arrangements are shown in the diagrams on the following page.

Internal details of ICs are of little importance to anyone except the designer of the microcircuits. Other than the circuit designer, the service technician is concerned that a microcircuit chip contains a number of basic circuits. What is important to the technician is *what* these circuits are and *how* connections can be made to them. If the logic circuits operate as specified by the chip manufacturer, everything is fine. If any one of the logic circuits does not operate as specified, its individual circuit components cannot be replaced; the *entire* chip must be *replaced*.

Integrated microcircuit chips containing a number of identical digital gates are widely available at very low cost. For example, inverter chips are available with six inverters in a single IC; AND gates are available in groups of four on a chip. Because popular chips are inexpensive, electronics designers may choose to make use of extra circuit elements in an IC to make the ones they want. Boolean logic makes it possible.

It may happen that an AND circuit is required but only NAND and NOT gates are available on the chips being used. By connecting these two gates in series, an AND is produced. Similarly, if a NOR gate is required, it can be made from two NOT gates and one AND gate.

NAND circuits are quite popular in digital technology. There are many integrated microcircuit chips with a number of NAND gates on a single IC, so it is a good idea to know how to use them for various purposes. Boolean logic permits the three basic AND, OR, and NOT plus NOR to be assembled from combinations of NAND gates. These combinations are shown in the diagram on the following page.

Examples of Digital Logic Circuits—Combination Circuits (continued)

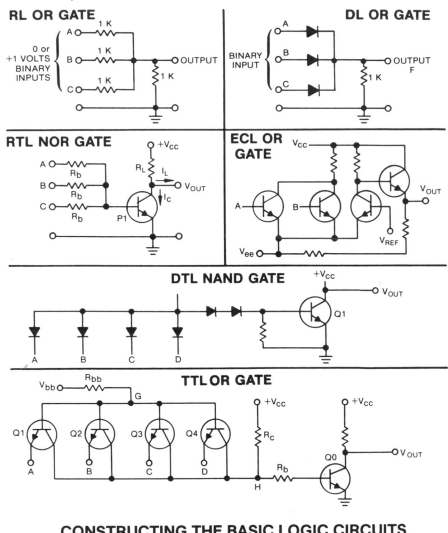

RL OR GATE

DL OR GATE

RTL NOR GATE

ECL OR GATE

DTL NAND GATE

TTL OR GATE

CONSTRUCTING THE BASIC LOGIC CIRCUITS WITH NAND AND NOT GATES
COMBINATION CIRCUITS

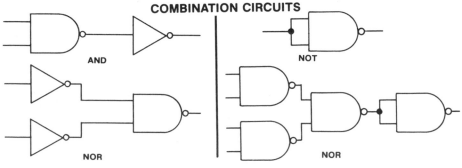

AND

NOT

NOR

NOR

Definition of High and Low Logic State

Most digital logic circuits operate from a +5-volt supply. Since we use cutoff and saturated transistors, respectively, to represent logic states 1 (high) and 0 (low), you might believe that the voltages provided as an output (or required as an input) must be 5 volts or 0 volts. Actually, this is not quite true, because a saturated transistor still has a residual voltage across it, and a cutoff transistor with a resistor in its collector circuit when coupled to a load resistor will, by divider action, produce an output of less than 5 volts. In addition, because of noise or transients that might trigger digital logic circuits, it is usually desirable that the threshold sensitivity to the low state be somewhat greater than 0 volts.

TRANSISTOR-TRANSISTOR LOGIC STATES

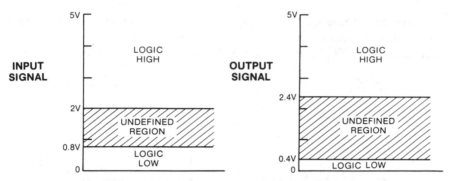

For transistor–transistor logic (TTL), the input/output (I/O) characteristics are as shown above. At a voltage below 0.8 volt, the driven circuit sees a low logic (0) state; above 2 volts at the input produces a high logic (1) state. Similarly, outputs are defined as 0.4 volt or less for low and 2.4 volts for high. As you can see, there is compatibility between the output of one device and the input of the next device. This compatibility is very important. Thus, functional circuits use combinations of compatible chips—e.g., all TTL or ECL. The maximum number of devices or the load current an output circuit of a device can deliver is usually specified. This is often called the *fanout capability* for the device, and it is a measure of the extent to which a single device can drive multiple devices. When a *fanout* exceeds the capability of the device, a *buffer* is used. A buffer is a separate noninverting (or inverting) amplifier that increases the available load current or fanout capability. If more output current is necessary, additional buffers can be used.

The problem of load capability and definition of high and low logic states are important to the digital designer. However, the *definition* of *high/ low logic states* is also important to the *troubleshooting of digital circuits*. For logic circuits a low input is defined as between 0 and 0.8 volt and a high input is above 2.0 volts. On the other hand, a low output is defined as between 0 and 0.4 volt, while a high output is anything greater than 2.4 volts.

A Simple Digital Control System

The table below shows how a combination of simple digital logic elements can be used to control many circuits with a limited set of control lines. For example, suppose eight lamps were to be operated using three control lines (three are necessary to get eight separate states—that is, $2^N = 2^3 = 8$). Set up a truth table to establish logic requirements:

TRUTH TABLE

LAMP NO.	CONTROL LINE LOGIC STATE		
	A	B	C
1	0	0	0
2	0	0	1
3	0	1	0
4	0	1	1
5	1	0	0
6	1	0	1
7	1	1	0
8	1	1	1

The necessary logic can be set up using a combination of three input AND gates to decode the logic state of the control lines and a set of three inverters (NOT) to provide $\overline{A}$, $\overline{B}$, and $\overline{C}$, to feed the AND circuits.

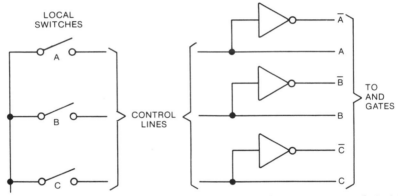

The three switches determine the states of lines A, B, and C (high or low). At the remote end, the states A, B, and C are available, as well as $\overline{A}$, $\overline{B}$, and $\overline{C}$.

The AND gates are connected to these outputs, each in a unique combination. For example, AND 3 is connected with $\overline{A}$, B, and $\overline{C}$ so that only when switches A and C are off and switch B is on does it light lamp 3. The final truth table is shown below.

TRUTH TABLE

SWITCH POSITION			LAMP NUMBER	LOGIC INPUT TO ANDs		
A	B	C		1	2	3
OFF	OFF	OFF	1	$\overline{A}$	$\overline{B}$	$\overline{C}$
OFF	OFF	ON	2	$\overline{A}$	$\overline{B}$	C
OFF	ON	OFF	3	$\overline{A}$	B	$\overline{C}$
OFF	ON	ON	4	$\overline{A}$	B	C
ON	OFF	OFF	5	A	$\overline{B}$	$\overline{C}$
ON	OFF	ON	6	A	$\overline{B}$	C
ON	ON	OFF	7	A	B	$\overline{C}$
ON	ON	ON	8	A	B	C

If you trace the circuit on the next page, keeping in mind what you know about NOT and AND circuits, you can see that different lamps will light as you manipulate the switches.

A Simple Digital Control System (continued)

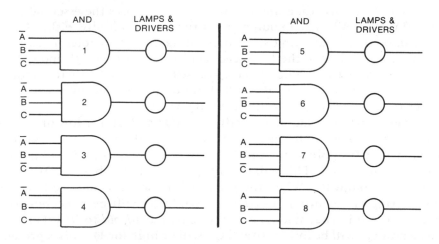

For example, if all switches are open (all lines low), then lamp 1 will light since AND gate 1 will have all 1s at its inputs—producing a 1 at the output. Similarly, if only the C switch is closed (making line C high), then AND circuit 2 will have all 1s at its inputs—and lamp 2 will light. If you follow the rest of the logic states at the AND inputs as the switches are operated, you will see that each lamp lights in response to a unique set of switch positions, allowing you to light any desired lamp by means of the three switches and the three control wires.

Flip-flop circuits are important digital-system elements because they can be used to *store* (*remember*) digital information and they can be used to count. These properties and their general usefulness in structuring digital systems makes them as important as the basic logic devices you have already learned about. While flip-flops can be constructed from discrete circuit elements, as shown in the diagram below, you will find that they are invariably ICs; in many cases, several flip-flops are arrayed on a single IC chip. The important thing is *how a flip-flop can be used to receive, store (remember), count, and feed out logic data.*

BASIC FLIP-FLOP SYMBOL AND CIRCUIT

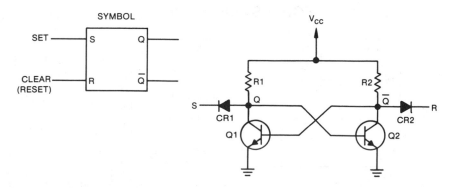

The Flip-Flop Circuit

The schematic diagram on page 5–29 illustrates the essential parts of a flip-flop. When the equipment is turned on initially, either Q1 or Q2 can be in a conducting state, depending on circuit tolerances, transients, etc. As shown, the base feedback paths between Q1 and Q2 are arranged so that only one of these transistors (Q1 *or* Q2) will conduct. This low collector output ($\overline{Q}$ or 0 state) occurs on one side, while the other transistor is cut off and is high (Q or 1 state).

Let us assume that Q2 is conducting (Q2 collector is low or $0 = \overline{Q}$) and that Q1 is nonconducting (Q1 collector is high or $1 = Q$). If a ground is momentarily applied at S, the collector of Q1 will be pulled to ground through diode CR1. This will cut off Q2, making the base of Q1 highly positive through R2. When the ground is removed from S, Q1 will be conducting and continue to conduct, while Q2 will be nonconducting and continue in that state. Thus, the logic has *flipped* to its other state since now Q1 will be low (or 0) and Q2 will be high (or 1). This represents the *set condition* of the flip-flop, and input port S is called the *set input*. Similarly, if a ground is momentarily applied to the R input, the circuit will be *flopped* to its original condition. This represents the *reset condition* of the flip-flop, and input R is called the *reset input*. Sometimes the reset input is called the *clear input* and is designated by a *C*. We also speak of a reset as *clearing* the flip-flop. You can understand now why this circuit is called a *flip-flop*. The symbolic drawing for a flip-flop is shown depicting the inputs and outputs, just as for the logic circuit elements you have already studied. As you can see from the schematic, the circuit is *symmetrical*, but by choosing the designation of the set and clear lines we have determined how and when the state of Q1 and Q2 collectors will change with the set and clear inputs.

The illustration on page 5–29 shows the symbol for a flip-flop. As you can see from the schematic of the flip-flop and the truth table shown below, if the circuit is set by a signal (low logic state), then further inputs to the set line will not affect its state. It can only be set if it is in the clear condition.

TRUTH TABLE
FOR R-S FLIP-FLOP

PRIOR STATE		INPUT STATE		OUTPUT		
Q	$\overline{Q}$	SET	(CLEAR) RESET	Q	$\overline{Q}$	
0	1	0	1	1	0	⎫
0	1	1	0	NO CHANGE		⎬ USEFUL
1	0	1	0	0	1	
1	0	0	1	NO CHANGE		⎭
0	1	1	1	0	0	⎫
1	0	1	1	0	0	⎬ AVOID
0	1	0	0	INDETERMINATE		THESE
1	0	0	0	INDETERMINATE		⎭

The Flip-Flop Circuit (continued)

Furthermore, if the flip-flop is clear, additional clear inputs will not affect its state. Thus, a set (or cleared) flip-flop *remains* in that condition until action is taken to change its state. Therefore, a flip-flop *remembers* its set (or clear) state until it is changed. This is a *highly significant property* that gives the flip-flop its important position in digital systems. In addition, the flip-flop provides not only the output Q, but also its complement $\overline{Q}$. In some applications, as you will see, this is very convenient.

When equipment containing flip-flops is turned on, the flip-flops are in an arbitrary state. If a particular state is desired, then an input is required to set the state of the device before the flip-flop outputs are used. Remember that the terms reset (R) and clear (C) are used interchangeably. The characteristics of the flip-flops described here are summarized in the truth table on page 5–30. Note that *indeterminate states can occur*. This means that for the input condition specified, the output *cannot* be specified. This uncontrolled operating state *must be avoided*.

DUAL NOR GATE (R-S) FLIP-FLOP

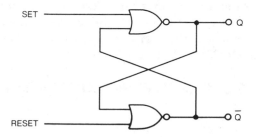

The simple flip-flops described above are called *R-S (reset-set) flip-flops*. Typically, they are made from a pair of NOR gates connected as shown above. From what you already know about NOR gates, you can see how they work as flip-flops when connected as shown. Check to see whether the results agree with the truth table shown above.

You have seen how the basic R-S flip-flop operates. It indicates at its output a prior action at its input and *retains this information* until its state is changed. There are many flip-flop configurations, but the most common is the R-S type.

In many cases, it is desirable to be able to change the state of a flip-flop at specific times only. These usually correspond to the times determined by a *clock* in any digital system that times and controls the sequence of events. For example, it may be desired to test the state of a signal at a specific time and set (or reset) a flip-flop based on the state of that signal at that time only and ignore changes between. This can be done by means of a *clocked flip-flop*. The figure on the top of page 5–32 illustrates this circuit.

Other Basic Flip-Flop Circuit Configurations

CLOCKED R-S FLIP-FLOP

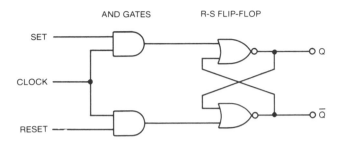

A typical clocked flip-flop circuit consists of a basic R-S flip-flop, except that the set-reset lines of the R-S flip-flop are fed by a pair of AND gates. The AND gates are fed by the clock and by the set-reset lines. Because both inputs to the AND gates must be high to produce a high output, the R-S flip-flop can only change its state when the clock line is high (1). At all other times, the set or reset cannot be propagated to the R-S flip-flop section because of the logical properties of the AND gates. Typically, clocked flip-flops are designed to change state on the transition of the clock to a high (or low) state.

R-S CLOCKED FLIP-FLOP SYMBOL

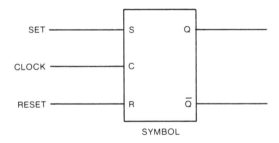

The symbol for the R-S clocked flip-flop is shown above. As you will notice, it is the same as that for the R-S flip-flop with the addition of the clock line. As with the R-S flip-flop, multiple units are available in single IC packages. The R-S flip-flop and the clocked R-S flip-flop have a problem if both inputs (R and S) are high (or low) at the same time, since the output is ambiguous (refer to the R-S flip-flop truth table). To prevent this, the D and J-K flip-flops have been developed.

Other Basic Flip-Flop Circuit Configurations (continued)

The D and J-K flip-flops are commonly used to avoid the ambiguity of the simple R-S flip-flop when both inputs are placed simultaneously in the same state. In the *D-type flip-flop*, an additional input (the D or data input) is connected to the set gate input, and an inverter is connected from the input (set) line to the reset. In this way, the set and reset levels can never be high (or low) simultaneously. Often, dc set and reset lines are provided on D-type flip-flops to provide for initialization. The D flip-flops are well suited for use as buffer registers, shift registers, and counters. (You will learn about these applications later in this volume.) As in the clocked R-S flip-flop, the D flip-flop changes state in response to a positive clock transition (when the clock goes high).

SYMBOLS OF OTHER BASIC TYPES OF FLIP-FLOPS

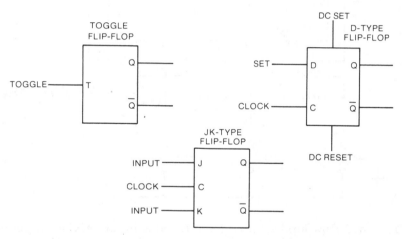

The *J-K flip-flop* is a *master-slave flip-flop*, or two flip-flops in series. Information is transferred into the master flip-flop on the positive transition of the clock. The stored data is then transferred to the slave (output) flip-flop on the negative transition of the clock pulse. The unique feature of the J-K flip-flop is that there will be *no change* in the output if both inputs are in the low logic state during the clock pulse. If both inputs are high, the output will complement (reversing Q and $\overline{Q}$).

The last of the common flip-flops is the *toggled flip-flop*. In this flip-flop, there is an additional (T) input. This input reverses output (complements) whenever a high (1) logic signal is provided to the T input.

Remember that the basic feature in all flip-flop designs is that the flip-flop output can be put into a given state when conditions are appropriate at the input, and this state remains at the output until it is changed by appropriate input signals (and clocks).

Experiment/Application—The Flip-Flop

You can see how a flip-flop works by setting one up and observing the outputs. To do this, suppose you set up the circuit shown below using a 7472 IC chip. The 7472 is a J-K flip-flop, which, as you remember, operates on the master-slave principle.

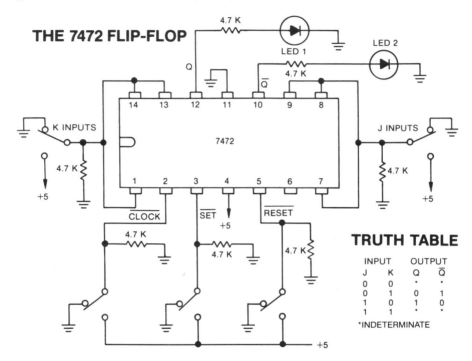

The 7472 flip-flop has three J and three K inputs ANDed together. For our experiment, we will strap these together for one input. If you set the J-K inputs as shown in the truth table, you will observe that nothing happens if the $\overline{set}$, $\overline{reset}$, and $\overline{clock}$ are at +5 volts. Momentary operation of the $\overline{reset}$ (connect to ground) will force the flip-flop into the state where $\overline{Q}$ = 1 and Q = 0. Therefore LED 2 is on and LED 1 is off. If you set the input section, for example, so that J = 1 and K = 0, nothing will happen at the output (LED 1 and LED 2 will not change) even though the input section has changed state. When you operate the clock switch (momentary ground), however, you will see the output change state, as evidenced by the fact that LED 1 will light and LED 2 will go out. This is because activation of the clock input transfers the data set at J-K to the slave flip-flop. Operation of $\overline{set}$ will reverse the outputs (LED 2 will be high and LED 1 low), since $\overline{set}$ and $\overline{reset}$ act directly on the slave output. If the J-K inputs are set at J = 0 and K = 1, with the $\overline{set}$, $\overline{clock}$, and $\overline{reset}$ on 1, after the set has been operated you will see that clocking does *not* change the output. If J and K are set at J = 1 and K = 0, then clocking will reverse the state. If both J and K are set on 1, clocking will produce an indeterminate state (either Q or $\overline{Q}$ may be 1).

Monostable (One-Shot) Multivibrator

In contrast to the flip-flop, the *monostable* or *one-shot multivibrator* has only one stable state. When a clock pulse or trigger pulse is applied, the state switches for a time—determined by component values (usually RC timing elements)—and then reverts to the original state. These multivibrators are useful in digital work for generating *pulse delays* and *gates*. Invariably today, multivibrator circuits are packaged as IC microcircuits; the RC timing elements are external, however, because of their size and the desire to be able to change RC values for different applications.

THE MONOSTABLE MULTIVIBRATOR (ONE SHOT)

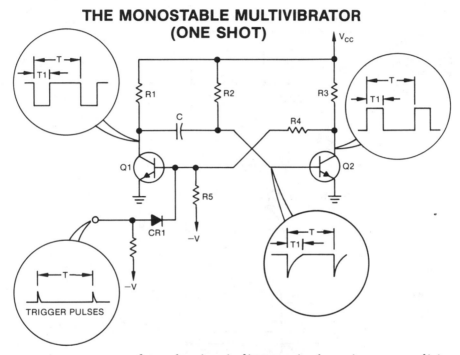

As you can see from the circuit diagram, in the quiescent condition, Q1 is biased off by divider R4–R5. Q2 is biased on by the current through R2 applied to its base. Thus, the Q2 collector is near ground potential (Q2 is saturated). Upon application of a positive trigger pulse at the Q1 base through CR1, Q1 is driven into conduction and the Q1 collector is pulled to near ground potential. This negative transition is propagated to the Q2 base via C, turning Q2 off. When the Q2 collector goes high, the voltage divider R3–R4–R5, keeps the Q1 base positive, keeping Q1 on. As the negative charge on C is neutralized by current through R2, the potential at the Q2 base gradually moves toward V_{cc} at a rate dependent basically on the value of R2 and C. When the Q2 base becomes slightly positive, Q2 conducts. This turns Q2 on and, in the process, the Q1 base becomes negative, which restores the circuit to its original state. When the next trigger is applied, the process repeats. Varying R2 and/ or C changes the duration of the pulse by varying the times during which Q2 is cut off. The pulse can be used directly as a gate.

The Squaring (Schmitt Trigger) Multivibrator

The *squaring multivibrator* (also known as a *Schmitt trigger*) generates a clean rectangular pulse when supplied with a sine, sawtooth, or poorly shaped rectangular pulse. It is therefore often used to clean up digital waveforms that have been corrupted by noise. The major difference between this circuit and the flip-flop is that the feedback path from Q2 to Q1 is replaced by an emitter resistor that is common to Q1 and Q2, as shown in the schematic. As with the multivibrator and other digital-circuit elements, Schmitt trigger devices are invariably IC devices today.

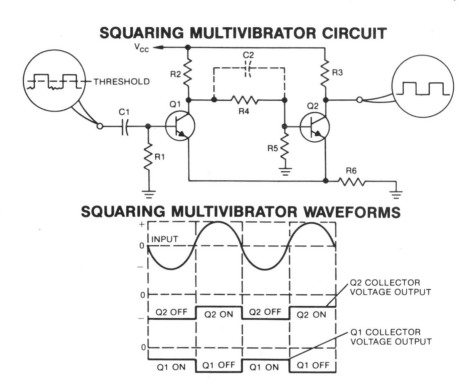

SQUARING MULTIVIBRATOR CIRCUIT

SQUARING MULTIVIBRATOR WAVEFORMS

Note that Q1 is normally off (the Q1 collector is high) and Q2 is biased on (the Q2 collector is low) under normal conditions. Transistor Q2 conducts because of the positive bias applied to its base via divider R2–R4–R5. Current through Q2 produces a positive voltage at the emitters, because of Q2 current flowing through common-emitter resistor R6. This positive bias is also applied to the Q1 emitter, which also helps keep Q1 off (positive emitter voltage on the npn transistor). When the base of Q1 is driven sufficiently positive by the input signal to make Q1 conduct, the base of Q2 goes essentially to ground, which cuts Q2 off. In addition, the current through Q1 flows through the common-emitter resistor R6, which ensures that Q2 is kept off (the Q2 base is grounded, with a positive bias on the npn Q2 emitter).

Experiment/Application—The Multivibrator

This condition remains until the potential on the Q1 base falls to the point where Q1 ceases to conduct. When this happens, Q2 conducts and the circuit is in its original state. Thus, as long as the Q1 base is *above* the threshold value, *Q1 conducts* and Q2 is off. When the Q1 base voltage falls *below* the threshold, Q1 is cut off and *Q2 conducts*. Capacitor C2 is a small capacitor often placed across R4 to speed up the switching action.

Below we have an experiment/application for the multivibrator.

Multivibrators are available as IC chips with external RC time constants to control the output pulse width. A simple experiment can be done to illustrate how multivibrators work. A typical multivibrator IC is the 74121, which is a monostable multivibrator. Suppose you set up the circuit shown:

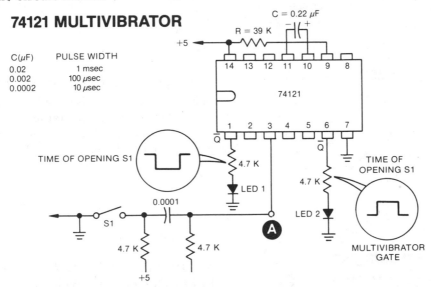

74121 MULTIVIBRATOR

C(μF)	PULSE WIDTH
0.02	1 msec
0.002	100 μsec
0.0002	10 μsec

The multivibrator will produce only relatively short pulses (10 milliseconds) so that you will only see a flicker of LED 2 and LED 1 when the switch is opened. However, if you have an oscilloscope you can look at the output pulse at pin 6 (or 1) directly. If you connect the oscilloscope up for either an internal trigger (or connect a trigger lead to point A), you can see the output pulse. With the RC value shown, it will be about 10 milliseconds long. Changing C (you could also change R) to values listed will give variable pulse widths. You can test this by trying the various values for C and observing the resulting pulse width on the oscilloscope. Check the value you get against those shown in the table. The 74123 is a retriggerable multivibrator. In this type of multivibrator, the return to the quiescent state is inhibited if the unit is retriggered before the active period is over. Retriggering is equivalent to restarting the time-delay interval. If you have a 74123, try it in the same circuit.

Review of Digital System Elements

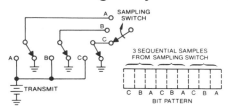

1. **BITS OF INFORMATION** can be transmitted in either serial or parallel form. In either case, the same information requires the same number of bits.

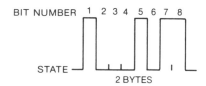

2. **A DIGITAL WORD** is a parallel or serial pattern of high or low (1s or 0s) state, with each element in the pattern called a *bit*.

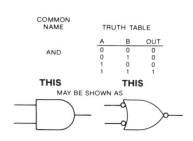

3. **BASIC DIGITAL LOGIC ELEMENTS** include AND (all inputs needed for output), OR (at least one input for output), NOT (reverses input), and the inverses NAND and NOR.

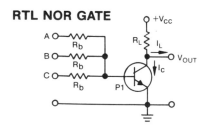

4. **DIGITAL LOGIC CIRCUIT ELEMENTS** can be made up of resistors or diodes (RL or DL), resistors and transistors (RTL), diodes and transistors (DTL), or transistors alone (TTL).

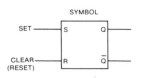

5. **A FLIP-FLOP** can store information, and, with others, can count. It has only two states and can "remember" its last state.

Self-Test—Review Questions

1. Define a bit, byte, logic state, digital word, serial transfer, parallel transfer, and bit patterns. What determines the information (intelligence) that is carried in a digital signal?
2. Draw the symbol for an AND circuit. Define its operation by showing the truth table for the AND circuit.
3. Draw the symbol for an OR circuit. Define its operation by showing the truth table for the OR circuit.
4. Draw the symbol for a NOT circuit. Define its operation by showing the truth table for the NOT circuit.
5. Draw the symbols for the NOR, NAND, and Exclusive OR circuits. Describe the functional operation of each; use truth tables to define their logic.
6. How are digital logic elements usually packaged?
7. Make up a simple digital control system to produce four separate outputs from one pair of input lines. Describe its operation and show, with a table, the logic states on the control line and the output line.
8. Draw a schematic diagram of a flip-flop. Describe how it works. What is the flip-flop used for? Why?
9. What is an R-S flip-flop? A clocked flip-flop? Why are these flip-flops useful? Define and describe the D and J-K flip-flops.
10. What is the difference between a multivibrator and a flip-flop? What is the Schmitt trigger useful for? How does it differ from the multivibrator or flip-flop?

Learning Objectives—Next Section

Overview—Now that you are familiar with the basic digital system elements, you will learn about digital arithmetic before you learn how these elements are used to perform these arithmetic functions.

$$10 + 5 = 15 \text{ (decimal)}$$

$$1010 + 0101 = 1111 \text{ (binary)}$$

$$10 \times 5 = 50 \text{ (decimal)}$$

$$1010 \times 0101 = 00110010 \text{ (binary)}$$

Numbers and Number Systems

Before you can proceed further in your study of digital systems, it will be useful to learn about *binary* numbers. However, before we study these you must know something generally about numbers and number systems. As you are probably aware, the decimal system is almost universally used because it is related directly to our 10-digit sets of fingers and toes. However, have you ever stopped to think about what a decimal number *means*? If you look at any decimal number you will find that while there may be many digits, none ever exceeds a value of 9 or is less than 0. (Of course we are ignoring the sign for the moment.) You also know that the *order* of the numbers is significant: 587 is different from 857 and 758. Therefore, not only is the *value* in any position never greater than 9, but the *sequence* of the numbers (their order) is vitally important.

Let's look at the decimal numbers 5, 50, and 500. Obviously they do not mean the same thing, even though they contain only the number 5 (and some zeros). What we really mean by these numbers is:

DECIMAL NUMBERS
USE POWERS OF 10

$$5 \times 10^0 = 5$$

$$5 \times 10^1 + 0 \times 10^0 = 50$$

$$5 \times 10^2 + 0 \times 10^1 + 0 \times 10^0 = 500$$

As you can see, the position of each digit in a number sequence implies that it is multiplied by some power of 10, with the exponent indicating that power determined by the position of the digit in the sequence. Furthermore, the number itself is the sum of the products in each position. The number 48,654 is really

$$4 \times 10^4 + 8 \times 10^3 + 6 \times 10^2 + 5 \times 10^1 + 4 \times 10^0 = 48,654$$

The number that is being raised to a power is called the *base*. As we shall soon see, this base can be any number, but the rules of arithmetic do not change. Notice that when you add decimal numbers, for example, if the sum is 10, we put a 0 in the 10^0 column and a 1 in the 10^1 column. If the sum is 27, we put a 7 (7×10^0) in the 10^0 column and a 2 (2×10^1) in the 10^1 column. Thus, we can form a *rule*: when values are equal to or greater than the base (in the case of decimals, the base if 10), we *shift* the *values greater* than the base *to the left* to get them into the proper column. This is called the *carry principle*, and it is characteristic of *any* number system. As you will see, numbers based on the binary (base 2), octal (base 8), and hexadecimal (base 16) systems are all used in digital systems.

Numbers and Number Systems—Binary Numbers

A generalization can be made that any number to any base is represented by the following set of digits (A4A3A2A1):

$$\text{Number} = A4A3A2A1$$
$$\text{Base} \quad = B$$

Therefore the number is

$$A4 \times B^3 + A3 \times B^2 + A2 \times B^1 + A1 \times B^0$$

In digital work, as you know, we use a binary sequence of 1 (high) and 0 (low) states to describe a number or bit pattern that conveys information. It is, therefore, very convenient to use a base of 2, since this means that only 1s and/or 0s will be present in any number sequence. "Base 2" requires that we use the powers of 2 (2^0, 2^1, 2^2, etc.) instead of powers of 10. If we follow the rule above, the number A4A3A2A1 with base 2 is

$$A4 \times 2^3 + A3 \times 2^2 + A2 \times 2^1 + A1 \times 2^0$$

Remember that a digit can have any value between 0 and 9 with a base of 10; a digit *must be* either 0 or 1 with a base of 2. Thus, the binary number represented by 1010 is equal to

$$1 \times 2^3 + 0 \times 2^2 + 1 \times 2^1 + 0 \times 2^0$$

The number 1010 therefore represents

$$
\begin{aligned}
1 \times 2^3 &= 8 \\
0 \times 2^2 &= 0 \\
1 \times 2^1 &= 2 \\
0 \times 2^0 &= \underline{0} \\
& 10
\end{aligned}
$$

Thus the binary number 1010_2 has a total value of 10_{10}. By convention, we use a subscript numeral to designate the base, just as a superscript (exponent) is used to designate a power. So we can write $1010_2 = 10_{10}$. The number 1010 in base 10 instead of base 2 would be

$$
\begin{aligned}
1 \times 10^3 &= 1000 \\
0 \times 10^2 &= 0000 \\
1 \times 10^1 &= 0010 \\
0 \times 10^0 &= \underline{0000} \\
& 1010 = 1{,}010
\end{aligned}
$$

As you can see, it takes more binary bits (base 2) to represent a value than it does to represent the same value in base 10. As you proceed through this section on digital arithmetic; remember the rules you have learned here, and you will have little difficulty.

Binary Number System—Decimal/Binary Conversion

It is useful to learn how to *convert* a *decimal* number to its *binary* equivalent. In doing this it is important to understand the terms *most significant digit (MSD)* and *least significant digit (LSD)*. The (MSD) is the one on the *extreme left*; it has the *highest* value (weighting) in determining the magnitude of the number. If all remaining digits were valued at zero, you would still have a good idea of the magnitude of the number. The LSD is the digit at the *extreme right* and plays the *least* importance in determining the magnitude of the number. In the number $53,872_{10}$, the 5 is the MSD and the 2 is the LSD. You can see that an error in the 5 would be very serious, while if the LSD were misread there would be only a negligible effect on the number. In most digital words, we are dealing with *bits* rather than digits; therefore we use the terms *MSB*, and *LSB*, where the B stands for bit.

THE MOST SIGNIFICANT DIGIT (MSD) AND LEAST (LSD)

IN THE BINARY NUMBERS THE LEFT-MOST 1 IS THE MOST SIGNIFICANT DIGIT

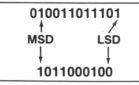

THE RIGHT-MOST 1 OR 0 IS THE LEAST SIGNIFICANT DIGIT

There are several methods of converting a decimal number to a binary equivalent. The simplest is the *divide-by-2* (divide by the base) method shown here. You divide the number by 2, writing down the quotient and remainder separately. The remainder will always be either a 0 or a 1, depending on whether the quotient is an even number or an odd number, respectively. The first remainder provided in this manner is the LSB. You then divide each new quotient successively by two and write down the quotient and remainder in the same manner. Repeat this process until the last remainder (or quotient) is formed, and that is the MSB. The calculation below shows the process for finding the binary equivalent of 173_{10} (decimal).

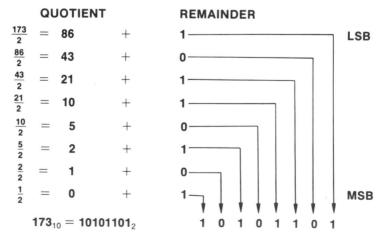

$$173_{10} = 10101101_2$$

Binary Number System—Decimal/Binary Conversion (continued)

As you saw on the previous page, $173_{10} = 10101101_2$. We can check this by a binary-to-decimal conversion using the definition for numbers established earlier. Thus,

$$10101101_2 = 1 \times 2^7 + 0 \times 2^6 + 1 \times 2^5 + 0 \times 2^4$$
$$+ 1 \times 2^3 + 1 \times 2^2 + 0 \times 2^1 + 1 \times 2^0$$
$$= 128 + 0 + 32 + 0 + 8 + 4 + 0 + 1 = 173_{10}$$

Thus, $10101101_2 = 173_{10}$. As you can see, conversion may be somewhat tedious, but it can be done. Fortunately, it isn't usually necessary to do such conversion by hand; but it is valuable to *know how* it is done since it also helps give you an insight into *how number systems work*. As another example, suppose we wished to convert 85_{10} to binary form and back again to check our result:

$$\frac{85}{2} = 42 + 1$$
$$\frac{42}{2} = 21 + 0$$
$$\frac{21}{2} = 10 + 1$$
$$\frac{10}{2} = 5 + 0$$
$$\frac{5}{2} = 2 + 1$$
$$\frac{2}{2} = 1 + 0$$
$$\frac{1}{2} = 0 + 1$$
$$85_{10} \qquad 1 \quad 0 \quad 1 \quad 0 \quad 1 \quad 0 \quad 1_2$$

$$1010101_2 =$$
$$1 \times 2^6 + 0 \times 2^5 + 1 \times 2^4 + 0 \times 2^3 +$$
$$1 \times 2^2 + 0 \times 2^1 + 1 \times 2^0$$
$$= 64 + 0 + 16 + 0 + 4 + 0 + 1$$
$$= 85_{10}$$

To check the result we can convert 1010101_2 back to base 10. As you know, we should get a value of 85_{10}. Using our definition,

$$1010101_2 = 1 \times 2^6 + 0 \times 2^5 + 1 \times 2^4$$
$$+ 0 \times 2^3 + 1 \times 2^2 + 0 \times 2^1 + 1 \times 2^0$$
$$= 64 + 0 + 16 + 0 + 4 + 0 + 1 = 85_{10}$$

Thus, it is easy to see how these conversions are done. For practice, prove to yourself that

$$236_{10} = 11101100_2$$
$$101_{10} = 1100101_2$$
$$10110110_2 = 182_{10}$$
$$11011_2 = 27_{10}$$

If you have difficulty with these, go back and read over the last few pages and try again. In any event, the important thing is that *a number in base 10 can be converted to a binary bit pattern suited to digital processing*.

Octal and Hexadecimal Number Systems

As you have already observed, the largest digit in any number system is one less than the base (9 is one less than 10, and 1 is one less than 2), which is why binary number systems can be *very cumbersome*. As a result, other number systems using base 8 (*octal*) and base 16 (*hexadecimal*) are commonly used in digital systems to provide a more compact notation. These have been chosen mainly as a *convenience* since 111 is equal to 7 in binary form and 1111 is equal to 15. By using these bases we have a convenient way to handle binary bit sequences, but with much shorter descriptors. Another convenient representation occasionally used is called *binary-coded decimal* (*BCD*). In this case 4-bit binary representations of the decimal digits 0 to 9 (0000 to 1001) are used.

The chart on the following page compares these numbers in different forms. The decimal number 932 has the hexadecimal representation $3A4_{16}$, the octal representation 1644_8, and the binary representation 1110100100. The decimal, hexadecimal, or octal representations are more convenient to use. Furthermore, the binary bit patterns 1001 0011 0010 (BCD), 0011 1010 0100 (hexadecimal), or 001 110 100 100 (octal) are also easier to deal with than the pure binary number. Three-bit coded groups are sufficient because in hexadecimal 7 is the largest digit.

The important thing to remember here, is not so much the number systems themselves, but that *any number* in any base *can be easily converted* to a *binary bit pattern* of 1s and 0s, and these bit patterns are easy to *store* and *manipulate*. In some cases, the *bit patterns can represent numbers* as we see here; in other cases, these bit patterns can represent *control data* or *information* of almost any kind. It is of some importance to know about octal and hexadecimal numbers, since they often are the form in which computer information is provided, so it is sometimes necessary to understand these numbers and also the bit patterns that they represent. Also, as you will see, arithmetic is very easy to do with binary numbers.

HEXADECIMAL NUMBERS

1	5	9	13D
2	6	10A	14E
3	7	11B	
4	8	12C	15F

HOW LETTERS ARE USED IN HEXADECIMAL

DEC.: 1, 2, 3, 4, 5, 6, 7, 8, 9, 10, 11, 12, 13, 14, 15, 16, 17
HEX.: 1, 2, 3, 4, 5, 6, 7, 8, 9, A, B, C, D, E, F, 10, 11

For hexadecimal numbers it is necessary to have single-digit descriptors between 0 and 15. To accomplish this, letters are used to describe the double-digit numbers from 10 to 15, as shown in the table above. However, the same rules apply—that is, FF_{16} is equal to 256_{10} ($F \times 16^1 + F \times 16^0 = 256$), and 3A is equal to 58_{10} ($3 \times 16^1 + 10 \times 16^0 = 58$).

Number Systems and Bit Patterns

DIGITAL ARITHMETIC

DECIMAL (BCD— 10^0)	BINARY REPRESENTATION
00	0000 0000
01	0000 0001
02	0000 0010
03	0000 0011
04	0000 0100
05	0000 0101
06	0000 0110
07	0000 0111
08	0000 1000
09	0000 1001
10 (10^1)	0001 0000
11	0001 0001
12	0001 0010
13	0001 0011
14	0001 0100
15	0001 0101
16	0001 0110
17	0001 0111
18	0001 1000
19	0001 1001
20	0010 0000
21	0010 0001
22	0010 0010
23	0010 0011
24	0010 0100
25	0010 0101
26	0010 0110
27	0010 0111
28	0010 1000
29	0010 1001
30	0011 0000
31	0011 0001
32	0011 0010

HEXADECIMAL (Base 16)	BINARY REPRESENTATION
00 16^0	0000 0000
01	0000 0001
02	0000 0010
03	0000 0011
04	0000 0100
05	0000 0101
06	0000 0110
07	0000 0111
08	0000 1000
09	0000 1001
0A	0000 1010
0B	0000 1011
0C	0000 1100
0D	0000 1101
0E	0000 1110
0F	0000 1111
10 16^1	0001 0000
11	0001 0001
12	0001 0010
13	0001 0011
14	0001 0100
15	0001 0101
16	0001 0110
17	0001 0111
18	0001 1000
19	0001 1001
1A	0001 1010
1B	0001 1011
1C	0001 1100
1D	0001 1101
1E	0001 1110
1F	0001 1111
20	0010 0000

OCTAL (Base 8)	BINARY REPRESENTATION
00 8^0	000 000
01	000 001
02	000 010
03	000 011
04	000 100
05	000 101
06	000 110
07	000 111
10 8^1	001 000
11	001 001
12	001 010
13	001 011
14	001 100
15	001 101
16	001 110
17	001 111
20	010 000
21	010 001
22	010 010
23	010 011
24	010 100
25	010 101
26	010 110
27	010 111
30	011 000
31	011 001
32	011 010
33	011 011
34	011 100
35	011 101
36	011 110
37	011 111
40	100 000

BINARY (Base 2)
000000 2^0
000001
000010 2^1
000011
000100 2^2
000101
000110
000111
001000 2^3
001001
001010
001011
001100
001101
001110
001111
010000 2^4
010001
010010
010011
010100
010101
010110
010111
011000
011001
011010
011011
011100
011101
011110
011111
100000 2^5

Converting Decimal Numbers to Octal and Hexadecimal Form

The technique for converting decimal numbers to octal or hexadecimal form is the same as for binary numbers. For example, the number 932_{10} is $3A4_{16}$ and 1644_8. These values are obtained by successively dividing by the base and retaining the remainders just as for binary (base 2) numbers:

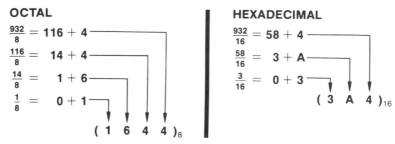

These can be reconverted to decimal form (base 10) by summing the products of the numbers times the power of the base, as before. Thus,

$$\text{Octal } 1644_8 = 1 \times 8^3 + 6 \times 8^2 + 4 \times 8^1 + 4 \times 8^0$$
$$= 512 + 384 + 32 + 4 = 932$$
$$\text{Hexadecimal } 3A4 = 3 \times 16^2 + A \times 16^1 + 4 \times 16^0$$
$$= 768 + 160 + 4 = 932$$

These numbers can be represented in binary form by forming a set of binary digits for each of the digits, as shown:

$$\overset{(1)\ \ (6)\ \ (4)\ \ (4)}{1644_8 = 001\ 110\ 100\ 100} \quad \text{and} \quad \overset{(3)\ \ \ (A)\ \ \ (4)}{3A4_{16} = 0011\ 1010\ 0100}$$

As you can see from the examples, it is only necessary to follow the rules, and converting numbers from one base to another will be easy. Furthermore, it is easy to convert from any base to binary patterns. If you remember, a byte usually means an 8-bit word. This means that each byte equals two hexadecimal digits. Since the BC group $1111_{16} = F$, then the largest number per byte is $255 = FF$. Obviously, bytes can be cascaded to provide any number desired. For example, the 2-byte $FF\ FF_{16}$ is equal to $65,536_{10}$ when the higher-order byte is considered an extension of the lower-order byte (1111 1111 upper byte, 1111 1111 lower byte). Similarly, a byte in octal is interpreted as 377_8 when all 8 bits are 1, and thus equal to 255_{10}. The binary bit pattern, as you know, is 11 111 111.

In addition to the usefulness of binary numbers in digital hardware, it is also very much easier to do binary arithmetic. To add binary numbers you proceed in very much the same manner as in adding decimal numbers, including the method of carrying. However, because there are only two digits the process is much simpler. The rules are:

$$0 \text{ plus } 0 = 0$$
$$0 \text{ plus } 1 = 1$$
$$1 \text{ plus } 0 = 1$$
$$1 \text{ plus } 1 = 0 \text{ and carry } 1$$

Binary Arithmetic—Addition and Subtraction

Thus, to add the binary numbers 10 and 11 you proceed as shown in the diagram below. In the 2^0 column $0 + 1 = 1$, so you write a 1 below the sum line. Then in the 2^1 column $1 + 1 = 0$ and carry 1, so you write a 0 below the sum line and carry the 1 to the next column to the left, the 2^2 column. Note that the binary numbers 10 and 11 are equal to 2_{10} and 3_{10} and that the binary sum 101 is equal to the sum 5_{10}. Thus, the final sums obtained with equivalent numbers in systems of different bases are always equal. The diagram also shows the details of adding three binary numbers and their decimal equivalents.

BINARY ADDITION

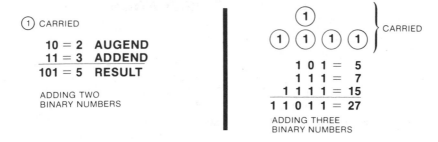

BINARY SUBTRACTION

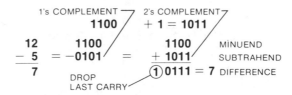

Digital subtraction is not done directly, but is accomplished by converting the subtraction to a problem in addition. In binary subtraction, you add to a number the *2's complement* of the number to be subtracted. The concept of such a complement is mathematically sophisticated, but the method of finding it is simple. To find the 2's complement of any binary number simply find the 1's complement (change 0s to 1s and 1s to 0s) and then add 1 to the result. Thus the 2's complement of 1001 is 0111.

Thus to subtract one binary number from another, find its 2's complement and add. There is one *special rule*: You cross out the final carry bit (if any) in the addition. The illustration above shows this subtraction process. The *sign* is determined by the leftmost bit. If it is a 1, it is dropped and the sign of the difference is positive. If it is a 0, the sign is negative (subtrahend greater than minuend).

Binary Arithmetic—Multiplication and Division

Binary multiplication is also easy to perform since it involves only a *shift* of columns and binary *addition*, similar to decimal multiplication except that we only have to deal with 1s and 0s. Since $1 \times 1 = 1$ and $1 \times 0 = 0$, the procedure requires that we both shift and add where the multiplier has a 1 and shift only (fill with 0s) when the multiplier shows a 0. In the simple illustration shown below, the multiplier is 0101_2 (5_{10}). The multiplier is written directly below the first bit (1) of the multiplicand. The second bit from the right of the multiplier is 0, so it is only necessary to shift with no adding (fill the line with 0s). The next bit in the multiplier is a 1, so the multiplicand is again shifted and written down in the proper position. After we again shift for the final multiplier bit (0), the partial products are added to produce the product, as shown:

BINARY MULTIPLICATION

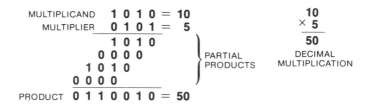

BINARY DIVISION

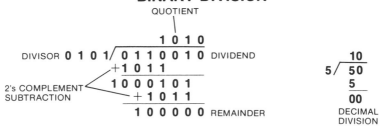

Binary division is also quite simple, since it can be done by a series of *subtractions* of the divisor. (Find the 2's complement and then add.) As shown in the illustration, a 1 is placed in the quotient whenever a 2's complement addition of the divisor is possible, and a 0 is placed in the quotient when the next binary bit brought down does not permit a 2's complement addition. Thus, the first addition of the 2's complement of 0101 (1011) results in a 1 in the quotient. When the next bit (0) in the dividend is brought down, the binary number is too small to perform a subtraction, so a 0 is put into the quotient and the next binary bit (1) is brought down. Since a subtraction is now possible, a 1 is placed in the quotient. When the final bit (0) is brought down, no further subtraction is possible, and this line is the remainder, while a 0 is placed in the quotient. Again, as you can see, binary division is very easy, since you must deal only with patterns of binary bits.

Review of Digital Arithmetic

DECIMAL NUMBERS USE POWERS OF 10

$$5 \times 10^0 = 5$$
$$5 \times 10^1 + 0 \times 10^0 = 50$$
$$5 \times 10^2 + 0 \times 10^1 + 0 \times 10^0 = 500$$

1. DECIMAL NUMBERS constitute our common number system, so named because it is based on the number 10.

HEXADECIMAL NUMBERS

1	5	9	13D
2	6	10A	14E
3	7	11B	
4	8	12C	15F

2. THE NUMBER SYSTEMS most commonly used in computers are *binary* (base 2), octal (base 8), and hexadecimal (base 16).

DECIMAL	932
BINARY CODED DECIMAL	1001 0011 0010
BINARY	1110100100
OCTAL	1644₈ or 001 110 100 100
HEXADECIMAL	3A4₁₆ OR 0011 1010 0100

3. CONVERSIONS FROM ONE SYSTEM TO ANOTHER can be made and each number in each system has an equivalent in each other system.

BINARY ADDITION

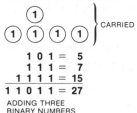

ADDING THREE
BINARY NUMBERS

4. BINARY NUMBERS MAY BE ADDED, remembering that $0 + 0 = 0$, $1 + 0$ or $0 + 1 = 1$, and $1 + 1 = 0$ with a 1 carry. Subtraction is accomplished using the 2's complement of the subtrahend.

BINARY MULTIPLICATION

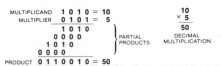

5. BINARY MULTIPLICATION AND DIVISION can be done in the same manner as for decimal, keeping in mind the rules of addition and subtraction.

Self-Test—Review Questions

1. Do the rules of arithmetic depend on the base used? Explain your answer.
2. What are the essential differences between binary and decimal arithmetic? What is the importance of binary numbers in digital systems?
3. What is an LSB? An MSB? Illustrate your answers with an example.
4. Convert the following numbers from decimal to binary form.

$$100_{10} = ? \quad 4,732_{10} = ? \quad 39,763_{10} = ? \quad 4_{10} = ? \quad 75_{10} = ? \quad 382_{10} = ?$$

5. What are the octal and hexadecimal number systems? Why are they used? Convert the decimal numbers of question 4 to octal and hexadecimal form.
6. Check your answers to question 6 by converting the octal and hexadecimal numbers obtained back to decimal form.
7. Perform the indicated arithmetic operations. Convert the numbers to binary form and perform the indicated operations. Check your results.

$$427_{10} + 256_{10} = ? \qquad FF_{16} + FF_{16} = ?$$
$$3,720_8 + 2,310_8 = ? \qquad 389_{10} + 573_{10} = ?$$
$$FF_{16} + IF_{16} = ? \qquad 1,210_{10} + 234_{10} = ?$$

8. Perform the indicated subtractions in decimal form and then do the same by 2's complement arithmetic after converting the decimal numbers to binary form. Check your results.

$$427_{10} - 256_{10} = ? \qquad 777_{10} - 777_{10} = ?$$
$$9,321_{10} - 8,302_{10} = ? \qquad 1,000_{10} - 100_{10} = ?$$
$$575_{10} - 410_{10} = ? \qquad 8,204_{10} - 273_{10} = ?$$

9. Perform the indicated multiplications and divisions in binary form and then check your results in decimal form.

$$101111_2 \times 011011_2 = ? \qquad 110110 \div 000101 = ?$$
$$1111_2 \times 0010_2 = ? \qquad 101010 \div 001101 = ?$$

Learning Objectives—Next Section

Overview—Now that you know about digital logic elements and binary arithmetic, you will learn in the next sections about digital system functions, timing and counting circuits, and some alphanumeric displays.

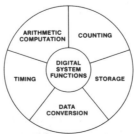

Arithmetic Computation by Digital Logic Circuits

One of the functions of digital logic circuits is to perform basic *arithmetic computation* or *number crunching*. You have already learned how binary 1s (high) and 0s (low) are added, subtracted, multiplied, and divided by using only *addition* and *shifting*. Now you will find out how these computations can be *performed electronically* by digital logic circuits.

The *serial adder circuits* have been almost totally replaced by *parallel adder circuits*. The operation of serial adders is explained here because it will simplify your understanding of parallel adders. The serial half-adder is a basic circuit for computer arithmetic. The basic arrangement consists of AND, OR, and NOT circuits. The NOT gate (or inverter) is a small circle at the input to the second AND gate.

In digital systems, the binary 0 and 1 bits are fed through circuits as a train of pulses. As described previously, one common arrangement is to operate on a series of bits in a group known as a *word*, which consists of 8-bit groups, known as *bytes*. For addition with a half-adder, the train of bits is fed into each of the two inputs concurrently; that is, the two serial trains are added in parallel.

ADDITION WITH SERIAL HALF-ADDERS

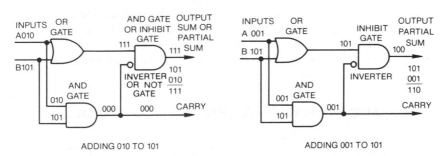

ADDING 010 TO 101   ADDING 001 TO 101

When you learned about binary addition you found out that if you add two 1s, the result is a 0 with a 1 carried to the next column. In the first example in this illustration, a 010_2 train is fed to the upper input, and a 101_2 train is fed to the lower input. These inputs are fed simultaneously to both inputs of the OR and AND gate circuits and timed so that the 010_2 is in step with the 101_2. Because only 0s and 1s are added at any one time for the numbers 010 and 101, only the OR circuit has any output, which is 111. The AND gate output is 000. Since there is no carry, the output of the OR gate is the true sum (111) of the binary numbers applied to inputs A and B. The second AND circuit with a NOT (inverter) at its lower input is an INHIBIT gate that is only used when there is a carry, as described below.

In the second example, 001 is added to 101. When 1s appear simultaneously at both the A and B inputs, the operation is more complex because $1 + 1 = 0$ with a carry. Now the OR gate produces a 101 (not a true sum of the two inputs) and the AND gate has an output of 001.

Arithmetic Computation by Digital Logic Circuits (continued)

When this 1 is applied to the lower input of the INHIBIT gate, the 1 output from the OR gate is inverted and thus blocked. Now the final output of the combination is 100, which is not a true sum. However, the half-adder has put out a 001 at its carry output, and this can be used in producing a true sum.

The serial full-adder circuit shown below produces a true sum by creating a *carry 1* when two 1s have to be added and then using that carry 1 in the addition. The circuit consists of two half-adders and a 1-bit delay device. The problem illustrated is the addition of 101 and 001, as in the second example on the previous page.

In the first half-adder (Half-Adder 1) the sequence of events is the same as that described on the previous page. The result is a 100 output from INHIBIT gate I1 and a 001 carry output from AND gate A1. To the upper input of OR gate G2 is fed the 100 input from I1.

What the delay device D1 does is to delay the carry output (001) for a time of exactly 1 bit (the cycle time of one binary pulse). What this means is that the 1 (carry) output from D1 is applied to the lower input of G2 (and A2) of the second half-adder, but displaced one position to the left. This puts the 1 in the column of the next higher order bit or 010. (This means that a 1 is given the weight of a 1 in the *next* column to the left—it is *carried* one column.) Thus the output from G2 is 110. Since there are no other carry operations in the A2 output (000) in this addition, the output of I2 is 110, which is the true sum. If A2 did produce a carry, it would be fed back to the input of D1 and result in another similar carry operation.

ADDITION WITH A SERIAL FULL-ADDER

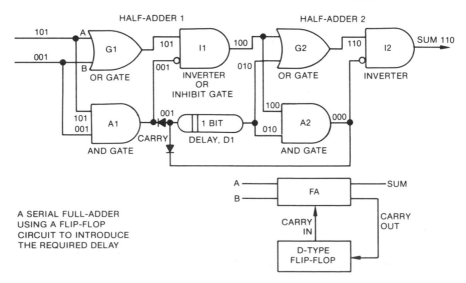

A SERIAL FULL-ADDER
USING A FLIP-FLOP
CIRCUIT TO INTRODUCE
THE REQUIRED DELAY

Arithmetic Computation by Digital Logic Circuits (continued)

A practical way to accomplish the function of the delay circuit is to use a D-type flip-flop circuit that is triggered by the same clock used to generate the pulses that are being fed into the adder. Thus, if the second half-adder produces a carry pulse, that pulse will be held in the flip-flop until the next clock pulse comes along. At that time, if there is a carry pulse in the flip-flop, it will be fed to the second adder and thus achieve a delay of one clock pulse.

Circuits similar to these adders are used to perform the other arithmetic computations. As you learned previously, subtraction is accomplished by adding the 2's complement of the number to be subtracted. Also, as described earlier, multiplication is performed as a series of additions and shifts, and division is accomplished as a series of additions of 2's complements and shifts. Later, you will learn how flip-flops are combined into devices called *shift registers* that allow the right and far-left shifting of stored data bits.

The adders that were just described are known as *serial adders* because they add each of a series of pairs of digits with one digit of each pair from each number to produce the final result, which consists of the result of these serial additions. The method known as *parallel addition* is much faster. This method uses more hardware to permit carrying to be extended to numbers with more significant figures. Examination of the half-adder circuits on the previous pages shows that they are the exact equivalent of an Exclusive OR gate plus an AND gate. As shown on the previous page, when you connect the two half-adders with an OR gate, you have a full-adder (FA) that adds two digits and also accepts a carry-in and produces a carry-out. For simplicity, this FA arrangement is drawn as a block.

A parallel adder consists of a number of such full-adders connected as shown in the diagram on the next page. Each FA is used to add two digits. Thus, to add two 4-bit numbers you need four FAs that are used to simultaneously add the number pairs in parallel. The diagram shows the addition of the two binary numbers 1110 and 1111. Each FA adds the bits of equal weight (in the same position) from each of the input numbers—that is, the FA on the extreme right adds the two least significant bits, and the FA on the extreme left adds the two most significant bits. If any of the FAs has a carry, it is sent to the next FA to its left.

Although the parallel adder requires more FAs than a serial adder, it offers a significant advantage in speed. Advances in integrated circuitry have made parallel adders capable of processing 8 or 16 bits widely available and low in cost. Therefore, serial adders are now obsolete and used only for special applications where the utmost in simplicity is necessary.

Arithmetic Computation by Digital Logic Circuits (continued)

DETAILS OF PARALLEL ADDITION

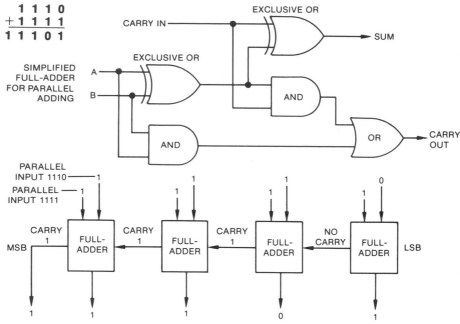

On the next several pages you will learn about how flip-flop circuits can be used to *count* and to form shift registers for *storage* and also for *serial-to-parallel conversion* and vice versa. Four of the toggle flip-flops described previously can be connected to each other and to lamps. The result is a *binary counter* that will *count* up to 15 input pulses ($2^N - 1 = 2^4 - 1 = 15$). Furthermore, the lighted or unlighted condition of the lamps will display the count at any point in binary code with a lit lamp equal to 1 and an unlit lamp is equal to 0.

Before any pulses are fed in, all the flip-flops are in the reset (0) condition and all the lamps are dark, indicating 0s in all the Q outputs. Now a single pulse is applied to the input. This causes FF1 to go into the set condition (Q1 = 1). This causes the lamp at the extreme right to light, showing a 0001 count. However, FF1 does not produce an output to FF2 and will not until the next pulse comes along. Each flip-flop must receive two input pulses to produce an output pulse to the next flip-flop.

When the next pulse is applied, FF1 returns to its reset condition. Its lamp goes out, but an output pulse is applied to FF2. This causes FF2's lamp to light, indicating a binary count of 0010, but no output pulse is applied to FF3.

When the third input pulse is applied (shown in row B of the diagram on next page), FF1 returns to its set (1) condition, and its lamp lights. No output pulse is sent to FF2 however, so it remains in its set (1) state and its lamp remains lighted. The two right-hand lamps are now lighted, as shown in the diagram, giving a binary readout of 0011.

Counting with Flip-Flop Circuits

COUNTING WITH FLIP-FLOPS

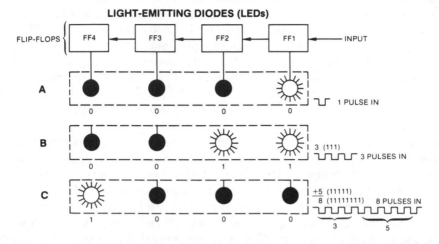

Now five more pulses are applied, as shown in row C. Because FF1 and FF2 are in the set state, the first pulse turns off both their lamps but it causes FF1 to produce an output to FF2 and FF2 to deliver a pulse to FF3. Now the lamps indicate a count of 0100. The next pulse only turns on FF1's lamp, producing a count of 0101 (equivalent to decimal 5).

Each successive pulse produces the same type of result until eight have been applied to the input. The lamp at the extreme left will light and all others will be out, indicating a count of 1000 (or decimal 8).

Additional input pulses can be applied; on a count of 15, all lamps will be lit and the bit pattern will be 1111. You can trace through the operation of the circuit using the rule that each flip-flop is set only on the reset of the flip-flop to the right. If you do this, you will arrive at the table shown below (0: lamp out; 1: lamp lit).

PULSE NO.	MSB LAMP 4	LAMP 3	LAMP 2	LSB LAMP 1	BIT PATTERN STORED
0	0	0	0	0	0000
1	0	0	0	1	0001
2	0	0	1	0	0010
3	0	0	1	1	0011
4	0	1	0	0	0100
5	0	1	0	1	0101
6	0	1	1	0	0110
7	0	1	1	1	0111
8	1	0	0	0	1000
9	1	0	0	1	1001
10	1	0	1	0	1010
11	1	0	1	1	1011
12	1	1	0	0	1100
13	1	1	0	1	1101
14	1	1	1	0	1110
15	1	1	1	1	1111
16(1) (RESET)	0	0	0	0	0000

Experiment/Application—Counting with Flip-Flops

As you can see in the table, the set of flip-flops acts as a counter and produces a parallel output that represents the count. This counter will only count to 15 (binary 1111), but if we added another four flip-flops as a register counter on the end fed by the output of FF4 the expanded circuit would count to (but not including) 256_{10} (1111 1111$_2$, FF_{16}, 377_8). Obviously, this could be extended with additional flip-flops. Several points are worth noting here: (1) The counter holds the number of counts at any point and therefore has the property of *remembering* what the count is. (2) The pulses were input in *serial* form but are available in *parallel* (more about this below). (3) If we wanted to *sense* any particular count, we could do so by adding the proper logic circuits (more about this in another few pages). (4) The bit pattern for each lamp is *predictable*—that is, lamp 1 changes state with each input pulse, lamp 2 changes state with each pair of input pulses, lamp 3 changes state with each fourth input pulse, and lamp 4 changes state with each eighth input pulse.

You could set up separate flip-flops to count, but it is much easier to use a single IC package that contains several flip-flops. For example, the SN7493 has four flip-flops set up as a 4-bit binary counter in a single chip. The internal configuration is shown below with the pin configuration and the truth table.

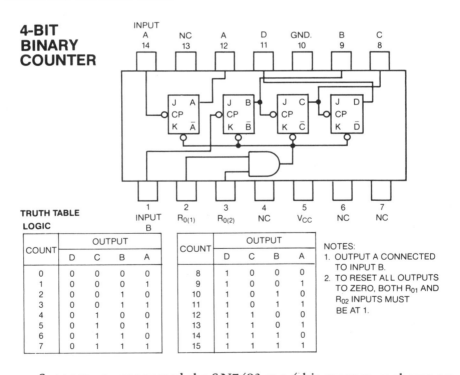

4-BIT BINARY COUNTER

TRUTH TABLE
LOGIC

COUNT	OUTPUT				COUNT	OUTPUT			
	D	C	B	A		D	C	B	A
0	0	0	0	0	8	1	0	0	0
1	0	0	0	1	9	1	0	0	1
2	0	0	1	0	10	1	0	1	0
3	0	0	1	1	11	1	0	1	1
4	0	1	0	0	12	1	1	0	0
5	0	1	0	1	13	1	1	0	1
6	0	1	1	0	14	1	1	1	0
7	0	1	1	1	15	1	1	1	1

NOTES:
1. OUTPUT A CONNECTED TO INPUT B.
2. TO RESET ALL OUTPUTS TO ZERO, BOTH R_{01} AND R_{02} INPUTS MUST BE AT 1.

Suppose you connected the SN7493 as a 4-bit counter, as shown on the next page, with LEDs to show the binary count. You will have to use a pulse generator such as the multivibrator you studied previously to generate pulses from the pushbutton switch to operate the counter.

Experiment/Application—Counting with Flip-Flops (continued)

The reset inputs $R_0(1)$ and $R_0(2)$ are set to a 1 condition to reset the counter. Since the A flip-flop is independent, it will have to be strapped as shown to work with flip-flops B, C, and D.

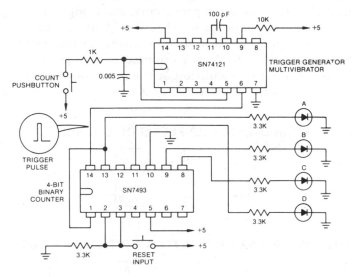

When you have connected the circuit as shown, you will see that the reset sets all the lamps to 0 (low) or off. As you depress the count pushbutton, the counter will advance one count for each input count. When 15 (binary 1111) is reached, the counter will stop. You can reset it with the reset pushbutton and begin to count again.

You can easily use this as an event counter by having the input events applied to the trigger multivibrator and a timing signal applied to the reset. For example, if a waveform of 5 volts for reset was applied to terminals 2 and 3 of the counter, and was at 0 volts for 1 minute, you could count the events at the input that occur in 1 minute.

Two counters can be connected in cascade to give a total count of 256 (0–255) or $1111\ 1111_2$ by connecting two SN7493 circuits.

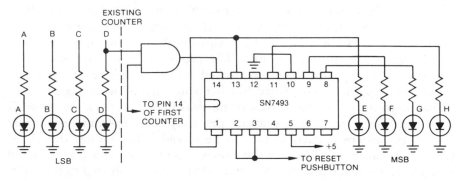

The circuit allows the second counter to be activated by the trigger pulses only after the first counter has reached maximum count. Draw a truth table for this counter. You will need a *long* piece of paper!

Storage with Register Circuits

A register is a digital circuit arrangement that *stores* a binary bit or word. Registers are typically composed of flip-flops and greatly resemble the counters described previously. The *shift register* is a very important circuit in digital systems, microprocessors, and computers. What it does is to *shift* a stored digital bit or word to the *left* or *right* each time a *shift (or clock) pulse* is applied. A basic shift register, one that shifts entries to the right, is shown below. It is composed of D-type flip-flops, all of which are in the reset (0) condition at the beginning. A binary word is entered serially, with the extreme-right (least significant) digit (LSB) first.

MAKING FOUR D-TYPE FLIP-FLOPS INTO A SHIFT REGISTER

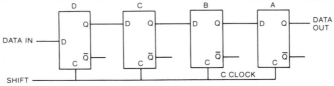

ENTERING BINARY 1101 INTO A 4-BIT SHIFT REGISTER

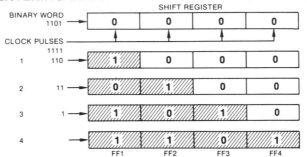

When the LSB (0 or 1) is fed to the input terminal, nothing happens until the next clock pulse comes along. At this time the LSB operates on the circuit in accordance with whether the LSB is a 0 or a 1. If it is a 0, flip-flop FF1, at the extreme left, remains in its 0 condition. If the LSB is a 1, FF1 changes to the set condition. Now the next significant digit is applied to the input. When the next clock pulse comes along, the 0 or 1 condition of FF1 is *shifted* to FF2, and the digit being applied to the input terminal is shifted into FF1. The procedure continues until each bit in the binary word has been shifted into the register. If the clock continues *after* the binary word has been stored, the shift register will continue to shift and the word will be shifted out. Thus, if the word is to be held for some later operation, precisely the same number of shift (clock) pulses are needed as bits to be stored. The counter described previously can be used to gate the precise number of clock pulses called for. Then, when the binary word is needed, it can be shifted to the right and out of the register by entering the next binary word or by applying clock pulses that continue the shifting operation.

Serial/Parallel Data Conversion with Shift Registers

Shift registers are extremely useful because they act as a conveniently available *temporary storage* of binary words. These *words* can represent *numbers, instructions*, or *data*. In addition to shifting to the right, as on the previous page, registers are available that will shift to the left or to both the right or left, in the direction selected by the operator or by logic.

Another important capability of shift registers is that of *converting serial data* to *parallel data*, and vice versa. The concept is similar to that of serial or parallel addition, discussed previously. The first illustration shows the conversion of parallel data input to serial data output. Here a binary number is fed in so that all the bits enter the register *at the same time* and set each register according to the input bit pattern, in parallel. The individual bits are arranged as shown, from left to right, and each is applied to an individual flip-flop input. Then a load (strobe) pulse is applied to all the flip-flops at the same time, and all the bits in the binary word are entered at the same time. Now four clock pulses can be applied, as described on the previous page, and the bits can be shifted to the right (or left) out of the register in serial fashion. While the binary word is stored in the register, the 0 or 1 state of each flip-flop can be sensed or *read* without affecting its condition.

SHIFT REGISTER SERIAL/PARALLEL CONVERSION

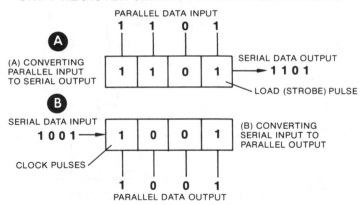

A similar procedure can be used to convert serial data to parallel data. Here, the binary word is fed in serially, using clock pulses and a pulse train. When it is stored in the register it can be read without affecting its content, or it can be shifted *down* in parallel fashion.

Serial-to-parallel and parallel-to-serial conversion are very useful capabilities in input/output circuits of digital devices. Internally, the system operates in parallel fashion for most operations; for ease of data handling, however, the external circuits are fed serially or the digital system itself is fed using serial data to minimize external connections. In many cases, there is serial-to-parallel conversion and parallel-to-serial conversion in the external device such as a video display or printer.

Timing by Digital Logic Circuits

You may be wondering how a set of clock pulses can be gated to operate a shift register or to execute a fixed sequence of events based on a known series of clock pulses. The diagram below shows how this can be done. Clock pulses are fed into the counter made up of four flip-flops (maximum count = 15). Suppose you wanted to gate in eight clock pulses, starting with a count of 1 and ending with a count of 8 (the desired bit pattern at the start is 0001 and at the stop is 1000). If you configure a set of AND gates with inverters to sense the outputs of the flip-flops, you will see that the output of U3 will go high only when the inputs to U1 and U2 are 0001. The output of U3 is used to set flip-flop U5. Before the set pulse is applied to the flip-flop, it is in the reset state so that its output is low. This low output is applied to AND gate U4 along with the clock pulses; however, no output is produced until flip-flop U5 is set. Then, the clock pulses appear at the output of U4.

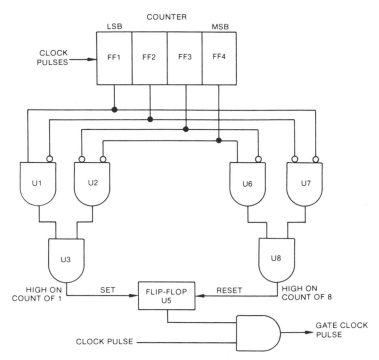

The AND gates and inverters U6, U7, and U8 function in a manner similar to U1, U2, and U3. However, as you can see, a high output from U8 is only obtained when the bit pattern in FF1–FF4 is 1000. The output of U8 is used to reset flip-flop U5 so that the clock pulse stream is inhibited, because one of the inputs to U4 is low and hence the output remains low. You can verify this by generating a truth table. This is not the easiest way to generate the desired count, although it does illustrate the point. Actually, the circuit can be constructed from three or four IC chips, as you will learn in an experiment/application.

Modern Digital Circuit Chips

Essentially all digital circuit elements are in IC chip form. As early as 1970, an entire microprocessor was successfully placed on a chip less than ½ inch square (1.27 cm²). The challenge for digital circuits today is not to cram more and more circuits onto a single chip but rather to package combinations of arithmetic and logic circuits that are generally useful to the equipment designer. The result is that there is now a vast variety of chips being produced in IC packages. In cases where there may be a large number of connections to be made to a small chip, the package may be made large enough to accommodate the additional terminals.

On preceding pages you have learned about a variety of digital circuit functions. These were generally described in discrete circuit terms because it was easier to understand. At present, there are so many digital IC packages, both of the MSI (medium-scale integration) and LSI (large-scale integration) type, that only reference to manufacturers' data can describe them all. They do share, however, a common scheme for sizes and pin designations. Some examples of ICs are shown below.

TYPICAL DIGITAL CIRCUIT CHIPS

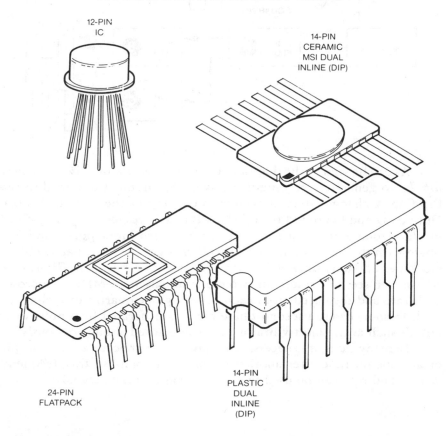

12-PIN IC

14-PIN CERAMIC MSI DUAL INLINE (DIP)

24-PIN FLATPACK

14-PIN PLASTIC DUAL INLINE (DIP)

Clocks

Most digital systems used in control and computer applications are under the control of a *clock* that *references* the *occurrence* of all signals and keeps the events in the system *synchronized*. This is necessary so that a *sequence* of events can occur in the *correct order* or at a *specified time*. Later you will learn how unsynchronized inputs are handled. For now, let's consider only signals synchronized by the clock.

Usually, the clock is crystal controlled at some frequency 2 or more times the operating frequency of the system. In principle, a 2-MHz crystal-controlled oscillator could be used for a digital device operating at a 1-microsecond (μsec) rate. For many applications, clocks with different phases but the same frequency are desired. This is done so that the data stream bits timed by clock phase I are fully in the *desired logic state* before the data bits are received using clock pulses of phase II.

TWO-PHASE CLOCK

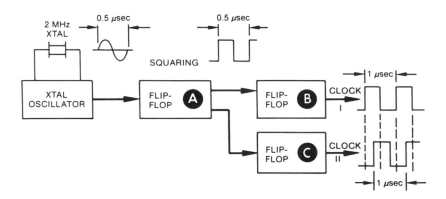

As can be seen in the diagram, the master oscillator drives a *squaring flip-flop* to generate a rectangular wave. The circuit is designed so that flip-flop A changes state at each zero crossing of the oscillator signal. Outputs Q and $\overline{Q}$ are fed to flip-flops B and C, respectively. Flip-flip B produces one output square wave for each two input pulses and is in phase with the input signal. Flip-flip C changes state one-half oscillator clock period later, since it is driven by the $\overline{Q}$ output from flip-flop A. As you can see, this results in two clock signals at a 1-MHz (1-microsecond) rate and displaced from each other by one-quarter of a clock period. In some cases, a second phase clock is generated by a pulse-delay circuit such as the multivibrator that you learned about previously.

In many cases, it is necessary to have an event occur at set multiples or at some set time after another event. To accomplish this, frequency *dividers* and *gates* can be used to create the appropriate signals.

Dividers and Gates

You have already studied an example of this in which a fixed set of clock pulses was gated through using a counter and the necessary digital logic. Any desired count can be obtained by the use of feedback and/or other logic. For example, a count of 10 (10:1 divide) can be obtained as shown below, using four flip-flops.

FLIP-FLOP CHAIN

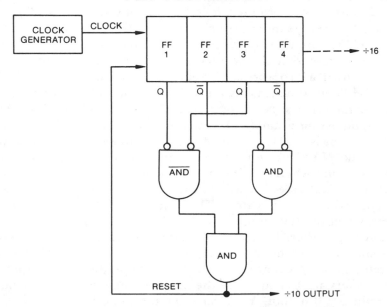

As shown, the clock is fed to four flip-flops that would normally produce a division of 16:1. When the flip-flops reach a count of 10 (1010), however, the logic senses this and resets the flip-flop chain and produces a single pulse output that occurs for every tenth clock pulse. The AND circuit is connected to the Q outputs of flip-flops 1 and 3 to sense the 1s in 1010. The other AND circuit is connected to the $\overline{Q}$ outputs of flip-flops 2 and 4 to sense when they are in the reset condition (the 0s of the 4-bit word). When these conditions are met, the two AND outputs are high, producing a high output from the third AND circuit. This AND produces an output for every tenth pulse. This output also resets the flip-flop string. As you know, the divider chain could be extended to sense another count and the two output pulses obtained could then be used to set and reset a flip-flop and thus produce a gate. Rearrangement of this hard-wired logic could obviously result in a different count. Conversely, a set of switches, appropriately wired, could be used to provide any count desired. In fact, such a switchable counter is used in the 40-channel CB radio frequency synthesizer to be described shortly.

Code Conversion

As you know, there are several different coding systems (pure binary, octal, hexadecimal, binary-coded decimal, etc.), and it is sometimes necessary to convert from one to another. As an example, we will discuss a binary-coded decimal (BCD) conversion to decimal format. This is useful for display purposes when it is desired to produce a decimal number on a display (LED or LCD) to indicate an output from the digital system. The diagram shows a Signetics 8251 *binary-coded-digital-to-pure-decimal* decoder. As you can see, it consists of NAND and NOT gates in an arrangement very similar to some you've studied previously. The input is a parallel four-digit binary code for the decimal number. The 10 output lines (0 to 9) represent the decimal number appearing at the input as BCD. As shown, a 0 appears on the output line to represent the decimal number while all other inputs are 1. The truth table shows the input/output characteristics. Note that the LSB is to the left and the MSB is to the right. As an example of how the converter works, suppose the BCD input code is 1000 (1_{10}); the 1 on line A is inverted twice and appears as a 1 at the NAND gate feeding output line 1. All the other inputs (BCD) are 0, and when these are inverted twice they appear as 0s at all the other NAND gates. However, they are inverted once only before being applied to NAND gate 1. Therefore, NAND gate 1 has all 1s at its input when the 1000 is applied. This leads to a 0 output for decimal line 1. All the others have a combination of 1 and 0 at their NAND inputs and hence produce a 1 output. According to the NAND gate truth table shown previously, the NAND gate produces a 0 output only if *all* its inputs are 1. Thus, only the gate feeding output line decimal 1 will produce a 0 output. You can trace the logic for the other input states to verify that the truth table is correct.

SIGNETICS 8251 DECODER

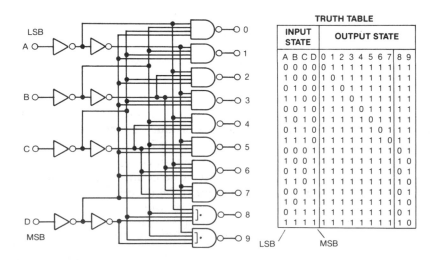

INPUT STATE	OUTPUT STATE									
A B C D	0	1	2	3	4	5	6	7	8	9
0 0 0 0	0	1	1	1	1	1	1	1	1	1
1 0 0 0	1	0	1	1	1	1	1	1	1	1
0 1 0 0	1	1	0	1	1	1	1	1	1	1
1 1 0 0	1	1	1	0	1	1	1	1	1	1
0 0 1 0	1	1	1	1	0	1	1	1	1	1
1 0 1 0	1	1	1	1	1	0	1	1	1	1
0 1 1 0	1	1	1	1	1	1	0	1	1	1
1 1 1 0	1	1	1	1	1	1	1	0	1	1
0 0 0 1	1	1	1	1	1	1	1	1	0	1
1 0 0 1	1	1	1	1	1	1	1	1	1	0
0 1 0 1	1	1	1	1	1	1	1	1	0	1
1 1 0 1	1	1	1	1	1	1	1	1	1	0
0 0 1 1	1	1	1	1	1	1	1	1	0	1
1 0 1 1	1	1	1	1	1	1	1	1	1	0
0 1 1 1	1	1	1	1	1	1	1	1	0	1
1 1 1 1	1	1	1	1	1	1	1	1	1	0

TRUTH TABLE

LED and LCD Devices

Forward-biased p-n semiconductor junctions made from the proper materials have the property of emitting photons. These junctions are the basis for the *light-emitting diode (LED)* displays with which you are probably familiar. LEDs available now are usually fabricated from either gallium arsenide phosphide (GaAsP) or gallium phosphide (GaP). GaAsP provides for red light output and is most common. GaP can provide red, yellow, or green light output, depending on the impurities added to the semiconductor material.

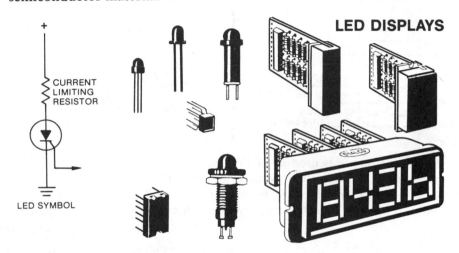

LEDs are used as indicators as well as in displays. Single LEDs are restricted in size and hence are used in groups to form characters. As you can see from the illustration above, the displays can be made from an array of diodes or can be arranged to light up a lens segment in groups to form a bar—groups of which are used to form a character, as will be described. The LED is basically a low-power device. Typically, the voltage drop across the diode junction is less than 1.5 volts. Continuous operating currents are usually less than 10 mA. Some units designed to be strobed (flashed at a relatively high rate to give the illusion of *continuous* illumination) can handle pulse currents of up to 150–200 mA.

The LED displays described above are very well suited to numeric or alphabetic character display but consume appreciable power to produce a light output. Because of their power consumption, their use in low-power devices like digital watches is limited to *intermittent* operation. To overcome this problem, the *liquid-crystal display* (LCD) was developed. The LCD operates on the basis of *variable reflectivity (light scattering)*. It provides *no* light of its own and, unlike the LED, requires ambient light (daylight, lamps, etc.) to be observed. However it consumes essentially *no power* and hence can be operated *continuously*. The most common LCDs are the *reflective* type, requiring front illumination, although other types using back illumination are also used. The principle of operation of the LCD involves the use of *special organic compounds* that are long cigar-shaped molecules charged at either end to form a dipole.

LED and LCD Devices (continued)

Although there are many variations of the LCD idea, most operate on the basis of *rotating* the *plane* of *polarized light* and the fact that the charged organic molecules can be aligned in an electrostatic field. If two polarizers are arrayed so that their plane of polarization is the same, they are essentially *transparent*; if, however, they are at 90 degrees, then *no* light is transmitted through them, as shown below. The arrows show the relative plane of polarization between the two polarizers. The dot shows that the plane of polarization is at 90 degrees.

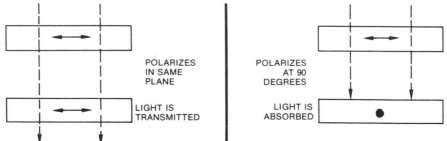

This principle is used in LCDs since the organic molecules can be aligned in an electric field because of their charge and hence can let polarized light through to the back plate or not, depending on whether a charge is present. An LCD display using this principle is shown below.

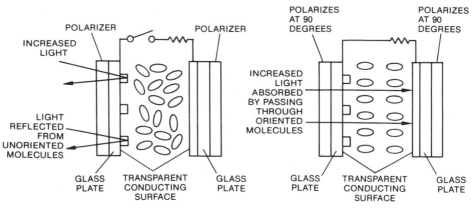

NO CHARGE ON CONDUCTING PLATES— LIGHT IS REFLECTED

As shown, light is reflected from the unoriented molecules when no power is applied, providing a uniform appearance. When power is applied to the transparent conducting layer, the molecules align so that the polarized light passes through and is absorbed by the second polarizer, which is crossed in polarization, leaving the area where the molecules are aligned darkened. Since the molecules are only aligned in the electric field, only those areas with an applied field will darken. If a segment pattern is plated on as a conductive surface and segments are properly energized, dark characters against a light background will appear.

Numerical Displays

Since we are dealing with electrostatic fields and molecular orientation, the only energy needed is that necessary to orient the molecular dipoles. Therefore, very little power is consumed by these displays. Thus, the individual segments can be driven in such a way as to produce different characters, as for the LED. Because of certain chemical reactions, dc is not normally used to excite LCD. Instead an ac square wave (typically 1–25 kHz) is used, since it doesn't matter what the polarity of the electric field is. A typical LCD in a watch operates at 3 volts and consumes about 0.5 μA to light up all numbers of a watch display simultaneously.

The code converters and the LED–LCD devices that you learned about on the previous pages are the key to forming the arabic numbers that you see on digital watches and calculators.

Take the simplest case, that of the arabic number readout on digital watches and calculators: These numbers are formed by energizing the individual elements of an array of seven *light-emitting diode (LED)* or *liquid-crystal display (LCD)* devices. The individual elements of the array are identified as A,B,C,D,E,F, and G. Applying a 1 to any of these elements makes it highly visible against its background. Thus if elements A,B, and C are energized, the display shows a number 7.

LED/LCD DEVICES

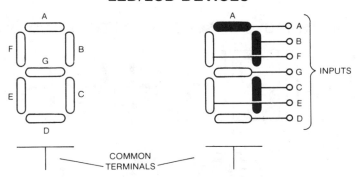

As you can see, by lighting the appropriate segments, the numbers 0 to 9 can be generated. Special symbols like − and + are generated by special displays or are included as special added elements in the seven-segment display. To properly illuminate the display, it is necessary to convert a binary code representing a number to the code necessary to illuminate the proper segments of the display. To do this, a BCD- (or other binary code) to-seven-segment decoder is used.

A BCD-to-seven-segment decoder operates in the same manner as the BCD to decimal decoder discussed earlier. The truth table shown on the next page illustrates the details. Numerical displays are rarely limited to a single digit. Even low-cost pocket calculators have a row of eight or more digits in their displays.

Numerical Displays (continued)

There are two general methods for doing this. One is to have a separate decoder for driving each array. More often, a *multiplexing technique* is used. Each element in each array has two terminals (input and ground). One of these terminals is connected in parallel to the corresponding terminal in each of the arrays. All the other terminals in each array are connected to a single common point, but this is shown as a single ground terminal to avoid complicating the diagram below.

The principle of operation should now be clear. Although the segments of all characters are excited in parallel, only the character whose common terminal is also grounded will be illuminated. Thus, to make a 7 appear on the array at the extreme right you apply a 1 across the A,B, and C inputs and the common terminal of that array. Although 7 is applied to one terminal of all arrays in the row, only the array at the extreme right is activated because only its common terminal is grounded.

To make the next array to the left show a particular number, you would apply the code of that number across the same A through G lines and the common terminal of that particular array. When you use this method to apply any given desired number in rapid succession to each display array in the row, *strobing* each array over and over again, all the numbers in the row *appear to be on* at the same time because of your persistence of vision.

TRUTH TABLE OF SEVEN-SEGMENT DECODER

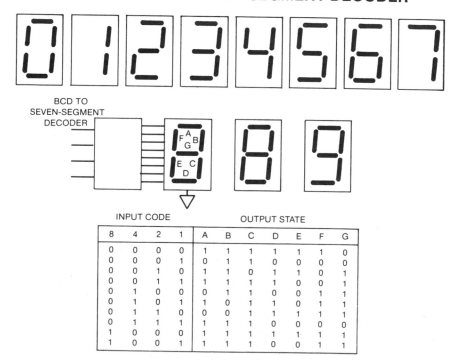

INPUT CODE				OUTPUT STATE						
8	4	2	1	A	B	C	D	E	F	G
0	0	0	0	1	1	1	1	1	1	0
0	0	0	1	0	1	1	0	0	0	0
0	0	1	0	1	1	0	1	1	0	1
0	0	1	1	1	1	1	1	0	0	1
0	1	0	0	0	1	1	0	0	1	1
0	1	0	1	1	0	1	1	0	1	1
0	1	1	0	0	0	1	1	1	1	1
0	1	1	1	1	1	1	0	0	0	0
1	0	0	0	1	1	1	1	1	1	1
1	0	0	1	1	1	1	0	0	1	1

Experiment/Application—Driving an LED Display

To illustrate how LEDs can be used in conjunction with a counter and a decoder, you can set up a simple experiment using a single seven-segment LED display driven by the counter you experimented with previously. To do this, you will also need a BCD- (binary-coded decimal) to-seven-segment decoder/driver to convert the binary output from the counter to the proper inputs to the LED display. As you might suspect, there is a single IC chip to do this. A typical one is the SN7447, which also includes the drivers necessary to operate the LEDs directly with a high current capability to give a bright image. The pin configuration, the truth table, and the output resultant display are shown in detail on page 5–70.

You can connect the A B C D outputs from the counter you built previously directly to the inputs of the decoder chip, as shown below.

CONNECTING OUTPUTS FROM COUNTER TO DECODER CHIP

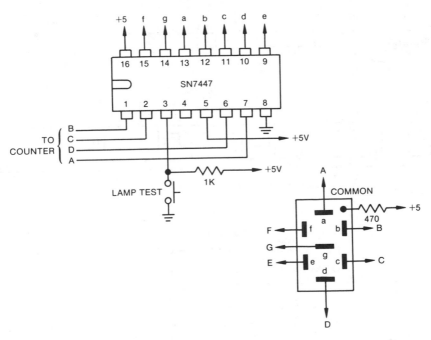

If you now push the lamp test to test all segments, you will see an 8 appear. You can now count by operating the counter described previously. When you get to 9, the remaining numbers (10–15) will produce the symbols shown earlier. When several digits are to be displayed, additional logic circuits between the counters and the LED displays allow only the numbers 0–9 to appear on each LED, and the count transfers to the succeeding LED as the count exceeds 9. In this way, a long number consisting of many digits can be displayed.

Experiment/Application—Driving an LED Display (continued)

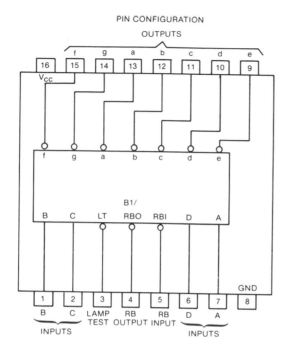

PIN CONFIGURATION

OUTPUTS

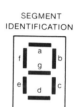

SEGMENT
IDENTIFICATION

0	1	2	3	4	5	6	7	8	9	10	11	12	13	14	15

NUMERICAL DESIGNATIONS—RESULTANT DISPLAYS

TRUTH TABLE

DECIMAL OR FUNCTION	INPUTS							OUTPUTS						
	LT	RBI	D	C	B	A	BI/RBO	a	b	c	d	e	f	g
0	1	1	0	0	0	0	1	0	0	0	0	0	0	1
1	1	x	0	0	0	1	1	1	0	0	1	1	1	1
2	1	x	0	0	1	0	1	0	0	1	0	0	1	0
3	1	x	0	0	1	1	1	0	0	0	0	1	1	0
4	1	x	0	1	0	0	1	1	0	0	1	1	0	0
5	1	x	0	1	0	1	1	0	1	0	0	1	0	0
6	1	x	0	1	1	0	1	1	1	0	0	0	0	0
7	1	x	0	1	1	1	1	0	0	0	1	1	1	1
8	1	x	1	0	0	0	1	0	0	0	0	0	0	0
9	1	x	1	0	0	1	1	0	0	0	1	1	0	0
10	1	x	1	0	1	0	1	1	1	1	0	0	1	0
11	1	x	1	0	1	1	1	1	1	0	0	1	1	0
12	1	x	1	1	0	0	1	1	0	1	1	1	0	0
13	1	x	1	1	0	1	1	0	1	1	0	1	0	0
14	1	x	1	1	1	0	1	1	1	1	0	0	0	0
15	1	x	1	1	1	1	1	1	1	1	1	1	1	1
BI	x	x	x	x	x	x	0	1	1	1	1	1	1	1
RBI	1	0	0	0	0	0	0	1	1	1	1	1	1	1
LT	0	x	x	x	x	x	1	0	0	0	0	0	0	0

Digital Tuning of CB Radio Transceivers

You have had some previous exposure to the use of digital processing techniques for tuning radio and TV transmitters and receivers in Volumes 3 and 4, where a phase-locked-loop (PLL) circuit was described. The Johnson Messenger models 4170 and 4175 40-channel CB transceivers, provided examples of how a PLL is used in digital tuning.

A block diagram of the PLL frequency synthesizer is shown in the illustration below. It serves as a part of the master oscillator, and its purpose is to put out a signal that is 4.3 MHz below the frequency of the selected channel. This output is used as the local oscillator (LO) for the receiver and is offset by a crystal oscillator/mixer to provide the carrier for transmission. The output signal from the voltage-controlled oscillator (VCO) Q201 is mixed with the output of the high-frequency crystal oscillator that generates a signal at 21.855 MHz. The resulting feedback frequency (VCO − 21.855) is divided by the digital programmable divider, which operates under the control of the channel-selection switch. The programmable divider output is a signal at about 10 kHz, which is fed into the phase detector.

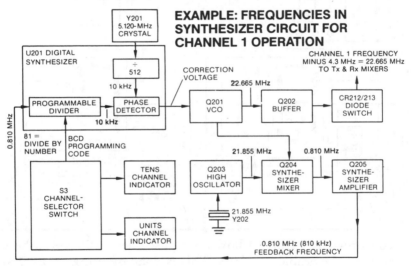

EXAMPLE: FREQUENCIES IN SYNTHESIZER CIRCUIT FOR CHANNEL 1 OPERATION

Another input to the phase detector is a 10-kHz reference signal generated by dividing the 5.120-MHz reference signal (÷ 512) generated by crystal Y201 and its associated circuit components in the digital frequency synthesizer integrated circuit U201. The frequency of the VCO is adjusted by the phase detector output from U201 to make the programmable divider output to the phase detector exactly equal to 10 kHz, as for the reference signal. When this condition occurs, the output of the VCO to the diode switch is equal to the selected channel frequency less 4.3 MHz. The diagram shows the circuit frequencies involved when tuned to channel 1, which has a transmit frequency of 26.965 MHz (22.665 MHz + 4.3 MHz).

Digital Tuning of CB Radio Transceivers (continued)

On the previous page it was noted that the channel-selector switch controls the programmable divider for dividing the feedback frequency to produce the 10-kHz output to the phase detector. The selector switch not only controls the programmable divider but also operates the display, which is made up from seven-segment LEDs.

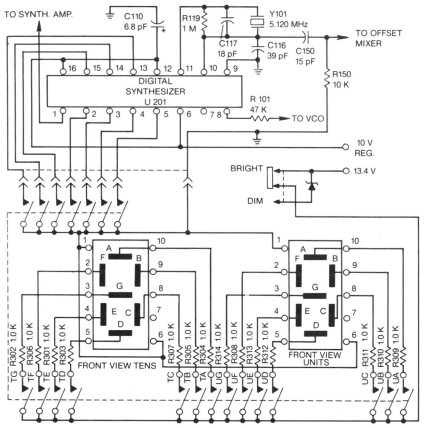

You previously learned how a seven-segment LED could be used to display all the arabic numbers. This is done by using an LED array having seven straight bars identified as A through G and a common terminal. By supplying voltage (1) or no voltage (0) to each of these LED segments, the desired number is formed. The CB transceiver has a rotating selector switch with 40 positions. The switch also has a number of contacts arranged as shown in the diagram. At each position of the selector switch is a set of *fingers* that can close any of the contacts, thus generating a 1 signal on the lead connected to that contact. Leads connected to unclosed contacts have no voltage applied to them and thus carry a 0 signal.

The table on page 5–74 shows the digital words from the three sections of the switch and their relative value (high or low). Note that this table shows a ground as 0 (switch closed) and an open as 1. The table also shows the synthesizer output frequency, divide-by number, and the feedback frequency to the divider.

Digital Tuning of CB Radio Transceivers (continued)

The first column shows the CB frequency channel number to which the transceiver is to be set by the selector switch. The next column gives the number by which the frequency of the signal from the synthesizer amplifier must be divided to become 10 kHz for the phase detector. Next is the programming code that is sent from the channel selector switch to the programmable divider, to indicate the frequency division necessary. The numbers 12, 13, 14, 15, 2, 3, 4, and 5 are the terminals of the digital synthesizer (U201), to which the switch sections supply the code.

The next column (VCO Output) is the frequency of the signal the VCO sends to the synthesizer mixer (Q204) to mix with the 21.855-MHz signal from the high oscillator (Q203). The mixing results in the signal to be divided to 10 kHz, whose frequency is different for each channel, thus requiring a different "divide by" number. As previously mentioned, the VCO frequency is 4.3 MHz below the transceiver's operating frequency. The next column, "Output Transmit," is the transceiver's operating frequency.

On the right side of the table are the channel indications sent from the channel selector switch to the LEDs so that they will indicate the number of the channel selected. Note the two LEDs, one for tens and one for units, with their elements indicated by the standard letters A through F. As previously stated, it is important to note that, in this equipment, an LED element is activated when the channel indicator digit is a zero, and not by a "1."

The switch contact elements in the diagram on the previous page are arranged into three groups. One group of seven is used to form the *units* digit, and its contacts are marked UA through UG. Another group of six contacts is used to form the *tens* digit, and its contacts are marked TA through TG. Finally, you will see that there are seven switch contacts connected to the input to the digital synthesizer. These feed a 7-bit binary code to the synthesizer. Thus, the sets of switch contacts are used not only to control the illumination of the seven-segment numeric indicators, but also to control the programmable divider.

As you can see from the table, for example, when the selector switch is set to channel 1, only the *units* LED segments B and C are illuminated to form a 1. The programming code, produced by the selector switch, puts a ground on pins 2,3,4,13,14, and 15 and an open on pins 5 and 12 of U201. This sets the programmable divider to divide the synthesizer mixer output by 81. The selected operating frequency is 26.965 MHz. For a receiver with an IF of 4.3 MHz, an LO frequency of 26.965 − 4.3 MHz or 22.665 MHz is required as the output of the VCO. As shown previously, the reference high-frequency oscillator (Q203) produces a constant output at 21.855 MHz. If the VCO is at 22.665 MHz, and is mixed with the 21.855-MHz fixed oscillator output, an 810-kHz signal results. This signal is fed to the programmable divider to produce the 10-kHz signal for the phase detector input. Thus, you can see that the operating frequency of the system is set by changing the programmable divider. By doing this, any operating channel in the entire band of frequencies can be selected.

DIGITAL SYNTHESIZER PROGRAMMING TABLE

CHANNEL NO.	DIVIDE BY NUMBER	PROGRAMMING CODE U202 PIN 12	13	14	15	2	3	4	5	VCO OUTPUT	OUTPUT TRANSMIT	TENS A	B	C	D	E	F	G	UNITS A	B	C	D	E	F	G
1	81	1	0	0	0	0	0	0	1	22.665	26.965	1	1	1	1	1	1	1	1	0	0	0	1	1	1
2	82	0	1	0	0	0	0	1	0	22.675	26.975	1	1	1	1	1	1	1	0	0	0	0	1	1	0
3	83	0	1	0	0	0	0	1	1	22.685	26.985	1	1	1	1	1	1	1	0	0	1	0	1	1	0
4	85	1	1	0	0	0	1	0	1	22.705	27.005	1	1	1	1	1	1	1	0	1	0	0	1	0	0
5	86	1	1	0	0	0	1	1	0	22.715	27.015	1	1	1	1	1	1	1	0	1	0	1	1	0	0
6	87	1	1	0	0	0	1	1	1	22.725	27.025	1	1	1	1	1	1	1	0	1	1	0	1	0	0
7	88	1	1	0	0	1	0	0	0	22.735	27.035	1	1	1	1	1	1	1	0	1	1	1	1	0	1
8	90	1	1	0	0	1	0	1	0	22.755	27.055	1	1	1	1	1	1	1	1	0	0	0	0	0	0
9	91	1	1	0	0	1	0	1	1	22.765	27.065	1	1	1	1	1	1	1	1	0	0	1	0	0	1
10	92	1	1	0	0	1	1	0	0	22.775	27.075	1	1	1	1	1	1	0	1	0	0	0	0	0	1
11	93	1	1	0	0	1	1	0	1	22.785	27.085	1	1	1	1	1	1	0	1	0	1	0	0	0	1
12	95	1	1	0	0	1	1	1	1	22.805	27.105	1	1	1	1	1	1	0	1	1	0	0	1	0	1
13	96	1	1	0	0	1	1	0	0	22.815	27.115	1	1	1	1	1	0	0	1	1	1	0	0	0	1
14	97	1	1	0	1	0	0	0	1	22.825	27.125	1	1	1	1	1	0	0	1	1	1	1	0	0	0
15	98	1	1	0	1	0	0	1	0	22.835	27.135	1	1	1	1	1	0	0	0	0	0	0	0	1	0
16	100	1	1	0	1	0	1	0	0	22.855	27.155	1	1	1	1	0	0	0	0	0	0	0	1	0	0
17	101	1	1	0	1	0	1	0	1	22.865	27.165	1	1	1	1	0	0	0	0	0	1	0	1	0	1
18	102	1	1	0	1	0	1	1	0	22.875	27.175	1	1	1	1	0	0	0	0	0	1	1	1	0	0
19	103	1	1	0	1	0	1	1	1	22.885	27.185	1	1	1	1	0	0	0	0	1	0	0	1	1	1
20	105	1	1	1	0	0	0	0	1	22.905	27.205	1	1	1	0	0	0	0	0	1	0	0	0	0	1
21	106	1	1	1	0	0	0	1	0	22.915	27.215	0	0	0	0	0	0	0	0	1	1	0	0	1	1
22	107	1	1	1	0	0	0	1	1	22.925	27.225	0	0	0	0	0	0	0	0	1	1	1	0	1	1
23	110	1	1	1	0	0	1	1	0	22.955	27.255	0	0	0	0	0	0	0	1	0	0	0	0	0	0
24	108	1	1	1	0	0	1	0	0	22.935	27.235	0	0	0	0	0	0	0	0	1	1	1	1	0	0
25	109	1	1	1	0	0	1	0	1	22.945	27.245	0	0	0	0	0	0	0	0	1	1	1	1	1	1
26	111	1	1	1	0	0	1	1	1	22.965	27.265	0	0	0	0	0	0	0	1	0	0	0	1	0	1
27	112	1	1	1	0	1	0	0	0	22.975	27.275	0	0	0	0	0	0	0	1	0	0	1	1	0	0
28	113	1	1	1	0	1	0	0	1	22.985	27.285	0	0	0	0	0	0	0	1	0	0	1	1	0	1
29	114	1	1	1	0	1	0	1	0	22.995	27.295	0	0	0	0	0	0	0	1	0	1	0	1	0	0
30	115	1	1	1	0	1	0	1	1	23.005	27.305	0	0	0	0	0	0	0	1	0	1	0	1	0	1
31	116	1	1	1	0	1	1	0	0	23.015	27.315	0	0	0	0	0	0	0	1	0	1	1	1	1	0
32	117	1	1	1	0	1	1	0	1	23.025	27.325	0	0	0	0	0	0	0	1	0	1	1	1	0	1
33	118	1	1	1	0	1	1	1	0	23.035	27.335	0	0	0	0	0	0	0	1	1	0	0	1	1	0
34	119	1	1	1	0	1	1	1	1	23.045	27.345	0	0	0	0	0	0	0	1	1	0	0	1	0	1
35	120	1	1	1	1	0	0	0	0	23.055	27.355	0	0	0	0	0	0	0	1	1	0	1	1	0	0
36	121	1	1	1	1	0	0	0	1	23.065	27.365	0	0	0	0	0	0	0	1	1	0	1	0	0	1
37	122	1	1	1	1	0	0	1	0	23.075	27.375	0	0	0	0	0	1	0	1	1	1	0	0	0	0
38	123	1	1	1	1	0	0	1	1	23.085	27.385	0	0	0	0	0	1	0	1	1	1	0	0	1	0
39	124	1	1	1	1	0	1	0	0	23.095	27.395	0	0	0	0	1	1	0	1	1	1	1	0	0	0
40	125	1	1	1	1	0	1	0	1	23.105	27.405	0	0	0	1	0	0	1	1	0	0	0	0	0	1

ALL FREQUENCIES IN MHz
"1" = SWITCH OPEN
"0" = SWITCH CLOSED

Digital Watch with Analog Display

A digital watch is essentially an electronic digital counter that counts time in terms of hours, minutes, and seconds (and sometimes in days and months) by means of the digital circuits you have already studied. Such a watch may be equipped with an *analog* display, which has the familiar *mechanical hands* seen on mechanical watches and clocks. Digital watches can also use LED and LCD digital displays formed by means of the seven-segment displays.

These watches are of special interest here because they illustrate how digital logic circuits can be used as timers and to control motor drives, relays, and other electromechanical components. In addition, the *time control* of an array of devices ranging from microwave ovens to traffic lights is based on the timing principles given here.

DIGITAL WATCH WITH ANALOG DISPLAY

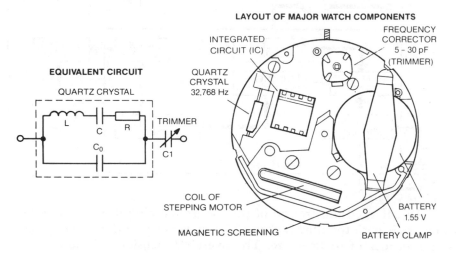

LAYOUT OF MAJOR WATCH COMPONENTS

The digital-analog watch used as an example here is manufactured by Ebauches S.A., known as ESA, and is typical. The basic timing element that enables most electronic watches to keep extremely accurate time is an electronic oscillator or clock using quartz-crystal control. As you know, quartz-crystal oscillators are very stable and produce a precisely controlled frequency output. The diagram above shows the equivalent circuit of the oscillator and the arrangement of the major parts in the watch.

As for most applications of this type, all of the circuit components except the crystal and frequency-trimmer capacitor (C1) are contained in a single IC package. The resonant frequency of the crystal oscillator is 32,768 Hz or exactly 2^{15} Hz. This frequency can be divided by 2^{15} to give the 1-Hz (1 cycle per second) frequency whose period is the basic timing interval. Adjustment of C1 permits time change of up to 4.5 seconds per day to account for manufacturing tolerance.

Digital Watch with Analog Display (continued)

All of the control and arithmetic/logic functions in the watch are performed in the single CMOS integrated circuit.

ENLARGED LAYOUT OF WATCH –
INTEGRATED CIRCUIT

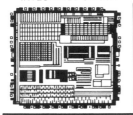

CURRENT PULSES TO STEPPER MOTOR

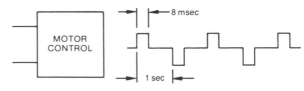

BLOCK DIAGRAM OF FREQUENCY DIVISION
PERFORMED BY WATCH INTEGRATED CIRCUIT

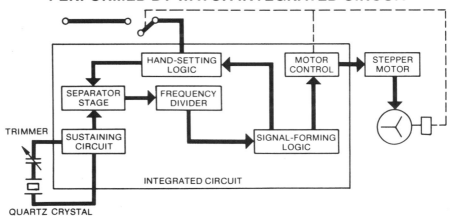

An enlarged illustration of the integrated circuit is shown above. Compare it with the illustration of the movement on the previous page to get an idea of its true size. The overall dimensions of the IC in its protective plastic capsule are 4.80 × 3.80 × 1.35 mm. The counter in the IC takes the 32,768-Hz output signal of the crystal oscillator, shapes it, and divides the frequency down to 8-millisecond (ms) ($^1/_{125}$ second) current pulses. These pulses are applied to the stepper motor, which rotates through half (180 degrees) of a complete rotation. The next current pulse of reverse polarity causes the motor to rotate another 180 degrees in the same direction. The direction of rotation is the same because the rotor is a permanent magnet with one North and one South pole inclined slightly to the field produced by the wire-wound stator. Each step of the rotor is connected through gears to the second, minute, and hour hands. The second hand (geared down from the motor by 60:1) takes half-second steps around the 60-second scale on the watch face.

In the block diagram the *sustaining circuit* provides the required feedback to keep the crystal oscillator going. The *separator stage* shapes the pulse from the oscillator and inhibits the counter when the hands are being set. The divider is a series of 15 divide-by-2 stages (flip-flops) that produce a final output at 0.5 Hz.

Digital Watch with Digital Display

The signal-forming circuit accepts this signal and shapes it to the required amplitude to drive the motor control section, which reverses the polarity of alternate pulses as shown and feeds the stepper motor. The hand-setting logic loop makes it easy to set the second hand to the exact second. It has a contact that opens the signal path in the separator stage and thus stops the motor until the hand-setting stem has been pushed back in.

ESA also produces a microprocessor-controlled digital watch using seven-segment LEDs to form arabic numbers. The layout of the watch is shown on this page, and the following page shows the block diagram. Two 1.55-volt silver oxide cells are mounted under the back of the case.

LAYOUT OF MAJOR COMPONENTS IN A DIGITAL MICROPROCESSOR WATCH

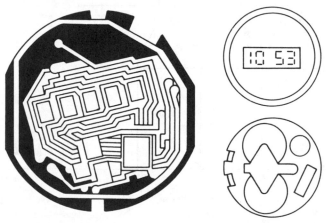

This watch also uses a quartz oscillator operating at a frequency of 32,768 Hz and adjusted by a trimmer capacitor as the basic timing element. The integrated circuit has wave-shaping and frequency-dividing circuits similar to the analog unit. The output of the frquency divider is at 768 Hz. Then a divider-decoder section uses this 768-Hz signal and additionally divides and encodes it to generate signals to the seven-segment LEDs that indicate seconds, minutes, hours, and date.

To minimize drain from the battery, power is not applied continuously to the LEDs, but is multiplexed to each digit at a frequency of about 32 Hz as a power-conservation measure. No display is shown until the user presses the demand switch. This activates a display of hours from 1 to 12 which is separated by two dots (:) from a minutes display of from 01 to 60. If the demand switch is held closed for more than a few seconds, the display changes and the seconds are displayed. If the demand switch is pressed twice, the number of the month is displayed from 1 to 12, separated by a dash (-) from days of the month from 01 to 31. Pressing the correction switch permits the user to operate each counter position (hours, seconds, day, date) at a fast rate for setting of the watch.

BLOCK DIAGRAM OF CONTROL AND DISPLAY
IN A DIGITAL MICROPROCESSOR WATCH

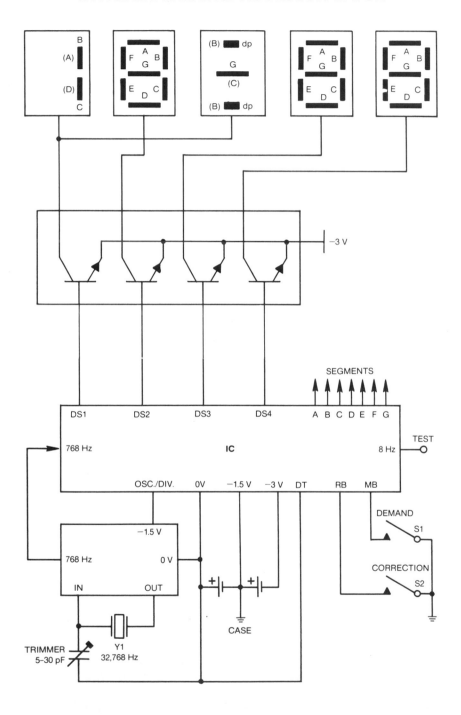

Digital Alarm Clocks and Clock Radios

The clock radios that are so popular today use single ICs that include all the logic needed to provide a complete control and timing system. Such ICs not only keep time but also control the radio, can be reset to provide variable alarm features, etc. These ICs are made by many manufacturers. Just as in the watches you learned about earlier, a simple divider chain is used to keep time. Since most of these devices work from the power line (60 or 50 Hz), they do not require the crystal clock for establishing the base operating frequency. Most of these clocks operate on a 24-hour cycle, so they count by 86,400 (seconds-to-a-day) for a total count of 5,184,000 (86,400 seconds × 60 per second). When the various alarms are set, a digital number is stored, and, typically, the count from the clock is compared to this stored time and the alarm is started (or radio turned on) when these counts coincide. Usually, an auxiliary counter is started at this time to allow for the turn off of the functions.

FEATURES
- 50- OR 60-Hz OPERATION
- SINGLE POWER SUPPLY
- LOW POWER DISSIPATION (32 mW AT 8 V)
- 12- OR 24-HOUR DISPLAY FORMAT
- AM/PM OUTPUTS
- LEADING-ZERO BLANKING
- 24-HOUR ALARM SETTING
- ALL COUNTERS ARE RESETTABLE
- FAST AND SLOW SET CONTROLS
- POWER-FAILURE INDICATION
- BLANKING/BRIGHTNESS CONTROL CAPABILITY
- ELIMINATION OF ILLEGAL TIME DISPLAY AT TURN-ON

APPLICATIONS
- ALARM CLOCKS
- DESK CLOCKS
- CLOCK RADIOS
- AUTOMOBILE CLOCKS
- STOPWATCHES
- INDUSTRIAL CLOCKS
- MILITARY CLOCKS
- PORTABLE CLOCKS
- PHOTOGRAPHY TIMERS
- INDUSTRIAL TIMERS
- APPLIANCE TIMERS
- SEQUENTIAL CONTROLLERS

Some features are shown above for a National Semiconductor MM5316 digital alarm clock chip—a typical unit.

The block diagram of the digital alarm IC is shown on page 5–80. As shown, the clock will work with a 50- or 60-Hz input. The incoming signal is shaped and operates a 50- or 60-Hz divider (determined by whether pin 36 is set high or low). The divider output (seconds) is divided again by 60 to count the minutes, and the minutes are divided by 12 or 24 to count hours. Whether the clock is on a 12- or 24-hour cycle is determined externally by whether pin 38 is strapped high or low.

The counter outputs (minutes/hours) are fed to the alarm comparator along with the stored information on alarm times. The outputs from the comparator and appropriate control signals are analyzed by the alarm and sleep circuits to provide for control of the radio and alarm. Display data (hours/minutes/seconds/AM–PM/etc.) are converted, and the IC chip output can be used to operate an LED, LCD, or similar digital segmented display. Additional features included are leading-zero blanking, power-failure indicator, and fast and slow setting. These fast/slow settings are accomplished by bypassing or decreasing the count in a given counter, which of course make the displayed time advance at a greater rate.

BLOCK AND CONNECTION DIAGRAMS

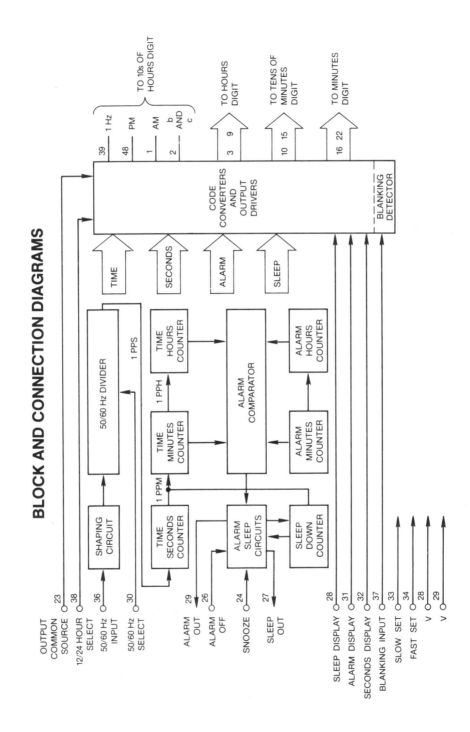

Experiment/Application—A Simple Clock

A simple clock can be built using the line frequency of 60 Hz as a reference. All you need is a counter that counts up to 60 and resets and counts to 60 again to give a 1-second timing or clock pulse. This can be done using the dual 4-bit counter you experimented with previously. Actually, the circuitry also exists to provide 1-minute pulses too, since the dual 4-bit counter can be cascaded with one more 4-bit counter to provide up to 4,096 counts ($16 \times 16 \times 16$). You will remember that in binary form the number 60 can be represented by 00111100. All you need to do is decode this sequence and reset the counter when it occurs to make a 1-second timer. To carry this operation to minutes would take a second counter working from the output of the first, with the reset coming at the end of a count of 60 in the second counter.

To illustrate the point, you could set up a pair of dual four-stage binary counters that you experimented with previously—with some modifications. The first modification is to trigger the first counter input with the line frequency by modifying the input circuit as shown below.

CLOCK PULSES

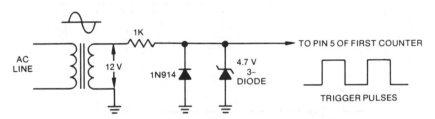

This circuit provides the 60-Hz input clock pulses from the power line to drive the counter. If no decoder and reset are used, the counter will count to 255 (FF_{16} or 377_8) and stop. We want to reset it at the end of a count of 60 and also count the seconds on the second counter to produce a counter of 60 seconds for a minute indication.

Now the first counter will work at a 60-Hz rate. You would like to have the second counter count once for each count of 60 on the first counter. To do that you must decode the output of the first counter at a count of 60. This can be done with the AND and NOT gates that you studied earlier. For this experiment we will use the SN7430 IC that has an eight-input positive NAND gate configuration. The diagram on the next page shows the configuration of the SN7430—that is, a 1 is required at all eight inputs to produce a 0 at the output and a 0 at any input produces 1 at the output. Since we want to decode the sequence 00111100, we will need inverters on the 0s to produce the necessary condition.

Experiment/Application—A Simple Clock (continued)

The diagram below shows the decoder configuration to provide a reset on a count of 60_{10} (00111100_2). As you will notice, not all the stages are necessary since we can count to 255, but only need to count to 60; however, since it is already available, we will use it.

The output of the SN7430 produces a low logic state (0) whenever the count reaches 60. At this point, the counter must be reset. To do this, we will use the output of the NAND gates to drive a multivibrator through an inverter as follows:

DECODER CONFIGURATION

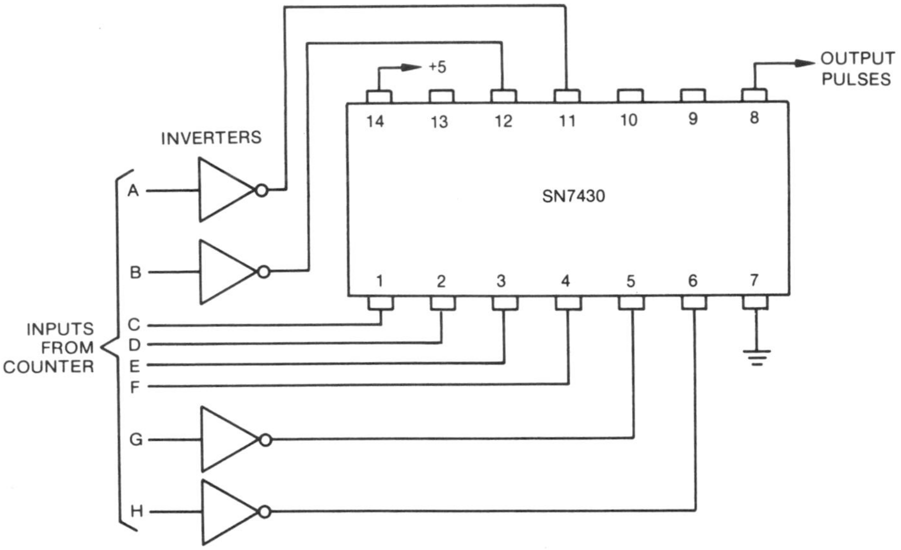

The multivibrator shown is the same one used earlier in your counter experiments, with the input and output reconnected as shown. To produce a signal that is visible (at 1-second intervals), a second multivibrator should be set up with a time constant of about 10 milliseconds. This should be connected to the reset line, as shown above. The configuration of this multivibrator is the same as that for the experimental material on multivibrators. Time the output and you will see that the LED flashes every second (60 counts). You can repeat the process to get 1-minute intervals with the second pair of counters driven by the reset for the first counter. In addition, a second reset to reset the second counter can be used to obtain 1-minute timing intervals. If you do this, you will see that the first counter output is at 1-second intervals, and the second counter output is at 1-minute intervals.

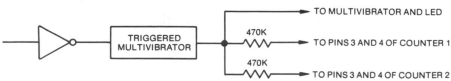

Review of Basic Digital System Functions and Applications

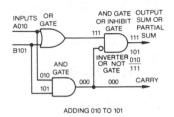

ADDING 010 TO 101

1. A HALF-ADDER is made up of an OR and two ANDs. It must be combined with another half-adder to do complete addition.

COUNTING WITH FLIP-FLOPS

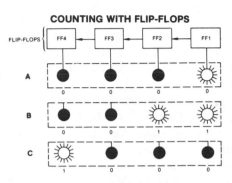

2. A FLIP-FLOP changes between its two states with each input pulse. Connected in series, flip-flops can perform counting.

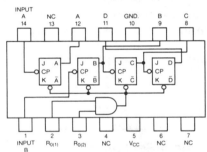

3. CLOCK AND REGISTER FUNCTIONS can be performed by counters by use of control circuits, binary to decimal converters, and indicators such as LEDs.

SIGNETICS 8251 DECODER

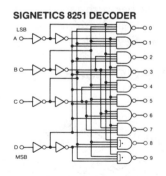

4. BINARY AND BINARY CODED NUMBERS must be converted for outputs in decimal numbers.

PULSES TO STEPPER MOTOR

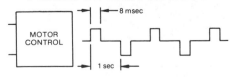

5. DIGITAL WATCHES use pulses derived from a quartz crystal oscillator, whose output frequency is divided down to operate a stepping motor or connected to LED circuits.

Self-Test—Review Questions

1. Draw a block diagram of serial half- and full-adders and describe how they add two numbers together. What is the fate of *carry* in a serial adder?

2. Draw a block diagram of a parallel adder and describe how it adds two numbers together. How is the *carry* handled?

3. Draw a block diagram of a counter using flip-flops. Describe how it is used for storage of digital words.

4. Create a table for a three-stage counter to show the state of each flip-flop as the count increments.

5. Describe via block diagram how serial-to-parallel and parallel-to-serial conversion is accomplished using digital storage elements (flip-flops).

6. Define clock and load strobe. Show by a block diagram how you can generate a set of four clock pulses from a pulse train using digital logic elements. Describe this operation.

7. How are flip-flops used as dividers? Show how counts of 3, 7, and 14 can be generated by appropriate logic and feedback. Describe briefly how a BCD-to-decimal decoder works using an appropriate truth table.

8. Briefly describe how LEDs work. How are they used in displays? Do the same for LCDs. What are the basic differences in their operation?

9. Describe the operation of a seven-segment LED (LCD) display. How are these used with BCD-to-seven-segment decoders to provide a numerical display. How is multiplexing used to minimize hardware in these displays?

10. Describe how digital tuning of a CB radio is accomplished. What is the role of the counter in these systems?

Learning Objectives—Next Section

Overview—You now have the necessary background in digital systems to begin your study of computers and microprocessors to see how they are organized and how they work.

COMPUTERS AND MICROPROCESSORS

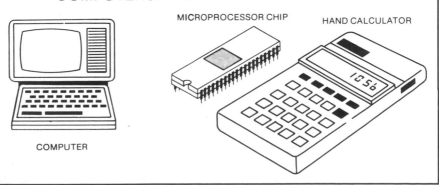

MICROPROCESSOR CHIP

HAND CALCULATOR

COMPUTER

Digital Computers

As you know, addition, subtraction, multiplication, and division are easy in binary form. A digital computer can perform these operations at very high speeds—and when properly directed by a *program* will perform not only calculations, but also make logical decisions on how to proceed as the calculations continue. A program is a set of logical instructions to the computer which are stored and used to tell the computer what to do, step-by-step. In addition to their role in processing great quantities of numerical data (or *number crunching*, as it is called), digital computers are very important in *control applications*. Here, these devices are programmed to accept data inputs, make necessary calculations, make logical decisions about what is to be done, and then proceed to issue instructions to the components under control.

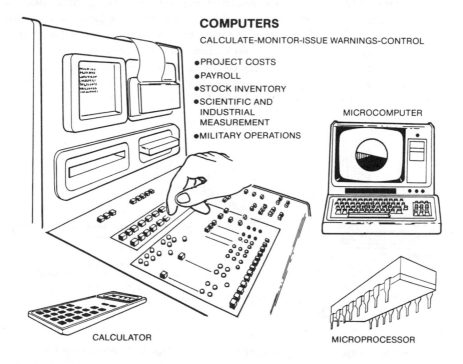

COMPUTERS

CALCULATE-MONITOR-ISSUE WARNINGS-CONTROL

- PROJECT COSTS
- PAYROLL
- STOCK INVENTORY
- SCIENTIFIC AND INDUSTRIAL MEASUREMENT
- MILITARY OPERATIONS

MICROCOMPUTER

CALCULATOR

MICROPROCESSOR

Digital computers may be built with a microprocessor or may have a very large *main frame* that can be very complex. Generally, however, digital computers can be divided into those that can be *flexibly programmed* or reprogrammed easily to carry out any desired set of instructions and those that are *fixed programmed* to do a specific task. Usually, the former are called *general-purpose computers* and the latter are called *dedicated computers* or *controllers*. As you might suspect, the computer elements may be identical. It is the *manner of programming* and, as you will see, *how* the *memory* of the computer *is organized* that determines whether it is a general-purpose or dedicated machine.

Development of Digital Computers

The differences between computers, minicomputers, microprocessors, calculators, etc. are not clearly definable. Early electronic digital computers used *vacuum tubes* and were big, extremely expensive, and unreliable. For example, the *Electronic Numerical Integrator And Computer (ENIAC)* built by the University of Pennsylvania in 1948 contained 18,000 vacuum tubes and was the size of a small house. In the 1950s, several types of digital computers—using vacuum tubes—were developed and used successfully for scientific work, military systems, and business. The advent of solid-state technology led to the development of solid-state systems with significantly improved capability while providing for enormous reduction in size, power consumption, and greatly increased reliability.

In 1965, the Digital Equipment Corporation (DEC) introduced a practical digital computer that would fit on a desk. Because of its comparatively small size, it became known as a *minicomputer*. Today, there are minicomputers available from many sources; they are quite inexpensive and used in a wide variety of applications. Minicomputers differ from large *main-frame-type* machines mainly in the *smaller amount* of *memory* incorporated into the minicomputer and in *operating speed*.

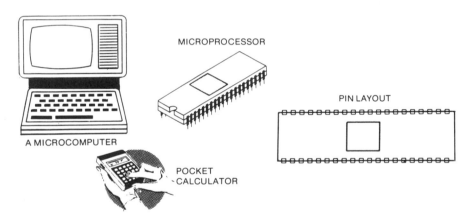

MICROPROCESSOR

PIN LAYOUT

A MICROCOMPUTER

POCKET CALCULATOR

The *microprocessor*—a broad term that can include anything from the CPU (central processing unit) in a minicomputer, to a complete CPU, to a complete computer incorporating memory and other capability—has come to dominate the small computer systems. In fact, microprocessors are also often used in large digital computers to support the CPU. Early microprocessors were of limited capability, but more modern ones can easily compete in both speed of operation and capacity with all but the largest machines. The low price of the microprocessor, coupled with the major reduction in price and size of associated memory and the expanded capacity of these microprocessors to handle more bits per cycle (faster speed of operation), has meant that they now have a prominent place in all digital computer systems as well as in control.

Major Elements of a Digital Computer

The major elements of a typical digital computer are the *arithmetic/ logic unit (ALU)*, the *control unit, memory*, and *input/output (I/O) buffers*. Usually, the control unit and arithmetic/logic units are combined into one assembly called the *central processing unit (CPU)*. Microprocessors usually include all the functions of the CPU *on a single chip*. In some cases, memory and I/O are also included in a single IC for specialized application, usually for specialized control functions.

ELEMENTS OF A DIGITAL COMPUTER

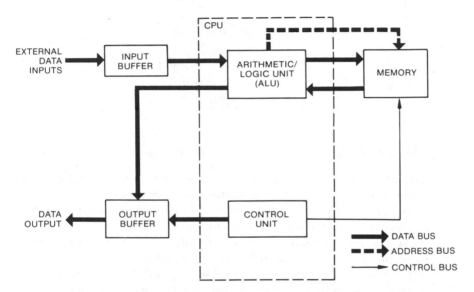

The input/output buffers (I/O) receive and act as a temporary store for external inputs or provide outputs to external devices. Typically, these could be a keyboard input and a printer output. The memory stores the results of various data manipulations, intermediate results, etc., and also contains the program or set of instructions that directs the computer's functions. The CPU contains the arithmetic/logic unit where data (bit patterns) are manipulated in accordance with the program, and the control unit controls the flow of information through the ALU, memory, and the I/O buffers. A set of lines carrying information is called a *Bus*. Most computers are organized in a two-bus structure—one for carrying and transfer of data (the *data bus*) and another that is used to control memory (the *address bus*).

The important thing to remember is that the basic organization of all digital computers is essentially the same and that they work by *controlled manipulation* of *bits* to convey specific information or produce a specific result.

The Central Processing Unit

All digital computers have a centralized processing unit called the *CPU* that contains the arithmetic, logic, and control functions. The CPU varies widely among computers. In a large system it may be very complicated and involve one or more circuit boards. On the other hand, what we usually call a *microprocessor* is really a *CPU* contained in a *single chip.* While the specific nature of a CPU can vary widely from computer to computer, there are certain fundamental requirements that must be met.

THE CPU IS A SINGLE CHIP
IN A MICROCOMPUTER

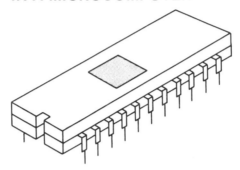

Basically all CPUs must have the following elements:

1. One or more *registers* called *accumulators* in which data taken from memory or to be returned to memory can be stored.
2. A *data counter*, the contents of which identify (point to) the location in memory where data are to be stored or retrieved.
3. An *instruction register*, the contents of which determine the operation to be performed as defined by the program or other logic.
4. A *program counter*, the contents of which point to the location in memory from which an instruction code has been obtained (*fetched*).
5. The necessary *logic, control,* and *arithmetic capability* to carry out the instructions of the program stored in memory.

Note that while the CPU performs the necessary bit manipulations, these bit patterns carrying the necessary information are generally fetched from and returned to memory. In addition, how the bit patterns are to be manipulated is determined by instructions obtained from memory in some defined sequence.

The bit manipulations within the CPU are performed by a group of logic components that are collectively called the *arithmetic/logic unit (ALU).* The ALU performs as a minimum the following functions:

1. Boolean operations
2. Binary addition
3. Complementation
4. Shift of data to right or left

The Central Processing Unit (continued)

Although additional logic elements are often included for speed of operation, the listed operations are all that are necessary. Boolean operations provide for logical decisions. Addition, complementation, and shifts allow for all addition, subtraction, multiplication, and division, which in turn allows for almost any desired bit manipulation.

THE ALU DOES THE DATA MANIPULATION

```
┌──────────── ALU ────────────┐
│                              │
│   • BOOLEAN LOGIC            │
│   • BOOLEAN ADDITION         │
│   • COMPLEMENTATION          │
│   • SHIFT RIGHT/LEFT         │
│                              │
└──────────────────────────────┘
```

All computer CPUs are designed to handle binary data in increments of a given number of bits. As you remember, an 8-bit word is usually called a *byte*. Large computers have word lengths of as much as 64 bits (8 bytes) or even 128 bits (16 bytes). Minicomputers typically handle 16- to 32-bit word lengths (2 to 4 bytes). Most inexpensive microprocessor CPUs use an 8-bit (1-byte) word length, although some of the newer microprocessor CPUs are using 16-bit (2-byte) word lengths. Some very simple microprocessor controllers use only a 4-bit word length. The memory, CPU registers, and ALU are all of *consistent word length*. If the processor is an 8-bit machine, all are compatible with the 8-bit word. The word length does not mean that the accuracy or precision are reduced or increased. It means that the data are processed in increments of 8 bits (or 1 byte). However, as many bytes as desired can be processed to give the desired accuracy.

The *control unit (CU)* of the CPU is responsible for determining the sequence of operations within the ALU. The CU is instructed by the contents of the instruction registers in the CPU, which is in turn driven by the contents of memory (program). The *instruction register* contents are decoded by the CU from the bit pattern in the instruction register, and on this basis the CU generates the sequence of signals to control the flow of data through the ALU and to control ALU logic at the right times.

THE CONTROL UNIT DETERMINES
OPERATION SEQUENCE

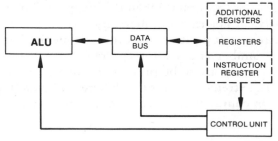

The Central Processing Unit (continued)

The ALU and the registers are interconnected by an *internal data bus*. When the control unit, operating in accordance with the bit pattern from the instruction register, perceives how data bits are to flow between the ALU and the registers via the data bus, and the ALU understands how it is to manipulate the data bits from the accumulators, instructions are sent from the control unit to the ALU and registers so that the appropriate signals appear on the data bus. In this way, data bits are transferred from the registers to the ALU, appropriately operated on, and shifted back via the data bus that is common between them. *These operations are all synchronized* via a common clock used to time all events.

Many CPUs have additional registers to store data from the ALU— or to permit the holding of intermediate results or to take in additional information from memory. These registers are useful for increasing speed of operation and make programming easier in some situations.

The CPU also generates a series of *status signals* called *flags*. These flags are usually 1-bit logic states that are set or reset to indicate the status and/or results of ALU operation. The most important of these status flags is the *carry flag*. This is essential when the total word being manipulated consists of several bytes. If the whole word cannot be handled in a single operation, the process must be done by a *series* of *operations*. For example, suppose we wanted to add two 16-bit (2-byte) words in a microcomputer that had only an 8-bit (single-byte) capability. We could do this by adding the lower-order bytes together and the higher-order bytes together, but we need to couple any carry from the lower-order byte to the higher-order byte operation. The carry bit allows us to do this because it is set when the lower-order bytes are added, if the result produces a carry. In this case, we would add 1 to the LSB of the higher-order byte addition.

STATUS FLAGS

- CARRY
- ZERO
- SIGN
- OVERFLOW
- INTERRUPT MASK

The *zero flag* tests a result and is set if the result is a 0. This is very helpful in logic schemes because it may be desired to go to a new part of the instructions (program) when this happens. The *sign flag* is used to test for positive or negative numbers. By convention, the sign flag is the bit at the extreme left position of a binary sequence, with a 0 interpreted as a positive number and a 1 interpreted as a negative number. When multiple-byte words are to be processed, some confusion arises as to the meaning of the extreme left bit of each byte. The sign flag and the *overflow flag* are used in conjunction to resolve this problem.

Experiment/Application—Arithmetic/Logic Unit

Lastly, the *interrupt flag* has meaning in programming with respect to getting data into and out of the computer without disrupting the sequence of events that are to take place within the computer. As you might suspect, there are many special flags and other features that vary among various CPUs.

You can see the operation of the ALU (arithmetic/logic unit) of the CPU by experimenting with a self-contained IC chip ALU, the SN74181. The SN74181 provides for two 4-bit word inputs, and provides for 16 combinations—not only arithmetic operation but Boolean logic operations as well. The chip has not only inputs for the two 4-bit digital words A0–A3 and B0–B3, but a 4-bit control function, S0–S3. Outputs are obtained at the outputs F0–F3. Control line M determines whether the ALU performs arithmetic or logic operations. When M is low, arithmetic operations are performed; when M is high, logic operations are performed. The table on the following page shows the arithmetic and logic functions available for either $\overline{A}$ and $\overline{B}$ or A and B (low-level active or high-level active) inputs for arithmetic operations and $\overline{A}$ and $\overline{B}$ or A and B (negative logic or positive logic) inputs for logic operations, depending on the state of the control inputs S0–S3.

The schematic on page 5–93 shows the pin connections for the SN74181 IC. For this experiment, we will not use the special carry features available at $\overline{C_1}$ Cb_{n+4}, G and P. A flag is available at pin 14 when A = B.

Suppose you connected up the SN74181 as shown. If you set the mode selector switch to the arithmetic position, you can, by manipulating the selector switches, specify the arithmetic function to be performed as indicated in the table. By using the appropriate interpretation of the input (high or low level), you can obtain the outputs specified. Repeat this with the mode selector switch in the logic position to obtain the outputs shown in the table. You will also note that when A = B, regardless of the mode or function selected, the A = B flag LED will come on. You can test the validity of the truth tables by manipulation of the mode and function selector for various arithmetic inputs. For example, when the mode selector is in the arithmetic position, S0–S3 set at 1001, and $\overline{A}$ set at 0110 with $\overline{B}$ set at 1001, the output at F0–F3 will be 1111. Try all the combinations to see how the SN74181 behaves as the mode and function selectors are changed and the $\overline{A}/\overline{B}$ inputs are changed.

When the ALU is part of the CPU, the mode and function selectors are under the control of the control unit, which is in turn controlled by the program (programming will be described a little later in this section). Although the ALU used in these experiments is very simple and handles only 4-bit words, it is *representative* of the ALU in any digital computer. Therefore you should experiment with the input/output relationships of this ALU so that you can have a good feel for how *any ALU operates*.

TABLE OF ARITHMETIC OPERATIONS

| FUNCTION SELECT | | | | OUTPUT FUNCTION | |
S3	S2	S1	S0	LOW LEVELS ACTIVE	HIGH LEVELS ACTIVE
L	L	L	L	F = A MINUS 1	F = A
L	L	L	H	F = AB MINUS 1	F = A + B
L	L	H	L	F = A$\overline{B}$ MINUS 1	F = A + $\overline{B}$
L	L	H	H	F = MINUS 1 (2's complement)	F = MINUS 1 (2's complement)
L	H	L	L	F = A PLUS (A + $\overline{B}$)	F = A PLUS A$\overline{B}$
L	H	L	H	F = AB PLUS (A + $\overline{B}$)	F = (A + B) PLUS A$\overline{B}$
L	H	H	L	F = A MINUS B MINUS 1	F = A MINUS B MINUS 1
L	H	H	H	F = A + $\overline{B}$	F = A$\overline{B}$ MINUS 1
H	L	L	L	F = A PLUS (A + B)	F = A PLUS AB
H	L	L	H	F = A PLUS B	F = A PLUS B
H	L	H	L	F = A$\overline{B}$ PLUS (A + B)	F = (A + $\overline{B}$) PLUS AB
H	L	H	H	F = A + B	F = AB MINUS 1
H	H	L	L	F = A PLUS A†	F = A PLUS A†
H	H	L	H	F = AB PLUS A	F = (A + B) PLUS A
H	H	H	L	F = A$\overline{B}$ PLUS A	F = (A + $\overline{B}$) PLUS A
H	H	H	H	F = A	F = A MINUS 1

TABLE OF LOGIC FUNCTIONS

| FUNCTION SELECT | | | | OUTPUT FUNCTION | |
S3	S2	S1	S0	NEGATIVE LOGIC	POSITIVE LOGIC
L	L	L	L	F = $\overline{A}$	F = $\overline{A}$
L	L	L	H	F = $\overline{AB}$	F = $\overline{A + B}$
L	L	H	L	F = $\overline{A}$ + B	F = $\overline{A}$B
L	L	H	H	F = LOGIC 1	F = LOGIC 0
L	H	L	L	F = $\overline{A + B}$	F = $\overline{AB}$
L	H	L	H	F = $\overline{B}$	F = $\overline{B}$
L	H	H	L	F = $\overline{A}$ + $\overline{B}$	F = A$\overline{B}$
L	H	H	H	F = A + $\overline{B}$	F = A$\overline{B}$
H	L	L	L	F = $\overline{A}$ + B	F = $\overline{A}$ + B
H	L	L	H	F = B	F = B
H	L	H	L	F = A + B	F = AB
H	L	H	H	F = LOGIC 0	F = LOGIC 1
H	H	L	L	F = $\overline{AB}$	F = A + $\overline{B}$
H	H	L	H	F = $\overline{A}$B	F = A + B
H	H	H	L	F = AB	F = A + B
H	H	H	H	F = A	F = A

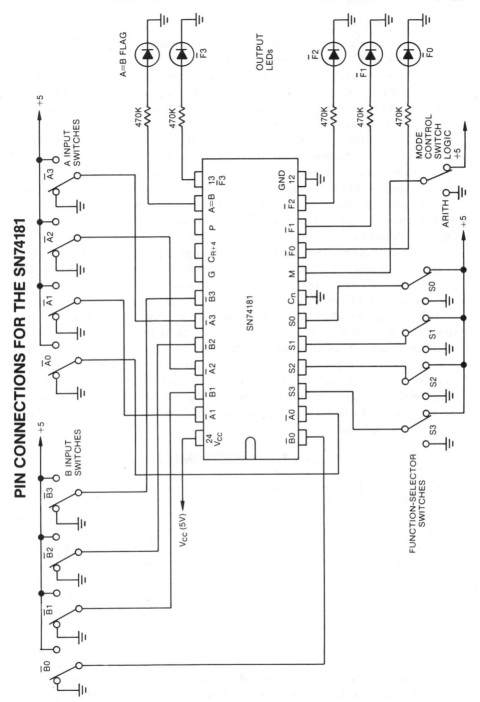

PIN CONNECTIONS FOR THE SN74181

Multiplexers

As you noticed, a *common bus* in a CPU is used to transfer data from registers to ALU. To make sure the correct interconnections are made, a *multiplexer* is used. Thus, multiplexers have important uses in the control section of a computer where they are used to conduct digital information from one place to another. Also, in a digital computer data bits are available at the outputs of keyboards, adders, registers, and counters, and for proper processing to take place these outputs must be connected to other adders, registers, counters, and output devices at the proper time. Multiplexers perform this interconnecting.

A simple way to picture a computer multiplexer is as a very fast selector switch that operates under the directions of the control unit. A multiplexer accepts a large number of inputs and by means of its very fast (electronic) switching *connects* the *right points together* at the *correct time.*

The inverse operation (demultiplexing) is performed at the output of the multiplexer. The *demultiplexer* receives its inputs from a single line that carries different types of data in sequence or serial form. First there may be a train of pulses from source 1, and then a train from source 4, and then a different pulse train from source 2, etc. By moving its electronic switching pole at the correct instant to the correct output terminal, the demultiplexer does its job of directing each component of the multiplexed train to its proper destination at the correct time. Multiplexers are particularly useful when several remote terminals are to be used with a computer. In this case, the output from the terminals can be multiplexed onto a single line, saving the considerable expense of having multiple sets of lines.

MULTIPLEXERS (MUX)

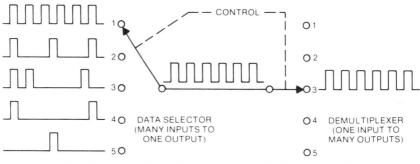

CONDUCT PULSE TRAINS FROM DIFFERENT SOURCES TO THEIR
CORRECT DESTINATIONS THROUGH A SINGLE CONDUCTOR

In its simplest form, a multiplexer can be made from one OR, four NAND, and four AND gates. When they are connected as shown below, any one of four data input lines can be connected to a single output line. The selection is made by connecting to the two data inputs a 0 or 1 in the proper combination.

Multiplexers (continued)

The diagram below shows how the basic approach can be expanded to produce an 8:1 multiplexer to select any 1 of 8 lines. This 8:1 device requires 3 binary bits to supply the input code. The 8:1 device also has an *inhibit* input to provide system flexibility. When the inhibit input is activated by applying a 1, a 1 at either output is forced to become 0.

The multiplexer can be expanded to enable any desired number of input lines to be connected to a single output line. Similarly, these methods can be inverted to distribute different pulse trains coming through the same line to as many different outputs as required.

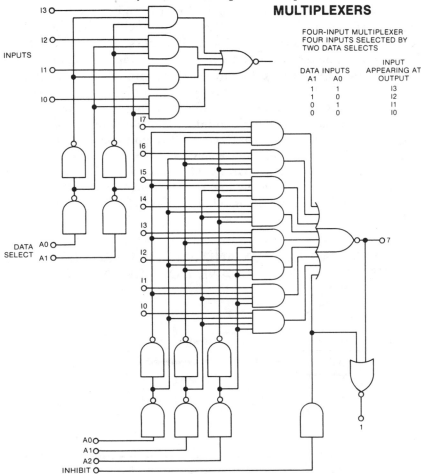

MULTIPLEXERS

FOUR-INPUT MULTIPLEXER
FOUR INPUTS SELECTED BY
TWO DATA SELECTS

DATA INPUTS		INPUT APPEARING AT OUTPUT
A1	A0	
1	1	I3
1	0	I2
0	1	I1
0	0	I0

As you can see, multiplexers are necessary in digital computers to allow the flow of data from one point to another. The common-bus structure used in computers provides for the multiplexing of the elements of the CPU as determined by the instruction as interpreted by the control unit. In this way, correct interconnections between the CPU elements occur at the proper times.

Buses and Tri-State Logic

As you know, a computer consists of a CPU (which can be in the form of a microprocessor) and various memory, input/output, and special-purpose devices. The CPU must be able to exchange information with and control the operation of these devices. Early computers used various types of multiplexer (MUX) to accomplish this. As more and more devices were added, multiplexing became too complicated.

Most modern computers are constructed around central data and address buses, which are arrays of parallel interconnecting lines. The CPU and all the devices are connected in parallel across these sets of lines. Information and control signals flow through these lines. Under the control of the CPU the various devices take out of the lines and feed back into the lines exactly what the CPU orders. Thus devices can be added, removed, or interchanged without involving changes to the system as a whole. Obviously, only one set of conditions can exist across a bus at one time. A disaster would occur if two units tried to drive a bus with one unit trying to put a 1 on a line and another trying to put a 0 on the line at the same time. Since most conventional digital devices are in either state, it is not practical to connect them together directly. *Tri-state logic* provides the answer since it has *three states* 1, 0, and *off* (disabled). Thus, when many devices are paralleled across a bus, all but one is held in the *disabled (off) state* while the one that is *enabled* drives the bus. Typically, the disabled state results in a high impedance at the output of the disabled device.

COMPARING MUX AND BUS METHODS OF DISTRIBUTING DATA AND CONTROL SIGNALS

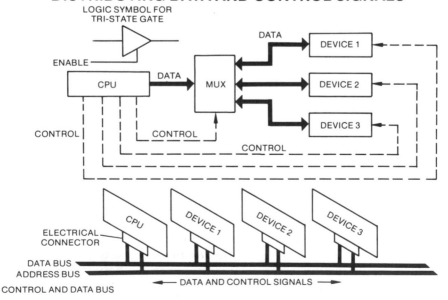

Memory

An essential element of a computer is its *electronic memory*. It is in memory that data and results are stored. It is in memory that the instructions for the operation of the CPU are stored. The memory of a computer is thus essential to its operation. In fact, the operation of a digital computer can be considered as the *manipulation of data* (bits) *in memory to produce the desired result*. As a result, computers can be *classified* on the basis of their memory word size, as shown:

System	Memory Word Size	Typical Direct Addressing Range
Microcomputer	8–16	16 bits
Minicomputer	16–32	16 bits
Main frame	32–64	16–32 bits or more

MEMORY GROUPS

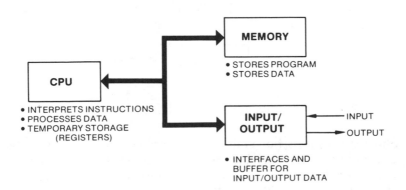

Computers use many different types of memory, either alone or in combination. Each has its advantages and disadvantages. Generally, the memory in a computer is divided into three general groups: (1) a *fixed memory*, containing fixed instructions that are used for *initializing* the system's loading programs and data, interpreting instructions from the input form to the form useful to the CPU control; (2) *working memory*, where working programs and data and other information are stored while a program is being run; and (3) *mass storage*, where large quantities of data, programs, and other information are stored. In contrast to fixed and working memory, mass memory usually has a slower access time and is generally external to the main part of the computer system. Thus memory is often defined as either internal (fixed and working memory) or external (mass memory).

Memory Types

Typically, fixed and dynamic memory are organized on an *addressable basis*. That is, each storage location can be reached by a suitable address. Thus, there are two buses associated with these memories. One establishes the location of the information (the *address bus*) and the other carries the information to and from memory (the *data bus*). Furthermore, the contents of memory (data) are changed only when new information is to be written into it. Thus, reading memory does not destroy the contents. This is done either by the nature of the memory device, which does not erase when read, or by replacing the contents with an identical word as each word is read out.

Memory is classified as either *volatile* or *nonvolatile*, depending on whether the contents are lost when power is removed. Obviously, for a fixed-program device, the program should remain. Furthermore, there are certain resident programming instructions (for loading memory, initializing the computer, etc.) that should remain regardless of the fact that the instructions to perform data manipulation may change on a short-term basis as in a general-purpose computer.

BUS TYPES

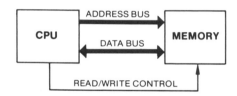

Since the address bus and the data are separate, it is possible to control numbers of memory locations that are not the same despite the size of the data word. For example, a microprocessor CPU typically has a word length of 1 byte (8 bits). However, the same CPU may have an addressing capability of 2 bytes (16 bits). Thus, there are 2^{16} (65,536) addressable locations, each containing one 8-bit (1-byte) word. For convenience, you will find memory capability described in rounded-off terms. Thus, a 65,536-word memory is referred to as 64K, a 32,768-word (2^{15}) memory is referred to as 32K, etc. Obviously, some means is necessary to determine whether memory is to be read or written into. This is done by a single-bit control line (read/write control). Below is a glossary of terms commonly used when referring to various types of memory.

Memory Glossary

Bubble memory

A recent nonvolatile random-access memory that provides for very dense packaging, thereby allowing for more memory per chip.

Memory Glossary (continued)

Core

A loosely used word, referring to a nonvolatile-type memory that stores data based on the direction of a magnetic field on a small magnetic ring.

Disk

A nonvolatile mass memory that can provide near-real-time access to multimegabytes of data. A very large disk system may store more than 100 million bytes of data and instructions.

Dynamic memory

Fresh input required; that is, the data will fade if not periodically refreshed in memory. This is usually done directly within the memory, so no external control is required.

Mass storage

Specialized storage media that are associated with computers and hold large quantities of data, programming, etc.—typically these media are tapes, disks, punched cards.

Nonvolatile memory

Data retained without power; data are changed when new data are written in.

PROM (EPROM)

Programmable read-only memory, a read-only memory device in which the contents can be changed under special conditions.

RAM

Random-access memory, a memory in which any location is directly addressable without regard to the previous adress.

ROM

Read-only memory, a memory that has a fixed bit pattern stored at each memory location—used for memory that will not be changed.

Static memory

No fresh input required; will hold data indefinitely.

Volatile memory

Data lost when power is removed.

Internal Memory-Core

The internal memory in a modern small digital computer usually consists of RAM and ROM. Larger and older computers use core memory, which has the advantage of being nonvolatile while providing random access. Core memory consists of tiny rings of magnetic oxide mixed with ceramic material for strength. When magnetized in one direction, the ring stores a 0; when magnetized in the other direction, it stores a 1. Cores are arranged in a rectangular pattern such that the number of rings in one dimension equals the word length for the computer, and as many as possible are placed in the other direction. Such a configuration is called a *plane*. Many planes can be stacked to increase core memory size.

MAGNETIC CORE MEMORY

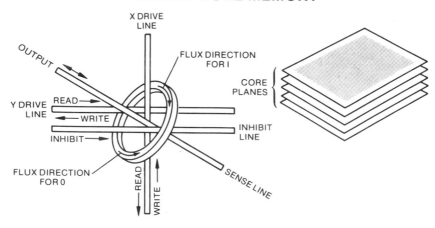

As shown, the cores are arranged in a rectangular pattern with wires going through the center of each core element to provide for read, write, and address functions. The directions of the currents in the X and Y lines determine whether the core is to be read or written into and provides the addressing capability. The sense and inhibit lines provide for output and control. Core memories require that the data be restored when they are read out, which complicates the design of core memory. However, this is done within the core plane so that externally it appears that the data have been retained. Cores are still used in larger computers where the internal memory may be greater than 1 million words. The method of addressing a core memory is similar to that for a semiconductor memory, which will be discussed next. The major disadvantage of core memory lies in its greater cost per bit coupled with greater circuit complexity and limitations of operating speed. On the other hand, modern core memories are reliable and can be made to hold large quantities of data. Their nonvolatile characteristics make them convenient to use, because the stored data are not lost when the computer is turned off.

The most popular internal memory today is based on semiconductor technology and consists of arrays of IC memory chips on circuit boards to provide random-access memory (RAM).

Internal Memory—Static/Dynamic Semiconductor RAM-CCDs

Microprocessors invariably use semiconductor internal memory. The two popular types are *static* and *dynamic RAM*. As you know, the difference lies in the fact that a static RAM holds the information without need for refreshing, whereas a dynamic RAM requires that the data be periodically refreshed. Typically, a static RAM can be considered an array of flip-flops set to 1 or 0 when written into and read by determining the state of the flip-flops. Obviously, if the power is turned off the flip-flops will have to be reset when power is restored. Recent developments have made available static RAM that can hold the bit patterns under very low power conditions. Thus, a battery or other low-power source will allow retention of bit patterns even when the main power is removed. Dynamic RAMs use a storage cell approach to memory: the state (1 or 0) is essentially stored on a small capacitor. Since the capacitor is imperfect, there is a need for refreshing the data at intervals (typically 1–2 μsec per refresh). *Charge-coupled devices (CCDs)* are becoming popular also as RAMs because of their low power requirements and small size. A CCD behaves like a continuously circulating shift register. Charges stored beneath metal plates for one clock pulse are circulated to the next plate on the next pulse and hence essentially rotate through the line representing a particular word. Their method of address is similar to that of other RAMs. RAM chips (static or dynamic) come in many many configurations, such as 128×4, 256×1, 256×4, $1,024 \times 1$, $1,024 \times 8$—$64,536 \times 1$, etc. These are usually arrayed so that a group of chips will form a memory of appropriate word size; for example, eight $16,384 \times 1$ chips will allow addressing of 16,384 locations at an 8-bit word length.

TYPICAL DYNAMIC RAM

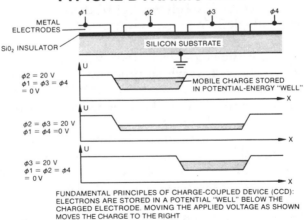

FUNDAMENTAL PRINCIPLES OF CHARGE-COUPLED DEVICE (CCD): ELECTRONS ARE STORED IN A POTENTIAL "WELL" BELOW THE CHARGED ELECTRODE. MOVING THE APPLIED VOLTAGE AS SHOWN MOVES THE CHARGE TO THE RIGHT

The block diagram above shows a typical dynamic RAM. Dynamic RAMs are popular, particularly for larger memories because, in spite of their need for refreshing they can provide more storage in a given space at a lower price and with lower power consumption than static RAMs. In smaller sizes, the static RAM is preferable because of its simplicity and the lack of the associated refresh circuitry.

Internal Memory—Semiconductor RAM

The illustration below shows the basic characteristics of a typical static RAM for a 4,096×1 bit memory. The block diagram and other data on this RAM are shown on the following page. As shown, the memory matrix is a 64 × 64 element array (64 × 64 = 4,096) that is addressed by two sets of address buses—A0–A5 for the row decoding and A6–A11 for the column decoding. Application of a high logic state to these inputs defines the row or column addressed.

MC4847

4096 x 1-bit static NMOS RAM

FEATURES
- SINGLE 5-VOLT SUPPLY AND HIGH-DENSITY STANDARD 18-PIN PACKAGE
- FAST ACCESS TIME.. (45 ns MAX FOR MC4847-2)
  (55 ns MAX FOR MC4847-3)
- DIRECTLY TTL COMPATIBLE................................... ALL INPUTS AND OUTPUTS
- COMPLETELY STATIC... NO CLOCK OR TIMING STROBE REQUIRED
- THREE-STATE OUTPUT
- AUTOMATIC POWER DOWN

For example, to access element 1,1 lines A0 and A6 are high and all others are low. To access element 32,32 line A5 and line A11 are high. In similar fashion, any element can be addressed of the 4,096 bits. The output is formed by the column I/O so that the output appears at D_{out}. By sequentially defining locations, a serial word can be formed at D_{out}.

The $\overline{CS}$ input is the chip-select input, which is high until the data from this chip are desired, at which time the $\overline{CS}$ is low to permit data readout. Input $\overline{WE}$ is the write-enable input. When $\overline{WE}$ is low, data on the D_{in} line can be written into the location specified by A0–A11. Thus, the appropriate binary code on inputs A0–A11 will select a particular location. If $\overline{CS}$ is low, the data stored will be available at D_{out}. If data are to be written in, then $\overline{WE}$ will be made low and the state on D_{in} will be stored in the location specified by A0–A11. The data input/output lines for this RAM are under tri-state logic control so that the lines can be paralleled with other memory chips. For example, a 4,096 × 8 bit RAM suitable for use with an 8-bit microprocessor can be made with an array of eight of these chips. Only 32 of these chips can be used to form a 16K × 8 memory. On the other hand 16K × 1 chips and even 64K × 1 chips are readily available, so a complete 64K × 8 RAM can now be easily made using only eight small chips.

FUNCTIONAL BLOCK DIAGRAM

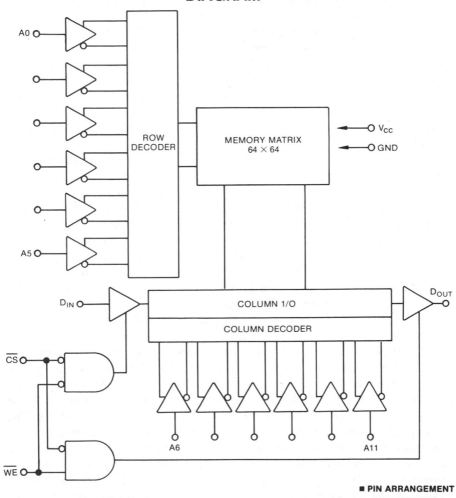

■ PIN ARRANGEMENT

A0 — 1   18 — V_CC
A1 — 2   17 — A6
A2 — 3   16 — A7
A3 — 4   15 — A8
A4 — 5   14 — A9
A5 — 6   13 — A10
D_OUT — 7   12 — A11
$\overline{WE}$ — 8   11 — D_IN
GND — 9   10 — $\overline{CS}$

(TOP VIEW)

■ ABSOLUTE MAXIMUM RATINGS

ITEM	SYMBOL	RATING	UNIT
RELATIVE TO GND VOLTAGE ON ANY PIN	V_{IN}	+0.5 TO 7.0	V
DC OUTPUT CURRENT	I_{OUT}	20	mA
OPERATING TEMPERATURE	T_{opr}	0 TO 70	°C
STORAGE TEMPERATURE	T_{stg}	−65 TO 150	°C

■ TRUTH TABLE

$\overline{CS}$	$\overline{WE}$	MODE	OUTPUT	POWER
H	X	NOT SELECTED	HIGH Z	STANDBY
L	L	WRITE	HIGH Z	ACTIVE
L	H	READ	D_{OUT}	ACTIVE

Experiment/Application—RAM

The Signetics 8225 IC chip is a 64-bit RAM capable of handling 16 words of 4 bits. While the memory capability is small, comparatively, it is very useful to illustrate how RAM works. In computers, it is often used as a *scratch-pad memory*; that is, it is used to hold temporarily some data that will be used shortly. The 8225 chip has 4 address lines (A0–A3). Lines I1–I4 are used for input control, and lines D1–D4 are data lines. Two additional inputs are R/W enable and chip enable that select the mode (read or write) and activate the chip for either mode. The following page shows the block diagram and the pin arrangement for this chip.

You can see how RAM works by assuming inputs and observing what the outputs would be.

The address lines can enable any of 16 addresses (0000 to 1111). The input data bus provides the data to be stored at the specified address, while the output lines provide output of the stored data for the given address. When power is applied, you should clear all memory. You can do this by setting all inputs to 0 (binary 0000) and then sequencing to each address (0000 to 1111). To write the data in, you must set the inputs and address switches and then momentarily put the RE switch in the write position. Alternatively, you can leave in the memory location arbitrary words that occur when power is applied, since you can change the words to the desired value in any of the memory locations to be used.

To illustrate how the RAM works, suppose you set the $\overline{CE}$ to ground (chip enabled). Then set the address code to the fifth address (0101_2) and set a 15 (binary 1111) at the input. If you now set the RE momentarily to write (0), the data will be written into RAM at location 1111. To verify this, leave the address switches as is and set RE to read (1). The data word is now stored. If you repeat the process with another data word input (for example, 0110) and operate the RE switch to the write position momentarily, then place it back to the read position, the data output will be 0110. By manipulating the address lines from 0000 to 1111 and the RE switch (make sure CE is at 0), you can store the same word in all 16 locations. If you read all addresses, you will find the same word output at all addresses. Obviously, you can store any set of 4-bit patterns in any of the memory locations and read them out as desired. You can also note that when the $\overline{CE}$ switch is set at 1, the outputs are all 1, and data cannot be entered or removed.

These examples illustrate how RAM operates. Of course, all of these functions occur at very high speeds in a digital computer (fractions of microseconds). However, the operation is *exactly the same*. Also note that if power is removed and then reapplied, the stored data bits are destroyed. Hence, RAMs are *volatile memories*.

RAM — EXPERIMENT/APPLICATION

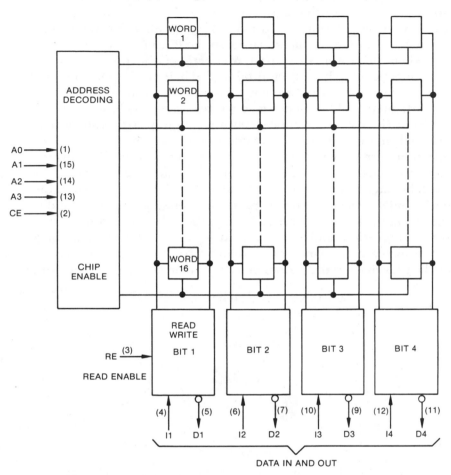

V_{CC}-16
GND-8
(X) DENOTES PIN NUMBER

TRUTH TABLE

RE	CE (CHIP ENABLE)	MODE	OUTPUTS
0	0	WRITE	"1"
1	0	READ	INFORMATION
X	1	CHIP DISABLE	"1"

X = EITHER STATE

Internal Memory—Magnetic Bubble Memory

Magnetic bubble memories have recently become available and hold promise to replace many existing static and dynamic RAMs. Not only are they capable of containing a very large amount of storage in a small space, but they are also nonvolatile, so the information in storage is not lost when the power is removed. Magnetic bubble memories with over 256K bits per chip have been produced. Their main disadvantage at the present time is that they are not particularly fast in operation.

Magnetic bubble memories operate on the principle that a localized magnetic field will form concentrations of highly magnetically polarized material in a chip made from gadolinium gallium garnet (GGG). The magnetic film grown on the substrate is such that it will allow the formation of magnetic domains (bubbles) in a direction perpendicular to the substrate. Normally, these domains are arbitrarily placed in the substrate and are free to move. A permanent biasing magnetic field causes the magnetic domains to concentrate into small compact magnetic strictures. Magnetic bubbles are created by a microscopic loop above the locale for each bit. Once the bubbles have been created, they need a method for moving them along a path, so that they can be read out or in.

MAGNETIC BUBBLE MEMORY

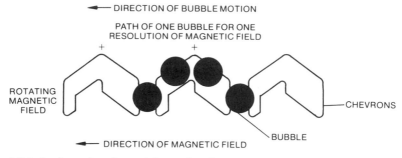

This is done by deposition of soft magnetic material in the form of chevron-shaped patterns on the chip above the magnetized field. Application of a rotating magnetic field shifts the bubbles along the chevron path just as in a rotating shift register. Magnetic bubbles created are read out as they are shifted; when a bubble is read out, it is reconstituted and passed back into the loop. Data bits are read out by a magnetoresistive detector that senses the presence of a bubble (or no bubble) for each bit of a data word. In this way, the original data are preserved. When new data are to be written in, the bubbles at the old locations are destroyed by leading them to a magnetic ground during the circulation. The five basic functions that take place in bubble memories therefore are: (1) generate, (2) replicate, (3) annihilate, (4) transfer data in; and (5) transfer data out.

The basic difference between RAM and ROM is that in ROM the bit patterns are permanent, and since the patterns placed in the device at time of manufacture *cannot be changed*, it is a *read-only memory (ROM)*.

Internal Memory—ROM

There are many solid-state memory devices for this purpose. They all consist of a row and column array of memory cells such as the simple one shown here. Each cell contains one bit of fixed data. At present, ROMs are available with over 64K memory bits per IC unit.

The exact nature of the memory cell varies with the type of solid-state technology used. However, for purposes of simplification each cell can be considered to be a combination of a diode and a switch. If the switch is closed, the circuit is complete, and an electrical signal can get through (a high or 1 state). If the switch is open, no electrical signal can get through (a low or 0 state). The switch pole is a link that is deposited by the manufacturer during the fabrication process. Thus, the 0 or 1 state of each memory element is *fixed* during manufacture according to the customer's order to do some specific job in the computer.

You can read out the state of any selected element. You *cannot change* the contents of this type of memory, and the contents are not lost if the computer power is turned off.

BASIC STRUCTURE OF A READ-ONLY MEMORY (ROM)

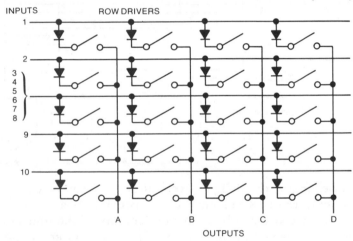

In the example shown here, there are 10 rows with four memory cells in each. To read out the contents of any row, you apply a high (1) state to that row; in other words, you *address* that row. The contents of the addressed row are seen by the outputs that appear at the 4 lines at the bottom. Thus, if the second row is addressed, the contents of the four elements appear at the 4 output lines, unaffected by the conditions in any of the other rows. If switches on row 2 to elements B and D are closed, the word 0101 would appear on the ABCD output lines.

To learn the condition of any specific element, you determine its row and column (X and Y) coordinates. Then you apply a high (1) state to that row and examine the output line for that column. If a high state appears, that memory element contains a 1. If a low state appears, that element contains a 0.

Internal Memory—ROM (continued)

The trouble with the arrangement on the previous page is that 14 connections are required to gain access to a memory containing either 10 words of 4 bits each or 40 individual memory elements. If you had an ordinary ROM containing 512 × 8 elements, it would seem that you would need 512 + 8 electrical connections to use this memory, and that would be an impractical number of wires to work with. Imagine how difficult it would be for a ROM containing 64K elements!

The answer is to *use decoders*, as described previously. You have already learned how a four-digit binary code can be used to select 1 of 10 output lines in a BCD-to-decimal decoder. Actually, a 4-bit input can be used to select any of up to 16 lines.

CONVERTERS PERMIT FEWER CONNECTIONS TO ROM ELEMENTS

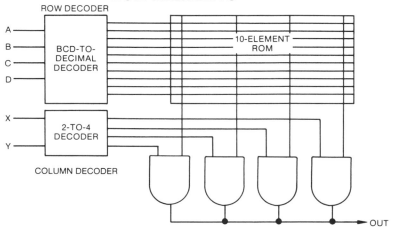

For example, the BCD-to-decimal decoder can be connected to the input lines of the 10 × 4 memory of the previous page. In the same manner, a 2-bit decoder can be used to select any of the four outputs. This decoder is connected to the output of the memory array together with AND circuits, as shown in the diagram. It is now only necessary to apply to each decoder the four-digit and two-digit binary codes that will produce a 1 at the desired row and column. Applying a 1 to the selected row produces an output of 1 or 0 at each output line, depending on the stored pattern. Applying a 1 to the selected-column AND gate energizes only the AND gate for that column. If the output is a 1, then a 1 is produced at output. If the matrix output is 0, a 0 is produced at output.

Thus, only six connections are required to address any of 40 memory elements to learn its content. By extending the techniques, it becomes practical to work with ROMs containing as many memory cells as desired.

The same methodology is used with RAM to address thousands of storage locations with a relatively small number of lines. For example, 16 lines can decode 65,536 storage locations.

Internal Memory—PROM/EPROM

Programmable read-only memory (PROM) are more economical for users who only want a limited number of ROM units with the same memory. The method of depositing the bit patterns by photographic mask during manufacture is very low in cost only if there are thousands of ROMs required with the same memory pattern. PROMs are electrically programmable with relatively simple equipment, but once programmed for a specific bit pattern are the same as ROM in their properties. One type employs a switch contact consisting of a thin layer (like a fuse) of nickel and chrome alloy, and all the switch contacts are manufactured in the closed condition, which is equivalent to a 1 in each memory element. The manufacturer, the distributor, or the user can then apply a large current through any selected contact by addressing its row and column leads and *blow* the fuse element, thereby creating a 0 in that memory element. All elements with undisturbed (unblown) contacts thus have a permanent 1 in their memories; those that have been disturbed (blown) have a permanent 0 in that location.

This type of fusible switch contact leaves unblown parts of the alloy layer in the vicinity, and sometimes these link up again and restore the undesired 1 condition. In some cases, a polycrystalline silicon material is used as the fusible link, and this makes a clean, permanent break (0) when blown.

EPROMS (erasable PROMs) do not use a permanently opened switch pole. The memory pattern or program can be removed electrically or by exposing the chip to ultraviolet light, and another bit pattern can be replaced on the chip. One type, developed by Intel, uses the *Floating-gate Avalanche injection MOS charge-storage device (FAMOS)*. This arrangement is the equivalent of a silicon-gate MOSFET, but there is no connection to the silicon gate. Applying more than -30 volts across the junction of a p-channel device of this type injects high-energy electrons from the pn junction surface avalanche region to the floating silicon gate. The presence or absence of such a charge indicates a 1 or 0 condition in the memory element. The presence of charge in the floating gate permits easy electrical conduction between source and drain, and represents the 1 condition. The charge can be sensed without destroying it for at least 10 years at operating temperatures of 125°C. The program can be removed at any time by exposing the device to strong ultraviolet light through a quartz window over the chip. The ultraviolet light neutralizes the charge under the floating gate, and the PROM can then be reprogrammed with another bit pattern.

FUSIBLE-LINK PROM   **FAMOS EPROM**

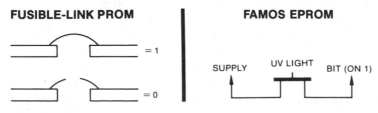

External Memory—Mass Storage

In addition to the internal memory of a computer, there may be an *external memory*. These external memories are usually for mass storage of data, programs, or other information that can be accessed with ease. Typically, the information stored in an external mass memory consists of data that are not needed at the moment but must be retained for some indefinite period without having to be reconstructed. All but the smallest computer systems make use of external mass memory. Today, the most popular mass memories are *disk* systems. As you will see, there are many disk configurations available capable of storing between 50K and 300 million or more bytes of information. Disk systems have the real advantage of providing almost immediate (millisecond) access to almost any data stored and can store data in the same time frame. Thus, in addition to providing for long-term storage of data, disk systems are used in conjunction with the internal memory to allow for storage of intermediate results, parts of programs, output data from working programs, data for input into working programs, etc. Simple computers—such as home computing systems—use a conventional cassette tape recorder for external storage of information.

EXTERNAL MEMORY

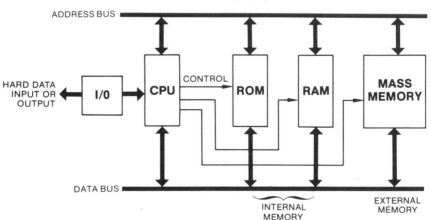

Digital tape is sometimes used as a mass memory, directly involved with the computer bus structure; however, generally because of problems with rapid access, tape is used for long-term storage of data and information that is read into internal or external memory. Punched cards and punched paper tape store information as bit patterns of punched holes and are used for mass storage in some applications.

Generally magnetic tape, punched cards, and punched paper tape are considered hard data inputs and memory or outputs from the computer and are accessed via a number of various I/O (input/output) ports. When information from these sources is needed, it is transferred into the computer via the I/O and stored in either RAM or in the mass memory. The reasons for this will become apparent as we discuss these media.

External Memory—Disk Systems

The most common forms of external mass memory in use today are disk systems. These basically consist of a disk (or several disks) coated with magnetic material rotated beneath read/write heads that pick up stored digital data or write data onto the disk. Within this basic framework, there are several classifications of disk memories varying from simple inexpensive *floppy disks* to very large *hard disks* used with either minicomputers or large main-frame computers, although their use is increasing in microcomputers.

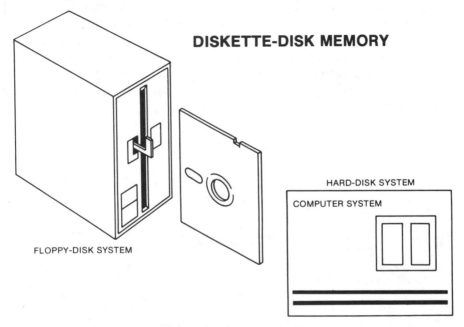

DISKETTE-DISK MEMORY

HARD-DISK SYSTEM

COMPUTER SYSTEM

FLOPPY-DISK SYSTEM

Disk systems are nonvolatile, and therefore can provide for semi-permanent storage. In addition, most have provisions for easy interchange of disks, so that different data and programming packages can be readily interchanged. The ease with which disks can be interchanged, the relatively rapid access to information, and the ease with which data can be read or written account for the popularity of disk systems.

As you will learn, most computers are programmed with either a *high-level programming language*, such as *FORTRAN* or *BASIC*, or a *mnemonic language*, usually referred to as *assembly language*. These must be translated into the binary instructions for instructing the computer CPU. These translators (or assemblers), which are themselves programs, are usually resident in disk as well as other fixed instruction data. Thus, when the computer is turned on, the computer memory is loaded with appropriate instructions and assemblers to allow for system operation from the disk. These housekeeping and translating functions—along with stored data of any type, from numerical data, customer accounts, financial records, or any other data desired—are commonly stored via a disk system.

External Memory—Disk Systems (continued)

Hard disks are usually associated with minicomputers and large main-frame computers. Typically, storage capacity is available from 2 million to as much as 300 million bytes. The disk pack may have one or several rigid disks on a single spindle, and is usually removable to allow for utilization of different disk packs. Often a specific disk (or a portion of one) is reserved for fixed programming, while most of the disks are available for other storage. The read/write/erase heads of a hard disk system do not touch the disk but are placed a very short distance (thousandths of an inch or centimeter) away from the disk. The data are laid down on the disk in sectors and in tracks. (One track may contain several sectors.) Usually the sectors are arranged to hold a specific number of bytes of information—typically, 256 bytes. The heads are moved by a precision stepping motor to read the different tracks. Typically, disks rotate at speeds of 300–600 rpm (revolutions per minute) and are usually 16–18 inches (40.64–45.72 cm) in diameter.

TYPICAL DISK SYSTEM

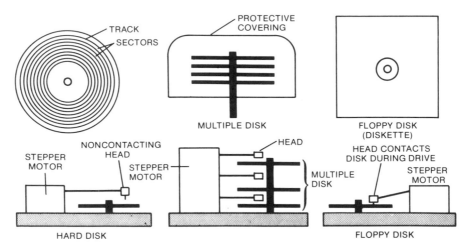

Floppy disks are very widely used with microprocessor/microcomputer applications. They use a flexible medium (vinyl or Mylar®) covered with magnetic oxide material. Typically these come in 5- or 8-inch (12.7- or 20.32-cm) *diskettes*. These are rotated, like the hard disks, except that the various heads are arranged to make contact with the disk during read/write. These systems may use either a single head with a stepper motor or multiple heads. The sectoring and track arrangement is similar to that for hard disks, except that fewer sectors and tracks are required. When more than one disk is required, separate drive units are used for each floppy disk. Floppy disks can store about 50K per side in the smallest units up to more than 500K on both sides in the more expensive units. By using several floppy disk drives, several million bytes can be stored.

External Memory—Magnetic Tape/Tape Cassettes

Both floppy disks and hard disks are controlled by a *disk controller* that handles the flow of data into or out of the disk system. One controller can often be used to control several disks. Disk systems allow for rapid access to data stored and for rapid storage of data, with access time to any data typically measured in milliseconds. They have been a primary factor in the proliferation of computers for applications requiring the storage and retrieval of large amounts of data.

One of the earliest reusable forms of computer mass memory was *magnetic tape*. Special tape transports using wide tape (0.5 inch to 2 inches, 1.27 to 5.08 cm) record data in a set of parallel tracks. Tape provides a relatively inexpensive form of storage, but suffers from the serious defect that data are accessible only by winding through the tape until the desired data are located. Even with very complicated transports, the access time to specific data may require many minutes. On the other hand, magnetic tape has the virtue of being storable in little space and is relatively permanent. Today, magnetic tape is used to store old data records and information that need be accessed only infrequently. Magnetic tapes are also used to back up disk systems. This is done because occasionally disk systems fail by having the disk itself break (crash). If this happens, the data stored on a disk can be lost. To prevent the loss of data, disks are periodically (for example, daily or weekly) copied onto tape. As you can imagine, the loss of business records or personnel data by a disk failure would be catastrophic. Thus, the backup of a disk system by tape is quite common. In some cases, the disk is copied onto another disk which is stored in a different location as a backup for the operating disk system.

MAGNETIC TAPE—TAPE CASSETTES

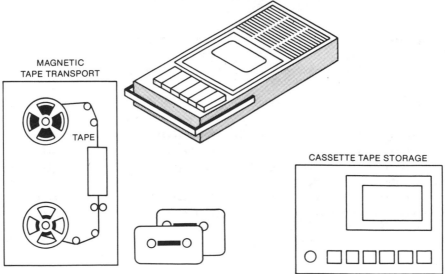

MAGNETIC
TAPE TRANSPORT

TAPE

CASSETTE TAPE STORAGE

External Storage—Punched Cards

Cassette magnetic tape storage is very popular for inexpensive home computer systems using conventional cassette tape recorder/player systems. More elaborate cassette tape transports are used with minicomputers or some more elaborate microcomputers. The cassette recording is used mainly to load programs and assemblers into computer internal memory and for data storage on a more limited basis. Inexpensive home computer systems record data by using two tones—one representing a mark (1) and one a space (0). You will learn more about this approach later when you study *MODEMs* for communication with computer systems. Cassette tape for digital data storage must be of good quality (as for all magnetic tape) to avoid dropout of data. This is usually caused by minor tape irregularities that can lift the tape from contact with the heads, causing weak or lost recording for that short portion.

The oldest form of external computer storage is the *punched card*. These are still used extensively since they provide permanence. In addition, data on punched cards can be easily sorted and tabulated by very simple sorting machines. Cards are punched by a special card-punching machine. Typically, cards contain 80 punch positions (columns) and have 12 rows at each punch position. Each card carries information in the form of punched rectangular or round holes, with each column equal to a single character or number (alphanumeric) or other symbol. The text across the top of the card is the interpretation of the punched pattern as it will be read by the computer. Numbers are represented by a single punch in any column 0–9. Alphabetic characters and other symbols are represented by punches in the number fields (0–9) and additional punches (superpunches) along the top of the card, as shown.

PUNCHED CARD

External Storage—Punched Paper Tape

Punched cards are read by sensing the sequence of punched holes for each column in a computer input device called a *card reader*. These are usually found associated with larger computer systems. You are familiar with punched cards for billing purposes. The punched card has printed material for you to understand in addition to your account and billing information stored as punched holes. When you return your bill, the payments are recorded as additional punches, and then the card is read into the computer to update your account.

Punched paper tape is an inexpensive permanent storage medium. Characters are punched in a special code, designated ASCII (American Standard Code for Information Interchange). The tape shown below illustrates the ASCII punching of the alphabet (A–Z), two spaces, and the numbers 1–9 and 0. Actually, only seven punched holes (vertically) convey information. The eighth hole is used as a *parity check*. Whenever a character in the first seven rows from the bottom has an odd number of punched holes, a hole is punched in the eighth column; if the number of punches in the first seven rows is even, no hole is punched in the eighth row. In this way, the computer checks to see that there is an even number of punches in all eight rows and rejects as incorrect those characters that do not meet the parity test. Parity checking will be discussed later in this volume. The seven positions representing data can code for 128 characters (2^7), which is sufficient to code for all of the characters on a standard typewriter-type keyboard.

PUNCHED PAPER TAPE

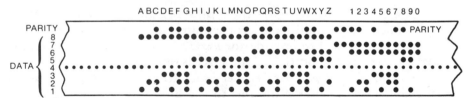

The ASCII system utilizes a 7-bit code represented by the punches on the paper tape. As you can see from the punched tape, an A is represented in the ASCII system by a binary 1000001 ($= 65_{10}$). The complete ASCII system is discussed in the next section. Paper tape is punched and read in a paper-tape reader and/or punch, driven by code data from the keyboard, computer, or other device. Punched paper tape is usually quite slow in operation and is therefore used to store limited quantities of information. In more modern, inexpensive systems, paper tape is replaced by magnetic cassette tape—which is more compact, is reusable, and can transfer data much more quickly.

Memory Organization

A computer, as you know, consists of the essential elements—CPU, memory, input/output (I/O). Also, as you have heard several times before, a digital computer can be described in terms of *bit manipulation* and *memory management*. Memory is used to *store data, results, intermediate calculations,* etc. In addition, memory holds the *instructions* necessary to have the computer accomplish the tasks required. This set of instructions is called the *program*. Actually, *many programs* may be involved in the programming process, as you shall see.

TYPICAL ORGANIZATION OF A 64K RAM

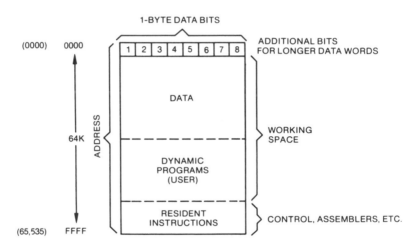

The diagram above shows the typical organization of a 64K RAM such as one would find in a microcomputer. The space is divided (not necessarily as shown) to provide for *working space*, where user programs and data are stored. A portion of the space is reserved for so-called *resident programs* and *instructions* that are necessary to operate I/O, control the CPU, and translate user programs to the binary codes for use by the CPU control. In many cases, some of the resident instructions that are basic to getting the system in operation (for example, the loading routine and initialization that gets the computer in operation) are stored in ROM since they do not change. Many modern computers may have several programming languages or operating modes available. Most of these instructions are stored on disk or other media and are transferred into the resident memory area when a program is to be translated to binary code form.

Since the computer memory generally has the capability for random access, the precise locations may vary. Actually, except for a few special cases, the computer user or programmer has only limited interest in the actual location of various items in memory—the computer worries about that! However, it must be directly specified by the user in some cases when programming at levels close to or at the machine level.

Memory Organization—Direct Memory Access (DMA)

To speed the transfer of data between memory and I/O devices, *direct memory access (DMA)* techniques are often used, particularly in microprocessor-based computers. Data stored in memory or external data to be transferred into memory can be moved independently of the CPU. In these systems, tri-state logic interfaces all elements of the computer tied to the address and data buses.

FUNDAMENTALS OF DIRECT MEMORY ACCESS (DMA)

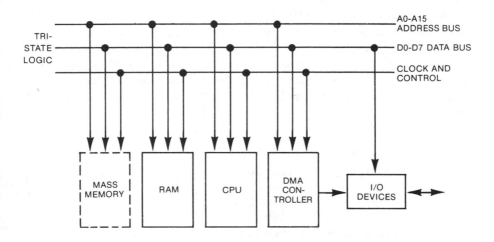

In operation, when there is to be a direct transfer of data in or out, the DMA controller is given control of the system control and data and address buses. The DMA controller, using stored address information, starts the readout or input of data stored in the RAM locations specified. Similar operations can take place with mass memory as well. The data transfer can take place in a block—block data transfer—during times when the CPU is not active, or when CPU operations are deliberately suspended to permit data transfer. Other modes involve *cycle stealing* (short CPU interrupts to allow transfer of single bytes or words of data and memory sharing), that is, transfer of data during periods when the CPU is busy internally and is not using the bus structure.

DMA is quite common in modern computing systems of all sizes because it allows for the efficient dumping of information in and out of memory with minimal involvement of the CPU. In this way input/output, which generally results in slowing of the computer, can be accomplished with minimal computer slowdown. In addition, DMA allows for much more rapid data transfer, which is very useful when very large amounts of data are to be transferred.

Programming—Introduction

Fundamentally, a computer takes digital words (bits from memory) loads these into the CPU, operates on the bit patterns in accordance with some instruction, and returns the modified bit patterns to memory. The *instructions*, which are themselves *binary bit patterns stored in memory*, determine what is to be done to the contents of memory. These instructions are called the *program*. The program, then, is a series of instructions organized in some logical sequence to produce the desired results in memory. As you know, programs are often referred to as *software*.

There are at least *two instructions per operation* in a computer: the *location* to be *addressed to find* the *information*, and the *operation* to be *performed on* the *information*. For example, if it is desired to load a data word or number into the accumulator, the instructions consist of an instruction to the controller in the CPU to load data into the accumulator of the ALU of the CPU and an instruction giving the address at which the information to be loaded is located (1). If the next operation is to add the loaded information to some number in memory, another instruction consisting of two pieces of information is required—the instruction to add the information and the location of the number where it should be added (2).

TO ADD TWO NUMBERS FROM MEMORY :

1 LOAD FIRST NUMBER IN ACCUMULATOR FROM MEMORY LOCATION A

2 ADD TO THIS A NUMBER STORED IN MEMORY LOCATION B

3 STORE THE RESULT IN MEMORY LOCATION C

. . . ALL COMPUTERS USE SEQUENCES OF THIS SORT

Finally, if the result is to be stored at a third location (3), this again requires two pieces of information—the instruction to store the information and the location at which it is to be stored. Note that the *program* consists of a *sequence* of *logical steps* to accomplish the result desired. The program itself is in memory and must be addressed. This is done by a register in the CPU, called *the program counter*, which starts at the first line of the program when the program is to be run and then *automatically sequences (points)* to the subsequent steps as each step is executed. If the result of a logical operation requires that control be transferred to another part of the program for a particular operation, the value of the program counter is stored temporarily so that when control returns to the main sequence in the program, the CPU will know where to continue from.

Programming—Introduction (continued)

The instruction to the control unit regarding the function to be implemented (executed) is called the *OP code*. The data to be operated on is called the *operand*. These are in binary form as stored in memory. For example, suppose you had an 8-bit microcomputer with 256 words of RAM (memory). All computers have an *instruction set* that describes the binary OP codes and the format of the data to be stored in memory. For example, the MC6800 CPU (which will be described in more detail later) has the following OP codes for performing the simple addition described on the previous page:

OP CODES FOR THE MC6800 CPU

OP CODE	FUNCTION
1001 0110	LOAD ACCUMULATOR A OF CPU WITH CONTENTS OF THE MEMORY LOCATION SPECIFIED BY THE NUMBER IN THE NEXT MEMORY LOCATION AFTER THE LOAD INSTRUCTIONS.
1001 1011	ADD TO THE NUMBER STORED IN ACCUMULATOR A THE NUMBER TO BE FOUND IN THE MEMORY LOCATION SPECIFIED BY NUMBER IN THE NEXT MEMORY LOCATION AFTER THE ADD INSTRUCTION.
1001 0111	STORE THE RESULT (IN ACCUMULATOR A) IN THE MEMORY LOCATION SPECIFIED BY THE NUMBER IN THE NEXT MEMORY LOCATION AFTER THE STORE INSTRUCTION.

As you can see, each OP code in memory (in this example) is followed by a word in memory that specifies the place where the information is to be found or delivered.

The program counter of the CPU is initialized at the location of the first instruction (OP code). This is transferred to the CPU, which then proceeds to carry out the instruction using the memory location information in the next one or more words, depending on the memory size. The program counter proceeds through the program, loading each instruction in turn into the CPU along with the memory location data (the data are taken from the memory location specified). The CPU executes the instruction, transferring results to memory location specified. You can now see the basis for the earlier statement that digital computers operate by memory management.

Suppose you wanted to add the numbers 83_{10} (01010011) and 138_{10} (10001010), which are stored in memory locations 10,000,000 (128) and 10,000,001 (129), with the result to be stored in location 10,000,010 (130). The program is executed by having the program counter in the CPU point to each successive memory location of the program, loading these into the CPU, and having the CPU react accordingly. For this example, the contents of memory and the location of the program counter before and after execution are shown in the figure on the top of page 5–120.

Programming—Introduction (continued)

MEMORY MANAGEMENT
BY CPU

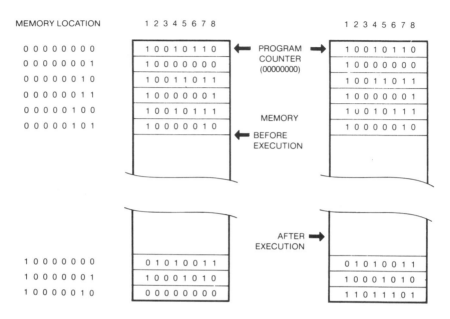

As you can see, the program counter (PC) points initially to the OP code in location 00000000 of memory. This OP code instructs the CPU to load the number in memory location 10000000 (128) into the accumulator. At this point, the PC is at location 00000010, and the CPU executes the instruction 1001 1011—*ADD* the contents of the memory location specified in program memory location 00000011. This proceeds for the final instruction to *STORE*. The result after execution is that the answer 221_{10} (1101 1101) is now found in memory location 10000010.

This is obviously a very cumbersome procedure even for simple programs, and, in general, programming is *not* done at the *binary machine level* but at a *higher* level—more easily understood language—that is converted into the binary code for use by the computer using a translator called a *compiler* to convert the higher-level language to the binary code that the computer understands.

Assembly language uses easily remembered and understood mnemonics to do the programming. (*Mnemonics* are word devices to aid in your memorization of a code, etc. and are made up from the initials of the successive steps for a procedure.) These mnemonic codes are then interpreted in a *compiler* that *translates* the *mnemonic codes* into the *binary codes* for use by the computer. In addition, compilers often do a good deal to help in assigning memory locations.

Programming—Assembly/High-Level Languages

To illustrate how the simple program on the previous page can be coded in assembly language, reference is made to the mnemonic instruction set for the MC6800 (detailed later in this volume). In this instruction set, the instruction to load accumulator A with the contents of memory location 1000 0000 (128) is simply written as shown below:

```
LDA, 128    LOAD ACCUMULATOR A WITH CONTENTS OF MEMORY LOCATE 128₁₀
ADDA, 129   ADD CONTENTS OF MEMORY LOCATION 129₁₀ TO CONTENTS OF ACCUMULATOR A
STA, 130    STORE THE RESULT IN ACCUMULATOR A IN MEMORY LOCATION 130
```

As you see, this is a very easy way to generate programs and gets you away from complicated binary bit patterns. Furthermore, it is much easier to *troubleshoot* this sort of program because it is easy to understand. In microcomputers, as in larger machines, the assembler compiler that converts the assembly language to the binary codes is resident in either ROM or is loaded into RAM from a disk or externally.

Even assembly language programming is cumbersome for the non-professional programmer. In many cases, scientists, engineers, or others using computers wish to do their own programming. Higher-level languages such as *FORTRAN* and *BASIC* were developed to allow for instructing or programming computers to do a specific job with instructions that closely resemble the constructing as done by paper and pencil. For example, our simple earlier example is easily programmed in FORTRAN or BASIC as shown below:

```
FORTRAN            BASIC
READ (3,6) A,B     10 READ A,B
Y = A + B          20 LET Y = A + B
WRITE (5,7) Y      30 PRINT Y
FORMAT (2I3)       40 DATA A,B
FORMAT (I3)        50 END
END
```

In these cases, a FORTRAN or BASIC compiler (or interpreter) converts the almost plain language instructions into the binary instruction set. As you can see, the process of writing high-level language programs can be quite easy because these programs are familiar looking. One major advantage of high-level language programming is that all of the tedious effort required to designate storage locations is also taken care of by the compiler. The *development* of *compilers* to allow programming in high-level languages has been a *major factor* in the *extended use* of computers. FORTRAN is used widely in general scientific work. BASIC is very popular with home microcomputer enthusiasts.

Today there are many specialized high-level languages such as *PASCAL* and *COBOL* that have features making them valuable for specific applications. The proliferation of programming languages and the various dialects available in many languages has led to a great deal of controversy. Most of this is pointless—you use the language that gets the job done most economically and efficiently.

Programming—Flow Charts

Flow charts are aids to programming. They allow the programmer to see the logical flow of calculations and decisions necessary to accomplish the objectives of the program. In this way, the programmer can see clearly—step-by-step—what needs to be done. One of the most important points in the flow chart is that it highlights decision points. While there is some standardization, it really doesn't matter what symbology is used, so long as it is consistent. For example, consider the flow charts written below for getting up and ready for work, and for multiplying two binary 8-bit numbers together (which, as you know, is only a series of shifts and additions).

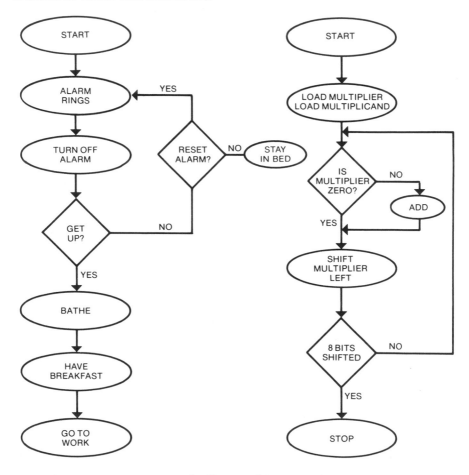

As you can see, practically anything can be charted. The major symbols are those that involve steps in the flow chart necessary to accomplish an objective (ovals or squares, usually) and decision-making points (diamonds). If you follow the flow charts above, you will see that they simply show the necessary steps and decision-making points required to achieve a result.

Review of Computers and Microprocessors—Background Information

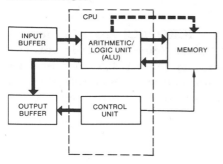

1. THE ARITHMETIC/LOGIC UNIT (ALU), The control unit (CU), memory, and input/output buffers are the major elements of a digital computer. The central processor unit (CPU) comprises the CU and ALU.

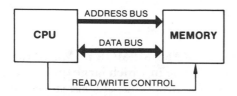

ALU

- **BOOLEAN LOGIC**
- **BOOLEAN ADDITION**
- **COMPLEMENTATION**
- **SHIFT RIGHT/LEFT**

2. The CU determines sequence of operations for the ALU, which manipulates data to produce desired results of computations and decisions.

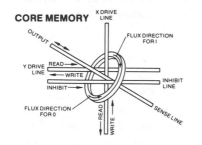

3. MEMORY can be divided into two main types: internal and external. Internal memory is more integrated with the CPU and more rapidly accessible.

CORE MEMORY

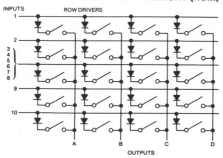

4. MEMORIES are also classified as volatile or nonvolatile. Volatile memories lose their data when power is removed; nonvolatile memories do not. Core memories are nonvolatile.

STRUCTURE OF A READ-ONLY MEMORY (ROM)

5. A RANDOM ACCESS MEMORY (RAM) is one that is directly addressable regardless of the previous address. A read-only memory (ROM) is one whose contents don't change and can be used only for readout.

Review of Computers and Microprocessors—Background Information (continued)

MC4847

4096 x 1-bit static NMOS RAM

FEATURES
- SINGLE 5-VOLT SUPPLY AND HIGH-DENSITY STANDARD 18-PIN PACKAGE
- FAST ACCESS TIME
- DIRECTLY TTL COMPATIBLE
- COMPLETELY STATIC
- THREE-STATE OUTPUT
- AUTOMATIC POWER DOWN

6. MICROPROCESSORS invariably use semiconductor internal memory for RAM. It may be static (information needs no refreshing) or dynamic (needs refreshing).

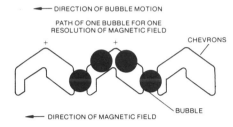

7. MAGNETIC BUBBLE MEMORIES have the advantages of space-saving and nonvolatility. They are, however, not as fast as some other types.

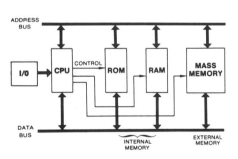

8. A MASS MEMORY (Mass Storage) is one in which large quantities of data, programs, etc. are stored. It is usually external to the main part of the computer system.

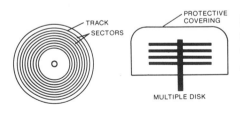

9. DISK SYSTEMS are today the most popular mass storage type. Advantages are: almost immediate access and same-time frame data storage.

```
>100 PRINT "HERE IS AN
     EXAMPLE"
>110 PRINT "OF AN APPLE"
>120 PRINT "INTEGER BASIC"
>130 PRINT "PROGRAM."
>130 END
>
```

10. A PROGRAM is a set of instructions that tell the computer how to take words out of memory, operate on them, and return them to memory.

Self-Test—Review Questions

1. Define the major elements of a digital computer and briefly describe their function. How are these elements tied together?
2. What functional elements are contained in a typical CPU?
3. What operations are typically carried out in the ALU?
4. What operations are typically carried out in the control unit? What specifically are the roles of the instruction register, the program counter, and the status flag?
5. How can a single-byte (8-bit) microprocessor handle multiple-byte words?
6. Describe the use of multiplexers and tri-state logic in computer systems. Why are they necessary?
7. How is memory organized in digital computers? How does it interface with the rest of the system? What are the basic differences between internal and external memory?
8. Define or describe volatile and nonvolatile memory, static and dynamic memory, core, RAM, ROM, PROM, and disk. Describe the operation of a static RAM memory.
9. Describe some typical external memory devices. How are they generally used?
10. What is a program? What is the minimum number of instructions per operation in a program. Define OP code and operand. How do the program counter and instruction set interact to execute a program?

Learning Objectives—Next Section

Overview—Now that you have learned about computers and microprocessors and how they are organized and operate, you are ready to learn how these systems interface with the outside world to receive information or to send it out.

COMPUTER I/O DEVICES

PRINTER

MODEM
TELEPHONE

CRT/KEYBOARD

Input/Output Devices

Obviously, it is necessary to communicate with any digital computer. This is necessary to give the computer information, programs, etc., and to receive output information. This is done via the I/O devices that *interface* between the computer CPU and memory and the outside world. Typically, input devices are keyboards for immediate input of instructions or data, disk data and punched cards, and magnetic or punched paper tape. Output devices usually consist of printers (to produce *hard copy*), CRT terminals that produce an alphanumeric display (which is temporary), and, in some cases, plotters. For some applications, outputs are produced on punched cards, magnetic tape, or punched paper tape.

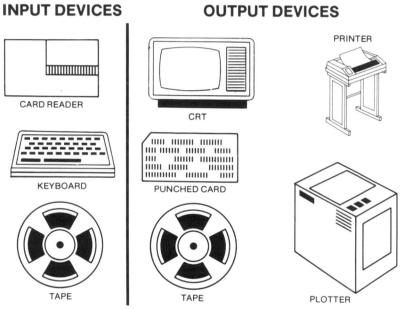

INPUT DEVICES — CARD READER, KEYBOARD, TAPE
OUTPUT DEVICES — CRT, PUNCHED CARD, TAPE, PRINTER, PLOTTER

In our discussion here, we will confine our attention to keyboard inputs and CRTs and printers (often called *line printers* if printing is done a line at a time) as output devices because these are quite common. In most cases, a keyboard and CRT are combined into a single communication unit. Except for systems where the I/O devices are directly connected to the computer, communication is via devices called *MODEMs* (*mo*dulator—*dem*odulator). These MODEMs allow for transmission and reception of information without difficulty over longer lines, telephone lines, etc. As you will see on the next few pages, standards for computer interface have been developed, and almost all data interface systems operate on the same principles. To save wiring, most interfaces are serial in format—data input or output takes place 1 bit at a time. However, data can also be transferred into and out of memory in parallel form. When a key is depressed on a keyboard, the output will usually be in parallel form. Typically, I/O interfaces have a parallel-to-serial conversion feature.

Data Rates—The Baud Rate

The rate of data transfer (in or out) is determined by the bandwidth of the system, as described in Volume 3 of this series. Of course, another restriction is the rate at which a device can generate or accept data. For a serial system, the data rate is defined in terms of the *baud* rate. The baud rate is the number of bits transmitted per second. Thus, a system that operates at a baud rate of 100 can transmit 100 bits per second (bps). Some common standard serial interface baud rates are shown below.

THE BAUD RATE IS THE NUMBER OF BITS PER SECOND TRANSMITTED

BAUD RATE (bps)	TYPE	TELEPHONE COMPATIBLE
110	ASYNCHRONOUS	TWO WAY
300	ASYNCHRONOUS	TWO WAY
600	ASYNCHRONOUS	ONE WAY
1200	ASYNCHRONOUS	ONE WAY
2400	SYNCHRONOUS/ASYNCHRONOUS	ONE WAY
4800	SYNCHRONOUS/ASYNCHRONOUS	SPECIAL
9600	SYNCHRONOUS/ASYNCHRONOUS	NOT COMPATIBLE

As you can see, the choice of baud rate determines data transfer speed, but the baud rate may be limited by other system elements such as telephone lines. You will notice that only the slower modes of operation permit two-way operation for telephone line use. That means that data can be transmitted and received on the same line *simultaneously*. Such operation is called *full-duplex*. Systems where operation is essentially a one-way mode are called *half-duplex*.

Synchronous and *asynchronous operation* refer to the *timing relationship* between the transmitting device and the computer. *Asynchronous* operation means that the timing of the data stream is *independent* of the system timing. In these systems, synchronizing pulses are provided at the beginning and end of the data word. In *synchronous* systems, the data bits are *precisely synchronized* to a *system clock*. Synchronous systems are more complex, but as you can see are also capable of high-speed operation. Our discussion will mainly cover asynchronous transmission since this is most common, particularly for microcomputers.

Most digital data transmission and reception systems use a standard character set. This standard character set is the *ASCII (American Standard Code for Information Interchange)* system. The system is described in detail on page 5–128.

ASCII Character Set

The basic characters are developed as a 7-bit code (2^7 characters). An eighth bit is used as a parity check bit. In the illustration shown below, even parity is used (the eighth bit is added as a 1 when sum of the first 7 bits is odd). The code configurations are shown on the following page. The ASCII system is rather complete and includes many special symbols (see the table of definition at the bottom of the following page) that are not generally used for computer applications. Those that are usually found are marked with an asterisk.

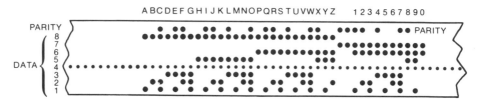

The illustration above shows an eight-level paper tape punched with ASCII characters, with the data in parallel form. In accordance with the tradition set by the early code-transmission systems, the quiescent state is in the mark (1) condition. This was done so that any break in the line could be observed immediately. Thus, the line is in the mark (1) state quiescently. Many of the ASCII characters start with a mark, however, and this could lead to difficulty. Therefore a complete ASCII character transmission consists of 11 bits. The first bit is a space (0) that is used to start the reception of data. The next 8 bits (2–9) are the ASCII character (marks, spaces, and parity check) and the final 2 bits (10–11) are two marks (1s) that always terminate a character.

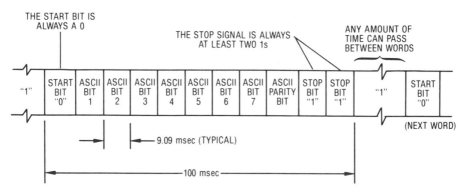

As you can see from the figure, this effectively eliminates any ambiguity in the reception of the ASCII system. Since 11 bits are needed per character, a 110-baud system can send (or receive) 10 characters per second.

AMERICAN STANDARD CODE FOR INFORMATION INTERCHANGE (ASCII)

BIT POSITION

7 6 5 0 0 0	7 6 5 0 0 1	7 6 5 0 1 0	7 6 5 0 1 1	7 6 5 1 0 0	7 6 5 1 0 1	7 6 5 1 1 0	7 6 5 1 1 1	4 3 2 1
*NUL	DLE	*SP	0	@	P		p	0 0 0 0
SOH	DC1	!	1	A	Q	a	q	0 0 0 1
STX	DC2	"	2	B	R	b	r	0 0 1 0
ETX	DC3	#	3	C	S	c	s	0 0 1 1
EOT	DC4	$	4	D	T	d	t	0 1 0 0
ENQ	NAK	%	5	E	U	e	u	0 1 0 1
ACK	SYN	&	6	F	V	f	v	0 1 1 0
BEL	ETB	'	7	G	W	g	w	0 1 1 1
*BS	CAN	(	8	H	X	h	x	1 0 0 0
HT	EM	)	9	I	Y	i	y	1 0 0 1
*LF	SUB	*	:	J	Z	j	z	1 0 1 0
VT	*ESC	+	;	K	[	k	{	1 0 1 1
*FF	FS	,	<	L	\	l	:	1 1 0 1
*CR	GS	—	=	M	]	m	}	1 1 0 1
SO	RS	.	>	N	^	n	~	1 1 1 0
SI	US	/	?	O	_	o	*DEL	1 1 1 1

*NUL	NULL, OR ALL 0s		DC1	DEVICE CONTROL 1
SOH	START OF HEADING		DC2	DEVICE CONTROL 2
STX	START OF TEXT		DC3	DEVICE CONTROL 3
ETX	END OF TEXT		DC4	DEVICE CONTROL 4
EOT	END OF TRANSMISSION		NAK	NEGATIVE ACKNOWLEDGE
ENQ	ENQUIRY		SYN	SYNCHRONOUS IDLE
ACK	ACKNOWLEDGE		ETB	END OF TRANSMISSION BLOCK
BEL	BELL, OR ALARM		CAN	CANCEL
*BS	BACKSPACE		EM	END OF MEDIUM
HT	HORIZONTAL TABULATION		SUB	SUBSTITUTE
*LF	LINE FEED		*ESC	ESCAPE
VT	VERTICAL TABULATION		FS	FILE SEPARATOR
*FF	FORM FEED		GS	GROUP SEPARATOR
*CR	CARRIAGE RETURN		RS	RECORD SEPARATOR
SO	SHIFT OUT		US	UNIT SEPARATOR
SI	SHIFT IN		*SP	SPACE
DLE	DATA LINK ESCAPE		*DEL	DELETE

*MOST COMMONLY USED SPECIAL CHARACTERS

MODEMs/Acoustical Couplers

Data transmission and reception systems, particularly those using telephone lines, cannot easily operate on a simple mark (on) and space (off) because a long time between transmissions would be essentially a dc signal, which will not pass through the system. Furthermore, it would not be possible to operate in full-duplex (simultaneous transmission and reception). To solve these problems, tone signals (bursts of audio) are used. These are carefully selected to avoid interference, but low enough to allow for easy passage over a telephone line. The devices that generate these tones in response to an input signal and produce an output when tones are received are called *MODEMs* (*Mo*dulator-*Dem*odulator).

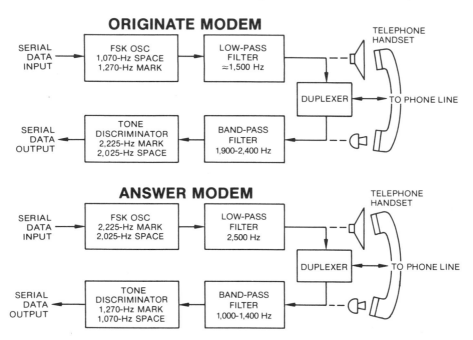

The illustration shows the configuration of the originate MODEM (at the keyboard) and the answer MODEM (at the computer). As you can see from the block diagram, two tones provide mark-and-space transmission information, and two tones provide mark-and-space reception information. The filters isolate the tone pairs in the transmit and receive channels. The FSK (frequency-shift keyed) oscillator generates approximate tones depending on whether the input signal is a mark or a space. The tone discriminator demodulates the received signal into the two levels (mark and space) in the receive channels. For acoustical coupling, the duplexer is not used, but the transmitted signal is sent to a small loudspeaker and the receive terminal is fed by a microphone. When a telephone handset is placed adjacent to these (loudspeaker to telephone microphone and microphone to telephone headphone), the signals are coupled into a conventional telephone, which makes it possible to use any telephone to communicate with a computer in a remote location.

The Teletype®

One of the earliest I/O devices for computers was the Teletype®. The most common Teletype is the Model ASR-33. (A newer unit, the ASR-43, is shown below and consists of a keyboard and matrix line printer, to be described.) The Teletype is a slow electromechanical device that operates at 110 baud. While the ASR-33 is an old device, and is slow, it has the advantage of combining a keyboard, printer, paper tape punch, and paper tape reader in one unit. Timing is accomplished by a constant-speed motor making a single revolution in 100 milliseconds, coupled to a commutator (11-segment) via a single-turn clutch. When the start bit (transition from 1 to 0) occurs, the clutch engages the motor to the commutator for a single turn. The incoming serial word is distributed to a bank of scanning solenoids that set up the ASCII code in parallel form, and these activate a bank of push rods that drive a typehead so that the appropriate character is set to face the printing paper. When the typehead is properly positioned, a hammer drives the typehead against the paper to print the appropriate character. Line feed, carriage return, etc. operate the appropriate functions in the typehead carriage.

TYPICAL TELETYPE

The operation is reversed when a keyboard key is struck. In this case, a parallel word is formed that energizes the appropriate segment of the commutator; the single turn of the motor then converts the parallel word to serial form by scanning the settings of the electromagnets while adding appropriate start and stop bits.

The push rods that set the typehead also set a group of eight (including parity) hole punches for the paper tape punch. When these are set, a hammer operates the punch, putting the appropriate hole pattern in the tape. A sprocket advances the tape between punches. When a tape is to be read, a set of fingers senses the bit pattern (through the tape holes). This parallel word is then read and converted to serial form for transmission, using the motor/commutator described above. This operation can also set the typehead bars to print the character locally. Generally, a string of characters are input from a Teletype (and a keyboard) and are stored at the computer in an input buffer; the information is fed to the computer memory only upon receipt of a carriage-return signal, indicating the end of a line of input data.

Keyboards

Most computer systems now use an electronic keyboard as one of the inputs. These keyboards strongly resemble conventional typewriter keyboards, except that they often have additional special characters peculiar to computer use. When a key is pushed on the keyboard, a corresponding ASCII character is fed out, usually in serial form. The basic keyboard usually is in the form of a matrix, as shown below. When a key is pushed, a pair of wires (one in a column, and one in a row) is energized. A unique wire pair is energized for each key pressed. These lines are fed to a decoding device that converts the unique wire pair signals into an ASCII character. In modern equipment, this is done in an LSI chip, but older keyboards use a diode matrix for decoding.

COMPUTER WITH KEYBOARD

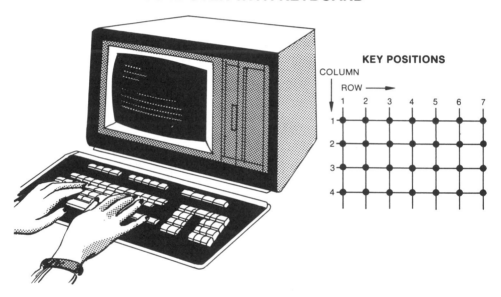

Typically, the output from the matrix-to-ASCII conversion is in parallel form. If the data are to be transferred out, they generally must be converted to serial form and the proper timing relationship set up so that the proper transmission speed and timing are maintained. Usually, this is done by a device called a *UART* (*U*niversal *A*synchronous *R*eceiver *T*ransmitter), covered on page 5-133. These devices (an IC chip) provide all of the functions necessary to convert parallel data to serial form (on transmit or send) or from serial form to parallel (on receive).

In addition to the basic ASCII data, many keyboards also provide a flag or flags (single bits) to indicate that a key has been depressed (or released), and a lockout device prevents other keys from functioning until the key is released. As with other manual input devices, the words output from the keyboard are stored in a buffer in the system until the carriage return or other suitable end-of-line signal is transmitted.

The UART

The UART (*Universal Asynchronous Receiver Transmitter*) is an IC that is used with electronic keyboards and display devices to provide for parallel-to-serial conversion for ASCII character transmission, for serial-to-parallel conversion in reception, and for timing of data.

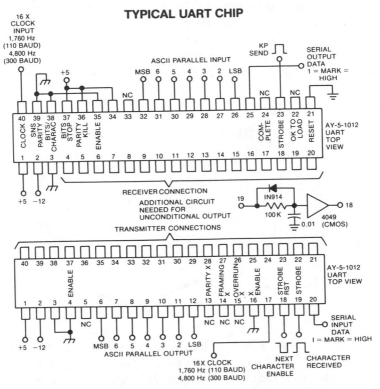

TYPICAL UART CHIP

As you can see from the illustration of a typical UART chip, a clock input is needed to properly time the data rate. The clock frequency depends on the baud rate. By proper setting of a high or low input on pins 34–39, the word length, parity bits, and stop bits can be hard-wired to provide the appropriate message length and format.

Besides providing for the basic character conversion and message formats, the UART has additional I/O ports that are used to control the reading out or writing in of data. These take the form of strobe, enable, and inhibit signals that provide for data readout only after loading is complete; character readout only after the entire character has been received; and holding when the UART is loaded but data have not been accepted into the computer (or, conversely, when the character has not yet been sent). These are necessary to control the orderly flow of data. For example, if the computer CPU is busy with computation when a character is received, the input of the character must wait until the computational cycle is complete before it can be interrupted to accept new inputs. Most UARTs are tri-state devices so that they may be directly connected into the data bus structure of the computer.

Printers

Printers are output devices that provide *hard copy*. Low-cost, low-speed printers that provide printing rates between 10 and 120 characters per second usually print character-by-character (like a typewriter) while high-speed printers print line-by-line (an entire line at a time). In this case, high-speed data transmission fills a *buffer* (short-term store) that will hold an entire line. *Line printers* are generally used with large high-speed computer systems where *throughput* (the input/output speed) is important (for example, billings for customers). Many minicomputers and essentially all microcomputers use the character-by-character printer.

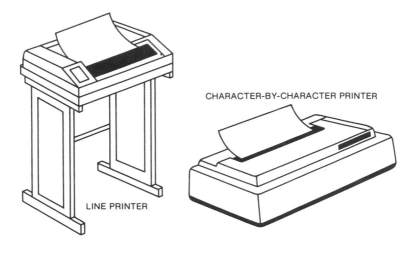

CHARACTER-BY-CHARACTER PRINTER

LINE PRINTER

Some printers use standard hard-face type with a ribbon. Many use special papers and print characters by chemical changes in the paper that are electrically induced. Most modern printers use a dot-matrix printing head. This consists of a 5×7 (or 7×9) array of wires driven by a series of electromagnets. These are driven by an interpreter that pushes the appropriate set of wires forward to form a character when driven against the ribbon/paper.

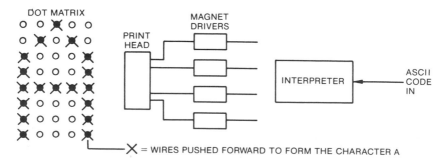

As you can see from the above, when the right wires are pushed forward, they will print any desired alphanumeric character. It may seem surprising, but printers using this principle can print in a minute more than 100 lines of 80 characters.

CRT Terminals

The most common and inexpensive data display in use today is the CRT (cathode-ray tube) terminal. This is essentially identical to the CRT in any TV receiver system, except that the bandwidth of the video circuits is greater (typically, 10–15 MHz) to give better definition. Characters are formed by the same 5 × 7 (or sometimes 7 × 9) dot matrix, except that spots are brightened to form the character. The CRT is scanned from left to right, sequentially, just like the TV receivers you learned about in Volume 4. For a 5 × 7 matrix, 7 TV lines are used to produce 1 line of characters. Most CRT terminals also include a keyboard for data input to the computer as an integral part of the unit. The dot matrix is generated by the same type of IC decoder as is used to produce printed dot-matrix characters.

CRT DOT MATRIX

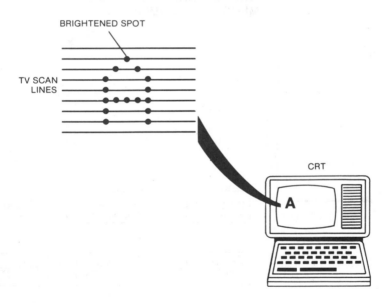

CRTs are very flexible display devices and can produce not only upper-case characters, but also lower-case as well as special characters not usually found on typewriter keyboards. Typically, displays show about 20–30 lines of information, with 80 characters to a line. However, for special purposes greater numbers of characters and/or lines can be displayed. For example, the displays used with the word processors that you will study shortly often have an elongated CRT screen so that an entire page of text can be displayed at one time.

The TV raster is usually scanned at a 60-Hz rate (noninterlaced) so the display screen is refreshed every 1/60 second to provide a flicker free display. Special precautions are used in the design of the deflection and focus circuits of CRT terminals to maintain a sharp, well-focused image over the entire CRT face.

CRT Terminals (continued)

The data rate to refresh the CRT is too high to be obtained directly from the computer. For an 80-character × 20-line display, 1,600 characters are generated every 1/60 second (96,000 characters must be read every second). To solve this problem, CRT terminals have an *internal memory* (typically, 8-bit words) which contains the data to be displayed. This memory is read at a 60-Hz rate to provide the data for the dot-matrix decoder. Simpler CRT terminals only have enough memory to hold what is to be displayed directly. When a new line is added, the old lines move up the screen, and the line at the top of the display is moved off screen and *lost* from memory. More sophisticated CRT terminals provide the capability for *scrolling*. In this case, there is additional memory provided to retain some number of lines not displayed, but these can be displayed by moving a cursor control.

**FOR SCANNING MORE THAN MINIMUM
MEMORY IS REQUIRED**

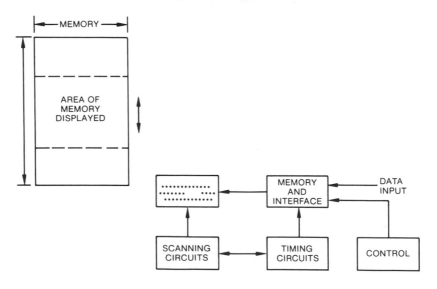

CRT terminals are either *dumb* or *smart* (or somewhere in between). *Dumb terminals* depend on the *main computer for control* of memory and the displayed data. *Smart terminals* are usually *controlled* from an *internal microprocessor*. These often have extensive memory and can be programmed to provide many varied display modes and do additional tasks. In fact, some smart terminals are essentially complete computing systems that can stand alone or provide for preprocessing and formatting of I/O data for most efficient interface with a computing system. Some smart terminals operate in conjunction with a printer to stand alone as a complete system, as you will see shortly. Many modern CRT terminals have a very flexible format so that *computer graphics* can be implemented. Computer graphics allow for drawing graphs, bar charts, or any figure desired.

Interfaces

As you learned previously, communication between a terminal and a computer can take place via a MODEM (*mo*dulator-*dem*odulator). Computer terminals can also be connected directly to the computer system. Usually this is done by a standard *interface* that inputs/outputs the MODEM or the computer. Usually, the input device (keyboard) and output devices (CRT) are independent. When a terminal is connected directly to a computer, two pairs of wires are usually used, since no MODEM is provided to offer continuous communication. When you strike a key on the keyboard, the character that appears on the screen usually does not come from inside the terminal but comes in *via the computer*. Thus, the computer shows an *echo* on the display of the information sent from the keyboard. This allows the operator to *verify* that the character sent has been received and acknowledged by the computer.

THE DISPLAY ECHOES
THE TRANSMITTED CHARACTERISTICS

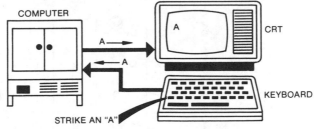

Two standard interfaces commonly in use currently are the *RS-232 standard* and the *20-mA current loop*. The 20-mA current loop is really obsolete, but is still used on some equipment, particularly Teletype® terminals. The 20-mA current loop has a source of constant current at the transmitting end that delivers 20 mA to the line regardless (within limits) of the line length and resistance. A 1 (or mark) is sensed as the current flow, a 0 (or space) as an open line (no current).

RS-232 INTERFACE

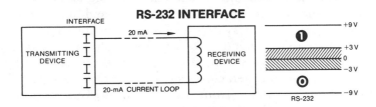

The RS-232 interface generates a 9-volt positive signal for a 1 (mark) and a 9-volt negative signal as a 0 (space). The region between ± 3 volts is undefined so that noise in the line (at lower levels) does not generate a false bit.

Other standards have been proposed and are used, as it is important that the interfaces between terminals and computers and/or MODEM be compatible; however, the RS-232 is most commonly used.

Analog-to-Digital/Digital-to-Analog Conversion

Analog-to-digital (A/D) converters are used between analog-output devices and digital devices. Their job is to convert the analog signal to an appropriate digital signal that can be used in digital-signal processing. For example, quantities like temperature, pressure, light intensity, or liquid level are typically sensed by analog devices. In digital processing, an A/D conversion is required. Important characteristics of A/D converters are the accuracy of conversion (number of bits generated) and the speed of conversion (how long it takes to generate the digital word).

SUCCESSIVE APPROXIMATION A/D CONVERTER

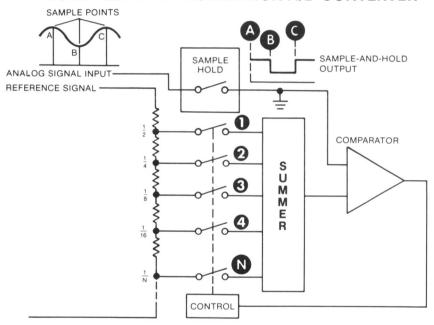

The block diagram above shows a *successive-approximation A/D converter*, the type most commonly used. The input signal is sampled at the beginning of the conversion cycle. The sample-and-hold is essentially a switch and capacitor. The switch is closed momentarily (at points A,B, and C) to charge the capacitor to the input voltage, and it holds this voltage until conversion is complete. Then another sample is taken and converted. The control then closes switch 1, and the comparator determines whether the voltage from the *precision divider/summer* is above or below the signal sample. If it is above, the switch is opened again; if it is below, the switch is kept closed and switch 2 is closed and the same procedure is followed. As you can see, eventually after all (N) switches are tried, a combination of open and closed switches will be arrived at to make the comparator output approximately zero. The switch positions (open = 0, closed = 1) now represent the *digital equivalent* of the *analog signal*. The digital output, then, is simply the switch settings.

Analog-to-Digital/Digital-to-Analog Conversion (continued)

Obviously, the switches are electronic, and the entire process proceeds very rapidly (typically, in fractions of a millisecond or even fractions of a microsecond). The number of switches (bits) determines the accuracy of conversion. For example, a 10-volt peak analog-signal level in an 8-bit A/D converter is represented digitally to 1 part in 256. If the A/D converter has 16 bits, the accuracy is 1 part in 65,536. A/D converters are usually available as IC devices in a single package, although high-speed, high-accuracy units may consist of several ICs.

Digital-to-analog (D/A) converters are used when it is desired to produce an analog-signal output for display or interface to a device that requires an analog-signal input. They are comparatively simple devices, as shown below.

D/A CONVERSION

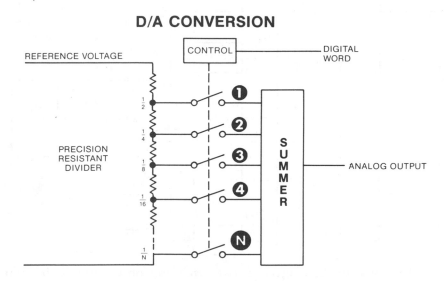

The digital word is applied to the control unit, which closes the switches in accordance with the bit pattern of the digital word. The switched outputs from the precision divider are fed into the summer, and the summed output is the analog representation of the digital signal. Thus, the output is the sum of the bits. For example, if the bit pattern on an 8-bit D/A converter is 10110111, the output voltage will be $\frac{1}{2} + 0 + \frac{1}{8} + \frac{1}{16} + 0 + \frac{1}{64} + \frac{1}{128} + \frac{1}{256}$ of the reference voltage, or 0.7148 times the reference voltage.

For both A/D and D/A conversion, you can see the importance of having a stable, well-defined reference voltage and accurate divider chain. For many applications, the reference voltage is internal. However, it can be supplied externally as well. Maximum voltages of ± 5 and ± 10 volts are common standards. Obviously, it is necessary to know the range and calibration of the input signal so that it has a known relationship to the reference for the digital and/or analog signals to have meaningful values.

Time-Sharing of Computers

Many computer systems, both large and small, are arranged to service multiple users simultaneously. This is possible because for many uses, the computer itself can operate much faster than necessary to service a single user. Basically, the time-sharing computer is just what you have already learned about, except that there are modifications to the software to permit multiple users or *time-sharing* operation. Users can be either local (directly tied in) or remote (with access via telephone lines) or a combination of both. Usually, to be effective, computers used for time-sharing should have a large dynamic memory and disk storage.

INTERFACE/LINES

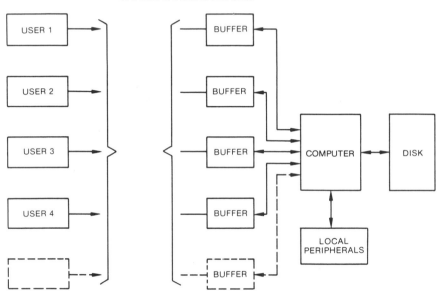

As shown in the diagram above, each user on the line is tied into a buffer that interfaces with the computer. The computer polls each buffer for information (or delivers information to it) sequentially. The computer then *dedicates* itself to that user for a fixed period of time (typically) and executes whatever portion of the program it can in the time alloted. At the end of the specific time, the program execution for that user is interrupted, and the next user is serviced. This process repeats continuously in a loop so that all users obtain service. Because, the input/output of the user's terminal is much slower than the rate at which data can be entered, removed, or manipulated by the computer, the user is usually unaware that there are other users on-line. Practical limits as to the number of users on-line are reached when the speed of operation per user slows to an unacceptable value. Typically, 16 to more than 100 users can tie into a medium or large computer system in this way. The time-sharing approach allows for high-power computing service at minimum cost, since the cost is shared among multiple users.

Review of Input/Output Devices

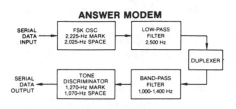

ANSWER MODEM

1. THE ASCII CHARACTER SYSTEM, which uses a 7-bit code, is used in most digital I/O systems. Ten characters per second can be handled with a 110-baud (bits-per-second) system.

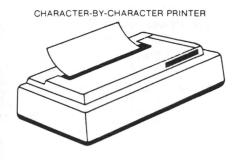

CHARACTER-BY-CHARACTER PRINTER

2. TELETYPE EQUIPMENT, KEYBOARDS, AND PRINTERS are important, widely used I/O devices.

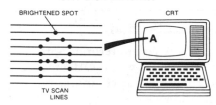

CRT DOT MATRIX

3. THE CRT TERMINAL is the most common and inexpensive data display in use today. It is like the CRT circuit in a TV receiver but has greater bandwidth.

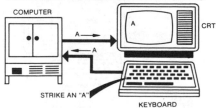

THE DISPLAY ECHOES TRANSMITTED CHARACTERISTICS

4. INTERFACES in a computer include display systems that "echo" what is put into the computer and A/D and D/A converters.

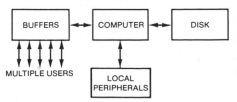

5. TIME-SHARING COMPUTERS serve multiple users simultaneously. Such computers have a large dynamic memory and disk storage.

Self-Test—Review Questions

1. What is a baud? What is baud rate? Explain synchronous and asynchronous operation.
2. What is ASCII? What is the format of an ASCII character? Why are 11 bits used to transmit what is basically a 7-bit word?
3. What is a MODEM? Why are they used?
4. What I/O devices are contained within the ASR-33 Teletype®?
5. How are keyboards organized? Is a separate wire required for each key? Explain.
6. What is a UART? Why is it used? How is it used?
7. Describe a typical CRT terminal. How is the alphanumeric display on a CRT terminal formed? What are the essential elements of a CRT/keyboard display? What is the difference between a "smart" and a "dumb" terminal?
8. Describe the nature of the 20-mA current loop and RS232 interfaces. Give details of their use.
9. Describe how A/D and D/A converters work. Where would they be used?
10. Sketch the organization of a time-sharing computer system and describe its operation.

Learning Objectives—Next Section

Overview—Thus far, you have studied how computer/microprocessor systems work and how they are interfaced with I/O devices. In the next section, we will look at some typical computer/microprocessor systems and their application.

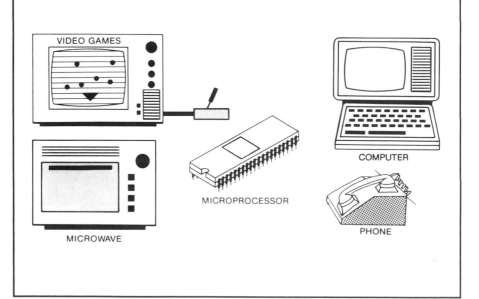

A Typical Microprocessor/Microcomputer—MC6800 Series

The Motorola MC6800 is a typical inexpensive microprocessor that has been in production for some time but is still widely used. It has an 8-bit (1-byte) word length and can address 65,536 memory locations in RAM or ROM (16-bit address structure). The CPU (MC6800) is supported by a wide selection of I/O, interface, ROM, and RAM ICs. ROM is also available to provide for assembly language programming. Extensive software is available to support the MC6800 microprocessor. The 6800 operates with a clock rate of 1 MHz. Improved versions (for example, the MC6806) operate at higher speeds, and a 16-bit version (MC68000) is available to provide for faster processing or *throughput*.

The CPU is made as a 40-pin package. As shown in the block diagram on the following page, the basic design is similar to that for all computers and microprocessors. Programming of the 6800 system is relatively easy in assembly language; however, high-level language such as BASIC and even a version of FORTRAN are available as software. A floppy disk controller is also available to expand the memory capacity and provide for program/data storage, as described earlier. As you can see by examining the Appendix, an instruction set of 72 instructions is available for assembly language (or machine-level) programming. It also features DMA and flexible interrupts and address modes for convenient I/O and data management.

The MC6800 CPU chip, like other microprocessor CPU chips, is used widely for both microcomputer and microprocessor applications. Some microprocessors have a much more extensive instruction set and additional storage registers available in the CPU. These extended instructions allow for more convenience in programming, but do not really give the microprocessor more capability. Remember, the essential things done by all computers (including microprocessors) are binary addition, Boolean operations, complement, shift of data to the right or left, movement of data in and out of memory.

The MC6800 CPU is organized as shown on the following page. The two 8-bit (1-byte) accumulators are the locations where arithmetic and logic functions are performed. Either or both can be used. As you will see in an example shortly, the two accumulators can be chained (via the carry bit) to form a 16-bit data word. The program counter (16 bits) is used to address the program stored in memory. It is moved sequentially through the program for execution. The 2-byte (16-bit) index register is used for addressing memory with respect to data. The stack pointer (2 bytes) is used in conjunction with the temporary storage of data in memory. It sets up and addresses an area in memory for this purpose. The control register accepts the OP code (instruction) from the appropriate location in memory, and the condition code register is used to indicate carry, zero (in an accumulator), and similar data. Special instructions (like RESET, HALT, MEMORY READ/WRITE, INTERRUPT, etc.) are available on individual pins of the microprocessor. All of these functions are contained in the CPU or microprocessor IC chip.

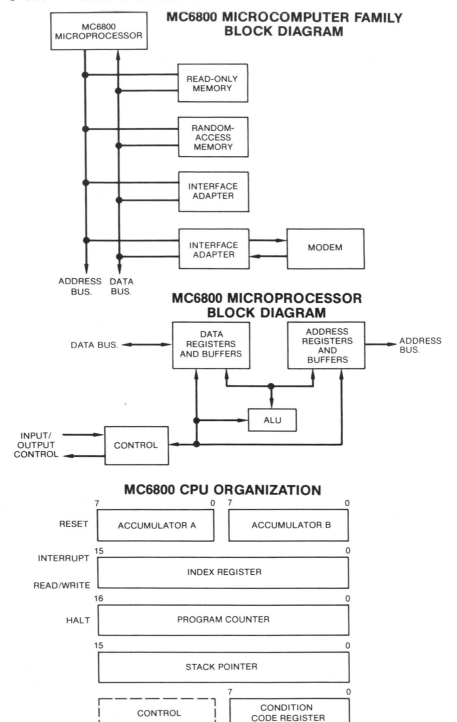

MC6800 MICROCOMPUTER FAMILY BLOCK DIAGRAM

MC6800 MICROPROCESSOR

READ-ONLY MEMORY

RANDOM-ACCESS MEMORY

INTERFACE ADAPTER

INTERFACE ADAPTER

MODEM

ADDRESS BUS.   DATA BUS.

MC6800 MICROPROCESSOR BLOCK DIAGRAM

DATA BUS.

DATA REGISTERS AND BUFFERS

ADDRESS REGISTERS AND BUFFERS

ADDRESS BUS.

ALU

INPUT/ OUTPUT CONTROL

CONTROL

MC6800 CPU ORGANIZATION

RESET

7   ACCUMULATOR A   0   7   ACCUMULATOR B   0

INTERRUPT

READ/WRITE

15   INDEX REGISTER   0

HALT

16   PROGRAM COUNTER   0

15   STACK POINTER   0

CONTROL

7   CONDITION CODE REGISTER   0

Experiment/Application—Programming the 6800 Microprocessor

While you cannot expect to become a programmer without further study, you can get a feel for how programming is done for the 6800 microprocessor system by trying to write some simple programs of your own. Refer to the Appendix for details of the complete instruction set. For our purposes, we will use assembly language to program a simple routine. For example, suppose you wanted to add two numbers located in memory at locations $00F0_{16}$ and $00F3_{16}$ and return the result to location $00F8_{16}$. You could proceed as follows:

LDAA 00F0   (LOADS ACCUMULATOR A WITH CONTENTS OF MEMORY LOCATION 00F0).

LDAB 00F3   (LOADS ACCUMULATOR B WITH CONTENTS OF MEMORY LOCATION 00F3).

ABA   (ADDS CONTENTS OF ACCUMULATORS A AND B AND PUTS RESULT IN
                ACCUMULATOR A).

STAA 00F8   (STORE CONTENTS OF ACCUMULATOR A INTO MEMORY LOCATION 00F8).

Try to program a subtraction of the contents of the data in the memory locations shown above. As you can see, it only requires that ABA be changed to SBA to achieve subtraction. Try to write some other simple programs to get a feel for programming. You should, of course, prepare a flow chart before you start, so that you can see the steps involved in performing the necessary operations, as shown below.

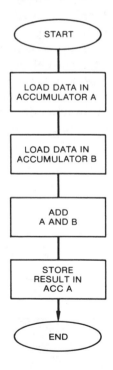

Introduction to Computer/Microprocessor Systems

Early computer applications invariably involved the use of a relatively elaborate computer with CPU, memory, and I/O at a single location. This central computer could service remotely stationed users, as described previously. While these systems are still important for many applications, the development of the microprocessor, floppy disks, and all of the hardware for use with microprocessors has made the use of *distributed processing* quite practical, and today many systems are configured as *distributed processors*.

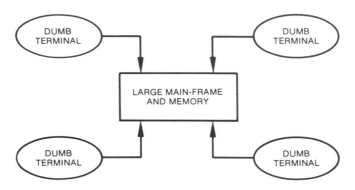

The availability of *smart terminals* with self-contained microprocessors (sometimes more than one) and convenient inexpensive mass media like floppy disks has made distributed processing quite practical. Here, there is a central main (mass) memory into which all users contribute data. Each user has access (assuming a need to know) to the central memory, but with the smart terminal you can load programs at the terminal to perform desired operations on or use the records in the main memory. In this way, *all* users have access to the same information files, but are capable of using it in independent ways because of the microprocessing power in their individual terminals.

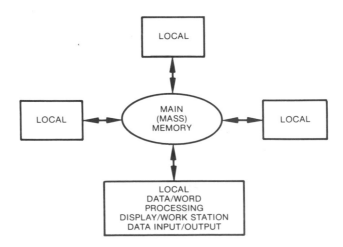

Microprocessor Applications

There is only enough space here to review a very few of the many applications of microprocessors and microcomputers. Both large and small computer systems are extensively used in business, industry, and government for record keeping, data processing, and data storage and retrieval. As you probably know, they are also widely used in research, scientific, and engineering work. Microprocessors/microcomputers have become more and more widely used and important in these areas as their capability has increased. Because of their small size and cost, the applications for which they are especially useful are in control systems, word processing, office information systems, small business computers, home appliances, home computing systems, TV games, etc.

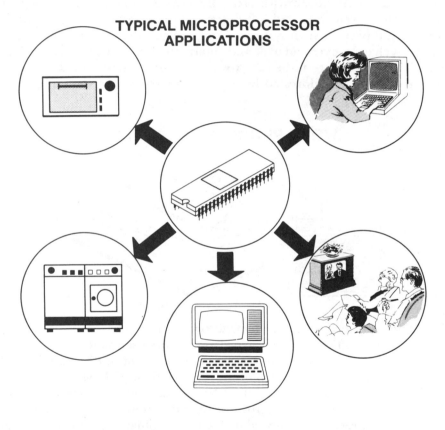

TYPICAL MICROPROCESSOR APPLICATIONS

There is no way to keep up with the increasing use of these devices in all aspects of our daily life. As the price decreases and utility increases, you will find them everywhere. While there are efforts at standardization, many different microprocessors are being made, in many cases for a single dedicated application. The important thing to remember is that all of these devices work in basically the *same* way; so if you know about one, you can easily understand others.

Word- and Information-Processing Systems

In the early days of computer use, *word processing*—as opposed to data processing—was limited to large systems with extensive mass memory and expensive printers. The availability of microprocessors, inexpensive memory (RAM, floppy disk), and high-quality, fast printers with a full keyboard of characters has resulted in the rapid development of word-processing systems. These systems are commonplace in business offices and in publishing companies. This is because word-processing systems allow for the rapid, easy editing of manuscript to provide clean copy, requiring only the typing of the first draft and the necessary editing. Adding, deleting, and shifting of words, sentences, and paragraphs are easily done. Intermediate copies for review and editing and final copy are easily produced with virtually *no human effort*. Features such as justification (even right and left margins), centering, and page margins are integral parts of such systems. Compare this to the usual method of typing each draft over and over and finally the final version. In addition, if multiple letters are to be sent to many people, these can all be individually typed and addressed by a word processor once the initial data are entered.

WORD-PROCESSING SYSTEM

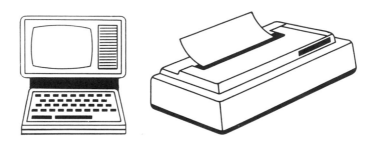

The word-processing systems allow for the production of perfect copy without difficulty. Early word processors used tape cartridges or magnetic cards and were quite limited in terms of ability to edit and make corrections and the size of storage available. Modern microprocessor systems have changed all this. Now, a manuscript is typed in (from dictation or notes), a draft is produced and edited, the corrections are made, and the results are viewed on a CRT terminal. A simple command then results in the letter or manuscript being typed out. An autofeed mechanism on the printer allows for the continuous typing of material page by page, if desired. Paper for these printers can be preprinted with fixed information, such as a letterhead.

Many word-processing systems are so compact that they are contained in little more space than a CRT terminal would occupy. In more elaborate installations, CRTs are multiplexed into a central mass storage system so that many operators can use these facilities simultaneously.

Word- and Information-Processing Systems (continued)

In word- and information-processing systems, the term *memory management* becomes very evident. An operator at a CRT/keyboard types the letter or manuscript, which is stored in RAM. The microprocessor controls the data input/output from RAM and the external I/O devices. Longer manuscripts may be stored partially in disk. Material to be used over again (like standard form letters or lists of clients) are also stored on disk. The initial writing into RAM is also presented on the CRT. In most cases, special CRTs allow for the display of more than the usual 20 lines. Since the writing is via RAM, the written words can be manipulated, moved, deleted, or written over at will via command from the keyboard and the microprocessor control. The final result, in a specified area of RAM, contains the final version. It is only necessary to have the RAM data read out via I/O to the printer. The CPU controls the printer as well as the RAM, so that justification and other print formatting desired will be obtained.

WORD PROCESSING

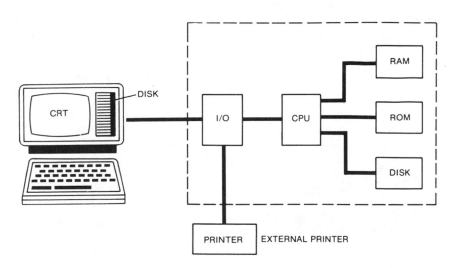

Some modern word- and information-processing systems even include dictionaries to correct spelling or to take care of proper consistent spelling and correct hyphenation throughout the text. Usually, draft copies are returned to the writer for editing, with the changes then incorporated into the text. Further draft copies are then easily prepared before the final copy is approved. The microprocessor can be coupled to a wide range of printers to produce text-quality type. Printers using either a replaceable ball or rotating character wheel allow for a wide choice of typefaces. In some cases, the CRT keyboard system is remote from the mass memory. This allows for several authors to contribute to a particular document without the need to work through a single word-processing work station.

Word- and Information-Processing Systems (continued)

As you can see, the system is not very different from other micro-processor systems including I/O, ROM, RAM, and mass storage, but the emphasis is on *memory manipulation* or *management*. As portrayed in the illustration, which shows what happens when a paragraph is to be inserted, space is made available by moving the information below the insertion point X to another unused portion of RAM. The new paragraph is typed in, and the information that was temporarily moved is replaced, immediately below the new paragraph. In a similar way, letters, words, or punctuation can be inserted, deleted, or changed. A *cursor* (usually a blinking rectangle about the size of a single character) is shown on the CRT face. The location of the cursor is controlled by a special set of keys added to the keyboard. The location of the cursor tells the operator *where* in memory the address bus is set. The operator then can set the cursor to a character and press an appropriate function key to change the character, delete it, insert something at this point, etc. For example, to insert a word, the cursor is positioned at the proper place, the insert key is depressed, and the operator proceeds to type the new word. As the typing proceeds, the material in memory below that point is moved as the new characters are inputted. Thus, when the change is complete, the corrected information is sequential in RAM—ready for output.

RAM MANIPULATION IN WORD PROCESSING

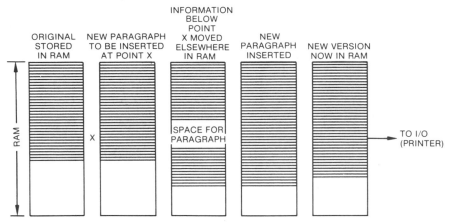

There is a great deal of work currently going on to eliminate even the need for initial typing of the manuscript! *Microprocessor-controlled speech-recognition systems* that take speech input directly via a microphone and produce written text have been developed with some success. There are ample indications that these devices will soon allow for direct input from the spoken word into written text. There currently are available machines that will convert written text into spoken words with a high degree of intelligibility. These devices use a general-purpose microcomputer as the basis for operation.

On-Line Data Storage and Retrieval Systems (Data Bases)

The explosive growth of information in this century has led to problems with locating information and data on particular topics. Conventional library searches and bibliographic indexes are becoming very cumbersome because of the volume of data to be handled. As a result, computer service firms offer on-line bibliographic service to handle inquiries on specific topics. Basically, these systems have massive memories (disk) into which the necessary bibliographic data have been inserted. These bibliographic files are updated continuously with data obtained from current literature.

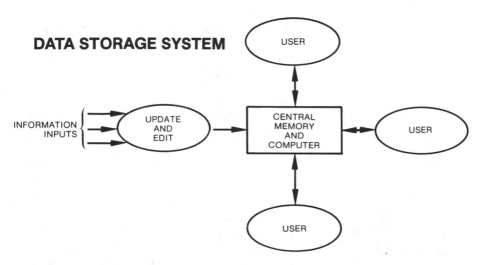

DATA STORAGE SYSTEM

In operation, users have access to the central memory and computer—usually called the *data base*. The items stored are characterized by key words or *descriptors*. Thus, data on a particular subject can be called by using these terms. Thesauri and dictionaries are available to help in choosing appropriate descriptors. The data base is then queried, and an initial estimate of the number of records that fit the description is given. Usually the number is too large, and additional descriptors are added. When the list is appropriate, a printout can be obtained. Requests can be made for any desired abstracts after examination of the lists. At this point, the specific references desired can be obtained from usual sources. Usually many data bases or data base services are accessible from a single terminal. Some of these are quite general in nature, although most are confined to relatively well-defined subjects.

One of the fastest growing areas of computer usage is in providing on-line information services (data bases). While these systems are organized in a similar manner to the bibliographic services described, they differ in that timely information on a broad range of subjects is available. These services provide specific information to businesses, the government, and other institutions. For example, information is available on the stock market and various commodity markets, as well as detailed data on the U.S., Japanese, and other economies.

On-Line Information Services (Data Bases)

Modern information services are very helpful to the legal profession; federal, state, and local regulations and laws are readily available in the most up-to-date form along with listings of court cases that have involved these regulations or laws. In addition, the status of pending legislation is readily available.

Another area of rapidly expanding service lies in the availability of on-line information services to individual households. Here, the consumer can find out what stores carry a particular item, what the cost is, and even where the best buy is.

ON-LINE INFORMATION SERVICES (DATA BASES)

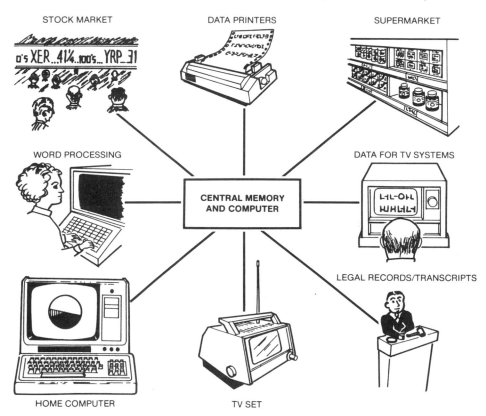

Other aspects of these information services are timely delivery of important news relative to specialized areas of interest or of only local interest, when conventional news sources might not provide adequate information. For example, local road conditions or the status of local rapid transit can be obtained from such systems.

On-Line Household Banking and Buying Services

There are currently experimental systems being developed to do banking and buying from your home. Using these services, it will be possible to transfer funds between accounts, pay goods and services bills (in full or in part), and purchase almost anything desired from airplane tickets to a new pair of shoes. In addition, the *electronic fund transfer (EFT)* is being widely used so that an employer or other source deposits money as needed directly into your account, and no checks or personal visits are required. Even now, electronic banking to make withdrawals using an appropriate identification card is quite common in some parts of the country. Another advantage to electronic banking/buying is that the account balance is available up to date at any time desired.

ELECTRONIC BANKING

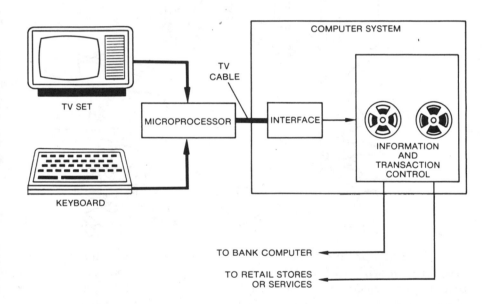

One advantage to doing all transactions (buying, bill paying, etc.) by EFT is that a complete record is made of the transaction. These records can be kept so that they provide a basis for preparing income tax forms as well as a summary of how money was spent. While one can keep records using a home microcomputer at the present time, such record keeping can be tedious and often overlooked or missed. In addition, there is a question as to the legality of records kept by computer that are not backed up by actual documents. On-line household banking and buying services may make life much more convenient for all of us in the near future.

Video Games

One of the most popular uses of microprocessors and controllers is in *video games*. These can vary from small, hand-held, self-contained systems having a limited choice of games to the large, elaborate, flexible microprocessor-based systems that can be programmed for a wide range of games using a standard TV set as the display terminal. With large systems, the number of games that can be played and their complexity are limited only by practical cost considerations.

TYPICAL VIDEO GAME

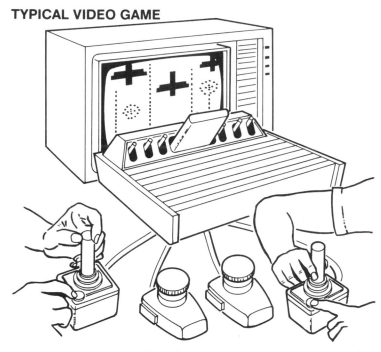

The fixed-program, hand-held games generally use a dedicated microprocessor with a ROM memory. Generally, the output drives a LED display. The electronic educational toys, like the QUIZ game shown above, have program cartridges. These cartridges supply a data input to the fixed preprogrammed microprocessor and thus allow for different data sets to be used.

At the other end of the spectrum are the flexibly preprogrammed video games. These use separate program software for each game, retaining in the microprocessor only the elements of firmware common to all games (for example, I/O). Here, a program cartridge (typically a cassette cartridge or ROM) is used to program the microcomputer to play a particular game. One of the most astonishing aspects of these video games is their low price compared to their capability. These low prices are directly the result of high-volume LSI production. It is this fact that allows us to predict that some of the low-cost communication facilities described earlier will indeed come into being.

Automotive Microcomputers

Microprocessors are now being used in automobiles for engine and emission control. These devices are integrated into the control systems and monitor engine parameters to establish the optimum settings for fuel, spark settings, air/fuel mixtures, etc. Data such as engine load and temperature are input by appropriate sensors. In addition, these microprocessors monitor the status of headlights, parking lights, and other parameters to warn the driver of problems with any of these.

A new application of microprocessors in automobiles is the *on-board computer*. These systems provide data on engine performance and gas consumption, as well as distance, speed, and time-to-destination.

INTERNAL-MICROPROCESSOR ENGINE CONTROL

Basic inputs such as engine speed, engine (oil) temperature, remaining gasoline, gasoline flow, and engine torque are provided to the microprocessor by external sensors. Fixed data on tire size, gear ratios, etc. are programmed in on installation (but can be changed if desired). The driver can not only monitor fuel consumption (miles/kilometers per gallon/liter), but can also calculate how far available gas will take the car, the time of arrival at a given point, and average speed; other options include an elapsed timer and accurate digital clock. In some systems, warning indicators also flag problems of overheating, low gas, etc.

The basic problem with the use of microcomputers for automotive application lies primarily in providing adequate input sensors. Most of the sensors on current automobiles are relatively inaccurate and produce analog outputs. As new sensors are developed that can interface directly with the microprocessor, systems will become more available. Another problem lies in servicing. The electronic systems on present automobiles are difficult to service. The introduction of microprocessors into automobiles will increase this problem, which must be resolved before the public will accept microprocessor-controlled automobiles.

Microwave Ovens

One of the first home appliances to use microprocessors is the microwave oven. The microwave oven provides a very flexible, rapid method for cooking and heating of foods and liquids. Because of its speed and flexibility, precise control and timing are necessary for best results. The microprocessor in a typical microwave oven is programmable in a limited way from the front panel. A digital clock (microprocessor-based) is included as part of the control system. Typical parameters that can be entered into the microprocessor are present time, time to start cooking, time to end cooking, microwave power level, and food temperature.

MICROWAVE OVEN—OPERATION

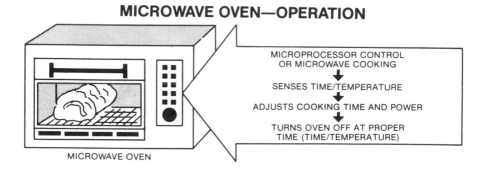

In operation, the control panel provides inputs on the time to start cooking (based on comparing input time to clock time). The microwave power can be adjusted by keyboard input or automatically controlled in a variable sequence under microprocessor control based on temperature. Different power settings are required for different cooking modes and for different foods. A temperature probe senses the food temperature and then (based on keyboard input) terminates and/or adjusts the microwave power. As an alternative, the system can be programmed manually for a fixed cooking time. For frozen foods, a two-step process can be programmed to defrost and then cook a food. Some models can even be programmed to keep food hot or to reheat it at a specific time to have hot food ready when desired. Microprocessors in applications like this essentially sense the environment (in this case, a foodstuff) and react appropriately. In the same way, a temperature controller for a home heating/cooling system can be made to provide for optimal temperature at all times to maximize comfort and minimize cost.

We are all familiar with the fact that a growing number of people are kept alive or in reasonable health by the use of a cardiac (heart) pacemaker. These are currently implanted so that the entire device, including the power source, is inside the patient's chest. Newer units are less than 2 inches × 2½ inches × ⅓ inch (5.08 cm × 6.35 cm × 0.85 cm), including the power source for operation for several years.

Programmable Cardiac Pacemaker

From time to time, physicians have found it necessary to vary the parameters of the pacemaker. Originally, it was necessary to *remove* the implant and substitute another with different characteristics. In addition, the physician found it difficult to evaluate the patient's condition because of the presence of the pacemaker itself.

CARDIAC PACEMAKER SYSTEM

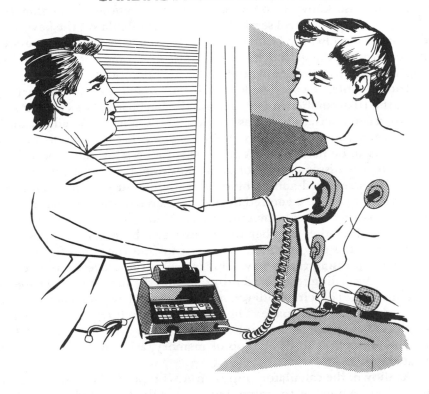

The Spectrax™ developed and manufactured by Medtronic is a cardiac pacemaker using a microprocessor that can be programmed while implanted. The programming can be temporary or permanent (but alterable). An external program is impressed on the internal microprocessor by a magnetic pulser that generates code information as a stream of magnetic pulses. These are picked up by the pacemaker and, according to the instructions supplied, make appropriate changes in the internal program that can vary pulse rate, pulse duration, pulse amplitude, and other parameters. In addition, the physician can exercise control over the pacemaker to evaluate the condition of a patient's heart. The magnetic device also functions as an output device, since the parameters programmed can be detected and displayed on an external display unit.

Basic Pocket Calculator

As you know, it is possible to buy a relatively inexpensive electronic pocket calculator that performs addition, subtraction, multiplication, division, and percentage calculations with sequential algebraic operations. More complex and expensive engineering calculators can do extensive mathematical operations and many modern engineering calculators can be simply programmed to perform repetitive calculations. In all cases, the operating principles are the same.

As you also know, a computer contains input/output, arithmetic/logic, memory, and control sections. In a pocket calculator, the keyboard is the input; the LED (or LCD) display is the output; and the calculator chip contains all the arithmetic/logic control and memory required to perform arithmetic and sequential algebraic operations.

For example, if you decide to add 867.3 and 452.6, you start by placing the first number in memory via the keyboard. The calculator is ready to receive serial entries as part of the number to be added until you tell it what the fate of the number is to be—that is, what arithmetic operation is to be performed. You do this by pressing the + key. The numbers on the display blink for an instant to indicate that the instruction has been received and memory is ready to add your next entry to the first. After you have entered the second number, the calculator waits until you instruct it to do something else. You can instruct the control system to add another number to the first two by pressing the + key and inserting the new number. Or, you can find the sum of your two entries by depressing the = key. This instructs the processor to add the two entries according to a stored program, and the sum of 1319.9 appears in the display. The circuit diagram for a typical pocket calculator is shown on the following page. The bulk of the circuitry is contained in a single LSI/IC chip. This single chip typically will have over 5,000 semiconductors in it to do the job of memory, arithmetic, control, and formatting of I/O.

As shown, the calculator chip is an AMI type S9411. The keyboard is simple, but embodies the same structure as the keyboard you studied earlier—that is, a code is set up when a key is depressed, exciting a wire pair defined by the key that was depressed. For example, if a 0 is depressed, lead 28 and lead 4 from the IC are connected together to define a unique address, etc. The decoding for the display segments is also contained in the IC. There are some special outputs for special symbols like the decimal point. The display uses a strobe to operate the LED characters to conserve power, which you learned about previously.

Since there are few accessible elements of the calculator, not much can be done to repair these calculators when they fail. The entire unit is usually assembled on a single mother board printed circuit with essentially no hand operations for manufacture. This *minimal-touch labor*, coupled with volume production of the components for low cost, is what makes these calculators so inexpensive.

TYPICAL SCHEMATIC DIAGRAM
WITH LED DISPLAY

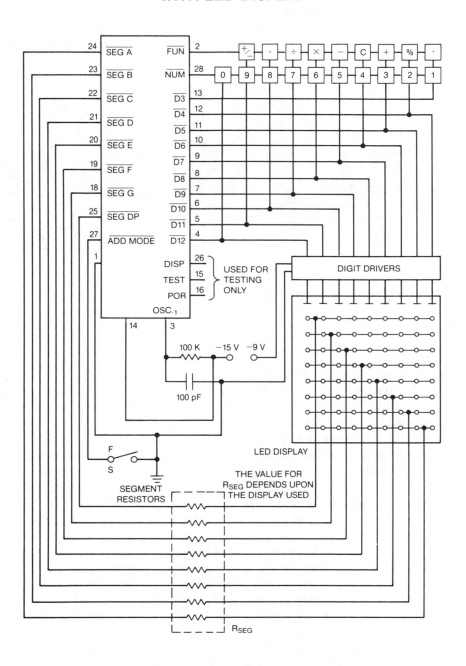

Microprocessor-Based Scales

You undoubtedly have seen *microprocessor-based scales* and *balances* that are commonly used in markets to weigh and exactly calculate the price based on the unit cost. In these devices, the scale tray (which is connected to an ordinary scale mechanism) is also connected to a *digital strain gauge* that measures the scale tray deflection obtained when an item is placed on the tray. This digital output is sent to a microprocessor that is programmed to accept this input.

MICROPROCESSOR-BASED SCALE

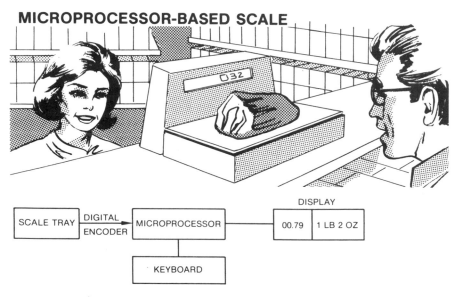

The microprocessor also has a keyboard input, which the clerk uses to manually insert the unit price (for example, the price per pound or per kilogram). The microprocessor then does the necessary calculation based on these two inputs to produce the total price. Microprocessor digital output is then used to operate a display that shows the total price and the weight of the item being purchased. In some supermarket systems the item's weight and price are recorded on the customer receipt.

As you can see, the calculation is very simple and can be done with essentially a single chip similar to a hand calculator chip; one quantity is inserted manually, and the other input is inserted automatically via the digital strain gauge. The weight is multiplied by the price per unit, and the result is displayed just as if a hand calculator were used to do the multiplication. The electronic part of the scale is inexpensive and contains essentially a single IC. The digital encoder is more complex, since it must be coupled to the scale mechanism with high accuracy. Some scales are more elaborate and have provision for compensating for the weight of the package, etc. This is done by weighing the package alone, recording it via the microprocessor, adding the item to be weighed, subtracting the container weight, and proceeding to do the necessary calculation and display, as described above.

Point-of-Sale Terminals

Microprocessors and microcomputers are becoming increasingly important in *point-of-sale terminals*, credit card processing of purchases, and supermarket and department store checkout and billing. You have undoubtedly seen some of these devices in operation. A point-of-sale terminal is a microprocessor-based terminal that is used to record and format transactions so that as charges are rung up, a tally is kept of the total while an itemized list is printed out or the sale automatically recorded in a central computer system. In many cases, the prices of items for sale are stored in the system, so that only the item number need be read in. In many of these systems, the central computer keeps inventory records as well, so that as each item is sold, the stored inventory record is reduced. When new stock is added, the additions are recorded at the central computer so that the quantity on hand is always readily available.

POINT-OF-SALE TERMINAL

In the simplest system, the data are entered manually. Credit cards with a magnetic strip on which a customer's number is recorded as a magnetic digital pattern are often used with credit systems to input customer data. In these systems, the digital pattern is read when a credit card is inserted in a slot. The customer's account is checked and, if the purchase is allowable, the transaction is recorded.

In many cases (particularly in supermarkets), the point-of-sale terminal is coupled to the scale described on the previous page and a *universal product code (UPC)* reader (to be described on the next page). When this is done, checkout is very easy. For items with a UPC, checkout is done simply by passing the item over the reader. The price and item are tallied automatically. For other items, like meats, the data are inserted manually. For items to be weighed, the item is placed on an electronic scale, and the unit price is keyed in manually. The total price is calculated and printed out on the customer's list.

The Universal Product Code

The *universal product code (UPC)* is now found on virtually all packaged materials and foods in the United States. It appears as a *bar code* somewhere on the package. The code contains 11 numbers in digital code and an additional number that is used to check that the code has been read properly.

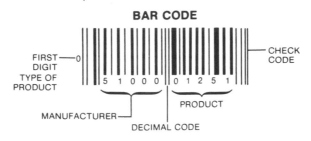

BAR CODE

The Universal Product Code

The check code is used to tell the optical scanning machine whether the bar code is to be read from right to left or left to right (it can be done either way). The first digit of the bar code tells the computer what type of product is involved. The next five digits identify the manufacturer, and the last five identify the product. Thus, the bar code provides all the necessary information for the microprocessor or computer. Bar codes also are used for many other applications, such as identification of books and patrons in libraries with computerized circulation-control systems; some credit card systems also use bar codes.

The bar codes are read *optically*. Typically, a light or laser beam is used with a *wand*, or a *laser reader* is contained under a transparent opening in the counter top of the optical scanner.

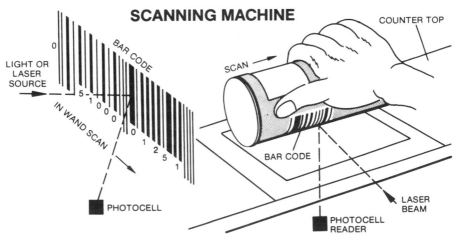

SCANNING MACHINE

Light from the laser or lamp source illuminates the bar code, and the reflected light is picked up by a photocell. The light and dark areas of the bar code are read and the data transferred into the microcomputer.

The UPC Optical Scanners

The information from the optical scanner/reader is compared with stored digital data in the microcomputer to identify the product. Price data for all items are input via a manual keyboard at the time any price is established or changed by management. Once it has located the item in its memory by means of the bar code, the microcomputer produces the price and the item description for display and printout. In addition, the prices are tallied by the microcomputer to provide a total price for all purchases. Thus, not only are patrons checked out more rapidly, but sales slips are produced with an itemized listing of purchases and the price for each item as well as a total of the bill.

OPTICAL SCANNER/READER

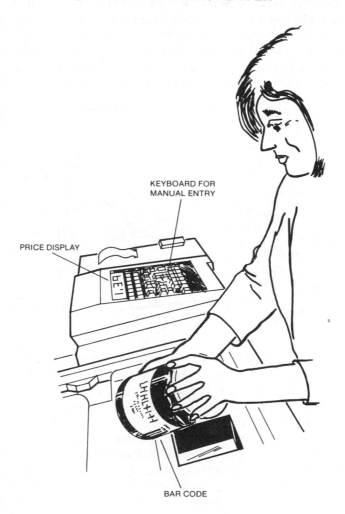

KEYBOARD FOR
MANUAL ENTRY

PRICE DISPLAY

BAR CODE

Data-Processing Systems—IBM Business Computer System/38

The computers studied earlier are considered *small* systems. Here, we will describe a more traditional *large* (main-frame) computer designed to store and process hundreds of megabytes of data and to permit simultaneous operation by up to 40 work stations. While this computer system may be much larger than the systems studied earlier, its operating principle is *exactly the same*. This computer is designed to provide information storage and retrieval as well as perform calculations in the many varied activities conducted by an industrial concern. People with duties within the company as varied as sales, manufacturing, payroll, shipping, and science and research can *all* use the system at the *same time*. Each can add to the information that is stored, and each can learn what is needed from this accumulation of information or program the computer to do any specific desired task.

The example to be described is the IBM System/38 produced by the International Business Machines Corporation, General Systems Division, of Atlanta, Georgia. The central control assembly of this computer system is shown here. While this computer can be made to do any computer job, it is basically organized for business use. Thus, it is organized basically for mass data storage and retrieval of information.

DISKETTE MAGAZINE DRIVE SYSTEM CONSOLE

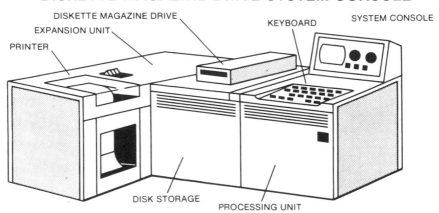

The processing unit (CPU) contains the working memory (RAM) and controls the use of this memory plus the external memories contained in the disk storage and diskette magazine drive. The processing unit also controls the flow of data to and from 12 to 40 remote work stations, to be described on the next page.

Depending on the needs of the purchaser, the disk storage unit can be supplied with from 1 to 6 magnetic disk units. These provide storage for from 64.5 to 129.0 megabytes of data. The expansion disk unit provides additional storage for up to an additional 258.0 megabytes.

The diskette magazine drive is used as an input/output unit that permits this system to make semipermanent records that can be stored off-line when data must be retained but are not needed immediately.

Data-Processing Systems—IBM Business Computer System/38 (continued)

Communication between the processing unit and the various memories and the operator are provided by the system console. This includes a keyboard for the operator to make entries plus a CRT display that shows entries and computer responses. A choice of 300-lines-per-minute or 600-lines-per-minute printers is available and provide a means for the computer to provide a large amount of output data in printed form.

From 12 to 40 individual remote work stations (CRT displays, keyboards, and printers) can be connected to the processing unit. These work stations are located as required, either hard-wired to the computer locally or remotely via MODEMs. The IBM 5251 display station includes a keyboard and a CRT with an alphanumeric display. Up to 1,920 characters can be shown at the same time, and these permit the user to see input data and responses from the computer. The keyboard and CRT also permit communication with other work stations and the system console. As required, printers are available to print out up to 120 characters per second.

THE IBM 5251 DISPLAY STATION

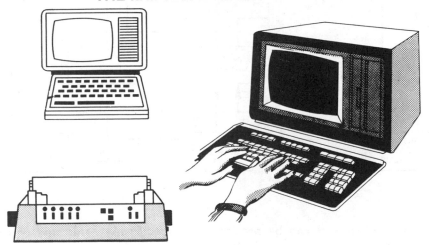

Optional accessories include two models of a multifunction card unit that reads, punches, or prints 96-column cards. The higher-speed model can, in 1 minute, read 500 cards and punch or print 120 cards. Also available is a choice of two magnetic tape units that can process data at rates up to 80K per second.

A family of IBM Licensed Programs (software) designed to perform specific general computer functions is available so that the general business operation does not require specialized programming except for these. Also available are 11 software program packages that perform a variety of commercial standardized activities that are common to most commercial operations.

Data-Processing Systems—IBM Business Computer System/38 (continued)

The System/38 computer can be supplied with software designed to do 11 different functions common to business operations. These operations can be performed concurrently. See the figure below.

THE PROGRAM PACKAGES WORK TOGETHER TO GET THE JOBS DONE EFFICIENTLY

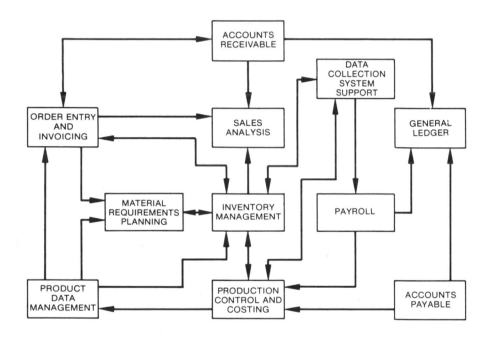

In addition, it can be programmed specifically to do any of the things that general-purpose digital computers can do.

You can see how the use of multiple interactive programs and multiple work stations permits each company department and section to use the computer as required to help individuals to do their jobs with the latest information and with greatest efficiency. Once information is entered, it becomes available to all who require it and in the form that they need it to best do their jobs. Simplified programming techniques enable programmers to easily create new programs that satisfy special needs.

Digital Communication

Digital transmission systems do not differ markedly from AM and FM analog systems. What is different is the way in which the information is transferred. In an analog system, the input signals are *continuous* and *can have all possible values* between the minimum and maximum value. In digital systems, *only discrete values* are possible. However, the precision of the data can be improved by expanding the number of digits in the message words.

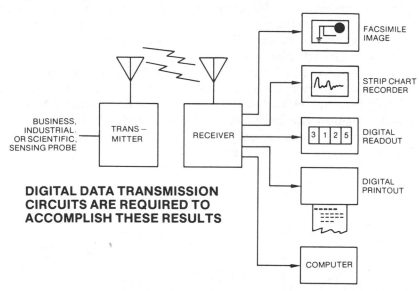

DIGITAL DATA TRANSMISSION CIRCUITS ARE REQUIRED TO ACCOMPLISH THESE RESULTS

Digital transmission systems take many forms. However, in all cases we are dealing with *pulses*. The arrangement of the pulse patterns carries the information. You have already learned something about wire data transmission systems from your study of the way terminals and other peripheral devices talk back and forth to computers. Some of the basic ideas apply. The main point is that a radio frequency or optical frequency link, rather than a wire link, is used. Of course, a two-way system requires a receiver and transmitter at each end.

A SIMPLE DIGITAL COMMUNICATIONS SYSTEM

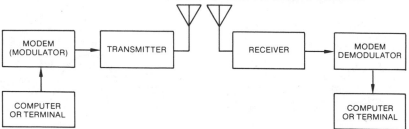

Digital Communication—PAM

One simple way to make a digital-data transmission system is to take the tones generated by a MODEM, as you learned about previously, and use these to modulate either an FM or AM transmitter. A receiver then would recover these tones to be demodulated by another MODEM at the receiving end. This approach is easy to implement, since the signals are already in a form well suited to this. If an analog signal is to be transmitted digitally, A/D and D/A converters can be used at the transmitting and receiving ends, respectively. As you might suspect, essentially all digital transmission systems use serial data, since a single communication channel can really handle only serial data. This problem can be overcome by multiplexing.

Digital data as serialized words formatted for data transmission can, of course, be transmitted directly on a suitable carrier using the mark/space approach for either an FM-FSK or tone-modulated AM carrier. Regardless of how the information is transferred, it is important to remember that we are dealing only with *bit patterns* arranged in some *defined way* so that they can be *interpreted*.

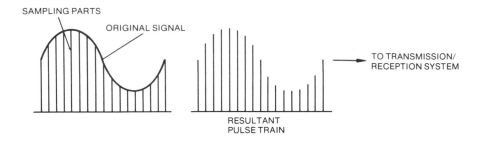

A simple approach to transmitting information in pulse form is *pulse-amplitude modulation (PAM)*. Here, an analog signal is *sampled*, and a pulse train is produced for transmission with amplitudes *proportional* to the signal amplitude, as shown. At the receiving end, the original signal can be recovered by a sample-and-hold (*boxcar detector*) and the resultant passed through a low-pass filter.

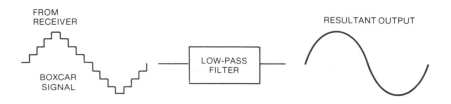

If the sampling rate is high enough, the output from the receiver/processor can be a very accurate representation of the input signal.

Digital Communication—PAM/TDM

As you also might suspect, other signals can be inserted into the *dead time* between pulses. In this way, more than one message can be transmitted at the same time. A suitable decoder at the other end can then separate the output signals into separate channels.

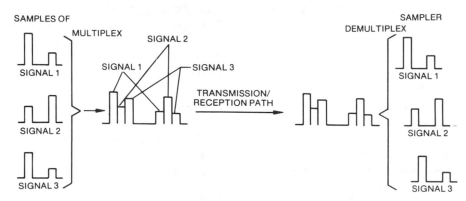

A data- or information-transfer system of this type, using multiplexers, is called *time-division multiplexing (TDM)*. Since the system described uses PAM and TDM, it is referred to as a *PAM/TDM system*. These systems are commonly used for the transmission of multiple conversations on a telephone system. If the sampling frequency is high enough (greater than about 5 kHz), the sampling is not apparent, and the conversations are perfectly clear.

In a similar way, the individual channel data can be placed on subcarriers (as in FM stereo or SCA, which you learned about in Volumes 3 and 4). These subcarriers are added together to modulate the transmitter (usually FM) and are separated by approximate filtering of the subcarriers at the receiving end. Such systems are called *PAM/FM systems*.

DIGITAL DATA TRANSMISSION VIA SUBCARRIERS

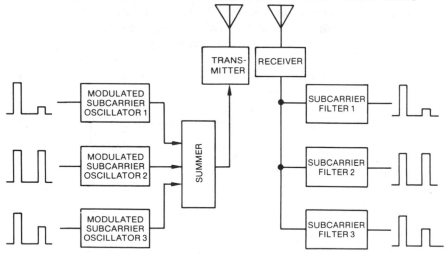

Digital Communication—PCM/PPM/PWM

PCM, PPM, and PWM refer to *pulse-code modulation, pulse-position modulation*, and *pulse-width modulation*, respectively. The only one that is widely used today is PCM. If the transmission is by FM, these are often referred to as PCM/FM, PPM/FM, or PWM/FM. If AM transmission is used, then the designations PCM/AM, PPM/AM, or PWM/AM are used.

In pulse-position modulation (PPM), the starting point is a train of equal-amplitude pulses. The spacing is equal between all pulses under conditions of no signal. When a signal is being transmitted, the more positive the modulating wave is, the earlier the pulse is generated with respect to the reference pulse. The less positive (or more negative) the amplitude of the modulating wave, the later the pulse is generated with respect to the reference pulse. When a sine wave is transmitted, the pulses are displaced as shown in the diagram. To demodulate this pulse train, it is fed into a circuit whose output voltage is proportional to the distance between the reference pulse and the modulated pulse.

PULSE-POSITION (WIDTH) MODULATION (PPM)

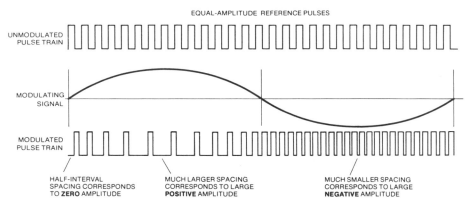

Pulse-width modulation (PWM) is also known as *pulse-length modulation (PLM)* or *pulse-duration modulation (PDM)*. In this method, the amplitude and spacing of the pulse train is kept constant. Pulse width, however, is expanded or compressed according to the amplitude of the signal being transmitted. The maximum voltage in the sine wave expands the width of the corresponding pulse to the greatest amount. Similarly, minimum voltage in the sine wave produces minimum pulse width. To recover the original signal from the PWM train, the train is fed into an interpreter circuit whose output voltage is proportional to pulse width.

PWM PATTERN

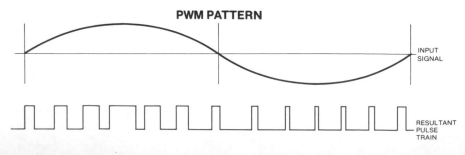

Troubleshooting Digital Circuits

Essentially all digital circuits use ICs. Therefore, troubleshooting consists of finding the defective IC and replacing it. Generally, the IC chips are soldered in, and removal is difficult. In some equipment, the ICs are in sockets, which makes replacement easy. Unlike linear analog systems, the simple *presence* or *absence* of a *signal cannot usually be used* as a basis for test. This is because, as you know, *no signal* (a *low* or *0 condition*) *is just as valid as a signal* (a *high* or *1 condition*). In addition, it is the *bit pattern* that is important. Therefore, you must approach the troubleshooting of digital circuits from a *slightly different* viewpoint, even though the basic ideas of knowing the circuit function and using your head still apply.

In order to troubleshoot a digital circuit effectively, you must *understand it*. In addition you must know the *input/output* (truth tables) of the chips in the unit under test. Using this information, you can determine whether the output is consistent with the input conditions. While an oscilloscope can be used to evaluate these conditions, it is often more convenient to use a *digital logic probe*, as shown here.

TROUBLESHOOTING

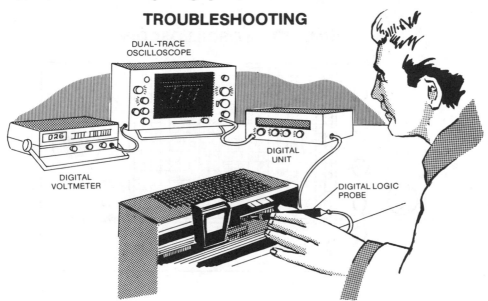

DUAL-TRACE
OSCILLOSCOPE

DIGITAL
UNIT

DIGITAL
VOLTMETER

DIGITAL LOGIC
PROBE

Assuming that the input signal is correct, the output can now be verified using either the scope probe or by using the logic probe. Some special digital oscilloscopes will allow the simultaneous storage and observation of signals for both input and output. This *dual-trace feature* is particularly useful for systems where these signals are changing dynamically on a short-term basis.

As you can see, troubleshooting digital circuits requires somewhat unique test hardware. However, the basic ideas of point-by-point analysis (making sure that the correct signals are present at each point) coupled with using your head (to be sure that what you observe is correct) are *no different* than for troubleshooting any other electronic circuit.

Troubleshooting Digital Circuits (continued)

When troubleshooting a digital circuit, it is important that you know the *function* of the ICs under examination and that you understand their behavior. For example, if you are checking an AND circuit and find that the two inputs are high, you would expect the output also to be high. If it is not, you might suspect that the chip either is defective or is being heavily loaded by the next chip or device in line. By using a digital logic probe and/or an oscilloscope, individual chips and circuit elements can be examined.

Things becomes a little more complicated as the circuit becomes more complex. For example, if a shift-register output in a serial-to-parallel conversion is not functioning correctly, it will be necessary to ascertain whether or not the input is correct, and, if it is, to determine why the ouput is not correct. Determining whether a serial signal is correct can be difficult because it is *transient* in nature. In addition, the clock timing may be important. Special oscilloscopes are available with the capability of storing and displaying a serial pulse train so that the input signal can be verified.

DUAL-TRACE OSCILLOSCOPE

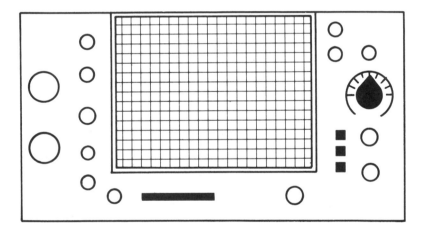

The logic probe is an inexpensive tool—easy to use on crowded printed-circuit boards. It gives an immediate picture of the signal on any given pin of an IC. A typical logic probe has two or three LEDs. For a three-LED unit, the LEDs indicate whether the pin under test is *high*, *low*, or *toggling* (switching rapidly between states). The two-LED units have both LEDs on when the signal is toggling. Unless the toggling is very slow, it is not possible to see it by eye, so an oscilloscope is required to see the time sequence of high/low states.

Review of Computer/Microprocessor Applications, Digital Communications, and Troubleshooting

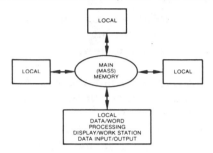

1. SMART TERMINALS, used with central computers, have made the use of *distributed processing* practical. All users have access to the same information files, but can do some processing locally.

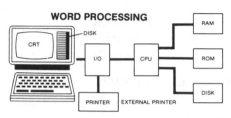

2. WORD PROCESSING SYSTEMS allow rapid, easy editing of manuscripts. The manuscript is typed in, and corrections and format adjustments made while viewing results on a CRT.

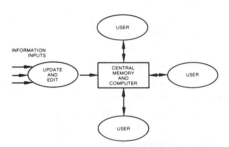

3. ON-LINE DATA SYSTEMS allow access to stored information on such things as books, banking, stock market activity, and general economic information.

4. MICROPROCESSORS have made possible such things as video games, auto microcomputers, microwave oven controls, and medical device controls.

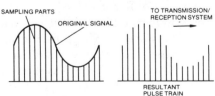

5. DIGITAL COMMUNICATION makes use of digitized information for greater reliability and precision. Pulse-code modulation (PCM), pulse-position modulation (PPM), and pulse-width modulation (PWM) are methods used.

Self-Test—Review Questions

1. What is distributed processing?
2. What is a "smart terminal" and how is it helpful to a computer user?
3. Explain problems that are made easier to solve through use of word-processor systems. What must be done first with a manuscript before it can undergo computer processing?
4. How does an on-line data storage and retrieval system operate? Name at least two fields of work in which such a system might be used.
5. In what ways is an automobile microprocessor useful?
6. What is the function of a microprocessor in a microwave oven?
7. In a "pocket calculator," what constitutes the input device? The output device?
8. Explain the principle and application of the universal product code (UPC).
9. What is the difference between an analog communications system and a digital communications system? Which is potentially more accurate? Why?

Overall Information Management System Epilogue

In this volume of *Basic Solid-State Electronics* you have learned about the principles and applications of digital systems. You have seen that these are merely extensions of what you learned previously about basic circuits, and you have now studied digital timing and logic circuits. As you have found in all of your studies in *Basic Solid-State Electronics*, the basic principle is the *control and management of information*. The source of information may vary, and it may be in either digital or analog form, but the basic idea of control and management remains the same. Of course, within this basic idea is embedded the need for the transmission and reception of information as well. This concept of the control and management of information is shown diagramatically on pages 5–176 and 5–177.

In Volume 1, you learned about semiconductor fundamentals and how simple discrete circuits and integrated circuits can be used to build power supplies that provide the necessary operating potentials for solid-state electronic equipment. In this initial volume, the concepts of passive resistive and capacitive components and the active diode and transistor components for rectification and amplification were studied. These practical and theoretical concepts laid the groundwork for later.

In Volume 2, you learned about amplifiers and oscillators. You found that the underlying principle in amplifiers is always the same; the configuration changes, however, to suit a particular need such as RF, IF, video, audio, etc. You also found that an oscillator is basically only a variation of the amplifier circuit. Then you learned how what you had studied about power supplies and amplifiers is used in the construction of audio amplifiers and hi-fi and stereo systems. And you learned about your first I/O devices—microphones and loudspeakers. That was the first step in information management and control, since these basic functions are incorporated into audio systems.

Overall Information Management System Epilogue (continued)

Volume 3 introduced the concept of information transfer and, in particular, transmission—an extension of Volume 2—since information transfer is not limited to wired systems, but can be radiated over great distances. The concept of modulation was discussed as a means for impressing information on a carrier for broadcast, using the oscillator-amplifier concepts learned previously, coupled with the study of antennas for the radiation of the modulated carrier. You were introduced to some aspects of pulsed systems when you studied how a TV image is formed by the TV camera (a transducer) and subsequently used to modulate a carrier for transmission. Here you also were exposed to basic timing concepts, since TV synchronization is of fundamental importance in the reconstruction of the transmitted signal.

Volume 4 provided you with an understanding of how broadcast signals containing information are received, appropriately demodulated (detected), and prepared for use by suitable transducers such as loudspeakers or TV picture tubes. The concepts of mixers coupled with amplifiers, demodulators, and detectors were expanded here to show how these circuit elements are used to provide for information reception. Both Volumes 3 and 4 emphasized that transmission and reception are only mechanisms for the transfer of information—audio, video, data, or sensory—from one place to another.

Finally, in Volume 5 you have concentrated on digital systems. Here, again, you have found that the basic concepts do not change, even when they are discussed in terms of on/off or high/low logic states. Furthermore, we have shown that digital information management, while different in detail, really uses the basic ideas set forth in the earlier volumes. The extraordinary flexibility of digital systems, coupled with the ability to pack many semiconductor elements into a very small IC package, has made the digital approach to information management and control fairly commonplace in many aspects of our lives. Surprisingly few building blocks—like timing circuits and logic elements —provide, in various combinations, almost limitless possibilities for special and general information acquisition, management, and control.

In summary, you can see that the essential building blocks of solid-state electronic systems are relatively simple and embody relatively straightforward concepts. The way in which these building blocks are configured determines *how* a particular function is to be performed. If you remember these ideas of simple building blocks, functional manipulation of information, and information transfer and control, then you will find it relatively easy to understand even systems that seem rather complex at first glance. You now have the necessary background to understand *any* solid-state system, if you keep these building blocks and functional concepts in mind and remember how they fit into the Overall Information Management System shown on the next two pages.

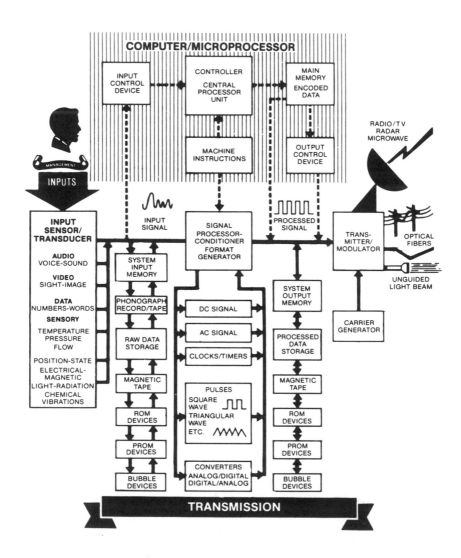

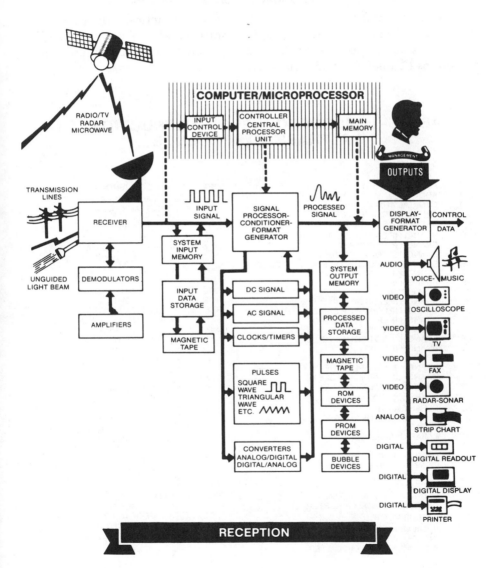

The MC6800 Microprocessor Instruction Set

While a complete description of the details of programming for a microprocessor is not possible here, the following material, presented in highly condensed form, will give you some idea of what a microprocessor instruction set is like. You can use these instructions (adapted from material kindly supplied by Motorola Semiconductor Products, Inc.) to try your hand at writing simple programs. If a microcomputer with a 6800 microprocessor is available, you can try out the programs you write.

MPU Instruction Set

The MC6800 has a set of 72 different instructions. Included are binary and decimal arithmetic, logical, shift, rotate, load, store, conditional or unconditional branch, interrupt, and stack-manipulation instructions (Tables 5–1 through 5–4).

TABLE 5-1: MICROPROCESSOR INSTRUCTION SET—
ALPHABETIC SEQUENCE

ABA	ADD ACCUMULATORS	JMP	JUMP
ADC	ADD WITH CARRY	JSR	JUMP TO SUBROUTINE
ADD	ADD	LDA	LOAD ACCUMULATOR
AND	LOGICAL AND	LDS	LOAD STACK POINTER
ASL	ARITHMETIC SHIFT LEFT	LDX	LOAD INDEX REGISTER
ASR	ARITHMETIC SHIFT RIGHT	LSR	LOGICAL SHIFT RIGHT
BCC	BRANCH IF CARRY CLEAR	NEG	NEGATE
BCS	BRANCH IF CARRY SET	NOP	NO OPERATION
BEQ	BRANCH IF EQUAL TO ZERO	ORA	INCLUSIVE OR ACCUMULATOR
BGE	BRANCH IF GREATER OR EQUAL TO ZERO		
BGT	BRANCH IF GREATER THAN ZERO	PSH	PUSH DATA
BHI	BRANCH IF HIGHER	PUL	PULL DATA
BIT	BIT TEST	ROL	ROTATE LEFT
BLE	BRANCH IF LESS OR EQUAL	ROR	ROTATE RIGHT
BLS	BRANCH IF LOWER OR SAME	RTI	RETURN FROM INTERRUPT
BLT	BRANCH IF LESS THAN ZERO	RTS	RETURN FROM SUBROUTINE
BMI	BRANCH IF MINUS	SBA	SUBTRACT ACCUMULATORS
BNE	BRANCH IF NOT EQUAL TO ZERO	SBC	SUBTRACT WITH CARRY
BPL	BRANCH IF PLUS	SEC	SET CARRY
BRA	BRANCH ALWAYS	SEI	SET INTERRUPT MASK
BSR	BRANCH TO SUBROUTINE	SEV	SET OVERFLOW
BVC	BRANCH IF OVERFLOW CLEAR	STA	STORE ACCUMULATOR
BVS	BRANCH IF OVERFLOW SET	STS	STORE STACK REGISTER
CBA	COMPARE ACCUMULATORS	STX	STORE INDEX REGISTER
CLC	CLEAR CARRY	SUB	SUBTRACT
CLI	CLEAR INTERRUPT MASK	SWI	SOFTWARE INTERRUPT
CLR	CLEAR	TAB	TRANSFER ACCUMULATORS
CLV	CLEAR OVERFLOW	TAP	TRANSFER ACCUMULATORS
CMP	COMPARE		TO CONDITION CODE REGISTER
COM	COMPLEMENT	TBA	TRANSFER ACCUMULATORS
CPX	COMPARE INDEX REGISTER	TPA	TRANSFER CONDITION CODE
DAA	DECIMAL ADJUST		REGISTER TO ACCUMULATOR
DEC	DECREMENT	TST	TEST
DES	DECREMENT STACK POINTER	TSX	TRANSFER STACK POINTER TO
DEX	DECREMENT INDEX REGISTER		INDEX REGISTER
EOR	EXCLUSIVE OR	TXS	TRANSFER INDEX REGISTER
INC	INCREMENT		TO STACK POINTER
INS	INCREMENT STACK POINTER	WAI	WAIT FOR INTERRUPT
INX	INCREMENT INDEX REGISTER		

TABLE 5-2: ACCUMULATOR AND MEMORY INSTRUCTIONS

OPERATIONS	MNEMONIC	IMMED OP ~ #	DIRECT OP ~ #	INDEX OP ~ #	EXTND OP ~ #	IMPLIED OP ~ #	BOOLEAN/ARITHMETIC OPERATION (ALL REGISTER LABELS REFER TO CONTENTS)	H	I	N	Z	V	C
ADD	ADDA	8B 2 2	9B 3 2	AB 5 2	BB 4 3		$A + M \to A$	↕	•	↕	↕	↕	↕
	ADDB	CB 2 2	DB 3 2	EB 5 2	FB 4 3		$B + M \to B$	↕	•	↕	↕	↕	↕
ADD ACCUMULATORS	ABA					1B 2	$A + B \to A$	↕	•	↕	↕	↕	↕
ADD WITH CARRY	ADCA	89 2 2	99 3 2	A9 5 2	B9 4 3		$A + M + C \to A$	↕	•	↕	↕	↕	↕
	ADCB	C9 2 2	D9 3 2	E9 5 2	F9 4 3		$B + M + C \to B$	↕	•	↕	↕	↕	↕
AND	ANDA	84 2 2	94 3 2	A4 5 2	B4 4 3		$A \cdot M \to A$	•	•	↕	↕	R	•
	ANDB	C4 2 2	D4 3 2	E4 5 2	F4 4 3		$B \cdot M \to B$	•	•	↕	↕	R	•
BIT TEST	BITA	85 2 2	95 3 2	A5 5 2	B5 4 3		$A \cdot M$	•	•	↕	↕	R	•
	BITB	C5 2 2	D5 3 2	E5 5 2	F5 4 3		$B \cdot M$	•	•	↕	↕	R	•
CLEAR	CLR			6F 7 2	7F 6 3		$00 \to M$	•	•	R	S	R	R
	CLRA					4F 2	$00 \to A$	•	•	R	S	R	R
	CLRB					5F 2	$00 \to B$	•	•	R	S	R	R
COMPARE	CMPA	81 2 2	91 3 2	A1 5 2	B1 4 3		$A - M$	•	•	↕	↕	↕	↕
	CMPB	C1 2 2	D1 3 2	E1 5 2	F1 4 3		$B - M$	•	•	↕	↕	↕	↕
COMPARE ACCUMULATORS	CBA					11 2	$A - B$	•	•	↕	↕	↕	↕
COMPLEMENT, 1'S	COM			63 7 2	73 6 3		$\overline{M} \to M$	•	•	↕	↕	R	S
	COMA					43 2	$\overline{A} \to A$	•	•	↕	↕	R	S
	COMB					53 2	$\overline{B} \to B$	•	•	↕	↕	R	S
COMPLEMENT, 2'S (NEGATE)	NEG			60 7 2	70 6 3		$00 - M \to M$	•	•	↕	↕	①	②
	NEGA					40 2	$00 - A \to A$	•	•	↕	↕	①	②
	NEGB					50 2	$00 - B \to B$	•	•	↕	↕	①	②
DECIMAL ADJUST, A	DAA					19 2	CONVERTS BINARY ADD. OF BCD CHARACTERS INTO BCD FORMAT	•	•	↕	↕	↕	③
DECREMENT	DEC			6A 7 2	7A 6 3		$M - 1 \to M$	•	•	↕	↕	④	•
	DECA					4A 2	$A - 1 \to A$	•	•	↕	↕	④	•
	DECB					5A 2	$B - 1 \to B$	•	•	↕	↕	④	•
EXCLUSIVE OR	EORA	88 2 2	98 3 2	A8 5 2	B8 4 3		$A \oplus M \to A$	•	•	↕	↕	R	•
	EORB	C8 2 2	D8 3 2	E8 5 2	F8 4 3		$B \oplus M \to B$	•	•	↕	↕	R	•
INCREMENT	INC			6C 7 2	7C 6 3		$M + 1 \to M$	•	•	↕	↕	⑤	•
	INCA					4C 2	$A + 1 \to A$	•	•	↕	↕	⑤	•
	INCB					5C 2	$B + 1 \to B$	•	•	↕	↕	⑤	•
LOAD ACCUMULATOR	LDAA	86 2 2	96 3 2	A6 5 2	B6 4 3		$M \to A$	•	•	↕	↕	R	•
	LDAB	C6 2 2	D6 3 2	E6 5 2	F6 4 3		$M \to B$	•	•	↕	↕	R	•
OR, INCLUSIVE	ORAA	8A 2 2	9A 3 2	AA 5 2	BA 4 3		$A + M \to A$	•	•	↕	↕	R	•
	ORAB	CA 2 2	DA 3 2	EA 5 2	FA 4 3		$B + M \to B$	•	•	↕	↕	R	•
PUSH DATA	PSHA					36 4	$A \to M_{SP}, SP - 1 \to SP$	•	•	•	•	•	•
	PSHB					37 4	$B \to M_{SP}, SP - 1 \to SP$	•	•	•	•	•	•
PULL DATA	PULA					32 4	$SP + 1 \to SP, M_{SP} \to A$	•	•	•	•	•	•
	PULB					33 4	$SP + 1 \to SP, M_{SP} \to B$	•	•	•	•	•	•

TABLE 5-2 (CONTINUED)

OPERATIONS	MNEMONIC	IMMED OP	~	#	DIRECT OP	~	#	INDEX OP	~	#	EXTND OP	~	#	IMPLIED OP	~	#	BOOLEAN/ARITHMETIC OPERATION	H	I	N	Z	V	C
ROTATE LEFT	ROL							69	7	2	79	6	3				M)	•	•	↕	↕	⑥	↕
	ROLA													49	2	1	A } C ← [b7…b0]	•	•	↕	↕	⑥	↕
	ROLB													59	2	1	B)	•	•	↕	↕	⑥	↕
ROTATE RIGHT	ROR							66	7	2	76	6	3				M)	•	•	↕	↕	⑥	↕
	RORA													46	2	1	A } C → [b7…b0]	•	•	↕	↕	⑥	↕
	RORB													56	2	1	B)	•	•	↕	↕	⑥	↕
SHIFT LEFT, ARITHMETIC	ASL							68	7	2	78	6	3				M)	•	•	↕	↕	⑥	↕
	ASLA													48	2	1	A } C ← [b7…b0] ← 0	•	•	↕	↕	⑥	↕
	ASLB													58	2	1	B)	•	•	↕	↕	⑥	↕
SHIFT RIGHT, ARITHMETIC	ASR							67	7	2	77	6	3				M)	•	•	↕	↕	⑥	↕
	ASRA													47	2	1	A } [b7…b0] → C	•	•	↕	↕	⑥	↕
	ASRB													57	2	1	B)	•	•	↕	↕	⑥	↕
SHIFT RIGHT, LOGIC	LSR							64	7	2	74	6	3				M)	•	•	R	↕	⑥	↕
	LSRA													44	2	1	A } 0 → [b7…b0] → C	•	•	R	↕	⑥	↕
	LSRB													54	2	1	B)	•	•	R	↕	⑥	↕
STORE ACCUMULATOR	STAA				97	4	2	A7	6	2	B7	5	3				A → M	•	•	↕	↕	R	•
	STAB				D7	4	2	E7	6	2	F7	5	3				B → M	•	•	↕	↕	R	•
SUBTRACT	SUBA	80	2	2	90	3	2	A0	5	2	B0	4	3				A − M → A	•	•	↕	↕	↕	↕
	SUBB	C0	2	2	D0	3	2	E0	5	2	F0	4	3				B − M → B	•	•	↕	↕	↕	↕
SUBTRACT ACCUMULATORS	SBA													10	2	1	A − B → A	•	•	↕	↕	↕	↕
SUBTRACT WITH CARRY	SBCA	82	2	2	92	3	2	A2	5	2	B2	4	3				A − M − C → A	•	•	↕	↕	↕	↕
	SBCB	C2	2	2	D2	3	2	E2	5	2	F2	4	3				B − M − C → B	•	•	↕	↕	↕	↕
TRANSFER ACCUMULATORS	TAB													16	2	1	A → B	•	•	↕	↕	R	•
	TBA													17	2	1	B → A	•	•	↕	↕	R	•
TEST, ZERO OR MINUS	TST							6D	7	2	7D	6	3				M − 00	•	•	↕	↕	R	R
	TSTA													4D	2	1	A − 00	•	•	↕	↕	R	R
	TSTB													5D	2	1	B − 00	•	•	↕	↕	R	R

LEGEND:
OP   OPERATION CODE (HEXADECIMAL)
~   NUMBER OF MPU CYCLES
#   NUMBER OF PROGRAM BYTES
+   ARITHMETIC PLUS
−   ARITHMETIC MINUS
·   BOOLEAN AND
M_{SP} CONTENTS OF MEMORY LOCATION POINTED TO STACK POINTER
+   BOOLEAN INCLUSIVE OR
⊕   BOOLEAN EXCLUSIVE OR
$\overline{M}$ COMPLEMENT OF M
→   TRANSFER INTO
0   BIT = ZERO
00   BYTE = ZERO

CONDITION CODE SYMBOLS:
H   HALF-CARRY FROM BIT 3
I   INTERRUPT MASK
N   NEGATIVE (SIGN BIT)
Z   ZERO (BYTE)
V   OVERFLOW, 2'S COMPLEMENT
C   CARRY FROM BIT 7
R   RESET ALWAYS
S   SET ALWAYS
↕   TEST AND SET IF TRUE, CLEARED OTHERWISE
•   NOT AFFECTED

NOTE—ACCUMULATOR ADDRESSING MODE INSTRUCTIONS ARE INCLUDED IN THE COLUMN FOR IMPLIED ADDRESSING

TABLE 5-3: INDEX REGISTER AND STACK MANIPULATION INSTRUCTIONS

POINTER OPERATIONS	MNEMONIC	IMMED OP	~	#	DIRECT OP	~	#	INDEX OP	~	#	EXTND OP	~	#	IMPLIED OP	~	#	BOOLEAN/ARITHMETIC OPERATION	H 5	I 4	N 3	Z 2	V 1	C 0
COMPARE INDEX REGISTER	CPX	8C	3	3	9C	4	2	AC	6	2	BC	5	3				$X_H - M, X_L - (M + 1)$	•	•	⑦	↕	⑧	•
DECREMENT INDEX REGISTER	DEX													09	4	1	$X - 1 \rightarrow X$	•	•	•	↕	•	•
DECREMENT STACK POINTER	DES													34	4	1	$SP - 1 \rightarrow SP$	•	•	•	•	•	•
INCREMENT INDEX REGISTER	INX													08	4	1	$X + 1 \rightarrow X$	•	•	•	↕	•	•
INCREMENT STACK POINTER	INS													31	4	1	$SP + 1 \rightarrow SP$	•	•	•	•	•	•
LOAD INDEX REGISTER	LDX	CE	3	3	DE	4	2	EE	6	2	FE	5	3				$M \rightarrow X_H, (M + 1) \rightarrow X_L$	•	•	⑨	↕	R	•
LOAD STACK POINTER	LDS	8E	3	3	9E	4	2	AE	6	2	BE	5	3				$M \rightarrow SP_H, (M + 1) \rightarrow SP_L$	•	•	⑨	↕	R	•
STORE INDEX REGISTER	STX				DF	5	2	EF	7	2	FF	6	3				$X_H \rightarrow M, X_L \rightarrow (M + 1)$	•	•	⑨	↕	R	•
STORE STACK POINTER	STS				9F	5	2	AF	7	2	BF	6	3				$SP_H \rightarrow M, SP_L \rightarrow (M + 1)$	•	•	⑨	↕	R	•
INDEX REGISTER → STACK POINTER	TXS													35	4	1	$X - 1 \rightarrow SP$	•	•	•	•	•	•
STACK POINTER → INDEX REGISTER	TSX													30	4	1	$SP + 1 \rightarrow X$	•	•	•	•	•	•

TABLE 5-4: JUMP AND BRANCH INSTRUCTIONS

OPERATIONS	MNEMONIC	RELATIVE OP	~	#	INDEX OP	~	#	EXTND OP	~	#	IMPLIED OP	~	#	BRANCH TEST	H 5	I 4	N 3	Z 2	V 1	C 0
BRANCH ALWAYS	BRA	20	4	2										NONE	•	•	•	•	•	•
BRANCH IF CARRY CLEAR	BCC	24	4	2										$C = 0$	•	•	•	•	•	•
BRANCH IF CARRY SET	BCS	25	4	2										$C = 1$	•	•	•	•	•	•
BRANCH IF = ZERO	BEQ	27	4	2										$Z = 1$	•	•	•	•	•	•
BRANCH IF ≥ ZERO	BGE	2C	4	2										$N \oplus V = 0$	•	•	•	•	•	•
BRANCH IF > ZERO	BGT	2E	4	2										$Z + (N \oplus V) = 0$	•	•	•	•	•	•
BRANCH IF HIGHER	BHI	22	4	2										$C + Z = 0$	•	•	•	•	•	•
BRANCH IF ≤ ZERO	BLE	2F	4	2										$Z + (N \oplus V) = 1$	•	•	•	•	•	•
BRANCH IF LOWER OR SAME	BLS	23	4	2										$C + Z = 1$	•	•	•	•	•	•
BRANCH IF < ZERO	BLT	2D	4	2										$N \oplus V = 1$	•	•	•	•	•	•
BRANCH IF MINUS	BMI	2B	4	2										$N = 1$	•	•	•	•	•	•
BRANCH IF NOT EQUAL TO ZERO	BNE	26	4	2										$Z = 0$	•	•	•	•	•	•
BRANCH IF OVERFLOW CLEAR	BVC	28	4	2										$V = 0$	•	•	•	•	•	•
BRANCH IF OVERFLOW SET	BVS	29	4	2										$V = 1$	•	•	•	•	•	•
BRANCH IF PLUS	BPL	2A	4	2										$N = 0$	•	•	•	•	•	•
BRANCH TO SUBROUTINE	BSR	8D	8	2											•	•	•	•	•	•
JUMP	JMP				6E	4	2	7E	3	3				SEE SPECIAL OPERATIONS	•	•	•	•	•	•
JUMP TO SUBROUTINE	JSR				AD	8	2	BD	9	3					•	•	•	•	•	•
NO OPERATION	NOP										01	2	1	ADVANCES PROG. CNTR. ONLY	•	•	•	•	•	•
RETURN FROM INTERRUPT	RTI										3B	10	1		⑩					
RETURN FROM SUBROUTINE	RTS										39	5	1		•	•	•	•	•	•
SOFTWARE INTERRUPT	SWI										3F	12	1	SEE SPECIAL OPERATIONS	•	•	•	•	•	•
WAIT FOR INTERRUPT*	WAI										3E	9	1		•	⑪	•	•	•	•

*WAI PUTS ADDRESS BUS, R/W, AND DATA BUS IN THE THREE-STATE MODE WHILE VMA IS HELD LOW.

MPU Addressing Modes

The MC6800 8-bit microprocessing unit has seven address modes that can be used by a programmer, with the addressing mode a function of both the type of instruction and the coding within the instruction.

Accumulator (ACCX) Addressing—In accumulator-only addressing, either accumulator A or accumulator B is specified. These are 1-byte instructions.

Immediate Addressing—In immediate addressing, the operand is contained in the second byte of the instruction—except LDS and LDX, which have the operand in the second and third bytes of the instruction, respectively. The MPU addresses this location when it fetches the immediate instruction for execution. These are 2- or 3-byte instructions.

Direct Addressing—In direct addressing, the address of the operand is contained in the second byte of the instruction. Direct addressing allows the user to directly address the lowest 256 bytes in the machine—that is, locations 0 through 255. Enhanced execution times are achieved by storing data in these locations. In most configurations, it should be a random-access memory. These are 2-byte instructions.

Extended Addressing—In extended addressing, the address contained in the second byte of the instruction is used as the higher 8 bits of the address of the operand. The third byte of the instruction is used as the lower 8 bits of the address for the operand. This is an absolute address in memory. These are 3-byte instructions.

Indexed Addressing—In indexed addressing, the address contained in the second byte of the instruction is added to the index register's lowest 8 bits in the MPU. The carry is then added to the higher-order 8 bits of the index register. This result is then used to address memory. The modified address is held in a temporary address register so there is no change to the index register. These are 2-byte instructions.

Implied Addressing—In the implied addressing mode, the instruction gives the address (that is, stack pointer, index register, etc.). These are 1-byte instructions.

Relative Addressing—In relative addressing, the address contained in the second byte of the instruction is added to the program counter's lowest 8 bits plus 2. The carry or borrow is then added to the high 8 bits. This allows the user to address data within a range of -125 to $+129$ bytes of the present instruction. These are 2-byte instructions.

CUMULATIVE INDEX

CUMULATIVE INDEX

(Note: The first number in each entry identifies
the *volume* in which the information is to be found;
the second number identifies the *page.*)

:   1